Y0-BTA-524

Galois Theory, Hopf Algebras, and Semiabelian Categories

THE FIELDS INSTITUTE FOR RESEARCH IN MATHEMATICAL SCIENCES

Galois Theory, Hopf Algebras, and Semiabelian Categories

George Janelidze
Bodo Pareigis
Walter Tholen
Editors

American Mathematical Society
Providence, Rhode Island

The Fields Institute for Research in Mathematical Sciences

The Fields Institute is named in honour of the Canadian mathematician John Charles Fields (1863–1932). Fields was a visionary who received many honours for his scientific work, including election to the Royal Society of Canada in 1909 and to the Royal Society of London in 1913. Among other accomplishments in the service of the international mathematics community, Fields was responsible for establishing the world's most prestigious prize for mathematics research—the Fields Medal.

The Fields Institute for Research in Mathematical Sciences is supported by grants from the Ontario Ministry of Training, Colleges and Universities, and the Natural Sciences and Engineering Research Council of Canada. The Institute is sponsored by Carleton University, McMaster University, the University of Ottawa, the University of Toronto, the University of Waterloo, the University of Western Ontario, and York University. In addition there are several affiliated universities and corporate sponsors, from Canada and the United States.

2000 *Mathematics Subject Classification.* Primary 08Bxx, 12Hxx, 13Bxx, 14Lxx, 16Dxx, 17Bxx, 18-xx, 19Dxx, 22Axx.

Library of Congress Cataloging-in-Publication Data

Galois theory, Hopf algebras, and semiabelian categories / George Janelidze, Bodo Pareigis, Walter Tholen, editors.
p. cm. — (Fields Institute Communications; v. 43)
Includes bibliographical references.
ISBN 0-8218-3290-5 (acid-free paper)
1. Differential algebra. 2. Galois theory. 3. Hopf algebras. 4. Rings (Algebra) I. Janelidze, G. (George), 1952– II. Pareigis, Bodo. III. Tholen, W. (Walter), 1947– IV. Series.
QA247.4.G35 2004
512′.32–dc22 2004050271
CIP

∞ The paper used in this book is acid-free and falls within the guidelines
established to ensure permanence and durability.
This publication was prepared by The Fields Institute.
Visit the AMS home page at http://www.ams.org/

10 9 8 7 6 5 4 3 2 1 09 08 07 06 05 04

Contents

Preface

During the week of September 23–28, 2002, the editors of this volume organized a *Workshop on Categorical Structures for Descent and Galois Theory, Hopf Algebras, and Semiabelian Categories* at the Fields Institute for Research in Mathematical Sciences in Toronto. The goal of the Workshop was to bring together researchers working in the quite distinct but nevertheless interrelated and partly overlapping areas mentioned in its title. The meeting was attended by almost eighty mathematicians from various research communities and boasted twenty invited lectures and numerous contributed talks that led to an inspiring atmosphere of learning and scientific exchange.

This volume can only partially reflect the Workshop's themes but should nevertheless give the reader a good idea about the current connections among abstract Galois theories, Hopf algebras, and semiabelian categories. Here is a very brief indication of the origins of those connections. Hopf algebras arrived to the Galois theory of rings as early as the 1960s — independently of, but in fact similarly to, the way in which algebraic group schemes were introduced to the theory of étale coverings in algebraic geometry. Galois theory, in turn, was extended to elementary toposes and was then formulated in purely categorical contexts. Eventually it became general enough to even include abstractions of the theory of central extensions, to mention only one of various fairly recent developments. In fact, classically, central extensions of groups together with the homology functors $H_1(-,\mathbf{Z})$ and $H_2(-,\mathbf{Z})$ can be used to begin homological algebra, just like covering spaces together with the homotopy functors π_0 and π_1 are the starting gadgets of homotopy theory. Finally, during the past four years semiabelian categories have emerged as a very good environment in which to pursue not just basic modern algebra but in fact homological algebra of groups and other non-abelian structures categorically.

Given the diversity of the backgrounds of the presenters at the Workshop, this volume cannot be expected to contain a homogeneous sequence of chapters on its themes. Rather, the reader will find a collection of beautiful but fairly independent articles on selected topics in algebra, topology, and pure category theory that should seriously contribute to the categorical unification of the subjects in question. The survey articles contained in this volume should be particularly helpful in this regard.

A rough general "map" of the topics/articles presented in this volume may be displayed as follows, with the numbers referring to the (alphabetical) list of contributions contained in the volume. Most of the papers are mentioned more than once. Solid lines represent links explicitly discussed in this volume, while dotted lines indicate other known links.

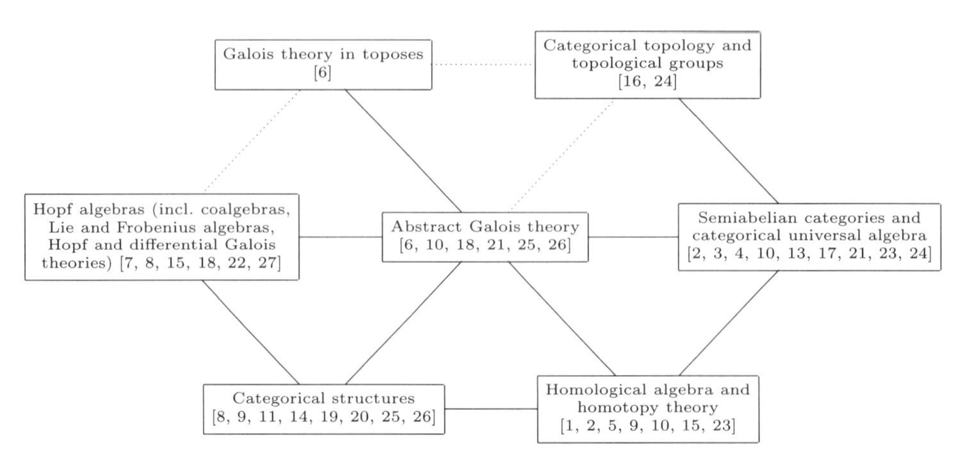

1. M. Barr, *Algebraic cohomology: the early days*
2. F. Borceux, *A survey of semi-abelian categories*
3. D. Bourn, *Commutator theory in regular Mal'cev categories*
4. D. Bourn and M. Gran, *Categorical aspects of modularity*
5. R. Brown, *Crossed complexes and homotopy groupoids as non commutative tools for higher dimensional local-to-global problems*
6. M. Bunge, *Galois groupoids and covering morphisms in topos theory*
7. S. Caenepeel, *Galois corings from the descent theory point of view*
8. B. Day and R. H. Street, *Quantum categories, star autonomy, and quantum groupoids*
9. J. W. Duskin, R. W. Kieboom, and E. M. Vitale, *Morphisms of 2-groupoids and low-dimensional cohomology of crossed modules*
10. M. Gran, *Applications of categorical Galois theory in universal algebra*
11. C. Hermida, *Fibrations for abstract multicategories*
12. J. Huebschmann, *Lie-Rinehart algebras, descent, and quantization*
13. P. T. Johnstone, *A note on the semiabelian variety of Heyting semilattices*
14. G. M. Kelly and S. Lack, *Monoidal functors generated by adjunctions, with applications to transport of structure*
15. M. Khalkhali and B. Rangipour, *On the cyclic homology of Hopf crossed products*
16. G. Lukács, *On sequentially h-complete groups*
17. J. L. MacDonald, *Embeddings of algebras*
18. A. R. Magid, *Universal covers and category theory in polynomial and differential Galois theory*
19. N. Martins-Ferreira, *Weak categories in additive 2-categories with kernels*
20. T. Palm, *Dendrotopic sets*
21. A. H. Roque, *On factorization systems and admissible Galois structures*
22. P. Schauenburg, *Hopf-Galois and bi-Galois extensions*
23. J. D. H. Smith, *Extension theory in Mal'tsev varieties*
24. L. Sousa, *On projective generators relative to coreflective classes*
25. J. J. Xarez, *The monotone-light factorization for categories via preorders*
26. J. J. Xarez, *Separable morphisms of categories via preordered sets*
27. S. Yamagami, *Frobenius algebras in tensor categories and bimodule extensions*

We express our sincere thanks to the Fields Institute for hosting and supporting the Workshop and publishing this volume. We are particularly grateful to Ms. Debbie Iscoe for her help in preparing the files. We also thank the Faculty of Arts of York University for additional financial assistance.

George Janelidze
Bodo Pareigis
Walter Tholen

Fields Institute Communications
Volume **43**, 2004

Algebraic Cohomology: The Early Days

Michael Barr
Department of Mathematics and Statistics
McGill University
805 Sherbrooke W.
Montreal, QC, H3A 2K6 Canada
barr@barrs.org

Abstract. This paper will survey the various definitions of homology theories from the first Eilenberg-Mac Lane theories for group cohomology through the Cartan-Eilenberg attempt at a uniform (co-)homology theory in algebra, cotriple cohomology theories and the various acyclic models theorems that tied them all together (as much as was possible).

1 Introduction

This paper is a mostly historical introduction to the topic of cohomology theories in algebra between 1940 and 1970 when my interests turned elsewhere. This is not to suggest that progress stopped that year, but that I did not keep up with things like crystalline cohomology, cyclic cohomology, etc., and therefore will have nothing to say about them. Much of what I report, I was directly involved in, but anything earlier than about 1962 is based either on the written record or on hearing such people as Samuel Eilenberg and Saunders Mac Lane reminisce about it. Therefore I report it as true to the best of my belief and knowledge.

I have also omitted any mention of sheaf cohomology. I had nothing to do with it and did not know the people most associated with it—Grothendieck, Godement and others. This was much more highly associated with developments in category theory which I knew little about until after the time frame I am dealing with here. The first development here was Mac Lane's paper [1950] which was the first paper to discover universal mapping properties and also attempt to define what we now call abelian categories, later given their full definition in [Buchsbaum, 1956] and [Grothendieck, 1957].

2000 *Mathematics Subject Classification.* Primary 01A60, 18-03; Secondary 18G10, 18G30, 18G35.

Key words and phrases. Homology and cohomology, triples, acyclic models.

This research has been supported by the NSERC of Canada.

1.1 Acknowledgment. I would like to thank the referee who took an inexcusably careless draft and read it with care and many—far too many than should have been necessary—valuable suggestions for improvements. Any remaining errors and obscurities are, of course, mine.

2 Eilenberg-Mac Lane cohomology of groups

2.1 The background. Eilenberg escaped from Poland in 1939 and spent the academic year 1939-40 at the University of Michigan. Mac Lane, then a junior fellow at Harvard, was invited to speak at Michigan. He was attempting to compute what I will call the Baer group of a group π with coefficients in a π-module A. This group, which I will denote $B(\pi, A)$, can be described as follows. Consider an exact sequence

$$0 \longrightarrow A \longrightarrow \Pi \longrightarrow \pi \longrightarrow 1$$

(Of course, that terminology did not exist in those days; Mac Lane—and Baer—would have said that A was a commutative normal subgroup of Π and $\pi = \Pi/A$.) Since A is normal subgroup of Π, it is a Π-module by conjugation. Since A is abelian, the action of A on itself is trivial, so that the Π-action induces a π-action on A, which may, but need not be the action we began with. Then $B(\pi, A)$ is the class of all such exact sequences that induce the π-action we started with. Say that the sequence above is equivalent to

$$0 \longrightarrow A \longrightarrow \Pi' \longrightarrow \pi \longrightarrow 1$$

if there is a homomorphism (necessarily an isomorphism) $f : \Pi \longrightarrow \Pi'$ such that

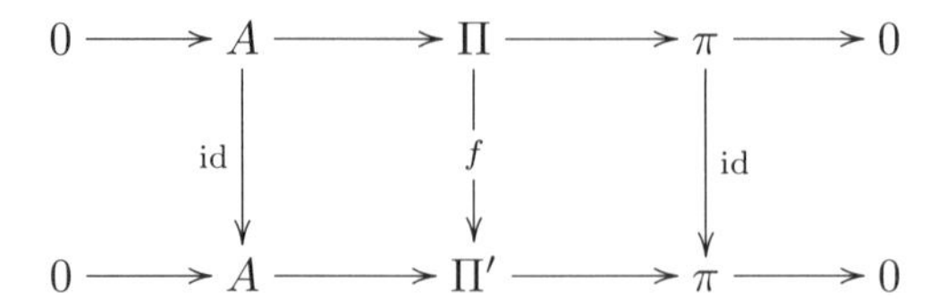

commutes. Then the equivalence classes of such sequences are a set. Moreover there is a way of adding two such equivalence classes that makes $B(\pi, A)$ into an abelian group. If $0 \longrightarrow A \longrightarrow \Pi \longrightarrow \pi \longrightarrow 0$ and $0 \longrightarrow A \longrightarrow \Pi' \longrightarrow \pi \longrightarrow 0$ are two such sequences, we first form the pullback $P = \Pi \times_\pi \Pi'$, let S be the subgroup of all elements of P of the form $(a, -a)$ for $a \in A$. It is immediate that S is a normal subgroup of P and then we can let $\Pi'' = P/S$. It is easy to see that the two maps $A \longrightarrow \Pi \longrightarrow P$ and $A \longrightarrow \Pi' \longrightarrow P$ are rendered equal by the projection $P \longrightarrow P/S$ so there is a canonical map $A \longrightarrow \Pi''$. Also the map $P \longrightarrow \pi$ vanishes on S and hence induces $\Pi'' \longrightarrow \pi$. A simple computation shows that $0 \longrightarrow A \longrightarrow \Pi'' \longrightarrow \pi \longrightarrow 0$ is exact. That defines the sum of the two sequences. The negative of the sequence $0 \longrightarrow A \longrightarrow \Pi \longrightarrow \pi \longrightarrow 0$ is simply the sequence in which the inclusion of $A \longrightarrow \Pi$ is negated. Since not every element has order 2 this also provides an example showing that equivalence of sequences is not simply isomorphism of the middle term.

Note that in a sequence the kernel A is not merely an abelian group, but also a π-module. This is because Π acts on the normal subgroup A by conjugation and A, being abelian, acts trivially on itself so that action extends to an action of π. The Baer group is the set of equivalence classes of extensions that induce the given module structure on A as a π-module.

Mac Lane's talk included the computations he was carrying out to compute the Baer group. It would have been along these lines: an extension determines and is determined by a function $\pi \times \pi \longrightarrow A$ satisfying a certain condition and two such functions determined equivalent extensions if there was a function $\pi \longrightarrow A$ that generated the difference of the two original functions. Details will be given later.

Apparently during Mac Lane's talk, Eilenberg noted that some of the calculations that Mac Lane was carrying out were the same as those he was doing in connection with certain cohomology groups. As they remarked in one of their first papers [Eilenberg and Mac Lane, 1942b], "This paper originated from an accidental observation that the groups obtained by Steenrod [1940] were identical with some groups that occur in the purely algebraic theory of *extension of groups*." The computations by Steenrod refer to some first cohomology classes (they are described as homology classes of infinite cycles, but I assume that they were really based on duals of chain groups since the dual of an infinite sum is an infinite product, whose elements can look like infinite sums of chains, especially if you do not have a clear idea of contravariant functors) of the complements of solenoids embedded in a 3-sphere. This is likely the second cohomology group of the solenoid itself. See [Eilenberg and Mac Lane, 1941] for more on infinite cycles.

Later in the same paper, in a footnote, they remark, "Group extensions are discussed by Baer [1934], Hall [1938], Turing [1938], Zassenhaus [1937] and elsewhere. Much of the discussion in the literature treats the case in which G but not H is assumed to be abelian and in which G is not necessarily in the centre of H." What the latter sentence means is that they were looking at extensions of the form

$$0 \longrightarrow G \longrightarrow H \longrightarrow H/G \longrightarrow 0$$

in which G and H are abelian, while the literature was mainly discussing the case that G was an abelian normal subgroup of H and the conjugation action of H (and therefore H/G) on G was non-trivial.

When I gave my talk, I said (and believed, relying on what I thought Eilenberg had said to me 30 years ago, from what he remembered from 30 years before that) that what Eilenberg was working on was calculating the cohomology groups of $K(\pi, 1)$, a space that had fundamental group π and no other homotopy. This was described as the main motivation in the later joint work [1947a, b]. But this goes back only to a paper of Hopf's from 1942 in which he showed that if X is a suitable space, then the fundamental group $\pi_1(X)$ determines the cokernel of the map from $\pi_2(X) \longrightarrow H_2(X)$. Thus in the particular case of a $K(\pi, 1)$, whose π_2 is 0, Hopf's result says that $\pi_1(X)$ determines $H_2(X)$ (as well, of course, as $H_1(X)$, which is the commutator quotient of π_1). It is not clear quite when it was realized that π determines all the homology and cohomology groups of a $K(\pi, 1)$, but it must have been before 1947. That paper came after Hochschild's cohomology theory for commutative algebras, which had, it would appear, no connection with topology. I am indebted to Johannes Hübschmann for pointing out some of the pre-history that I was previously unaware of.

2.2 The Eilenberg-Mac Lane groups. Here is a brief description of the Eilenberg-Mac Lane theory. For a non-negative integer n, an n-cochain on π with coefficients in A is a function $f : \pi^n \longrightarrow A$. When $n = 0$, this is simply an element of A. Denote the set of such functions by $C^n(\pi, A)$. Define a function $\delta_n : C^n(\pi, A)$

$\longrightarrow C^{n+1}(\pi, A)$ by

$$(\delta_n f)(x_1, \ldots, x_{n+1}) = x_1 f(x_2, \ldots, x_{n+1})$$

$$+ \sum_{i=1}^{n} (-1)^i f(x_1, \ldots, x_i x_{i+1}, \ldots, x_{n+1}) + (-1)^{n+1} f(x_1, \ldots, x_n)$$

The first term uses the action (which could be the trivial, or identity, action) of π on A. It is a simple exercise to show that $\delta_n \circ \delta_{n-1} = 0$ so that $\operatorname{im}(\delta_{n-1}) \subseteq \ker(\delta_n)$ so that we can define

$$H^n(\pi, A) = \ker(\delta_n)/\operatorname{im}(\delta_{n-1})$$

When $n = 0$, the cohomology is simply the kernel of δ_0. The elements of $\ker(\delta_n)$ are called n-cocycles and those of $\operatorname{im}(\delta_{n-1})$ are called n-coboundaries. Cochains that differ by a coboundary are said to be cohomologous.

2.3 Interpretation of $H^2(\pi, A)$. We will see how the group $H^2(\pi, A)$ is the same as $B(\pi, A)$. Given an extension

$$0 \longrightarrow A \xrightarrow{j} \Pi \xrightarrow{p} \pi \longrightarrow 1$$

let s be a set-theoretic section of p. This means that s is a function, not necessarily a homomorphism, $\pi \longrightarrow \Pi$ such that $p \circ s = \mathrm{id}$. We can and will suppose that $s(1) = 1$. If we write $j(a) = (a, 1)$ for $a \in A$, $s(x) = (0, x)$ for $x \in \pi$, and (a, x) for $(a, 1)(0, x)$ then the fact that every element of Π can be written in the form $j(a)s(x)$ for $a \in A$ and $x \in \Pi$ means that we can identify the underlying set of Π with $A \times \pi$. Since j is a homomorphism, we see that $(a, 0)(b, 0) = (a + b, 0)$. Although s is not a homomorphism, p is and that forces the second coordinate of $(0, x)(0, y)$ to be xy so that we can write $(0, x)(0, y) = (f(x, y), xy)$ with $f \in C^2(\pi, A)$. Note that $(0, x)(a, 1) = (xa, 1)(0, x)$ since the action of π on A is just conjugation. It follows that

$$\begin{aligned}(a, x)(b, y) &= (a, 1)(0, x)(b, 1)(0, y) = (a, 1)(xb, 1)(0, x)(0, y)\\ &= (a + xb, 1)(f(x, y), xy) = (a + xb, 1)(f(x, y), 1)(0, xy)\\ &= (a + xb + f(x, y), 1)(0, xy) = (a + xb + f(x, y), xy)\end{aligned}$$

Next I claim that the associative law of group multiplication forces f to be a 2-cocyle. In fact, we have for a, b, $c \in A$ and x, y, $z \in \pi$,

$$\begin{aligned}((a, x)(b, y))(c, z) &= (a + xb + f(x, y), xy)(c, z)\\ &= (a + xb + f(x, y) + xyc + f(xy, z), xyz)\end{aligned}$$

while

$$\begin{aligned}(a, x)((b, y)(c, z)) &= (a, x)(b + yc + f(y, z), yz)\\ &= (a + xb + xyc + xf(y, z) + f(x, yz), xyz)\end{aligned}$$

and comparing them we see that

$$f(x, y) + f(xy, z) = xf(y, z) + f(x, yz)$$

which is equivalent to $\delta_2 f = 0$.

The same computation shows that if we begin with a 2-cocycle f and define multiplication on $A \times \pi$ by the formula

$$(a, x)(b, y) = (a + xb + f(x, y), xy)$$

we have an associative multiplication. To see it is a group first suppose that $f(x, 1) = 0$ for all $x \in \pi$. It is then an easy computation to show that $(0, 1)$ is a right identity and then that $(-x^{-1}a - x^{-1}f(x, x^{-1}), x^{-1})$ is a right inverse for (a, x). We will show below that the group extensions constructed using cohomologous cocycles are equivalent. We use that fact here by showing that any cocycle is cohomologous to one for which $f(x, 1) = 0$. From

$$\delta f(x, 1, 1) = xf(1, 1) - f(x, 1) - f(x, 1) + f(x, 1) = 0$$

we see that $f(x, 1) = xf(1, 1)$. Then for $g(x) = f(x, 1)$,

$$(f - \delta g)(x, 1) = f(x, 1) - xg(1) + g(x) - g(x) = f(x, 1) - xf(1, 1) = 0$$

Finally, we claim that equivalent extensions correspond to cohomologous cocycles. For suppose that f and f' are 2-cocycles and g is a 1-cochain such that $f - f' = \delta g$. Suppose we denote the two possible multiplications by $*$ and $*'$ so that

$$(a, x) * (b, y) = (a + xb + f(x, y), xy)$$

$$(a, x) *' (b, y) = (a + xb + f'(x, y), xy)$$

Define $\alpha : A \times \pi \longrightarrow A \times \pi$ by $\alpha(a, x) = (a + g(x), x)$. Then

$$\begin{aligned}
\alpha((a, x) * (b, y)) &= \alpha(a + xb + f(x, y), xy) \\
&= (a + xb + f(x, y) + g(xy), xy) \\
&= (a + xb + f'(x, y) + xg(y) + g(x), xy) \\
&= (a + g(x), x) *' (b + g(y), y) \\
&= \alpha(a, x) *' \alpha(b, y)
\end{aligned}$$

which shows that α is a homomorphism, while it is obviously invertible. Clearly the isomorphism commutes with the inclusion of A and the projection on π.

Conversely, suppose f and f' are cocycles that give equivalent extensions. Again, we will denote the two multiplications by $*$ and $*'$. Since we are trying to show that f and f' are cohomologous, we can replace them by cohomologous cocycles that satisfy $f(x, 1) = f'(x, 1) = 0$ for all $x \in \pi$. The fact that the extensions are actually equivalent (not merely isomorphic) implies that the isomorphism α has the property that elements of A are fixed and that the second coordinate of $\alpha(a, x)$ is x. In particular, we can write $\alpha(0, x) = (g(x), x)$. Then

$$\begin{aligned}
\alpha(a, x) = \alpha((0, x) * (x^{-1}a, 1)) &= \alpha(0, x) *' \alpha(x^{-1}a, 1) \\
&= (g(x), x) *' (x^{-1}a, 1) = (a + g(x), x)
\end{aligned}$$

Then from

$$\begin{aligned}(a + xb + f(x, y) + g(xy), xy) &= \alpha(a + xb + f(x, y), xy) \\ &= \alpha((a, x) * (b, y)) \\ &= \alpha(a, x) *' \alpha(b, y) \\ &= (a + g(x), x) *' (b + g(y), y) \\ &= (a + g(x) + xb + xg(y) + f'(xy), xy)\end{aligned}$$

we conclude that $f - f' = \delta g$.

2.4 Interpretations in other low dimensions. The definition of $H^0(\pi, A)$ as the kernel of δ_0 makes it obvious that it is simply

$$\{a \in A \mid xa = a \text{ for all } x \in \pi\}$$

otherwise known as the group of fixed elements of A and denoted A^π.

The kernel of δ_1 consists of those maps $d : \pi \longrightarrow A$ such that $d(xy) = xd(y) + d(x)$. In the case that π acts as the identity on A, this is exactly a homomorphism of π to the abelian group A. For this reason, such a function is sometimes called a crossed homomorphism. We prefer to call it a derivation for compatibility with other examples. The image of δ_0 consists of those derivations of the form $d(x) = xa - a$ for some element $a \in A$. These are called the inner derivations and $H^1(\pi, A)$ is simply the quotient group of derivations modulo inner derivations.

There is also an interpretation of H^3 that gave information (more limited than H^2) for extensions with non-abelian kernels. We give no proofs here, but content ourselves with a brief description.

Suppose that

$$1 \longrightarrow G \longrightarrow \Pi \longrightarrow \pi \longrightarrow 1 \qquad (*)$$

is an exact sequence of groups. This means that G is a normal subgroup of Π and the quotient is π. Since G is normal, Π acts on G by conjugation. This gives a map $\Pi \longrightarrow \operatorname{Aut}(G)$ the group of automorphisms of G. Unless G is commutative, this does not vanish on G and hence does not give a natural map $\pi \longrightarrow \operatorname{Aut}(G)$. However, if $\operatorname{In}(G)$ denotes the (normal) subgroup of $\operatorname{Aut}(G)$ consisting of the inner automorphisms, then the composite map $\Pi \longrightarrow \operatorname{Aut}(G) \longrightarrow \operatorname{Aut}(G)/\operatorname{In}(G)$ does vanish on G and hence induces a natural map $\theta : \pi \longrightarrow \operatorname{Aut}(G)/\operatorname{In}(G)$. We will say that θ is induced by $(*)$.

Now let $\mathscr{Z}(G)$ denote the centre of G. One may check that the centre of a normal subgroup is also a normal subgroup so that $\mathscr{Z}(G)$ is also a Π-module, but one that G acts trivially on, so that $\mathscr{Z}(G)$ is a π-module. Now it turns out that any homomorphism $\theta : \pi \longrightarrow \operatorname{Aut}(G)/\operatorname{In}(G)$ induces, after a number of choices, a cocycle in $C^3(\pi, \mathscr{Z}(G))$ whose cohomology class we denote $[\theta]$.

A homomorphism $\theta : \pi \longrightarrow \operatorname{Aut}(G)/\operatorname{In}(G)$ induces a π-module structure on $\mathscr{Z}(G)$. Simply choose, for each $x \in \pi$, an element $\overline{\theta}(x) \in \operatorname{Aut}(G)$ whose class mod $\operatorname{In}(G)$ is $\theta(x)$ and define $xz = \overline{\theta}(x)(z)$ for $z \in \mathscr{Z}(G)$. This is well-defined since inner automorphisms are trivial on the centre.

The main result is contained in the following.

Theorem 2.1 *Suppose G is a group whose centre $\mathscr{Z}(G)$ is a π-module and that $\theta : \pi \longrightarrow \mathrm{Aut}(G)/\mathrm{In}(G)$ is a homomorphism that induces the given action of π on $\mathscr{Z}(G)$. Then*

1. *the cohomology class $[\theta]$ in $H^3(\pi, \mathscr{Z}(G))$ does not depend on the arbitrary choices made;*
2. *the cohomology class $[\theta] = 0$ if and only if θ comes from an extension of the form $(*)$;*
3. *the equivalence class of extensions $(*)$ that give rise to a given cohomology class $[\theta]$ are in 1-1 correspondence with the elements of $H^2(\pi, \mathscr{Z}(G))$; and*
4. *given a π-module A, every element of $H^3(\pi, A)$ has the form $[\theta]$ for some group G and some homomorphism $\theta : \pi \longrightarrow \mathrm{Aut}(G)/\mathrm{In}(G)$ such that $A \cong \mathscr{Z}(G)$ as π-modules, the latter with the π-action induced by θ.*

The 1-1 correspondence in the third point above is actually mediated by a principal homogeneous action of $H^2(\pi, \mathscr{Z}(G))$ on extensions $(*)$ ([Barr, 1969]). The class $[\theta]$ is called the **obstruction** of $[\theta]$ (to arising from an extension) and the last clause says that every element of $H^3(\pi, A)$ is the obstruction to some homomorphism's coming from an extension.

The most striking application of that theory was that if $\mathscr{Z}(G) = 1$, then the equivalence classes of extensions

$$1 \longrightarrow G \longrightarrow \Pi \longrightarrow \pi \longrightarrow 1$$

is in 1-1 correspondence with the homomorphisms

$$\theta : \pi \longrightarrow \mathrm{Aut}(G)/G$$

3 Hochschild cohomology of associative algebras

Gerhard Hochschild defined a cohomology theory for associative algebras in [1945, 1946]. Formally, his definitions look almost identical to those of Eilenberg and Mac Lane.

The setting of this theory is that of an associative algebra Λ over a (commutative) field K. It was later observed that the definitions can be given when K is any commutative ring, but the resultant extension theory is limited to those extensions that split as K-modules. Let A be a two-sided Λ-module. Define $C^n(\Lambda, A)$ to be the set of all n-linear functions $\Lambda^n \longrightarrow A$. Define $\delta_n : C^n(\Lambda, A) \longrightarrow C^{n+1}(\Lambda, A)$ by the formula

$$\begin{aligned}(\delta_n f)(x_1, \ldots, x_{n+1}) = x_1 f(x_2, \ldots, x_{n+1}) \\ + \sum_{i=1}^{n} (-1)^i f(x_1, \ldots, x_i x_{i+1}, \ldots, x_{n+1}) \\ + (-1)^{n+1} f(x_1, \ldots, x_n) x_{n+1}\end{aligned}$$

This differs from the formula for group cohomology only in that the last term is multiplied on the right by x_{n+1}. One can make the formulas formally the same by making any left π-module into a two-sided π-module with the identity action on the right. Alternatively, you can use two-sided modules, using the same coboundary formula as for the Hochschild cohomology. This does not really give a different theory, since you can make a two-sided π-module into a left π-module by the formula $x * a = xax^{-1}$ without changing the cohomology.

Much the same interpretation of the low dimensional cohomology holds for the Hochschild cohomology as for groups. Derivations are defined slightly differently: $d(xy) = xd(y) + d(x)y$, that is the Leibniz formula for differentials. Commutative normal subgroups are replaced by ideals of square 0.

The limitation to extensions that split as modules was avoided by having two kinds of cocycles (in degree 2; n in degree n), one to express the failure of additive splitting and one for the multiplication. This was first done by Mac Lane [1958] and then generalized to algebras in [Shukla 1961].

4 Chevalley-Eilenberg cohomology of Lie algebras

The third cohomology theory that was created during the half decade after the war was the Chevalley-Eilenberg cohomology of Lie algebras [1948]. The formulas are a little different, although the conclusions are much the same. If $\mathfrak{g}$ is a Lie algebra over the field K, then a $\mathfrak{g}$-module A is abelian group with an action of $\mathfrak{g}$ that satisfies $[x,y]a = x(ya) - y(xa)$. It can be considered a two-sided module by defining $ax = -xa$, a fact we will make use of when discussing the Cartan-Eilenberg cohomology. A derivation $d : \mathfrak{g} \longrightarrow A$ satisfies $d[xy] = xd(y) - yd(x)$. An n-cochain is still an n-linear map $f : \mathfrak{g}^n \longrightarrow M$, but it is required to alternate. That is $f(x_1, \ldots, x_n) = 0$ as soon as two arguments are equal. The coboundary formula is also quite different. In fact the coboundary formulas used for groups and associative algebras would not map alternating functions to alternating functions. The definition found by Chevalley and Eilenberg is given by the following formula, in which the hat, $\hat{}$, is used to denote omitted arguments.

$$\delta f(x_0, \ldots, x_n) = \sum_{i=1}^{n} (-1)^i x_i f(x_0, \ldots, \hat{x}_i, \ldots, x_n)$$

$$+ \sum_{0 \le i < j \le n} (-1)^{i+j} f([x_i, x_j], x_0, \ldots, \hat{x}_i, \ldots, \hat{x}_j, \ldots, x_n)$$

The formula arose naturally as the infinitesimal version of the de Rham coboundary formula on Lie groups.

The same kinds of interpretations in low dimensions hold as in the other two cases.

4.1 Comments on these definitions. Two of these three definitions arose from topology in which homology and cohomology were, by the 1940s, coming to be well understood. The cohomology of π is the cohomology of the space $K(\pi, 1)$ and the cohomology of a real or complex Lie algebra is that of the corresponding Lie group. As far as I know, there was no topological motivation behind Hochschild's theory. Of course, the interpretations in dimensions ≤ 3 presumably gave some confidence that the basic theory was correct. Nonetheless, the definitions, viewed as purely algebraic formulas, were inexplicable and *ad hoc*. Among the questions one might raise was the obvious, "What is a module?" To this question, at least, we will give Beck's surprising and entirely convincing answer.

5 Cartan-Eilenberg cohomology

When, as a graduate student in 1959, I took a course from David Harrison on homological algebra, the definitions of cohomology of groups and associative algebras given above were the definitions I learned. (If Lie algebras were mentioned,

I do not recall it.) The course was mainly concerned with Ext, Tor and the like, leading to a proof of the Auslander-Buchsbaum theorem. I imagine the book of Cartan-Eilenberg [1956] was mentioned, but I do not think I ever looked at it. The purchase date inscribed in own copy is April, 1962, just after I had finished typing my thesis. My thesis was on commutative algebra cohomology, which I will discuss later, and the methods of that book do not work in that case for reasons I will explain.

When I got to Columbia as a new instructor, the Cartan-Eilenberg book was the bible of homological algebra and I gradually learned its methods. Basically, they had found a uniform method for defining cohomology theories that included the three theories I have described. Only later did it become apparent their methods applied only to the three cohomology theories it was based on and not on any other.

The way that Cartan and Eilenberg had proceeded was based on the observation that in all three cases, for each object X of the category of interest, there was an enveloping algebra X^e that had the property that the coefficient modules for the cohomology were exactly the X^e-modules. And in all three cases, there was some module—call it X_J—for which the cohomology groups could be described as $\mathrm{Ext}_{X^e}(X_J, M)$.

5.1 Comparison of Cartan-Eilenberg with earlier theories. Here is how you see that the Cartan-Eilenberg (CE) cohomology of a group is the same as the Eilenberg-Mac Lane (EM) cohomology. To compute the EM cohomology of a group π with coefficients in a π-module A, you let $C^n(\pi, A) = \mathrm{Hom}_{\mathsf{Set}}(\pi^n, A)$. By using various adjunctions, we see that

$$\begin{aligned} C^n(\pi, A) = \mathrm{Hom}_{\mathsf{Set}}(\pi^n, A) &\cong \mathrm{Hom}_{\mathsf{Ab}}(\mathbf{Z}(\pi^n), A) \\ &\cong \mathrm{Hom}_{\mathsf{Ab}}(\mathbf{Z}(\pi)^{\otimes n}, A) \cong \mathrm{Hom}_{\mathbf{Z}(\pi)}(\mathbf{Z}(\pi)^{\otimes(n+1)}, A) \end{aligned}$$

Here $M^{\otimes n}$ stands for the nth tensor power of a module M. It is easy to compute that the boundary operator $\delta : C^n(\pi, A) \longrightarrow C^{n+1}(\pi, A)$ is induced by the map $\partial : \mathbf{Z}(\pi)^{\otimes(n+2)} \longrightarrow \mathbf{Z}(\pi)^{\otimes(n+1)}$ defined by

$$\begin{aligned} \partial(x_0 \otimes \cdots \otimes x_{n+1}) = \sum_{i=0}^{n} (-1)^i x_0 \otimes \cdots \otimes x_i x_{i+1} \otimes \cdots \otimes x_{n+1} \\ + (-1)^{n+1} x_0 \otimes \cdots \otimes x_n \end{aligned}$$

I claim that if you set $C_n(\pi) = \mathbf{Z}(\pi)^{\otimes(n+1)}$ with the boundary operator ∂, the resultant complex is a projective resolution of $\mathbf{Z}$ made into a π-module by having each element of π act as the identity. Let us note that the action of π on $\mathbf{Z}(G)^{\otimes(n+1)}$ is on the first coordinate, in accordance with the adjunction isomorphism above. Thus $\mathbf{Z}(\pi)^{\otimes(n+1)} = \mathbf{Z}(\pi) \otimes \mathbf{Z}(\pi)^{\otimes n}$ and $\mathbf{Z}(\pi)^{\otimes n}$ is just the free abelian group generated by π^n and hence $\mathbf{Z}(\pi)^{\otimes(n+1)}$ is the free π-module generated by π^n. The augmentation $\mathbf{Z}(\pi) \longrightarrow \mathbf{Z}$ is the linear map that takes the elements of π to the integer 1. In order to show that $\mathrm{Ext}_\pi(\mathbf{Z}, M)$ is the Eilenberg-Mac Lane cohomology of π with coefficients in M, it is sufficient to show that the augmented chain complex

$$\cdots \longrightarrow \mathbf{Z}(\pi)^{\otimes(n+1)} \longrightarrow \mathbf{Z}(\pi)^{\otimes n} \longrightarrow \cdots \longrightarrow \mathbf{Z}(\pi) \longrightarrow \mathbf{Z} \longrightarrow 0$$

is acyclic (that is, is an exact sequence), which implies that the unaugmented complex is a projective resolution of $\mathbf{Z}$ as a $\mathbf{Z}(\pi)$-module. We will actually show that the augmented complex is contractible as a complex of $\mathbf{Z}$-modules, although not as a complex of $\mathbf{Z}(\pi)$-modules. Define $s_{-1} : \mathbf{Z} \longrightarrow \mathbf{Z}(\pi)$ to be the linear map that sends 1 to $1 \in \pi$ (that is, the identity of π). For $n > 0$, define $s_{n-1} : \mathbf{Z}(\pi)^{\otimes n} \longrightarrow \mathbf{Z}(\pi)^{\otimes(n+1)}$ to be the linear map that takes $x_1 \otimes \cdots \otimes x_n$ to $1 \otimes x_1 \otimes \cdots \otimes x_n$. Then $\partial_0 \circ s_{-1} = \mathrm{id}$ clearly. For $n = 1$, we have

$$(\partial_1 \circ s_0 + s_{-1}\partial_0)(x) = \partial_1(1 \otimes x) + s_{-1}(1) = x - 1 + 1 = x$$

For $n > 1$,

$$\begin{aligned} \partial_n \circ s_{n-1}(x_1 \otimes \cdots \otimes x_n) &= \partial_n(1 \otimes x_1 \otimes \cdots \otimes x_n) \\ &= x_1 \otimes \cdots \otimes x_n \\ &\quad + \sum_{i=1}^{n-1} (-1)^i (1 \otimes x_1 \otimes \cdots \otimes x_i x_{i+1} \otimes \cdots \otimes x_n) \\ &\quad + (-1)^n 1 \otimes x_1 \otimes \cdots \otimes x_{n-1} \end{aligned}$$

while

$$\begin{aligned} s_{n-2} \circ \partial_{n-1}(x_1 \otimes \cdots \otimes x_n) &= s_{n-2}\left(\sum_{i=1}^{n-1} (-1)^{i-1} (x_1 \otimes \cdots \otimes x_i x_{i+1} \otimes \cdots \otimes x_n) \right) \\ &\quad + (-1)^{n-1} s_{n-2}(x_1 \otimes \cdots \otimes x_{n-1}) \\ &= \sum_{i=1}^{n-1} (-1)^{i-1} (1 \otimes x_1 \otimes \cdots \otimes x_i x_{i+1} \otimes \cdots x_n) \\ &\quad + (-1)^{n-1} 1 \otimes x_1 \otimes \cdots \otimes x_{n-1} \end{aligned}$$

and adding them up, we see that $\partial_n \circ s_{n-1} + s_{n-2} \circ \partial_{n-1} = \mathrm{id}$.

Thus it follows that the Eilenberg-Mac Lane cohomology groups of a group π with coefficients in a π-module A are just $\mathrm{Ext}_{\mathbf{Z}(\pi)}(\mathbf{Z}, A)$.

In the case of associative K-algebras, the development is similar. In this paragraph, we will denote by $\otimes$, the tensor product $\otimes_K$. A two-sided Λ-module is a left $\Lambda^{\mathrm{e}} = \Lambda \otimes \Lambda^{\mathrm{op}}$-module. The Hochschild cochain complex with coefficients in a module A is shown to be the complex $\mathrm{Hom}_{\Lambda^{\mathrm{e}}}(C_\bullet, A)$ in which $C_n = \Lambda^{\otimes(n+2)}$ with boundary $\partial = \partial_n : C_n \longrightarrow C_{n-1}$ given by

$$\partial(x_0 \otimes \cdots \otimes x_{n+1}) = \sum_{i=0}^{n} (-1)^i x_0 \otimes \cdots \otimes x_i x_{i+1} \otimes \cdots \otimes x_{n+1}$$

By a similar formula to the group case, one shows that this chain complex augmented over Λ is linearly (actually even right Λ-linearly) contractible and therefore a projective resolution of Λ as a Λ^{e}-module. It should be observed that $\Lambda \otimes \Lambda \cong \Lambda \otimes \Lambda^{\mathrm{op}}$ as a Λ^{e}-module, although not, of course, as an algebra. The upshot is that the Hochschild cohomology of Λ with coefficients in a two-sided Λ-module A is simply $\mathrm{Ext}_{\Lambda^{\mathrm{e}}}(\Lambda, A)$.

The analysis of the Lie algebra case is somewhat more complicated (because the coboundary operator is more complicated) but the outcome is the same. If $\mathfrak{g}$ is a K-Lie algebra, the category of $\mathfrak{g}$-modules is equivalent to the category of left modules over the enveloping associative algebra $\mathfrak{g}^{\mathrm{e}}$. This is the quotient of the tensor algebra over the K-module underlying $\mathfrak{g}$ modulo the ideal generated by all

$x \otimes y - y \otimes x - [x, y]$, for $x, y \in \mathfrak{g}$. Then one can show that the cochain complex comes from a projective resolution of K—with trivial $\mathfrak{g}$-action—so that the cohomology with coefficients in the $\mathfrak{g}$-module A is $\mathrm{Ext}_{\mathfrak{g}^e}(K, A)$.

5.2 Comments. It was certainly elegant that Cartan and Eilenberg were able to find a single definition that gave cohomology theories of the previous decade in all three cases. Nonetheless there were three *ad hoc* features of their definitions that rendered their answer less than satisfactory:

1. what is a module;
2. what is X^e; and
3. what is X_J?

But the least satisfactory aspect of their definition did not become clear right away. It was those three cohomology theories ended up as the only ones for which the Cartan-Eilenberg theory was correct. This first showed up in Harrison's theory for commutative algebras.

6 The Harrison cohomology theory

Around 1960, Dave Harrison [1962] defined a cohomology theory for commutative algebras over a field. His original definition was rather obscure. As modified by the referee (who identified himself to Harrison as Mac Lane) it became somewhat less obscure, but still not obvious. Let R ba commutative K-algebra with K a field. If $0 < i < n$ define a linear map $* : R^{\otimes i} \otimes R^{\otimes(n-i)} \longrightarrow R^{\otimes n}$ by

$$(x_1 \otimes \cdots \otimes x_i) * (x_{i+1} \otimes \cdots \otimes x_n) = (x_1 \otimes (x_2 \otimes \cdots \otimes x_i) * (x_{i+1} \otimes \cdots \otimes x_n))$$
$$+ (-1)^i (x_{i+1} \otimes (x_1 \otimes \cdots \otimes x_i) * (x_{i+2} \otimes \cdots \otimes x_n))$$

which together with

$$(x_1 \otimes \cdots \otimes x_n) * () = () * (x_1 \otimes \cdots \otimes x_n) = (x_1 \otimes \cdots \otimes x_n)$$

defines inductively an operation on strings, called the shuffle.[1] If A a left R-module that is made into a two-sided R-module by having the same operation on both sides. Harrison defined a commutative n-cochain as an f in the Hochschild cochain group $C^n(R, A)$ such that

$$f((x_1 \otimes \cdots \otimes x_i) * (x_{i+1} \otimes \cdots \otimes x_n)) = 0$$

for all $0 < i < n$ and all $(x_1 \otimes \cdots \otimes x_n) \in R^n$. In other words, a commutative cochain was one that vanished on all proper shuffles. The commutative coboundary formula was the same as Hochschild's and it required a non-trivial argument to show that Harrison's cochains are invariant under it. The details are in his 1962 paper and can also be found in [Barr, 2002].

The coboundary of a 0-chain in $C(R, A)$—an element of $a \in A$ is the one chain f for which $f(x) = ax - xa$. But if A has the same action on both sides, this is 0. Thus the cochain complex breaks up into two pieces, the degree 0 piece and the rest. Hence for Harrison, $H^0(R, A) = A$ and $H^1(R, A) = \mathrm{Der}(R, A)$, the

[1] The reason it is called a shuffle is that it is the alternating sum of all possible ways of shuffling the first i cards in an n card deck with the remaining $n - i$. The inductive formula corresponds to the obvious fact that any shuffle can be thought of as consisting of taking the top card from one of the two decks, shuffling the remaining cards and then replacing that top card on top.

group of derivations of R to A. The groups $H^2(R, A)$ and $H^3(R, A)$ have the same interpretations as in the Hochschild theory, but restricted to the case that everything is commutative.

Harrison observed moreover that, provided the characteristic of the underlying ground field is not 2, the group of 2-cochains can be written as the direct sum of the commutative cochains and a complementary summand in such a way that the Hochschild cohomology splits as the direct sum of two groups, one of which is the commutative cohomology and the other is a complementary term. In my Ph.D. thesis, [1962], I pushed this splitting up to degree 4, except that now characteristic 3 also had to be excluded. What I actually showed was the Hochschild cochain complex, truncated to degree 4, could be written as a direct sum of the Harrison cochain subcomplex and a complementary subcomplex and this splits the cohomology up to degree 4. (It might be thought that you need a splitting of the cochain complex up to degree 5 to do this, but that turns out to be unnecessary.) This was used to show various facts about the Harrison cohomology group of which the perhaps the most interesting was that when R is a polynomial ring, then $H^n(R, A) = 0$ for all R-modules A and $n = 2, 3, 4$. Actually, Harrison had shown this for $n = 2, 3$ in a different way, but it was not obvious how to extend his argument to higher dimension.

My proofs were highly computational and it was unclear how to extend any of it to higher dimensions. But I gnawed at it for five years and did eventually [Barr, 1968] find a relatively simple and non-computational construction of an idempotent in the rational group rings of symmetric groups that when applied to the Hochschild chain complex of an algebra over a field of characteristic 0 splits it into a commutative part and a complement, each invariant under the coboundary. This showed that when the ground field has characteristic 0, the Hochschild cohomology splits into two parts; one of the two is the Harrison cohomology and the other is a complement. Later, Gerstenhaber and Shack [1987], by discovering and factoring the characteristic polynomial of these idempotents, discovered a "Hodge decomposition" of the Hochschild groups:

$$H^n(R, A) = \sum_{i=1}^{n} H^{n\,i}(R, A)$$

where $H^{n\,1}(R, A)$ is the Harrison group and the other summands are part of an infinite series whose nth piece vanishes in dimensions below n, so that in each degree the sum is finite.

Harrison's original paper also contained an appendix—written by me—that used the same ideas from Mac Lane's 1958 paper to deal with the case of a commutative algebra over a general coefficient ring. However, the referee—who identified himself to Harrison as Mac Lane—insisted on an appendectomy and the paper is now lost. From this, I conclude that Mac Lane felt the approach was likely a dead end and I tend to agree.

6.1 Does the Harrison cohomology fit the Cartan-Eilenberg model? As I have already hinted, Harrison was not a fan of the Cartan-Eilenberg model of what a cohomology theory was. I do not recall that he ever mentioned it, I do not think I was even aware of it before arriving at Columbia in the fall of 1962 and I do not know if Harrison had ever considered whether his theory fit into it. It is a bit odd, since it seems clear that Cartan and Eilenberg believed that their book defined what a cohomology theory was. They scarcely mentioned the older definitions in their book and, as far as I can tell (the book lacks a bibliography), they did not even cite either Eilenberg-Mac Lane paper [1947a,b]. But Harrison's definition was in the style of the older definitions. To put his theory into the Cartan-Eilenberg framework, there would have to be, for each commutative ring R, an enveloping algebra R^e such that left R^e-modules were the same as left R-modules and a module R_J for which $H^n(R, A) = \mathrm{Ext}^n_{R^e}(R_J, A)$. Since left R-modules were the same as left R^e-modules, we would have to take R^e to be R or something Morita equivalent to it. But Ext is invariant under Morita equivalence so nothing can be gained by using anything but R. As for R_J, we leave that aside for the moment and point out that one consequence of the cohomology being an Ext is that it vanishes whenever A is injective. This is true for cohomology theories of groups, associative algebras, and Lie algebras, but is not true for commutative algebras. The easiest example is the following. Let K be any field. It is easy to see (and well-known) that the algebra of dual numbers $R = K[x]/(x^2)$ is self-injective. On the other hand, we have the non-split exact sequence

$$0 \longrightarrow x^2 \cdot K[x]/x^4 \longrightarrow K[x]/(x^4) \longrightarrow K[x]/(x^2) \longrightarrow 0$$

of commutative rings, whose kernel is, as an R-module, isomorphic to R and hence is injective. Thus $H^2(R, R) \neq 0$. See [Barr, 1968a]. This example doomed the attempted redefinition of commutative cohomology that appeared in [Barr, 1965a, 1965b].

Here is an explanation of what goes wrong. If R is a commutative K-algebra, then the chain complex that has $R^{\otimes(n+2)}$ with the Hochschild boundary operator, is a projective resolution of R as an $R \otimes R$-module. If A is a symmetric module and B is any two-sided R-module, it is evident that

$$\mathrm{Hom}_{R\otimes R}(B, A) \cong \mathrm{Hom}_R(R \otimes_{R\otimes R} B, A)$$

since $R \otimes_{R\otimes R} B$ is just the symmetrization of B. Since $R \otimes_{R\otimes R} R^{\otimes(n+2)} = R^{\otimes(n+1)}$, the Hochschild cochain complex of R with coefficients in A is

$$\mathrm{Hom}_{R\otimes R}(R^{\otimes(n+2)}, A) \cong \mathrm{Hom}_R(R^{\otimes(n+1)}, A)$$

It follows that the symmetrized chain complex is not generally acyclic; its homology is $\mathrm{Tor}^{R\otimes R}(R, R)$ and that is trivial if and only if R is separable. Even if it is, the quotient complex modulo the shuffles will not be generally acyclic.

7 Cohomology as a functor in the first argument

Although Harrison had an 8 term exact sequence involving the cohomology of R, that of R/I for an ideal I and $\mathrm{Ext}_R(I, -)$, the functoriality of cohomology in its first argument had been mostly ignored. If a connected series of functors is to look like a derived functor in a contravariant variable, it should vanish in all positive dimensions when that variable is free. But cohomology did not vanish when its first variable was free. In fact, when the group, algebra, or Lie algebra is free, the cohomology vanishes in dimensions greater than 1. This is also the case for commutative algebras, but only in characteristic 0. This, as well as other indications, suggested that if one wanted to view cohomology as a functor in the first variable, it would be best to drop the lowest degree term, renumber all the rest by -1 and also change the new lowest degree term from derivations modulo inner derivations to simply derivations. Thus was born the idea of cohomology as the derived functor of derivations, see [Barr and Rinehart, 1964].

Let us call this cohomology with the lowest term dropped and the next one modified the dimension-shifted cohomology. For the commutative cohomology, as already noted, there are no inner derivations and the lowest degree term is merely the coefficient module, so dropping it entails no loss of information. For the others, there is some cost. For associative and Lie algebras the cohomology vanishes in all dimensions if and only if the algebra is (finite-dimensional and) separable. For other purposes, the dimension-shifted cohomology seems better. For instance, the Stallings-Swan theorem states that a group is free if and only if its cohomological dimension is 1. Using the dimension-shifted theory, that number changes to 0, which is what you might expect for projectives (of course, every projective group is free).

So the first definition of the dimension-shifted is simply apply the functor $\mathrm{Der}(X,-)$ (here X is the group or algebra or whatever) to the category of X-modules and form the derived functor. This means, given a module A find an injective resolution

$$0 \longrightarrow Q_0 \longrightarrow Q_1 \longrightarrow \cdots \longrightarrow Q_n \longrightarrow \cdots$$

apply $\mathrm{Der}(X,-)$ to get the chain complex

$$0 \longrightarrow \mathrm{Der}(X,Q_0) \longrightarrow \mathrm{Der}(X,Q_1) \longrightarrow \cdots \longrightarrow \mathrm{Der}(X,Q_n) \longrightarrow \cdots$$

and define the cohomology groups to be the cohomology of that cochain complex. The trouble with this definition is that it automatically makes the cohomology vanish when the coefficients are injective and therefore can represent the cohomology only for theories that vanish on injective coefficients.

Before turning to other theories, there is one more point to be made. Since Der is like a homfunctor, it is quite easy to see that it preserves limits. The special adjoint functor theorem implies that it is representable by an object we call Diff (for module of differentials, since differentials are dual to derivations). The map $A \longrightarrow \mathrm{Der}(X,A)$ that sends an element of A to the inner derivation at A is the component at A of a natural transformation $\mathrm{Hom}_{X^e}(X^e,-) \longrightarrow \mathrm{Hom}_{X^e}(\mathrm{Diff},-)$ and is thereby induced by a homomorphism $\mathrm{Diff} \longrightarrow X^e$. In the three classic cases of groups, associative algebras and Lie algebras, this homomorphism is injective and the quotient module X^e/Diff is the heretofore mysterious X_J. Thus in those three cases, not only is the cohomology $\mathrm{Ext}_{X^e}(X_J,A)$, but the dimension-shifted cohomology is $\mathrm{Ext}_{X^e}(\mathrm{Diff},A)$. Thus we have removed one of the *ad hoc* features from the Cartan-Eilenberg theory. It can be replaced by a question as to why the map from $\mathrm{Diff} \longrightarrow X^e$ is injective in those cases, but at least we know where X_J comes from. The second special item is X^e but X^e is determined, up to Morita equivalence, by the fact that X^e-modules are the same as X-modules. Since Ext is invariant under Morita equivalence, that ambiguity is not important. The third *ad hoc* feature of the Cartan-Eilenberg definition is the definition of module and we will explain Jon Beck's surprising and elegant answer to that question next.

8 Beck modules

8.1 The definition of Beck modules. We begin Beck's theory by looking at an example in some detail. Denote by Grp the category of groups. If π is a group, the category Grp/π has as objects group homomorphisms $p : \Pi \longrightarrow \pi$. If $p' : \Pi' \longrightarrow \pi$ is another object, a map $f : p' \longrightarrow p$ is a commutative triangle

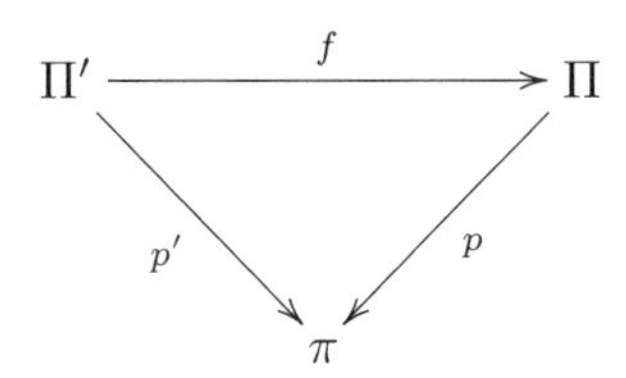

Given a π-module A, denote by $A \rtimes \pi$ the extension corresponding to the 0 element of $H^2(\pi, A)$. This means the underlying set is $A \times \pi$ and the multiplication is given by $(a, x)(b, y) = (a + xb, xy)$. The second coordinate projection makes $A \rtimes \pi$ into an object of Grp/π.

Let us calculate the set $\mathrm{Hom}_{\mathsf{Grp}/\pi}(\Pi \xrightarrow{p} \pi, A \rtimes \pi \to \pi)$. A homomorphism $f : \Pi \longrightarrow A \rtimes \pi$ is a function $\Pi \longrightarrow A \times \pi$. But in order to be a morphism in the category, the second coordinate must be p. If we call the first coordinate d, then $f = (d, p)$. In order to be a group homomorphism, we must have, for any $x, y \in \Pi$ that

$$(d(xy), p(xy)) = (d(x), p(x))(d(y), p(y)) = (d(x) + p(x)d(y), p(x)p(y))$$

Since p is already a homomorphism, the second coordinates are equal. As for the first coordinates, the required condition is that $d(xy) = p(x)d(y) + d(x)$. Except for the p on the right hand side, this is just the formula that defines a derivation. But we can use p to make any π-module into a Π-module and then this condition is just that d is a derivation of Π to A. Thus we have shown that

$$\mathrm{Hom}_{\mathsf{Grp}/\pi}(\Pi \xrightarrow{p} \pi, A \rtimes \pi \to \pi) = \mathrm{Der}(\Pi, A)$$

Evidently, $\mathrm{Der}(\Pi, A)$ is an abelian group from the additive structure of A and it is easy to see that a morphism $(\Pi' \to \pi) \longrightarrow (\Pi \to \pi)$ induces not merely a function but a group homomorphism $\mathrm{Der}(\Pi, A) \longrightarrow \mathrm{Der}(\Pi', A)$. This means that $A \rtimes \pi \to \pi$ is actually an abelian group object in Grp/π.

The converse is also true. It is a standard fact about abelian group objects in categories with finite products that a group object G is given by a global section $1 \longrightarrow G$, an inverse map $G \longrightarrow G$ and a multiplication $G \times G \longrightarrow G$ and that the multiplication map is just the product of the two projections. If $p : \Pi \longrightarrow \pi$ is a group object in Grp/π, the terminal object of the category is $\mathrm{id} : \pi \longrightarrow \pi$ and thus the zero map is just a homomorphism $\pi \longrightarrow \Pi$ that splits p. If K is the kernel of p, it is a normal subgroup and we will write its group operation as $+$ and its identity element as 0 even though we have not yet proved it commutative. By extending the notation to the non-commutative case, we can write $\Pi = K \rtimes \pi$. In this notation, the zero morphism is given by $z(x) = (0, x)$. The product in the category Grp/π is just the fibered product over π. Since $\Pi \times_\pi \Pi = K \times K \rtimes \pi$, the multiplication is a homomorphism

$$m : K \times K \rtimes \pi \longrightarrow K \rtimes \pi$$

over π. We can write this as $m(a, b, x) = (f(a, b, x), x)$ for $a, b \in K$ and $x \in K$. Since m preserves the identity of the group object, $m(0, 0, x) = (0, x)$ from which it follows that $f(0,0,x) = 0$ for any $x \in \pi$. But then

$$m(a,b,x) = m(a,b,1)m(0,0,1) - (f(a,b,1),1)(f(0,0,x),x)$$

$$= (f(a,b,1),1)(0,x) = (f(a,b,1),x)$$

so that f does not depend on x. Write $f(a,b,1) = a * b$. Then since $(0,x)$ is the identity in the fiber over x, $m(a,0,x) = (a,x) = m(0,a,x)$ so that $a*0 = 0*a = a$. But then

$$m(a,b,x) = m(a,0,1)m(0,b,1)m(0,0,x) = (a*0,1)(0*b,1)(0,x)$$

$$= (a*0+0*b,x) = (a+b,x)$$

while at the same time

$$m(a,b,x) = m(0,b,1)m(a,0,1)m(0,0,x) = (0*b,1)(a*0,1)(0,x)$$

$$= (0*b+a*0,x) = (b+a,x)$$

from which we conclude that $* = +$ and is commutative. The action of Π on K is by conjugation and since K is commutative, this induces an action of π on K. If $q : \Gamma \longrightarrow \pi$ is a group over π, then a morphism $f : (\Gamma \to \pi) \longrightarrow (\Pi \to \pi)$ has the form $f(y) = (dy, qy)$ and to be a homomorphism we must have

$$(d(yy'), q(yy')) = f(yy') = f(y)f(y') = (dy, qy)(dy', qy')$$

$$= (dy + (qy)(dy'), (qy)(qy'))$$

which means that $d(yy') = dy + (qy)(dy')$ which is the definition of a derivation (with respect to q).

This is one case of:

Theorem 8.1 (Beck) *Let $\mathscr{X}$ be one of the familiar categories (groups, algebras and rings, Lie algebras, commutative algebras and rings, Jordan algebras, ...). Then the category of modules over an object X of $\mathscr{X}$ is (equivalent to) the category of abelian group objects in the category $\mathscr{X}/X$. Moreover, let A be an X-module with $Y \longrightarrow X$ the corresponding abelian group object in $\mathscr{X}/X$. Then for an object $Z \longrightarrow X$ of that category,* $\mathrm{Hom}(Z \to X, Y \to X)$ *is canonically isomorphic to* $\mathrm{Der}(Z, A)$.

The upshot of this result is that not only do we know what a module is for an object of any category, we also know what a derivation into that module is. For example, a module over the set I is just an I-indexed family $A = \{A_i \mid i \in I\}$ of abelian groups and $\mathrm{Der}(I, A) = \prod_{i \in I} A_i$. For any $X \longrightarrow I$, which is just an I-indexed family $\{X_i \mid i \in I\}$, one can see that $\mathrm{Der}(X, A) = \prod_{i \in I} A_i^{X_i}$.

After I gave my talk, I had a private discussion with Myles Tierney and Alex Heller. Heller remarked that he first heard the definition of Beck module from Eilenberg and wondered whether the definition was originally his. Tierney, who was a student of Eilenberg's at the same time as Beck, recalled that it taken a couple months for Beck to convince Eilenberg of the correctness of his definition, but that once convinced, Eilenberg embraced it enthusiastically.

As an aside, let me say that this is the definition of bimodule. One-sided modules cannot be got as special cases of this and are apparently a different animal. Apparently one-sided modules are part of representation theory and two-sided modules are part of extension theory.

9 Enter triples

Triples (also known as monads) have been used in many places and for many reasons in category theory (as well as in theoretical computer science, especially in the theory of datatypes). They were originally invented (by Godement in [1958]) for the purpose of describing standard flabby resolutions of sheaves. They were being used for this purpose by Eckmann and his students around 1960 and by Eilenberg and Moore in their Memoir [1965a]. But this was always in additive categories (although the triples were not always additive, but Eilenberg and Moore assumed not only that but that they preserved kernels, although neither hypothesis was necessary for their purposes).

Beck was the first person who used a triple (more precisely, a cotriple) in a non-additive category to define a cohomology theory. Once more, I will illustrate what he did in the category of groups.

9.1 Triples and Eilenberg-Moore algebras. The definitions are available in many places and so I will just give a rapid sketch. If $\mathscr{X}$ is a category, a triple $\mathbf{T}$ on $\mathscr{X}$ consists of (T, η, μ) where T is an endofunctor on $\mathscr{X}$ and η and μ are natural transformations: $\eta : \mathrm{Id} \longrightarrow T$ and $\mu : T^2 \longrightarrow T$ such that the diagrams

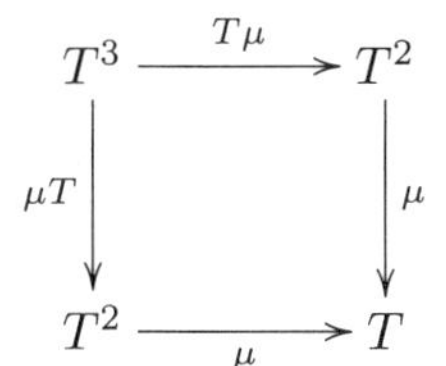

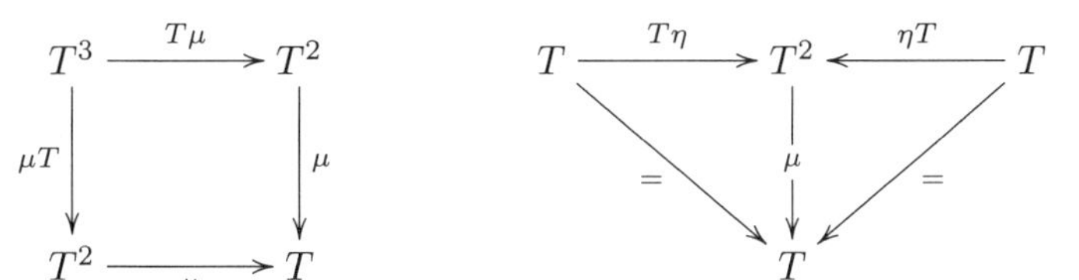

commute. A cotriple is a triple in the dual category. Hence a cotriple $\mathbf{G}$ consists of (G, ϵ, δ) with G an endofunctor and $\epsilon : G \longrightarrow \mathrm{Id}$ and $\delta : G \longrightarrow G^2$ natural transformations such that the dual of the above diagrams commute. Given an adjoint pair $F : \mathscr{X} \longrightarrow \mathscr{Y}$ left adjoint to $U : \mathscr{Y} \longrightarrow \mathscr{X}$, with adjunction arrows $\eta : \mathrm{Id} \longrightarrow UF$ and $\epsilon : FU \longrightarrow \mathrm{Id}$, then $(UF, \eta, U\epsilon F)$ is a triple on $\mathscr{X}$ and $(FU, \epsilon, F\eta U)$ is a cotriple on $\mathscr{Y}$. Peter Huber (a student of Eckmann's) told me once that they had proved this theorem—a simple argument with naturality—because they were having so much trouble verifying the commutations in the diagram above and noticed that all their triples and cotriples were associated with adjoint pairs (see [Huber, 1961]). It is much easier to verify an adjunction than a triple. The converse is also true; every (co-)triple does arise in this way from an adjoint pair. There are two distinct proofs of this; due to Kleisli [1964, 1965] and Eilenberg-Moore [1965b]. For most purposes in algebra, the Eilenberg-Moore algebras are more interesting. But triples have also entered theoretical computer science and there the Kleisli construction is the main one.

9.2 Beck cohomology. As above, it is convenient to illustrate Beck's results in the category of groups. The adjoint pair $F \dashv U$ where $F : \mathsf{Set} \longrightarrow \mathsf{Grp}$ and $U : \mathsf{Grp} \longrightarrow \mathsf{Set}$ is the underlying set functor, gives rise to a cotriple on Grp as described. If π is a group, then $G\pi$ is the free group on the underlying set of π. As with any cotriple, there results a functor that assigns to each group π a simplicial group that has $G^{n+1}\pi$ in degree n. The face operators are constructed from ϵ by $d^i = d^i_n = G^i \epsilon G^{n-i} : G^{n+1} \longrightarrow G^n$ and $s^i = s^i_n = G^i \delta G^{n-i} : G^{n+1} \longrightarrow G^{n+2}$. The naturality and the commuting diagrams satisfied by a cotriple imply the simplicial identities:

$$
\begin{aligned}
d_n^i \circ d_{n+1}^j &= d_n^{j-1} \circ d_{n+1}^i && \text{if } 0 \le i < j \le n+1\\
s_n^j \circ s_{n-1}^i &= s_n^i \circ s_{n-1}^{j-1} && \text{if } 0 \le i < j \le n\\
d_{n+1}^i \circ s_n^j &= \begin{cases} s_{n-1}^{j-1} \circ d_n^i & \text{if } 0 \le i < j \le n\\ 1 & \text{if } 0 \le i = j \le n \text{ or } 0 \le i-1 = j < n\\ s_{n-1}^j \circ d_n^{i-1} & \text{if } 0 < j < i-1 \le n \end{cases}
\end{aligned}
$$

We now form, for each object X and each X-module A, the cochain complex

$$0 \longrightarrow \operatorname{Der}(GX, A) \longrightarrow \cdots \longrightarrow \operatorname{Der}(G^n X, A) \longrightarrow \operatorname{Der}(G^{n+1}X, A) \longrightarrow \cdots$$

with the coboundary map

$$\sum_{i=0}^{n} (-1)^i \operatorname{Der}(d^i X, A) : \operatorname{Der}(G^n X, A) \longrightarrow \operatorname{Der}(G^{n+1}X, A)$$

Curiously, the δ of the cotriple plays no part in this, but it is absolutely necessary in applying the acyclic models theorem mentioned below.

This gives a uniform treatment for cohomology in any equational category. Beck also showed that the group $H^1(X, A)$ did classify extensions of X with kernel A, as expected. He did not explore the second cohomology that in known cases was relevant to extensions with non-abelian kernels. Later I did in a couple cases and discovered some problems in the general case that showed that the interpretation of the dimension-shifted H^2 could not be quite the same as in the known cases (Jordan algebras supplied the first counter-example).

10 Comparison theorems

Jon Beck and I started thinking in early 1964 about the connection between the various cohomology theories that had been invented in an *ad hoc* way and the Beck (or cotriple) cohomology theories. Let us assume that we are talking about the dimension-shifted versions of the former. So we know that $H^0(X, A) = \operatorname{Der}(X, A)$ for both theories and that both versions of $H^1(X, A)$ classify the same set of extensions and hence that they are at least isomorphic (although naturality was still an issue). But we had no idea whatever about H^2 or any higher dimension. The complexes look entirely different. The Eilenberg-Mac Lane complex has, in degree n, functions of $n+1$ (because of the dimension shift) variables from π to A, while the cotriple complex has functions of one variable from $G^{n+1}\pi$ to A. We spent the fall term of 1964 working on this problem and getting exactly nowhere. For group cohomology, one approach would have been to show that the chain complex

$$\cdots \longrightarrow \operatorname{Diff} G^{n+1}\pi \longrightarrow \operatorname{Diff} G^n\pi \longrightarrow \cdots \longrightarrow \operatorname{Diff} G\pi \longrightarrow \operatorname{Diff} \pi \longrightarrow 0$$

with boundary operator $\sum_{i=1}^{n} (-1)^n \operatorname{Diff} d^i$, is exact; for then the positive part of that complex would have been a projective resolution of $\operatorname{Diff} \pi$. In retrospect, we now know how to do that directly. Fortunately for us we did not find that argument for it would have been a dead end.

Instead, Beck spoke to Harry Appelgate during the term break between 1964 and 1965. Appelgate suggested trying to solve the problem by using acyclic models, one of the subjects of his thesis [1965]. We did and within a few days we had solved the comparison problem for groups and associative algebras. For some reason, we never looked at Lie algebras, while commutative algebras, for the time being, resisted our methods.

The case of commutative algebras was settled in the late 1960s. In finite characteristic, Harrison's definition was not equivalent to the cotriple cohomology, as shown by an example due to Michel André. In characteristic 0, it was shown in [Barr, 1968] that the two theories coincided. It is shown in [Barr, 1996] that the same is true for Lie algebras.

10.1 Acyclic models. Here is a brief description of the acyclic models theorem that we proved and used to compare the Cartan-Eilenberg cohomology with the cotriple cohomology. The context of the theorem is a category $\mathscr{X}$ equipped with a cotriple $\mathbf{G} = (G, \epsilon, \delta)$ and an abelian category $\mathscr{A}$. We are given augmented chain complex functors $K_\bullet = \{K_n\}$, $L_\bullet = \{L_n\} : \mathscr{X} \longrightarrow \mathscr{A}$, for $n \geq -1$. We say that $K_\bullet$ is G-presentable if there is a natural transformation $\theta_n : K_n \longrightarrow K_nG$ for all $n \geq 0$ (note: not for $n = -1$) such that $K_n\epsilon \circ \theta_n = \mathrm{id}$ for all $n \geq 0$. We say that $L_\bullet$ is G-contractible if the complex $L_\bullet G \longrightarrow 0$ has a natural contracting homotopy (which is called s below).

Theorem 10.1 *Suppose that $K_\bullet$ is G-presentable and $L_\bullet$ is G-contractible. Then any natural transformation $f_{-1} : K_{-1} \longrightarrow L_{-1}$ can be extended to a natural transformation $f_\bullet : K_\bullet \longrightarrow L_\bullet$. Any two extensions of f_{-1} are naturally homotopic.*

Proof Here are the two diagrams required for the proof. The computations are straightforward. The map f_n is defined as the composite,

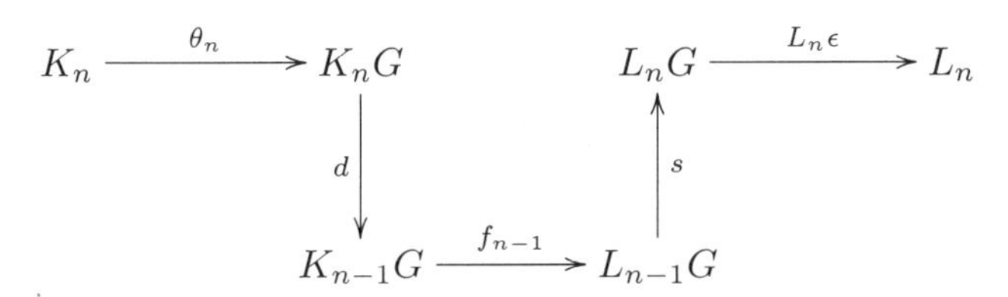

while the nth homotopy is defined as the difference of the upper and lower composite in,

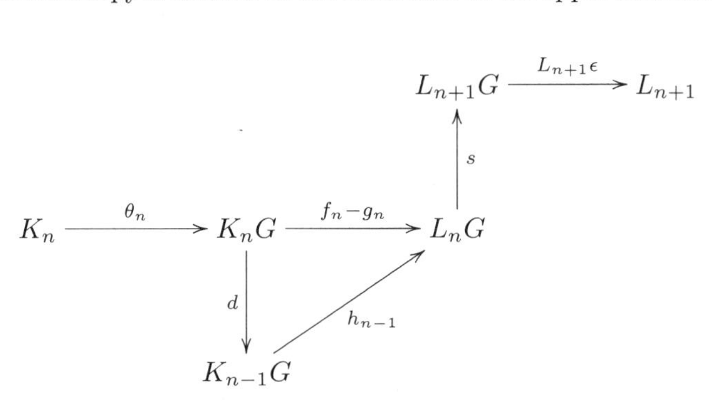

□

Of course, an immediate consequence of this is that if $K_\bullet$ and $L_\bullet$ each satisfy both hypotheses and K_{-1} is naturally isomorphic to L_{-1}, then $K_\bullet$ is naturally homotopic to $L_\bullet$. This result is augmented by

Proposition 10.2 *Suppose that $L_n = L_{-1}G^{n+1}$ with coboundary*

$$\sum_{i=0}^{n}(-1)^i G^i \epsilon G^{n-i}$$

Then $L_\bullet$ is both G-presentable and G-contractible.

In fact, δ is used to show both claims.

Before describing the modern acyclic models theorem that can be used to prove this, we will describe the issues involved. Basically, two things are required to show that a cochain complex functor is cohomologous to the cotriple complex. The first seems somewhat odd at first glance, but is satisfied in examples. That is that the terms of the complex do not depend on the structure of the object in question, but only on the underlying set (or module). It is a curious fact, but obvious from the definition that the structure is used only in the definition of the coboundary homomorphism. The second condition required is that the cohomology of a free object should vanish in positive degrees. This condition is automatic for the cotriple cohomology; hence this is a necessary condition if the two theories are to be equivalent. This is where the problem arises for Harrison's commutative cohomology in finite characteristic. Michel André showed that in characteristic p there was a non-zero cohomology class in degree $2p-1$.

Here is the argument for acyclicity in the case of free groups. We begin by observing that a free group is also free as far as derivations is concerned.

Proposition 10.3 *Suppose that Π is free on basis X and M is a Π-module. Then any function $\tau : X \longrightarrow M$ extends to a unique derivation $\Pi \longrightarrow M$.*

Proof Let $U : \mathsf{Grp} \longrightarrow \mathsf{Set}$ denote the underlying functor. This result follows from the sequence of isomorphisms

$$\begin{aligned} \mathrm{Der}(\Pi, M) &\cong \mathrm{Hom}_{\mathsf{Grp}/\pi}(\Pi \to \pi, M \rtimes \pi \to \pi) \\ &\cong \mathrm{Hom}_{\mathsf{Set}/U\pi}(X \to U\pi, UM \times U\pi \to U\pi) \\ &\cong \mathrm{Hom}_{\mathsf{Set}}(X, UM) \end{aligned}$$

□

This implies that $\mathrm{Diff}(\Pi)$ is the free π-module generated by S.

It is not hard to show that $C_\bullet$ is an exact chain complex and hence $C_\bullet(\Pi)$ is a free resolution of $\mathrm{Diff}(\Pi)$. In the case that Π is free, this is then a free resolution of a free module and hence necessarily split. However, we would rather get the extra information available if we know that the splitting is natural, namely that we then get a homotopy equivalence between the two chain complex functors.

We start by defining a homomorphism $\partial : C_0(\Pi) \longrightarrow \mathrm{Diff}(\Pi)$. There is a function $\tau : X \longrightarrow \mathrm{Diff}(\Pi)$ which is the inclusion of the basis. This extends to a derivation $\tau : \Pi \longrightarrow \mathrm{Diff}(\Pi)$ as above. Since $C_0(\Pi)$ is freely generated by the elements of Π, this derivation τ extends to a π-linear function $\partial : C_0(\Pi) \longrightarrow \mathrm{Diff}(\Pi)$. In accordance with the recipe above, ∂ is defined on elements of Π recursively as follows. We will denote by $\langle w \rangle$ the basis element of $C_0(\Pi)$ corresponding to $w \in \Pi$. As above, either $w = 1$ or $w = xv$ or $w = x^{-1}v$ for some $x \in X$ and some $v \in \Pi$ shorter than w. Then

$$\partial\langle w \rangle = \begin{cases} x\partial\langle v \rangle + x & \text{if } w = xv \\ x^{-1}\partial\langle v \rangle - x^{-1}x & \text{if } w = x^{-1}v \\ 0 & \text{if } w = 1 \end{cases}$$

Now define $s : \mathrm{Diff}(\Pi) \longrightarrow C_0(\Pi)$ to be the unique π-linear map such that $s(dx) = \langle x \rangle$ for $x \in X$. Since $\mathrm{Diff}(\Pi)$ is freely generated by all dx for $x \in X$, this does define a unique homomorphism. For $x \in X$, we have that $\partial \circ s(dx) = \partial\langle x \rangle = dx$ and so $\partial \circ s = \mathrm{id}$.

For each $n \geq 0$ we define a homomorphism $s : C_n \longrightarrow C_{n+1}$ as follows. We know that C_n is the free π-module generated by Π^{n+1}. We will denote a generator by $\langle w_0, \cdots, w_n \rangle$ where $w_0, \ldots, w_n$ are words in elements of X and their inverses. Then we define $s : C_n \longrightarrow C_{n+1}$ by induction on the length of the first word:

$$s\langle w_0, \ldots, w_n \rangle = \begin{cases} xs\langle w, w_1, \ldots, w_n \rangle - \langle x, w, w_1, \ldots \rangle & \text{if } w_0 = xw \\ x^{-1}s\langle w, w_1, \ldots, w_n \rangle + x^{-1}\langle x, w_0, w_1, \ldots \rangle & \text{if } w_0 = x^{-1}w \\ \langle 1, 1, w_1, \ldots, w_n \rangle & \text{if } w_0 = 1 \end{cases}$$

Proposition 10.4 *For any word w and any $x \in X$*

$$s\langle xw, w_1, \ldots, w_n \rangle = xs\langle w, w_1, \ldots, w_n \rangle - \langle x, w, w_1, \ldots, w_n \rangle$$
$$s\langle x^{-1}w, w_1, \ldots, w_n \rangle = x^{-1}s\langle w, w_1, \ldots, w_n \rangle + \langle x^{-1}, w_0, w_1, \ldots, w_n \rangle$$

Proof These are just the recursive definitions unless w begins with x^{-1} for the first equation or with x for the second. Suppose $w = x^{-1}v$. Then from the definition of s,

$$s\langle w, w_1 \ldots, w_n \rangle = x^{-1}s\langle v, w_1, \ldots, w_n \rangle + x^{-1}\langle x, w, w_1, \ldots, w_n \rangle$$

so that

$$s\langle xw, w_1, \ldots, w_n \rangle = s\langle v, w_1, \ldots, w_n \rangle$$
$$= xs\langle w, w_1, \ldots, w_n \rangle - \langle x, w, w_1, \ldots, w_n \rangle$$

The second one is proved similarly. □

Now we can prove that s is a contraction. First we will do this in dimension 0, then, by way of example, in dimension 2; nothing significant changes in any higher dimension. In dimension 0, suppose w is a word and we suppose that for any shorter word v, we have that $s \circ \partial\langle v \rangle + \partial \circ s\langle v \rangle = \langle v \rangle$. If $x = 1$, then

$$s \circ \partial\langle 1 \rangle + \partial \circ s\langle 1 \rangle = \partial\langle 1, 1 \rangle = 1\langle 1 \rangle - \langle 1 \rangle + \langle 1 \rangle = \langle 1 \rangle$$

If $w = xv$, with $x \in X$, then

$$\partial \circ s\langle w \rangle + s \circ \partial\langle w \rangle = \partial(xs\langle v \rangle - \partial\langle x, v \rangle) + s(dw)$$
$$= x\partial \circ s\langle v \rangle - x\langle v \rangle + \langle xv \rangle - \langle x \rangle + s(x\partial(v) + dx)$$
$$= \langle w \rangle + x(\partial \circ s + s \circ \partial - 1)\langle v \rangle - \langle x \rangle + \langle x \rangle = \langle w \rangle$$

A similar argument takes care of the case that $w = x^{-1}v$. In dimension 2, the chain group $C_2(\Pi)$ is freely generated by Π^3. If we denote a generator by $\langle w_0, w_1, w_2 \rangle$, we argue by induction on the length of w_0. If $w_0 = 1$, then

$$s \circ \partial\langle 1, w_1, w_2 \rangle = s(\langle w_1, w_2 \rangle - \langle w_1, w_2 \rangle + \langle 1, w_1w_2 \rangle - \langle 1, w_1 \rangle)$$
$$= \langle 1, 1, w_1w_2 \rangle - \langle 1, 1, w_1 \rangle$$

while

$$\partial \circ s\langle 1, w_1, w_2 \rangle = \partial(\langle 1, 1, w_1, w_2 \rangle)$$
$$= \langle 1, w_1, w_2 \rangle - \langle 1, w_1, w_2 \rangle + \langle 1, w_1, w_2 \rangle - \langle 1, 1, w_1w_2 \rangle + \langle 1, 1, w_1 \rangle$$

and these add up to $\langle 1, w_1, w_2\rangle$. Assume that $(\partial \circ s + s \circ \partial)\langle w\rangle = \langle w\rangle$ when w is shorter than w_0. Then for $w_0 = xw$,

$$\begin{aligned}\partial \circ s\langle xw, w_1, w_2\rangle &= x\partial \circ s\langle w, w_1, w_2\rangle - \partial\langle x, w, w_1, w_2\rangle \\ &= x\partial \circ s\langle w, w_1, w_2\rangle - x\langle w, w_1, w_2\rangle + \langle xw, w_1, w_2\rangle \\ &\quad - \langle x, ww_1, w_2\rangle + \langle x, w, w_1w_2\rangle - \langle x, w, w_1\rangle\end{aligned}$$

while

$$\begin{aligned}s \circ \partial\langle x, w, w_1, w_2\rangle &= xws\langle w_1, w_2\rangle - s\langle xww_1, w_2\rangle + s\langle xw, w_1w_2\rangle - s\langle xw, w_1\rangle \\ &= xws\langle w_1, w_2\rangle - xs\langle ww_1, w_2\rangle + \langle x, ww_1, w_2\rangle \\ &\quad + xs\langle w, w_1w_2\rangle - \langle x, w, w_1w_2\rangle - xs\langle w, w_1\rangle + \langle x, w, w_1\rangle \\ &= xs \circ \partial\langle w, w_1, w_2\rangle + \langle x, ww_1, w_2\rangle - \langle x, w, w_1w_2\rangle + \langle x, w, w_1\rangle\end{aligned}$$

Then,

$$\begin{aligned}(\partial \circ s + s \circ \partial)\langle xw, w_1, w_2\rangle &= x(\partial \circ s + s \circ \partial)\langle w, w_1, w_2\rangle - x\langle w, w_1, w_2\rangle \\ &\quad + \langle xw, w_1, w_2\rangle - \langle x, ww_1, w_2\rangle + \langle x, w, w_1w_2\rangle \\ &\quad - \langle x, w, w_1\rangle + \langle x, ww_1, w_2\rangle - \langle x, w, w_1w_2\rangle + \langle x, w, w_1\rangle\end{aligned}$$

Using the inductive assumption, the first two terms cancel and all the rest cancel in pairs, except for $\langle xw, w_1, w_2\rangle$, which shows that $s \circ \partial + \partial \circ s = 1$ in this case. The second case, that w_0 begins with the inverse of a letter is similar.

This completes the proof of the homotopy equivalence for group cohomology. For associative algebras the argument is quite similar. For Lie algebras and, especially for commutative algebras, it is a good deal more complicated because it is less obvious that the cochain complex of a free algebras splits. In fact, for Harrison's cochain complex, it splits only in characteristic 0 and the equivalence fails in finite characteristic. As a result, the cohomology that results from the cotriple resolution has been seen as primary.

11 Acyclic models now

Besides the acyclic models theorem quoted above, there was a weaker form due to Michel André [André, 1967, 1974] in which the conclusion was the weaker homology isomorphism and one could not infer naturality, at least as it was stated and proved. In the process of trying to settle the naturality question I discovered an acyclic models theorem that included both the version above and André's as special cases, along with at least one other interesting version. I outline the definitions and theorems. For proofs, I refer to my recent book [Barr, 2002].

In this definition, $\mathscr{C} = \mathsf{CC}(\mathscr{A})$ is the category of chain complexes of an abelian category $\mathscr{A}$.

11.1 Acyclic classes. A class Γ of objects of $\mathscr{C}$ will be called an *acyclic class* provided:

AC–1. The 0 complex is in Γ.
AC–2. The complex $C_\bullet$ belongs to Γ if and only if $SC_\bullet$ does.
AC–3. If the complexes $K_\bullet$ and $L_\bullet$ are homotopic and $K_\bullet \in \Gamma$, then $L_\bullet \in \Gamma$.

AC–4. Every complex in Γ is acyclic.

AC–5. If $K_{\bullet\bullet}$ is a double complex, all of whose rows are in Γ, then the total complex of $C_\bullet$ belongs to Γ.

Given an acyclic class Γ, let Σ denote the class of arrows f whose mapping cone is in Γ. It can be shown that this class lies between the class of homotopy equivalences and that of homology equivalences.

Suppose that $G : \mathscr{X} \longrightarrow \mathscr{X}$ is an endofunctor and that $\epsilon : G \longrightarrow \mathrm{Id}$ is a natural transformation. If $F : \mathscr{X} \longrightarrow \mathscr{A}$ is a functor, we define an augmented chain complex functor we will denote $FG^{\bullet+1} \longrightarrow F$ as the functor that has FG^{n+1} in degree n, for $n \geq -1$. Let $\partial^i = FG^i \epsilon G^{n-i} : FG^{n+1} \longrightarrow FG^n$. Then the boundary operator is $\partial = \sum_{i=0}^{n}(-1)^i \partial^i$. If, as usually happens in practice, G and ϵ are 2/3 of a cotriple, then this chain complex is the chain complex associated to a simplicial set built using the comultiplication δ to define the degeneracies. Next suppose that $K_\bullet \longrightarrow K_{-1}$ is an augmented chain complex functor. Then there is a double chain complex functor that has in bidegree (n, m) the term $K_n G^{m+1}$. This will actually commute since

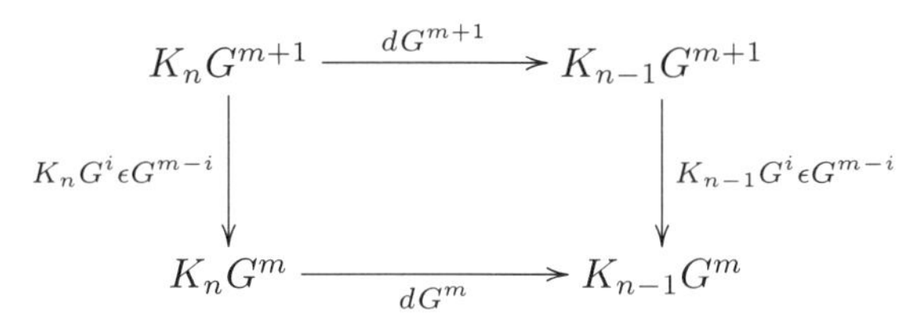

commutes by naturality for $0 \leq i \leq m$ hence so does

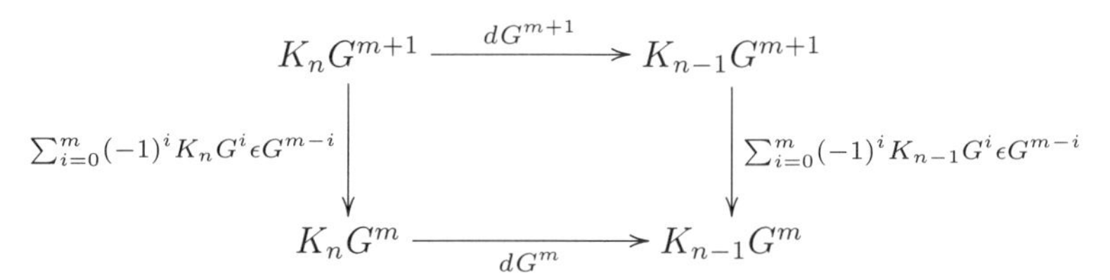

However, the usual trick of negating every second column produces an anticommuting double complex.

This is augmented in both directions, once, using ϵ, over the single complex $K_\bullet$ and second, using the augmentation of $K_\bullet$, over the complex $K_{-1}G^{\bullet+1}$. We say that $K_\bullet$ is *ϵ-presentable with respect to* Γ if for each $n \geq 0$, the augmented chain complex $K_n G^{\bullet+1} \longrightarrow K_n \longrightarrow 0$ belongs to Γ. We say that $K_\bullet$ is *G-acyclic with respect to* Γ if the augmented complex $K_\bullet G \longrightarrow K_{-1}G \longrightarrow 0$ belongs to Γ.

It could be notes that an augmented chain complex $C_\bullet \longrightarrow C_{-1} \longrightarrow 0$ is the desuspension of the mapping cone of the map of chain complexes $C_\bullet \longrightarrow C_{-1}$ in which the latter is considered as a chain complex with C_{-1} in degree 0 and 0 in all other degrees. Thus an equivalent formulation of the two definitions above is that $K_\bullet$ is *ϵ-presentable with respect to* Γ if for each $n \geq 0$, the chain map $K_n G^{\bullet+1} \longrightarrow K_n$ belongs to Σ and that $K_\bullet$ is *G-acyclic with respect to* Γ if the chain map $K_\bullet G \longrightarrow K_{-1}G \longrightarrow 0$ belongs to Σ.

Theorem 11.1 *Let Γ be an acyclic class and Σ be the associated class of arrows. Suppose $\alpha : K_\bullet \longrightarrow K_{-1}$ and $\beta : L_\bullet \longrightarrow L_{-1}$ are augmented chain complex functors. Suppose G is an endofunctor on $\mathscr{X}$ and $\epsilon : G \longrightarrow \mathrm{Id}$ a natural transformation for which $K_\bullet$ is ϵ-presentable and $L_\bullet \longrightarrow L_{-1} \longrightarrow 0$ is G-acyclic,* *both with respect to* Γ. *Then given any natural transformation* $f_{-1} : K_{-1} \longrightarrow L_{-1}$ *there is, in* $\Sigma^{-1}\mathscr{C}$, *a unique arrow* $f_\bullet : K_\bullet \longrightarrow L_\bullet$ *that extends* f_{-1}.

Although we give no proof, here is the "magic diagram" from which it all follows:

$$\begin{array}{ccccc}
K_{-1}G^{\bullet+1} & \xleftarrow{\alpha G^{\bullet+1}} & K_\bullet G^{\bullet+1} & \xrightarrow{K_\bullet \epsilon} & K_\bullet \\
{\scriptstyle f_{-1}G^{\bullet+1}}\Big\downarrow & & & & \\
L_{-1}G^{\bullet+1} & \xleftarrow[\beta G^{\bullet+1}]{} & L_\bullet G^{\bullet+1} & \xrightarrow[L_\bullet \epsilon]{} & L_\bullet
\end{array}$$

Note that each of the two hypotheses implies that one of the "wrong way" arrows belongs to Γ. When these are inverted, you get a map $K_\bullet \longrightarrow L_\bullet$.

11.2 Examples. We mention three examples of acyclic classes.

Let Γ be the class of contractible complexes. In that case, Σ is the class of homotopy equivalences. It is trivial to show that when the rows and columns of a double complex are contractible so is the total complex. It is a little more surprising—but still true—that if the rows *or* the columns are contractible, so is the total complex. The only other point to make is that in this case, the functor $\mathscr{C} \longrightarrow \Sigma^{-1}\mathscr{C}$ is surjective on arrows, so that you actually get a homotopy equivalence as conclusion.

Let Γ be the class of acyclic complexes. In that case, Σ is the class of chain maps that induce an isomorphism on homology (called homology isomorphisms). One can use a trivial spectral sequence argument to show that if all the rows or all the columns of a double complex are acyclic, then so is the total complex, but it is not hard to give a direct argument using a filtration. Since the functor $\mathscr{C} \longrightarrow \Sigma^{-1}\mathscr{C}$ is not surjective on arrows, the arrows you get are not induced by arrows between the chain complexes, but they are natural.

Let Γ be the class of chain complexes that are, at each object of $\mathscr{X}$, contractible, but not naturally so. In that case, Σ consists of arrows that induce, at each X, homotopy equivalences. This means that the arrows are natural, but the homotopy inverse and the homotopies involved are not necessarily natural. This situation arises quite naturally in topology. For example, on the category of C^∞ manifolds, the inclusion of the C^∞ chains into all the chains can be shown to induce a homotopy equivalence on any space, but there is no obviously natural way of doing this. This example answered a question raised by Rob Milson who was working with these homology groups.

References

[1] M. André (1967), *Méthode simpliciale en algèbre homologique et algèbre commutative.* Lecture Notes in Mathematics **43**, Springer-Verlag, Berlin, Heidelberg, New York.

[2] M. André (1974), *Homologie des algèbres commutatives.* Springer-Verlag, Berlin, Heidelberg, New York.

[3] H. Appelgate (1965), *Categories with models.* Dissertation, Columbia University.

[4] R. Baer (1934), *Erweiterungen von Gruppen und ihren Isomorphismem.* Math. Zeit. **38**, 375–416.

[5] M. Barr (1962), *On the cohomology of commutative algebras.* Dissertation, University of Pennsylvania.

[6] M. Barr (1965a), *A cohomology theory for commutative algebras, I.* Proc. Amer. Math. Soc. **16**, 1379–1384.

[7] M. Barr (1965b), *A cohomology theory for commutative algebras, II.* Proc. Amer. Math. Soc. **16**, 1385–1391.
[8] M. Barr (1966), *Cohomology in tensored categories. Proceedings of the Conference on Categorical Algebra*, Springer-Verlag, Berlin, Heidelberg, New York, 344–354.
[9] M. Barr (1967), *Shukla cohomology and triples.* J. Algebra **5**, 222–231.
[10] M. Barr (1968a), *A note on commutative algebra cohomology.* Bull. Amer. Math. Soc. **74**, 310–313.
[11] M. Barr (1968b), *Harrison homology, Hochschild homology and triples.* J. Algebra, **8**, 314–323.
[12] M. Barr (1969), Cohomology and obstructions: commutative algebras. In B. Eckmann, ed. *Seminar on Triples and Categorical Homology Theory*, Lecture Notes in Mathematics **80**, 357–375, Springer-Verlag, Berlin, Heidelberg, New York.
[13] M. Barr (1996a), *Cartan-Eilenberg cohomology and triples.* J. Pure Applied Algebra, **112**, 219–238.
[14] M. Barr (1996b), *Acyclic models.* Canadian J. Math., **48**, 258-273.
[15] M. Barr (2002), *Acyclic Models.* Centre de Recherche Mathématiques.
[16] M. Barr (forthcoming), *Resolutions and derived functors.* ftp.math.mcgill.ca/pub/barr/derfun.{ps,pdf}.
[17] M. Barr and J. Beck (1966), *Acyclic models and triples.* Proceedings of the Conference on Categorical Algebra, Springer, 336–343.
[18] M. Barr and J. Beck (1969), Homology and standard constructions. In M. Tierney and A. Heller, eds., *Seminar on Triples and Categorical Homology Theory*, Lecture Notes in Mathematics **80**, Springer-Verlag, Berlin, Heidelberg, New York, 245–535.
[19] M. Barr and M.-A. Knus (1971), *Extensions of Derivations.* Proc. Amer. Math. Soc. **28**, 313–314.
[20] M. Barr, and G. S. Rinehart (1966), *Cohomology as the derived functor of derivations.* Trans. Amer. Math. Soc., **122**, 416–426.
[21] D. Buchsbaum (1956), *Exact categories.* Appendix to [Cartan and Eilenberg, 1956].
[22] H. Cartan and S. Eilenberg (1956), *Homological Algebra.* Princeton University Press.
[23] C. Chevalley and S. Eilenberg (1948), *Cohomology theory of Lie groups and Lie algebras.* Trans. Amer. Math. Soc., **63**, 85–124.
[24] S. Eilenberg and S. MacLane (1941), *Infinite cycles and homologies.* Proc. Nat. Acad. Sci. U. S. A., **27**, 535–539.
[25] S. Eilenberg and S. MacLane (1942a), *Natural isomorphisms in group theory.* Proc. Nat. Acad. Sci. U. S. A., **28**, 537–543.
[26] S. Eilenberg and S. MacLane (1942b), *Group extensions and homology.* Ann. of Math., **43**, 757–831.
[27] S. Eilenberg and S. MacLane (1943), *Relations between homology and homotopy groups.* Proc. Nat. Acad. Sci. U. S. A., **29**, 155–158.
[28] S. Eilenberg and S. MacLane (1945a), *General theory of natural equivalences.* Trans. Amer. Math. Soc., **58**, 231–294.
[29] S. Eilenberg and S. MacLane (1945b), *Relations between homology and homotopy groups of spaces.* Ann. of Math., **46**, 480–509.
[30] S. Eilenberg and S. MacLane (1947a), *Cohomology theory in abstract groups, I.* Ann. of Math., **48**, 51–78.
[31] S. Eilenberg and S. MacLane (1947b), *Cohomology theory in abstract groups, II. Group extensions with a non-Abelian kernel.* Ann. of Math., **48**, 326–341.
[32] S. Eilenberg and S. MacLane (1951), *Homology theories for multiplicative systems.* Trans. Amer. Math. Soc., **71**, 294–330.
[33] S. Eilenberg and S. MacLane (1953), *Acyclic models.* Amer. J. Math., **75**, 189–199.
[34] S. Eilenberg and J. C. Moore (1965a), *Foundations of Relative Homological Algebra*, Memoirs Amer. Math. Soc. **55**.
[35] S. Eilenberg and J. C. Moore (1965b), *Adjoint functors and triples* Illinois J. Math., **9**, 381–398.
[36] Murray Gerstenhaber and S. D. Schack (1987), *A Hodge-type decomposition for commutative algebra cohomology.* J. Pure Appl. Algebra **48**, 229–247.
[37] R. Godement (1958), *Topologie algébrique et théorie des faisceaux.* Actualités Sci. Ind. No. 1252. Publ. Math. Univ. Strasbourg, **13**, Hermann, Paris.

[38] A. Grothendieck (1957), *Sur quelques points d'algèbre homologique.* Tohôku Math. Journal **2**, 119–221.

[39] M. Hall (1938), *Group rings and extensions, I.* Ann. of Math. **39**, 220–234.

[40] D. K. Harrison (1962), *Commutative algebras and cohomology.* Trans. Amer. Math. Soc., **104**, 191–204.

[41] G. Hochschild (1945), *On the cohomology groups of an associative algebra.* Ann. of Math. (2), **46**, 58–67.

[42] G. Hochschild (1946), *On the cohomology theory for associative algebras.* Ann. of Math. (2), **47**, 568–579.

[43] H. Hopf (1942, 1943), *Fundamentalgruppe und zweite Bettische Gruppen* Comment. Math. Helv. **14**, 257–309. *Nachtrag,* **15**, 27–32.

[44] H. Hopf (1945), *Ueber die Bettischen Gruppen die zu einem beliebigen Gruppe gehören.* Comment. Math. Helv. **17**, 39–79.

[45] P. Huber (1961), *Homotopy theory in general categories.* Math. Ann., **144**, 361–385.

[46] H. Kleisli (1964), *Comparaison de la résolution simpliciale à la bar-résolution.* In Catégories Non-Abéliennes, L'Université de Montréal, 85–99.

[47] H. Kleisli (1965), *Every standard construction is induced by a pair of adjoint functors.* Proc. Amer. Math. Soc., **16**, 544–546.

[48] S. Mac Lane (1950), *Duality for groups.* Bull. Amer. Math. Soc. **56**, 485–516.

[49] S. Mac Lane (1958), *Extensions and obstructions for rings.* Ill. J. Math. **2**, 316–345.

[50] S. Mac Lane (1963), *Homology.* Springer-Verlag, Berlin, Heidelberg, New York.

[51] U. Shukla (1961), *Cohomologie des algèbres associatives.* Ann. Sci. École Norm. Sup. (3), **78**, 163–209.

[52] A. M. Turing (1938), *The extensions of a group.* Compositio Math. **5**, 357–367.

[53] H. Zassenhaus (1937), *Lehrbuch der Gruppentheorie.* Hamburg Math. Einzelschriften **21**, Leipzig.

Received October 14, 2002; in revised form May 14, 2003

Fields Institute Communications
Volume **43**, 2004

A Survey of Semi-abelian Categories

Francis Borceux
Département de Mathématiques
Université de Louvain
2 chemin du Cyclotron
1348 Louvain-la-Neuve, Belgium
borceux@math.ucl.ac.be

Abstract. We define the notion of semi-abelian category and discuss its basic properties. Such a category is in particular a Mal'cev category. The five lemma, the nine lemma and the snake lemma hold true. Theories of semi-direct products, commutators and centrality can be developed. Every semi-abelian category contains an abelian core.

Introduction

An elementary introduction to the theory of abelian categories culminates generally with the proof of the basic diagram lemmas of homological algebra: the five lemma, the nine lemma, the snake lemma, and so on. This gives evidence of the power of the theory, but leaves the reader with the misleading impression that abelian categories constitute the most natural and general context where these results hold. This is indeed misleading, since all those lemmas are valid as well – for example – in the category of all groups, which is highly non-abelian.

This survey paper intends to give evidence, among other things, that a natural and more general context in which the diagram lemmas are valid is that of a semi-abelian category. More precisely, a finitely cocomplete exact category (for example, an algebraic variety) turns out to be semi-abelian precisely when it admits a zero object and when all the diagram lemmas of homological algebra hold true. The categories of groups, rings without unit, sheaves or presheaves of these, and so on, are semi-abelian. And a category $\mathcal{E}$ is abelian precisely when both $\mathcal{E}$ and its dual $\mathcal{E}^{\mathsf{op}}$ are semi-abelian.

We pay a special attention to the normal subobjects (=the kernels) which suffice, in a semi-abelian category, to characterize all the quotients by an equivalence relation. We prove in particular that a semi-abelian category is a Mal'cev category,

---

2000 *Mathematics Subject Classification.* Primary: 18E10, 18C10, 18G50; Secondary: 08B05, 08C05.

This survey paper has been written upon invitation of the editors of this volume; the author aknowledges support of FNRS grant 1.5.096.02.

that is, a category in which every reflexive relation is at once an equivalence relation. We show also the existence of semi-direct products and commutators in a semi-abelian category, extending so classical results in the case of groups. Finally, we prove that every semi-abelian category contains an abelian full subcategory of "abelian objects". Many of these results do not require the full strength of the axiomatics of semi-abelian categories: but it is a major interest of semi-abelian categories to allow recapturing all these important results from a unique and simple list of axioms.

Of course since more than 50 years, various mathematicians have tried to settle down an axiomatic context forcing the validity of the diagram lemmas of non-abelian homology. If none of these attempts imposed itself as a natural categorical context in which to work, it is probably because of their heavy technical nature, in contrast with the elegance of the theory of abelian categories. We refer the reader to the introduction of [23] for a reliable historical discussion of these pioneering works. It is in this same paper of G. Janelidze, L. Márki and W. Tholen that the definition of a semi-abelian category appears for the first time and is put in relation with the more general notion of protomodular category, due to D. Bourn (see [8]). This is the presentation adopted in this survey paper. Many other results of the paper are borrowed from [23] and subsequent work of D. Bourn, whose ideas are everywhere present in this paper. The presentation is often borrowed from [5].

1 Protomodular and semi-abelian categories

In the category Set of sets, a morphism $p\colon A \longrightarrow I$ can be seen as an I-indexed family of sets $(A_i)_{i\in I}$, simply by putting $A_i = p^{-1}(i)$. When p admits a section s, the pair (p,s) can now be seen as an I-indexed family of pointed sets $(A_i, a_i)_{i\in I}$, by putting further $a_i = s(i) \in A_i$.

More generally, given a category $\mathcal{E}$ and an object $I \in \mathcal{E}$, the category $\mathsf{Pt}_I(\mathcal{E})$ of pointed objects over I has for objects the triples

$$\big(A, p\colon A \to I, s\colon I \to A\big),\ \ p\circ s = \mathsf{id}_I.$$

A morphism

$$f\colon \big(p,s\colon A \leftrightarrows I\big) \longrightarrow \big(p',s'\colon A' \leftrightarrows I\big)$$

is a morphism $f\colon A \longrightarrow A'$ making both triangles commutative, that is, $p'\circ f = p$ and $f\circ s = s'$.

When the category $\mathcal{E}$ admits pullbacks, every morphism $v\colon J \longrightarrow I$ in $\mathcal{E}$ induces by pullback a functor

$$v^*\colon \mathsf{Pt}_I(\mathcal{E}) \longrightarrow \mathsf{Pt}_J(\mathcal{E}).$$

The reader familiar with the theory of fibrations will notice that we have defined a pseudo-functor to the "category" of categories ("pseudo" meaning that $(v\circ w)^*$ is isomorphic to $w^*\circ v^*$, not equal)

$$\mathcal{E} \longrightarrow \mathsf{Cat},\ \ I \mapsto \mathsf{Pt}_I(\mathcal{E}),\ \ v \mapsto v^*,$$

thus a corresponding fibration $\mathsf{Pt}(\mathcal{E}) \longrightarrow \mathcal{E}$ of pointed objects in $\mathcal{E}$ (see [3]). This fibration has been extensively studied by D. Bourn (see [6] and [7]) and has very strong classifying properties. Its most powerful application is certainly the theory of protomodular categories (see [6], [7], [8], [10]), which allows in particular an elegant treatment of normal subobjects in categories without (necessarily) a zero object.

Definition 1.1 A category $\mathcal{E}$ is protomodular when

1. $\mathcal{E}$ admits pullbacks;
2. for every morphism $v\colon J \longrightarrow I$ in $\mathcal{E}$, the pullback functor

$$v^* : \mathsf{Pt}_I(\mathcal{E}) \longrightarrow \mathsf{Pt}_J(\mathcal{E})$$

reflects isomorphisms.

A zero object $\mathbf{0}$ in a category $\mathcal{E}$ is an object which is both initial and terminal. Given objects A, B in $\mathcal{E}$, we write $0\colon A \longrightarrow B$ for the unique morphism factoring through $\mathbf{0}$. The kernel $\mathsf{Ker}\, f$ of an arrow $f\colon A \longrightarrow B$ is the equalizer of f and 0, and dually for the cokernel. Equivalently, the kernel k of f is given by the following pullback:

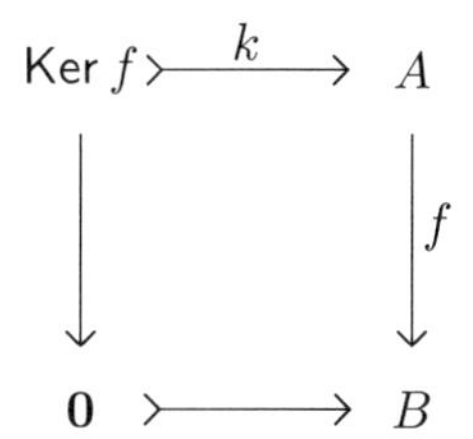

We recall that a category $\mathcal{E}$ with pullbacks and a terminal object is finitely complete (see [4]).

Proposition 1.2 *Let $\mathcal{E}$ be a category with pullbacks and a zero object. The following conditions are equivalent:*

1. *$\mathcal{E}$ is protomodular;*
2. *the split short five lemma holds in $\mathcal{E}$, that is: given a diagram where all squares are commutative*

$$\begin{array}{ccccccccc}
\mathbf{0} & \longrightarrow & A & \stackrel{k}{\rightarrowtail} & B & \underset{p}{\overset{s}{\leftrightarrows}} & C & \longrightarrow & \mathbf{0} \\
 & & {\scriptstyle f}\downarrow{\scriptstyle \cong} & & {\scriptstyle g}\downarrow & & {\scriptstyle \cong}\downarrow{\scriptstyle h} & & \\
\mathbf{0} & \longrightarrow & A' & \stackrel{k'}{\rightarrowtail} & B' & \underset{p'}{\overset{s'}{\leftrightarrows}} & C' & \longrightarrow & \mathbf{0}
\end{array}$$

$$k' \circ f = g \circ k, \; p' \circ g = h \circ p, \; g \circ s = s' \circ h,$$

and where moreover

$$p \circ s = \mathsf{id}_C, \; p' \circ s' = \mathsf{id}_{C'}, \; k = \mathsf{Ker}\, p, \; k' = \mathsf{Ker}\, p',$$

if f and h are isomorphisms, then g is an isomorphism as well.

Proof The category $\mathsf{Pt}_{\mathbf{0}}(\mathcal{E})$ is isomorphic to the category $\mathcal{E}$. Pulling back an arrow along the morphism $0_I\colon \mathbf{0} \longrightarrow I$ is taking its kernel. Thus the split short five lemma means that the functor

$$0_I^*\colon \mathsf{Pt}_I(\mathcal{E}) \longrightarrow \mathsf{Pt}_{\mathbf{0}}(\mathcal{E}) \cong \mathcal{E}$$

reflects isomorphisms. This proves already (1) $\Rightarrow$ (2). Conversely, given an arbitrary morphism $v\colon J \longrightarrow I$, the equality $v \circ 0_J = 0_I$ implies $0_J^* \circ v^* = 0_I^*$. Since 0_J^* preserves isomorphisms and 0_I^* reflects them, v^* reflects isomorphisms. $\square$

Anticipating on 4.3, it should be observed that in opposition to the usual formulation of the short five lemma which involves normal epimorphisms p, p' and their kernels k, k', the split short five lemma is a statement involving only finite limits.

A category $\mathcal{E}$ is regular when it admits pullbacks, coequalizers of kernel pairs (= the two projections of the pullback of an arrow with itself), and when every pullback of a regular epimorphism (= a coequalizer) is still a regular epimorphism. Obviously, every regular epimorphism is then the coequalizer of its kernel pair. The category $\mathcal{E}$ is exact when it is regular and every equivalence relation in $\mathcal{E}$ is a kernel pair (see [1]).

The following definition is borrowed from G. Janelidze, L. Márki and W. Tholen (see [23]):

Definition 1.3 A category $\mathcal{E}$ is semi-abelian when

1. $\mathcal{E}$ has a zero object $\mathbf{0}$;
2. $\mathcal{E}$ has binary coproducts $A \amalg B$;
3. $\mathcal{E}$ is exact;
4. $\mathcal{E}$ is protomodular.

A semi-abelian category is thus in particular finitely complete. As the terminology suggests, the first example must be (see also 7.6 for an even more convincing result):

Example 1.4 Every abelian category is semi-abelian.

Proof An abelian category is additive, exact, finitely complete and finitely cocomplete and satisfies the short five lemma (see [19]). One concludes by 1.2. □

To produce a whole bunch of other examples, let us characterize the semi-abelian varieties of universal algebra (or equivalently, the semi-abelian Lawvere-algebraic categories). This characterization is due to D. Bourn and G. Janelidze (see [15]).

Theorem 1.5 *Let $\mathcal{E}$ be an algebraic variety. The following conditions are equivalent:*

1. *$\mathcal{E}$ is semi-abelian;*
2. *$\mathcal{E}$ is protomodular with a zero object;*
3. *the corresponding theory contains:*
   - *a unique constant 0;*
   - *n binary operations $\alpha_i(x, y)$ such that $\alpha_i(x, x) = 0$;*
   - *an $(n+1)$-ary operation θ such that $\theta\big(\alpha_1(x, y), \ldots, \alpha_n(x, y), y\big) = x$.*

Proof An algebraic variety is always complete, cocomplete and exact, proving (1) ⇔ (2). Moreover, the existence of a unique constant in the theory is equivalent to the existence of a zero object in $\mathcal{E}$ (see [26]).

Let us first assume that $\mathcal{E}$ is semi-abelian. Writing F for the free algebra functor, consider the following diagram

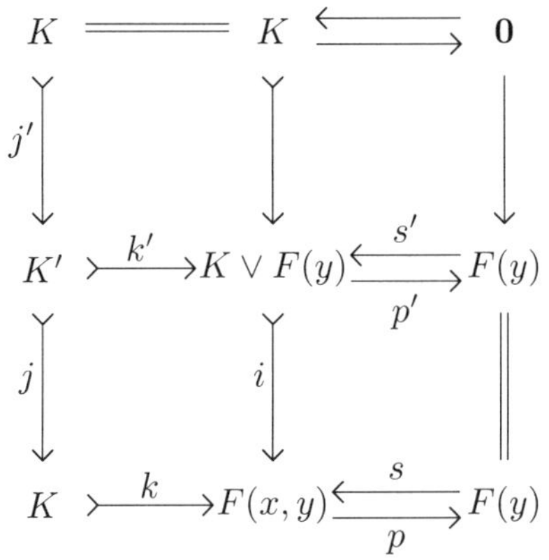

$$
\begin{array}{ccccc}
K & = & K & \underset{\longrightarrow}{\overset{\longleftarrow}{}} & \mathbf{0} \\
{\scriptstyle j'}\downarrow & & \downarrow & & \downarrow \\
K' & \overset{k'}{\rightarrowtail} & K \vee F(y) & \underset{p'}{\overset{s'}{\rightleftarrows}} & F(y) \\
{\scriptstyle j}\downarrow & & \downarrow{\scriptstyle i} & & \| \\
K & \overset{k}{\rightarrowtail} & F(x,y) & \underset{p}{\overset{s}{\rightleftarrows}} & F(y)
\end{array}
$$

where $p(x) = y = p(y)$, $s(y) = y$ and $k = \mathsf{Ker}\, p$. Consider further the union $K \vee F(y)$ in the variety; p' is the restriction of p which is thus zero on K, while s factors through $F(y)$, yielding the section s'. Put $k' = \mathsf{Ker}\, p'$ and consider the corresponding factorization j. Complete this commutative diagram with the first line, where obviously $\mathsf{id}_K = \mathsf{Ker}\, 0$; this yields the factorization j'. Since k is a monomorphism, $j \circ j' = \mathsf{id}_K$ and the monomorphism j is an isomorphism. By the split short five lemma (see 1.2), the inclusion i is an isomorphism as well.

Since $F(x,y) = K \vee F(y)$, we get $x \in K \vee F(y)$. This proves the existence of n elements $\alpha_i(x,y) \in K$ and a $(n+1)$-ary operation θ such that

$$x = \theta\big(\alpha_1(x,y), \ldots, \alpha_n(x,y), y\big).$$

It remains to notice that by definition of K, $\alpha_i(x,x) = 0$ for each index i.

Conversely, let us prove first that given two elements a, b in an algebra A

$$\big(\forall i\ \alpha_i(a,b) = 0\big) \Rightarrow \big(a = b\big).$$

Indeed

$$b = \theta(\alpha_1(b,b), \ldots, \alpha_n(b,b), b\big) = \theta(0, \ldots, 0, b) = \theta(\alpha_1(a,b), \ldots, \alpha_n(a,b), b\big) = a.$$

It remains to prove the split short five lemma; we use the notation of 1.2. For the injectivity of g, consider $x, y \in B$ such that $g(x) = g(y)$. This yields

$$(h \circ p)\big(\alpha_i(x,y)\big) = (p' \circ g)\big(\alpha_i(x,y)\big) = p'\Big(\alpha_i\big(g(x), g(y)\big)\Big) = p'(0) = 0.$$

Since h is injective, $p\big(\alpha_i(x,y)\big) = 0$ and viewing k as a canonical inclusion, this yields $\alpha_i(x,y) \in A$. But then,

$$(k' \circ f)\big(\alpha_i(x,y)\big) = (g \circ k)\big(\alpha_i(x,y)\big) = \alpha_i\big(g(x), g(y)\big) = 0.$$

Since k' and f are injective, $\alpha_i(x,y) = 0$. This implies $x = y$, thus the injectivity of g.

For the surjectivity, choose now $x \in B'$ and put $y = (s \circ h^{-1} \circ p')(x) \in B$. This yields

$$p'\Big(\alpha_i\big(x, g(y)\big)\Big) = \alpha_i\big(p'(x), (p' \circ g)(y)\big) = \alpha_i\big(p'(x), p'(x)\big) = 0.$$

This proves the existence of elements $z_i \in A'$ such that $k'(z_i) = \alpha_i\big(x, g(y)\big)$. Viewing still k as a canonical inclusion, this forces

$$\begin{aligned} g\Big(\theta\big(f^{-1}(z_1),\dots,f^{-1}(z_n),y\big)\Big) &= \theta\big(k'(z_1),\dots,k'(z_n),g(y)\big) \\ &= \theta\Big(\alpha_1\big(x,g(y)\big),\dots,\alpha_n\big(x,g(y)\big),g(y)\Big) \\ &= x \end{aligned}$$

and proves the surjectivity of g. □

It should be noticed that condition 3 of theorem 1.5 appeared already in [33] and later in the appendix of [34]. To my best knowledge, it has been used only to investigate properties of congruences.

The following corollary emphasizes the case $n = 1$ in the characterization given by theorem 1.5. This choice $n = 1$ cannot always be done, as proved in [25].

Corollary 1.6 *Let $\mathbb{T}$ be an algebraic theory which possesses a unique constant 0 and two binary operations $x + y$ and $x - y$ which satisfy the axioms*

$$(x - y) + y = x, \ \ x - x = 0.$$

The corresponding variety $\mathcal{V}$ is semi-abelian. In particular, when the theory $\mathbb{T}$ contains a group operation, the variety $\mathcal{V}$ is semi-abelian: this contains the cases of groups, Ω-groups, rings without unit, R-algebras, and so on.

Proof Put $n = 1$, $\alpha_1(x, y) = x - y$ and $\theta(x, y) = x + y$ in 1.5. □

Example 1.7 Let $\mathbb{T}$ be an algebraic theory as in theorem 1.5. The models of $\mathbb{T}$ in every Grothendieck topos constitute a semi-abelian category. When moreover the theory $\mathbb{T}$ admits a finite presentation, its models in every topos with Natural Number Object still constitute a semi-abelian category.

Proof When $\mathbb{T}$ admits a finite presentation, its models in an elementary topos $\mathcal{E}$ with Natural Number Object constitute a finitely complete and finitely cocomplete exact category, monadic over $\mathcal{E}$ (see [24], section D.5.3). This takes care of axioms 1, 2, 3 in definition 1.3. This allows also repeating the proof of the split short five lemma, as in 1.5, in the internal logic of the topos $\mathcal{E}$. The assumption of finite presentability can be dropped as soon as the topos is finitely complete and cocomplete. □

An abelian category admits a full and faithful exact embedding in a category of modules, yielding a highly useful corresponding metatheorem which allows developing most proofs in terms of elements. Up to now, no such theorem has been proved for semi-abelian categories. Nevertheless, Barr's metatheorem for regular categories (see [1]) can already be adapted to the present context.

Metatheorem 1.8 *Let $\mathcal{P}$ be a statement of the form $\varphi \Rightarrow \psi$, where φ and ψ can be expressed as conjunctions of properties in the following list:*

1. *some arrow is a zero arrow;*
2. *some finite diagram is commutative;*
3. *some morphism is a monomorphism;*
4. *some morphism is a regular epimorphism;*
5. *some morphism is an isomorphism;*
6. *some finite diagram is a limit diagram;*

7. *some arrow* $f\colon A \longrightarrow B$ *factors through some monomorphism* $s\colon S \rightarrowtail B$.

If this statement $\mathcal{P}$ *is valid in the category* Set_* *of pointed sets, it is valid in every regular category* $\mathcal{E}$ *with a zero object, thus in particular in every semi-abelian category.*

Proof If $\mathcal{E}$ is a finitely complete regular category, Barr's theorem (see [1]) indicates the existence of a full and faithful embedding

$$Z\colon \mathcal{E} \longrightarrow [\mathcal{C}^{\mathsf{op}}, \mathsf{Set}]$$

in a category of presheaves, where Z preserves and reflects finite limits and regular epimorphisms.

When $\mathcal{E}$ has a zero object, $Z(\mathbf{0})$ is the constant functor on the singleton. The natural transformation $Z(\mathbf{0}) \Rightarrow Z(A)$ defines a base point in each set $Z(A)(X)$, for all $X \in \mathcal{C}$. Thus each functor $Z(A)$ factors naturally through the category Set_* of pointed sets. This proves that the Barr embedding factors as

$$Z'\colon \mathcal{E} \longrightarrow [\mathcal{C}^{\mathsf{op}}, \mathsf{Set}_*].$$

This functor Z' is faithful since so is Z. It is also full since every natural transformation over Set_* is in particular a natural transformation over Set. Since the forgetful functor $\mathsf{Set}_* \longrightarrow \mathsf{Set}$ preserves and reflects finite limits and regular epimorphisms, Z' preserves and reflects finite limits and regular epimorphisms, since so does Z. And by construction, Z' preserves and reflects zero morphisms.

Since Z is full and faithful, it preserves and reflects isomorphisms and the commutativity of diagrams. A morphism f is a monomorphism when the kernel pair (u, v) of f is a pair of isomorphisms. Condition 7 means that the pullback of s along f is an isomorphism. So if one of the statements 1 to 7 is valid in the category of pointed sets, it is valid pointwise in $[\mathcal{C}^{\mathsf{op}}, \mathsf{Set}_*]$ and thus holds true in $\mathcal{E}$. $\square$

We shall implicitly use this metatheorem in several proofs, when a property falls under its scope and is trivial in the case of pointed sets via an elementary "chase on the diagram".

2 Subobjects, quotients and pullbacks

First, we show that in a semi-abelian category, kernels and cokernels behave like in abelian categories.

Proposition 2.1 *In a semi-abelian category* $\mathcal{E}$*, pulling back along an arrow reflects monomorphisms.*

Proof Consider the following commutative diagram

$$\begin{array}{ccccc}
K[f'] & \underset{p'_1}{\overset{\Delta'}{\leftrightarrows}} & A' & \overset{f'}{\rightarrowtail} & X' \\
{\scriptstyle\gamma}\downarrow & & \downarrow{\scriptstyle\beta} & & \downarrow{\scriptstyle\alpha} \\
K[f] & \underset{p_1}{\overset{\Delta}{\leftrightarrows}} & A & \underset{f}{\longrightarrow} & X
\end{array}$$

where $K[f]$ is the kernel pair of f with first projection p_1 and diagonal Δ; analogously for f'. When the right hand square is a pullback, so is the left hand square involving the projections. When f' is a monomorphism, $p'_1 = \mathsf{id}_{A'} = \Delta'$, yielding the isomorphism

$$\beta^*(p_1, \Delta\colon K[f] \leftrightarrows A) \cong (\mathsf{id}_{A'}, \mathsf{id}_{A'}\colon A' \leftrightarrows A') \cong \beta^*(\mathsf{id}_A, \mathsf{id}_A\colon A \leftrightarrows A)$$

between pointed objects. By protomodularity, (p_1, Δ) is isomorphic to $(\mathsf{id}_A, \mathsf{id}_A)$ and p_1 is an isomorphism. Thus f is a monomorphism. □

Theorem 2.2 *In a semi-abelian category $\mathcal{E}$, the following conditions are equivalent for a morphism $f\colon A \longrightarrow B$:*

1. *f is a monomorphism:*
2. $\mathsf{Ker}\, f = 0$.

Proof Computing the kernel is pulling back along $0\colon \mathbf{0} \longrightarrow B$; one concludes by 2.1. □

The dual result does not hold in a semi-abelian category. Of course an epimorphism has zero cokernel, but the converse is not true, not even in the category of groups. If G is a non trivial simple group with non trivial subgroup H, the cokernel of the inclusion $i\colon H \rightarrowtail G$ must be zero while i is not surjective.

We recall that in a category with finite limits, a family $(f_i\colon X_i \longrightarrow Y)_{i\in I}$ of morphisms is strongly epimorphic (and in particular epimorphic) when it does not factor through any proper subobject of Y. In a finitely complete regular category, strong epimorphisms coincide with regular ones (see [4]).

Lemma 2.3 *In a semi-abelian category $\mathcal{E}$, given a pullback diagram $p \circ u = v \circ q$ and a section $p \circ s = \mathsf{id}_Y$,*

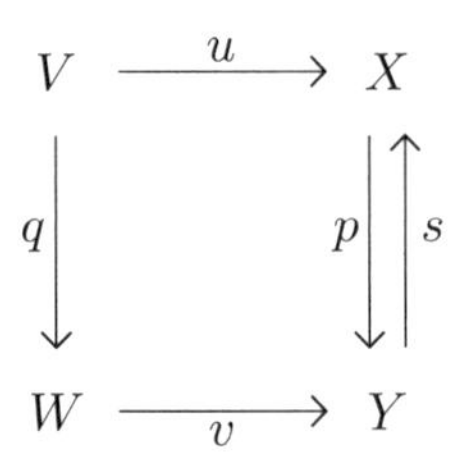

the pair (u, s) is strongly epimorphic.

Proof Consider the diagram

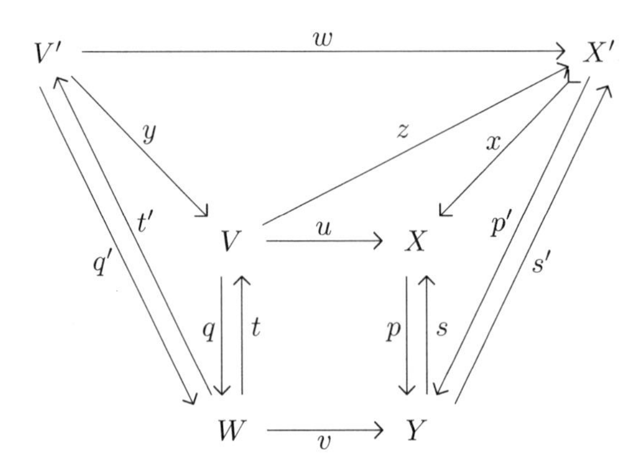

where t is the factorization of the pair $(\mathsf{id}_W, s \circ v)$ through the pullback. If u and s factor through some monomorphism x as $u = x \circ z$ and $s = x \circ s'$, we put $p' = p \circ x$ and view x as a morphism in $\mathsf{Pt}_Y(\mathcal{E})$. Pulling back this situation along v yields a corresponding morphism

$$y\colon (q', t'\colon V' \leftrightarrows W) \longrightarrow (q, t\colon V \leftrightarrows W)$$

in $\mathsf{Pt}_W(\mathcal{E})$. Since the downward directed square and the downward directed outer trapezium are pullbacks, the upper trapezium $u \circ y = x \circ w$ is a pullback as well, that is, $V' = u^{-1}(X')$. But since u factors through the subobject X', $u^{-1}(X') = V$ and y is an isomorphism. By protomodularity, x is an isomorphism as well. □

Corollary 2.4 *Every semi-abelian category $\mathcal{E}$ is unital, meaning that given two objects A and B, the pair*

$$A \xrightarrow{(\mathsf{id}_A, 0)} A \times B \xleftarrow{(0, \mathsf{id}_B)} B$$

is strongly epimorphic.

Proof In 2.3, choose $V = A$, $Y = B$, $X = A \times B$, $W = \mathbf{0}$, $u = (\mathsf{id}_A, 0)$, $p = p_B$, $s = (0, \mathsf{id}_B)$. □

Theorem 2.5 *Let $\mathcal{E}$ be a semi-abelian category $\mathcal{E}$. Every regular epimorphism $f\colon A \twoheadrightarrow B$ is the cokernel of its kernel: $f = \mathsf{Coker}\,(\mathsf{Ker}\, f)$.*

Proof Consider the following diagram, with f a regular epimorphism and k, (d_1, d_2) respectively, the kernel and the kernel pair of f.

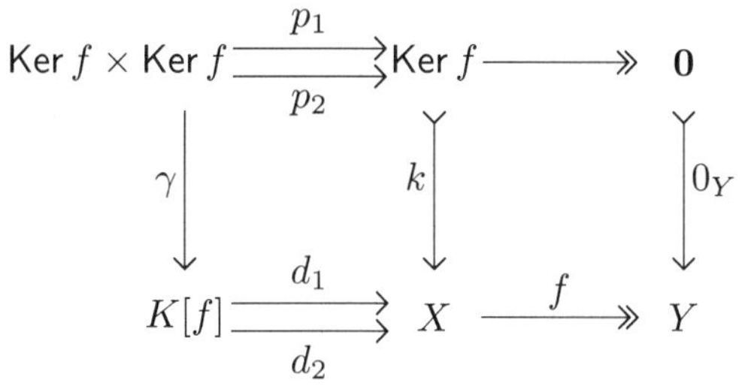

$$\begin{array}{ccccc}
\mathsf{Ker}\, f \times \mathsf{Ker}\, f & \overset{p_1}{\underset{p_2}{\rightrightarrows}} & \mathsf{Ker}\, f & \twoheadrightarrow & \mathbf{0} \\
\gamma \downarrow & & k \downarrow & & \downarrow 0_Y \\
K[f] & \overset{d_1}{\underset{d_2}{\rightrightarrows}} & X & \overset{f}{\twoheadrightarrow} & Y
\end{array}$$

Both left hand squares are pullbacks. Since each morphism d_i admits the diagonal $\Delta\colon X \longrightarrow K[f]$ as a section, lemma 2.3 implies that the pair (γ, Δ) is strongly epimorphic.

Choose now $h\colon X \longrightarrow Z$ such that $h \circ k = 0$; we must prove that h factors uniquely through f. Since $f = \mathsf{Coeq}(d_1, d_2)$, it suffices to prove that $h \circ d_1 = h \circ d_2$. We have indeed

$$h \circ d_1 \circ \gamma = h \circ k \circ p_1 = 0 = h \circ k \circ p_2 = h \circ d_2 \circ \gamma;$$

$$h \circ d_1 \circ \Delta = h = h \circ d_2 \circ \Delta.$$

This forces the conclusion, since the pair (γ, Δ) is epimorphic. □

Next we focus on an unusual cancellation property of pullbacks.

Lemma 2.6 *In a semi-abelian category $\mathcal{E}$, consider a diagram*

$$\begin{array}{ccccc}
V & \xrightarrow{u} & X & \xrightarrow{h} & A \\
{\scriptstyle q}\downarrow & (1) & {\scriptstyle p}\downarrow \uparrow{\scriptstyle s} & (2) & \downarrow{\scriptstyle g} \\
W & \xrightarrow[v]{} & Y & \xrightarrow[f]{} & B
\end{array}$$

where $p \circ s = \mathsf{id}_Y$ and the downward directed squares are commutative. If the square (1) and the outer rectangle (1)+(2) are pullbacks, the square (2) is a pullback as well.

Proof Since the square (1) is a pullback, there is a morphism $t\colon W \longrightarrow V$ such that $q \circ t = \mathsf{id}_W$ and $u \circ t = s \circ v$. Consider the diagram

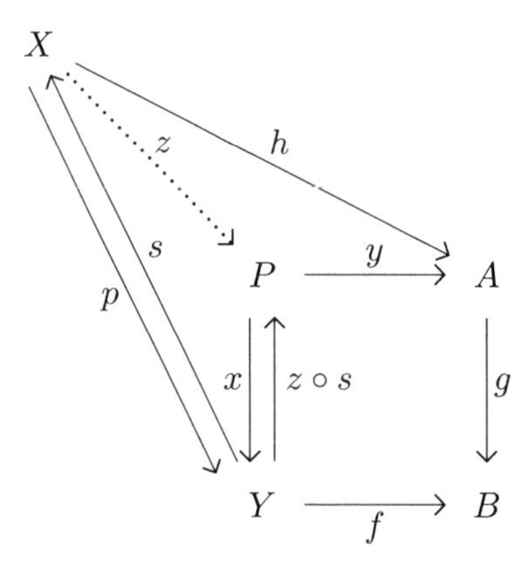

where the downward directed square is a pullback and z is the unique factorization of the outer downward directed quadrilateral through this pullback. This yields a morphism

$$z\colon (p, s\colon X \leftrightarrows Y) \longrightarrow (x, z \circ s\colon P \leftrightarrows Y)$$

in the category $\mathsf{Pt}_Y(\mathcal{E})$. The pullback assumptions in the statement imply

$$v^*(p, s\colon X \leftrightarrows Y) \cong (q, t\colon V \leftrightarrows W) \cong v^*(x, z \circ s\colon P \leftrightarrows Y).$$

Thus $v^*(z)$ is an isomorphism and therefore z is an isomorphism, by protomodularity. By construction of P, this proves that the square (2) is a pullback. □

Theorem 2.7 *In a semi-abelian category $\mathcal{E}$, consider a commutative diagram,*

$$\begin{array}{ccccc}
A & \xrightarrow{u} & B & \xrightarrow{v} & C \\
{\scriptstyle f_1}\downarrow & (1) & {\scriptstyle f_2}\downarrow\!\!\downarrow & (2) & \downarrow{\scriptstyle f_3} \\
X & \xrightarrow[s]{} & Y & \xrightarrow[t]{} & Z
\end{array}$$

where f_2 is a regular epimorphism. If the outer rectangle (1)+(2) and the square (1) are both pullbacks, the square (2) is a pullback as well.

Proof Let us extend the diagram of the statement with the kernel pairs (α_i, β_i) of the morphisms f_i, the corresponding diagonals Δ_i and the obvious factorizations u', v' through these.

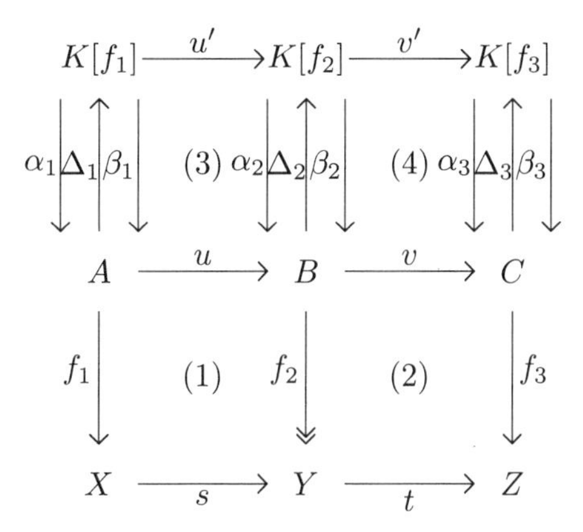

Using our metatheorem 1.8, it is easy to observe that since (1) and (1)+(2) are pullbacks, both downward directed squares (3) are pullbacks and both downward directed rectangles (3)+(4) are pullbacks. The morphisms α_2 and β_2 admit the section Δ_2. By 2.6, both downward directed squares (4) are pullbacks as well. By 1.8 again, this implies at once that the square (2) is a pullback. □

3 Normal subobjects and equivalence relations

As the example of groups suggests, we define:

Definition 3.1 In a semi-abelian category $\mathcal{E}$, a subobject $k\colon K \rightarrowtail A$ is normal when it is the kernel of some morphism.

Proposition 3.2 *Let $\mathcal{E}$ be a semi-abelian category $\mathcal{E}$. Every normal subobject $k\colon K \rightarrowtail A$ is the kernel of its cokernel: $k = \mathsf{Ker}\,(\mathsf{Coker}\,k)$.*

Proof Given $f\colon A \longrightarrow B$ and $k = \mathsf{Ker}\,f$, consider the image factorization $f = i \circ p$ of f. We have still $k = \mathsf{Ker}\,p$, with p a regular epimorphism. By 2.5, $p = \mathsf{Coker}\,k$. □

Proposition 3.3 *In a semi-abelian category $\mathcal{E}$, the pullback of a normal subobject along an arbitrary arrow is again a normal subobject.*

Proof Consider $f\colon A \longrightarrow B$ and a subobject $s\colon S \rightarrowtail B$ which is the kernel of $g\colon B \longrightarrow C$. Then $f^{-1}(S)$ is the kernel of $g \circ f$. □

The crucial property of normal subobjects is then:

Theorem 3.4 *In a semi-abelian category $\mathcal{E}$, the normal subobjects of a fixed object A correspond bijectively with the equivalence relations on A.*

Proof Given an equivalence relation R on A, we associate with it the kernel $\mathsf{Ker}\,q_R$ of the corresponding quotient $q_R\colon A \twoheadrightarrow A/R$.

This correspondence is surjective: every normal subobject $i\colon B \rightarrowtail A$ is the kernel of its cokernel q (see 3.2) and q itself is the quotient $q = q_R$, where R is the kernel pair of q.

By exactness of $\mathcal{E}$, every equivalence relation R on an object A is the kernel pair of its quotient $q_R \colon A \twoheadrightarrow A/R$. Pulling back the first (or the second) projection $p_1^R \colon R \longrightarrow A$ over $\mathbf{0}$ yields the kernel of p_1^R, which coincides thus with the kernel of q_R.

$$
\begin{array}{ccccc}
K & \overset{k}{\rightarrowtail} & R & \overset{p_2^R}{\longrightarrow} & A \\
\downarrow & \text{p.b.} & \downarrow{\scriptstyle p_1^R} & \text{p.b.} & \downarrow{\scriptstyle q_R} \\
\mathbf{0} & \longrightarrow & A & \underset{q_R}{\twoheadrightarrow} & A/R
\end{array}
$$

To prove the injectivity, consider two equivalence relations R and S on A such that the corresponding quotients $q_R \colon A \twoheadrightarrow A/R$ and $q_S \colon A \twoheadrightarrow A/S$ admit the same kernel $k \colon K \rightarrowtail A$. It follows at once that the intersection $R \cap S$ is an equivalence relation whose quotient $q_{R\cap S} \colon A \twoheadrightarrow A/(R\cap S)$ still admits k as kernel. The split short five lemma applied to the diagram

$$
\begin{array}{ccccc}
K & \overset{k}{\rightarrowtail} & R\cap S & \underset{p_1^{R\cap S}}{\overset{\Delta_{R\cap S}}{\leftrightarrows}} & A \\
\| & & \downarrow & & \| \\
K & \underset{k}{\rightarrowtail} & R & \underset{p_1^R}{\overset{\Delta_R}{\leftrightarrows}} & A
\end{array}
$$

where $\Delta_{R\cap S}$, Δ_R denote the diagonals, proves that $R \cap S \cong R$, that is , $R \subseteq S$. Analogously, $S \subseteq R$. □

The reader is probably familiar with the property in theorem 3.4, but he should convince himself that the property fails for non semi-abelian categories, like the category of pointed sets or that of monoids. For example in the case of pointed sets, the equivalence class of 0 determines ... the equivalence class of 0, nothing else, and certainly not the whole congruence. The next corollary, whose precise meaning is recalled in the proof, emphasizes further the "good behaviour" of equivalence relations in a semi-abelian category.

Corollary 3.5 *In a semi-abelian category $\mathcal{E}$, consider an equivalence relation R on an object A. If a subobject $i \colon B \rightarrowtail A$ contains the R-equivalence class of* 0*, it is saturated for R.*

Proof The equivalence class of 0 is given by the left hand pullback below.

$$
\begin{array}{ccc}
[0]_R & \overset{\rho}{\longrightarrow} & R \\
{\scriptstyle r_0}\downarrow & & \downarrow{\scriptstyle r} \\
A & \underset{(0,\mathsf{id}_A)}{\longrightarrow} & A \times A
\end{array}
\qquad
\begin{array}{ccc}
B_1 & \overset{i_1}{\rightarrowtail} & R \\
{\scriptstyle q_1}\downarrow & & \downarrow{\scriptstyle p_1^R} \\
B & \underset{i}{\rightarrowtail} & A
\end{array}
\qquad
\begin{array}{ccc}
B_2 & \overset{i_2}{\rightarrowtail} & R \\
{\scriptstyle q_2}\downarrow & & \downarrow{\scriptstyle p_2^R} \\
B & \underset{i}{\rightarrowtail} & A
\end{array}
$$

Notice that the monomorphism $(\mathsf{id}_A, 0)$ is the inverse image of $\mathbf{0} \rightarrowtail A$ along the second projection; by composition of pullbacks, $[0]_R$ is thus also the kernel of $p_2 \circ r = p_2^R$, that is the kernel of the quotient $A \twoheadrightarrow A/R$, as observed in the proof of 3.4. The saturation of B for R means that in the other two pullbacks, the monomorphisms i_1 and i_2 determine the same subobject of R. In the language of 1.8,

$$B_1 = \{(x, y) \in R \mid x \in B\}, \; B_2 = \{(x, y) \in R \mid y \in B\}.$$

The saturation of B for R means thus

$$\text{If } (x, y) \in R, \text{ then } x \in B \Leftrightarrow y \in B,$$

which is the usual notion of R-saturation.

To prove the corollary, consider the diagram

$$\begin{array}{ccccc}
[0]_R \cap B & \rightarrowtail & B_1 \cap B_2 & \overset{\delta}{\underset{\pi}{\leftrightarrows}} & B \\
\Vert & & \downarrow \beta_1 & & \Vert \\
[0]_R & \rightarrowtail & B_1 & \overset{\delta_1}{\underset{\pi_1}{\leftrightarrows}} & B
\end{array}$$

where π, π_1 are the first projections and δ, δ_1 are the diagonals. An obvious diagram chase proves $\mathsf{Ker}\,\pi_1 = [0]_R$ and $\mathsf{Ker}\,\pi = [0]_R \cap B$. Since $[0]_R \subseteq B$ by assumption, $[0]_R \cap B = [0]_R$ and the split short five lemma implies that β_1 is an isomorphism. Thus $B_1 \subseteq B_2$ and analogously, $B_2 \subseteq B_1$. □

Now let us switch to another celebrated concept: the Mal'cev categories, generalizing Mal'cev varieties (see [30] and [32]). A Mal'cev operation is a ternary operation $p(x, y, z)$ which satisfies the axioms

$$p(x, y, y) = x, \; p(y, y, z) = z.$$

The most celebrated example is that of groups, where $p(x, y, z) = x - y + z$. A Mal'cev algebraic theory is one which contains a Mal'cev operation; the corresponding variety is then called a Mal'cev variety. This notion admits an elegant categorical formalization (see [17] and [18]):

Definition 3.6 A Mal'cev category $\mathcal{E}$ is a category with finite limits in which every reflexive relation is an equivalence relation.

D. Bourn has given a very elegant characterization of Mal'cev categories: they are those finitely complete categories $\mathcal{E}$ such that each category $\mathsf{Pt}_I(\mathcal{E})$ is unital in the sense of 2.4 (see [7]). Let us prove directly that:

Theorem 3.7 *Every semi-abelian category $\mathcal{E}$ is a Mal'cev category.*

Proof Let us use our metatheorem 1.8. Let $p_1, p_2 \colon R \rightrightarrows A$ be a reflexive relation on A in $\mathcal{E}$. We write $R \times_A R$ for the pullback of p_2 and p_1, which in Set_* is given by

$$R \times_A R = \{(x, y, z) | (x, y) \in R, \; (y, z) \in R\}.$$

The first and the third projections $R \times_A R \rightrightarrows A$ induce a factorization to $A \times A$ and we write T for the inverse image of R along this factorization. In Set_* this yields

$$T = \{(x, y, z) | (x, y) \in R, \; (y, z) \in R, \; (x, z) \in R\}$$

with an obvious factorization $i\colon T \rightarrowtail R\times_A R$. The kernel of $(p_1^T, p_3^T)\colon T \longrightarrow A\times A$ is given elementwise by

$$\{(0,y,0)\,|\,(0,y)\in R,\ (y,0)\in R,\ (0,0)\in R\}$$

while the kernel of $(p_1^{R\times_A R}, p_3^{R\times_A R})\colon R\times_A R \longrightarrow A\times A$ is

$$\{(0,y,0)\,|\,(0,y)\in R,\ (y,0)\in R\}.$$

Both kernels coincide since R is reflexive; let us write K for these kernels. This yields the diagram

$$\begin{array}{ccccc}
K & \rightarrowtail & T & \underset{(p_1^T,p_3^T)}{\overset{s^T}{\leftrightarrows}} & A\times A \\
\| & & \downarrow{\scriptstyle i} & & \| \\
K & \rightarrowtail & R\times_A R & \underset{(p_1,p_3)}{\overset{s}{\leftrightarrows}} & A\times A
\end{array}$$

where elementwise, $s_T(x,z) = (x,x,z) = s(x,z)$; this makes sense again because R is reflexive. The split short five lemma applied to this diagram indicates that i is an isomorphism. This proves the transitivity of R.

We know thus already that every reflexive relation in $\mathcal{E}$ is transitive. Given a reflexive relation R on A, we construct now a reflexive relation S on R. The object $R\times R$ is provided with four projections to A; S is the inverse image of R along the morphism $(p_1^{R\times R}, p_4^{R\times R})\colon R\times R \longrightarrow A\times A$. Elementwise, this means

$$S = \Big\{\big((a,b),(c,d)\big)\Big|(a,b)\in R,\ (c,d)\in R,\ (a,d)\in R\Big\}.$$

This relation S on R is trivially reflexive, thus it is transitive as well by the first part of the proof. Now given $(x,y)\in R$, we have both $\big((y,y),(x,y)\big)\in S$ and $\big((x,y),(x,x)\big)\in S$ because R is reflexive. By transitivity of S, $\big((y,y),(x,x)\big)\in S$, proving $(y,x)\in R$. □

Proposition 3.8 *In a semi-abelian category $\mathcal{E}$, the normal subobjects of every object A constitute a modular lattice.*

Proof By 3.4, it is equivalent to work with equivalence relations. If R and S are equivalence relations on A, their composite $R\circ S$ contains both R and S, but also the diagonal of A. By 3.7, $R\circ S$ is an equivalence relation. If T is another equivalence relation containing R and S, then $R\circ S \subseteq T\circ T = T$. Thus $R\circ S = R\vee S$ in the poset of equivalence relations. Notice that this implies $R\circ S = S\circ R$, since $R\vee S = S\vee R$. On the other hand the intersection $R\cap S$ is an equivalence relation as well, the diagonal Δ_A is the smallest equivalence relation on A and $A\times A$ is the biggest one. This proves that the equivalence relations constitute a lattice with top and bottom element.

The modularity of this lattice means the restricted distributivity law

$$R\leq T \;\Rightarrow\; T\wedge(R\vee S) = (T\wedge R)\vee(T\wedge S) = R\vee(T\wedge S),$$

given three equivalence relations R, S, T on A. One inequality is obvious. To prove the converse inequality, we use our metatheorem 1.8. Take a pair (x,y) in

$T \wedge (R \vee S)$. Since $R \vee S = R \circ S = S \circ R$,

$$\exists u \in A \ (x,u) \in R \ (u,y) \in S, \ \exists v \in A \ (x,v) \in S \ (v,y) \in R.$$

Since $R \subseteq T$, $(x,y) \in T$ and $(x,u) \in R$ imply $(u,y) \in T$. But $(u,y) \in S$, thus $(u,y) \in T \wedge S$. On the other hand $(x,u) \in R$, thus $(x,y) \in R \circ (T \wedge S)$. □

Proposition 3.9 *Given an object A in a semi-abelian category $\mathcal{E}$, write* $\mathsf{Sub}(A)$ *for its poset of subobjects.*

1. *The intersection of two normal subobjects of A in* $\mathsf{Sub}(A)$ *is still a normal subobject of A.*
2. *The union of two normal subobjects of A exists in* $\mathsf{Sub}(A)$ *and is a normal subobject of A.*
3. *Given a epimorphism $f\colon A \twoheadrightarrow B$ and a normal subobject $U \subseteq A$, its image $f(U) \subseteq B$ is a normal subobject of $f(A)$.*

Proof Consider two morphisms $f\colon A \longrightarrow X$, $g\colon A \longrightarrow Y$ and the corresponding factorization $(f,g)\colon A \longrightarrow X \times Y$. It is immediate that $\mathsf{Ker}\, f \cap \mathsf{Ker}\, g = \mathsf{Ker}\,(f,g)$. This proves the first assertion.

To prove the second assertion, consider two equivalence relations R, S on A and the corresponding normal subobjects $u\colon U \rightarrowtail A$, $v\colon V \rightarrowtail A$ given by theorem 3.4. By 3.8, the union of U and V in the lattice of normal subobjects of A exists: it is the normal subobject $w\colon W \rightarrowtail A$ corresponding to the equivalence relation $R \circ S$. We shall prove that $W = U \cup V$ as ordinary subobjects of A. This means that given a subobject $z\colon Z \rightarrowtail A$ containing U and V, Z contains W as well. By 3.5 we know already that Z is saturated for both R and S. Using 1.8, $x \in W$ means $(0,x) \in R \circ S$, that is the existence of y such that $(0,y) \in R$ and $(y,x) \in S$. But $(0,y) \in R$ implies $y \in Z$, from which $(y,x) \in S$ implies $x \in Z$.

Consider now a regular epimorphism $f\colon A \twoheadrightarrow B$ and an equivalence relation R on A. The subobject $(f \times f)(R) \subseteq B \times B$ contains the diagonal of B, thus it is an equivalence relation on B by 3.7. For short, we write it simply $f(R)$. To prove assertion 3, it suffices by 3.1 to show that the image of the R-equivalence class of $0 \in A$ is the $f(R)$-equivalence class of 0 in B. We apply again our metatheorem 1.8. Trivially $(a,0) \in R$ implies $\big(f(a),0\big) \in f(R)$. Conversely given $(b,0) \in f(R)$, we have $(a,a') \in R$ such that $f(a) = b$ and $f(a') = 0$, that is, $(a',0) \in S$. This implies $(a,0) \in R \circ S = S \circ R$, thus the existence of $a'' \in A$ with $(a,a'') \in S$ and $(a'',0) \in R$. This yields finally

$$(b,0) = \big(f(a),0\big) = \big(f(a''),0\big) = (f \times f)(a'',0) \in f(R).$$

When f is an arbitrary morphism, it suffices to consider its image factorization $f = i \circ p$ and apply the argument above to the regular epimorphism g. □

Proposition 3.10 *A semi-abelian category $\mathcal{E}$ is finitely complete and finitely cocomplete.*

Proof We know already that $\mathcal{E}$ is finitely complete, since it has pullbacks and a terminal object. Since $\mathcal{E}$ has an initial object and binary coproducts, it has all finite coproducts. It remains to prove the existence of the coequalizer of a pair $f,g\colon A \rightrightarrows B$ of morphisms.

Consider the morphism $(f,g)\colon A \longrightarrow B \times B$ and its image factorization $(f,g) = r \circ q$.

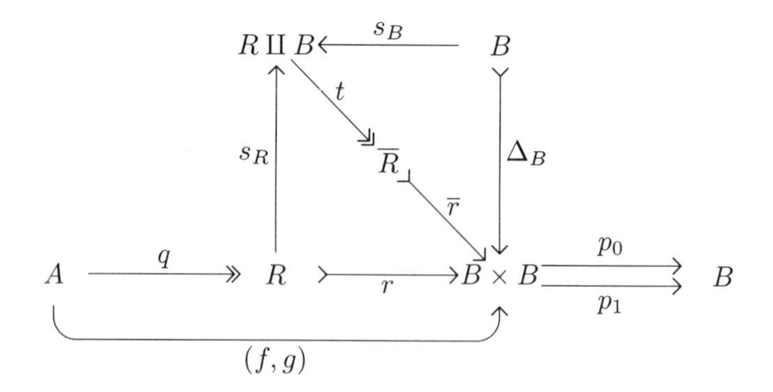

Since q is an epimorphism, this diagram shows at once that $\mathsf{Coeq}(f, g)$ exists if and only if $\mathsf{Coeq}(p_0 \circ r, p_1 \circ r)$ exists and these coequalizers are then equal. Thus it suffices to prove that the relation R on B has a coequalizer.

Consider now the coproduct of R and B, the factorization τ of the pair (r, Δ_B) through this coproduct and its image factorization $\tau = \overline{r} \circ t$. An arrow $x \colon B \longrightarrow X$ coequalizes $(p_0 \circ \overline{r}, p_1 \circ \overline{r})$ precisely when it coequalizes their composites with the epimorphism t. Since (s_R, s_B) are the injections of a coproduct, this is further equivalent to

$$x \circ p_0 \circ \overline{r} \circ t \circ s_R = x \circ p_1 \circ \overline{r} \circ t \circ s_R, \; x \circ p_0 \circ \overline{r} \circ t \circ s_B = x \circ p_1 \circ \overline{r} \circ t \circ s_B$$

that is

$$x \circ p_0 \circ r = x \circ p_1 \circ r, \; x \circ p_0 \circ \Delta_B = x \circ p_1 \circ \Delta_B.$$

But since $p_0 \circ \Delta_B = \mathsf{id}_B = p_1 \circ \Delta_B$, we conclude that x coequalizes $(p_0 \circ \overline{r}, p_1 \circ \overline{r})$ precisely when it coequalizes $(p_0 \circ r, p_1 \circ r)$. This proves that the relation R has a coequalizer if and only if the relation $\overline{R}$ has a coequalizer, in which case those coequalizers are equal. So it suffices to prove that the relation $\overline{R}$ has a coequalizer.

By construction, the relation $\overline{R}$ contains the diagonal of B, that is, is reflexive. Since $\mathcal{E}$ is a Mal'cev category (see 3.7), that reflexive relation $\overline{R}$ is an equivalence relation. Since the category $\mathcal{E}$ is exact, this equivalence relation is a kernel pair and therefore has a coequalizer. □

4 Exact sequences and diagram lemmas

In a semi-abelian category, the notion of exact sequence is the classical one:

Definition 4.1 Let $\mathcal{E}$ be a semi-abelian category. A pair (f, g) of composable morphisms is an exact sequence when the mono-part i of the image factorization of f

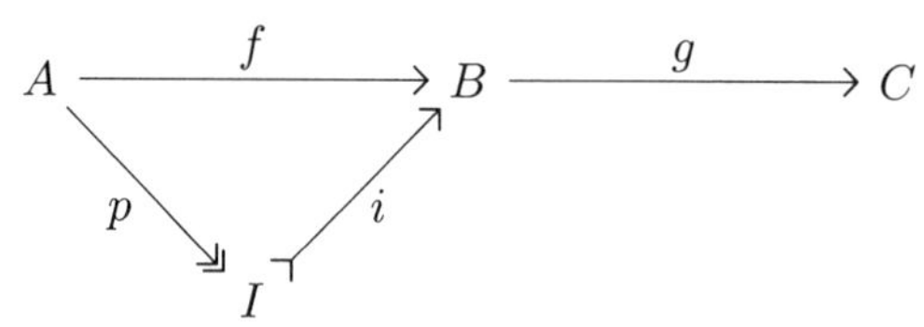

is also the kernel of g.

As usual, a long exact sequence is a sequence of composable morphisms such that each pair of consecutive morphisms is exact. And a short exact sequence is a long exact sequence of the form

$$\mathbf{0} \longrightarrow A \xrightarrow{\ f\ } B \xrightarrow{\ g\ } C \longrightarrow \mathbf{0}.$$

Notice nevertheless a major difference with the abelian case: not every morphism f can appear as first morphism in an exact sequence (f, g): indeed, the exactness property forces the image of f to be a normal subobject.

To avoid any ambiguity, it is thus useful to make explicit some special cases of interest:

Proposition 4.2 *Let $\mathcal{E}$ be a semi-abelian category.*

1. *a morphism $f\colon A \longrightarrow B$ is a monomorphism precisely when the sequence*

$$\mathbf{0} \longrightarrow A \xrightarrow{\ f\ } B$$

is exact;

2. *$k = \mathsf{Ker}\, f$ precisely when the sequence*

$$\mathbf{0} \longrightarrow K \xrightarrow{\ k\ } A \xrightarrow{\ f\ } B$$

is exact;

3. *a morphism $f\colon A \longrightarrow B$ is a regular epimorphism precisely when the sequence*

$$A \xrightarrow{\ f\ } B \longrightarrow \mathbf{0}$$

is exact;

4. *$q = \mathsf{Coker}\, f$ and the image of f is a normal subobject, precisely when the sequence*

$$A \xrightarrow{\ f\ } B \xrightarrow{\ q\ } Q \longrightarrow \mathbf{0}$$

is exact;

5. *a morphism $f\colon A \longrightarrow B$ is an isomorphism precisely when the sequence*

$$\mathbf{0} \longrightarrow A \xrightarrow{\ f\ } B \longrightarrow \mathbf{0}$$

is exact.

Proof Conditions 1 and 2 follow at once from 2.2, since the morphism $\mathbf{0} \longrightarrow A$ is its own image. Condition 3 holds because f is a regular epimorphism precisely when its image is B. The case of isomorphisms follows from assertions 1 and 3.

To prove condition 4, let us write $f = i \circ p$ for the image factorization of f. Since p is an epimorphism, $\mathsf{Coker}\, f = \mathsf{Coker}\,(i \circ p) = \mathsf{Coker}\, i$. The exactness of the given sequence means thus first that q is a regular epimorphism and second that $i = \mathsf{Im}\, f = \mathsf{Ker}\, q$. By 2.5, this implies $q = \mathsf{Coker}\, i = \mathsf{Coker}\, f$.

Conversely, let $q = \mathsf{Coker}\, f$ and $i = \mathsf{Ker}\, g$, for some morphism g. There is no restriction to suppose that g is a regular epimorphism (if not, factor it through its image). By 2.5 again, this implies

$$g = \mathsf{Coker}\, i = \mathsf{Coker}\, f = q.$$

Thus $i = \mathsf{Ker}\, g = \mathsf{Ker}\, q$ and the sequence (f, q) is exact. □

Let us now turn our attention to the classical diagram lemmas involving exact sequences (see [9]).

Theorem 4.3 (Short five lemma) *In a semi-abelian category $\mathcal{E}$, consider the following commutative diagram where the rows are exact sequences.*

$$\begin{array}{ccccccccc} 0 & \longrightarrow & A & \xrightarrow{u} & B & \xrightarrow{p} & C & \longrightarrow & 0\\ & & \Big\downarrow{\scriptstyle a} & & \Big\downarrow{\scriptstyle b} & & \Big\downarrow{\scriptstyle c} & & \\ 0 & \longrightarrow & A' & \xrightarrow{u'} & B' & \xrightarrow{p'} & C' & \longrightarrow & 0 \end{array}$$

If a and c are isomorphisms (respectively, monomorphisms, regular epimorphisms), b is an isomorphism (respectively, monomorphism, regular epimorphism) as well.

Proof We handle first the case of isomorphisms. Consider the commutative diagram

$$\begin{array}{ccccccc} A' & \xrightarrow[\cong]{a^{-1}} & A & \rightarrowtail^{u} & B & \xrightarrow{b} & B' \\ \Big\downarrow & (1) & \Big\downarrow & (2) & \Big\downarrow{\scriptstyle p} & (3) & \Big\downarrow{\scriptstyle p'} \\ 0 & = & 0 & \longrightarrow & C & \xrightarrow[c]{\cong} & C' \end{array}$$

(with the outer arrow $u' \colon A' \to B'$ along the top)

The square (2) is a pullback because $u = \mathsf{Ker}\, p$. The outer part of the diagram is a pullback because $u' = \mathsf{Ker}\, p'$. Since the left hand horizontal morphisms are isomorphisms, the rectangle (2)+(3) is a pullback. By theorem 2.7, the square (3) is a pullback, thus b is an isomorphism, since so is c.

Next let us consider the factorization $b = b_1 \circ b_2$ of b through its image. When a and c are monomorphisms, p factors through b_2 and we get the following diagram.

$$\begin{array}{ccccccccc} 0 & \longrightarrow & A & \xrightarrow{u} & B & \xrightarrow{p} & C & \longrightarrow & 0\\ & & \Big\| & & \Big\downarrow{\scriptstyle b_2} & & \Big\| & & \\ 0 & \longrightarrow & A & \xrightarrow{b_2 \circ u} & B'' & \xrightarrow{p''} & C & \longrightarrow & 0 \end{array}$$

An easy diagram chase, using our metatheorem 1.8, shows that the second line is still exact. By the first part of the proof, b_2 is an isomorphism. Thus $b \cong b_1$ is a monomorphism.

When a and c are regular epimorphisms, u' factors through b_1 and we obtain the diagram

$$\begin{array}{ccccccccc} 0 & \longrightarrow & A' & \xrightarrow{u''} & B'' & \xrightarrow{p' \circ b_1} & C' & \longrightarrow & 0\\ & & \Big\| & & \Big\downarrow{\scriptstyle b_1} & & \Big\| & & \\ 0 & \longrightarrow & A' & \xrightarrow{u'} & B' & \xrightarrow{p''} & C' & \longrightarrow & 0 \end{array}$$

Again an easy diagram chase proves that the first line is still an exact sequence. The first part of the proof implies that b_1 is now an isomorphism. Thus $b \cong b_2$ is a regular epimorphism. $\square$

Theorem 4.4 (Five lemma) *In a semi-abelian category $\mathcal{E}$, consider the commutative diagram*

$$\begin{array}{ccccccccc}
A & \xrightarrow{k} & B & \xrightarrow{h} & C & \xrightarrow{g} & D & \xrightarrow{f} & E \\
\alpha \downarrow \cong & & \beta \downarrow \cong & & \gamma \downarrow & & \cong \downarrow \delta & & \cong \downarrow \varepsilon \\
A' & \xrightarrow{k'} & B' & \xrightarrow{h'} & C' & \xrightarrow{g'} & D' & \xrightarrow{f'} & E'
\end{array}$$

where the rows are exact sequences and the morphisms α, β, δ, ε are isomorphisms. Then γ is an isomorphism as well.

Proof Considering the left hand square, the images of k and k' are isomorphic, thus also the coequalizers of these images. But these coequalizers are the epimorphic parts of the image factorizations of h and h'. Working "dually" from the right, the images of g and g' are isomorphic as well. Factoring all the horizontal morphisms through their images, it remains to apply the short five lemma (see 4.3) to the central part of this new diagram, delimited by the image objects of h, h', g, g'. $\square$

Theorem 4.5 (Nine lemma) *In a semi-abelian category $\mathcal{E}$, consider the commutative diagram 1 where the three columns and the middle row are short exact*

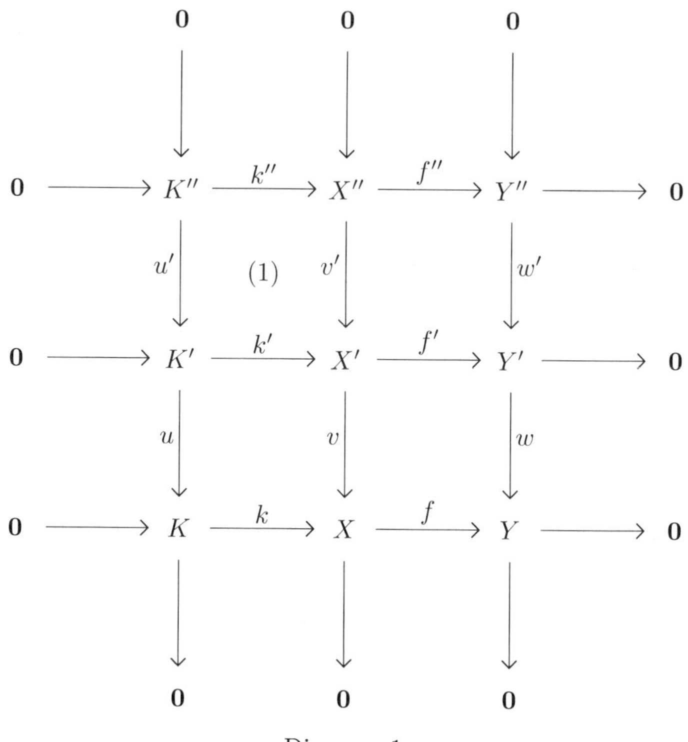

Diagram 1

sequences. The first row is a short exact sequence if and only if the last row is a short exact sequence.

Proof Notice at once that, unlike the case of abelian categories, no duality principle can be used to interchange the roles of the first and the last row. Indeed, the notion of semi-abelian category is not self-dual (see 7.6 for a convincing argument). We use freely 4.2 and our metatheorem 1.8, without recalling it every time.

Let us first assume that the last row is exact. The morphisms k' and u' are monomorphisms, thus k'' is a monomorphism and the first row is exact at K''.

A straightforward diagram chase on the first two columns indicates that the square (1) is a pullback. Another diagram chase on the first two lines concludes that $k'' = \operatorname{Ker} f''$. So the first row is exact at X''.

It remains to prove that f'' is a regular epimorphism. For this we consider the following diagram, where the square (2) is a pullback.

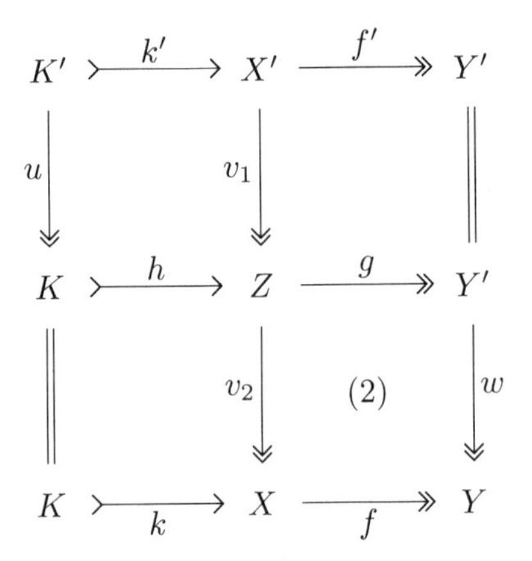

The equality $f \circ v = w \circ f'$ yields the factorization v_1, with $v_2 \circ v_1 = v$ and $g \circ v_1 = f'$. The equality $f \circ k = 0 = w \circ 0$ yields h such that $g \circ h = 0$ and $v_2 \circ h = k$. The morphisms v_2 and g are regular epimorphisms, since so are w and f. The morphism h is a monomorphism, since so is k. An easy diagram chase, using $k = \operatorname{Ker} f$ and next the pullback (2), proves that $h = \operatorname{Ker} g$. Since g is a regular epimorphism, (h, g) is a short exact sequence. By the short five lemma (see 4.3), v_1 is a regular epimorphism.

Since $w \circ w' = 0 = f \circ 0$, we get a factorization $s \colon Y'' \longrightarrow Z$ such that $v_2 \circ s = 0$ and $g \circ s = w'$. Since w' is a monomorphism, so is s. A simple diagram chase proves $s = \operatorname{Ker} v_2$, because the square (2) is a pullback. We obtain in this way the following commutative diagram

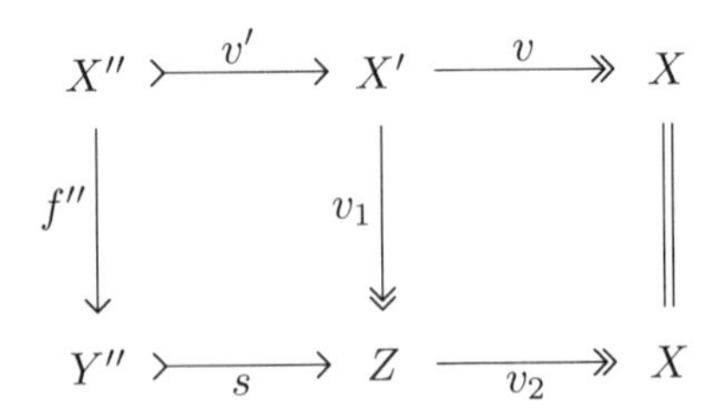

where $v' = \mathsf{Ker}\, v$ and $s = \mathsf{Ker}\, v_2$. The left hand square is pullback. Indeed, if $s \circ x = v_1 \circ y$, then $v_2 \circ v_1 \circ y = 0$, from which the expected factorization through the kernel v' of $v = v_2 \circ v_1$. Since v_1 is a regular epimorphism, so is f''.

Next let us assume that the first row is an exact sequence. Since w and f' are regular epimorphisms, f is a regular epimorphism. Notice that we can "dualize" a little bit further the proof of the first part: the square $f \circ v = w \circ f'$ is a pushout and $f = \mathsf{Coker}\, k$. But this does not imply the exactness of the last row at X (see 4.2.4).

Instead, let us prove first that k is a monomorphism. Using 1.8, if $k(x) = 0$, choose $y \in K'$ such that $u(y) = x$. Then $(v \circ k')(y) = 0$, thus $k'(y) = v'(z)$ for some $z \in X''$. Since the square (1) is a pullback, we have $(y, z) \in K''$. Finally, $x = (u \circ u')(y, z) = 0$. By 2.2, k is a monomorphism.

Now in diagram 1, replace the morphism $k\colon K \longrightarrow X$ by the kernel of f, which we write $k_f\colon \mathsf{Ker}\, f \rightarrowtail X$. We get accordingly a factorization $u_f\colon K' \longrightarrow \mathsf{Ker}\, f$. The three rows (k'', f''), (k', f'), (k_f, f) and the last two columns (v', v), (w', w) of this new diagram are now exact. Thus the first column (u', u_f) is exact as well, by the first part of the proof.

Since (u', u) and (u', u_f) are short exact sequences, $\overline{k}$ is an isomorphism and $k \cong k_f = \mathsf{Ker}\, f$, which concludes the proof. □

Let us mention that the "middle nine lemma" holds as well: when the three columns, the first and the last row are exact sequences and when moreover $f' \circ k' = 0$, then the middle row is an exact sequence (see [5]).

Theorem 4.6 (Snake lemma) *In a semi-abelian category $\mathcal{E}$, consider diagram 2, where all squares of solid arrows are commutative and all sequences of solid arrows are exact. There exists an exact sequence of dotted arrows still making the diagram commutative.*

Proof Since the proof uses techniques analogous to those developed for proving the nine lemma, we simply give the construction of the various morphisms. An explicit proof can be found in [9], in the special case where f is a monomorphism and g' is a regular epimorphism. It is routine to adapt it to the general case, by factoring f and g' through their images, as we do it below.

Of course f_K and g_K are the obvious factorizations through the kernels k_u of u, k_v of v and k_w of w. And dually for f'_Q, g'_Q.

To define the diagonal morphism d, we consider first the factorizations $f = f_2 \circ f_1$ and $g' = g'_2 \circ g'_1$ of f and g' through their images. The pairs (f_2, g) and (f', g'_1) are then short exact sequences.

Next, we construct diagram 3, whose various ingredients will now be described.

The square (1) is a pullback by definition. Since k_w is a monomorphism, h is a monomorphism. Since g is a regular epimorphism, φ is a regular epimorphism.

From $g \circ k_v = k_w \circ g_k$, we obtain the factorization γ through the pullback, yielding $h \circ \gamma = k_v$ and $\varphi \circ \gamma = g_k$. Since k_v is a monomorphism, so is γ.

From the equality $g \circ f_2 = 0 = k_w \circ 0$ we obtain a factorization θ through the pullback such that $h \circ \theta = f_2$ and $\varphi \circ \theta = 0$. Since f_2 is a monomorphism, θ is a monomorphism as well. It follows at once that $\theta = \mathsf{Ker}\, \varphi$, thus (θ, φ) is a short exact sequence.

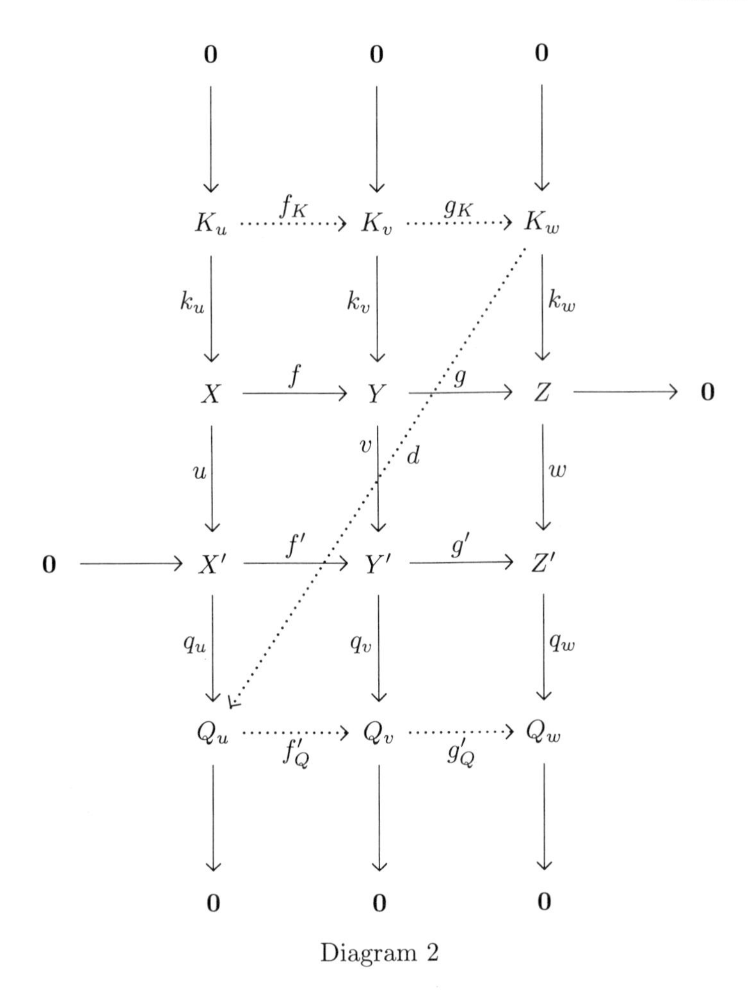

Diagram 2

The commutative squares $f' \circ u = v \circ f$ and $g' \circ v = w \circ g$ yield corresponding factorizations u', w' through the images of f and g'. Since f_1 is a regular epimorphism and g'_2 is a monomorphism, we have still $q_u = \mathsf{Coker}\, u'$ and $k_w = \mathsf{Ker}\, w'$.

Observe that $h = \mathsf{Ker}\,(w' \circ g) = \mathsf{Ker}\,(g'_1 \circ v)$. Since $f' = \mathsf{Ker}\, g'_1$, this yields the factorization ψ and the "square" $f' \circ \psi = v \circ h$ is a pullback. It follows at once that $\gamma = \mathsf{Ker}\, \psi$. Finally the equalities $f' \circ \psi \circ \theta = v \circ h \circ \theta = v \circ f_2 = f' \circ u'$ imply $\psi \circ \theta = u'$, because f' is a monomorphism. Therefore $q_u \circ \psi \circ \theta = q_u \circ u' = 0$, which yields the factorization d through the cokernel of θ, such that $d \circ \varphi = q_u \circ \psi$. This morphism d is the connecting morphism that we wanted to construct. □

Various other diagram lemmas can be proved in a semi-abelian category. This is in particular the case for the two Noether isomorphism theorems. We refer the interested reader to [5] where detailed proofs of these results are given.

The observant reader will have noticed that the results of this section did not use the existence of binary coproducts nor the effectiveness of equivalence relations. They are thus valid in a regular protomodular category with a zero object. They are even valid in the slightly more general context of γ-categories (see [16]).

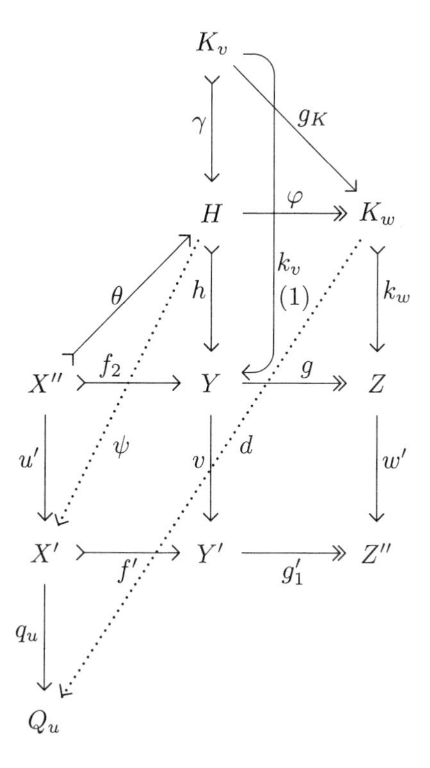

Diagram 3

5 G-actions and semi-direct product

Another important property that semi-abelian categories share with the category of groups is the existence of semi-direct products. For the sake of clarity, we define first that notion and show afterwards that it reduces to the classical notion in the case of groups.

The crucial result is the following one; it is due to D. Bourn and G. Janelidze and uses the full strength of the axioms of a semi-abelian category (see [14]).

Theorem 5.1 *For every morphism* $v\colon J \longrightarrow I$ *of a semi-abelian category* $\mathcal{E}$, *the pullback functor*

$$v^*\colon \mathsf{Pt}_I(\mathcal{E}) \longrightarrow \mathsf{Pt}_J(\mathcal{E})$$

is monadic.

Proof We refer to [2], [28] or [4] for the theory of monads and in particular the Beck criterion for monadicity.

First of all, the functor v^* admits a left adjoint functor

$$v_!\colon \mathsf{Pt}_J(\mathcal{E}) \longrightarrow \mathsf{Pt}_I(\mathcal{E})$$

which is given by the pushout along v (see 3.10).

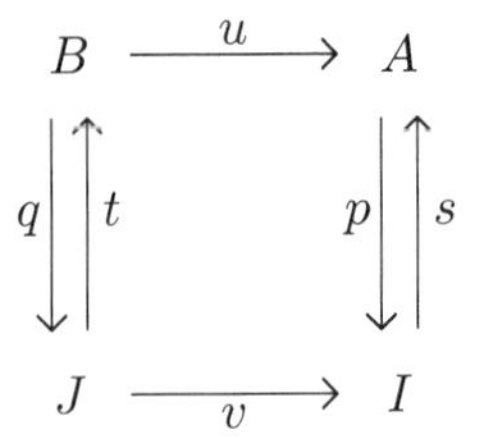

Indeed, if t admits a retraction q, its pushout s along v admits a retraction p such that $s \circ v = u \circ t$. The commutativity of this diagram implies the existence of a unique factorization

$$z\colon (q,t) \longrightarrow (h,r) = v^* v_!(q,t)$$

through the pullback (h,r) of (p,s) along v. This yields the first natural transformation of the expected adjunction. Analogously, using the same diagram but with different assumptions, let us start from $(p,s) \in \mathsf{Pt}_I(\mathcal{E})$. We can compute the pullback (q,t) of (p,s) over J; the commutativity of the diagram implies now the existence of a unique factorization

$$z'\colon (h',r') = v_! v^*(p,s) \longrightarrow (p,s),$$

from the pushout (h',r') of (q,s) along v. This yields the second natural transformation of the expected adjunction. It is routine to check the triangular identities.

Next, by protomodularity of $\mathcal{E}$, the functor v^* reflects isomorphisms (see 1.1). Thus to apply the Beck criterion, it remains to check the condition on coequalizers of reflexive pairs.

First, it is obvious to observe that pullbacks, pushouts, equalizers and coequalizers exist in $\mathsf{Pt}_I(\mathcal{E})$ and are computed as in $\mathcal{E}$. Since $\mathcal{E}$ is exact, it follows at once that each category $\mathsf{Pt}_I(\mathcal{E})$ is exact and each pullback functor v^* preserves pullbacks and regular epimorphisms. This implies that each functor v^* preserves coequalizers of kernel pairs, thus by exactness, coequalizers of equivalence relations.

Observe next that each category $\mathsf{Pt}_I(\mathcal{E})$ has binary coproducts, computed as pushouts in $\mathcal{E}$ (see 3.10). It has also a zero object, namely the pair $(\mathsf{id}_I, \mathsf{id}_I)$; each functor v^* preserves obviously these zero objects. Since pullbacks and the terminal object generate all finite limits, each category $\mathsf{Pt}_I(\mathcal{E})$ is finitely complete and each functor v^* preserves finite limits.

If $\big(p, s\colon A \leftrightarrows I\big)$ is an object in $\mathsf{Pt}_I(\mathcal{E})$, a pointed object over (p,s) in $\mathsf{Pt}_I(\mathcal{E})$ is simply a pointed object over A in $\mathcal{E}$. Since pullbacks and isomorphisms in $\mathsf{Pt}_I(\mathcal{E})$ are computed as in $\mathcal{E}$, $\mathsf{Pt}_I(\mathcal{E})$ is protomodular, since so is $\mathcal{E}$. This concludes the proof that $\mathsf{Pt}_I(\mathcal{E})$ is semi-abelian. In particular, $\mathsf{Pt}_I(\mathcal{E})$ is a Mal'cev category, by 3.7.

Consider now a reflexive pair in $\mathsf{Pt}_I(\mathcal{E})$, that is

$$\big(q, t\colon B \leftrightarrows I\big) \underset{v}{\overset{u}{\underset{w}{\rightrightarrows}}} \big(p, s\colon A \leftrightarrows I\big), \quad u \circ w = \mathsf{id}_A = v \circ w.$$

Consider the morphism (u,v) to the product $(p,s) \times (p,s)$ in $\mathsf{Pt}_I(\mathcal{E})$ and its image factorization $(u,v) = r \circ \pi$ still in $\mathsf{Pt}_I(\mathcal{E})$.

$$
\begin{array}{c}
\overset{(u,v)}{\frown} \\
(p,s) \xrightarrow{\;w\;} (q,t) \xrightarrow{\;\pi\;} R \xrightarrow{\;r\;} (p,s)\times(p,s) \underset{p_1}{\overset{p_0}{\rightrightarrows}} (p,s) \\
\underset{\Delta}{\smile}
\end{array}
$$

By assumption, the composite $(u,v) \circ w$ is the diagonal Δ. Thus R is a reflexive relation and therefore an equivalence relation, by the Mal'cev property. Since π is an epimorphism, $\mathsf{Coeq}(u,v) = \mathsf{Coeq}(p_0 \circ r, p_1 \circ r)$ and the coequalizer of (u,v) is that of the equivalence relation R. We know already that the coequalizer of R is preserved by the functor v^*. But since v^* preserves all the ingredients of the diagram, the coequalizer of $v^*(R)$ is also that of $\big(v^*(u), v^*(v)\big)$. $\square$

Corollary 5.2 *For each object I of a semi-abelian category $\mathcal{E}$, the functor*

$$\mathsf{Ker} : \mathsf{Pt}_I(\mathcal{E}) \longrightarrow \mathcal{E}, \;\; \big(p, s\colon A \leftrightarrows I\big) \mapsto \mathsf{Ker}\, p$$

is monadic and admits the functor

$$\sigma_I\colon \mathcal{E} \longrightarrow \mathsf{Pt}_I(\mathcal{E}), \;\; B \mapsto \big((0, \mathsf{id}_I), s_I\colon B \amalg I \leftrightarrows I\big)$$

as left adjoint.

Proof In 5.1, put $J = \mathbf{0}$. One has $\mathsf{Pt}_{\mathbf{0}}(\mathcal{E}) \cong \mathcal{E}$, while the pushout over $\mathbf{0}$ is the coproduct. $\square$

Here is now the definition of the semi-direct product:

Definition 5.3 Let $\mathcal{E}$ be a semi-abelian category and $G \in \mathcal{E}$ an object of $\mathcal{E}$.

1. A G-algebra is an algebra for the monad $\mathbb{T}_G$ corresponding to the monadic functor $\mathsf{Ker} : \mathsf{Pt}_G(\mathcal{E}) \longrightarrow \mathcal{E}$ (see 5.2).
2. The semi-direct product $(X,\xi) \rtimes G$ of a G-algebra (X,ξ) and the object $G \in \mathcal{E}$ is the domain part H of the pointed object $\big(p, s\colon H \leftrightarrows G\big)$ corresponding to (X,ξ) via the equivalence $\mathsf{Pt}_G(\mathcal{E}) \cong \mathcal{E}^{\mathbb{T}_G}$.

Let us emphasize at once an important property of the semi-direct product:

Theorem 5.4 *Let $\mathcal{E}$ be a semi-abelian category, $G \in \mathcal{E}$ an object of $\mathcal{E}$ and (X,ξ) a G-algebra. There exists a split short exact sequence*

$$\mathbf{0} \longrightarrow X \xrightarrow{\;k\;} (X,\xi) \rtimes G \underset{p}{\overset{s}{\leftrightarrows}} G \longrightarrow \mathbf{0}.$$

Proof The morphisms p and s are those of definition 5.3, thus $p \circ s = 0$. By commutativity of the triangle

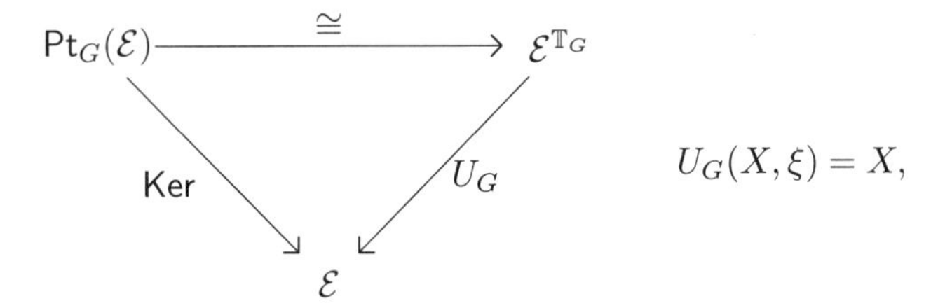

$$U_G(X,\xi) = X,$$

and again with the notation of 5.3, $\mathsf{Ker}\, p \cong X$. $\square$

It remains now to convince the reader that in the category Gp of groups, this definition recaptures the classical notions of a G-group and the semi-direct product of groups. Let us first recall the basic elements about semi-direct products of groups.

Definition 5.5 Let $(G, \cdot)$ be a group. A G-group is a triple $(X, +, m)$ consisting of a group $(X, +)$ and an action

$$m\colon G \times X \longrightarrow X,\ (g, x) \mapsto gx$$

which satisfies the axioms

$$1x = x,\ g'(gx) = (g \cdot g')x,\ g(x + x') = gx + gx'$$

for all elements $g, g' \in G$ and $x, x' \in X$.

It is immediate to observe that in the conditions of definition 5.5, the equalities $g0 = 0$ and $g(-x) = -(gx)$ hold for all elements $g \in G$ and $x \in X$.

The semi-direct product is then classically defined via the following well-known proposition:

Proposition 5.6 *Let G be a group and $(X, +, m)$ a G-group. The set $X \times G$, provided with the multiplication*

$$(x, g) \star (x', g') = (x + gx', g \cdot g')$$

is a group. This group is called the semi-direct product of $(X, +, m)$ and $(G, \cdot)$ and is written $(X, m) \rtimes G$.

Proof See [27] or any classical text on group theory. □

The key result to exhibit the link with definition 5.3 is then:

Proposition 5.7 *Let* Gp *be the category of groups and group homomorphisms and let G be a fixed group. The category of pointed objects $\mathsf{Pt}_G(\mathsf{Gp})$ is equivalent to the category of G-groups and their morphisms.*

Proof A morphism $f\colon (X, +, m) \longrightarrow (Y, +, n)$ of G-groups is, of course, a group homomorphism $f\colon X \longrightarrow Y$ which commutes with the actions of G, that is, $f(gx) = gf(x)$ for all elements $x \in X$ and $g \in G$.

Every G-group $(X, +, m)$ yields a pointed object in $\mathsf{Pt}_G(\mathsf{Gp})$

$$p_G, i_G\colon (X, m) \rtimes G \underset{\longrightarrow}{\longleftarrow} G,\ p_G(x, g) = g,\ i_G(g) = (0, g).$$

This construction extends easily in a functor

$$\Pi\colon G\text{-}\mathsf{Gp} \longrightarrow \mathsf{Pt}_G(\mathsf{Gp}).$$

Conversely, given a group $(H, \star)$ and a pointed object

$$p, s\colon H \underset{\longrightarrow}{\longleftarrow} G,\ p \circ s = \mathsf{id}_G$$

in $\mathsf{Pt}_G(\mathsf{Gp})$, we define the group $(X, +)$ to be the kernel of p. We provide X with a G-action by defining

$$m\colon G \times X \longrightarrow X,\ (g, x) \mapsto gx = s(g) \star x \star s(g)^{-1};$$

checking the axioms of definition 5.5 is routine. Again this construction extends easily in a functor

$$\Gamma\colon \mathsf{Pt}_G(\mathsf{Gp}) \longrightarrow G\text{-}\mathsf{Gp}.$$

It remains to observe that both constructions are mutually inverse, which is straightforward computation left to the reader. □

We are now ready to conclude. By 5.7 and 5.2, we have equivalences of categories

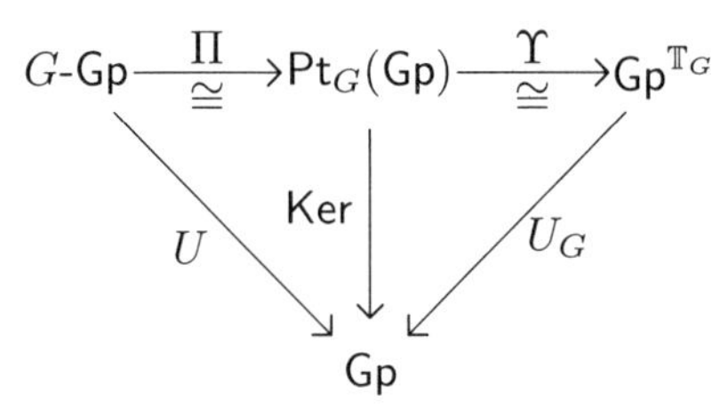

where

$$U(X,m) = X,\ \Pi(X,m) = \big(p, s\colon (X,m) \rtimes G \leftrightarrows G\big),\ (U_G \circ \Upsilon)(p,s) = \mathsf{Ker}\, p.$$

In particular the $\mathbb{T}$-algebra $(X', \xi) = (\Upsilon \circ \Pi)(X, m)$ corresponding to the G-group (X, m) is such that $X' = \mathsf{Ker}\, p = X$. This proves that providing a group X with a G-action is equivalent to provide it with a $\mathbb{T}_G$-action. Moreover the object part of the corresponding pointed object over G is the classical semi-direct product $(X, m) \rtimes G$.

6 Commutators and centrality

Let us now switch to the study of the "commutative aspects" of a semi-abelian category.

Definition 6.1 Let $\mathcal{E}$ be a semi-abelian category. Two morphisms f, g with the same codomain commute when there exists a (necessarily unique) morphism φ making the following diagram commutative,

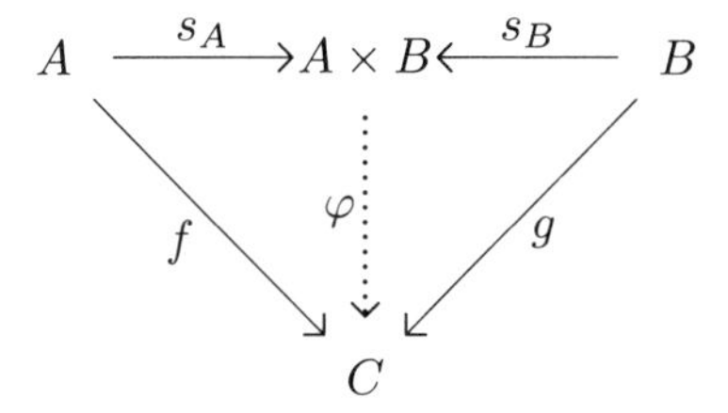

where $s_A = (\mathsf{id}_A, 0)$ and $s_B = (0, \mathsf{id}_B)$. We call φ the *connector* of f and g.

By 2.4, the pair (s_A, s_B) is strongly epimorphic, from which we have the uniqueness of φ, when it exists. The terminology varies according to the authors (see [10], [12], [21], [22], [31]); ours is justified by the following example.

Example 6.2 In the category of groups, two morphisms f, g as in 6.1 commute when

$$\forall a \in A\ \forall b \in B\ f(a) + g(b) = g(b) + f(a).$$

Proof If the condition of the statement holds, the formula $\varphi(a, b) = f(a) + g(b)$ defines a group homomorphism such that $\varphi(a, 0) = f(a)$ and $\varphi(0, b) = g(b)$. Conversely if a connector φ exists, then

$$f(a) + g(b) = \varphi(a, 0) + \varphi(0, b) = \varphi(a, b) = \varphi(0, b) + \varphi(a, 0) = g(b) + f(a).$$

□

The basic result concerning the commutation of morphisms is the possibility to force it universally. The very recent and not yet published construction in the proof of 6.3 is due to D. Bourn (see [11]) and yields a definition of the commutator (see 6.4) which gains in elegance with respect to anterior definitions (see [31]).

Theorem 6.3 *In a semi-abelian category $\mathcal{E}$, consider two morphisms*

$$A \xrightarrow{f} C \xleftarrow{g} B$$

with the same codomain. There exists a morphism $\psi\colon C \longrightarrow D$, universal for making the composites $\psi \circ f$ and $\psi \circ g$ commute. This morphism ψ is a regular epimorphism.

Proof We define D to be the colimit of the outer part of the following diagram, with the dotted arrows as colimit cocone.

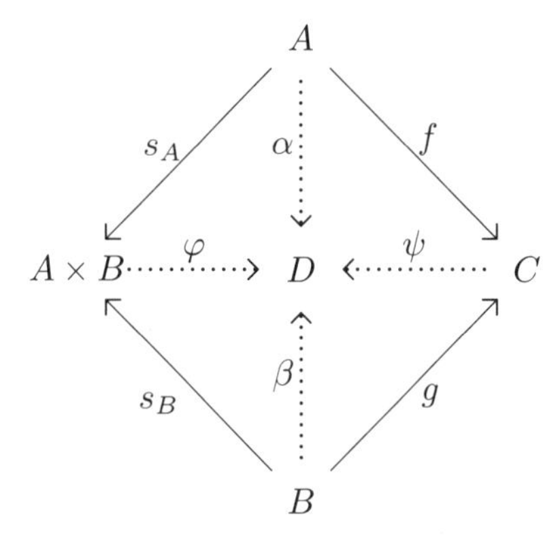

Of course the composites $\psi \circ f$ and $\psi \circ g$ commute, with connector φ.

We must prove that given another morphism $\psi'\colon C \longrightarrow D'$ such that $\psi' \circ f$ and $\psi' \circ g$ commute (let us say, with connector φ'), then ψ' factors uniquely through ψ. Putting $\alpha' = \psi' \circ f$ and $\beta' = \psi' \circ g$, we obtain a new cocone $(\alpha', \beta', \varphi', \psi')$ with vertex D' on the same diagram, from which a unique factorization $d\colon D \longrightarrow D'$ through the colimit. In particular, $d \circ \psi = \psi'$.

Next we prove that ψ is a strong (thus regular) epimorphism, from which the uniqueness of the factorization. Indeed if ψ factors through a subobject $s\colon S \rightarrowtail D$, then by commutativity of the diagram, α, β, $\varphi \circ s_A$ and $\varphi \circ s_B$ factor through s as well. Since the colimit cocone and the pair (s_A, s_B) are strongly epimorphic (see 2.4), this forces s to be an isomorphism. Thus ψ is a strong epimorphism. □

Definition 6.4 Let $\mathcal{E}$ be a semi-abelian category. Given two morphisms f, g with the same codomain, their commutator $[f, g]$ is the kernel of the universal morphism ψ of theorem 6.3.

Example 6.5 In the category Gp of groups, consider two normal subgroups $h\colon H \rightarrowtail G$ and $k\colon K \rightarrowtail G$. The commutator $[h, k]$ in the sense of definition 6.4 is the usual group $[H, K]$ of commutators.

Proof The group $[H, K]$ of commutators is the subgroup of G generated by all the elements of the form $x + y - x - y$, with $x \in H$ and $y \in K$. This is a normal subgroup, by normality of H and K. The quotient $G \longrightarrow G/[H, K]$ is well-known to be the universal morphism making H and K commute. □

Centrality is a special case of interest. Still with example 6.2 in mind, we define

Definition 6.6 Let $\mathcal{E}$ be a semi-abelian category. A morphism $f\colon A \longrightarrow C$ is central when it commutes with the identity on C.

Proposition 6.7 *Let $\mathcal{E}$ be a semi-abelian category. A central morphism $f\colon A \longrightarrow C$ commutes with every morphism $g\colon B \longrightarrow C$.*

Proof If $\varphi\colon A \times C \longrightarrow C$ is the connector of f and id_C, the composite $\varphi \circ (\mathsf{id}_A \times g)$ is the connector of f and g. □

Finally, again in view of example 6.2, we define:

Definition 6.8 An object C of a semi-abelian category $\mathcal{E}$ is abelian when the identity on C commutes with itself (i.e. is central).

Of course example 6.2 shows that the abelian objects of the category of groups are exactly the abelian groups. This is a much more general fact, as the following theorem proves:

Theorem 6.9 *For an object C in a semi-abelian category $\mathcal{E}$, the following conditions are equivalent:*

1. *C is abelian;*
2. *the diagonal $\Delta_C\colon C \rightarrowtail C \times C$ of C is a normal subobject;*
3. *C can be provided with the structure of an internal abelian group.*

Moreover, the structure of abelian group on C is necessarily unique.

Proof An abelian group addition $+\colon C \times C \longrightarrow C$ on C must in particular satisfy the axioms $+ \circ (\mathsf{id}_C, 0) = \mathsf{id}_C$ and $+ \circ (0, \mathsf{id}_C) = \mathsf{id}_C$. By 2.4, this proves the uniqueness of $+$.

If C is abelian, write $+\colon C \times C \longrightarrow C$ for the connector of id_C with itself. We must prove that $+$ is an abelian group structure on C. Since this reduces to prove the commutativity of diagrams involving finite limits, this can be done elementwise using our metatheorem 1.8. The letters x, y, z denote "elements" of C. By definition of $+$, $x + 0 = x = 0 + x$.

Let us write $\mathsf{tw}\colon C \times C \longrightarrow C \times C$ for the twisting isomorphism. The commutativity of $+$ is the equality $+ \circ \mathsf{tw} = +$. Since the morphisms $(\mathsf{id}_C, 0)$, $(0, \mathsf{id}_C)$ constitute an epimorphic pair (see 2.4), it suffices to prove the equality after composition with them. This reduces to proving $x + 0 = 0 + x$, which holds by definition of $+$. Analogously, iterating 2.4, it suffices to prove the associativity axiom $x + (y + z) = (x + y) + z$ when two of the variables are equal to zero, which reduces again to the definition of $+$.

Thus $+$ provides already C with the structure of a commutative monoid. To prove that it is a group, consider the diagram below, where $k = \mathsf{Ker}\,+$ and $\gamma = p_2 \circ k$. The three columns and the first two rows are short exact sequences; by the nine lemma, the last row is a short exact sequence.

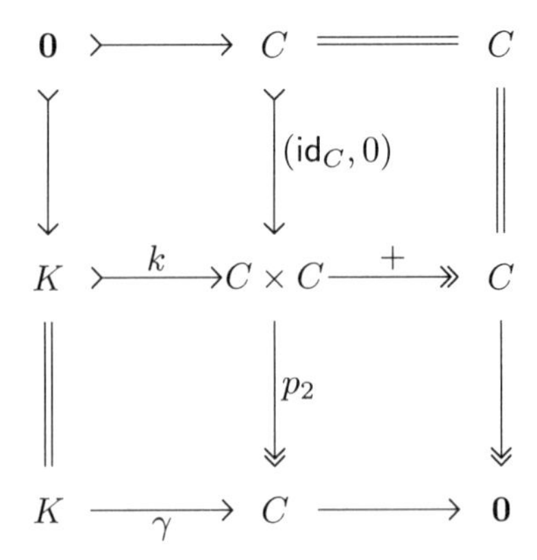

This proves that γ is an isomorphism (see 4.2). The composite

$$\iota\colon C \xrightarrow{\gamma^{-1}} K \xrightarrow{k} C \times C \xrightarrow{p_1} C$$

yields the opposite for the addition. Indeed by commutativity of the diagram, $(k \circ \gamma^{-1})(x)$ is a pair (y, z) such that $y = \iota(x)$, $z = x$ and $y + z = 0$.

Assuming now that C is provided with a group addition $+$, the diagonal of C is the kernel of the corresponding subtraction $-\colon C \times C \longrightarrow C$.

Finally assume that the diagonal of C is the kernel of some morphism $q\colon C \times C \longrightarrow Q$. Factoring q through its image, there is no restriction to assume that q is a regular epimorphism. Consider the diagram below where $\xi = q \circ (\mathsf{id}_C, 0)$. The three rows and the first two columns are short exact sequences; by the nine lemma, the last column is a short exact sequence as well. By 4.2, this proves that ξ is an isomorphism. We define then the subtraction "$-$" of C as the composite $\xi^{-1} \circ q$.

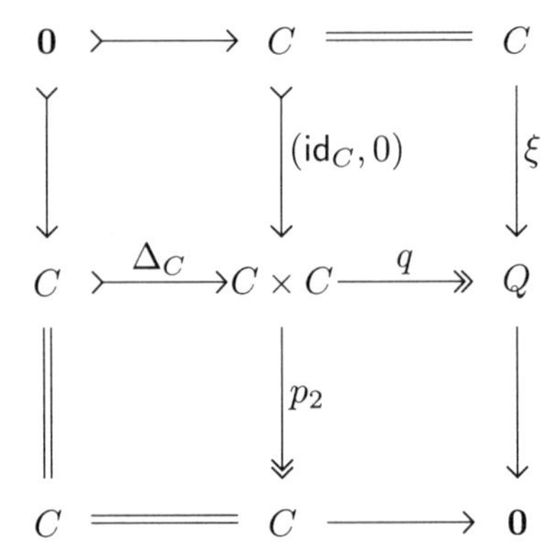

By commutativity of the diagram, we have at once $x - x = 0$ and $x - 0 = x$. From these properties, it follows at once that the equality

$$(y - z) - (x - z) = y - x$$

is satisfied when two of the variables are equal to 0. By an iterated application of 2.4, this proves the equality in full generality. It is well-known (or routine to check) that these three axioms are those defining a group in terms of its subtraction. In particular, the group addition of C is the connector of id_C with itself. $\square$

7 Semi-abelian versus abelian

This section intends to "measure the distance" between abelian categories and semi-abelian categories. Our first comparison (see [21], [20], [13]) between both notions focuses on the construction of the "abelian core" of a semi-abelian category.

Proposition 7.1 *Let $\mathcal{E}$ be a semi-abelian category. The full subcategory $\mathsf{Ab}(\mathcal{E})$ of abelian objects is closed in $\mathcal{E}$ under finite limits, subobjects and regular quotients. In particular, this category $\mathsf{Ab}(\mathcal{E})$ is exact.*

Proof By 6.9, an object is abelian when its diagonal is a normal subobject. Given a finite limit $L = \mathsf{lim}_{i\in I}\, G_i$ in $\mathcal{E}$, the diagonal Δ_L of L is the corresponding limit of the diagonals Δ_i of the various G_i. If each G_i is abelian, its diagonal is the kernel of its cokernel $q_i\colon G_i \twoheadrightarrow Q_i$ (see 3.2). The original diagram induces factorizations between these cokernels Q_i. The limit of these morphisms q_i admits as kernel the limit of their kernels Δ_i, that is, the diagonal Δ_L of L. This proves the normality of Δ_L. Thus the limit L in $\mathcal{E}$ is an abelian object.

If G is abelian and $i\colon H \rightarrowtail G$ is an arbitrary subobject, the diagonal of H is the pullback along i of the diagonal of G. Thus H is abelian by 3.3.

Finally consider a regular epimorphism $p\colon A \twoheadrightarrow B$ with kernel $k\colon K \rightarrowtail A$ and kernel pair $R \rightrightarrows A$. Write $q\colon A\times A \twoheadrightarrow Q$ for the cokernel of $R \rightarrowtail A\times A$. This yields the following commutative diagram

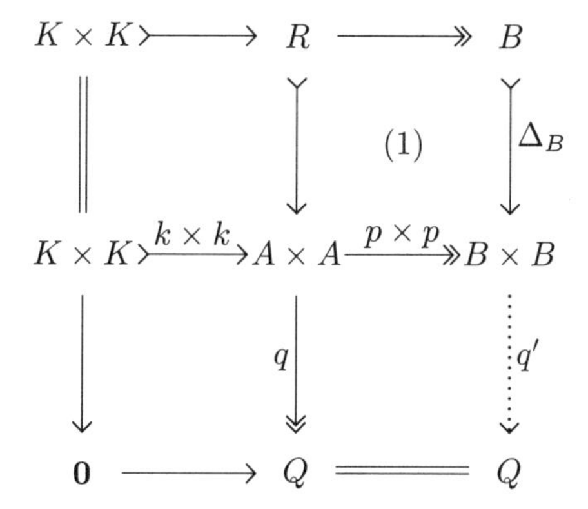

where Δ_B is the diagonal of B, the square (1) is a pullback, the three rows and the first two columns are short exact sequences. This forces the existence of the factorization q' and $\Delta_B = \mathsf{Ker}\, q'$, by the nine lemma (see 4.5). Thus B is abelian. □

Theorem 7.2 *Given a semi-abelian category $\mathcal{E}$, the full subcategory $\mathsf{Ab}(\mathcal{E})$ of abelian objects is abelian and regular-epi-reflective in $\mathcal{E}$.*

Proof The category $\mathsf{Ab}(\mathcal{E})$ is exact by 7.1. By 6.9, the objects of $\mathsf{Ab}(\mathcal{E})$ are the internal abelian groups. But an arbitrary morphism $f\colon A \longrightarrow B$ between abelian objects commutes with their abelian group structures. Indeed, by 2.4 this reduces to the elementwise properties

$$f(x+0) = f(x) + f(0),\ \ f(0+x) = f(0) + f(x)$$

which hold trivially since $f(0) = 0$. Thus $\mathsf{Ab}(\mathcal{E})$ is equivalent to the category of abelian groups in $\mathcal{E}$, which is additive. Therefore $\mathsf{Ab}(\mathcal{E})$ is abelian, because it is both additive and exact (see [1]).

By theorem 6.3, every object $C \in \mathcal{E}$ has an abelian reflection D given by the universal morphism $\psi\colon C \twoheadrightarrow D$ forcing the commutativity of id_C with itself. Moreover, ψ is a strong epimorphism. Observe that in this special case, the colimit diagram in the proof of 6.3 reduces to the coequalizer

$$C \underset{(0,\mathsf{id}_C)}{\overset{(\mathsf{id}_C,0)}{\rightrightarrows}} C \times C \xrightarrow{\;\psi\;} D.$$

□

Corollary 7.3 *A semi-abelian category $\mathcal{E}$ is abelian if and only if every object of $\mathcal{E}$ is abelian.*

Let us now investigate the distance between semi-abelian and abelian categories in terms of normal subobjects.

Proposition 7.4 *In a semi-abelian category $\mathcal{E}$, every subobject $s\colon S \rightarrowtail A$ of an abelian object A is normal.*

Proof By 7.1, s is a monomorphism in $\mathsf{Ab}(\mathcal{E})$. By 7.2, this category is abelian, thus s is a kernel in $\mathsf{Ab}(\mathcal{E})$. By 7.1 again, s is the same kernel in $\mathcal{E}$. □

Theorem 7.5 *A semi-abelian category $\mathcal{E}$ is abelian if and only if every subobject is normal in $\mathcal{E}$.*

Proof In an abelian category, every subobject is normal. Conversely, if every subobject is normal, every object is abelian (see 6.9); one concludes by 7.2. □

Let us now conclude this paper with a further comparison between the notions of semi-abelian and abelian categories (see [23]) ... a comparison which justifies fully the terminology!

Theorem 7.6 *A category $\mathcal{E}$ is abelian if and only if both $\mathcal{E}$ and its dual $\mathcal{E}^{\mathsf{op}}$ are semi-abelian.*

Proof The notion of abelian category is self-dual, thus one implication follows at once from 1.4.

If $\mathcal{E}$ is semi-abelian, the two morphisms in 2.4 yield a strongly epimorphic factorization $A \amalg B \longrightarrow A \times B$. If $\mathcal{E}^{\mathsf{op}}$ is semi-abelian as well, by duality, this same morphism is also a regular monomorphism, thus an isomorphism. Therefore the codiagonal of the coproduct yields on every object A an "addition" making the following diagram commutative:

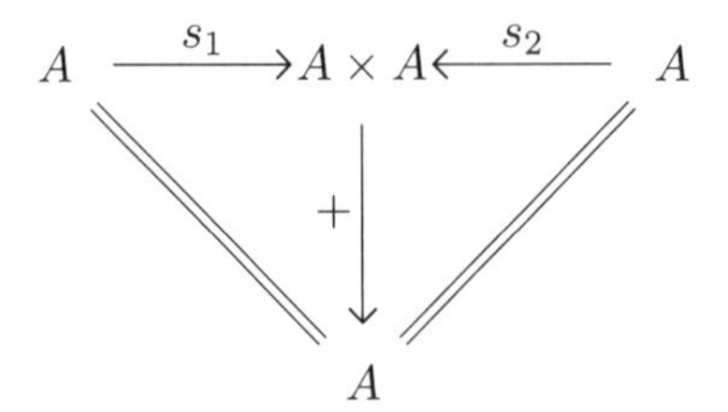

This shows that every object $A \in \mathcal{E}$ is abelian (see 6.8. By 7.3, the category $\mathcal{E}$ is abelian. □

References

[1] M. Barr, *Exact categories*, Springer Lect. Notes in Math. **236** (1971) 1–120
[2] M. Barr and J. Beck, *Homology and standard constructions*, Springer Lect. Notes in Math. **80** (1969) 245–335
[3] J. Bénabou, *Fibered categories and the foundations of naive category theory*, J. of Symbolic Logic **50** (1985) 10–37
[4] F. Borceux, *Handbook of Categorical Algebra*, Cambridge Univ. Press **vol. 1–3** (1994)
[5] F. Borceux and D. Bourn, *Mal'cev, protomodular, homological and semi-abelian categories*, Mathematics and its Applications, Kluwer (2004)
[6] D. Bourn, *Normalization equivalence, kernel equivalence and affine categories*, Springer Lect. Notes in Math. **1448** (1991) 43–62
[7] D. Bourn, *Mal'cev categories and fibrations of pointed objects*, Appl. Categorical Structures **4** (1996) 302–327
[8] D. Bourn, *Normal subobjects and abelian objects in protomodular categories*, J. Algebra **228** (2000) 143–164
[9] D. Bourn, 3×3 *lemma and protomodularity*, J. Algebra **236** (2001) 778–795
[10] D. Bourn, *Intrinsic centrality and associated classifying properties*, Cahiers LMPA, Univ. du Littoral **153** (2001)
[11] D. Bourn, *Commutator theory in regular Mal'cev categories*, this volume
[12] D. Bourn and M. Gran, *Centrality and normality in protomodular categories*, Theory Appl. Categories **9** (2002) 151–165
[13] D. Bourn and M. Gran, *Central extensions in semi-abelian categories* J. Pure Appl. Alg. **175** (2002) 31–44
[14] D. Bourn and G. Janelidze, *Protomodularity, descent and semi-direct product*, Theory Appl. Categories **4** (1998) 37–46
[15] D. Bourn and G. Janelidze, *Characterization of protomodular varieties of universal algebra*, preprint (2002)
[16] M.S. Burgin, *Categories with involution and correspondences in γ-categories*, Trans. Moscow Math. Soc. **22** (1970) 181–257
[17] A. Carboni, G. M. Kelly and M. C. Pedicchio, *Some remarks on Maltsev and Goursat categories*, Appl. Categorical Structures **1** (1993) 385–421
[18] A. Carboni, J. Lambek and M. C. Pedicchio, *Diagram chasing in Mal'cev categories*, J. Pure Appl. Alg. **69** (1990) 271–284
[19] P. Freyd, *Abelian categories*, Harper and Row (1964)
[20] M. Gran, *Central extensions and internal groupoids in Maltsev categories*, J. Pure Appl. Alg. **155** (2001) 139–166
[21] S.A. Huq, *Commutator, nilpotency and solvability in categories*, Quart. J. Math. Oxford (2)**19** (1968) 363–389
[22] G. Janelidze and M. C. Pedicchio, *Pseudogroupoids and commutator theory*, Theory Appl. Categories **8** (2001) 405–456
[23] G. Janelidze, L. Márki and W. Tholen, *Semi-abelian categories*, J. Pure Appl. Alg. **168** (2002) 367–386
[24] P.T. Johnstone, *Sketches of an elephant: a topos theory compendium*, Oxford Science Publications, vol 1–2, 2002
[25] P.T. Johnstone, *A Note on the Semiabelian Variety of Heyting Semilattices*, This volume (2002)
[26] F.W. Lawvere, *Functorial semantics of algebraic theories*, Proc. Nat. Acad. Sci. U.S.A. **50** (1963) 869–873
[27] S. Mac Lane, *Homology*, Springer Verlag (1963)
[28] S. Mac Lane, *Categories for the working mathematician*, 2nd edition, Springer Verlag (1998)
[29] S. Mac Lane and I. Moerdijk, *Sheaves in geometry and logic*, Universitext, Springer Verlag (1992)
[30] A. I. Mal'cev, *On the general theory of algebraic systems*, Mat. Sbornik N. S. **35** (1954) 3–20
[31] M. C. Pedicchio, *A categorical approach to commutator theory*, J. Algebra **177** (1995) 647–657
[32] J. D. H. Smith, *Mal'cev varieties*, Springer Lect. Notes in Math. **554** (1976)

[33] A. Ursini *Osservazioni sulla varietá BIT*, Boll. U. M. I. **(4) 7** (1973) 205–211
[34] A. Ursini, *On subtractive varieties, I*, Algebra Universalis **31** (1994) 204–222

Received October 24, 2002; in revised form December 18, 2002

Fields Institute Communications
Volume **43**, 2004

Commutator Theory in Regular Mal'cev Categories

Dominique Bourn
Université du Littoral
Laboratoire de Mathématiques Pures et Appliquées
Bât. H. Poincaré
50 Rue F. Buisson BP 699
62228 Calais, France
bourn@lmpa.univ-littoral.fr

Abstract. In the categorical Mal'cev setting, we give a new construction of the commutator $[R, S]$ of two equivalence relations which extends the existence of commutators from the exact to the regular context and includes the case of topological groups. All the classical properties are satisfied, except $[R, S] \leq R \wedge S$ which requires the category $\mathbb{C}$ to be exact.

Introduction

The notion of commutator in Mal'cev varieties has been introduced by J.D.H. Smith [22]. Once the concept of Mal'cev category ([11], [12])had been established, it was quite natural to investigate the notion of commutator in this more general setting. This was first done by Pedicchio ([20] and [21]) for exact Mal'cev categories with cokernels, in a way mimicking the varietal construction of Hagemann-Herrmann [14] which, in the Mal'cev varietal context, was proved to coincide with the original construction of Smith ([22]; see also [13]).

We present here a new construction of different nature. The idea behind this comes from two directions. The first one deals with the notion of connector ([8] and [9]), which, although expressed in a seemingly easier language, is equivalent to the notion of centralizing relation and produces a means to assert that the commutator $[R, S] = 0$ in a Mal'cev categorical setting, freed of any right exactness condition. The second one deals with the notion of unital category where there is an intrinsic notion of commutativity and centrality [6], a setting in which there is a natural categorical way to force a pair of morphisms to commute, provided that regularity holds.

2000 *Mathematics Subject Classification.* Primary 18C10, 08B05, 18D35; Secondary 08A30, 18D40, 08G15.

Key words and phrases. Unital category, central morphism, Mal'cev category, commutator, abelian object.

The relationship between these two directions is that a category $\mathbb{C}$ is Mal'cev if and only if the associated fibration $\pi : Pt\mathbb{C} \to \mathbb{C}$ of pointed objects (see below) has unital fibres [4]. Consequently it is possible to translate the unital means to force commutation from the unital setting to the Mal'cev one.

One of the advantages of this new construction is to extend the notion of commutator from the exact Mal'cev context to the regular Mal'cev one, enlarging the range of examples to the Mal'cev (quasi-varieties) and the category $Gp(Top)$ of topological groups for instance. Of course, as a pointed, finitely complete, exact and protomodular category, any semi-abelian category ([16], [2]) is included in this construction. Many of the classical properties of commutators hold with our new definition (see sections 3 and 4), and some new applications are given. In particular, we obtain a left adjoint to the inclusion of the abelian objects. However the important property $[R, S] \leq R \wedge S$ requires, here, the category $\mathbb{C}$ to be exact.

1 Unital categories

We shall suppose $\mathbb{C}$ a pointed category, i.e. a finitely complete category with a zero object. We shall denote by $\alpha_X : 1 \to X$ and $\tau_X : X \to 1$ the initial and terminal maps. Clearly the class $\Omega(\mathbb{C}) = \{0_{X,Y} = \alpha_Y.\tau_X\}$ of null maps is an ideal of $\mathbb{C}$.

1.1. Definition. A *punctual span* in the pointed category $\mathcal{C}$ is a diagram of the form

$$X \underset{f}{\overset{s}{\leftrightarrows}} Z \underset{g}{\overset{t}{\rightleftarrows}} Y$$

with $f.s = 1_X, g.t = 1_Y, g.s = 0, f.t = 0$ (where 0 is the zero arrow). A punctual relation is a punctual span such that the pair of maps (f, g) is jointly monic.

1.2. Examples. For any pair (X, Y) of objects in $\mathbb{C}$, there is a canonical punctual relation which is called the coarse relation:

$$X \underset{p_X}{\overset{l_X}{\leftrightarrows}} X \times Y \underset{p_Y}{\overset{r_Y}{\rightleftarrows}} Y$$

where $l_X = (1_X, 0)$ and $r_Y = (0, 1_Y)$.

1.3. Definition. A pointed category $\mathbb{C}$ is called *unital*, see [6], when for each pair (X, Y) of objects in $\mathbb{C}$, the pair of maps (l_X, r_Y) is jointly strongly epic.

In any unital category, there are no other punctual relations but the coarse ones.

1.4. Examples. A variety $\mathcal{V}$ is unital if and only if it is Jonsson-Tarski, see [3], i.e such that its theory contains a unique contant 0 and a binary term + satisfying $x + 0 = x = 0 + x$. In particular, the categories Mag, Mon, CoM, Gp, Ab, Rg of respectively unitary magmas, monoids, commutative monoids, groups, abelian groups, rings are unital.

By obvious pointwise arguments, any category $\mathbb{C}^{\mathbb{E}}$ with $\mathbb{C}$ any of the previous examples are unital.

1.5. Examples. Let $U : \mathbb{C} \to \mathbb{C}'$ be any left exact conservative (i.e. reflecting isomorphisms) functor. Then if $\mathbb{C}$ is pointed, $\mathbb{C}'$ unital implies $\mathbb{C}$ unital.

When the category $\mathbb{E}$ has products, then the Yoneda embedding: $Y : \mathbb{E} \to Set^{\mathbb{E}^{op}}$ has a natural extension $Y : Mag(\mathbb{E}) \to Mag^{\mathbb{E}^{op}}$ to the category $Mag(\mathbb{E})$ of internal unitary magmas in $\mathbb{E}$ which is still left exact and conservative. Consequently $Mag(\mathbb{E})$ is unital when $\mathbb{E}$ is finitely complete. For similar reasons, the categories $Mon(\mathbb{E})$, $CoM(\mathbb{E})$, $Gp(\mathbb{E})$, $Ab(\mathbb{E})$, $Rg(\mathbb{E})$ of respectively internal monoids, commutative monoids, groups, abelian groups, rings in $\mathbb{E}$ are unital. In particular the categories $Mon(Top)$ and $Gp(Top)$ of topological monoids and topological groups are unital.

1.6. Examples. We have a non syntactical example with the dual Set_*^{op} of the category of pointed sets, and more generally with the dual of the category of pointed objects in any topos $\mathbb{E}$.

One of the main consequences of unitality is the fact that there is an intrinsic notion of commutativity and centrality. Indeed, given a unital category $\mathbb{C}$, the pair (l_X, r_Y), being jointly strongly epic, is actually jointly epic. Therefore a map $\varphi : X \times Y \to Z$ is uniquely determined by the pair of maps (f, g), $f : X \to Z$ and $g : Y \to Z$, with $f = \varphi.l_X$ and $g = \varphi.r_Y$. Accordingly the existence of such a map φ becomes a property in respect of the pair (f, g). Whence the following definitions, see [6] and also [15]:

1.7. Definition. Given a pair (f, g) of morphisms in a unital category $\mathbb{C}$, when such a map φ exists, we say that the maps f and g *cooperate* and that the map φ is the *cooperator* of the pair (f, g). A map $f : X \to Y$ is *central* when f and 1_Y cooperate. An object X is called *commutative* when the map $1_X : X \to X$ is central.

We eventually preferred the terminology "cooperate" to "commute" because of the non syntactic examples.

1.8. Examples. 1) In the category Mag, a map $f : X \to Y$ is central if and only if the following identities hold:

i) $f(x).y = y.f(x)$, for each pair $(x, y) \in X \times Y$,

ii) $f(x).(y.y') = (f(x).y).y'$ and $(y.y').f(x) = y.(y'.f(x))$ for any $(x, y, y') \in X \times Y \times Y$.

2) In the categories Mon and Gp, a map $f : X \to Y$ is central if and only if it take its values in the center of Y. In the categories CoM and Ab, any map $f : X \to Y$ is central.

3) In the category Rg, a map $f : X \to Y$ is central if and only if $f(x).y = 0$, for each pair $(x, y) \in X \times Y$.

4) In the category Set_*^{op} the only central maps are the null maps.

5) In the category Mon of monoids, two submonoids: $H \rightarrowtail G \leftarrowtail K$ cooperate precisely when they commute, that is: $\forall x \in H \ \forall y \in K \ x.y = y.x$.

6) More generally the subcategory $Com(\mathbb{C})$ of the commutative objects in the unital category $\mathbb{C}$ is always a full subcategory of $\mathbb{C}$.

We shall suppose now that the unital category $\mathbb{C}$ is moreover finitely cocomplete and regular [1], i.e. such that the regular epimorphisms are stable by pullback and any effective equivalence relation admits a quotient. This is the case of any Jonsson-Tarski variety $\mathcal{V}$ and also of the category $Gp(Top)$ of topological groups for instance. Moreover, in a regular category, the strong epis are exactly the regular epis.

In this context, we shall construct, from any pair $f : X \to Z$, $g : Y \to Z$ of coterminal maps, a map which universally makes them cooperate. Indeed consider the following diagram, where T is the colimit of the diagram made of the plain arrows:

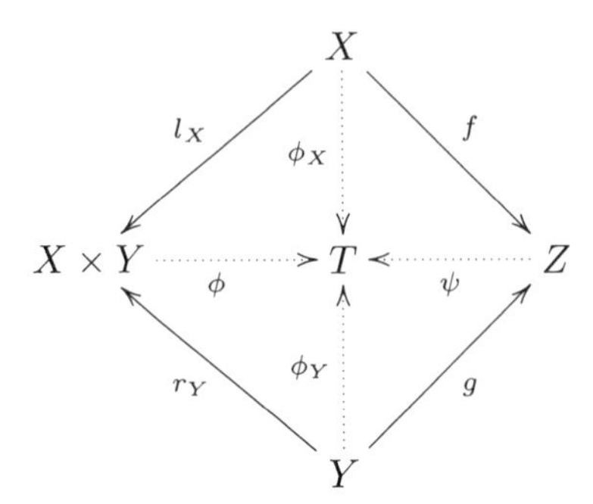

Clearly the maps ϕ_X and ϕ_Y are completely determined by the pair (ϕ, ψ), and clearly the map ϕ is the cooperator of the pair $(\psi.f, \psi.g)$. This map ψ measures the lack of cooperation between f and g.

1.9. Proposition. *Suppose $\mathbb{C}$ unital, finitely cocomplete. Then ψ is the universal arrow which, by composition, makes the pair (f, g) cooperate. The map ψ is an iso if and only if the pair (f, g) cooperates.*

Proof First let us show that the map ψ is a strong epimorphism. If $j : U \rightarrowtail T$ is a mono which, pulled back along ψ, is an isomorphism, this is also the case along ϕ_X and ϕ_Y, and thus along $\phi.l_X$ and $\phi.r_Y$. But the pair (l_X, r_Y) is jointly epic, consequently the pullback of j along ϕ is also an iso. Now the four dotted arrows form a colimit cone, and thus j is itself an iso.

Secondly suppose given a map $\chi : Z \to W$ which makes, by means of a cooperator θ, cooperate the pair (f, g) by composition. This determines a cocone on the previous diagram:

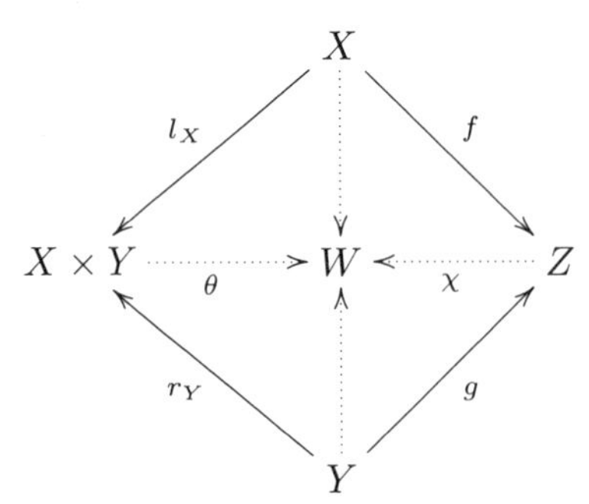

and consequently a factorization $\bar{\chi}$ such that $\bar{\chi}.\psi = \chi$. The uniqueness of this factorization is a consequence of the fact that ψ is a strong epimorphism. The end of the lemma is straightforward. $\square$

Whence two easy applications:

1.10. Corollary. *Suppose $\mathbb{C}$ unital, finitely cocomplete and regular. Let $f : X \to Z$ be a map. Consider the following coequalizer:*

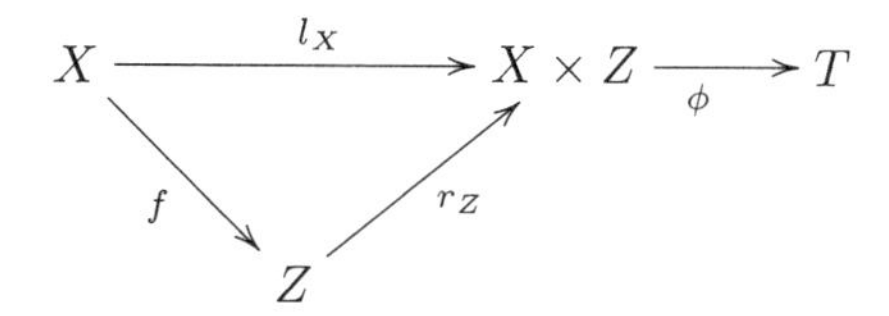

then $\psi = \phi.r_Z$ is the universal map which makes f central by composition.

Let X be any object of $\mathbb{C}$. Then the associated commutative object $\gamma(X)$ is given by the following coequalizer:

$$X \underset{r_X}{\overset{l_X}{\rightrightarrows}} X \times X \xrightarrow[\phi]{} \gamma(X)$$

In other words, the full inclusion of the commutative objects $Com(\mathbb{C}) \hookrightarrow \mathbb{C}$ admits a left adjoint γ.

Proof 1) When $g = 1_Z$ the previous colimit T is reduced to that coequalizer, and $\psi = \phi.r_Z$. Accordingly, the pair $(\psi.f, \psi)$ cooperate. But ψ, as a strong epi in a regular category, is a regular epi, and then the pair $(\psi.f, 1_Z)$ cooperate, which means that $\psi.f$ is central.

2) When $f = g = 1_X$ the previous colimit T is reduced to the coequalizer in question, and the pair (ψ, ψ) cooperates. But, again, ψ is a regular epi, and then the pair $(1_X, 1_X)$ cooperates, which means that the object X is commutative. □

2 Mal'cev categories

Let us recall that a category $\mathbb{C}$ is Mal'cev when it is finitely complete and such that every reflexive relation is an equivalence relation, see [12] and [11]. Following the famous Mal'cev argument, a variety $\mathcal{V}$ is Mal'cev if and only if its theory contains a ternary term p, satisfying : $p(x, y, y) = x = p(y, y, x)$ (called a Mal'cev operation).

There is a strong connection with unital categories which is given by the following observation. Let $\mathbb{C}$ be a finitely complete category. We denote by $Pt\mathbb{C}$ the category whose objects are the split epimorphisms in $\mathbb{C}$ with a given splitting and morphisms the commutative squares between these data. We denote by $\pi : Pt\mathbb{C} \to \mathbb{C}$ the functor associating its codomain with any split epimorphism. Since the category $\mathbb{C}$ has pullbacks, the functor π is a fibration which is called the *fibration of pointed objects.*

A finitely complete category $\mathbb{C}$ is Mal'cev if and only if the fibres of the fibration π are unital, see [4].

Now consider $(d_0, d_1) : R \rightrightarrows X$ an equivalence relation on the object X in $\mathbb{C}$. We shall denote by $s_0 : X \to R$ the inclusion arising from the reflexivity of the relation, and we shall write ΔX and ∇X respectively for the smallest $(1_X, 1_X) : X \rightrightarrows X$ and the largest $(p_0, p_1) : X \times X \rightrightarrows X$ equivalence relations on X.

Since the category $\mathbb{C}$ is Mal'cev, to give the equivalence relation $(d_0, d_1) : R \rightrightarrows X$ on the object X in $\mathbb{C}$ is equivalent to give the following inclusion of the object

$((d_0, s_0) : R \rightleftarrows X)$ into the object $((p_0, s_0) : X \times X \rightleftarrows X)$ in the fibre $Pt_X(\mathbb{C})$ above X:

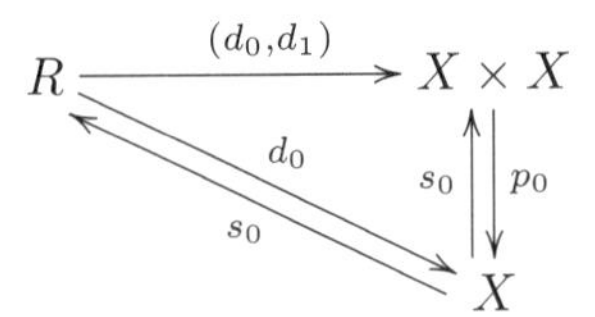

So, by abuse of notation, we shall often identify the equivalence relation R with the object $((d_0, s_0) : R \rightleftarrows X)$ of the fibre $Pt_X(\mathbb{C})$, and conversely.

Now consider $(d_0, d_1) : R \rightrightarrows X$ and $(d_0, d_1) : S \rightrightarrows X$ two equivalence relations on the same object X in $\mathbb{C}$. Then take the following pullback:

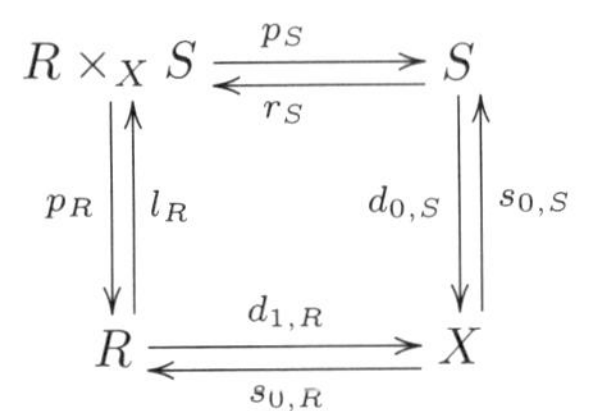

where l_R and r_S are the sections induced by the maps $s_{0,R}$ and $s_{0,S}$.

Let us recall the following definition, see [8], and also [19], [18], [12], [20]:

2.1. Definition. In a Mal'cev category $\mathbb{C}$, a connector on the pair (R, S) in a Mal'cev category is a morphism

$$p : R \times_X S \to X, \ (xRySz) \mapsto p(x, y, z)$$

which satisfies the identities : $p(x, y, y) = x$ and $p(y, y, z) = z$.

This notion actually makes sense in any finitely complete category provided that some further conditions are satisfied [8], which are always fulfilled in a Mal'cev category. Moreover, in a Mal'cev category, a connector is necessarily unique when it exists (since the pair (l_R, r_S) is jointly epic), and thus the existence of a connector becomes a property. We say then that R and S are *connected.*

2.2. Examples. By Proposition 3.6, Proposition 2.12 and definition 3.1 in [20], two relations R and S in a Mal'cev variety $\mathcal{V}$ are connected if and only if $[R, S] = 0$ in the sense of Smith [22].

Accordingly we shall denote a connected pair of equivalence relations by $[R, S] = 0$. One of the advantage of the notion of connectors is that it keeps a meaning in a context freed of any right exactness condition. So, in the same way as in the varietal case, we can say that an object X in a Mal'cev category is *abelian* when $[\nabla X, \nabla X] = 0$. Similarly, a map $f : X \to Y$ is said to have *central kernel* when its kernel equivalence $R[f]$ is such that $[R[f], \nabla X] = 0$. It is said to have *abelian kernel* when $[R[f], R[f]] = 0$ (this is clearly equivalent to saying that the object $f : X \to Y$ in the slice category $\mathbb{C}/Y$ is abelian). Consequently, for instance, a naturally Mal'cev category [17], is merely a Mal'cev category in which any object is abelian or any map has a central kernel.

On the other hand, the fibre $Pt_X(\mathbb{C})$ being unital, it is natural to ask when two subobjects R and S of ∇X cooperate in this fibre.

2.3. Proposition. *Let $\mathbb{C}$ be a Mal'cev category, the subobjects R and S of ∇X cooperate in the fibre $Pt_X(\mathbb{C})$ if and only if the equivalence relations R and S are connected in $\mathbb{C}$.*

Proof Let us consider the product of R and S in $Pt_X(\mathbb{C})$. It is given by the following pullback in $\mathbb{C}$:

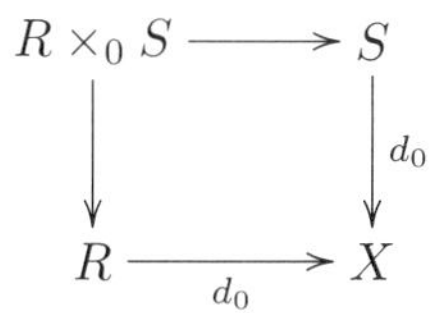

A cooperator between R and S in $Pt_X(\mathbb{C})$ is thus a map $\phi : R \times_0 S \to X \times X$ such that $\phi(x, y, x) = (x, y)$ and $\phi(x, x, z) = (x, z)$. But ϕ is a morphism in the fibre and necessarily is of the form $\phi(x, y, z) = (x, q(x, y, z))$. Accordingly a cooperator between R and S is just a map $q : R \times_0 S \to X$ such that $q(x, y, x) = y$ and $q(x, x, z) = z$. Consequently to set $p(u, v, w) = q(v, u, w)$ is to define a bijection between the cooperators and the connectors. □

So the two points of view are equivalent. We shall prefer the one of connectors because the guiding results of [9] concerning the commutators are given in these terms.

From this observation, and the universal construction of the first section, we shall derive a new construction of the commutator. We shall suppose from now on the category $\mathbb{C}$ finitely cocomplete, Mal'cev and regular, as is the case for the Mal'cev (quasi-varieties) and for the category $Gp(Top)$ of topological groups for instance. Regular Mal'cev categories have stronger stability properties than unital regular ones since, in particular, they are stable by slice and coslice. We shall appreciate below the advantages of these stability properties.

In a regular Mal'cev category, given a regular epi $f : X \to Y$, any equivalence relation R on X has a direct image $f(R)$ along f on Y. It is given by the regular epi/mono factorization of the map $(f.d_0, f.d_1) : R \twoheadrightarrow f(R) \rightarrowtail Y \times Y$. Clearly in any regular category $\mathbb{C}$, the relation $f(R)$ is reflexive and symmetric. When moreover $\mathbb{C}$ is Mal'cev, $f(R)$ is an equivalence relation.

Now let us consider the following diagram (which we shall denote $D(R, S)$) where T is the colimit of the plain arrows:

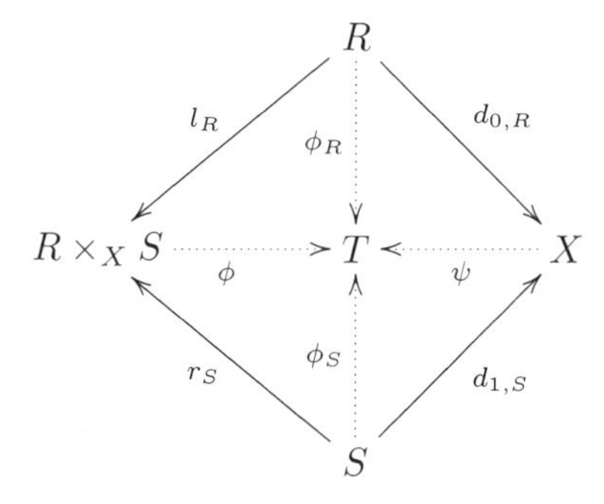

Notice that, in consideration of the pullback defining $R \times_X S$, the roles of the projections d_0 and d_1 have been interchanged. As above, the maps ϕ_R and ϕ_S are completely determined by the pair (ϕ, ψ). We shall see that this map ψ measures the lack of connection between R and S.

2.4. Theorem. *Let the category $\mathbb{C}$ be finitely cocomplete, Mal'cev and regular. Then the map ψ is the universal regular epimorphism which makes the images $\psi(R)$ and $\psi(S)$ connected. The equivalence relations R and S are connected (i.e. $[R, S] = 0$) if and only if ψ is an isomorphism.*

Proof First, the map ψ is a strong epi (and thus a regular epi) for exactly the same reasons as in the previous construction in unital categories. Secondly let us denote by $\psi_R : R \to \psi(R)$ and $\psi_S : S \to \psi(S)$ the respective regular factorizations. Thanks to Proposition 4.1 in [7], the induced factorization $\bar{\psi} : R \times_X S \to \psi(R) \times_T \psi(S)$ is itself a regular epi. The aim now will be to show that $\phi : R \times_X S \to T$ factors through $\psi(R) \times_T \psi(S)$, producing a connector $p : \psi(R) \times_T \psi(S) \to T$ on $\psi(R)$ and $\psi(S)$. For that we must show that ϕ coequalizes the kernel relation $R[\bar{\psi}]$ of $\bar{\psi}$. Clearly $R[\bar{\psi}]$ is obtained by the following pullback:

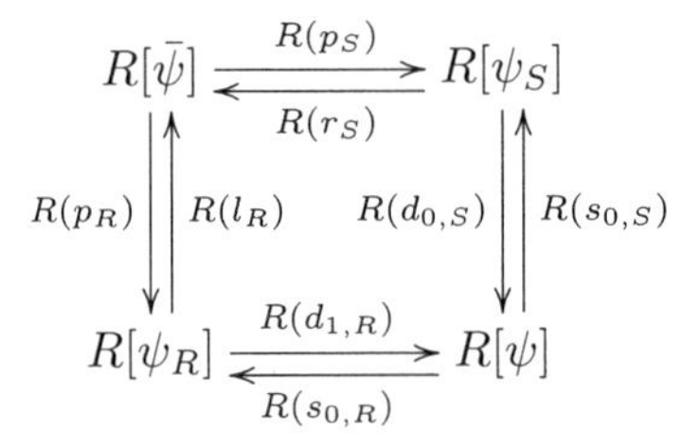

But the fibre $Pt_{R[\psi]}(\mathbb{C})$ is unital, and consequently the pair $(R(l_R), R(r_S))$ is jointly (strongly) epic. So it is sufficient to check the coequalization in question by composition with this pair of maps, which is straightforward. The axioms asserting that p is a connector for the pair of equivalence relations $\psi(R)$ and $\psi(S)$ are a consequence of the form of the colimit T.

Now assume that we have a regular epi $\chi : X \to Y$ such that the images $\chi(R)$ and $\chi(S)$ are connected by a map $\theta : \chi(R) \times_Y \chi(S) \to Y$. Consider the induced factorization $\bar{\chi} : R \times_X S \to \chi(R) \times_Y \chi(S)$. Then the following cocone produces the required factorization $\chi' : T \to Y$:

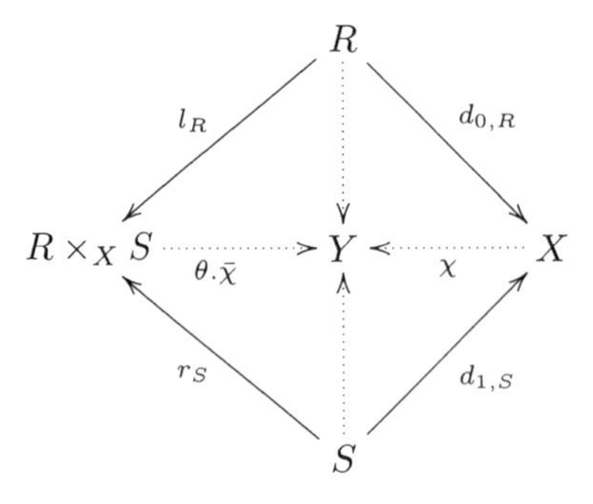

The last point of the theorem is straightforward. □

The map ψ being a regular epi, its distance from being an isomorphism is its distance from being a monomorphism, which is exactly measured by its kernel relation $R[\psi]$. Accordingly it is meaningful to introduce the following definition:

2.5. Definition. Let the category $\mathbb{C}$ be finitely cocomplete, Mal'cev and regular. Let two equivalence relations $(d_0, d_1) : R \rightrightarrows X$ and $(d_0, d_1) : S \rightrightarrows X$ be given on the same object X in $\mathbb{C}$. The kernel relation $R[\psi]$ of the map ψ is called the commutator of R and S. It is classically denoted by $[R, S]$.

2.6. Examples. If we suppose moreover the category $\mathbb{C}$ exact [1], namely such that any equivalence relation is effective, i.e. the kernel relation of some map, then thanks to Theorem 3.9 in [20], the previous definition is equivalent to the definition of [20], and accordingly to the definition of Smith [22] in the Mal'cev varietal context.

So one of the advantage of this definition is that it extends the meaning of commutator from the exact Mal'cev context to the regular Mal'cev one, enlarging the range of examples to the Mal'cev (quasi-varieties), or to the case of the topological groups for instance.

3 Universal constructions

Let us come back to a finitely cocomplete regular Mal'cev category $\mathbb{C}$. Let us write $Rel(\mathbb{C})$ for the following category:

1. the objects are the pairs (X, R) where X is an object of $\mathbb{C}$ and R is an equivalence relation of X;
2. a morphism $f : (X, R) \to (X', R')$ is a morphism $f : X \to X'$ in $\mathbb{C}$ such that $R \leq f^{-1}(R')$.

We shall write further $Rel^2(\mathbb{C})$ for the category:

1. whose objects are the triples (X, R, S) of an object X and two equivalence relations R, S on X;
2. the morphisms $f : (X, R, S) \to (X', R', S')$ are the morphisms f such that $R \leq f^{-1}(R')$ and $S \leq f^{-1}(S')$.

Finally we write $ZRel(\mathbb{C})$ for the subcategory (actually full subcategory saturated for subobjects, see [9], and [3] for the details) of $Rel^2(\mathbb{C})$ whose objects are the triples (X, R, S) such that R and S admit a connector, and we denote by $j : ZRel(\mathbb{C}) \to Rel^2(\mathbb{C})$ the full inclusion. Then the construction T easily extends to a left adjoint $T : Rel^2(\mathbb{C}) \to ZRel(\mathbb{C})$ of j. This remark has a number of interesting consequences.

3.1. Corollary. *Let $\mathbb{C}$ be a finitely cocomplete regular Mal'cev category. Given an object X, the object $T(\nabla_X, \nabla_X)$ is the abelian object universally associated with X. This yields the left adjoint to the full inclusion $Ab(\mathbb{C}) \hookrightarrow \mathbb{C}$ of the abelian objects of $\mathbb{C}$.*

Proof Straightforward, since the object X is abelian if and only if the pair (∇_X, ∇_X) is connected. □

If we "localize" the previous result in the slice category $\mathbb{C}/Y$, we obtain:

3.2. Corollary. *Let $\mathbb{C}$ be a pointed finitely cocomplete regular Mal'cev category. Every map $f : X \to Y$ in $\mathbb{C}$ factorizes universally through a map with abelian kernel.*

Proof The slice category $\mathbb{C}/Y$ is finitely cocomplete, regular and Mal'cev. We noticed that a map $f : X \to Y$ is an abelian object in $\mathbb{C}/Y$ if and only if it has abelian kernel. By the previous corollary, the object $f : X \to Y$ has a universally associated abelian object $\tilde{X} : \tilde{f} \to Y$ in $\mathbb{C}/Y$ and the universal morphism $h : (X, f) \to (\tilde{X}, \tilde{f})$ in $\mathbb{C}/Y$ yields the factorization of f through the map $\tilde{f}$ with abelian kernel. □

3.3. Corollary. *Let $\mathbb{C}$ be a finitely cocomplete, regular Mal'cev category. Every map $f : X \to Y$ factorizes universally through a map with central kernel.*

Proof Clearly the map f in the category $\mathbb{C}$ determines a map $f : (X, R[f], \nabla_X) \to (Y, \Delta_Y, \nabla_Y)$ in $Rel^2(\mathbb{C})$. Moreover, (Y, Δ_Y, ∇_Y) lies in $ZRel(\mathbb{C})$. So, we get a factorization $f' : T(R[f], \nabla_X) \to Y$ such that $f = f'.\psi$, where $\psi : X \to T(R[f], \nabla_X)$ is the canonical strong epimorphism. Since ψ is a regular epimorphism, $\psi(\nabla_X) = \nabla_T$; moreover, the factorization $\overline{\psi} : R[f] \to R[f']$ induced by ψ is itself a regular epimorphism, proving that $\psi(R[f]) = R[f']$. By the universal property of ψ, the relations $\psi(R[f]) = R[f']$ and $\psi(\nabla_X) = \nabla_T$ are connected, proving that f' has central kernel. □

We shall also be able to associate a groupoid with any reflexive graph. Let us recall, see [9] for instance, that, given a reflexive graph on an object X in a Mal'cev category:

$$G \underset{d_0}{\overset{d_1}{\underset{\longleftarrow s_0}{\rightrightarrows}}} X$$

this graph is underlying a groupoid if and only if the commutator $[R[d_0], R[d_1]] = 0$. Consequently there is at most one structure of groupoid on a given reflexive graph. Moreover the inclusion $j : Grd(\mathbb{C}) \hookrightarrow Gph(\mathbb{C})$ of the internal groupoids into the reflexive graphs is a full inclusion saturated for subobjects, see [4], and also [12].

3.4. Proposition. *Let $\mathbb{C}$ be a finitely cocomplete, regular Mal'cev category. The full inclusion $j : Grd(\mathbb{C}) \hookrightarrow Gph(\mathbb{C})$ admits a left adjoint.*

Proof Let us start with a reflexive graph. Then consider the map $\psi : G \to X_1 = T(R[d_0], R[d_1])$. The morphism d_0 is a split epimorphism, with section s_0. The image $d_0(R[d_0])$ is simply ΔX. Thus it is connected with every equivalence relation on X, and in particular with $d_0(R[d_1])$. So we get a factorization $\overline{d_0} : X_1 \to X$ such that $\overline{d_0}.\psi = d_0$. In the same way, there is a morphism $\overline{d_1} : X_1 \to X$ such that $\overline{d_1}.\psi = d_1$. Set $\overline{s_0} = \psi.s_0$. It is then obvious that

$$X_1 \underset{\overline{d_0}}{\overset{\overline{d_1}}{\underset{\longleftarrow \overline{s_0}}{\rightrightarrows}}} X$$

is a reflexive graph and that $\psi : G \to X_1$ is a morphism of reflexive graphs. Since ψ is a regular epimorphism, we have $\psi(R[d_i]) = R[\overline{d_i}]$. Thus $R[\overline{d_0}]$ and $R[\overline{d_1}]$ are connected by definition of $X_1 = T(R[d_0], R[d_1])$. So the previous graph is provided with a structure of a groupoid. The universal property of this construction follows easily from the universal property of T. □

4 Commutator theory

We are now going to investigate the properties of the commutator.

4.1 The regular context. Let us begin with the regular context. We shall later study the exact one.

4.2. Proposition. *Let two equivalence relations* $(d_0, d_1) : R \rightrightarrows X$ *and* $(d_0, d_1) : S \rightrightarrows X$ *be given on the same object* X *in* $\mathbb{C}$. *Then* $[R, S] = [S, R]$.

Proof The symmetry of the relations interchanges the d_0 and the d_1; it induces a natural isomorphism between the diagrams $D(R, S)$ and $D(S, R)$. Accordingly the colimits $T(R, S)$ and $T(S, R)$ are isomorphic, and $[R, S] = [S, R]$. □

4.3. Proposition. *Let three equivalence relations* R, R' *and* S *be given on* X. *Then* $R' \leq R$ *implies* $[R', S] \leq [R, S]$.

Proof Clearly the inclusion $R' \leq R$ determines a natural transformation from $D(R', S)$ to $D(R, S)$. Accordingly there is a factorization $T(R', S) \to T(R, S)$ which implies that $[R', S] \leq [R, S]$. □

We have also:

4.4. Proposition. *Suppose given pairs of equivalence relations* (R, R') *on* X *and* (S, S') *on* X'. *Then* $[R \times S, R' \times S'] \leq [R, R'] \times [S, S']$.

Proof This comes from the functoriality of the construction T. The projection $p_X : X \times X' \to X$ determines a map $p_X : (X \times X', R \times S, R' \times S') \to (X, R, R')$ in $Rel^2(\mathbb{C})$ which gives an image $T(p_X) : T(R \times S, R' \times S') \to T(R, R')$. The factorization $(T(p_X), T(p_{X'})) : T(R \times S, R' \times S') \to T(R, R') \times T(S, S')$ provides the required inequality. □

In a regular Mal'cev category, the equivalence relations on X can be composed, and we have moreover $R \circ S = S \circ R = R \vee S$, see [11]. The composite relation $R \circ S$ is defined in the following way: consider the following pullback:

$$\begin{array}{ccc} R \times_X S & \xrightarrow{p_S} & S \\ {\scriptstyle p_R}\downarrow & & \downarrow{\scriptstyle d_{0,S}} \\ R & \xrightarrow[d_{1,R}]{} & X \end{array}$$

and take the canonical regular-epi/mono factorization of the map $(d_0.p_0, d_1.p_1) : R \times_X S \to X \times X$, $(x, y, z) \mapsto (x, z)$, namely:

$$R \times_X S \twoheadrightarrow R \circ S \rightarrowtail X \times X$$

When R and S are two equivalence relations on X, then, in a regular category, $R \circ S$ is a reflexive relation which is not necessarily symmetric nor transitive. But this is clearly the case in a regular Mal'cev category.

4.5. Proposition. *Let three equivalence relations* R, S_1 *and* S_2 *be given on* X. *Then* $[R, S_1 \vee S_2] = [R, S_1] \vee_{eff} [R, S_2]$, *that is the supremum of* $[R, S_1]$ *and* $[R, S_2]$ *in the poset of effective equivalence relations on* X.

Proof The following square is trivially a pushout in $Rel^2(\mathbb{C})$:

$$\begin{array}{ccc} (X,R,\Delta X) & \xrightarrow{1_X} & (X,R,S_1) \\ {\scriptstyle 1_X}\downarrow & & \downarrow{\scriptstyle 1_X} \\ (X,R,S_2) & \xrightarrow[1_X]{} & (X,R,S_1 \vee S_2) \end{array}$$

Accordingly its image by the left adjoint functor T is still a pushout in $ZRel(\mathbb{C})$:

$$\begin{array}{ccc} T(R,\Delta X) & \xrightarrow{\psi_1} & T(R,S_1) \\ {\scriptstyle \psi_2}\downarrow & \searrow^{\psi} & \downarrow{\scriptstyle \gamma_1} \\ T(R,S_2) & \xrightarrow[\gamma_2]{} & T(R,S_1 \vee S_2) \end{array}$$

But clearly R and ΔX are connected, so that $T(R,\Delta X) = X$, and, finally, the following diagram is a pushout in $\mathbb{C}$:

$$\begin{array}{ccc} X & \xrightarrow{\psi_1} & T(R,S_1) \\ {\scriptstyle \psi_2}\downarrow & \searrow^{\psi} & \downarrow{\scriptstyle \gamma_1} \\ T(R,S_2) & \xrightarrow[\gamma_2]{} & T(R,S_1 \vee S_2) \end{array}$$

This precisely means that:

$$[R,S_1 \vee S_2] = R[\psi] = R[\psi_1] \vee_{eff} R[\psi_2] = [R,S_1] \vee_{eff} [R,S_2].$$

□

4.6. Proposition. *Let $f : X \to Y$ be a regular epi, and (R,S) a pair of equivalence relations on X. Then $[f(R), f(S)] = f(R[f] \vee_{eff} [R,S])$, and $f([R,S]) \leq [f(R), f(S)]$.*

Proof The following diagram is trivially a pushout in $Rel^2(\mathbb{C})$:

$$\begin{array}{ccc} (X,R,\Delta X) & \xrightarrow{f} & (Y,f(R),\Delta Y) \\ {\scriptstyle 1_X}\downarrow & & \downarrow{\scriptstyle 1_Y} \\ (X,R,S) & \xrightarrow[f]{} & (Y,f(R),f(S)) \end{array}$$

Accordingly its image by the left adjoint functor T is still a pushout in $ZRel(\mathbb{C})$:

$$\begin{array}{ccc} X = T(R,\Delta X) & \xrightarrow{f} & T(f(R),\Delta Y) = Y \\ {\scriptstyle \psi}\downarrow & & \downarrow{\scriptstyle \psi'} \\ T(R,S) & \xrightarrow[T(f)]{} & T(f(R),f(S)) \end{array}$$

Therefore we have:

$$f^{-1}([f(R),f(S)]) = f^{-1}(R[\psi']) = R[\psi'.f] = R[f] \vee_{eff} R[\psi] = R[f] \vee_{eff} [R,S]$$

But in a regular Mal'cev category, we have always $f(f^{-1}(\Sigma)) = \Sigma$ for any equivalence relation Σ on Y. Accordingly:

$$[f(R), f(S)] = f(f^{-1}([f(R), f(S)])) = f(R[f] \vee_{eff} [R, S])$$

On the other hand:

$$f([R, S]) = \Delta Y \vee f([R, S]) = f(R[f]) \vee f([R, S]) = f(R[f]) \vee [R, S])$$

But $R[f] \vee [R, S] \leq R[f] \vee_{eff} [R, S]$, and $f([R, S]) \leq [f(R), f(S)]$. □

4.7. Corollary. *Let $\mathbb{C}$ be a cocomplete regular Mal'cev category and $f : X \to Y$ a regular epi. Given (U, V) any pair of equivalence relations on Y, we have:*

$$f^{-1}[U, V] = R[f] \vee_{eff} [f^{-1}(U), f^{-1}(V)]$$

Proof Taking $R = f^{-1}(U)$ and $S = f^{-1}(V)$, the following square is a pushout as above:

$$\begin{array}{ccc} X & \xrightarrow{f} & Y \\ \psi \downarrow & & \downarrow \psi' \\ T(f^{-1}(U), f^{-1}(V)) & \xrightarrow[T(f)]{} & T(U, V) \end{array}$$

Accordingly, $R[\psi'.f] = R[f] \vee_{eff} R[\psi] = R[f] \vee_{eff} [f^{-1}(U), f^{-1}(V)]$, while $R[\psi'.f] = f^{-1}(R[\psi']) = f^{-1}[U, V]$. □

4.8 The exact context. We shall suppose from now on the category $\mathbb{C}$ finitely cocomplete, exact and Mal'cev, as is the case for the Mal'cev varieties. We already noticed that, in the exact context, our construction of the commutator coincides with the one previously given by Pedicchio, and consequently coincides in the Mal'cev varietal context with Smith's one. We shall then recover easily all the main identities.

4.9. Proposition. *We have always $[R, S_1 \vee S_2] = [R, S_1] \vee [R, S_2]$.*

Proof Straightforward, since any equivalence relation being effective, we have "$\vee = \vee_{eff}$". □

4.10. Proposition. *The commutators are stable by regular direct images.*

Proof Let $f : X \to Y$ be a regular epi. We noticed that the following square is always a pushout:

$$\begin{array}{ccc} X = T(R, \Delta X) & \xrightarrow{f} & T(f(R), \Delta Y) = Y \\ \psi \downarrow & & \downarrow \psi' \\ T(R, S) & \xrightarrow[T(f)]{} & T(f(R), f(S)) \end{array}$$

But in an exact Mal'cev category, any pushout of regular epimorphisms is a regular pushout, which means that the factorization through the induced pullback is a regular epi, see [7] and also [10]. Then according to Lemma 2.1 in [7], the induced factorization

$$\tilde{f} : R[\psi] = [R, S] \to R[\psi'] = [f(R), f(S)]$$

is a regular epi, and consequently $f([R,S]) = [f(R), f(S)]$. □

Similarly:

4.11. Corollary. *Let $f : X \to Y$ be a regular epi. Given (U,V) any pair of equivalence relations on Y, we have:*

$$f^{-1}[U,V] = R[f] \vee [f^{-1}(U), f^{-1}(V)]$$

And finally:

4.12. Theorem. *We have always $[R,S] \leq R \wedge S$.*

Proof The result will follow from $[R,S] \leq R$. Take the quotient $q : X \twoheadrightarrow X/R$. Then the direct image $q(R) = \Delta(X/R)$. Thus $q(R)$ is connected to any relation, and in particular to $q(S)$. Therefore there is a factorization $T(R,S) \to X/R$ which gives $[R,S] \leq R$. □

References

[1] M. Barr, *Exact categories*, Lecture Notes in Math., **236**, Springer, 1971, 1-120.

[2] F. Borceux, *A survey of semi-abelian categories*, this volume.

[3] F. Borceux and D. Bourn, *Mal'cev, protomodular, homological and semi-abelian categories*, Kluwer, to appear 2004.

[4] D. Bourn, *Mal'cev categories and fibration of pointed objects*, Appl. Categorical Structures, **4**, 1996, 307-327.

[5] D. Bourn, *Normal subobjects and abelian objects in protomodular categories*, Journal of Algebra, **228**, 2000, 143-164.

[6] D. Bourn, *Intrinsic centrality and associated classifying properties*, Journal of Algebra, **256**, 2002, 126-145.

[7] D. Bourn, *The denormalized* 3×3 *lemma*, J. Pure Appl. Algebra, **177**, 2003, 113-129.

[8] D. Bourn and M. Gran, *Centrality and normality in protomodular categories*, Th. and Applications of Categories, **9**, 2002, 151-165.

[9] D. Bourn and M. Gran, *Centrality and connectors in Maltsev categories*, Algebra Univers., **48**, 2002, 309-331.

[10] A. Carboni, G.M. Kelly and M.C. Pedicchio, *Some remarks on Maltsev and Goursat categories*, Appl. Categorical Structures, **1**, 1993, 385-421.

[11] A. Carboni, J. Lambek and M.C. Pedicchio, *Diagram chasing in Mal'cev categories*, J. Pure Appl. Algebra, **69**, 1991, 271-284.

[12] A. Carboni, M.C. Pedicchio and N. Pirovano, *Internal graphs and internal groupoids in Mal'cev categories*, CMS Conference Proceedings, **13**, 1992, 97-109.

[13] R. Freese and R. McKenzie, *Commutator theory for congruence modular varieties*, LMS Lecture Notes, **125**, 1987.

[14] J. Hagemann and C. Hermann, *A concrete ideal multiplication for algebraic systems and its relation to congruence distributivity*, Arch. Math. , **32**, 1979, 234-245.

[15] S.A. Huq, *Commutator, nilpotency and solvability in categories*, Quart. J. Oxford, **19**, 1968, 363-389.

[16] G. Janelidze, L. Marki and W. Tholen, *Semi-abelian categories*, J. Pure Appl. Algebra, **168**, 2002, 367-386.

[17] P.T. Johnstone, *Affine categories and naturally Mal'cev categories*, J. Pure Appl. Algebra, **61**, 1989, 251-256.

[18] P.T. Johnstone, *The closed subgroup theorem for localic herds and pregroupoids*, J. Pure Appl. Algebra, **70**, 1989, 97-106.

[19] A. Kock, *The algebraic theory of moving frames*, Cahiers Top. Géom. Diff. Cat., **23**, 1982, 347-362.

[20] M.C. Pedicchio, *A categorical approach to commutator theory*, Journal of Algebra, **177**, 1995, 647-657.

[21] M.C. Pedicchio, *Arithmetical categories and commutator theory*, Appl. Categorical Structures, **4**, 1996, 297-305.
[22] J.D.H. Smith, *Mal'cev varieties*, Lecture Notes in Math., **554**, Springer, 1976.

Received December 9, 2002; in revised form February 26, 2003

Fields Institute Communications
Volume **43**, 2004

Categorical Aspects of Modularity

Dominique Bourn
Lab. Mathématiques Pures et Appliquées
Université du Littoral Côte d'Opale
50 Rue F. Buisson BP 699
62228 Calais, France
bourn@lmpa.univ-littoral.fr

Marino Gran
Lab. Mathématiques Pures et Appliquées
Université du Littoral Côte d'Opale
50 Rue F. Buisson BP 699
62228 Calais, France
gran@lmpa.univ-littoral.fr

Abstract. We investigate internal categorical structures in varieties of universal algebras and in more general categories. Special attention is paid to pseudogroupoids, which play a central role in commutator theory. We work in the general context of categories satisfying the shifting property, given by a categorical formulation of Gumm's Shifting Lemma. This general context includes in particular any congruence modular variety as well as any regular Maltsev category. Several characterizations of internal categories and internal groupoids in these categories are given.

Introduction

The purpose of this paper is to present some new results in the categorical approach to commutator theory and centrality.

The theory of commutators, first developed by Smith [22] in the context of Maltsev varieties, was then extended by Hagemann and Hermann to congruence modular varieties [14] [13] [10]. This theory can be considered as an extension of the classical notion of commutator for groups to more general varieties of universal algebras.

The categorical approach to commutator theory [20] [21] [15] [5] focuses on the connection between the properties of some internal categorical structures in varieties of universal algebras and the properties of the commutators. In many important

2000 *Mathematics Subject Classification.* Primary 08B10, 08C05, 18C05; Secondary 08A30,18D05.

Key words and phrases. Modular varieties, commutators, abelian objects, internal categories and groupoids, pseudogroupoids, shifting lemma.

algebraic varieties these internal structures are themselves of special mathematical interest: for instance, the category of internal groupoids in the category of groups, which is itself a variety of universal algebras, is important in homotopy theory [19], since it is equivalent to the variety of crossed modules [6].

Various descriptions of internal categories and internal groupoids were obtained by Janelidze and Pedicchio in the general context of congruence modular varieties [15]. These results clarified how deep the relationship between commutators and internal groupoids in modular varieties is. A good illustration of this fact is also provided by the following characterization of internal groupoids [12]: given an internal reflexive graph X in a modular variety

$$X_1 \underset{d_1}{\overset{d_0}{\underset{\longleftarrow}{\overset{\longrightarrow}{\longrightarrow}}}} X_0 \quad (s_0 : X_0 \to X_1)$$

X has a (unique) internal groupoid structure if and only if

$$(1) \quad [R[d_0], R[d_1]] = \Delta_{X_1} \qquad \text{and} \qquad (2) \quad R[d_0] \circ R[d_1] = R[d_1] \circ R[d_0]$$

(where $R[d_0]$ and $R[d_1]$ are the congruences arising as the kernel pairs of d_0 and d_1, respectively, and Δ_{X_1} is the smallest congruence on X_1).

Important progress was made when the internal structure of *pseudogroupoid* was introduced [16]. In any modular variety it was proved that two congruences R and S on an algebra X have trivial commutator $[R, S] = \Delta_X$ if and only if there is a (unique) pseudogroupoid structure on R and S.

In this paper we investigate internal pseudogroupoids, internal categories and internal groupoids in a very general categorical context. This context includes, on the one hand, any congruence modular variety and, on the other hand, any regular Maltsev category [8], so in particular also the categories of topological groups, Hausdorff groups and torsion-free abelian groups [7].

The main axiom we require, that we call the *shifting property*, is a categorical formulation of the well-known Shifting Lemma for modular varieties: it was Gumm who proved that, for a variety of universal algebras, the validity of the Shifting Lemma is equivalent to congruence modularity [13]. In categorical language, the shifting property can be expressed very simply by the requirement that a certain kind of internal functors between internal equivalence relations are discrete fibrations (see Lemma 2.2). In the presence of this weak assumption, the internal categorical structures behave surprisingly well: the notion of pseudogroupoid can be significantly simplified, and a pseudogroupoid on two equivalence relations is necessarily unique, when it exists. Accordingly, for two given equivalence relations, having a pseudogroupoid structure becomes a property. Many other results follow from this simple axiom, and some of them improve previously known ones in the special case of modular varieties. Accordingly, this axiomatic approach allows one not only to cover a wider range of examples, but also to sharpen the results. Under many respects, the categories which satisfy the shifting property are in the same relationship with modular varieties, as Maltsev categories are with Maltsev varieties.

The paper is structured as follows:

1. Pseudogroupoids
2. The Shifting Property
3. Connectors and Abelian Objects

4. Internal Categories
5. Internal Groupoids

In the first section we recall the definition of a pseudogroupoid, and we show that the "diamond associativity" condition in its definition can be replaced by two natural and simpler conditions. In the second section we introduce the shifting property, we give some examples, and we then establish our main results on pseudogroupoids in a category satisfying the shifting property. In the third section we apply the previous results to study internal connectors [4] [5] and abelian objects in these categories. In the last two sections several characterizations of the internal categories and of the internal groupoids are given. In particular, the above-mentioned characterization of internal groupoids in modular varieties is extended to any regular category satisfying the shifting property.

Acknowledgement: The authors would like to thank M.C. Pedicchio for having drawn their attention to Gumm's work.

1 Pseudogroupoids

In this section we recall the definition of an internal pseudogroupoid introduced by Janelidze and Pedicchio in [16]. We then show that the "diamond associativity" requirement in its definition is equivalent to two simpler and natural axioms.

Let us first fix some notations. $\mathcal{C}$ will always denote a category with finite limits. Thanks to the Yoneda embedding, in order to prove finite limit properties in $\mathcal{C}$ it is sufficient to do it "elementwise" in the category of sets. We shall freely use this well-known technique through the whole paper.

For any object X in $\mathcal{C}$, we write Δ_X for the smallest equivalence relation on X, and ∇_X for the largest equivalence relation on X. For any arrow $f \colon A \to B$ let $R[f]$ denote its *kernel relation*, that is the equivalence relation arising as its kernel pair. If R and S are equivalence relations on X, we write $R\square S$ for the double equivalence relation on R and S obtained by the following pullback:

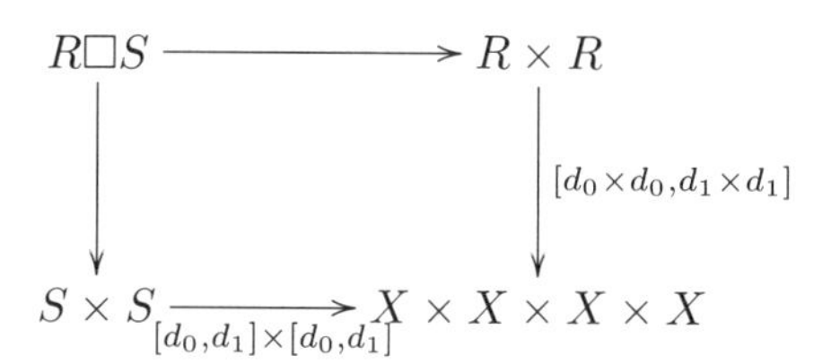

where d_0 and d_1 represent, respectively, the first and the second projection of the equivalence relation R or S. An element in $R\square S$ is called an R-S rectangle: it consists of four elements x, y, t, z in X with the property that xRy, tRz, xSt and ySz. An R-S rectangle will be represented equivalently by a diagram of the form

$$\begin{array}{ccc} x & \overset{S}{\text{———}} & t \\ {\scriptstyle R}\Big| & & \Big|{\scriptstyle R} \\ y & \underset{S}{\text{———}} & z, \end{array}$$

by a matrix $\begin{pmatrix} x & t \\ y & z \end{pmatrix}$, or by (x, y, t, z). This last "linear" notation for an element in $R\Box S$ explains the following notations for the projections of $R\Box S$ on R and S:

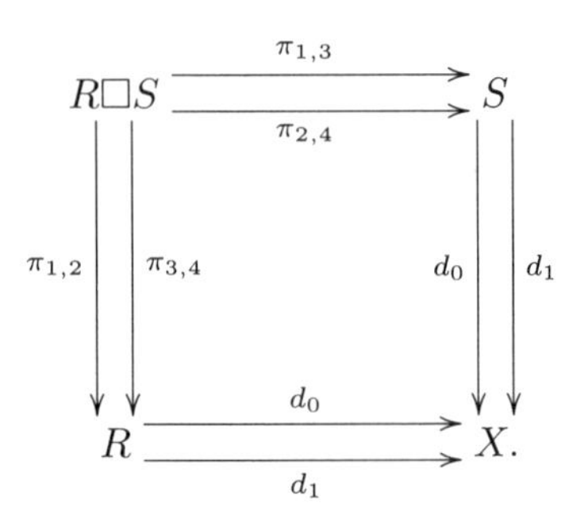

Let us then recall the definition of a pseudogroupoid [16]:

1.1. Definition. A *pseudogroupoid* on R and S is an arrow $m\colon R\Box S \to X$ in $\mathcal{C}$, written as

$$m\begin{pmatrix} x & t \\ y & z \end{pmatrix} = m(x, y, t, z),$$

with the following properties:

1. $xSm(x, y, t, z)Rz$

$$\begin{array}{ccc} x & \overset{S}{\text{———}} & m(x,y,t,z) \\ {\scriptstyle R}\,\Big| & & \Big|\,{\scriptstyle R} \\ y & \underset{S}{\text{———}} & z \end{array}$$

2. $m(x, y, t, z) = m(x, y, t', z)$ (i.e. m does not depend on the third variable)

3A. $m(x, x, t, y) = y$ 3B. $m(x, y, t, y) = x$

4. $m(m(x_1, x_2, y, x_3), x_4, t, x_5) = m(x_1, x_2, t, m(x_3, x_4, z, x_5))$ for every diagram of the following form:

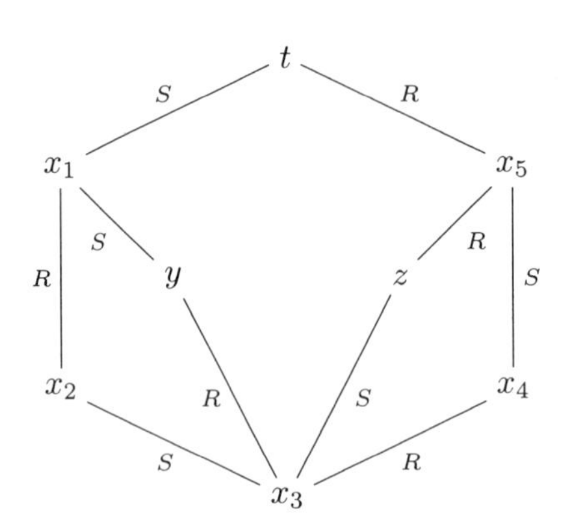

1.2. Examples. If R and S are two equivalence relations on X, we denote by $R \times_X S$ the pullback

$$\begin{array}{ccc} R \times_X S & \xrightarrow{p_1} & S \\ {\scriptstyle p_0}\downarrow & & \downarrow{\scriptstyle d_0} \\ R & \xrightarrow[d_1]{} & X. \end{array}$$

A *connector* between R and S [4] (see also [20] [21]) is an arrow $p : R \times_X S \to X$ in $\mathcal{C}$ such that

$I.\ xSp(x,y,z)Rz$

$IIA.\ p(x,x,y) = y$ $\qquad$ $IIB.\ p(x,y,y) = x$

$IIIA.\ p(x,y,p(y,u,v)) = p(x,u,v)$ $\qquad$ $IIIB.\ p(p(x,y,u),u,v) = p(x,y,v)$

Any connector p determines a pseudogroupoid: for any (x,y,t,z) in $R\Box S$ one defines $m(x,y,t,z) = p(x,y,z)$.

1.3. Examples. Let X be an *internal groupoid* in $\mathcal{C}$, represented by the diagram

$$X_1 \times_{X_0} X_1 \overset{p_1}{\underset{p_0}{\overset{m}{\rightrightarrows}}} X_1 \overset{d_0}{\underset{d_1}{\overset{s_0}{\leftrightarrows}}} X_0$$

where X_0 represents the "object of objects", X_1 is the "object of arrows", $X_1 \times_{X_0} X_1$ is the "object of composable arrows", d_0 is the domain, d_1 is the codomain and m is the groupoid composition. It determines a (unique) connector, and then a pseudogroupoid, on the kernel relations $R[d_0]$ and $R[d_1]$: the connector structure is internally defined by $p(x,y,z) = z \circ y^{-1} \circ x$. Conversely, a connector on $R[d_0]$ and $R[d_1]$ determines a groupoid structure on the underlying reflexive graph.

1.4. Examples. If R and S are two equivalence relations on X with the property that $R \cap S = \Delta_X$, then R and S have a unique pseudogroupoid structure [16]: for any (x,y,t,z) in $R\Box S$ this structure is given by $m(x,y,t,z) = t$. Indeed, the condition $R \cap S = \Delta_X$ guarantees the uniqueness of such a t, when it exists.

1.5. Proposition. *An arrow $m \colon R\Box S \to X$ is a pseudogroupoid on R and S if and only if it satisfies the axioms $1, 2, 3A$, $3B$ and the following axioms $4A$ and $4B$:*

$4A.$ $m(x,y,m(x,y,x',y'),m(y,z,y',z')) = m(x,z,x',z')$ *for every diagram*

$$\begin{array}{ccc} x & \overset{S}{\text{———}} & x' \\ {\scriptstyle R}\Big| & & \Big|{\scriptstyle R} \\ y & \underset{S}{\text{———}} & y' \\ {\scriptstyle R}\Big| & & \Big|{\scriptstyle R} \\ z & \overset{S}{\text{———}} & z' \end{array}$$

$4B.$ $m(m(x,y,x',y'),y',m(x',y',x'',y''),y'') = m(x,y,x'',y'')$ *for every diagram*

$$\begin{array}{ccccc} x & \overset{S}{\text{———}} & x' & \overset{S}{\text{———}} & x'' \\ R\Big| & & \Big|R & & \Big|R \\ y & \underset{S}{\text{———}} & y' & \underset{S}{\text{———}} & y'' \end{array}$$

Proof Let us begin by proving that the axioms $1, 2, 3A, 3B$ and 4 implies $4A$ and $4B$. We consider the situation

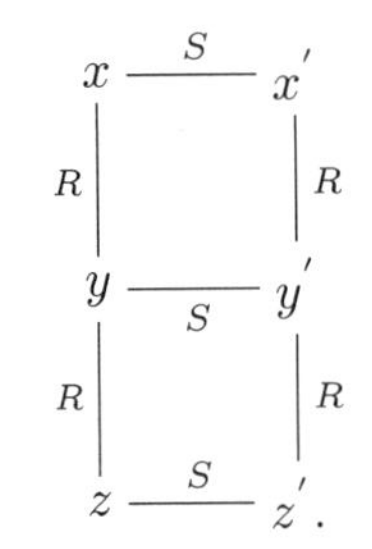

Then, thanks to axiom 1 we can form the diagram

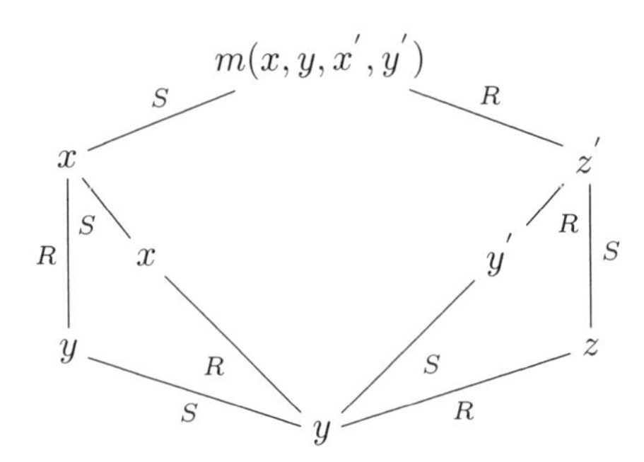

and then

$$\begin{aligned} & m(x, y, m(x, y, x', y'), m(y, z, y', z')) \\ = \;& m(m(x, y, x, y), z, m(x, y, x', y'), z') \\ = \;& m(x, z, m(x, y, x', y'), z') \\ = \;& m(x, z, x', z'). \end{aligned}$$

where the first equality follows from axiom 4, the second one from axiom 3B and the last one from axiom 2.

Similarly, if we start from the situation

$$\begin{array}{ccccc} x & \overset{S}{\text{———}} & x' & \overset{S}{\text{———}} & x'' \\ R\Big| & & \Big|R & & \Big|R \\ y & \underset{S}{\text{———}} & y' & \underset{S}{\text{———}} & y'', \end{array}$$

we can form the diagram

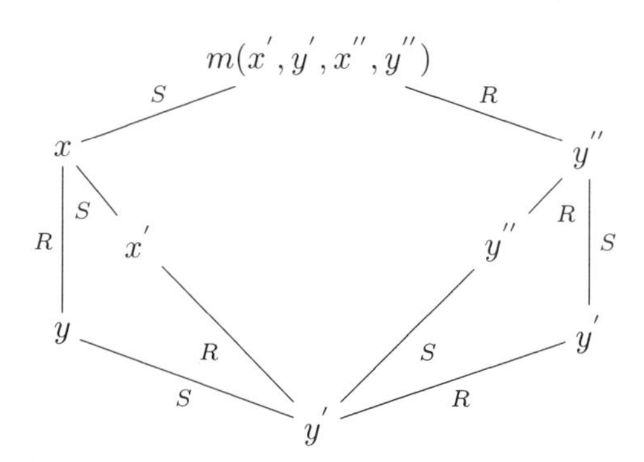

Then

$$
\begin{aligned}
& m(m(x,y,x',y'),y',m(x',y',x'',y''),y'') \\
= \; & m(x,y,m(x',y',x'',y''),m(y',y',y'',y'')) \\
= \; & m(x,y,m(x',y',x'',y''),y'') \\
= \; & m(x,y,x'',y''),
\end{aligned}
$$

by axioms 4, 3A and 2.

Conversely, if $m\colon R\square S \to X$ satisfies $1, 2, 3A, 3B, 4A$ and $4B$, let us then prove that m also satisfies axiom 4. We begin with the diagram

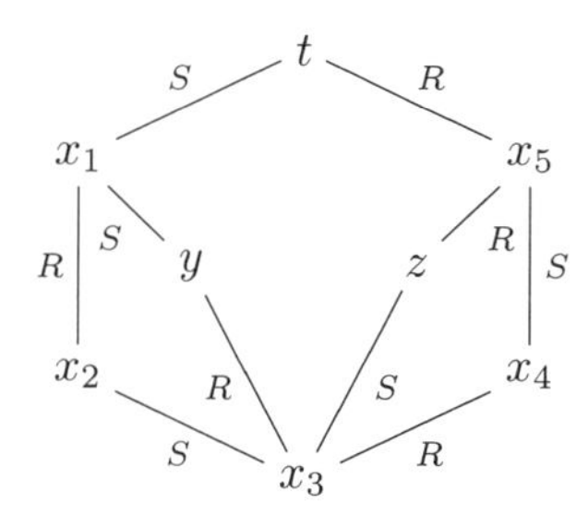

Axiom 1 allows us to form the diagram

$$
\begin{array}{ccccc}
x_1 & \overset{S}{\text{---}} & y & \overset{S}{\text{------}} & t \\
R \,\big| & & \big|\, R & & \big|\, R \\
x_2 & \underset{S}{\text{---}} & x_3 & \underset{S}{\text{---}} & m(x_3,x_4,z,x_5).
\end{array}
$$

Thanks to axiom $4B$ and to the fact that m does not depend on the third variable we get

$$
\begin{aligned}
& m(x_1,x_2,t,m(x_3,x_4,z,x_5)) \\
= \; & m(m(x_1,x_2,y,x_3),x_3,m(y,x_3,t,m(x_3,x_4,z,x_5)),m(x_3,x_4,z,x_5)) \\
= \; & m(m(x_1,x_2,y,x_3),x_3,m(m(x_1,x_2,y,x_3),x_3,t,z),m(x_3,x_4,z,x_5)) \\
= \; & m(m(x_1,x_2,y,x_3),x_4,t,x_5),
\end{aligned}
$$

where the last equality comes from axiom $4A$ applied to

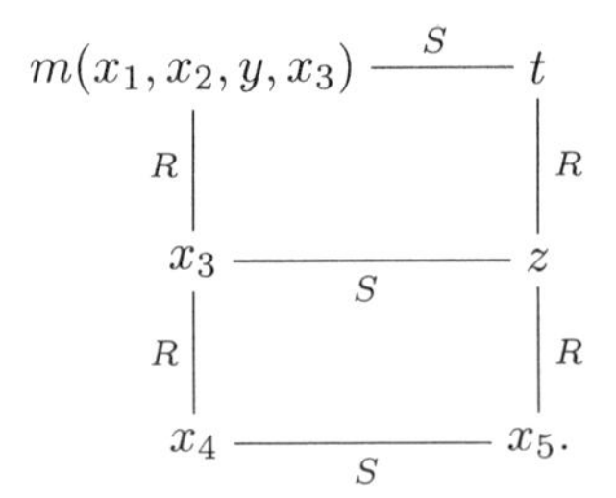

□

2 The shifting property

In this section we introduce the main axiom of this paper, namely the shifting property. Under this rather weak assumption we are going to show that a pseudogroupoid structure on two equivalence relations is necessarily unique, when it exists. Consequently, in this context, having a pseudogroupoid structure for two equivalence relations becomes a property.

Let $Eq_X(\mathcal{C})$ be the poset of equivalence relations in $\mathcal{C}$ on a fixed object X.

Let us recall the well-known *Shifting Lemma* due to Gumm [13]: a variety $\mathcal{V}$ of universal algebras satisfies the Shifting Lemma if for any R, S, $T \in Eq_X(\mathcal{V})$ with $R \cap S \leq T$ the situation

$$\begin{array}{ccc} x & \overset{S}{—} & t \\ R\,| & & |\,R \\ y & \underset{S}{—} & z, \end{array} \qquad (x\,T\,y)$$

implies that tTz.

A remarkable fact concerning the Shifting Lemma is that it characterizes congruence modular varieties, namely the varieties with the property that the lattice $Eq_X(\mathcal{V})$ is modular for any algebra X in $\mathcal{V}$.

2.1. Theorem. [13] *A variety $\mathcal{V}$ of universal algebras is congruence modular if and only if $\mathcal{V}$ satisfies the Shifting Lemma.*

In any finitely complete category the validity of the Shifting Lemma is equivalent to the following categorical property:

2.2. Lemma. *Let $\mathcal{C}$ be a finitely complete category. Then $\mathcal{C}$ satisfies the Shifting Lemma if and only if for any X in $\mathcal{C}$, for any R, S, $U \in Eq_X(\mathcal{C})$ with $R \cap S \leq U \leq R$ the canonical inclusion of equivalence relations*

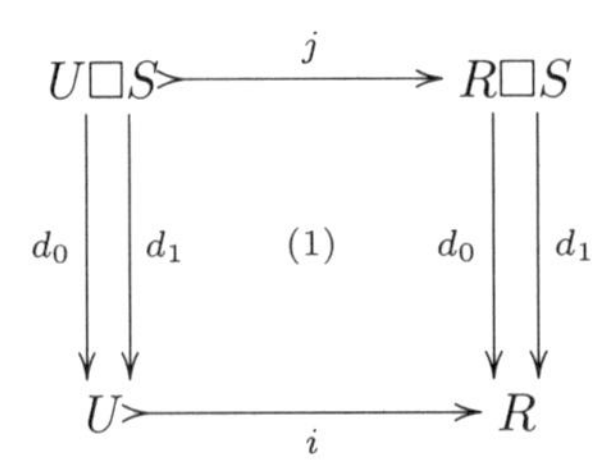

is a discrete fibration.

Proof Let us first recall that an internal functor (1) as above is a discrete fibration when the commutative square involving the arrows d_1 (or, equivalently, the arrows d_0) is a pullback.

If we assume that the property here above holds, one can define $R \cap S = U$ in the assumption of the Shifting Lemma and one immediately concludes that tUz, i.e. $t(R \cap S)z$. Conversely, when the Shifting Lemma holds and $R \cap S \leq U \leq R$ the diagram

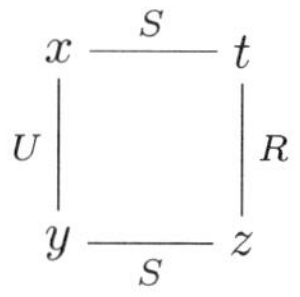

$$\begin{array}{ccc} x & \overset{S}{\text{———}} & t \\ {\scriptstyle U}\Big| & & \Big|{\scriptstyle R} \\ y & \underset{S}{\text{———}} & z \end{array}$$

allows one to form the diagram

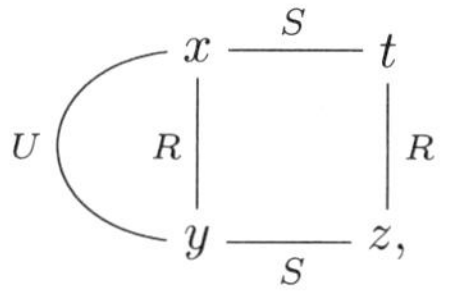

$$\begin{array}{ccc} x & \overset{S}{\text{———}} & t \\ {\scriptstyle U}\;{\scriptstyle R}\Big| & & \Big|{\scriptstyle R} \\ y & \underset{S}{\text{———}} & z, \end{array}$$

so that the assumption $R \cap S \leq U$ yields tUz, proving that the internal functor (1) is a discrete fibration. □

It follows from the previous result that the Shifting Lemma is a finite limit statement:

2.3. Definition. A finitely complete category $\mathcal{C}$ satisfies the *shifting property* when it satisfies any of the two equivalent conditions of the previous Lemma.

2.4. Examples.

1. *Congruence modular varieties*: in particular any Maltsev variety [22] (groups, rings, associative algebras, Lie algebras, quasi-groups, Heyting algebras) and any distributive variety (lattices, right complemented semigroups) satisfy the shifting property.

2. Any *regular Maltsev category* satisfies the shifting property, as well as, more generally, any regular Goursat category. This fact follows easily from Proposition 3.2 in [7]. In particular the categories of topological groups, Hausdorff groups, torsion-free abelian groups, or the dual category of an elementary topos satisfy the shifting property.

Since the *shifting property* is a finite limit statement, any functor category $\mathcal{C}^{\mathcal{A}}$ of functors from any small category $\mathcal{A}$ to $\mathcal{C}$ satisfies the shifting property, provided $\mathcal{C}$ satisfies the shifting property. A useful criterion to check whether a category satisfies the shifting property is given by the following

2.5. Lemma. *Let $F\colon \mathcal{C} \to \mathcal{D}$ be a finite limit preserving functor between finitely complete categories. If F is conservative (i.e. it reflects isomorphisms) and $\mathcal{D}$ satisfies the shifting property, then $\mathcal{C}$ satisfies the shifting property.*

Proof For any X in $\mathcal{C}$, for any R, S, $U \in Eq_X(\mathcal{C})$ with $R \cap S \leq U \leq R$ the internal functor

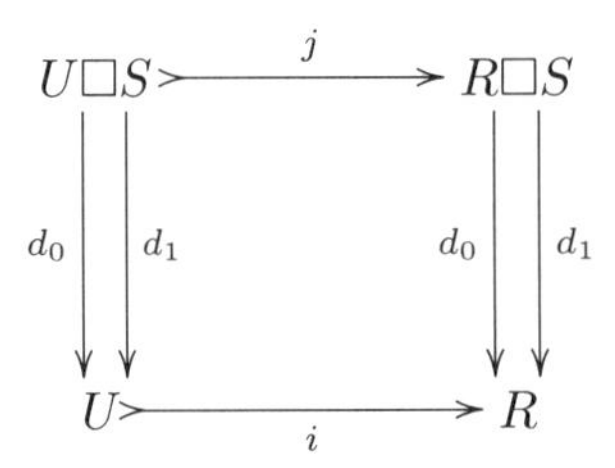

is sent to the internal functor

$$\begin{array}{ccc} F(U)\square F(S) & \xrightarrow{F(j)} & F(R)\square F(S) \\ {\scriptstyle F(d_0)} \downarrow\downarrow {\scriptstyle F(d_1)} & & {\scriptstyle F(d_0)} \downarrow\downarrow {\scriptstyle F(d_1)} \\ F(U) & \xrightarrow[F(i)]{} & F(R), \end{array}$$

which is necessarily a discrete fibration because $\mathcal{D}$ satisfies the shifting property. The functor F also reflects pullbacks, and then the first internal functor (the one in $\mathcal{C}$) is a discrete fibration as well. $\square$

The previous result easily gives some more examples of categories satisfying the shifting property: if $\mathcal{C}$ satisfies the shifting property, so do the categories $Cat(\mathcal{C})$ of internal categories in $\mathcal{C}$ and the category $Grpd(\mathcal{C})$ of internal groupoids in $\mathcal{C}$. Indeed, if one considers the forgetful functor associating its object of morphisms with any internal category or groupoid, one easily sees that it is conservative and it preserves finite limits.

On the other hand, if $\mathcal{C}$ satisfies the shifting property, so does the comma category $\mathcal{C}/Y$ for any object Y in $\mathcal{C}$. This follows from the fact that if R and S are two equivalence relations on an object $f: X \to Y$ in $\mathcal{C}/Y$, then the object $R\square S$ in the category $\mathcal{C}/Y$ is the same as $R\square S$ in the category $\mathcal{C}$. Accordingly, the argument given in the proof of Lemma 2.5 and the fact that the forgetful functor from $\mathcal{C}/Y$ to $\mathcal{C}$ is conservative and preserves pullbacks allows one to conclude.

As we shall see in Proposition 2.12, the notion of pseudogroupoid can be simplified in any category satisfying the shifting property. In order to present this simplification, we first need to introduce the following definitions:

2.6. Definition. A *double zero sequence* in a pointed category $\mathcal{C}$ is a diagram of the form

$$X \underset{s}{\overset{f}{\leftrightarrows}} Z \underset{g}{\overset{t}{\leftrightarrows}} Y$$

with $f \circ s = 1_X, g \circ t = 1_Y, g \circ s = 0, f \circ t = 0$ (where 0 is the zero arrow).

Let us then define the category $DZS(X, Y)$ of double zero sequences between X and Y, with arrows $\alpha: Z \to Z'$ making the following diagram commutative:

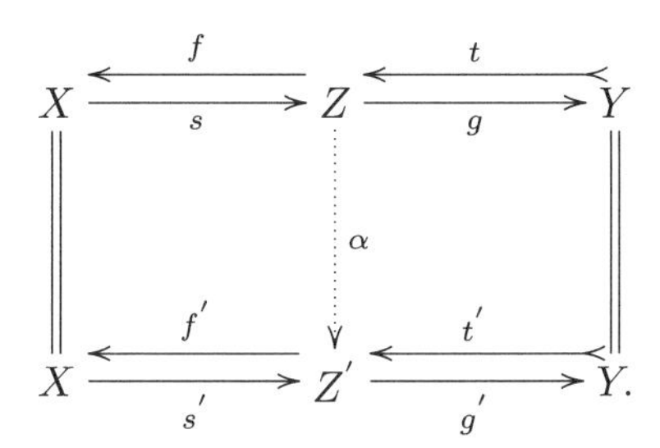

If $\mathcal{C}$ is any finitely complete category (not necessarily pointed) we denote by $Pt(\mathcal{C})$ the category whose objects (X, p, s) are the split epimorphisms $p\colon X \to Y$ in $\mathcal{C}$ with a given splitting $s\colon Y \to X$, with arrows the commutative squares between these data

$$\begin{array}{ccc} X & \xrightarrow{f_1} & X' \\ p \downarrow \uparrow s & & q \downarrow \uparrow t \\ Y & \xrightarrow[f_0]{} & Y' \end{array}$$

where $f_0 \circ p = q \circ f_1$ and $f_1 \circ s = t \circ f_0$.

The functor associating its codomain with any split epimorphism is denoted by $\pi\colon Pt(\mathcal{C}) \to \mathcal{C}$. This functor π is a fibration, called the *fibration of pointed objects* [3]. For any X in $\mathcal{C}$, we write $Pt_X(\mathcal{C})$ for the fiber above X, which is clearly pointed (by the trivial split epi $1_X\colon X \to X$).

Given R and S in $Eq_X(\mathcal{C})$, there is a canonical comparison arrow $\alpha\colon R\square S \to R \times_X S$ induced by the universal property of the pullback

$$\begin{array}{ccc} R \times_X S & \xrightarrow{p_1} & S \\ p_0 \downarrow & & \downarrow d_0 \\ R & \xrightarrow[d_1]{} & X, \end{array}$$

which associates the element (x, y, z) in $R \times_X S$ with any element (x, y, t, z) in $R\square S$. This arrow α is an arrow of double zero sequences in $DZS((R, d_1, s_0), (S, d_0, s_0))$ in the pointed category $Pt_X(\mathcal{C})$, where we write s_0 for the arrows giving the reflexivity of the relations R or S. The arrow α actually is the terminal arrow in this category:

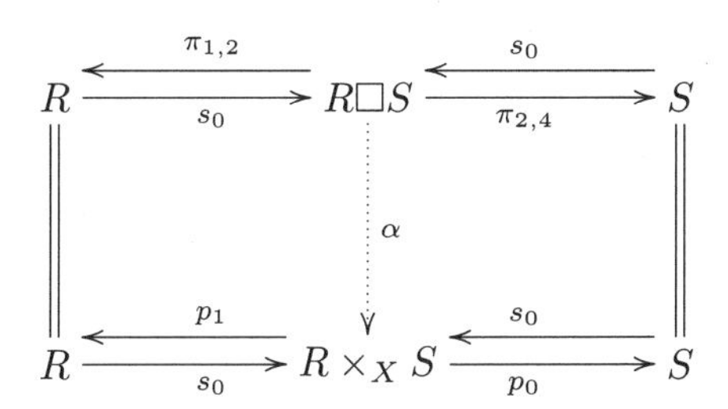

2.7. Definition. A *quasiconnector* on R and S is an arrow $\sigma\colon R\Box S \to R\Box S$ of double zero sequences in $Pt_X(\mathcal{C})$

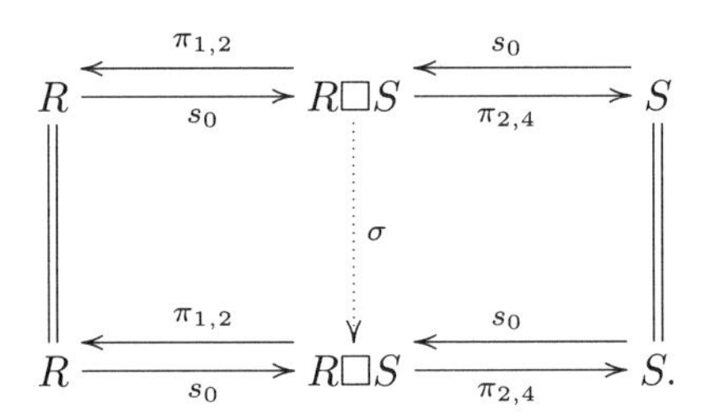

such that the kernel relation $R[\sigma]$ is $R[\pi_{1,2}] \cap R[\pi_{2,4}] = R[\alpha]$ (where $\alpha\colon R\Box S \to R \times_X S$ is the canonical terminal arrow defined above).

2.8. Remark. The axiom $R[\sigma] = R[\alpha]$ exactly says that σ does not depend on the third variable.

A quasiconnector $\sigma\colon R\Box S \to R\Box S$ is always idempotent: indeed, this follows from the following simple

2.9. Lemma. *Let $\sigma\colon Z \to Z$ be an arrow in a category with pullbacks $\mathcal{C}$ such that $R[\sigma] = \nabla_Z$ (where $\nabla_Z = R[\tau_Z]$ and $\tau_Z\colon Z \to 1$ is the terminal arrow). Then $\sigma \circ \sigma = \sigma$.*

Proof Consider the kernel relation of the terminal arrow:

$$R[\sigma] = Z \times Z \underset{p_1}{\overset{p_0}{\rightrightarrows}} Z \longrightarrow 1.$$

The arrow $\sigma\colon Z \to Z$ induces the arrow $(\sigma, 1_Z)\colon Z \to Z \times Z$. Then the equality $\sigma \circ p_0 = \sigma \circ p_1$ gives

$$\sigma \circ p_0 \circ (\sigma, 1_Z) = \sigma \circ p_1 \circ (\sigma, 1_Z).$$

It follows that

$$\sigma \circ \sigma = \sigma \circ 1_Z = \sigma.$$

□

2.10. Corollary. *Any quasiconnector satisfies $\sigma \circ \sigma = \sigma$.*

Proof One just needs to notice that in the category $DZS((R, d_1, s_0), (S, d_0, s_0))$ of double zero sequences between R and S in $Pt_X(\mathcal{C})$, the diagram

$$R \underset{s_0}{\overset{p_0}{\rightleftarrows}} R \times_X S \underset{p_1}{\overset{s_0}{\leftrightarrows}} S$$

is the terminal object, and the arrow $\alpha\colon R\Box S \to R \times_X S$ is the terminal arrow. Then the requirement $R[\sigma] = R[\alpha]$ allows one to apply the previous Lemma to σ. □

2.11. Remark. Given a quasiconnector σ, the arrow $m = d_1 \circ \pi_{1,3} \circ \sigma\colon R\Box S \to R\Box S \to S \to X$ defines an arrow $m\colon R\Box S \to X$ satisfying the axioms 1 and 2 in the definition of pseudogroupoid, and is such that

$3C \quad m(x,x,y,y) = y$

$3D \quad m(x,y,x,y) = x.$

Conversely, an arrow $m\colon R\Box S \to X$ satisfying axioms $1, 2, 3C$ and $3D$ gives rise to a quasiconnector. Accordingly, we shall refer equivalently to σ or to m when speaking of a quasiconnector.

Now, when the shifting property holds in $\mathcal{C}$, the notion of quasiconnector is equivalent to the one of pseudogroupoid:

2.12. Proposition. *If $\mathcal{C}$ is finitely complete and satisfies the shifting property, then any quasiconnector is a pseudogroupoid.*

Proof Clearly the axioms 2, $3C$ and $3D$ give

$$m(x,x,t,y) = m(x,x,y,y) = y$$

and

$$m(x,y,t,y) = m(x,y,x,y) = x,$$

so that axioms $3A$ and $3B$ always hold.

Let us then prove axiom $4B$. We write $\pi_{1,2}$ also for the kernel relation of the projection $\pi_{1,2}\colon R\Box S \to R$ and we write π_4 for the kernel relation of the projection $\pi_4\colon R\Box S \to X$ sending an element $\begin{pmatrix} x & t \\ y & z \end{pmatrix}$ in $R\Box S$ to z in X. The fact that $R[\sigma] = R[\alpha]$, so that $m\colon R\Box S \to X$ is independent of the third variable, implies that the kernel relation of m, also denoted by m, contains $R[\alpha] = \pi_{1,2} \cap \pi_4$. This remark allows us to apply the shifting property to the diagram

$$\begin{array}{ccc}
\begin{pmatrix} m(x,y,x',y') & m(x,y,x',y') \\ y' & y' \end{pmatrix} & \overset{\pi_{1,2}}{\text{———}} & \begin{pmatrix} m(x,y,x',y') & m(x',y',x'',y'') \\ y' & y'' \end{pmatrix} \\
m \Big(\ \pi_4 \Big| & & \Big| \pi_4 \\
\begin{pmatrix} x & x' \\ y & y' \end{pmatrix} & \underset{\pi_{1,2}}{\text{—————}} & \begin{pmatrix} x & x'' \\ y & y'' \end{pmatrix}
\end{array}$$

for any situation as in the following diagram

$$\begin{array}{ccccc}
x & \overset{S}{\text{——}} & x' & \overset{S}{\text{——}} & x'' \\
R \Big| & & \Big| R & & \Big| R \\
y & \underset{S}{\text{——}} & y' & \underset{S}{\text{——}} & y''.
\end{array}$$

This implies that $m(m(x,y,x',y'),y',m(x',y',x'',y''),y'') = m(x,y,x'',y'')$ (axiom $4B$). Similarly one can prove axiom $4A$. $\Box$

Given a finitely complete category $\mathcal{C}$, we write 2-$Eq(\mathcal{C})$ for the category whose objects are pairs of equivalence relations (R, S, X) on the same object X:

$$R \underset{d_0}{\overset{d_1}{\rightrightarrows}} X \underset{d_0}{\overset{d_1}{\leftleftarrows}} S \, ,$$

and arrows in 2-$Eq(\mathcal{C})$ are triples of arrows (f_R, f_S, f) making the following diagram commutative:

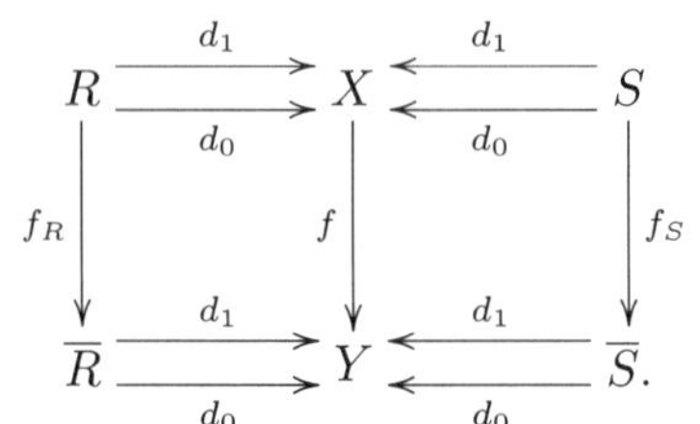

Let $Psdgrd(\mathcal{C})$ be the category whose objects are pairs of equivalence relations on a given object (R, S, X, m) equipped with a pseudogroupoid structure $m\colon R\square S \to X$. An arrow $(f_R, f_S, f)\colon (R, S, X, m) \to (\overline{R}, \overline{S}, Y, \mu)$ in $Psdgrd(\mathcal{C})$ is an arrow in 2-$Eq(\mathcal{C})$ that preserves the pseudogroupoid structure, i.e. such that the diagram

$$\begin{array}{ccc} R\square S & \xrightarrow{\tilde{f}} & \overline{R}\square\overline{S} \\ {\scriptstyle m}\downarrow & & \downarrow{\scriptstyle \mu} \\ X & \xrightarrow[f]{} & Y \end{array}$$

commutes. There is a forgetful functor $U\colon Psdgrd(\mathcal{C}) \to$ 2-$Eq(\mathcal{C})$ defined on objects by $U(R, S, X, m) = (R, S, X)$. This functor U is clearly faithful.

A remarkable consequence of the shifting property is that it forces a pseudogroupoid structure to be unique, when it exists. Moreover, the forgetful functor $U\colon Psdgrd(\mathcal{C}) \to$ 2-$Eq(\mathcal{C})$ is full:

2.13. Theorem. *Let $\mathcal{C}$ be a finitely complete category in which the shifting property holds. Then*

1. *a pseudogroupoid structure on two equivalence relations is unique (when it exists)*
2. *the forgetful functor $U\colon Psdgrd(\mathcal{C}) \to$ 2-$Eq(\mathcal{C})$ is a full and faithful inclusion*

Proof Let us first show that any pseudogroupoid $m\colon R\square S \to X$ satisfies the implication

$$m(x, y, t, z) = m(x, y, t', z') \quad \Rightarrow \quad z = z'.$$

We know that $\pi_{1,2} \cap \pi_4 \leq m$, hence the shifting property applied to the situation

$$\begin{array}{ccc} \begin{pmatrix} x & t \\ y & z \end{pmatrix} & \overset{\pi_4}{\rule{2em}{0.4pt}} & \begin{pmatrix} y & z \\ y & z \end{pmatrix} \\ m\left(\pi_{1,2} \right| & & \left| \pi_{1,2} \right. \\ \begin{pmatrix} x & t' \\ y & z' \end{pmatrix} & \underset{\pi_4}{\rule{2em}{0.4pt}} & \begin{pmatrix} y & z' \\ y & z' \end{pmatrix} \end{array}$$

shows that

$$z = m(y, y, z, z) = m(y, y, z', z') = z'.$$

This means that $m \cap \pi_{1,2} \leq \pi_4$.

Now, let $m \colon R \square S \to X$ and $m' : R \square S \to X$ be two pseudogroupoid structures on R and S. For any $\begin{pmatrix} x & t \\ y & z \end{pmatrix}$ in $R \square S$, the shifting property applied to

$$\begin{array}{ccc}
\begin{pmatrix} x & x \\ y & y \end{pmatrix} & \overset{\pi_{1,2}}{\rule{3cm}{0.4pt}} & \begin{pmatrix} x & t \\ y & z \end{pmatrix} \\
m' \Big(\quad m \Big| & & \Big| \, m \\
\begin{pmatrix} x & x \\ x & x \end{pmatrix} & \underset{\pi_{1,2}}{\rule{1.5cm}{0.4pt}} & \begin{pmatrix} x & m(x,y,t,z) \\ x & m(x,y,t,z) \end{pmatrix}
\end{array}$$

yields

$$m'(x, y, t, z) = m'(x, x, m(x, y, t, z), m(x, y, t, z)) = m(x, y, t, z).$$

Of course, the shifting property can be applied because $m \cap \pi_{1,2} = m \cap \pi_{1,2} \cap \pi_4 \leq m'$.

Let us then prove that the inclusion $U \colon Psdgrd(\mathcal{C}) \to 2\text{-}Eq(\mathcal{C})$ is full. Let $\tilde{f} \colon R \square S \to \overline{R} \square \overline{S}$ be the arrow induced by $(f, f_R, f_S) \colon (R, S, X, m) \to (\overline{R}, \overline{S}, Y, \mu)$. Consider any $\begin{pmatrix} x & t \\ y & z \end{pmatrix}$ in $R \square S$ and the following situation

$$\begin{array}{ccc}
\begin{pmatrix} x & x \\ y & y \end{pmatrix} & \overset{\pi_{1,2}}{\rule{3cm}{0.4pt}} & \begin{pmatrix} x & t \\ y & z \end{pmatrix} \\
\mu \circ \tilde{f} \Big(\quad m \Big| & & \Big| \, m \\
\begin{pmatrix} x & x \\ x & x \end{pmatrix} & \underset{\pi_{1,2}}{\rule{1.5cm}{0.4pt}} & \begin{pmatrix} x & m(x,y,t,z) \\ x & m(x,y,t,z) \end{pmatrix}
\end{array}$$

which clearly holds since we have:

$$(\mu \circ \tilde{f}) \begin{pmatrix} x & x \\ y & y \end{pmatrix} = \mu \begin{pmatrix} f(x) & f(x) \\ f(y) & f(y) \end{pmatrix} = f(x) = (\mu \circ \tilde{f}) \begin{pmatrix} x & x \\ x & x \end{pmatrix}.$$

Moreover,

$$m \cap \pi_{1,2} = m \cap \pi_{1,2} \cap \pi_4 \leq \pi_{1,2} \cap \pi_4 \leq \mu \circ \tilde{f},$$

so that the shifting property gives

$$(\mu \circ \tilde{f}) \begin{pmatrix} x & t \\ y & z \end{pmatrix} = (\mu \circ \tilde{f}) \begin{pmatrix} x & m(x,y,t,z) \\ x & m(x,y,t,z) \end{pmatrix} = \mu \begin{pmatrix} f(x) & (f \circ m)(x,y,t,z) \\ f(x) & (f \circ m)(x,y,t,z) \end{pmatrix}$$

and then

$$(\mu \circ \tilde{f}) \begin{pmatrix} x & t \\ y & z \end{pmatrix} = (f \circ m) \begin{pmatrix} x & t \\ y & z \end{pmatrix}.$$

$\square$

3 Connectors and Abelian objects

In this section we apply the previous results to the category of connectors ([4] [5]) and to the category of abelian objects in a category satisfying the shifting property. In particular it turns out that, for a given objet in such a category, being abelian becomes a property.

We begin with the following

3.1. Corollary. *Let R and S be two equivalence relations on X in a finitely complete category satisfying the shifting property. Then any arrow $p\colon R\times_X S\to X$ satisfying the axioms*
I $\quad xSp(x,y,z)Rz$
IIA $\quad p(x,x,y)=y$
IIB $\quad p(x,y,y)=x$
is a connector between R and S. Moreover, such a structure is necessarily unique.

Proof We are going to check only the axiom $IIIB$

$$p(p(x,y,u),u,v)=p(x,y,v)$$

the verification of the axiom $IIIA$ being similar.

Clearly, any arrow p satisfying the axioms I, IIA and IIB is a quasiconnector, by defining $m\colon R\square S\to X$ as $m(x,y,t,z)=p(x,y,z)$, which is then a pseudogroupoid, thanks to Proposition 2.12. Now, for any x,y,u,v with $xRySuSv$ one forms the diagram

$$\begin{array}{ccccc} x & \overset{S}{\text{———}} & p(x,y,u) & \overset{S}{\text{———}} & p(p(x,y,u),u,v) \\ {\scriptstyle R}\vert & & \vert{\scriptstyle R} & & \vert{\scriptstyle R} \\ y & \underset{S}{\text{———}} & u & \underset{S}{\text{———}} & v. \end{array}$$

Then the validity of axiom $4B$ for the pseudogroupoid structure gives:

$$\begin{aligned} p(p(x,y,u),u,v) &= m(m(x,y,p(x,y,u),u),u,m(p(x,y,u),u,p(p(x,y,u),u,v),v),v) \\ &= m(x,y,p(p(x,y,u),u,v),v) \\ &= p(x,y,v). \end{aligned}$$

The uniqueness of the structure follows from the uniqueness of a pseudogroupoid structure in $\mathcal{C}$. □

Let $Conn(\mathcal{C})$ be the category (R,S,X,p) of pairs of equivalence relations on a given object equipped with a connector, with arrows $(f_R,f_S,f)\colon(R,S,X,p)\to(\overline{R},\overline{S},Y,\overline{p})$ those arrows in $2\text{-}Eq(\mathcal{C})$ that preserve the connector, i.e. such that the diagram

$$\begin{array}{ccc} R\times_X S & \xrightarrow{\tilde{f}} & \overline{R}\times_Y\overline{S} \\ {\scriptstyle p}\downarrow & & \downarrow{\scriptstyle \overline{p}} \\ X & \xrightarrow[f]{} & Y \end{array}$$

commutes. Any connector being a pseudogroupoid, it follows from Theorem 2.13 that:

3.2. Corollary. *When $\mathcal{C}$ satisfies the shifting property, then the forgetful functor $V\colon Conn(\mathcal{C}) \to 2\text{-}Eq(\mathcal{C})$ is a full and faithful inclusion.*

It is worth mentioning that this forgetful functor V has been used to characterize Maltsev categories. As shown in [5], a finitely complete category $\mathcal{C}$ is Maltsev exactly when V is closed under subobjects: this means that given any monomorphism $i\colon X \to Y$ in $2\text{-}Eq(\mathcal{C})$, if Y is in $Conn(\mathcal{C})$, then also X is in $Conn(\mathcal{C})$.

3.3. Definition. An object X in $\mathcal{C}$ is *abelian* if it is endowed with an internal Maltsev operation, i.e. there is an arrow $p\colon X \times X \times X \to X$ with $p(x, x, y) = y$ and $p(x, y, y) = x$.

In the presence of the shifting property being abelian becomes a property:

3.4. Corollary. *Any internal Maltsev operation $p\colon X \times X \times X \to X$ in a finitely complete category satisfying the shifting property is associative and commutative. This operation is necessarily unique, when it exists.*

Proof The classical associativity $p(x, y, p(z, u, v)) = p(p(x, y, z), u, v)$ follows from Corollary 3.1:

$$p(x, y, p(z, u, v)) = p(p(x, y, z), z, p(z, u, v)) = p(p(x, y, z), u, v).$$

As far as commutativity is concerned, let us write $\pi_2\colon X \times X \times X \to X$ for the projection sending the element (x, y, z) to y, and $\pi_{1,3}\colon X \times X \times X \to X \times X$ for the projection sending the element (x, y, z) to (x, z). Clearly $\Delta_{X \times X \times X} = \pi_{1,3} \cap \pi_2 \leq p$, and the shifting property applied to the diagram

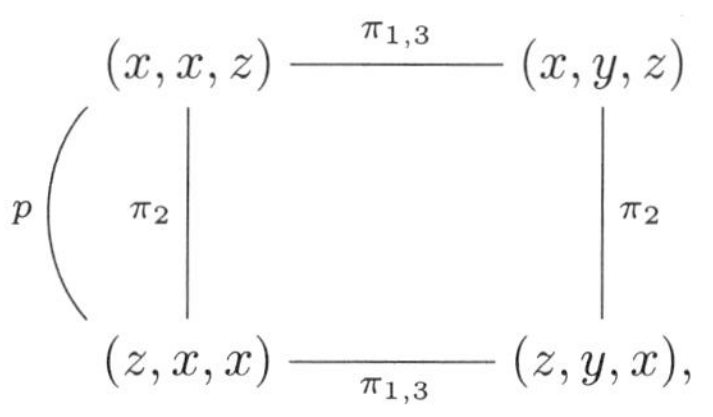

$$\begin{array}{ccc}
(x, x, z) & \overset{\pi_{1,3}}{\text{———}} & (x, y, z) \\
{\scriptstyle p}\Big(\;\; \Big| {\scriptstyle \pi_2} & & \Big| {\scriptstyle \pi_2} \\
(z, x, x) & \underset{\pi_{1,3}}{\text{———}} & (z, y, x),
\end{array}$$

gives $p(x, y, z) = p(z, y, x)$, as desired. The uniqueness of p follows from Theorem 2.13. □

The previous Corollary shows that in a category $\mathcal{C}$ satisfying the shifting property an object X is abelian precisely when there is a connector between ∇_X and ∇_X. Let $Mal(\mathcal{C})$ be the subcategory of the abelian objects of $\mathcal{C}$, with arrows those arrows in $\mathcal{C}$ which also respect the Maltsev operation. Now, the assumption that $\mathcal{C}$ satisfies the shifting property implies that any arrow in $\mathcal{C}$ respects the Maltsev operation (Corollary 3.2). Consequently:

3.5. Corollary. *When $\mathcal{C}$ satisfies the shifting property, the category of abelian objects $Mal(\mathcal{C})$ is naturally Maltsev.*

Proof Recall that a finitely complete category $\mathcal{A}$ is naturally Maltsev [17] if, in the category $[\mathcal{A}, \mathcal{A}]$ of endofunctors and natural transformations, the identity functor $Id_{\mathcal{A}}$ is provided with a Maltsev operation. By definition of $Mal(\mathcal{C})$ any object X in $Mal(\mathcal{C})$ is provided with a Maltsev operation $p_X\colon X \times X \times X \to X$ and the naturality of p follows from the fullness of $Mal(\mathcal{C})$ in $\mathcal{C}$. □

4 Internal categories

A description of internal categories in congruence modular varieties was obtained in [15]. We are now going to extend and improve some of the results proved in [15] thanks to a systematic use of the shifting property. Remark that an internal category, unlike an internal groupoid (Example 1.3), does not necessarily give rise to an internal pseudogroupoid. Consequently, the results of this section can not be deduced directly from the preceding ones concerning pseudogroupoids.

An internal reflexive graph X in $\mathcal{C}$

$$X_1 \underset{d_1}{\overset{d_0}{\underset{\longleftarrow s_0}{\rightrightarrows}}} X_0$$

has the property that $d_0 \circ s_0 = 1_{X_0} = d_1 \circ s_0$. X is endowed with an *internal category* structure

$$X_1 \times_{X_0} X_1 \overset{\pi_1}{\underset{\pi_0}{\overset{m}{\rightrightarrows}}} X_1 \underset{d_1}{\overset{d_0}{\underset{\longleftarrow s_0}{\rightrightarrows}}} X_0$$

when there is a multiplication $m\colon X_1 \times_{X_0} X_1 \to X_1$, satisfying the axioms

1. $m(1_{d_0(x)}, x) = x = m(x, 1_{d_1(x)})$
2. $d_0(m(x,y)) = d_0(x)$
3. $d_1(m(x,y)) = d_1(y)$
4. $m(x, m(y,z)) = m(m(x,y), z)$

(where we write 1_X for $s_0(X)$).

4.1. Lemma. *Let X be an internal reflexive graph in a finitely complete category $\mathcal{C}$ satisfying the shifting property. If there is a multiplication $m\colon X_1 \times_{X_0} X_1 \to X_1$ on X satisfying the axiom 1, then $m \cap \pi_0 = \Delta_{X_1 \times_{X_0} X_1}$ (right cancellation property).*

Proof If we assume that $m(a,x) = m(a,y)$, then we can form the diagram

$$\begin{array}{ccc} (a,x) & \overset{\pi_1}{\text{———}} & (1_{d_1(a)}, x) \\ \pi_0 \Big| & & \Big| \pi_0 \\ (a,y) & \underset{\pi_1}{\text{———}} & (1_{d_1(a)}, y), \end{array}$$

(with m relating (a,x) and (a,y)), where each represented element is in $X_1 \times_{X_0} X_1$. Of course $\pi_0 \cap \pi_1 = \Delta_{X_1 \times_{X_0} X_1}$, and the shifting property gives

$$x = m(1_{d_1(a)}, x) = m(1_{d_1(a)}, y) = y.$$

□

4.2. Proposition. *Let X be an internal reflexive graph in a finitely complete category $\mathcal{C}$ satisfying the shifting property. Then there is at most one multiplication $m\colon X_1 \times_{X_0} X_1 \to X_1$ on X satisfying the axioms 1 and 2.*

Proof Let us consider two multiplications m and m' on X satisfying the axioms 1 and 2. For any $(x, y) \in X_1 \times_{X_0} X_1$ we can form the diagram

$$\begin{array}{ccc} (x, 1_{d_1(x)}) & \overset{\pi_0}{\text{———}} & (x, y) \\ m' \left(\quad m \right| & & \Big| m \\ (1_{d_0(x)}, x) & \underset{\pi_0}{\text{———}} & (1_{d_0(m(x,y))}, m(x, y)), \end{array}$$

where $d_0(x) = d_0(m(x, y))$ by axiom 2. Then the previous Lemma gives

$$\Delta_{X_1 \times_{X_0} X_1} = m \cap \pi_0 \leq m',$$

and clearly one has that

$$m'(x, 1_{d_1(x)}) = x = m'(1_{d_0(x)}, x).$$

Then the shifting property gives

$$m'(x, y) = m'(1_{d_0(m(x,y))}, m(x, y)) = m(x, y).$$

□

4.3. Proposition. *Let $\mathcal{C}$ be a finitely complete category satisfying the shifting property. Then any multiplication m on a reflexive graph satisfying the axioms 1, 2 and 3 is associative.*

Proof Given three composable arrows x, y, z in X_1

$$A \xrightarrow{x} B \xrightarrow{y} C \xrightarrow{z} D,$$

we can form the following diagram of four elements in the equivalence relation π_1 on $X_1 \times_{X_0} X_1$ (we write ∇ instead of $\nabla_{X_1 \times_{X_0} X_1}$ to save space):

$$\begin{array}{ccc} [(m(x, y), 1_C), (y, 1_C)] & \overset{\pi_0 \Box \pi_1}{\text{————}} & [(m(x, y), z), (y, z)] \\ m\Box\pi_1 \left(\quad \nabla \times m \right| & & \Big| \nabla \times m \\ [(x, y), (1_B, y)] & \underset{\pi_0 \Box \pi_1}{\text{————}} & [(x, m(y, z)), (1_B, m(y, z))], \end{array}$$

where $\pi_0 \Box \pi_1$ and $m \Box \pi_1$ are thought as equivalence relations on π_1, and $\nabla \times m$ is the equivalence relation on π_1 defined by $[(a, b), (c, b)](\nabla \times m)[(d, e), (f, e)]$ if and only if $m(c, b) = m(f, e)$. Then, if $(\nabla \times m) \cap (\pi_0 \Box \pi_1) \leq m \Box \pi_1$, the shifting property implies that

$$[(m(x, y), z), (y, z)](m \Box \pi_1)[(x, m(y, z)), (1_B, m(y, z))],$$

which in particular gives

$$m(m(x, y), z) = m(x, m(y, z)),$$

as desired.

The reason why the inequality $(\nabla \times m) \cap (\pi_0 \Box \pi_1) \leq m \Box \pi_1$ holds is that $(\nabla \times m) \cap (\pi_0 \Box \pi_1) = \Delta_{\pi_1}$. For this, let us consider

$$[(a, b), (c, b)](\nabla \times m) \cap (\pi_0 \Box \pi_1)[(a, e), (c, e)].$$

The fact that $(c,b)(m \cap \pi_0)(c,e)$ implies that $b = e$ because $m \cap \pi_0 = \Delta_{X_1 \times_{X_0} X_1}$ (see Lemma 4.1). Accordingly, $[(a,b),(c,b)] = [(a,e),(c,e)]$. $\square$

4.4. Corollary. *For a reflexive graph X in a finitely complete category $\mathcal{C}$ satisfying the shifting property the following conditions are equivalent:*

1. *there is an internal category structure whose underlying reflexive graph is X*
2. *there is a unique internal category structure whose underlying reflexive graph is X*
3. *there is a multiplication m satisfying the axioms $1, 2$ and 3*
4. *there is a unique multiplication m satisfying the axioms $1, 2$ and 3*

4.5. Remark. This corollary extends the equivalence between conditions a, b, c, d of Theorem 4.1 in [15] from modular varieties to any category $\mathcal{C}$ satisfying the shifting property. Moreover, we improve that result by dropping the assumption that $R[d_0]$ and $R[d_1]$ have trivial commutator.

Let $RG(\mathcal{C})$ denote the category of internal reflexive graphs in $\mathcal{C}$.

4.6. Proposition. *Let $\mathcal{C}$ be a finitely complete category satisfying the shifting property. Then the forgetful functor*

$$Cat(\mathcal{C}) \to RG(\mathcal{C})$$

is a full inclusion.

Proof Thanks to Proposition 4.2 we only need to prove that any arrow (ϕ_0, ϕ_1) of internal reflexive graphs

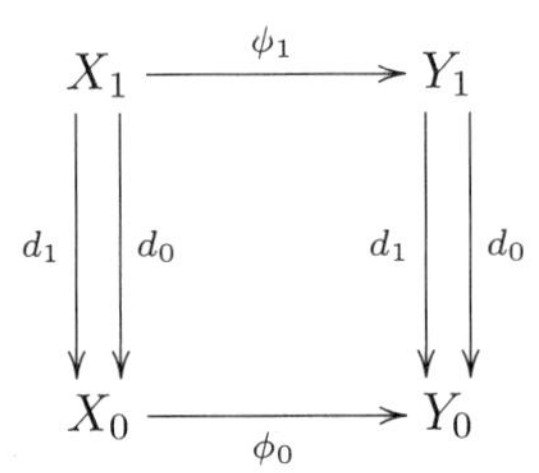

between two internal categories (X, m) and (Y, μ) always preserves the (unique) multiplication. If $\phi \colon X_1 \times_{X_0} X_1 \to Y_1 \times_{Y_0} Y_1$ is the arrow induced by the universal property of the pullbacks, then, given any pair of composable arrows in X_1

$$X \xrightarrow{f} Y \xrightarrow{g} Z,$$

one has that

$$(\mu \circ \phi)(f, 1_Y) = \phi_1(f) = (\mu \circ \phi)(1_X, f).$$

Since $m \cap \pi_0 = \Delta_{X_1 \times_{X_0} X_1}$ by Lemma 4.1, we can apply the shifting property to

$$\begin{array}{ccc} (f, 1_Y) & \overset{\pi_0}{\text{———}} & (f, g) \\ {\scriptstyle m} \Big| & & \Big| {\scriptstyle m} \\ (1_X, f) & \underset{\pi_0}{\text{———}} & (1_X, m(f,g)). \end{array}$$

(with $\mu \circ \phi$ relating $(f, 1_Y)$ and $(1_X, f)$)

This implies that

$$\begin{aligned}(\mu \circ \phi)(f,g) &= (\mu \circ \phi)(1_X, m(f,g)) \\ &= \mu(1_{\phi_1(X)}, (\phi_1 \circ m)(f,g)) \\ &= (\phi_1 \circ m)(f,g).\end{aligned}$$

□

Another result can be obtained when the finitely complete category $\mathcal{C}$ is assumed to be also *regular*, i.e. $\mathcal{C}$ has coequalizers of effective equivalence relations and the regular epis are pullback stable.

4.7. Proposition. *Let $\mathcal{C}$ be a regular category satisfying the shifting property. Let X be an internal reflexive graph in $\mathcal{C}$*

$$X_1 \underset{d_1}{\overset{d_0}{\underset{\longleftarrow}{\overset{\longrightarrow}{\longrightarrow}}}} X_0 \quad (s_0 : X_0 \to X_1)$$

equipped with a pseudogroupoid structure μ on $R[d_0]$ and $R[d_1]$. Then the following conditions are equivalent:

1. *there is a (unique) category structure on X*
2. *X_1 determines a preorder on X_0. This means that the regular image $I = \frac{X_1}{R[d_0] \cap R[d_1]}$ of the reflexive graph X is also a transitive relation on X_0.*

Proof In this proof we shall use Barr's metatheorem for regular categories. [1]. Thanks to this result, the elementwise procedure we used so far to prove finite limit properties can be augmented, in a regular category, by the fact that one may treat regular epimorphisms as if they were surjections.
It is clear that any internal category structure on X has the property that the image I is transitive. Indeed, for any (X,Y) and (Y,Z) in I there exist in X_1 at least one arrow $f\colon X \to Y$ and one arrow $g\colon Y \to Z$, and consequently the arrow $m(f,g)\colon X \to Z$ shows that (X,Z) is also in I.

Conversely, let us assume that I is transitive. Then, for any $f\colon X \to Y$ and $g\colon Y \to Z$, there exists an arrow $\phi\colon X \to Z$ with the property that $\begin{pmatrix} g & \phi \\ 1_Y & f \end{pmatrix}$ is in $R[d_0]\square R[d_1]$. We can then define $m(f,g) = \mu(g, 1_Y, \phi, f)$ (μ does not depend on the third variable). We then have that

$$f R[d_0] m(f,g) R[d_1] g$$

and, moreover,

$$m(f, 1_Y) = \mu(1_Y, 1_Y, f, f) = f = \mu(f, 1_X, f, 1_X) = m(1_X, f).$$

By Corollary 4.4 the proof is then complete. □

Corollary 4.4 and Proposition 4.7 here above then extend Theorem 4.1 in [15] to any regular category satisfying the shifting property.

5 Internal groupoids

Since any internal groupoid is an internal category, the previous results give in particular a description of the internal reflexive graphs underlying a (necessarily unique) groupoid structure. Indeed, it is sufficient to require also the existence of an arrow $i\colon X_1 \to X_1$ satisfying the usual conditions of an inverse.

However, a very simple and neat description of the internal groupoids in a category $\mathcal{C}$ satisfying the shifting property can be obtained by using the fact that a groupoid always gives rise to a pseudogroupoid.

We first establish a more general result, which clarifies the relationship between the notions of pseudogroupoid and of connector:

5.1. Proposition. *Let $\mathcal{C}$ be a regular category satisfying the shifting property, and let R and S be two equivalence relations on X, for X in $\mathcal{C}$. Then the following conditions are equivalent:*

1. *there is a (unique) connector between R and S*
2. *there is a (unique) pseudogroupoid on R and S and $R \circ S = S \circ R$*
3. *there is a (unique) quasiconnector on R and S and $R \circ S = S \circ R$*
4. *the canonical arrow $\alpha\colon R\square S \to R \times_X S$ is split by a (unique) arrow $i\colon R \times_X S \to R\square S$ of double zero sequences in $Pt_X(\mathcal{C})$*

Proof The uniqueness of the various structures follows from Theorem 2.13.
$1 \Rightarrow 2$ For any $xRySz$ there exists $p(x,y,z)$ such that $xSp(x,y,z)Rz$, hence $R\circ S \leq S \circ R$, and then $R \circ S = S \circ R$.
$2 \Rightarrow 3$ Trivial.
$3 \Rightarrow 4$ By using once again Barr's metatheorem for regular categories one can prove that the canonical arrow $\alpha\colon R\square S \to R \times_X S$ is a regular epi precisely when $R \circ S = S \circ R$. Now, if $\sigma\colon R\square S \to R\square S$ is the quasiconnector on R and S, the fact that $R[\alpha] = R[\sigma]$ implies that α is the coequalizer of σ and $1_{R\square S}$. It follows that there is a unique $i\colon R \times_X S \to R\square S$ with $i \circ \alpha = \sigma$ and $\alpha \circ i = 1_{R\times_X S}$. It is then easy to check that i is an arrow of double zero sequences in $Pt_X(\mathcal{C})$.
$4 \Rightarrow 1$ By defining $p = d_1 \circ \pi_{1,3} \circ i$ one obtains an internal partial Maltsev operation satisfying also axiom I in the definition of a connector. By Corollary 3.1 this is sufficient to conclude that $p\colon R \times_X S \to X$ is a connector. □

5.2. Corollary. *Let $\mathcal{C}$ be a regular category satisfying the shifting property. For an internal reflexive graph X*

$$X_1 \underset{d_1}{\overset{d_0}{\underset{\longleftarrow s_0}{\rightrightarrows}}} X_0$$

in $\mathcal{C}$ the following conditions are equivalent:

1. *there is a (unique) groupoid structure on X*
2. *there is a (unique) connector between $R[d_0]$ and $R[d_1]$*
3. *there is a (unique) quasiconnector on $R[d_0]$ and $R[d_1]$ and*

$$R[d_0] \circ R[d_1] = R[d_1] \circ R[d_0]$$

Proof It follows by the previous Proposition. □

We can now extend Corollary 4.3 in [15] characterizing internal groupoids in modular varieties:

5.3. Corollary. *Let $\mathcal{C}$ be a regular category satisfying the shifting property. For an internal reflexive graph X in $\mathcal{C}$*

$$X_1 \underset{d_1}{\overset{d_0}{\underset{\longrightarrow}{\overset{\longrightarrow}{\xleftarrow{s_0}}}}} X_0$$

the following conditions are equivalent:

1. *there is a unique groupoid structure on X*
2. *there is a unique pseudogroupoid structure on $R[d_0]$ and $R[d_1]$, and X determines an equivalence relation $I = \frac{X_1}{R[d_0] \cap R[d_1]}$ on X_0.*

Proof $1 \Rightarrow 2$ Since any groupoid X gives rise to a pseudogroupoid structure on $R[d_0]$ and $R[d_1]$, thanks to Proposition 4.7 we only need to prove that I is a symmetric relation on X_0. This is clear, because when X is a groupoid, then for any $f\colon X \to Y$ there is an arrow $f^{-1}\colon Y \to X$, so that if (X, Y) is in I, so is (Y, X). $2 \Rightarrow 1$ We already know that X is an internal category (Proposition 4.7). On the other hand, if for any (X, Y) in I, then there is also (Y, X) in I, then for any $f\colon X \to Y$ there is at least one arrow $g\colon Y \to X$. By setting $f^{-1} = \mu(1_X, f, g, 1_Y)$ and by axiom $4B$ in the definition of a pseudogroupoid one has that

$$m(f, \mu(1_X, f, g, 1_Y)) = \mu(\mu(1_X, f, g, 1_Y), 1_Y, \mu(g, 1_Y, 1_X, f), f) = \mu(1_X, f, 1_X, f),$$

so that $m(f, \mu(1_X, f, g, 1_Y)) = 1_X$. Similarly, by $4A$ one has that

$$m(\mu(1_X, f, g, 1_Y), f) = \mu(f, 1_X, \mu(f, 1_X, 1_Y, g), \mu(1_X, f, g, 1_Y)) = \mu(f, f, 1_Y, 1_Y),$$

hence $m(\mu(1_X, f, g, 1_Y), f) = 1_Y$. This proves that $\mu(1_X, f, g, 1_Y)$ is the inverse of f, and X is then an internal groupoid. □

References

[1] M. Barr, *Exact Categories*, LNM 236, Springer-Verlag, 1971, 1-120.

[2] F. Borceux, *A survey of semi-abelian categories*, to appear in this volume.

[3] D. Bourn, *Normalization, equivalence, kernel equivalence and affine categories,* LNM 1488, Springer-Verlag, 1991, 43–62.

[4] D. Bourn - M. Gran, *Centrality and normality in protomodular categories*, Theory Appl. Cat., Vol. 9, No. 8, 2002, 151-165.

[5] D. Bourn - M. Gran, *Centrality and connectors in Maltsev categories*, Algebra Universalis, 48, 2002, 309-331.

[6] R. Brown - C. Spencer, *G-Groupoids, crossed modules and the fundamental groupoid of a topological group,* Proc. Konn. Ned. Akad. Wet, 79, 1976, 296-302.

[7] A. Carboni - G.M. Kelly - M.C. Pedicchio, *Some remarks on Maltsev and Goursat categories* Appl. Categorical Structures, 1, 1993, 385-421.

[8] A. Carboni - J. Lambek - M.C. Pedicchio, *Diagram chasing in Mal'cev categories,* J. Pure Appl. Algebra, 69, 1990, 271-284.

[9] A. Carboni - M.C. Pedicchio - N. Pirovano, *Internal graphs and internal groupoids in Mal'cev categories,* Proc. Conference Montreal 1991, 1992, 97-109.

[10] R. Freese - R. McKenzie, *Commutator theory for congruence modular varieties*, Lond. Math. Soc. Lect. Notes Series, 125, Cambr. Univ. Press, 1987.

[11] M. Gran, *Internal categories in Mal'cev categories,* J. Pure Appl. Algebra, 143, 1999, 221-229.

[12] M. Gran, *Commutators and central extensions in universal algebra,* J. Pure Appl. Algebra, 174, 2002, 249-261.

[13] H. P. Gumm, *Geometrical Methods in Congruence Modular Algebras*, Mem. Amer. Math. Soc., 45 286, 1983.

[14] J. Hagemann - C. Herrmann, *A concrete ideal multiplication for algebraic systems and its relation to congruence distributivity*, Arch. Math. (Basel), 32, 1979, 234-245.
[15] G. Janelidze - M.C. Pedicchio, *Internal categories and groupoids in congruence modular varieties*, Journal of Algebra 193, 1997, 552-570.
[16] G. Janelidze - M.C. Pedicchio, *Pseudogroupoids and commutators*, Theory Appl. Cat., Vol. 8, No. 15, 2001, 408-456.
[17] P.T. Johnstone, *Affine categories and naturally Mal'cev categories,* J. Pure Appl. Algebra, 61, 1989, 251-256.
[18] E.W. Kiss, *Three remarks on the modular commutator,* Alg. Universalis 29, 1992, 455-476.
[19] J.-L. Loday, *Spaces with finitely many non-trivial homotopy groups,* J. Pure Appl. Algebra, 24, 1982, 179-202.
[20] M.C. Pedicchio, *A categorical approach to commutator theory*, Journal of Algebra, 177, 1995, 647-657.
[21] M.C. Pedicchio, *Arithmetical categories and commutator theory*, Appl. Categorical Structures, 4, 1996, 297-305.
[22] J.D.H. Smith, *Mal'cev Varieties*, LNM 554, Springer-Verlag, 1976.

Received November 28, 2002; in revised form March 14, 2003

Fields Institute Communications
Volume **43**, 2004

Crossed Complexes and Homotopy Groupoids as Non Commutative Tools for Higher Dimensional Local-to-global Problems

Ronald Brown
Department of Mathematics
University of Wales, Bangor
Gwynedd LL57 1UT, U.K.
r.brown@bangor.ac.uk

With thanks to many colleagues and Bangor research students.

Abstract. We outline the main features of the definitions and applications of crossed complexes and cubical ω-groupoids with connections. These give forms of higher homotopy groupoids, and new views of basic algebraic topology and the cohomology of groups, with the ability to obtain some non commutative results and compute some homotopy types.

Contents

2000 *Mathematics Subject Classification.* Primary 01-01, 16E05, 18D05, 18D35, 55P15, 55Q05;

This is an extended account of a lecture given at the meeting on 'Categorical Structures for Descent, Galois Theory, Hopf algebras and semiabelian categories', Fields Institute, September 23-28, 2002, for which the author is grateful for support from the Fields Institute and a Leverhulme Emeritus Research Fellowship. Other support is indicated later.

Introduction

An aim is to give a survey of results obtained by R. Brown and P.J. Higgins and others over the years 1974-2002, and to point to applications and related areas. This work gives an account of some basic algebraic topology which differs from the standard account through the use of *crossed complexes*, rather than chain complexes, as a fundamental notion. In this way one obtains comparatively quickly not only classical results such as the Brouwer degree and the relative Hurewicz theorem, but also non commutative results on second relative homotopy groups, as well as higher dimensional results involving the action of and also presentations of the fundamental group. For example, the fundamental crossed complex ΠX_* of the skeletal filtration of a CW-complex X is a useful generalisation of the usual cellular chains of the universal cover of X. It also gives a replacement for singular chains by taking X to be the geometric realisation of a singular complex of a space.

A replacement for the excision theorem in homology is obtained by using cubical methods to prove a colimit theorem for the fundamental crossed complex functor on filtered spaces. This colimit theorem is a higher dimensional version of a classical example of a *non commutative local-to-global theorem*, which itself was the initial motivation for the work described here. This Seifert-Van Kampen Theorem (SVKT) determines completely the fundamental group $\pi_1(X, x)$ of a space X with base point which is the union of open sets U, V whose intersection is path connected and contains the base point x; the 'local information' is on the morphisms of fundamental groups induced by the inclusions $U \cap V \to U, U \cap V \to V$. The importance of this result reflects the importance of the fundamental group in algebraic topology, algebraic geometry, complex analysis, and many other subjects. Indeed the origin of the fundamental group was in Poincaré's work on monodromy for complex variable theory.

Essential to this use of crossed complexes, particularly for conjecturing and proving local-to-global theorems, is a construction of *higher homotopy groupoids*, with properties described by *an algebra of cubes*. There are applications to local-to-global problems in homotopy theory which are more powerful than purely classical tools, while shedding light on those tools. It is hoped that this account will increase the interest in the possibility of wider applications of these methods and results, since homotopical methods play a key role in many areas. It is relevant that Atiyah in [8] identifies some major themes in 20th century mathematics as: *local to global, from commutative to non commutative, increase in dimensions, homology and K-theory.* The higher categorical methods described here allow the combining of these themes, and yield steps towards a *non commutative algebraic topology.*

Higher homotopy groups

Topologists in the early part of the 20th century were well aware that: the non commutativity of the fundamental group was useful in geometric applications; for path connected X there was an isomorphism

$$H_1(X) \cong \pi_1(X, x)^{\mathrm{ab}};$$

and the abelian homology groups existed in all dimensions. Consequently there was a desire to generalise the non commutative fundamental group to all dimensions.

In 1932 Čech submitted a paper on higher homotopy groups $\pi_n(X, x)$ to the ICM at Zurich, but it was quickly proved that these groups were abelian for $n \geqslant 2$, and on these grounds Čech was persuaded to withdraw his paper, so that only a small paragraph appeared in the Proceedings [55]. We now see the reason for this commutativity as the result (Eckmann-Hilton) that a group internal to the category of groups is just an abelian group. Thus the vision of a non commutative higher dimensional version of the fundamental group has since 1932 been generally considered to be a mirage. Before we go back to the SVKT, we explain in the next section how nevertheless work on crossed modules did introduce non commutative structures relevant to topology in dimension 2.

Of course higher homotopy groups were strongly developed following on from the work of Hurewicz (1935). The fundamental group still came into the picture with its action on the higher homotopy groups, which J.H.C. Whitehead once remarked (1957) was especially fascinating for the early workers in homotopy theory. Much of Whitehead's work was intended to extend to higher dimensions the methods of combinatorial group theory of the 1930s – hence the title of his papers: 'Combinatorial homotopy, I, II' [104, 105]. The first of these two papers has been very influential and is part of the basic structure of algebraic topology. It is the development of work of the second paper which we explain here.

Whitehead's paper on 'Simple homotopy types', [106], which deals with higher dimensional analogues of Tietze transformations, has a final section using crossed complexes. We refer to this again in section 15. Related work is by R.A. Brown [15].

It is hoped also that this survey will be useful background to work on the Van Kampen Theorem for diagrams of spaces in [42], which uses a form of homotopy groupoid which is in one sense much more powerful than that given here, since it encompasses n-adic information, but in which current expositions are still restricted to the reduced (one base point) case.

1 Crossed modules

In the years 1941-50, Whitehead developed work on crossed modules to represent the structure of the boundary map of the relative homotopy group

$$\pi_2(X, A, x) \to \pi_1(A, x) \qquad (*)$$

in which both groups can be non commutative. Here is the definition.

A *crossed module* is a morphism of groups $\mu : M \to P$ together with an action $(m, p) \mapsto m^p$ of the group P on the group M satisfying the two axioms

CM1) $\mu(m^p) = p^{-1}(\mu m)p$

CM2) $n^{-1}mn = m^{\mu n}$

for all $m, n \in M, p \in P$.

Standard algebraic examples of crossed modules are:

(i) an inclusion of a normal subgroup, with action given by conjugation;

(ii) the inner automorphism map $\chi : M \to \mathrm{Aut}\, M$, in which χm is the automorphism $n \mapsto m^{-1}nm$;

(iii) the zero map $M \to P$ where M is a P-module;

(iv) an epimorphism $M \to P$ with kernel contained in the centre of M.

Simple consequences of the axioms for a crossed module $\mu : M \to P$ are:

1.1 $\mathrm{Im}\, \mu$ is normal in P.

1.2 $\mathrm{Ker}\, \mu$ is central in M and is acted on trivially by $\mathrm{Im}\, \mu$, so that $\mathrm{Ker}\, \mu$ inherits an action of $M/\mathrm{Im}\, \mu$.

Another important construction is the *free crossed P-module*

$$\partial : C(\omega) \to P$$

determined by a function $\omega : R \to P$, where R is a set. The group $C(\omega)$ is generated by $R \times P$ with the relations

$$(r,p)^{-1}(s,q)^{-1}(r,p)(s,qp^{-1}(\omega r)p)$$

the action is given by $(r,p)^q = (r,pq)$ and the boundary morphism is given by $\partial(r,p) = p^{-1}(\omega r)p$, for all $(r,p),(s,q) \in R \times P$.

A major result of Whitehead was:

Theorem W [105] *If the space $X = A \cup \{e_r^2\}_{r \in R}$ is obtained from A by attaching 2-cells by maps $f_r : (S^1, 1) \to (A, x)$, then the crossed module of (*) is isomorphic to the free crossed $\pi_1(A,x)$-module on the classes of the attaching maps of the 2-cells.*

Whitehead's proof, which stretched over three papers, 1941-1949, used transversality and knot theory – an exposition is given in [18]. Mac Lane and Whitehead [90] used this result as part of their proof that crossed modules capture all homotopy 2-types (they used the term '3-types').

The title of the paper in which the first intimation of Theorem W appeared was 'On adding relations to homotopy groups' [103]. This indicates a search for higher dimensional SVKTs.

The concept of free crossed module gives a non commutative context for *chains of syzygies.* The latter idea, in the case of modules over polynomial rings, is one of the origins of homological algebra through the notion of *free resolution.* Here is how similar ideas can be applied to groups. Pioneering work here, independent of Whitehead, was by Peiffer [94] and Reidemeister [97]. See [37] for an exposition of these ideas.

Suppose $\mathcal{P} = \langle X \mid \omega \rangle$ is a presentation of a group G, where $\omega : R \to F(X)$ is a function, allowing for repeated relators. Then we have an exact sequence

$$1 \xrightarrow{i} N(\omega R) \xrightarrow{\phi} F(X) \longrightarrow G \longrightarrow 1$$

where $N(\omega R)$ is the normal closure in $F(X)$ of the set ωR of relations. The above work of Reidemeister, Peiffer, and Whitehead showed that to obtain the next level of syzygies one should consider the free crossed $F(X)$-module $\partial : C(\omega) \to F(X)$, since this takes into account the operations of $F(X)$ on its normal subgroup $N(\omega R)$. Elements of $C(\omega)$ are a kind of 'formal consequence of the relators', so that the relation between the elements of $C(\omega)$ and those of $N(\omega R)$ is analogous to the relation between the elements of $F(X)$ and those of G. The kernel $\pi(\mathcal{P})$ of ∂ is a G-module, called the module of *identities among relations*, and there is considerable work on computing it [37, 96, 78, 67, 47]. By splicing to ∂ a free G-module resolution of $\pi(\mathcal{P})$ one obtains what is called a *free crossed resolution* of the group G. These resolutions have better realisation properties than the usual resolutions by chain complexes of G-modules, as explained later.

This notion of using crossed modules as the first stage of syzygies in fact represents a wider tradition in homological algebra, in the work of Frölich and Lue [69, 87].

Crossed modules also occurred in other contexts, notably in representing elements of the cohomology group $H^3(G, M)$ of a group G with coefficients in a G-module M [89], and as coefficients in Dedecker's theory of non abelian cohomology [60]. The notion of free crossed resolution has been exploited by Huebschmann [79, 81, 80] to represent cohomology classes in $H^n(G, M)$ of a group G with coefficients in a G-module M, and also to calculate with these.

Our results can make it easier to compute a crossed module arising from some topological situation, such as an induced crossed module [50, 51], or a coproduct crossed module [19], than the cohomology element in $H^3(G, M)$ it represents. To obtain information on such an element it is useful to work with a small free crossed resolution of G, and this is one motivation for developing methods for calculating such resolutions. However, it is not so clear what a *calculation* of such a cohomology element would amount to, although it is interesting to know whether the element is non zero, or what is its order. Thus the use of algebraic models of cohomology classes may yield easier computations than the use of cocycles, and this somewhat inverts traditional approaches.

Since crossed modules are algebraic objects generalising groups, it is natural to consider the problem of explicit calculations by extending techniques of computational group theory. Substantial work on this has been done by C.D. Wensley using the program GAP [71, 52].

2 The fundamental groupoid on a set of base points

A change in prospects for higher order non commutative invariants was derived from work of the writer published in 1967 [16], and influenced by Higgins' paper [75]. This showed that the Van Kampen Theorem could be formulated for the *fundamental groupoid* $\pi_1(X, X_0)$ *on a set* X_0 *of base points*, thus enabling computations in the non-connected case, including, as explained in [17, p.319], those in Van Kampen's original paper [83]. This use of groupoids in dimension 1 suggested the question of the use of groupoids in higher homotopy theory, and in particular the question of the existence of *higher homotopy groupoids*.

In order to see how this research programme could go it is useful to consider the statement and special features of this generalised Van Kampen Theorem for the fundamental groupoid. First, if X_0 is a set , and X is a space, then $\pi_1(X, X_0)$ denotes the fundamental groupoid on the set $X \cap X_0$ of base points. This allows the set X_0 to be chosen in a way appropriate to the geometry. For example, if the circle S^1 is written as the union of two semicircles $E_+ \cup E_-$, then the intersection $\{-1, 1\}$ of the semicircles is not connected, so it is not clear where to take the base point. Instead one takes $X_0 = \{-1, 1\}$, and so has two base points. This flexibility is very important in computations, and this example of S^1 was a motivating example for this development. As another example, you might like to consider the difference between the quotients of the actions of $\mathbb{Z}_2$ on the group $\pi_1(S^1, 1)$ and on the groupoid $\pi_1(S^1, \{-1, 1\})$ where the action is induced by complex conjugation on S^1. Relevant work on orbit groupoids has been developed by Higgins and Taylor [77, 98].

Consideration of a set of base points leads to the theorem:

Theorem 2.1 [16] *Let the space X be the union of open sets U, V with intersection W, and let X_0 be a subset of X meeting each path component of U, V, W. Then*
(C) (connectivity) X_0 *meets each path component of* X *and*
(I) (isomorphism) *the diagram of groupoid morphisms induced by inclusions*

$$\begin{array}{ccc} \pi_1(W, X_0) & \xrightarrow{i} & \pi_1(U, X_0) \\ {\scriptstyle j}\downarrow & & \downarrow \\ \pi_1(V, X_0) & \longrightarrow & \pi_1(X, X_0) \end{array} \tag{2.1}$$

is a pushout of groupoids.

From this theorem, one can compute a particular fundamental group $\pi_1(X, x_0)$ using combinatorial information on the graph of intersections of path components of U, V, W, but for this it is useful to develop the algebra of groupoids. Notice two special features of this result.
(i) The computation of the invariant you want, the fundamental group, is obtained from the computation of a larger structure, and so part of the work is to give methods for computing the smaller structure from the larger one. This usually involves non canonical choices, e.g. that of a maximal tree in a connected graph. The work on applying groupoids to groups gives many examples of this [75, 76, 17].
(ii) The fact that the computation can be done is surprising in two ways: (a) The fundamental group is computed *precisely*, even though the information for it uses input in two dimensions, namely 0 and 1. This is contrary to the experience in homological algebra and algebraic topology, where the interaction of several dimensions involves exact sequences or spectral sequences, which give information only up to extension, and (b) the result is a non commutative invariant, which is usually even more difficult to compute precisely.

The reason for the success seems to be that the fundamental groupoid $\pi_1(X, X_0)$ contains information in dimensions 0 and 1, and so can adequately reflect the geometry of the intersections of the path components of U, V, W and the morphisms induced by the inclusions of W in U and V.

This suggested the question of whether these methods could be extended successfully to higher dimensions.

Part of the initial evidence for this quest was the intuitions in the proof of this groupoid SVKT, which seemed to use three main ideas in order to verify the universal property of a pushout for diagram (2.1) and given morphisms f_U, f_V from $\pi_1(U, X_0), \pi_1(V, X_0)$ to a groupoid G, satisfying $f_U i = f_V j$:

• A deformation or filling argument. Given a path $a : (I, \dot{I}) \to (X, X_0)$ one can write $a = a_1 + \cdots + a_n$ where each a_i maps into U or V, but a_i will not necessarily have end points in X_0. So one has to deform each a_i to a_i' in U, V or W, using the connectivity condition, so that each a_i' has end points in X_0, and $a' = a_1' + \cdots + a_n'$

is well defined. Then one can construct using f_U or f_V an image of each a'_i in G and hence of the composite, called $F(a) \in G$, of these images. Note that we subdivide in X and then put together again in G (this uses the condition $f_U i = f_V j$ to prove that the elements of G are composable), and this part can be summarised as:

- Groupoids provided a convenient *algebraic inverse to subdivision.*

Next one has to prove that $F(a)$ depends only on the class of a in the fundamental groupoid. This involves a homotopy rel end points $h : a \simeq b$, considered as a map $I^2 \to X$; subdivide h as $h = [h_{ij}]$ so that each h_{ij} maps into U, V or W; deform h to $h' = [h'_{ij}]$ (keeping in U, V, W) so that each h'_{ij} maps the vertices to X_0 and so determines a commutative square in one of $\pi_1(Q, X_0)$ for $Q = U, V, W$. Move these commutative squares over to G using f_U, f_V and recompose them (this is possible again because of the condition $f_U i = f_V j$), noting that:

- in a groupoid, *any composition of commutative squares is commutative.*

Two opposite sides of the composite commutative square in G so obtained are identities, because h was a homotopy rel end points, and the other two are $F(a)$, $F(b)$. This proves that $F(a) = F(b)$ in G.

Thus the argument can be summarised: a path or homotopy is divided into small pieces, then deformed so that these pieces can be packaged and moved over to G, where they are reassembled. There seems to be an analogy with the processing of an email.

Notable applications of the groupoid theorem were: (i) to give a proof of a formula in Van Kampen's paper of the fundamental group of a space which is the union of two connected spaces with non connected intersection, see [17, 8.4.9]; and (ii) to show the topological utility of the construction by Higgins [76] of the groupoid $f_*(G)$ over Y_0 induced from a groupoid G over X_0 by a function $f : X_0 \to Y_0$. (Accounts of these with the notation $U_f(G)$ rather than $f_*(G)$ are given in [76, 17].) This latter construction is regarded as a 'change of base', and analogues in higher dimensions yielded generalisations of the Relative Hurewicz Theorem and of Theorem W, using induced modules and crossed modules.

There is another approach to the Van Kampen Theorem which goes via the theory of covering spaces, and the equivalence between covering spaces of a reasonable space X and functors $\pi_1(X) \to \mathsf{Set}$ [17]. See for example [61, 13] for an exposition of the relation with Galois theory. The paper [39] gives a general formulation of conditions for the theorem to hold in the case $X_0 = X$ in terms of the map $U \sqcup V \to X$ being an 'effective global descent morphism' (the theorem is given in the generality of lextensive categories). This work has been developed for toposes [53]. Analogous interpretations for higher dimensional Van Kampen theorems are not known.

The justification of the breaking of a paradigm in changing from groups to groupoids is several fold: the elegance and power of the results; the increased linking with other uses of groupoids [20, 102]; and the opening out of new possibilities in higher dimensions, which allowed for new results and calculations in homotopy theory, and suggested new algebraic constructions.

3 The search for higher homotopy groupoids

Contemplation of the proof of the SVGKT in the last section suggested that a higher dimensional version should exist, though this version amounted to an idea of a proof in search of a theorem. In the end, the results exactly encapsulated this intuition.

One intuition was that in groupoids we are dealing with a partial algebraic structure, in which composition is defined for two arrows if and only if the source of one arrow is the target of the other. This seems easily to generalise to directed squares, in which two are composable horizontally if and only if the left hand side of one is the right hand side of the other (and similarly vertically).

However the formulation of a theorem in higher dimensions required specification of the three elements of a functor

$$\Pi : (\text{topological data}) \to (\text{higher order groupoids})$$

which would allow the expression of these ideas for the proof.

C. Ehresmann had defined double categories in [64].

Experiments were made in the years 1967-1973 to define some functor Π from spaces to some kind of double groupoid, using compositions of squares in two directions, but these proved abortive. However considerable progress was made in work with Chris Spencer in 1971-3 on investigating the algebra of double groupoids [48, 49], and showing a relation to crossed modules. Further evidence was provided when it was found [49] that group objects in the category of groupoids are not necessarily commutative objects, since they are equivalent to crossed modules. See [46] for an application of this equivalence to covering spaces of non-connected topological groups. (It turned out that this equivalence was known to the Grothendieck school in the 1960s, but not published. The equivalence should be regarded as a generalisation of the fact that congruences on a group correspond to normal subgroups of the group.)

A key discovery was that a category of double groupoids with one vertex and what we called 'special connections' [48] is equivalent to the category of crossed modules. Using these connections we could define what we called a 'commutative cube' in such a double groupoid. The key equation for this was:

$$c_1 = \begin{bmatrix} \ulcorner & a_0^{-1} & \urcorner \\ -b_0 & c_0 & b_1 \\ \llcorner & a_1 & \lrcorner \end{bmatrix}$$

which corresponded to folding flat five faces of a cube and filling in the corners with new 'canonical' elements which we called 'connections', because of a crucial 'transport law' which was borrowed from a paper on path connections in differential geometry, and can be written

$$\begin{bmatrix} \ulcorner & = \\ \| & \ulcorner \end{bmatrix} = \ulcorner .$$

The connections provide, additionally to the usual compositions, identities, and inverses, a structure which can be expressed intuitively by saying that in the 2-dimensional algebra of squares not only can you move forward and backwards, and stay still, but you can also turn left and right. For more details see for example [45, 2]. As you might imagine, there are problems in finding a formula in still higher dimensions. In the groupoid case, this is handled by a homotopy addition lemma and thin elements [30], but in the category case a formula for just a commutative 4-cube is complicated [70].

The blockage of defining a functor Π to double groupoids was resolved in 1974 in discussions with Higgins, by considering Whitehead's Theorem W. This showed that a 2-dimensional universal property was available in homotopy theory, which was encouraging; it also suggested that a theory to be any good should recover Theorem W. But this theorem was about *relative* homotopy groups. This suggested studying a relative situation $X_* : X_0 \subseteq X_1 \subseteq X$. On looking for the simplest way to get a homotopy functor from this situation using squares, the 'obvious' answer came up: consider maps $(I^2, \partial I^2, \partial\partial I^2) \to (X, X_1, X_0)$, i.e. maps of the square into X which take the edges into X_1 and the vertices into X_0, and then take homotopy classes rel vertices of such maps to form a set $\varrho_2(X_*)$. Of course this set will not inherit a group structure but the surprise is that it does inherit the structure of double groupoid with connections - the proof is not entirely trivial, and is given in [28] and the expository article [22]. In the case X_0 is a singleton, the equivalence of such double groupoids to crossed modules takes $\varrho(X_*)$ to the usual relative homotopy crossed module.

Thus a search for a *higher homotopy groupoid* was realised in dimension 2. It might be that a tendency of mathematicians to despise the notion of groupoid, as suggested in [56], contributed to such a construction not being found earlier.

Finding a good homotopy double groupoid led rather quickly, in view of the previous experience, to a substantial account of a 2-dimensional SVKT [28]. This recovers Theorem W, and also leads to new calculations in 2-dimensional homotopy theory, and in fact to some new calculations of 2-types. For a recent summary of some results and some new ones, see the paper in the J. Symbolic Computation [52] – publication in this journal illustrates that we are interested in using general methods in order to obtain specific calculations, and ones to which there seems no other route. For purely algebraic calculations, see [4].

Once the 2-dimensional case had been completed in 1975, it was easy to conjecture the form of general results for dimensions > 2, and announcements were made in [29] with full details in [30, 31]. However, these results needed a number of new ideas, even just to construct the higher dimensional compositions, and the proof of the Generalised SVKT was quite hard and intricate. Further, for applications, such as to explain how the general Π behaved on homotopies, we also needed a theory of tensor products, so that the resulting theory is quite complex. In the next section we give a summary.

4 Main results

Major features of the work over the years with Philip Higgins and others can be summarised in the following diagram of categories and functors:

Diagram 4.1

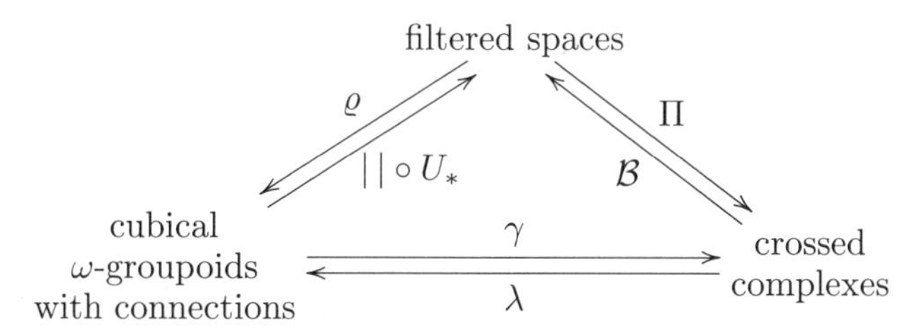

in which

4.1.1 the categories FTop of filtered spaces, ω-Gpd of ω-groupoids, and Crs of crossed complexes are monoidal closed, and have a notion of homotopy using $\otimes$ and a unit interval object;
4.1.2 ϱ, Π are homotopical functors (that is they are defined in terms of homotopy classes of certain maps), and preserve homotopies;
4.1.3 λ, γ are inverse adjoint equivalences of monoidal closed categories;
4.1.4 there is a natural equivalence $\gamma\varrho \simeq \Pi$, so that either ϱ or Π can be used as appropriate;
4.1.5 ϱ, Π preserve certain colimits and certain tensor products;
4.1.6 by definition, the *cubical filtered classifying space* is $\mathcal{B}^{\square} = |\,| \circ U_*$ where U_* is the forgetful functor to filtered cubical sets using the filtration of an ω-groupoid by skeleta, and $|\,|$ is geometric realisation of a cubical set;
4.1.7 there is a natural equivalence $\Pi \circ \mathcal{B}^{\square} \simeq 1$;
4.1.8 if C is a crossed complex and its cubical classifying space is defined as $B^{\square}C = (\mathcal{B}^{\square}C)_{\infty}$, then for a CW-complex X, and using homotopy as in 4.1.1 for crossed complexes, there is a natural bijection of sets of homotopy classes

$$[X, B^{\square}C] \cong [\Pi(X_*), C].$$

Here a *filtered space* consists of a (compactly generated) space X_{∞} and an increasing sequence of subspaces

$$X_* : X_0 \subseteq X_1 \subseteq X_2 \subseteq \cdots \subseteq X_{\infty}.$$

With the obvious morphisms, this gives the category FTop. The tensor product in this category is the usual

$$(X_* \otimes Y_*)_n = \bigcup_{p+q=n} X_p \times Y_q.$$

The closed structure is easy to construct from the law

$$\mathsf{FTop}(X_* \otimes Y_*, Z_*) \cong \mathsf{FTop}(X_*, \mathsf{FTOP}(Y_*, Z_*)).$$

An advantage of this monoidal closed structure is that it allows an enrichment of the category FTop over either crossed complexes or ω-Gpd using Π or ϱ applied to $\mathsf{FTOP}(Y_*, Z_*)$.

The structure of *crossed complex* is suggested by the canonical example, the *fundamental crossed complex* $\Pi(X_*)$ of the filtered space X_*. So it is given by a diagram

Diagram 4.2

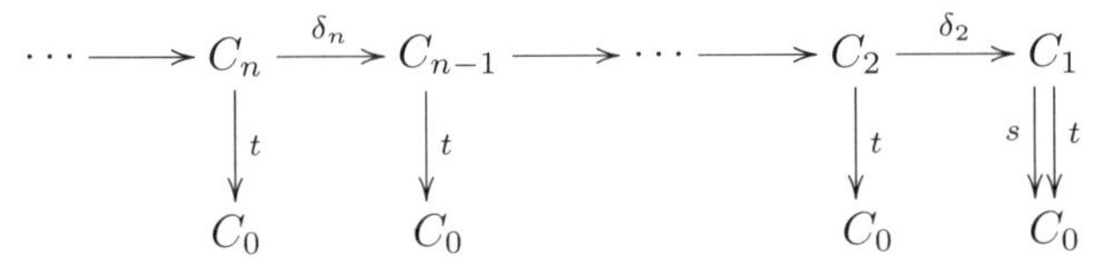

in which in this example C_1 is the fundamental groupoid $\pi_1(X_1, X_0)$ of X_1 on the 'set of base points' $C_0 = X_0$, while for $n \geqslant 2$, C_n is the family of relative homotopy groups $\{C_n(x)\} = \{\pi_n(X_n, X_{n-1}, x) \mid x \in X_0\}$. The boundary maps are those standard in homotopy theory. There is for $n \geqslant 2$ an action of the groupoid C_1 on C_n (and of C_1 on the groups $C_1(x)$, $x \in X_0$ by conjugation), the boundary morphisms are operator morphisms, $\delta_{n-1}\delta_n = 0$, $n \geqslant 3$, and the additional axioms are satisfied that

4.3 $b^{-1}cb = c^{\delta_2 b}$, $b, c \in C_2$, so that $\delta_2 : C_2 \to C_1$ is a crossed module (of groupoids),

4.4 for $n \geqslant 3$, each group $C_n(x)$ is abelian, and with trivial action by $\delta_2(C_2(x))$, so that the family C_n becomes a $C_1/\delta_2(C_2)$-module.

Clearly we obtain a category Crs of crossed complexes; this category is not so familiar and so we give arguments for using it in the next section.

As algebraic examples of crossed complexes we have: $C = \mathbb{C}(G, n)$ where G is a group, commutative if $n \geqslant 2$, and C is G in dimension n and trivial elsewhere; $C = \mathbb{C}(G : M, n)$, where G is a group, M is a G-module, $n \geqslant 2$, and C is G in dimension 1, M in dimension n, trivial elsewhere, and zero boundary if $n = 2$; C is a crossed module (of groups) in dimensions 1 and 2 and trivial elsewhere.

A crossed complex C has a fundamental groupoid $\pi_1 C = C_1 / \operatorname{Im} \delta_2$, and also for $n \geqslant 2$ a family $\{H_n(C, p) \mid p \in C_0\}$ of homology groups.

5 Why crossed complexes?

• They generalise groupoids and crossed modules to all dimensions. Note that the natural context for second relative homotopy groups is crossed modules of groupoids, rather than groups.

• They are good for modelling CW-complexes.

• Free crossed resolutions enable calculations with small CW-models of $K(G, 1)$s *and their maps* (Whitehead, Wall, Baues).

• Crossed complexes give a kind of 'linear model' of homotopy types which includes all 2-types. Thus although they are not the most general model by any means (they do not contain quadratic information such as Whitehead products), this simplicity makes them easier to handle and to relate to classical tools. The new methods and results obtained for crossed complexes can be used as a model for more complicated situations. This is how a general n-adic Hurewicz Theorem was found [41].

• They are convenient for *calculation*, and the functor Π is classical, involving relative homotopy groups. We explain some results in this form later.

• They are close to chain complexes with a group(oid) of operators, and related to some classical homological algebra (e.g. *chains of syzygies*). In fact if SX is the simplicial singular complex of a space, with its skeletal filtration, then the crossed complex $\Pi(SX)$ can be considered as a slightly non commutative version of the singular chains of a space.

• The monoidal structure is suggestive of further developments (e.g. *crossed differential algebras*) see [12, 11, 10]. It is used in [23] to give an algebraic model of homotopy 3-types, and to discuss automorphisms of crossed modules.

• Crossed complexes have a good homotopy theory, with a *cylinder object, and homotopy colimits.* The homotopy classification result 4.1.8 generalises a classical theorem of Eilenberg-Mac Lane.

• They have an interesting relation with the Moore complex of simplicial groups and of simplicial groupoids (see section 18).

6 Why cubical ω-groupoids with connections?

The definition of these objects is more difficult to give, but will be indicated later. Here we explain why we need to introduce such new structures.

• The functor ϱ gives a form of *higher homotopy groupoid*, thus realising the dreams of the early topologists.

• They are equivalent to crossed complexes.

• They have a clear *monoidal closed structure*, and a notion of homotopy, from which one can deduce those on crossed complexes, using the equivalence of categories.

• It is easy to relate the functor ϱ to tensor products, but quite difficult to do this directly for Π.

• Cubical methods, unlike globular or simplicial methods, allow for a simple *algebraic inverse to subdivision*, which is crucial for our local-to-global theorems.

• The additional structure of 'connections', and the equivalence with crossed complexes, allows for the sophisticated notion of *commutative cube*, and the proof that *multiple compositions of commutative cubes are commutative.* The last fact is a key component of the proof of the GSVKT.

• They yield a construction of a *(cubical) classifying space* $B^{\square}C = (\mathcal{B}^{\square}C)_{\infty}$ of a crossed complex C, which generalises (cubical) versions of Eilenberg-Mac Lane spaces, including the local coefficient case. This has convenient relation to homotopies.

• There is a current *resurgence of the use of cubes* in for example combinatorics, algebraic topology, and concurrency. There is a Dold-Kan type theorem for cubical abelian groups with connections [36].

7 The equivalence of categories

Let Crs, ω-Gpd denote respectively the categories of crossed complexes and ω-groupoids. A major part of the work in [30] consists in defining these categories and proving their equivalence, which thus gives an example of two algebraically defined categories whose equivalence is non trivial. It is even more subtle than that because the functors $\gamma : \mathsf{Crs} \to \omega{-}\mathsf{Gpd}$, $\lambda : \omega{-}\mathsf{Gpd} \to \mathsf{Crs}$ are not hard to define, and it is easy to prove $\gamma\lambda \simeq 1$. The hard part is to prove $\lambda\gamma \simeq 1$, which shows that an ω-groupoid G may be reconstructed from the crossed complex $\gamma(G)$ it contains. The proof involves using the connections to construct a 'folding map' $\Phi : G_n \to G_n$, and establishing its major properties, including the relations with the compositions. This gives an algebraic form of some old intuitions of several ways of defining relative homotopy groups, for example using cubes or cells.

On the way we establish properties of *thin elements*, as those which fold down to 1, and show that G satisfies a strong Kan condition, namely that every box has a unique thin filler. This result plays a key role in the proof of the GSVKT for ϱ, since it is used to show an independence of choice. That part of the proof goes by showing that the two choices can be seen, since we start with a homotopy, as given by the two ends $\partial^{\pm}_{n+1}x$ of an $(n+1)$-cube x. It is then shown by induction, using the method of construction and the above result, that x is degenerate in direction $n+1$. Hence the two ends in that direction coincide.

Properties of the folding map are used also in showing in [31] that $\Pi(X_*)$ is actually included in $\varrho(X_*)$; in relating two types of thinness for elements of $\varrho(X_*)$; and in proving a *homotopy addition lemma* in $\varrho(X_*)$.

Any ω-Gpd G has an underlying cubical set UG. If C is a crossed complex, then the cubical set $U(\lambda C)$ is called the *cubical nerve* $N^{\square}C$ of C. It is a conclusion of the theory that we can also obtain $N^\square C$ as

$$(N^\square C)_n = \mathsf{Crs}(\Pi(I^n_*), C)$$

where I^n_* is the usual geometric cube with its standard skeletal filtration. The (cubical) geometric realisation $|N^\square C|$ is also called the *cubical classifying space* $B^\square C$ of the crossed complex C. The filtration C^* of C by skeleta gives a filtration $\mathbb{B}C = B^\square C^*$ of $B^\square C$ and there is (as in 4.1.7) a natural isomorphism $\Pi(B^\square C^*) \cong C$. Thus the properties of a crossed complex are those that are universally satisfied by $\Pi(X_*)$. These proofs use the equivalence of the homotopy categories of cubical Kan complexes and of CW-complexes. We originally took this from the Warwick Masters thesis of S. Hintze, but it is now available with a different proof from R. Antolini [5, 6].

As said above, by taking particular values for C, the classifying space $B^\square C$ gives cubical versions of Eilenberg-Mac Lane spaces $K(G,n)$, including the case $n = 1$ and G non commutative. If C is essentially a crossed module, then $B^\square C$ is called the *cubical classifying space* of the crossed module, and in fact realises the k-invariant of the crossed module.

Another useful result is that if K is a cubical set, then $\varrho(|K|_*)$ may be identified with $\varrho(K)$, the *free* ω-Gpd *on the cubical set* K, where here $|K|_*$ is the usual filtration by skeleta. On the other hand, our proof that $\Pi(|K|_*)$ is the free crossed complex on the non-degenerate cubes of K uses the generalised SVKT of the next section.

It is also possible to give simplicial and globular versions of some of the above results, because the category of crossed complexes is equivalent also to those of simplicial T-complexes [7] and of globular ∞-groupoids [32]. In fact the published paper on the classifying space of a crossed complex [35] is given in simplicial terms, in order to link more easily with well known theories. This is helpful in the equivariant results in [25, 26].

8 Main aim of the work: colimit, or local-to-global, theorems

These theorems give *non commutative tools for higher dimensional local-to-global problems* yielding a variety of new, often non commutative, calculations, which *prove* (i.e. test) the theory. We now explain these theorems in a way which strengthens the relation with descent.

We suppose given an open cover $\mathcal{U} = \{U^\lambda\}_{\lambda\in\Lambda}$ of X. This cover defines a map

$$q : E = \bigsqcup_{\lambda\in\Lambda} U^\lambda \to X$$

and so we can form an augmented simplicial space

$$\check{\mathrm{C}}(q): \quad \cdots \qquad E\times_X E\times_X E \;\substack{\longrightarrow\\[-3pt]\longrightarrow\\[-3pt]\longrightarrow}\; E\times_X E \;\substack{\longrightarrow\\[-3pt]\longrightarrow}\; E \xrightarrow{\;q\;} X$$

where the higher dimensional terms involve disjoint unions of multiple intersections U^ν of the U^λ.

We now suppose given a filtered space X_*, a cover $\mathcal{U}$ as above of $X = X_\infty$, and so an augmented simplicial filtered space $\check{\mathrm{C}}(q_*)$ involving multiple intersections U^ν_* of the induced filtered spaces.

We still need a connectivity condition.

Definition 8.1 A filtered space X_* is *connected* if and only if for all $n > 0$ the induced map $\pi_0 X_0 \to \pi_0 X_n$ is surjective and for all $r > n > 0$ and $x \in X_0, \pi_n(X_r, X_n, x) = 0$.

Theorem 8.2 (MAIN RESULT (GSVKT)) *If U^ν_* is connected for all finite intersections U^ν of the elements of the open cover, then*
(C) (connectivity) *X_* is connected, and*
(I) (isomorphism) *the following diagram as part of $\varrho(\check{C}(q_*))$*

$$\varrho(E_* \times_{X_*} E_*) \rightrightarrows \varrho(E_*) \xrightarrow{\varrho(q_*)} \varrho(X_*). \qquad (c\varrho)$$

is a coequaliser diagram. Hence the following diagram of crossed complexes

$$\Pi(E_* \times_{X_*} E_*) \rightrightarrows \Pi(E_*) \xrightarrow{\Pi(q_*)} \Pi(X_*). \qquad (c\Pi)$$

is also a coequaliser diagram.

So we get calculations of the fundamental crossed complex $\Pi(X_*)$.

It should be emphasised that to get to and apply this theorem takes the two papers [30, 31], of 58 pages together. With this we deduce in the first instance:

- the usual SVKT for the fundamental groupoid on a set of base points;
- the Brouwer degree theorem ($\pi_n S^n = \mathbb{Z}$);
- the relative Hurewicz theorem;
- Whitehead's theorem that $\pi_n(X \cup \{e^2_\lambda\}, X)$ is a free crossed module;
- a more general excision result on $\pi_n(A \cup B, A, x)$ as an induced module (crossed module if $n = 2$) when $(A, A \cap B)$ is $(n-1)$-connected.

The assumptions required of the reader are quite small, just some familiarity with CW-complexes. This contrasts with some expositions of basic homotopy theory, where the proof of say the relative Hurewicz theorem requires knowledge of singular homology theory. Of course it is surprising to get this last theorem without homology, but this is because it is seen as a statement on the morphism of relative homotopy groups

$$\pi_n(X, A, x) \to \pi_n(X \cup CA, CA, x) \cong \pi_n(X \cup CA, x)$$

and is obtained, like our proof of Theorem W, as a special case of an excision result. The reason for this success is that we use algebraic structures which model the geometry and underlying processes more closely than those in common use.

Note also that these results cope well with the action of the fundamental group on higher homotopy groups.

The calculational use of the GSVKT for $\Pi(X_*)$ is enhanced by the relation of Π with tensor products (see section 15 for more details).

9 The fundamental cubical ω–groupoid $\varrho(X_*)$ of a filtered space X_*

Here are the basic elements of the construction.

I^n_*: the n-cube with its skeletal filtration.

Set $R_n(X_*) = \mathsf{FTop}(I^n_*, X_*)$. This is a *cubical set with compositions, connections, and inversions.*

For $i = 1, \ldots, n$ there are standard:
face maps $\partial^\pm_i : R_n X_* \to R_{n-1} X_*$;
degeneracy maps $\varepsilon_i : R_{n-1} X_* \to R_n X_*$
connections $\Gamma^\pm_i : R_{n-1} X_* \to R_n X_*$

compositions $a \circ_i b$ defined for $a, b \in R_n X_*$ such that $\partial_i^+ a = \partial_i^- b$
inversions $-_i : R_n \to R_n$.

The connections are induced by $\gamma_i^\alpha : I^n \to I^{n-1}$ defined using the monoid structures $\max, \min : I^2 \to I$. They are essential for many reasons, e.g. to discuss the notion of *commutative cube*.

These operations have certain algebraic properties which are easily derived from the geometry and which we do not itemise here – see for example [2, 72]. These were listed first in the Bangor thesis of Al-Agl [1]. (In the paper [30] the only connections needed are the Γ_i^+, from which the Γ_i^- are derived using the inverses of the groupoid structures.)

Definition 9.1

$$p : R_n(X_*) \to \varrho_n(X_*) = (R_n(X_*)/ \equiv)$$

is the quotient map, where $f \equiv g \in R_n(X_*)$ means *filter homotopic (i.e. through filtered maps) rel vertices of* I^n.

The following results are proved in [31].

9.2 The compositions on R are inherited by ϱ to give $\varrho(X_*)$ the structure of cubical multiple groupoid with connections.

9.3 The map $p : R(X_*) \to \varrho(X_*)$ is a Kan fibration of cubical sets.

The proofs of both results use methods of collapsing which are indicated in the next section. The second result is almost unbelievable. Its proof has to give a systematic method of deforming a cube with the right faces 'up to homotopy' into a cube with exactly the right faces, using the given homotopies. In both cases, the assumption that the relation $\equiv$ uses homotopies rel vertices is essential to start the induction. (In fact the paper [31] does not use homotopy rel vertices, but imposes an extra condition J_0, that each loop in X_0 is contractible X_1. A full exposition of the whole story is in preparation.)

Here is an application which is essential in many proofs.

Theorem 9.4 (Lifting multiple compositions) *Let* $[\alpha_{(r)}]$ *be a multiple composition in* $\varrho_n(X_*)$. *Then representatives* $a_{(r)}$ *of the* $\alpha_{(r)}$ *may be chosen so that the multiple composition* $[a_{(r)}]$ *is well defined in* $R_n(X_*)$.

Proof: The multiple composition $[\alpha_{(r)}]$ determines a cubical map

$$A : K \to \varrho(X_*)$$

where the cubical set K corresponds to a representation of the multiple composition by a subdivision of the geometric cube, so that top cells $c_{(r)}$ of K are mapped by A to $\alpha_{(r)}$.

Consider the diagram, in which $*$ is a corner vertex of K,

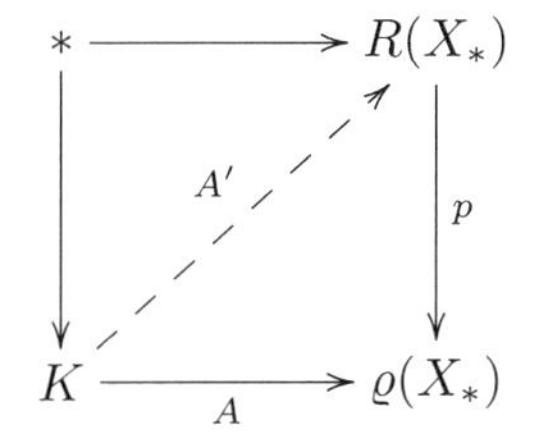

Then K collapses to $*$, written $K \searrow *$. By the fibration result, A lifts to A', which represents $[a_{(r)}]$, as required. □

So we have to explain collapsing.

10 Collapsing

We use a basic notion of collapsing and expanding due to J.H.C. Whitehead.

Let $C \subseteq B$ be subcomplexes of I^n. We say C is an *elementary collapse* of B, $B \searrow^e C$, if for some $s \geqslant 1$ there is an s-cell a of B and $(s-1)$-face b of a, the *free face*, such that

$$B = C \cup a, \qquad C \cap a = \partial a \setminus b$$

(where ∂a denotes the union of the proper faces of a).

We say $B_1 \searrow B_r$, B_1 *collapses* to B_r, if there is a sequence

$$B_1 \searrow^e B_2 \searrow^e \cdots \searrow^e B_r$$

of elementary collapses.

If C is a subcomplex of B then

$$B \times I \searrow (B \times \{0\} \cup C \times I)$$

(this is proved by induction on the dimension of $B \setminus C$).

Further I^n collapses to any one of its vertices (this may be proved by induction on n using the first example). These collapsing techniques are crucial for proving 9.2, that $\varrho(X_*)$ does obtain the structure of multiple groupoid, since it allows the construction of the extensions of filtered maps and filtered homotopies that are required.

However, more subtle collapsing techniques using partial boxes are required to prove the fibration theorem 9.3, as partly explained in the next section.

11 Partial boxes

Let C be an r-cell in the n-cube I^n. Two $(r-1)$-faces of C are called *opposite* if they do not meet.

A partial box in C is a subcomplex B of C generated by one $(r-1)$-face b of C (called a *base* of B) and a number, possibly zero, of other $(r-1)$-faces of C none of which is opposite to b.

The partial box is a box if its $(r-1)$-cells consist of all but one of the $(r-1)$-faces of C.

The proof of the fibration theorem uses a filter homotopy extension property and the following:

Proposition 11.1 Key Proposition: Let B, B' be partial boxes in an r-cell C of I^n such that $B' \subseteq B$. Then there is a chain

$$B = B_s \searrow B_{s-1} \searrow \cdots \searrow B_1 = B'$$

such that

(i) each B_i is a partial box in C

(ii) $B_{i+1} = B_i \cup a_i$ where a_i is an $(r-1)$-cell of C not in B_i;

(iii) $a_i \cap B_i$ is a partial box in a_i.

The proof is quite neat, and follows the pictures.

Induction up a such a chain of partial boxes is one of the steps in the proof of the fibration theorem 9.3.

The proof of the fibration theorem gives a program for carrying out the deformations needed to do the lifting. In some sense, it implies computing a multiple composition as in Theorem 9.4 can be done using collapsing as the guide.

Methods of collapsing are related to notions of shelling in [82], and of course any finite tree collapses to a point.

12 Thin elements

Another key concept is that of *thin element* $\alpha \in \varrho_n(X_*)$ for $n \geqslant 2$. The proofs here use strongly the algebraic results of [30].

We say α is *geometrically thin* if it has a *deficient* representative, i.e. an $a : I^n_* \to X_*$ such that $a(I^n) \subseteq X_{n-1}$.

We say α is *algebraically thin* if it is a multiple composition of degenerate elements or those coming from repeated negatives of connections. Clearly any composition of algebraically thin elements is thin.

Theorem 12.1 *(i) algebraically thin $\equiv$ geometrically thin.*
(ii) In a cubical ω–groupoid with connections, any box has a unique thin filler.

Proof The proof of the forward implication in (i) uses lifting of multiple compositions, in a stronger form than stated above.

The proofs of (ii) and the backward implication in (i) uses the full force of the algebraic relation between ω–groupoids and crossed complexes. □

These results allow one to replace arguments with commutative cubes by arguments with thin elements.

13 Sketch proof of the GSVKT

We go back to the following diagram whose top row is part of $\varrho(\check{\mathrm{C}}(q_*))$

$$\begin{array}{ccccc} \varrho(E_* \times_{X_*} E_*) & \overset{\partial_0}{\underset{\partial_1}{\rightrightarrows}} & \varrho(E_*) & \xrightarrow{\varrho(q_*)} & \varrho(X_*) \\ & & & \searrow^{f} & \downarrow f' \\ & & & & G \end{array} \qquad (c\varrho)$$

To prove this top row is a coequaliser diagram, we suppose given a morphism $f : \varrho(E_*) \to G$ of cubical ω-groupoids with connection such that $f \circ \partial_0 = f \circ \partial_1$, and prove that there is unique $f' : \varrho(X_*) \to G$ such that $f' \circ \varrho(q_*) = f$.

To define $f'(\alpha)$ for $\alpha \in \varrho(X_*)$, you subdivide a representative a of α to give $a = [a_{(r)}]$ so that each $a_{(r)}$ lies in an element $U^{(r)}$ of $\mathcal{U}$; use the connectivity conditions and this subdivision to deform a into $b = [b_{(r)}]$ so that

$$b_{(r)} \in R(U^{(r)}_*)$$

and so obtain

$$\beta_{(r)} \in \varrho(U^{(r)}_*).$$

The elements

$$f\beta_{(r)} \in G$$

may be composed in G (by the conditions on f), to give an element

$$\theta(\alpha) = [f\beta_{(r)}] \in G.$$

So the proof of the universal property has to use an *algebraic inverse to subdivision.* Again an analogy here is with sending an email: the element you start with is subdivided, deformed so that each part is correctly labelled, the separate parts are sent, and then recombined.

The proof that $\theta(\alpha)$ is independent of the choices involved makes crucial use of properties of thin elements. The key point is: *a filter homotopy* $h : \alpha \equiv \alpha'$ *in* $R_n(X_*)$ *gives a deficient element of* $R_{n+1}(X_*)$.

The method is to do the subdivision and deformation argument on such a homotopy, push the little bits in some

$$\varrho_{n+1}(U_*^\lambda)$$

(now thin) over to G, combine them and get a thin element

$$\tau \in G_{n+1}$$

all of whose faces not involving the direction $(n+1)$ are thin *because* h *was given to be a filter homotopy.* An inductive argument on unique thin fillers of boxes then shows that τ *is degenerate in direction* $(n+1)$, so that the two ends in direction $(n+1)$ are the same.

This ends a rough sketch of the proof of the GSVKT for ϱ.

Note that the theory of these forms of multiple groupoids is designed to make this last argument work. We replace a formula for saying a cube h has commutative boundary by a statement that h is thin. It would be very difficult to replace the above argument, on the composition of thin elements, by a higher dimensional manipulation of formulae such as that given in section 3 for a commutative 3-cube.

14 Tensor products and homotopies

The construction of the monoidal closed structure on the category ω-Gpd is based on rather formal properties of cubical sets, and the fact that for the cubical set $\mathbb{I}^n$ we have $\mathbb{I}^m \otimes \mathbb{I}^n \cong \mathbb{I}^{m+n}$. The details are given in [33]. The equivalence of categories implies then that the category Crs is also monoidal closed, with a natural isomorphism

$$\mathsf{Crs}(A \otimes B, C) \cong \mathsf{Crs}(A, \mathsf{CRS}(B, C)).$$

Here the elements of $\mathsf{CRS}(B, C)$ are in dimension 0 the morphisms $B \to C$, in dimension 1 the *left homotopies of morphisms*, and in higher dimensions are forms of higher homotopies. The precise description of these is obtained of course by tracing out in detail the equivalence of categories. It should be emphasised that certain choices are made in constructing this equivalence, and these choices are reflected in the final formulae that are obtained.

An important result is that if X_*, Y_* are filtered spaces, then there is a natural transformation

$$\eta : \varrho(X_*) \otimes \varrho(Y_*) \to \varrho(X_* \otimes Y_*)$$
$$[a] \otimes [b] \mapsto [a \otimes b]$$

where if $a : I^m_* \to X_*$, $b : I^n_* \to Y_*$ then $a \otimes b : I^{m+n}_* \to X_* \otimes Y_*$. It not hard to see, in this cubical setting, that η is well defined. It can also be shown using previous results that η is an isomorphism if X_*, Y_* are the geometric realisations of cubical sets with the usual skeletal filtration.

The equivalence of categories now gives a natural transformation of crossed complexes

$$\eta' \ : \ \Pi(X_*) \otimes \Pi(Y_*) \to \Pi(X_* \otimes Y_*). \tag{14.1}$$

It would be hard to construct this directly. It is proved in [35] that η' is an isomorphism if X_*, Y_* are the skeletal filtrations of CW-complexes. The proof uses the GSVKT, and the fact that $A \otimes -$ on crossed complexes has a right adjoint and so preserves colimits. It is proved in [10] that η is an isomorphism if X_*, Y_* are cofibred, connected filtered spaces. This applies in particular to the useful case of the filtration $B^\square C^*$ of the classifying space of a crossed complex.

It turns out that the defining rules for the tensor product of crossed complexes which follows from the above construction are obtained as follows. We first define a bimorphism of crossed complexes.

Definition 14.1 A *bimorphism* $\theta : (A, B) \to C$ of crossed complexes is a family of maps $\theta : A_m \times B_n \to C_{m+n}$ satisfying the following conditions, where $a \in A_m, b \in B_n, a_1 \in A_1, b_1 \in B_1$ (temporarily using additive notation throughout the definition):

(i)

$$\beta(\theta(a,b)) \;=\; \theta(\beta a, \beta b) \;\text{ for all }\; a \in A, b \in B\,.$$

(ii)

$$\begin{aligned}\theta(a, b^{b_1}) &= \theta(a,b)^{\theta(\beta a, b_1)} \;\text{ if }\; m \geqslant 0, n \geqslant 2\,,\\ \theta(a^{a_1}, b) &= \theta(a,b)^{\theta(a_1, \beta b)} \;\text{ if }\; m \geqslant 2, n \geqslant 0\,.\end{aligned}$$

(iii)

$$\theta(a, b+b') = \begin{cases} \theta(a,b) + \theta(a,b') & \text{if } m = 0, n \geqslant 1 \text{ or } m \geqslant 1, n \geqslant 2\,,\\ \theta(a,b)^{\theta(\beta a, b')} + \theta(a,b') & \text{if } m \geqslant 1, n = 1\,,\end{cases}$$

$$\theta(a+a', b) = \begin{cases} \theta(a,b) + \theta(a',b) & \text{if } m \geqslant 1, n = 0 \;\text{ or } m \geqslant 2, n \geqslant 1\,,\\ \theta(a',b) + \theta(a,b)^{\theta(a', \beta b)} & \text{if } m = 1, n \geqslant 1\,.\end{cases}$$

(iv)

$$\delta_{m+n}(\theta(a,b)) = \begin{cases} \theta(\delta_m a, b) + (-)^m \theta(a, \delta_n b) & \text{if } m \geqslant 2, n \geqslant 2\,,\\ -\theta(a, \delta_n b) - \theta(\beta a, b) + \theta(\alpha a, b)^{\theta(a, \beta b)} & \text{if } m = 1, n \geqslant 2\,,\\ (-)^{m+1}\theta(a, \beta b) + (-)^m \theta(a, \alpha b)^{\theta(\beta a, b)} + \theta(\delta_m a, b) & \text{if } m \geqslant 2, n = 1\,,\\ -\theta(\beta a, b) - \theta(a, \alpha b) + \theta(\alpha a, b) + \theta(a, \beta b) & \text{if } m = n = 1\,.\end{cases}$$

(v)

$$\delta_{m+n}(\theta(a,b)) = \begin{cases} \theta(a, \delta_n b) & \text{if } m = 0, n \geqslant 2\,,\\ \theta(\delta_m a, b) & \text{if } m \geqslant 2, n = 0\,.\end{cases}$$

(vi)

$$\alpha(\theta(a,b)) = \theta(a,\alpha b) \quad \text{and} \quad \beta(\theta(a,b)) = \theta(a,\beta b) \quad \text{if } m=0, n=1 \,,$$
$$\alpha(\theta(a,b)) = \theta(\alpha a,b) \quad \text{and} \quad \beta(\theta(a,b)) = \theta(\beta a,b) \quad \text{if } m=1, n=0 \,.$$

The *tensor product* of crossed complexes A, B is given by the universal bimorphism $(A,B) \to A \otimes B$, $(a,b) \mapsto a \otimes b$. The rules for the tensor product are obtained by replacing $\theta(a,b)$ by $a \otimes b$ in the above formulae.

The conventions for these formulae for the tensor product arise from the derivation of the tensor product via the category of cubical ω-groupoids with connections, and the formulae are forced by our conventions for the equivalence of the two categories [30, 33].

The complexity of these formulae is directly related to the complexities of the cell structure of the product $E^m \times E^n$ where the n-cell E^n has cell structure e^0 if $n=0$, $e^0_\pm \cup e^1$ if $n=1$, and $e^0 \cup e^{n-1} \cup e^n$ if $n \geqslant 2$.

It is proved in [33] that the bifunctor $- \otimes -$ is symmetric and that if a_0 is a vertex of A then the morphism $B \to A \otimes B$, $b \to a_0 \otimes b$, is injective.

There is a standard groupoid model I of the unit interval, namely the indiscrete groupoid on two objects $0, 1$. This is easily extended trivially to either a crossed complex or an ω-Gpd. So using $\otimes$ we can define a 'cylinder object' $\mathsf{I} \otimes -$ in these categories and so a homotopy theory (cf. [84]).

15 Free crossed complexes and free crossed resolutions

Let C be a crossed complex. A *free basis* B_* for C consists of the following:
B_0 is set which we take to be C_0;
B_1 is a graph with source and target maps $s, t : B_1 \to B_0$ and C_1 is the free groupoid on the graph B_1: that is B_1 is a subgraph of C_1 and any graph morphism $B_1 \to G$ to a groupoid G extends uniquely to a groupoid morphism $C_1 \to G$;
for $n \geqslant 2$, B_n is a totally disconnected subgraph of C_n with target map $t : B_n \to B_0$; for $n = 2$, C_2 is the free crossed C_1-module on B_2 while for $n > 2$, C_n is the free $(\pi_1 C)$-module on B_n.

It may be proved using the GSVKT that if X_* is a CW-complex with the skeletal filtration, then $\Pi(X_*)$ is the free crossed complex on the characteristic maps of the cells of X_*. It is proved in [35] that the tensor product of free crossed complexes is free.

A *free crossed resolution* F_* of a groupoid G is a free crossed complex which is aspherical together with an isomorphism $\phi : \pi_1(F_*) \to G$. Analogues of standard methods of homological algebra show that free crossed resolutions of a group are unique up to homotopy equivalence.

In order to apply this result to free crossed resolutions, we need to replace free crossed resolutions by CW-complexes. A fundamental result for this is the following, which goes back to Whitehead [106] and Wall [101], and which is discussed further by Baues in [9, Chapter VI, §7]:

Theorem 15.1 *Let X_* be a CW-filtered space, and let $\phi : \pi X_* \to C$ be a homotopy equivalence to a free crossed complex with a preferred free basis. Then there is a CW-filtered space Y_*, and an isomorphism $\pi Y_* \cong C$ of crossed complexes with preferred basis, such that ϕ is realised by a homotopy equivalence $X_* \to Y_*$.*

In fact, as pointed out by Baues, Wall states his result in terms of chain complexes, but the crossed complex formulation seems more natural, and avoids questions of realisability in dimension 2, which are unsolved for chain complexes.

Corollary 15.2 If A is a free crossed resolution of a group G, then A is realised as free crossed complex with preferred basis by some CW-filtered space Y_*.

Proof We only have to note that the group G has a classifying CW-space BG whose fundamental crossed complex $\Pi(BG)$ is homotopy equivalent to A. □

Baues also points out in [9, p.657] an extension of these results which we can apply to the realisation of morphisms of free crossed resolutions.

Proposition 15.3 Let $X = K(G,1)$, $Y = K(H,1)$ be CW-models of Eilenberg - Mac Lane spaces and let $h : \Pi(X_*) \to \Pi(Y_*)$ be a morphism of their fundamental crossed complexes with the preferred bases given by skeletal filtrations. Then $h = \Pi(g)$ for some cellular $g : X \to Y$.

Proof Certainly h is homotopic to $\Pi(f)$ for some $f : X \to Y$ since the set of pointed homotopy classes $X \to Y$ is bijective with the morphisms of groups $A \to B$. The result follows from [9, p.657,(**)] ('if f is Π-realisable, then each element in the homotopy class of f is Π-realisable'). □

These results are exploited in [91, 44], to calculate free crossed resolutions of the fundamental groupoid of a graph of groups.

An algorithmic approach to the calculation of free crossed resolutions for groups is given in [47], by constructing partial contracting homotopies for the universal cover at the same time as constructing this universal cover inductively. This theme leads to, indeed necessitates, new ideas in rewriting, as shown in [73, 74]. A practical application in group theory is: given a presentation $\langle X \mid R\rangle$ of a group, write a given element of $N(R)$ as a consequence of the relators.

16 Classifying spaces and homotopy classification of maps

The formal relations of cubical sets and of cubical ω-groupoids with connections and the relation of Kan cubical sets with topological spaces, allow the proof of a homotopy classification theorem:

Theorem 16.1 *If K is a cubical set, and G is an ω-groupoid, then there is a natural bijection of sets of homotopy classes*

$$[|K|, |UG|] \cong [\varrho(|K|_*), G],$$

where on the left hand side we work in the category of spaces, and on the right in ω-groupoids.

Here $|K|_*$ is the filtration by skeleta of the geometric realisation of the cubical set.

We explained earlier how to define a cubical classifying space say $B^\square(C)$ of a crossed complex C as $B^\square(C) = |UN^\square C| = |U\lambda C|$. The properties already stated now give the homotopy classification theorem 4.1.7.

It is shown in [31] that if Y is a connected CW-complex, then there is a map $p : Y \to B^\square\Pi Y_*$ whose homotopy fibre is n-connected if $\pi_i Y = 0$ for $2 \leqslant i \leqslant n-1$.

It follows that if also X is a connected CW-complex with $\dim X \leqslant n$, then p induces a bijection

$$[X, Y] \to [X, B\Pi Y_*].$$

So under these circumstances we get a bijection

$$[X, Y] \to [\Pi X_*, \Pi Y_*]. \tag{16.1}$$

This result, due to Whitehead [105], translates a topological homotopy classification problem to an algebraic one. We explain below how this result can be translated to a result on chain complexes with operators.

It is also possible to define a simplicial nerve $N^\Delta(C)$ of a crossed complex C by

$$N^\Delta(C)_n = \mathsf{Crs}(\Pi(\Delta^n), C).$$

The *simplicial classifying space* of C is then defined by

$$B^\Delta(C) = |N^\Delta(C)|.$$

The properties of this simplicial classifying space are developed in [35], and in particular an analogue of 4.1.7 is proved. An application of the classifying space of a crossed module is given in [65].

The simplicial nerve and an adjointness

$$\mathsf{Crs}(\Pi(L), C) \cong \mathsf{Simp}(L, N^\Delta(C))$$

are used in [25, 26] for an equivariant homotopy theory of crossed complexes and their classifying spaces. Important ingredients in this are notions of homotopy coherence from [57] and an Eilenberg-Zilber type theorem for crossed complexes proved in Tonks' Bangor thesis [99, 100].

Labesse in [85] defines a *crossed set.* In fact a crossed set is exactly a crossed module $\delta : C \to X \rtimes G$ where G is a group acting on the set X, and $X \rtimes G$ is the associated actor groupoid; thus the simplicial construction from a crossed set described by Larry Breen in [85] is exactly the nerve of the crossed module, regarded as a crossed complex. Hence the cohomology with coefficients in a crossed set used in [85] is a special case of cohomology with coefficients in a crossed complex, dealt with in [35]. (We are grateful to Breen for pointing this out to us in 1999.)

17 Relation with chain complexes with a groupoid of operators

Chain complexes with a group of operators are a well known tool in algebraic topology, where they arise naturally as the chain complex $C_*\widetilde{X}_*$ of cellular chains of the universal cover $\widetilde{X}_*$ of a reduced CW-complex X_*. The group of operators here is the fundamental group of the space X.

J.H.C. Whitehead in [105] gave an interesting relation between his free crossed complexes (he called them 'homotopy systems') and such chain complexes. We refer later to his important homotopy classification results in this area. Here we explain the relation with the Fox free differential calculus [68].

Let $\mu : M \to P$ be a crossed module of groups, let $G = \mathrm{Coker}\mu$, and let $\phi : P \to G$ be the quotient map. Then there is an associated diagram

$$\begin{array}{ccccc} M & \xrightarrow{\mu} & P & \xrightarrow{\phi} & G \\ {\scriptstyle h_2}\downarrow & & \downarrow{\scriptstyle h_1} & & \downarrow{\scriptstyle h_0} \\ M^{\mathrm{ab}} & \xrightarrow[\partial_2]{} & D_\phi & \xrightarrow[\partial_1]{} & \mathbb{Z}[G] \end{array} \tag{17.1}$$

in which the second row consists of (right) G-modules and module morphisms. Here h_2 is simply the abelianisation map; $h_1 : P \to D_\phi$ is the universal ϕ-derivation, that is it satisfies $h_1(pq) = h_1(p)^{\phi q} + h_1(q)$, for all $p, q \in P$, and is universal for this property; and h_0 is the usual derivation $g \mapsto g - 1$. Whitehead in his Theorem 8, p. 469, of [105] gives essentially this diagram in the case P is a free group, when he takes D_ϕ to be the free G-module on the same generators as the free generators of P. Our formulation (Proposition 3.1 of [34]), which uses the derived module due to Crowell [58], includes his case. It is remarkable that diagram (17.1) is a commutative diagram in which the vertical maps are operator morphisms, and that the bottom row is defined by this property. The proof in [34] follows essentially Whitehead's proof. The bottom row is exact: this follows from results in [58], and is a reflection of a classical fact on group cohomology, namely the relation between central extensions and the Ext functor, see [89]. In the case the crossed module is the crossed module $\delta : C(\omega) \to F(X)$ derived from a presentation of a group, then $C(\omega)^{\mathrm{ab}}$ is isomorphic to the free G-module on R, D_ϕ is the free G-module on X, and it is immediate from the above that ∂_2 is the usual derivative $(\partial r/\partial x)$ of Fox's free differential calculus [68]. Thus Whitehead's results anticipate those of Fox.

It is also proved in [105] that if the restriction $M \to \mu(M)$ of μ has a section which is a morphism but not necessarily a P-map, then h_2 maps $\mathrm{Ker}\,\mu$ isomorphically to $\mathrm{Ker}\,\partial_2$. This allows calculation of the module of identities among relations by using module methods, and this is commonly exploited, see for example [67] and the references there.

Whitehead introduced the categories CW of reduced CW-complexes, HS of homotopy systems, and FCC of free chain complexes with a group of operators, together with functors

$$\mathsf{CW} \xrightarrow{\Pi} \mathsf{HS} \xrightarrow{C} \mathsf{FCC}.$$

In each of these categories he introduced notions of homotopy and he proved that C induces an equivalence of the homotopy category of HS with a subcategory of the homotopy category of FCC. Further, $C\Pi X_*$ is isomorphic to the chain complex $C_*\widetilde{X}_*$ of cellular chains of the universal cover of X, so that under these circumstances there is a bijection of sets of homotopy classes

$$[\Pi X_*, \Pi Y_*] \to [C_*\widetilde{X}_*, C_*\widetilde{Y}_*]. \tag{17.2}$$

This with the bijection (16.1) can be interpreted as an operator version of the Hopf classification theorem. It is surprisingly little known. It includes results of Olum [93] published later, and it enables quite useful calculations to be done easily, such as the homotopy classification of maps from a surface to the projective plane [66], and other cases. Thus we see once again that this general theory leads to specific calculations.

All these results are generalised in [34] to the non free case and to the non reduced case, which requires a groupoid of operators, thus giving functors

$$\mathsf{FTop} \xrightarrow{\Pi} \mathsf{Crs} \xrightarrow{\Delta} \mathsf{Chain}.$$

One utility of the generalisation to groupoids is that the functor Δ then has a right adjoint, and so preserves colimits. An example of this preservation is given in [34, Example 2.10]. The construction of the right adjoint to Δ builds on a number of constructions used earlier in homological algebra.

The definitions of the categories under consideration in order to obtain a generalisation of the bijection (17.2) has to be quite careful, since it works in the groupoid case, and not all morphisms of the chain complex are realisable.

This analysis of the relations between these two categories is used in [35] to give an account of cohomology with local coefficients. See also [46] for a relation between extension theory, crossed complexes, and covering spaces of non-connected topological groups.

18 Crossed complexes and simplicial groups and groupoids

The Moore complex NG of a simplicial group G is in general not a (reduced) crossed complex. Let D_nG be the subgroup of G_n generated by degenerate elements. Ashley showed in his thesis [7] that NG is a crossed complex if and only if $(NG)_n \cap (DG)_n = \{1\}$ for all $n \geqslant 1$.

Ehlers and Porter in [62, 63] show that there is a functor C from simplicial groupoids to crossed complexes in which $C(G)_n$ is obtained from $N(G)_n$ by factoring out

$$(NG_n \cap D_n)d_{n+1}(NG_{n+1} \cap D_{n+1}),$$

where the Moore complex is defined so that its differential comes from the last simplicial face operator.

This is one part of an investigation into the Moore complex of a simplicial group, of which the most general investigation is by Carrasco and Cegarra in [54].

An important observation in [95] is that if $N \lhd G$ is an inclusion of a normal simplicial subgroup of a simplicial group, then the induced morphism on components $\pi_0(N) \to \pi_0(G)$ obtains the structure of crossed module. This is directly analogous to the fact that if $F \to E \to B$ is a fibration sequence then the induced morphism of fundamental groups $\pi_1(F, x) \to \pi_1(E, x)$ also obtains the structure of crossed module. This is relevant to algebraic K-theory, where for a ring R the homotopy fibration sequence is taken to be $F \to B(GL(R)) \to B(GL(R))^+$.

19 Other homotopy multiple groupoids

The proof of the GSVKT outlined earlier does seem to require cubical methods, so there is still a question of the place of globular and simplicial methods in this area. A simplicial analogue of the equivalence of categories is given in [7, 92], using Dakin's notion of *simplicial T-complex* [59]. However it is difficult to describe in detail the notion of tensor product of such structures, or to formulate a proof of the colimit theorem in that context. It may be that the polyhedral methods of [82] would help here.

It is easy to define a homotopy globular set $\varrho^{\bigcirc}(X_*)$ of a filtered space X_* but it is not quite so clear how to prove directly that the expected compositions are well defined. However there is a natural graded map

$$i : \varrho^{\bigcirc}(X_*) \to \varrho(X_*) \tag{19.1}$$

and applying the folding map of [1, 2] analogously to methods in [31] allows one to prove that i of (19.1) is injective. It follows that the compositions on $\varrho(X_*)$ are inherited by $\varrho^{\bigcirc}(X_*)$ to make the latter a globular ω-groupoid. A paper is in preparation on this.

Loday in 1982 [86] defined the fundamental cat^n-group of an n-cube of spaces, and showed that cat^n-groups model all reduced weak homotopy $(n+1)$-types. Joint work [42] formulated and proved a GSVKT for the cat^n-group functor from n-cubes of spaces and this allows new local to global calculations of certain homotopy n-types [21]. This work obtains more powerful results than the purely linear theory of crossed complexes, which has however other advantages. Porter in [95] gives an interpretation of Loday's results using methods of simplicial groups. There is clearly a lot to do in this area.

Recently some absolute homotopy 2-groupoids and double groupoids have been defined, see [27] and the references there, and it is significant that crossed modules have been used in a differential topology situation by Mackaay and Picken [88]. Reinterpretations of these ideas in terms of double groupoids are started in [24].

It seems reasonable to suggest that, in the most general case, double groupoids are still somewhat mysterious objects; a generalisation of the work in [48] is given in [43], but this does not apply to the homotopy double groupoid of a map constructed in [40].

20 Conclusion and questions

• The emphasis on filtered spaces rather than the absolute case is open to question.

• *Mirroring the geometry by the algebra* is crucial for conjecturing and proving universal properties.

• *Thin elements* are crucial as modelling commutative cubes, a concept not so easy to define or handle algebraically.

• *Colimit theorems* give, when they apply, exact information even in non commutative situations. The implications of this for homological algebra and its applications could be important.

• One construction inspired eventually by this work was the *non abelian tensor product of groups*, defined in full generality with Loday in 1985, and which now has a bibliography of 79 papers (http://www.bangor.ac.uk/~mas010/nonabtens.html).

• Globular methods do fit into the scheme of higher order categorical structures, but so far have not yielded new local-to-global results in the style of the current work.

• For computations we really need strict structures (although we do want to compute invariants of homotopy colimits, cf, [44]).

• In homotopy theory, identifications in low dimensions can affect high dimensional homotopy. So we need structure in a range of dimensions to model homotopical identifications algebraically. The idea of identifications in low dimensions is reflected in the algebra by 'induced constructions'.

• In this way we calculate some crossed modules modelling homotopy 2-types, whereas the corresponding k-invariant is often difficult to calculate.

• The use of crossed complexes in Čech theory is a current project with Tim Porter.

• **Question:** Are there applications of higher homotopy groupoids in other contexts where the fundamental groupoid is currently used, such as algebraic geometry?

• **Question:** There are uses of double groupoids in differential geometry, for example in Poisson geometry, and in 2-dimensional holonomy [38]. Is there a non abelian De Rham Theory, using an analogue of crossed complexes?

• **Question:** Is there a truly non commutative integration theory based on limits of multiple compositions of elements of multiple groupoids?

Acknowledgements

This work has been partially supported by the SRC, EPSRC, British Council, ARC, and INTAS, the latter under grants INTAS 97-31961 'Algebraic Homotopy, Galois Theory and Descent'; 93-367 'Algebraic K-theory, groups and categories'; 'Algebraic K-theory, Groups and Algebraic Homotopy Theory'.

References

[1] F. Al-Agl, 1989, *Aspects of multiple categories*, Ph.D. thesis, University of Wales, Bangor.

[2] F. Al-Agl, R. Brown and R. Steiner, *Multiple categories: the equivalence between a globular and cubical approach*, Advances in Math., 170 (2002) 71–118.

[3] M. Alp, *GAP, Crossed modules, Cat^1-groups: Applications of computational group theory* Ph.D. thesis, University of Wales, Bangor, (1997), `http://www.informatics.bangor.ac.uk/public/math/research/ftp/theses/alp.ps.gz`.

[4] M. Alp and C.D. Wensley, *Enumeration of cat^1-groups of low order*, Int. J. Math. Computation, 10 (2000) 407–424.

[5] R. Antolini, *Cubical structures and homotopy theory*, Ph.D. thesis, Univ. Warwick, Coventry, (1996).

[6] R. Antolini, *Cubical structures, homotopy theory*, Ann. Mat. Pura Appl., 178 (2000) 317–324.

[7] N. Ashley, *Simplicial T-complexes*, University of Wales, Bangor, Ph.D. thesis (1978), published as *Simplicial T-Complexes: a non abelian version of a theorem of Dold-Kan*, Dissertationes Math., 165 (1988) 11–58.

[8] M.F. Atiyah, *Mathematics in the 20th Century*, Bull. London Math. Soc., 34 (2002) 1–15.

[9] H.J. Baues, *Algebraic Homotopy*, volume 15 of *Cambridge Studies in Advanced Mathematics*, Cambridge University Press (1989).

[10] H.J. Baues and R. Brown, *On relative homotopy groups of the product filtration, the James construction, and a formula of Hopf*, J. Pure Appl. Algebra, 89 (1993) 49–61.

[11] H.J. Baues and D. Conduché, *On the tensor algebra of a non-abelian group and applications*, K-theory, 5 (1992) 531–554.

[12] H.-J. Baues and A. Tonks, *On the twisted cobar construction*, Math. Proc. Cambridge Philos. Soc., 121 (1997) 229–245.

[13] F. Borceux and G. Janelidze, *Galois theories*, Cambridge Studies in Advanced Mathematics, 72, Cambridge University Press, Cambridge, 2001.

[14] L. Breen, *Théorie de Schreier supérieure*, Ann. Sci. École. Norm. Sup., 25 (1992) 465–514.

[15] R.A. Brown, *Generalized group presentation and formal deformations of CW complexes*, Trans. Amer. Math. Soc., 334 (1992) 519–549.

[16] R. Brown, *Groupoids and van Kampen's Theorem*, Proc. London Math. Soc., 3 (1967) 385–340.

[17] R. Brown, *Topology:a geometric account of general topology, homotopy types and the fundamental groupoid*, Ellis Horwood Series: Mathematics and its Applications, Ellis Horwood Ltd., Chichester, second edition (1988) (first edition, McGraw Hill, 1968).

[18] R. Brown, *On the second relative homotopy group of an adjunction space: an exposition of a theorem of J. H. C. Whitehead*, J. London Math. Soc. (2), 22 (1980) 146–152.

[19] R. Brown, *Coproducts of crossed P-modules: applications to second homotopy groups and to the homology of groups*, Topology, 23 (1984) 337–345.

[20] R. Brown, *From groups to groupoids: a brief survey*, Bull. London Math. Soc., 19 (1987) 113–134.

[21] R. Brown, 1992, *Computing homotopy types using crossed n-cubes of groups*, in *Adams Memorial Symposium on Algebraic Topology, 1 (Manchester, 1990)*, volume 175 of *London Math. Soc. Lecture Note Ser.*, 187–210, Cambridge University Press, Cambridge.

[22] R. Brown, *Groupoids and crossed objects in algebraic topology*, Homology, homotopy and applications, 1 (1999) 1–78.

[23] R. Brown and N. D. Gilbert, *Algebraic models of 3-types and automorphism structures for crossed modules*, Proc. London Math. Soc. (3), 59 (1989) 51–73.

[24] R. Brown and J. F. Glazebrook, *Connections, local subgroupoids, and a holonomy Lie groupoid of a line bundle gerbe*, Univ. Iagellonicae Acta Math. (to appear), math.DG/02/0322..

[25] R. Brown, M. Golasiński, T. Porter and A. Tonks, *Spaces of maps into classifying spaces for equivariant crossed complexes*, Indag. Math. (N.S.), 8 (1997) 157–172.

[26] R. Brown, M. Golasiński, T. Porter and A. Tonks, *Spaces of maps into classifying spaces for equivariant crossed complexes. II. The general topological group case*, K-Theory, 23 (2001) 129–155.

[27] R. Brown, K. A. Hardie, K. H. Kamps and T. Porter, *A homotopy double groupoid of a Hausdorff space*, Theory Appl. Categ., 10 (2002) 71–93.

[28] R. Brown and P. J. Higgins, *On the connection between the second relative homotopy groups of some related spaces*, Proc.London Math. Soc., (3) 36 (1978) 193–212.

[29] R. Brown and P. J. Higgins, *Sur les complexes croisés d'homotopie associés à quelques espaces filtrés*, C. R. Acad. Sci. Paris Sér. A-B, 286 (1978) A91–A93.

[30] R. Brown and P. J. Higgins, *The algebra of cubes*, J. Pure Appl. Algebra , 21 (1981) 233–260.

[31] R. Brown and P. J. Higgins, *Colimit theorems for relative homotopy groups*, J. Pure Appl. Alg, 22 (1981) 11–41.

[32] R. Brown and P. J. Higgins, *The equivalence of crossed complexes and ω-groupoids,*, Cahiers Top. Géom. Diff., 22 (1981) 370–386.

[33] R. Brown and P. J. Higgins, *Tensor products and homotopies for ω-groupoids and crossed complexes,*, J. Pure Appl. Alg, 47 (1987) 1–33.

[34] R. Brown and P. J. Higgins, *Crossed complexes and chain complexes with operators*, Math. Proc. Camb. Phil. Soc., 107 (1990) 33–57.

[35] R. Brown and P. J. Higgins, *The classifying space of a crossed complex*, Math. Proc. Cambridge Philos. Soc., 110 (1991) 95–120.

[36] R. Brown and P. J. Higgins, *Cubical abelian groups with connections are equivalent to chain complexes*, Homology, Homotopy and Applications, 5 (2003) 49–52.

[37] R. Brown and J. Huebschmann, *Identities among relations*, in R.Brown and T.L.Thickstun, eds., *Low Dimensional Topology*, London Math. Soc Lecture Notes, Cambridge University Press (1982) 153–202.

[38] R. Brown and I. Icen, *Towards two dimensional holonomy*, Advances in Math., 178 (2003) 141–175.

[39] R. Brown and G. Janelidze, *Van Kampen theorems for categories of covering morphisms in lextensive categories*, J. Pure Appl. Algebra, 119 (1997) 255–263.

[40] R. Brown and G. Janelidze, *A new homotopy double groupoid of a map of spaces*, Appl. Cat. Struct. (to appear).

[41] R. Brown and J.-L. Loday, *Homotopical excision, and Hurewicz theorems, for n-cubes of spaces*, Proc. London Math. Soc., (3) 54 (1987) 176–192.

[42] R. Brown and J.-L. Loday, *Van Kampen Theorems for diagrams of spaces*, Topology, 26 (1987) 311–337.

[43] R. Brown and K.C.H. Mackenzie, *Determination of a double Lie groupoid by its core diagram*, J. Pure Appl. Algebra, 80 (1992) 237–272.

[44] R. Brown, E. Moore, T. Porter and C. Wensley, *Crossed complexes, and free crossed resolutions for amalgamated sums and HNN-extensions of groups*,Georgian Math. J., 9 (2002) 623–644.

[45] R. Brown and G. H. Mosa, *Double categories, 2-categories, thin structures and connections*, Theory Appl. Categ., 5 (1999) 163–175.

[46] R. Brown and O. Mucuk, *Covering groups of non-connected topological groups revisited*, Math. Proc. Camb. Phil. Soc, 115 (1994) 97–110.

[47] R. Brown and A. Razak Salleh, *Free crossed resolutions of groups and presentations of modules of identities among relations*, LMS J. Comput. Math., 2 (1999) 28–61.

[48] R. Brown and C. B. Spencer, *Double groupoids and crossed modules*, Cah. Top. Géom. Diff., 17 (1976) 343–362.

[49] R. Brown and C. B. Spencer, *G-groupoids, crossed modules and the fundamental groupoid of a topological group*, Proc. Kon. Ned. Akad. v. Wet, 79 (1976) 296–302.

[50] R. Brown and C. D. Wensley, *On finite induced crossed modules, and the homotopy 2-type of mapping cones*, Theory Appl. Categ., 1 (1995) 54–70.

[51] R. Brown and C. D. Wensley, *Computing crossed modules induced by an inclusion of a normal subgroup, with applications to homotopy 2-types*, Theory Appl. Categ., 2 (1996) 3–16.

[52] R. Brown and C. D. Wensley, *Computations and homotopical applications of induced crossed modules*, J. Symb. Comp., 35 (2003) 59–72.

[53] M. Bunge and S. Lack, *Van Kampen theorems for toposes*, Advances in Math., 179 (2003) 291–317.

[54] P. Carrasco and A. M. Cegarra, *Group-theoretic algebraic models for homotopy types*, J. Pure Appl. Algebra, 75 (1991) 195–235.

[55] E. Ĉech, 1933, *Höherdimensionale Homotopiegruppen*, in *Verhandlungen des Internationalen Mathematiker-Kongresses Zurich 1932*, volume 2, 203, International Congress of Mathematicians (4th : 1932 : Zurich, Switzerland, Walter Saxer, Zurich, reprint Kraus, Nendeln, Liechtenstein, 1967.

[56] A. Connes, *Noncommutative geometry*, Academic Press Inc., San Diego, CA (1994).

[57] J.-M. Cordier and T. Porter, *Homotopy coherent category theory*, Trans. Amer. Math. Soc., 349 (1997) 1–54.

[58] R.H. Crowell, *The derived module of a homomorphism*, Advances in Math., 5 (1971) 210–238.

[59] K. Dakin, *Kan complexes and multiple groupoid structures*, Ph.D. thesis, University of Wales, Bangor (1977).

[60] P. Dedecker, *Sur la cohomologie non abélienne. II*, Canad. J. Math., 15 (1963) 84–93.

[61] A. Douady and R. Douady, *Algebres et théories Galoisiennes*, volume 2, CEDIC, Paris (1979).

[62] P. J. Ehlers and T. Porter, *Varieties of simplicial groupoids. I. Crossed complexes*, J. Pure Appl. Algebra, 120 (1997) 221–233.

[63] P. J. Ehlers and T. Porter, *Erratum to: "Varieties of simplicial groupoids. I. Crossed complexes" [J. Pure Appl. Algebra 120 (1997) no. 3, 221–233;* , J. Pure Appl. Algebra, 134 (1999) 207–209.

[64] C. Ehresmann, *Catégories et structures*, Dunod, Paris, (1964).

[65] G.J. Ellis, *Homology of 2-types*, J. London Math. Soc., (2) 46 (1992) 1–27.

[66] G. J. Ellis, *Homotopy classification the J. H. C. Whitehead way*, Exposition. Math., 6 (1988) 97–110.

[67] G. Ellis and I. Kholodna, *Three-dimensional presentations for the groups of order at most 30*, LMS J. Comput. Math., 2 (1999) 93–117 +2 appendixes (HTML and source code).

[68] R.H. Fox, *Free differential calculus I: Derivations in the group ring*, Ann. Math., 57 (1953) 547–560.

[69] A. Fröhlich, *Non-Abelian homological algebra. I. Derived functors and satellites.*, Proc. London Math. Soc. (3) 11 (1961) 239–275.

[70] P. Gaucher, *Combinatorics of branchings in higher dimensional automata*, Theory Appl. Categ., 8 (2001) 324–376.

[71] The GAP Group, 2002, *Groups, Algorithms, and Programming, version 4.3*, Technical report, http://www.gap-system.org.

[72] M. Grandis and L. Mauri, *Cubical sets and their site*, Theory and Apl. Categories, (to appear).

[73] A. Heyworth, *Applications of rewriting systems and Groebner bases to computing Kan extensions and identities among relations*, U.W. Bangor Ph.D. thesis (1998) (Preprint 98.23).

[74] A. Heyworth and C.D. Wensley, *Logged rewriting and identities among relators*, Groups St Andrews 2001 in Oxford Vol I, eds. C. Campbell, E. Robertson, G. S. Smith, C.U.P., (2003) 256–276.

[75] P.J. Higgins, *Presentations of groupoids, with applications to aroups*, Proc. Camb. Phil. Soc., 60 (1964) 7–20.

[76] P.J. Higgins, *Categories and Groupoids*, van Nostrand, New York (1971).

[77] P.J. Higgins and J. Taylor, *The fundamental groupoid and homotopy crossed complex of an orbit space*, in K. H. Kamps et al., ed., *Category Theory: Proceedings Gummersbach 1981*, SLNM 962 (1982) 115–122.

[78] C. Hog-Angeloni and W. Metzler, eds., *Two-dimensional homotopy and combinatorial group theory*, volume 197 of *London Mathematical Society Lecture Note Series*, Cambridge University Press, Cambridge (1993).

[79] J. Huebschmann, *Crossed n-fold extensions of groups and cohomology*, Comment. Math. Helv., 55 (1980) 302–313.

[80] J. Huebschmann, *Automorphisms of group extensions and differentials in the Lyndon-Hochschild-Serre spectral sequence*, J. Algebra 72 (1981) 296–334.

[81] J. Huebschmann, *Group extensions, crossed pairs and an eight term exact sequence*, Jour. fur. reine. u. ang. Math., 321 (1981) 150–172.

[82] D.W. Jones, *Polyhedral T-complexes*, Ph.D. thesis, University of Wales, Bangor, (1984). Published as *A general theory of polyhedral sets and their corresponding T-complexes*, Diss. Math., 266 (1988).

[83] E.H.v. Kampen, *On the connection between the fundamental groups of some related spaces*, Amer. J. Math., 55 (1933) 261–267.

[84] K.H. Kamps and T. Porter, *Abstract homotopy and simple homotopy theory*, World Scientific Publishing Co. Inc., River Edge, NJ, (1997).

[85] J.-P. Labesse, *Cohomologie, stabilisation et changement de base*, Astérisque, appendix A by Laurent Clozel and Labesse, and Appendix B by Lawrence Breen, 257 (1999).

[86] J.-L. Loday, *Spaces with finitely many homotopy groups*, J.Pure Appl. Algebra , 24 (1982) 179–202.

[87] A.S.T. Lue, *Cohomology of groups relative to a variety*, J. Algebra , 69 (1981) 155–174.

[88] M. Mackaay and R. Picken, *Holonomy and parallel transport for abelian gerbes*, Advances in Math., 170 (2002) 287–339.

[89] S. MacLane, *Homology*, number 114 in Grundlehren, Springer (1967).

[90] S. MacLane and J.H.C. Whitehead, *On the 3-type of a complex*, Proc. Nat. Acad. Sci. U.S.A., 36 (1950) 41–48.

[91] E.J. Moore, , *Graphs of Groups: Word Computations and Free Crossed Resolutions*, Ph.D. thesis, University of Wales, Bangor (2001) UWB Preprint 01.02.

[92] G. Nan Tie, *A Dold-Kan theorem for crossed complexes*, J. Pure Appl. Algebra, 56 (1989) 177–194.

[93] P. Olum, *Mappings of manifolds and the notion of degree*, Ann. of Math., 58 (1953) 458–480.

[94] R. Peiffer, *Über Identitäten zwischen Relationen*, Math. Ann., 121 (1949) 67 – 99.

[95] T. Porter, *n-types of simplicial groups and crossed n-cubes*, Topology, 32 (1993) 5–24.

[96] S. Pride, 1991, *Identities among relations*, in A. V. E.Ghys, A.Haefliger, ed., *Proc. Workshop on Group Theory from a Geometrical Viewpoint*, 687–716, International Centre of Theoretical Physics, Trieste, 1990, World Scientific.

[97] K. Reidemeister, *Über Identitäten von Relationen*, Abh. Math. Sem. Hamburg, 16 (1949) 114–118.

[98] J. Taylor, *Quotients of groupoids by the action of a group*, Math. Proc. Camb. Phil. Soc., 103 (1988) 239–249.

[99] A. P. Tonks, 1993, *Theory and applications of crossed complexes*, Ph.D. thesis, University of Wales, Bangor (UWB Math Preprint 93.17).

[100] A.P. Tonks, *On the Eilenberg-Zilber theorem for crossed complexes*, J. Pure Appl. Algebra, 179 (2003) 199–220.

[101] C.T.C. Wall, *Finiteness conditions for CW-complexes II*, Proc. Roy. Soc. Ser. A, 295 (1966) 149–166.

[102] A. Weinstein, *Groupoids: unifying internal and external symmetry*, Notices. Amer. Math. Soc., 43 (1996) 744–752. math.RT/9602220.

[103] J.H.C. Whitehead, *On adding relations to homotopy groups*, Ann. of Math., 42 (1941) 409–428.
[104] J.H.C. Whitehead, *Combinatorial Homotopy I*, Bull. Amer. Math. Soc., 55 (1949) 213–245.
[105] J.H.C. Whitehead, *Combinatorial Homotopy II*, Bull. Amer. Math. Soc., 55 (1949) 453–496.
[106] J.H.C. Whitehead, *Simple homotopy types*, Amer. J. Math., 72 (1950) 1–57.

Received December 16, 2002; in revised form May 10, 2003

Fields Institute Communications
Volume **43**, 2004

Galois Groupoids and Covering Morphisms in Topos Theory

Marta Bunge
Department of Mathematics and Statistics
McGill University,
805 Sherbrooke St West
Montréal QC, Canada H3A 2K6
bunge@math.mcgill.ca

Abstract. The goals of this paper are (1) to compare the Galois groupoid that appears naturally in the construction of the coverings fundamental groupoid topos given by Bunge (1992) with the formal Galois groupoid defined by Janelidze (1990) in a very general setting given by a pair of adjoint functors, and (2) to discuss a good notion of covering morphism of a topos that is general enough to include, in addition to the covering projections determined by the locally constant objects, also the unramified morphisms of topos theory given by those local homeomorphisms which are at the same time complete spreads in the sense of Bunge-Funk (1996, 1998). We also (3) introduce and study a notion of a Galois topos that generalizes in different ways those of Grothendieck (1971) and Moerdijk (1989), (4) explain the role of stack completions in distinguishing Galois groupoids from fundamental groupoids when the base topos is not Sets but arbitrary, (5) extend to the case of an arbitrary base topos results of Bunge-Moerdijk (1997) concerning the comparison between the coverings and the paths fundamental groupoid toposes, and (6) discuss pseudofunctorialy of the fundamental groupoid constructions and apply it to give a simple version of the van Kampen theorem for toposes of Bunge-Lack (2003).

1 Introduction

One of the purposes of this paper is to compare, given a locally connected and locally simply connected topos $\mathscr{E}$ over a base topos $\mathscr{S}$, the Galois groupoid of automorphisms of the canonical "point" (in effect, a "bag of points") of the

2000 *Mathematics Subject Classification.* Primary 18B, 18D, 18F; Secondary 57M.
Key words and phrases. topos, covering, locally constant object, Galois groupoid, fundamental groupoid, stack completion, Galois topos, complete spread, van Kampen theorems.
Invited lecture for the Workshop on Categorical Structures for Descent and Galois Theory, Hopf Algebras and Semiabelian Categories, The Fields Institute for Research in the Mathematical Sciences, University of Toronto, September 23-28, 2002. Research partially supported by the Natural Sciences and Engineering Research Council of Canada.

coverings fundamental groupoid topos of $\mathscr{E}$ as defined in [9], with the formal Galois groupoid of Janelidze [23, 24] in this setting.

The notion of a locally constant object is central to these considerations; we begin then by investigating in section 2 (Locally constant objects in toposes) the connection between (rather, identification of) the notion of a *locally constant object* given in [14] (in turn inspired by [9]), with that of Janelidze [24, 25] and, in passing, also with that of Barr and Diaconescu [3], with which both agree if the topos $\mathscr{E}$ is connected and the base topos $\mathscr{S}$ is Set.

In section 3 (Stack completions and the fundamental groupoid of a topos) we revisit the construction of the coverings fundamental groupoid of a locally connected topos $\mathscr{E}$ over a base topos $\mathscr{S}$ given in [9], an account of which is briefly given in [15] under the implicit assumption that the base topos $\mathscr{S}$ is Set. A new ingredient here is the observation, not previously made explicit in either [9] or [15], that there are *two* groupoids involved, to wit, for each cover U in $\mathscr{E}$, there is on the one hand the *internal* Galois groupoid G_U of automorphisms of the universal cover of the topos of U-split objects in $\mathscr{E}$, and on the other hand its stack completion π_U which classifies U-split torsors and which in principle need not be internal to $\mathscr{S}$ but only indexed (fibered) over $\mathscr{S}$. When the base topos is Set, this distinction dissapears and the two groupoids are usually identified in practice [21, 15]. This brings into consideration the possible advantages of assuming that the base (elementary) topos $\mathscr{S}$ satisfies an *axiom of stack completions* (ASC), which was suggested by Lawvere in 1974 and which is known to hold at least of all Grothendieck toposes $\mathscr{S}$, as shown in [18, 27]. This axiom guarantees that the (coverings) fundamental groupoid of a locally connected and locally simply connected topos $\mathscr{E}$ over $\mathscr{S}$, which is the stack completion of the internal Galois groupoid, is represented, as an $\mathscr{S}$-indexed category, by an internal groupoid, weakly equivalent to the Galois groupoid.

We prove in section 4 (Galois groupoids and Galois toposes) that there is an equivalence (and not just a Morita equivalence) between the Galois groupoid of automorphisms of a universal cover of a locally connected and locally simply connected topos $e : \mathscr{E} \to \mathscr{S}$, and the Galois groupoid of the pure theory associated with the first pair of adjoints in the 3-tuple $e_! \dashv e^* \dashv e_*$ given by the (locally connected) geometric morphism e. We also show that in topos theory, and just as in the pure Galois theory of [24, 5], the Galois theory is implicit in the construction of the Galois groupoid; in the case of toposes, we obtain it from the presence of the third adjoint in the sequel of three determined by e, but a different argument that employs only the cartesian closed structure is available, as shown by Janelidze.

Also in section 4 we introduce and study the notion of a *Galois topos* over an arbitrary base topos $\mathscr{S}$. Although these relative Galois toposes are not assumed to be either connected or pointed, they come naturally equipped with a bag of points indexed by the connected components of a (non-connected) universal cover; this is in line with the view advocated by Grothendieck [22] and Brown [6] that, rather than fixing a single base point, one ought work with a suitable "paquet des points", for instance one that is invariant under the symmetries in the given situation. This idea was naturally and independently incorporated in topos theory both by Kennison [29] and myself [9], by discussing the fundamental groupoid of an *unpointed* (and possibly pointless) locally connected topos.

Pointed connected Galois toposes over $\mathscr{S}$ have been investigated by Moerdijk [31] following Grothendieck [20] (see also [27]). We obtain here characterization theorems in a manner parallel to [31]; in particular we show that our Galois toposes over a base topos $\mathscr{S}$ satisfying (ASC) (and which correspond, modulo the intervention of locales in the subject, to the multi-Galois toposes of Grothendieck [21]) are precisely the classifying toposes of prodiscrete (localic) groupoids in $\mathscr{S}$.

In section 5 (Locally paths simply connected toposes over an arbitrary base) we recall the paths version of the fundamental group topos [33, 34], give a constructive version of the existence of a comparison map from the paths to the coverings fundamental groupoid toposes, and then prove the equivalence of the comparison map under an assumption of the locally paths simply connected type. We also investigate pseudonaturality of the fundamental groupoid topos constructions and their universal properties, and apply this to give a simplified version of the van Kampen theorem for locally paths simply connected toposes, in connection with [14].

In section 6 (Generalized covering morphisms and a van Kampen theorem) we turn to the topic of covering morphisms, also in connection with the Galois and fundamental groupoids. The notion we give here has good enough properties so that (a pushout version of) the van Kampen theorem from [14] holds for these coverings and is also general enough to include, not only the familiar covering projections determined by the locally constant objects, but also the unramified coverings introduced and studied in [11, 12, 19], namely those local homeomorphisms which are also complete spreads.

2 Locally constant objects in toposes

Let $\mathscr{E}$ be an elementary topos bounded over an elementary topos $\mathscr{S}$ by means of a geometric morphism $e : \mathscr{E} \to \mathscr{S}$.

Before proceeding, we explain this assumption roughly in simple terms. The notion of an *elementary topos* is at the same time a generalization of the notion of a topological space via sheaves, and a universe for set theory that is not necessarily a classical one. The base topos $\mathscr{S}$ above is interpreted as a choice of set theory, whereas the topos $\mathscr{E}$ is thought of as sheaves on a site in $\mathscr{S}$, that is, on a category in $\mathscr{S}$ equipped with a Lawvere-Tierney topology, explicitated by the *bounded* nature of the geometric morphism $e : \mathscr{E} \to \mathscr{S}$. If $\mathscr{S}$ is the category of sets, what this amounts to is simply that $\mathscr{E}$ is a *Grothendieck topos*, with $e : \mathscr{E} \to \mathscr{S}$ the unique geometric morphism whose direct image part $e_* : \mathscr{E} \to \mathscr{S}$ is the global sections functor, and whose inverse image part $e^* : \mathscr{S} \to \mathscr{E}$ assigns to a set S the associated sheaf the presheaf whose value is constantly S. In particular, we may think of $\mathscr{E}$ as a generalized space defined within the set theory $\mathscr{S}$. It is of course of interest, in the setting of generalized spaces within $\mathscr{S}$, to consider geometric morphisms $\mathscr{F} \to \mathscr{E}$ over $\mathscr{S}$, for instance covering morphisms.

One can also express, in the setting of elementary toposes and geometric morphisms, generalizations of familiar topological notions, such as that of a locally connected space, by making special assumptions about the given (in general not unique) "structure morphism" $e : \mathscr{E} \to \mathscr{S}$. For instance, one says that $\mathscr{E}$ is *locally connected* over $\mathscr{S}$ by means of e if there is given an $\mathscr{S}$-indexed left adjoint $e_! \dashv e^*$, where e is the geometric morphism given by the (necessarily $\mathscr{S}$-indexed) adjoint pair $e^* \dashv e_*$. We think of $e_! : \mathscr{E} \to \mathscr{S}$ as the (set of) connected components functor.

We now recall a *fundamental pushout* construction in the definition of the fundamental groupoid of $\mathscr{E}$ as given in [9]. We call it "fundamental" both on account of its role in the fundamental groupoid, and since this construction contains basically all of the information we use to develop Galois theory in the context of toposes.

For an epimorphism $U \to 1$ in $\mathscr{E}$ (a *cover* in $\mathscr{E}$) denote by $\mathscr{G}_U$ the topos defined by the following pushout in $\mathbf{Top}_{\mathscr{S}}$

$$\begin{array}{ccc} \mathscr{E}/U & \xrightarrow{\varphi_U} & \mathscr{E} \\ \downarrow{\scriptstyle \rho_U} & & \downarrow{\scriptstyle \sigma_U} \\ \mathscr{S}/e_!U & \xrightarrow[p_U]{} & \mathscr{G}_U \end{array} \qquad (*)$$

where $\varphi_U : \mathscr{E}/U \to \mathscr{E}$ is the canonical local homeomorphism and where ρ_U is the connected locally connected part in the (unique) factorization of the composite $e\ \varphi_U : \mathscr{E}/U \to \mathscr{S}$, which is the composite of two locally connected hence locally connected, into a connected locally connected morphism followed by a (surjective) local homeomorphism.

We point out that in the 2-category $\mathbf{Top}_{\mathscr{S}}$ of toposes bounded over $\mathscr{S}$, geometric morphisms over $\mathscr{S}$, and iso 2-cells, pushouts exist and are calculated simply as pullbacks of the inverse image parts in **CAT**, so that anything that depends solely on pushouts has, in principle, "nothing to do with toposes". However, the fact that, by performing a pullback in **CAT** of this kind to a diagram of toposes and (inverse images of) geometric morphisms, one obtains again toposes and (inverse images of) geometric morphisms is what makes this construction so powerful in topos theory. Without this general fact, one may need to show that a certain category (respectively, function) is a topos (respectively, geometric morphism) "by hand" in each such situation, which may be rather involved.

The fundamental pushout construction $(*)$ was what motivated the "family" definition of a locally constant object formally introduced in [14] and given below in the special context of a locally connected topos $\mathscr{E}$ over $\mathscr{S}$. It is, in fact, the natural notion to consider in topology in the case of a non-connected cover; in the context of Grothendieck toposes it also appears in [20]. This was seen independently also by Janelidze [23, 24]. There is, however, a non constructively given equivalence with a "single object" description when the topos $\mathscr{E}$ is connected and defined over *Set*, as shown by Barr and Diaconescu [3], but it is not clear to us at present whether this equivalence (which is of course reasonable to expect) can be made constructive.

Definition 2.1 An object A of (a locally connected topos) $\mathscr{E}$ (with structure map $e : \mathscr{E} \to \mathscr{S}$) is said to be *locally constant* if there exists a cover U in $\mathscr{E}$ which splits A in the sense that there is a morphism $\alpha : S \to e_!U$ in $\mathscr{S}$ and an isomorphism $\theta : e^*S \times_{e^*e_!U} U \to A \times U$ over U, where U is equipped with the morphism $\eta_U : U \to e^*e_!U$ given by the unit of the adjointness $e_! \dashv e^*$ evaluated at U.

The following result is easily proven and appears in [14].

Proposition 2.2 *If $e : \mathscr{E} \to \mathscr{S}$ is locally connected, A an object and U a cover in $\mathscr{E}$, the following are equivalent:*

1. *A is locally constant and split by U in the sense of Definition 2.1.*
2. *A is U-locally trivial with respect to the adjoint pair $e_! \dashv e^*$ in the sense of [24], that is, there is a morphism $\alpha : S \to e_!U$ and a morphism $\zeta : A \to$*

$U \to e^*S$ such that the square

$$\begin{array}{ccc} A \times U & \xrightarrow{\pi_2} & U \\ \downarrow{\scriptstyle \zeta} & & \downarrow{\scriptstyle \eta} \\ e^*S & \xrightarrow[e^*\alpha]{} & e^*e_!U \end{array}$$

is a pullback.

3. *There is a morphism $\alpha : J \to I$ in $\mathscr{S}$, a morphism $\eta : U \to e^*I$ in $\mathscr{E}$, and an isomorphism $\theta : e^*J \times_{e^*I} U \to A \times U$ over U.*

Remark 2.3 The equivalence between (1) and (2) in Proposition 2.2 shows that an object A of a locally connected topos $\mathscr{E}$ is locally constant (relative to its structure map $e : \mathscr{E} \to \mathscr{S}$) in the sense of Definition 2.1 if and only if the unique morphism $A \to 1$ is locally trivial with respect to the adjoint pair $e_! \dashv e^*$ in the sense of Janelidze [24]. The equivalence between (1) and (3) in Proposition 2.2 strips the notion to its barest terms and in particular has the desirable consequence of its *stability* under pullback along arbitrary geometric morphisms. Note that (3) is meaningful in an arbitrary topos $\mathscr{E}$ bounded over $\mathscr{S}$ and that it agrees with (1) (and so also with (2)) if $\mathscr{E}$ is locally connected over $\mathscr{S}$. In [14] we gave (3) as the notion of a locally constant object in a topos $\mathscr{E}$ defined over $\mathscr{S}$.

Corollary 2.4 *Let $\varphi : \mathscr{F} \to \mathscr{E}$ be a geometric morphism in* $\mathbf{Top}_{\mathscr{S}}$. *If A is a locally constant object in $\mathscr{E}$ split by the cover U in $\mathscr{E}$ in the sense of* (2) *in Proposition 2.2, then φ^*A is a locally constant object in $\mathscr{F}$, split by the cover φ^*U in $\mathscr{F}$.*

Proof The conclusion is easily checked. □

Denote by $\mathrm{Spl}(U)$ the full subcategory of $\mathscr{E}$ determined by the U-split objects of $\mathscr{E}$ in the sense of Definition 2.1.

Lemma 2.5 [9] *Let*

$$\begin{array}{ccc} \mathscr{C} & \xrightarrow{q} & \mathscr{B} \\ \downarrow{\scriptstyle p} & & \downarrow{\scriptstyle h} \\ \mathscr{A} & \xrightarrow[k]{} & \mathscr{D} \end{array}$$

be a pushout in $\mathbf{Top}_{\mathscr{S}}$ *where p and q are both locally connected. Then, each of h and k is locally connected. Furthermore, if p (respectively, q) is connected, then so is h (respectiveky k).*

Proposition 2.6 *Let $\mathscr{G}_U$ be the topos in the fundamental pushout* $(*)$ *applied to $\mathscr{E}$ and U. Then there exists an equivalence functor $\Phi : \mathscr{G}_U \to \mathrm{Spl}(U)$. Under this equivalence, the fully faithful functor ${\sigma_U}^* : \mathscr{G}_U \to \mathscr{E}$ corresponds to the inclusion of* $\mathrm{Spl}(U)$ *into $\mathscr{E}$.*

Proof An object of $\mathscr{G}_U$ in the fundamental pushout $(*)$ of toposes is precisely a locally constant object split by U in the sense of Definition 2.1. This is a consequence of the well-known fact that pushouts in $\mathbf{Top}_{\mathscr{S}}$ are calculated as pullbacks in **CAT** of the inverse image parts. A morphism from $\langle X\ , S \to e_!U\ , \theta\rangle$ to $\langle X'\ , S' \to e_!U\ , \theta'\rangle$, both objects in $\mathscr{G}_U$, is a pair of morphisms $f : X \to X'$ and $\alpha : S \to S'$, the latter over $e_!U$, compatible with the isomorphisms θ and θ'.

It follows from Lemma 2.5 that σ_U is connected and locally connected. Since the forgetful functor ${\sigma_U}^* : \mathscr{G}_U \to \mathscr{E}$ is then fully faithful and factors through the inclusion $\mathrm{Spl}(U) \hookrightarrow \mathscr{E}$, it follows from the equivalence of (1) and (3) in Proposition 2.2 that there is an equivalence $\Phi : \mathscr{G}_U \to \mathrm{Spl}(U)$. □

Remark 2.7 The equivalence functor $\Phi : \mathscr{G}_U \to \mathrm{Spl}(U)$ forgets the splitting of the U-split objects in the sense of Definition 2.1, as well as the compatibility with the splittings of any morphism between objects in $\mathscr{G}_U$. In the absence of the axiom of choice for $\mathscr{S}$, there is no inverse to Φ. The importance of this remark will emerge from the developments in the next section.

We give now a brief review of the pure Galois theory of Janelidze [23, 24, 5]. There is given a pair of adjoint functors $I \dashv H$ with $I : \mathscr{E} \to \mathscr{S}$ and $H : \mathscr{S} \to \mathscr{E}$ (here $\mathscr{E}$ and $\mathscr{S}$ are *categories*) such that for every object $E \in \mathscr{E}$ the counit $I^E H^E \to \mathrm{Id}_{\mathscr{S}/IE}$ of the induced adjoint pair $I^E \dashv H^E$ in the diagram

$$\mathscr{E}/E \underset{I^E}{\overset{H^E}{\rightleftarrows}} \mathscr{S}/IE$$

is an iso. In this context, a morphism $\alpha : A \to B$ in $\mathscr{E}$ is said to be *trivial* if the diagram

$$\begin{array}{ccc} A & \xrightarrow{\alpha} & B \\ \downarrow{\scriptstyle \eta_A} & & \downarrow{\scriptstyle \eta_B} \\ HIA & \xrightarrow[HI\alpha]{} & HIB \end{array}$$

is a pullback; $\alpha : A \to B$ is a *locally trivial* morphism relative to the adjoint pair $I \dashv H$ if there exists $p : E \to B$ of effective descent such that $p^*(A, \alpha)$ is trivial in $\mathscr{E}/E$.

The category $\mathrm{Spl}(p)$ of locally trivial morphisms split by p is easily seen to be given as the pullback

$$\begin{array}{ccc} \mathscr{E}/E & \xleftarrow{p^*} & \mathscr{E}/B \\ {\scriptstyle H^E}\uparrow & & \uparrow \\ \mathscr{S}/IE & \longleftarrow & \mathrm{Spl}(p) \end{array}$$

.

It is shown in [23] that if in the pure setting one lets $\mathrm{Gal}(p)$ be given by the canonical diagram

$$I((E \times_B E) \times_E (E \times_B E)) \;\substack{\longrightarrow \\ \longrightarrow \\ \longrightarrow}\; I(E \times_B E) \;\substack{\longrightarrow \\ \longleftarrow \\ \longrightarrow}\; I(E)$$

then, although $\mathrm{Gal}(p)$ is not in general an internal category in $\mathscr{S}$, it makes sense to consider its actions and to form a category $\mathscr{S}^{\mathrm{Gal}(p)}$. Furthermore, there is an equivalence

$$\mathrm{Spl}(p) \simeq \mathscr{S}^{\mathrm{Gal}(p)}$$

to be interpreted as the *fundamental theorem of Galois theory.*

In order to get an actual groupoid, it is assumed in [24]of the cover $p : E \to B$ considered in [23] that it is a normal (or regular) cover. As we show in the next section, the existence of a normal cover (actually of the *universal cover*) in the topos setting can be obtained using general properties of the fundamental pushout construction.

If $p : E \to B$ is an I-*normal* object of $\mathscr{E}$ in the sense that $p \in \mathrm{Spl}(p)$, then as shown in [24], the canonical morphism

$$I((E \times_B E) \times_E (E \times_B E)) \to I(E \times_B E) \times_{I(E)} I(E \times_B E)$$

is an isomorphism, from which the desired groupoid structure can be extracted. The reader ought to consult [23, 24, 5] for details.

We prove next that there is also a Galois groupoid in the topos context, associated with a cover $p : U \to 1$, which we obtain from the pushout topos $\mathscr{G}_U$ with the aid of the Joyal-Tierney theorem [28] and additional topos-theoretic considerations [9].

Proposition 2.8 *Let $\mathscr{E}$ be a locally connected topos over $\mathscr{S}$, and let U be a cover in $\mathscr{E}$. Then, the topos $\mathscr{G}_U$ in the fundamental pushout $(*)$ is the classifying topos $\mathscr{B}G_U$ of an etale complete discrete localic groupoid G_U in $\mathscr{S}$, by an equivalence*

$$\mathscr{G}_U \simeq \mathscr{B}G_U$$

which identifies p_U with the canonical bag of points of $\mathscr{B}G_U$.

Proof It follows from properties of the fundamental pushout $(*)$ that the geometric morphism $p_U : \mathscr{S}/e_!U \to \mathscr{G}_U$ is a local homeomorphism. Indeed, since both φ_U and ρ_U are locally connected, so is p_U by Lemma 2.5, and since p_U is also totally disconnected in the sense of [9], it must be a local homeomorphism; it is surjective since $U \to 1$ is an epi, hence φ_U is surjective. From general facts about descent toposes [28, 33] it is derived that there is determined an (etale complete) discrete localic groupoid G_U whose discrete locale of objects is $e_!U$ and is such that $\mathscr{G}_U$ is equivalent to the classifying topos $\mathscr{B}G_U$. □

Recall that one of our goals is to compare, in the case of a *locally simply connected* topos $\mathscr{E}$ over $\mathscr{S}$, the Galois groupoid of Janelidze [24] associated with the universal cover, with the Galois groupoid G_U that arises as in Proposition 2.8 above. In section 4 we shall prove their equivalence as groupoids, but prior to that we will, in section 3, make explicit its connection with the fundamental groupoid $\pi_1^c(\mathscr{E})$ of $\mathscr{E}$ as constructed in [9].

3 Stack completions and the fundamental groupoid of a topos

In this section we update the definition of the coverings fundamental groupoid $\pi_1^c(\mathscr{E})$ given in [9] (see also [15]), where $\mathscr{E}$ is a locally connected topos over $\mathscr{S}$ by means of a geometric morphism $e : \mathscr{E} \to \mathscr{S}$. A new ingredient here is making explicit what was only implicit in those sources, namely, the role of stacks (for the topology of regular epimorphisms) of groupoids in $\mathscr{S}$ in the definition of the fundamental groupoid of $\mathscr{E}$. To this end, we require that the (elementary) topos $\mathscr{S}$ satisfy an additional assumption, to wit, that the stack completions of groupoids in $\mathscr{S}$ be (the externalizations of) groupoids in $\mathscr{S}$. Without it, the main result about the fundamental groupoid is stated in the form "is weakly equivalent to a prodiscrete groupoid" (as in [9]) rather than more simply in the form "is a prodiscrete groupoid".

The notion of a stack ("champ" in the work of Grothendieck and Giraud) can be made relative to the regular epimorphism topology of a given topos $\mathscr{S}$. Stacks of category objects (for the regular epimorpisms topology of $\mathscr{S}$) are discussed in [16], and stack completions shown to exist (and described) in [17]. Roughly speaking, stacks are *good* ($\mathscr{S}$-indexed)categories in the sense that "locally in it" implies "in it".

We are here interested only in groupoids. We begin by making some remarks about stack completions in the case of groupoids in $\mathscr{S}$. For G a groupoid in $\mathscr{S}$ (regarded as a category object), its stack completion is given (up to equivalence) by any pair $< \mathscr{A}, F >$, with $\mathscr{A}$ an $\mathscr{S}$-indexed category, $F : [G] \to \mathscr{A}$ an $\mathscr{S}$-indexed functor, and where $\mathscr{A}$ is a stack (of groupoids) and $F : [G] \to \mathscr{A}$ a weak equivalence. The stack completion of a groupoid G in $\mathscr{S}$, as a category object in which idempotents split, is identified in [17] with the $\mathscr{S}$-indexed category $\mathrm{Pointess}_{\mathscr{S}}(\mathscr{B}G)$ of $\mathscr{S}$-essential points of $\mathscr{B}G$. Also the $\mathscr{S}$-indexed category $\mathrm{Tors}(G)$, given by assigning to an object I of $\mathscr{S}$ the category $\mathrm{Tors}(G)^I$ of G-torsors in $\mathscr{S}/I$, and letting transition morphisms given by pullback, is a stack completion of G. Diaconescu's theorem [27], stating that for a groupoid G the topos $\mathscr{B}G$ classifies G-torsors, suggests yet another identification of the stack completion of $[G]$, as in the following theorem.

Theorem 3.1 *Let G be a groupoid in $\mathscr{S}$ given by a diagram*

$$G_1 \times_{G_0} G_1 \;\overset{\longrightarrow}{\underset{\longrightarrow}{\longrightarrow}}\; G_1 \;\overset{\longrightarrow}{\underset{\longrightarrow}{\longleftarrow}}\; G_0 \,.$$

Then, the stack completion of G is given by the $\mathscr{S}$-indexed category of points of $\mathscr{B}G$.

Proof There is a canonical bag of points $p_G : \mathscr{S}/G_0 \to \mathscr{B}G$ such that composing with it defines an $\mathscr{S}$-indexed functor

$$F : [G] \to \mathrm{Points}_{\mathscr{S}}(\mathscr{B}G).$$

Explicitly, given an object I of $\mathscr{S}$, $F^I : [I, G] \to \mathbf{Top}_{\mathscr{S}}(\mathscr{S}/I, \mathscr{B}G)$ assigns, to an object $x : I \to G_0$ of the category $[I, G]$, the composite $\mathscr{S}/I \to \mathscr{S}/G_0 \to \mathscr{B}G$ of the geometric morphism induced by x with p_G, and to a morphism $f : x \to y$ in $[I, G]$, given by a morphism $f : I \to G_1$ in $\mathscr{S}$ such that $d_0 f = x$ and $d_1 f = y$, the natural transformation $f : p_G x \to p_G y$. This assignment is easily checked to be $\mathscr{S}$-indexed.

It follows just as in [8] (Proposition 3.1) that $\mathrm{Points}_{\mathscr{S}}(\mathscr{B}G)$ is a stack, using for this now that the $\mathscr{S}$-essential surjections (a particular case of open surjections) are of effective descent in $\mathbf{Top}_{\mathscr{S}}$ [28].

It remains to verify that F is a weak equivalence. That for each I the functor F^I is fully faithful is a direct consequence of the etale completeness of G, which means that the square in $\mathbf{Top}_{\mathscr{S}}$ given below

$$\begin{array}{ccc} \mathscr{S}/G_1 & \xrightarrow{d_1} & \mathscr{S}/G_0 \\ \downarrow{\scriptstyle d_0} & & \downarrow{\scriptstyle p_G} \\ \mathscr{S}/G_0 & \xrightarrow[p_G]{} & \mathscr{B}G \end{array} \qquad (1)$$

is a pullback, and which holds for any discrete (localic) groupoid G. We show next that F is essentially surjective. Let $q : \mathscr{S}/I \to \mathscr{B}G$ be a geometric morphism, that is, an object of $\mathrm{Points}_{\mathscr{S}}(\mathscr{B}G)^I$. Consider the pullback of q along p_G, which we claim looks like this:

$$\begin{array}{ccc} \mathscr{S}/J & \xrightarrow{r} & \mathscr{S}/G_0 \\ \Big\downarrow{\scriptstyle \alpha} & & \Big\downarrow{\scriptstyle p_G} \\ \mathscr{S}/I & \xrightarrow[q]{} & \mathscr{B}G \end{array} \tag{2}$$

where $\alpha : J \to I$ is an epimorphism in $\mathscr{S}$ and where $r : \mathscr{S}/J \to \mathscr{S}/G_0$ is induced by some $y : J \to G_0$, so that $F^J(y) \cong \alpha^*(q)$, or F is indeed essentially surjective. The claim is true because the canonical p_G is a surjective local homeomorphism, a fact already noticed in [9] and used in the proof of Proposition 2.8 in the previous section, and that therefore so is the geometric morphism opposite it in the pullback; it follows that the latter is induced by an epimorphism $\alpha : J \to I$ in $\mathscr{S}$, as claimed. Furthermore, the geometric morphism $r : \mathscr{S}/J \to \mathscr{S}/G_0$ opposite q in the pullback is necessarily induced by a morphism $y : J \to G_0$ in $\mathscr{S}$, hence the result. □

Corollary 3.2 *Let G be a groupoid in $\mathscr{S}$ and let $\mathscr{G} = \mathscr{B}G$ be its classifying topos. Then, the following hold:*

1. *Any geometric morphism $p : \mathscr{S}/I \to \mathscr{G}$ of $\mathscr{G}$ is $\mathscr{S}$-essential.*
2. *With p as above, the inverse image part $p^* : \mathscr{G} \to \mathscr{S}/I$ is naturally equivalent to a functor*

$$\mathrm{Hom}_{\mathscr{G}/g^*(I)}(A \to g^*(I), (-) \times g^*(I)) : \mathscr{G} \to \mathscr{S}/I$$

where $A \to g^*(I)$ has the structure of a G-torsor in $\mathscr{G}/g^*(I)$.

Proof 1. Since the $\mathscr{S}$-indexed category $\mathrm{Points}_{\mathscr{S}}(\mathscr{B}G)$ and its sub-indexed category $\mathrm{Pointess}_{\mathscr{S}}(\mathscr{B}G)$ are both stack completions of (the externalization $[G]$ of) G, they must be equivalent.

2. The $\mathscr{S}$-indexed category $\mathrm{Tors}(G)$ is also a stack completion of $[G]$ and a full subcategory of $\mathscr{B}G = \mathscr{S}^{G^{op}}$, hence equivalent to $\mathrm{Pointess}_{\mathscr{S}}(\mathscr{B}G)$ by an equivalence which gives the desired representability of inverse image parts of $\mathscr{S}$-essential (bags of) points by torsors.

□

Definition 3.3 A topos $\mathscr{S}$ is said to satisfy the *(ASC)* (or the axiom of stack completions) if for any groupoid G in $\mathscr{S}$ there exists (i) a groupoid $\tilde{G}$ in $\mathscr{S}$ whose externalization $[\tilde{G}]$ is a stack, and (ii) a weak equivalence internal functor (or homomorphism of groupoids) $G \to \tilde{G}$. (It is known that any Grothendieck topos $\mathscr{S}$ satisfies this axiom on account of the existence of a small generating class [18, 27].)

The *Morita equivalence theorem* from [17] takes on the following form in the case of groupoids.

Theorem 3.4 *Let $\mathscr{S}$ be a topos which satisfies the axiom of stack completions in the sense of Definition 3.3. Let G and H be groupoids in $\mathscr{S}$. Then there exists an equivalence of categories*

$$\mathrm{Hom}_{\mathscr{S}}(\tilde{G}, \tilde{H}) \simeq \mathbf{Top}_{\mathscr{S}}(\mathscr{B}G, \mathscr{B}H)$$

where $\tilde{G}$ and $\tilde{H}$ are the stack completions of G and H.

We now turn to *topos cohomology* of a locally connected topos $\mathscr{E}$ defined over a base topos $\mathscr{S}$, of which we shall furthermore assume satisfies (ASC) in the sense of Definition 3.3. For each discrete group K, the cohomology of $\mathscr{E}$ with coefficients in

K is denoted by $H^1(\mathscr{E};K)$ and defined as the object in $\mathscr{S}$ of connected components of the category $\mathrm{Tors}(\mathscr{E};K)$, that is, as the "set" of isomorphism classes of K-torsors in $\mathscr{E}$.

Consider now the equivalence

$$H^1(\mathscr{E};K) \simeq \Pi_0(\mathrm{colim}\{\mathrm{Tors}(\mathscr{E};K)^U\})$$

where the indexing is taken over a (cofinal set in a) generating poset of covers U in $\mathscr{E}$, and where $\mathrm{Tors}(\mathscr{E};K)^U$ is the groupoid of U-split K-torsors in $\mathscr{E}$. It is now *a priori* clear that the question of representing the functor $H^1(\mathscr{E};-)$ by some suitable (prodiscrete localic) groupoid $\pi_1^c(\mathscr{E})$ reduces to that of obtaining discrete groupoids π_U representing the functors $\mathrm{Tors}(\mathscr{E};-)^U : \mathrm{Gps}(\mathscr{S}) \to \mathrm{Gpds}(\mathscr{S})$, since then the desired groupoid $\pi_1^c(\mathscr{E})$ would be definable as a limit of a filtered system of the π_U.

The central idea in definining the fundamental groupoid of $\mathscr{E}$ is contained in the pushouts $(*)$ of the previous section and in the Galois groupoids G_U associated with them as in Proposition 2.8. The G_U are etale complete groupoids (hence Galois groupoids), and the toposes $\mathscr{B}(G_U)$ their classifying toposes.

Let $U \leq V$, where U and V are covers in $\mathscr{E}$, and let $\alpha : U \to V$ be a given morphism in $\mathscr{E}$ witnessing the relation. Then there is an induced geometric morphism $\varphi_\alpha : \mathscr{G}_U \to \mathscr{G}_V$ over $\mathscr{S}$ which commutes with the bags of points p_U and p_V, and which therefore may be interpreted as an object of $\mathbf{Top}_{\mathscr{S}}[\mathscr{B}G_U, \mathscr{B}G_V]_+$, the full subcategory of $\mathbf{Top}_{\mathscr{S}}[\mathscr{B}G_U, \mathscr{B}G_V]$ whose objects are geometric morpisms commuting with the canonical (bags of) points. The square brackets in both cases indicate that the morphisms in those Hom-categories are to be taken to be iso 2-cells in $\mathbf{Top}_{\mathscr{S}}$.

This is clear from the cube below, using the pushout property $(*)$ of the front face:

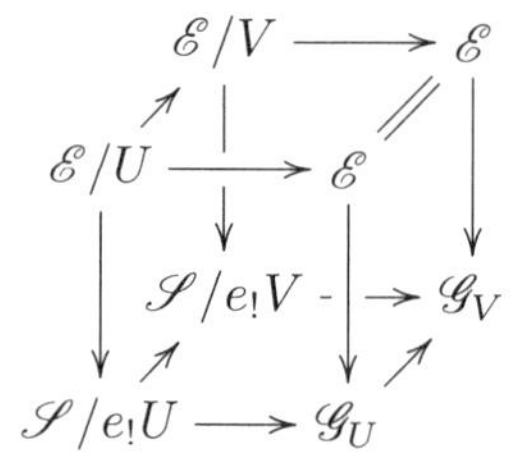

As shown in [8] there is a strong equivalence of categories:

$$\mathrm{Hom}(G_U, G_V) \simeq \mathbf{Top}_{\mathscr{S}}[\mathscr{B}(G_U), \mathscr{B}(G_V)]_+.$$

We derive from it that there are induced groupoid homomorphisms $g_\alpha : G_U \to G_V$, for each α. Furthermore, since the geometric morphisms φ_α are connected (and locally connected), the homomorphisms g_α are *full and essentially surjective* by [32].

From the above diagram follows, by virtue of the connectedness of $\sigma_U : \mathscr{E} \to \mathscr{G}_U$, that the geometric morphisms φ_α depend on α up to unique iso 2-cell. In particular, the groupoid homomorphisms g_α, which are determined by the geometric morphisms φ_α up to iso 2-cells, also depend on the α up to unique iso 2-cells. This and the cofinality of $\mathrm{Cov}(\mathscr{E})$ implies that the 2-system is bifiltered (and biordered) in the sense of the following definition [29].

Definition 3.5 A 2-system $\{G_i\}$ of discrete groupoids, groupoid homomorphisms and 2-cells between them, indexed by a category $\mathscr{C}$, is said to be *bifiltered and biordered* if

1. For any two groupoids G_i , G_j in the system, there is $k \in \mathscr{C}$ and morphisms $\alpha : i \to k$ and $\beta : j \to k$, inducing homomorphisms $g_\alpha : G_i \to G_k$, $g_\beta : G_j \to G_k$.
2. For any other morphism $\alpha' : i \to j$, and corresponding $g'_\alpha : G_i \to G_j$, there exists a unique (iso) 2-cell $\lambda_{i,i'} : g_\alpha \to g'_\alpha$.

We have also that for any group K in $\mathscr{S}$ there is an equivalence

$$\mathbf{Top}_{\mathscr{S}}(\mathscr{B}(G_U), \mathscr{B}(K)) \simeq \mathrm{Tors}(\mathscr{E}; K)^U$$

natural in K. Indeed, by Diaconescu's theorem [27], the topos $\mathscr{B}(K)$ classifies K-torsors in $\mathscr{E}$, and under this correspondence, the U-split K-torsors in $\mathscr{E}$ are in bijection with those geometric morphisms $\varphi : \mathscr{E} \to \mathscr{B}(K)$ whose inverse image part φ^* factors through the inclusion of $\mathrm{Spl}(U)$ in $\mathscr{E}$ and the latter is equivalent to $\mathscr{B}(G_U)$); notice that the canonical $p_G : \mathscr{S}/G_0 \to \mathscr{B}(G_U)$ has no bearing in this equivalence.

It follows that the Galois groupoids G_U only *weakly* represent cohomology. In order to get a strong representation, we must consider their stack completions.

Lemma 3.6 *Let $\mathscr{S}$ be a topos satisfying (ASC) in the sense of Definition 3.3. Let $\mathscr{E}$ be a locally connected topos over $\mathscr{S}$, U a cover in $\mathscr{E}$, and G_U the Galois groupoid obtained in Proposition 2.8. Then, if $F_U : G_U \to \tilde{G_U}$ is the stack completion of G_U, the discrete groupoid $\pi_U = \tilde{G_U}$ represents the functor*

$$\mathrm{Tors}(\mathscr{E}; -)^U : \mathrm{Gps}(\mathscr{S}) \to \mathrm{Gpds}(\mathscr{S}).$$

Further, the weak equivalence $F_U : G_U \to \tilde{G_U}$ induces an equivalence $\tilde{F_U} : \mathscr{B}(G_U) \to \mathscr{B}(\tilde{G_U})$.

Proof It follows from Theorem 3.4 that for the groupoid G_U and any discrete group K, there is a strong equivalence

$$\mathrm{Hom}(\tilde{G_U}, K) \simeq \mathbf{Top}_{\mathscr{S}}[\mathscr{B}(G_U), \mathscr{B}(K)]$$

and therefore, also a strong equivalence

$$\mathrm{Hom}(\tilde{G_U}, K) \to \mathrm{Tors}(\mathscr{E}; K)^U.$$

The equivalence of the classifying toposes is a consequence of the fact that $\mathscr{S}$ is a stack and therefore the functor $\mathscr{B}(-) \simeq \mathscr{S}^{(-)^{op}}$ carries weak equivalence functors into (strong) equivalence functors. $\square$

Definition 3.7 Denote by $\Pi_1^c(\mathscr{E})$ the limit in $\mathbf{Top}_{\mathscr{S}}$ of the filtered system of toposes $\mathrm{Spl}(U)$ and connected locally connected geometric morphisms $h_{UV} : \mathrm{Spl}(U) \to \mathrm{Spl}(V)$ for each pair of covers U, V with $U \leq V$, indexed by a cofinal generating poset of covers in $\mathscr{E}$. The topos $\Pi_1^c(\mathscr{E})$ is said to be the *coverings fundamental group topos* of $\mathscr{E}$ over $\mathscr{S}$. It is equivalent, as a category, to the full subcategory of $\mathscr{E}$ generated by the locally constant covers in $\mathscr{E}$, in the sense of Definition 2.1. There is a connected locally connected geometric morphism $\sigma_{\mathscr{E}} : \mathscr{E} \to \Pi_1^c(\mathscr{E})$.

The main theorem of [9] may now be stated as follows, under the assumption that the base topos $\mathscr{S}$ satisfies (ASC).

Theorem 3.8 *Let $\mathscr{E}$ be a locally connected topos over a base topos $\mathscr{S}$, where $\mathscr{S}$ is assumed to satisfy (ASC) in the sense of Definition 3.3. Then the coverings fundamental group topos $\Pi_1^c(\mathscr{E})$ defined in Definition 3.7 is the classifying topos of a prodiscrete localic groupoid $\pi_1^c(\mathscr{E})$ which represents first-degree cohomology of $\mathscr{E}$ with coefficients in discrete groups.*

Proof Let $\{\pi_U\}$ be system of groupoids obtained from the system of groupoids $\{G_U\}$ as in Lemma 3.6, that is, by taking stack completions, and where the transition homomorphisms are homomorphisms $\tilde{g_{UV}} : \pi_U \to \pi_V$ induced by the $g_\alpha : G_U \to G_V$ for *any* α witnessing $U \leq V$, as follows from the Morita equivalence theorem for discrete groupoids. Since the $g_\alpha : G_U \to G_V$ are full and essentially surjective homomorhisms, the same is true of the homomorphisms $g_{UV} : \tilde{\pi_U} \to \pi_V$. Furthermore, since the original 2-system is *bi*filtered and *bi*ordered solely on account of the dependency of the transition homomorphisms g_α on the witnessing morphisms $\alpha : U \to V$, the 1-system of the stack completions, whose transition homomorphisms $\tilde{g_{UV}}$ only depend on $U \leq V$, is simply filtered.

Let $\pi_1^c(\mathscr{E}) = \lim\{\pi_U\}$ be the limit of the filtered system of discrete groupoids, taken in the category of *localic groupoids*. We now prove that the prodiscrete groupoid $\pi_1^c(\mathscr{E})$ represents cohomology of $\mathscr{E}$ with coefficients in discrete groups. Explicitly, the claim is that for a discrete group K, there is an isomorphism

$$H^1(\mathscr{E}; K) \simeq [\pi_1^c(\mathscr{E}), K].$$

This follows from Lemma 3.6 and various canonical isomorphisms:

$$\begin{aligned}
[\pi_1^c(\mathscr{E}), K] &\simeq \Pi_0(\mathrm{Hom}(\pi(\mathscr{E}), K)) = \Pi_0(\mathrm{Hom}(\lim\{\pi_U\}, K)) \\
&\simeq \Pi_0(\mathrm{colim}\{\mathrm{Hom}(\pi_U, K)\}) \\
&\simeq \Pi_0(\mathrm{colim}\{\mathrm{Tors}(\mathscr{E}; K)^U\}) \simeq \Pi_0(\mathrm{Tors}(\mathscr{E}; K)) \simeq H^1(\mathscr{E}; K)
\end{aligned}$$

□

In the rest of the paper we shall be dealing for the most part with locally simply connected toposes. We recall the definition.

Definition 3.9 A locally connected topos $\mathscr{E}$ over $\mathscr{S}$ is said to be *locally simply connected* if there is a single cover U in $\mathscr{E}$ which splits all locally constant objects in $\mathscr{E}$ in the sense of Definition 2.1.

4 Galois groupoids and Galois toposes

In this section we compare, in the case of a (locally connected and) locally simply connected topos $\mathscr{E}$ over $\mathscr{S}$, the Galois groupoid of [24] given any $e_!$-normal cover in $\mathscr{E}$, with the fundamental groupoid $\pi_1^c(\mathscr{E})$; in particular, this applies to the universal cover in $\mathscr{E}$, easily seen to be $e_!$-normal cover in $\mathscr{E}$.

Recall (e.g., from [8]) the notion of a G-torsor in $\mathscr{E}$, for G a groupoid in $\mathscr{E}$ given by means of a diagram

$$G_1 \times_{G_0} G_1 \overset{\longrightarrow}{\underset{\longrightarrow}{\longrightarrow}} G_1 \overset{\longrightarrow}{\underset{\longrightarrow}{\longleftarrow}} G_0 \ .$$

A (right) G-object $\mathbf{T}$ is given by the data $< \gamma : T \to G_0 \ , \ \theta >$ with T an object of $\mathscr{E}$, $p : T \to 1$ is the unique morphism into the terminal object 1, and $\theta : T \times_{G_0} G_1 \to T$

an action (unitary and associative). We say that $\mathbf{T}$ is a *G-torsor* in $\mathscr{E}$ if it is a right G-object for which $p : T \to 1$ is an epimorphism and

$$< \mathrm{proj}_1 \ , \ \theta >: T \times_{G_0} G_1 \to T \times T$$

is an isomorphism.

Definition 4.1 Let $e : \mathscr{E} \to \mathscr{S}$ be a locally connected geometric morphism. By an *$\mathscr{S}$-Galois family* of $\mathscr{E}$ we mean an object X of $\mathscr{E}$ of global support, equipped with a morphism $\delta : X \to e^*I$, such that the pair $(X \ , \ \delta)$ is both connected and a $\mathrm{Aut}(\delta)$-torsor in $\mathscr{E}/e^*I$ over $\mathscr{S}$. Explicitly, for the the canonical action θ of $e^*(\mathrm{Aut}(\delta))$ (a discrete groupoid in $\mathscr{E}$ with object of objects e^*I) on δ, the induced

$$< \mathrm{proj}_1 \ , \ \theta >: X \times_{e^*I} e^*(Aut(\delta)) \to X \times X$$

is an isomorphism. In particular, X is a locally constant object in the sense of Definition 2.1 and a cover in $\mathscr{E}$.

We omit the subscripts U in what follows for the geometric morphisms σ_U and p_U which occur in the pushout diagram $(*)$ defining $\mathscr{G}_U$.

Denote by $g : \mathscr{G}_U \to \mathscr{S}$ the structure geometric morphism. There is a natural isomorphism $g\sigma \simeq e$. Since $\sigma : \mathscr{E} \to \mathscr{G}_U$ is connected and locally connected and since $e : \mathscr{E} \to \mathscr{S}$ is locally connected, it follows that $g : \mathscr{G}_U \to \mathscr{S}$ is locally connected. We will use this in what follows.

Theorem 4.2 *Let $\mathscr{E}$ be a locally connected topos bounded over $\mathscr{S}$ by means of a geometric morphism $e : \mathscr{E} \to \mathscr{S}$. Let U be a cover in $\mathscr{E}$. Let $p : \mathscr{S}/e_!U \to \mathscr{G}_U$ be the discrete point that arises in the pushout definition $(*)$. Then, The morphism $p : \mathscr{S}/e_!U \to \mathscr{G}_U$ is a surjective local homeomorphism, its inverse image part is represented (uniquely up to isomorphism) by an $\mathscr{S}$-Galois family $(\sigma^*A, \sigma^*\zeta)$ and the square*

$$\begin{array}{ccc} \mathscr{E}/\sigma^*A & \xrightarrow{\varphi_{\sigma^*A}} & \mathscr{E} \\ {\scriptstyle \rho_{\sigma^*\zeta}}\downarrow & & \downarrow{\scriptstyle \sigma} \\ \mathscr{S}/e_!U & \xrightarrow[p]{} & \mathscr{G}_U \end{array}$$

is a pullback. In addition, $\mathscr{G}_U$ is generated by a single $\mathscr{S}$-Galois family $(\sigma^*A, \sigma^*\zeta : \sigma^*A \to e^*e_!U)$ and is the classifying topos of the discrete localic groupoid $\mathrm{Aut}(\sigma^*\zeta)$.

Proof It follows from Lemma 2.5 that p is locally connected therefore $\mathscr{S}$-essential; it is also surjective. Therefore, by [7] Proposition 1.2, its inverse image p^* is represented by a family $\zeta\colon A \to g^*e_!U$ in the sense that there is a natural isomorphism

$$p^* \simeq \mathrm{hom}_{\mathscr{G}_U(\mathscr{E})/g^*e_!U}(\zeta \ , \ \Delta_{e_!(U)}(-))$$

where $\Delta_{e_!(U)}(X) = X \times g^*e_!(U)$, equipped with the appropriate projection. The morphism $\zeta\colon A \to g^*e_!U$ represents p^* which has a right adjoint; in particular, $\zeta\colon A \to g^*e_!U$ is connected in $\mathscr{G}_U/g^*e_!U$. From the remark that there is a natural isomorphism

$$\mathrm{hom}_{\mathscr{G}_U/g^*e_!U}(\zeta \ , \ \Delta_{e_!(U)}(-)) \simeq \mathrm{hom}_{\mathscr{G}_U}(A, -)$$

and the faithfulness of p follows that A has global support. An explicit description of the representing family is given in the proof of [8] Theorem 1.2; let $\zeta = (p^{e_!U})_!(\delta_{e_!U} : e_!U \to e_!U \times e_!U)$.

We shall also regard this family as one in $\mathscr{E}/e^*e_!U$ by applying the fully faithful σ^*.

That p is a local homeomorphism is proven in [9] by recourse to the notion of a totally disconnected geometric morphism. In particular, there is induced an equivalence $\mathscr{G}_U/A \simeq \mathscr{S}/e_!U$ which identifies the localic point p with the local homeomorphism $\varphi_A : \mathscr{G}_U/A \to \mathscr{G}_U$. The statement about the mentioned square being a pullback is a consequence of this.

Also a consequence of this is the claim that $(\sigma^*A, \sigma^*\zeta)$ is a torsor, as can be seen by evaluating the naturally isomorphic functors p^* and $\varphi^*\sigma^*$ at σ^*A; this gives the desired isomorphism

$$< \mathrm{proj}_1 \, , \, \theta >: \sigma^*A \times_{e^*e_!U} e^*(Aut(\sigma^*\zeta)) \to \sigma^*A \times \sigma^*A$$

as in Definition 4.1.

The family $\zeta : A \to g^*e_!U$ representing p^* is generating for $\mathscr{G}_U$ since p^* is faithful. Hence $\mathscr{G}_U$ is generated by a single $\mathscr{S}$-Galois family.

Finally, since p is a local homeomorphism hence open, and since it is surjective, it is of effective descent by [28]. It follows that there is an equivalence $\mathscr{G}_U \simeq \mathscr{B}(G_U)$ where $G_U = \mathrm{Aut}(p)$ is a localic groupoid, discrete since p is (not just open but) a local homeomorphism.

From the fact that ζ represents p^* now follows that $\mathrm{Aut}(p)$ and $\mathrm{Aut}(\sigma^*\zeta)$ are equivalent groupoids in $\mathscr{S}$. □

The following proposition states that the universal cover $\sigma^*\zeta : \sigma^*A \to e^*e_!U$ in the locally simply connected topos $\mathscr{E}$ is $e_!$-normal in the sense of [24]. This is immediate from the properties of universal covers; we give below an alternative and direct proof of the assertion in question using only part of the data supplied by the pushout square $(*)$ in $\mathbf{Top}_{\mathscr{S}}$.

Lemma 4.3 *The $\mathscr{S}$-Galois family ζ representing p^* as in Theorem 4.2 may be alternatively obtained as the morphism*

$$\sigma^*(u_{\sigma_!U}) : \sigma^*\sigma_!U \to \sigma^*g^*g_!\sigma_!U \simeq e^*e_!U,$$

where u is the unit of the adjointness $g_! \dashv g^*$.

Proof This follows readily from Theorem 4.2 and the commutativity of the pushout square $(*)$, in particular, from the natural isomorphism $p_!\rho_! \simeq \sigma_!\varphi_{U!}$. □

Proposition 4.4 *Let $\mathscr{E}$ be locally simply connected with U as a single cover splitting all locally constant objects. Let (A, ζ) be the $\mathscr{S}$-Galois family of $\mathscr{E}$ with ζ representing $p : \mathscr{S}/e_!U \to \mathscr{G}_U = \mathscr{B}(\pi_1^c(\mathscr{E}))$. Then (A, ζ), which we identify with $(\sigma^*A, \sigma^*\zeta)$ if regarded in $\mathscr{E}$, is a universal cover in the sense that σ^*A is a cover, $(\sigma^*A, \sigma^*\zeta)$ is a normal object for the adjoint pair $e_! \dashv e^*$, and σ^*A splits every locally constant object of $\mathscr{E}$.*

Proof It is shown in Theorem 4.2 that the discrete groupoid G_U is the groupoid of automorphisms of the localic point p and, since the latter is represented by an $\mathscr{S}$-Galois family $\zeta : A \to g^*e_!U$, G_U is also realized as the discrete groupoid $\mathrm{Aut}(\zeta)$. Furhermore, a concrete instance of such a family is $\sigma^*(u_{\sigma_!U}) : \sigma^*\sigma_!U \to e^*e_!U$.

We now show that $\sigma^*A = \sigma^*\sigma_!U$ has global support and is a normal object (for the adjoint pair $e_! \dashv e^*$). For the former, notice that since σ is connected and locally connected (in particular $\sigma_!1 \simeq 1$) and U has global support, $\sigma^*A = \sigma^*\sigma_!U \to \sigma^*\sigma_!1 \simeq 1$ is an epimorphism.

For the latter, observe that since σ^* is fully faithful and $A = \sigma_! U \in \mathrm{Spl}(U)$, $\sigma^* A \simeq \sigma^* \sigma_! U$ is U-split. Use now Lemma 5.3 of [14] applied to σ (since it is locally connected, and in particular the adjoint pair $\sigma_! \dashv \sigma^*$ is $\mathscr{S}$-indexed); it says in this case that for any object X of $\mathscr{G}_U$, U splits $\sigma^* X$ if and only if $\sigma_! U$ splits X. In particular, since U splits $\sigma^* A = \sigma^* \sigma_! U$, it follows that $\sigma_! U$ splits $\sigma_! U$, hence (by stability) $\sigma^* A = \sigma^* \sigma_! U$ splits $\sigma^* A = \sigma^* \sigma_! U$. □

Corollary 4.5 *Let $\mathscr{E}$ be a locally connected and locally simply connected topos over $\mathscr{S}$. Let U be a single cover which splits all locally constant objects of $\mathscr{E}$. Let A be the universal cover. Let $H(\mathscr{E}) = \mathrm{Gal}(\sigma^*\zeta)$ be the Galois groupoid obtained as in* [24] *using that $\sigma^*\zeta$ is an $e_!$-normal object, and let $G = \pi_1^c(\mathscr{E})$. Then, there is an equivalence $H \simeq G$, explicitly,*

$$\mathrm{Gal}(\sigma^*\zeta) \simeq \pi_1^c(\mathscr{E}).$$

Proof This follows readily from the isomorphism

$$\sigma^* A \times_{e^* e_! U} e^*(Aut(\sigma^*\zeta)) \to \sigma^* A \times \sigma^* A$$

applying the functor $e_!$ to both sides to get

$$e_!(\sigma^* A \times_{e^* e_! U} e^*(Aut(\sigma^*\zeta))) \simeq e_!(\sigma^* A \times \sigma^* A)$$

and then using that

$$e_! \sigma^* A \simeq g_! \sigma_! \sigma^* \sigma_! U \simeq g_! \sigma_! U \simeq e_! U$$

and *Frobenius Reciprocity* for the $\mathscr{S}$-indexed adjoint pair $e_! \dashv e^*$ to obtain

$$e_!(\sigma^* A \times_{e^* e_! U} e^*(Aut(\sigma^*\zeta))) \simeq e_! U \times_{e_! U} Aut(\sigma^*\zeta) \simeq Aut(\sigma^*\zeta)$$

hence the desired isomorphism

$$Aut(\sigma^*\zeta) \simeq e_!(\sigma^* A \times \sigma^* A).$$

□

Remark 4.6 We now sum up our argument leading to the Galois groupoid in the rich context of a locally connected topos, while comparing it with the purely categorical Galois theory of Janelidze.

In the case of a locally connected topos $e : \mathscr{E} \to \mathscr{S}$, we also start with a pure set-up given by an adjoint pair $e_! \dashv e^*$ satisfying certain conditions explicitated in [23]. In addition, we use, *primo*, that from the $\mathscr{S}$-indexedness of the adjoint pair $e_! \dashv e^*$ follows the existence of a normal cover (X, δ), and that in consequence one gets (as in [24]) a groupoid $\mathrm{Gal}(X, \delta)$ for which the "fundamental theorem of Galois theory" holds; *secundo*, that using the further right adjoint e_* and everything that derives from it in the pushout $(*)$, one has that (X, δ) is σ^* applied to a family which represents (the inverse image part of) the surjective $\mathscr{S}$-essential point p_U of the topos $\mathrm{Spl}(U)$ and is therefore an $\mathscr{S}$-ATO in the appropriate slice category, and *tertio*, from the additional observation that p_U is a local homeomorphism follows that (X, δ) is a Galois family (a torsor, not just an $\mathscr{S}$-ATO) in $\mathscr{E}$, and that $\mathrm{Gal}(X, \delta)$ *is* equivalent to the (Galois) groupoid of automorphisms of the universal cover.

In categorical Galois theory there is another way to justify these steps given the assumption of normality, as pointed out to me by Janelidze. To give a concrete situation consider the "most classical" example, namely the topos $\mathrm{Sh}(B)$ of sheaves on a connected and locally (simply) connected space B admitting a universal covering $p : E \to B$. According to the categorical Galois theory, the Galois group is defined as $e_!(E \times_B E)$, but classically (Chevalley-Grothendieck) it is to be defined as the automorphism group $\mathrm{Aut}_B(E)$. From the fundamental theorem of Galois theory, Janelidze obtains (i) the category equivalence $\mathrm{Spl}(E,p) \cong \mathrm{Set}^G$ for $G = e_!(E \times_B E)$; (ii) under this equivalence, (E,p) corresponds to G acting on itself via its multiplication; (iii) therefore the automorphism group $\mathrm{Aut}(E,p) = \mathrm{Aut}_B(E)$ is isomorphic to $\mathrm{Aut}(G)$, the automorphism group of G considered as a G-set as above, and (iv) since $\mathrm{Aut}(G)$ is isomorphic to the group G, it can be concluded that $\mathrm{Aut}_B(E) \cong e_!(E \times_B E)$. The same argument can be used when the base category $\mathcal{S}$ is not necessarily Set but is cartesian closed. in conclusion, the full topos setting (and in particular, the right adjoint e_* and the Joyal-Tierney theorem) is certainly not *needed*, contrary to what I had previously implied.

We will now consider an unpointed version of a notion of a Galois topos introduced by Moerdijk [31] following Grothendieck [1].

Definition 4.7 A topos $\mathcal{E}$ bounded over $\mathcal{S}$ will be said to be a *$\mathcal{S}$-Galois topos* if $\mathcal{E}$ is locally connected and has an internal site $\mathbf{C}$ of definition determined by objects A of $\mathcal{E}$ for which there exists a morphism $\zeta : A \to e^*I$ which is an $\mathcal{S}$-Galois family in $\mathcal{E}$ in the sense of Definition 4.1.

We set out to *characterize* $\mathcal{S}$-Galois toposes. For this we need to observe the behaviour of $\mathcal{S}$-Galois objects under the transition morphisms in the system whose limit gives the fundamental groupoid.

Lemma 4.8 *Let $\mathcal{E}$ and $\mathcal{F}$ be locally connected toposes over $\mathcal{S}$ with corresponding structure maps e and f. Let $\varphi : \mathcal{F} \to \mathcal{E}$ be a connected and locally connected geometric morphism over $\mathcal{S}$ and let $\zeta : A \to e^*I$ be any $\mathcal{S}$-Galoisfamily in $\mathcal{E}$. Then,*

$$\varphi^*A \xrightarrow{\varphi^*\zeta} \varphi^*e^*I \cong f^*I$$

is an $\mathcal{S}$-Galois family in $\mathcal{F}$.

Proof Observe that $\varphi^*(Aut(\zeta)) \simeq Aut(\varphi^*(\zeta))$ since φ^* is a fully faithful left exact functor which has a right adjoint. Observe also that, since φ is connected and locally connected, if $\zeta : A \to e^*I$ is connected in $\mathcal{E}/e^*I$, then $\varphi^*(\zeta) : \varphi^*(A) \to \varphi^*(e^*I) = f^*I$ is connected in $\mathcal{F}/f^*I$. □

Proposition 4.9 *The limit of an inversely filtered system of $\mathcal{S}$-Galois toposes of the form $\mathcal{B}(G_\alpha)$ with G_α a discrete groupoid in $\mathcal{S}$, and connected locally connected transition morphisms between them, is an $\mathcal{S}$-Galois topos.*

Proof For each $\mathcal{G}_\alpha$ there is a single generating $\mathcal{S}$-Galois family $\zeta_\alpha : A_\alpha \to e^*(I_\alpha)$ by assumption. Moreover, for each α one has the canonical internal site given by the discrete groupoid $\mathrm{Aut}(\zeta_\alpha)$. By the construction of filtered inverse limits of toposes in terms of (internal) sites [33], it follows that these families determine families in $\mathcal{G}$ which form an internal site for $\mathcal{G}$. Since the transition morphisms are connected locally connected, it follows from Lemma 4.8 that these are $\mathcal{S}$-Galois objects of $\mathcal{G}$. □

The following theorem is the analogue of [31] Theorem 3.2 in the unpointed case.

Theorem 4.10 *Let $\mathcal{E}$ be a locally connected bounded over $\mathcal{S}$. Then, (1) and (3) are equivalent. If furthermore the base topos $\mathcal{S}$ is assumed to satisfiy the* *axiom of stack completions in the sense of Definition 3.3, then all four conditions are equivalent.*

1. *$\mathscr{E}$ is an $\mathscr{S}$-Galois topos.*
2. *The canonical morphism $\sigma_{\mathscr{E}} : \mathscr{E} \to \Pi_1^c(\mathscr{E})$ is an equivalence.*
3. *$\mathscr{E}$ is generated by its locally constant covers.*
4. *$\mathscr{E}$ is equivalent to a topos $\mathscr{B}(G)$ for G a prodiscrete localic groupoid.*

Proof The implication (1) $\Rightarrow$ (3) follows from Proposition 4.4. The converse (3) $\Rightarrow$ (1) amounts to the existence of a Galois closure in this constructive setting and follows immediately from Theorem 4.2. The implication (2) $\Rightarrow$ (1) follows from the conjunction of Theorem 3.8 and Proposition 4.9. The converse (1) $\Rightarrow$ (2) holds because, from the construction of filtered inverse limits of toposes in terms of sites [33], it follows that the inverse image of $\sigma_{\mathscr{E}}$ establishes an equivalence of sites, hence of the respective toposes. The implications (2) $\Rightarrow$ (4) and (4) $\Rightarrow$ (2) are part of Theorem 3.8. □

Corollary 4.11 *Let $\mathscr{E}$ be a locally connected topos bounded over $\mathscr{S}$. Then, the following are equivalent*

1. *$\mathscr{E}$ is a locally simply connected $\mathscr{S}$-Galois topos.*
2. *$\mathscr{E}$ is locally simply connected and the canonical morphism $\sigma_{\mathscr{E}} : \mathscr{E} \to \Pi_1^c(\mathscr{E})$ is an equivalence.*
3. *$\mathscr{E}$ is locally connected and $\mathscr{E} \simeq \mathscr{E}_{lc}$ as a category, where $\mathscr{E}_{lc}$ is the full subcategory of $\mathscr{E}$ whose objects are the locally constant objects in $\mathscr{E}$.*
4. *$\mathscr{E}$ is the classifying topos $\mathscr{B}(G)$ for G a discrete groupoid in $\mathscr{S}$.*

Remark 4.12 Grothendieck [21] *defines* the fundamental groupoid of a (locally simply connected) Galois (Grothendieck) topos $\mathscr{G}$ as the groupoid of points of $\mathscr{G}$. This is consistent with the results of the last section, particularly Theorem 3.1 and Theorem 3.8, since Set satisfies the axiom of choice, hence trivially the axiom of stack completions. This definition is then extended in [21] to what are called therein "multi-Galois toposes" and which classify the actions of a progroup. In the case of $\mathscr{S}$-Galois toposes for any topos $\mathscr{S}$ (including Set), we are dealing with classifying toposes $\mathscr{B}(\pi)$ of prodiscrete groupoids π, which are localic but not discrete; as proven in [8], the stack completion of such a π (for the topology of open surjections of locales) is obtained by considering the *localic* points of $\mathscr{B}(\pi)$. In the case of $\pi_1^c(\mathscr{E}) = lim\{\pi_U\}$, there is a canonical localic point p of the classifying topos $\Pi_1^c(\mathscr{E}) = lim\{\mathscr{B}(\pi_U)\}$ and this localic point may be said to be a *prodiscrete localic point*, as it is obtained as a limit of (discrete) bags of points [9].

5 Locally paths simply connected toposes over an arbitrary base

Recall the definition of the paths fundamental group topos $\Pi_1^p(\mathscr{E})$ for $\mathscr{E}$ a topos bounded over $\mathscr{S}$ [31]. It is given as the colimit $\tau_{\mathscr{E}} : \mathscr{E} \to \Pi_1^p(\mathscr{E})$ in $\mathbf{Top}_{\mathscr{S}}$ for the descent diagram

$$\mathscr{E}^{\Delta} \;\substack{\longrightarrow\\ \longrightarrow\\ \longrightarrow}\; \mathscr{E}^{I} \;\substack{\xrightarrow{\varepsilon_0}\\ \xrightarrow[\varepsilon_1]{}}\; \mathscr{E}$$

obtained from the cosimplicial locale by taking its exponential into $\mathscr{E}$. The assignment of $\Pi_1^p(\mathscr{E})$ to $\mathscr{E}$ is evidently pseudofunctorial and the geometric morphisms $\tau_{\mathscr{E}} : \mathscr{E} \to \Pi_1^p(\mathscr{E})$ are the components of a natural transformation. In the case of a *connected* locally connected topos, $\tau_{\mathscr{E}} : \mathscr{E} \to \Pi_1^p(\mathscr{E})$ is an open surjection, hence

$\Pi_1^p(\mathscr{E})$ is the classifying topos of an open localic groupoid $\pi_1^p(\mathscr{E})$; we shall not make this assumption here.

The (unique) path-lifting property can be expressed in topos theory. Let Δ_n denote the standard n-simplex locale in the topos $\mathscr{S}$, constructed from the unit interval locale I in the usual manner.

Definition 5.1 (i) An object Y of $\mathscr{E}$ is said to have the *path-lifting property* if the induced geometric morphism

$$(\varphi_Y)^{\Delta_n} : (\mathscr{E}/Y)^{\Delta_n} \to (\mathscr{E})^{\Delta_n}$$

is an open surjection.

(ii) An object Y of $\mathscr{E}$ is said to have the *unique path-lifting property* if the commutative diagram

$$\begin{array}{ccc} (\mathscr{E}/Y)^{\Delta_n} & \xrightarrow{\varphi_Y^{\Delta_n}} & \mathscr{E}^{\Delta_n} \\ \big\downarrow \varepsilon_0 & & \big\downarrow \varepsilon_0 \\ \mathscr{E}/Y & \xrightarrow[\varphi_Y]{} & \mathscr{E} \end{array}$$

is a pullback.

Proposition 5.2 *Any cover Y in $\mathscr{E}$ with the unique path-lifting property also satisfies the path-lifting property.*

Proof Open surjections are pullback stable. □

Proposition 5.3 *Let A be a locally constant object in $\mathscr{E}$. Then A has the unique path-lifting property.*

Proof The Lemma 6.1 of [15], established for Grothendieck toposes, is valid over an arbitrary base topos. Indeed, locally constant sheaves in the sense of Definition 2.1 (or "constructible sheaves" in the sense of Grothendieck [1]) are constant on contractible locales, the argument is just as in topology. Furthermore, the required stability holds by Lemma 4.8. □

The *comparison lemma* between the paths and the coverings fundamental group toposes was established in [15] for a *connected* locally connected topos $\mathscr{E}$ over $\mathscr{S}$. In the non-connected case, we need to work with the locally constant objects of Definition 2.1 rather than with those of Barr and Diaconescu [3]. The proof we give here uses Galois closure and the fact that locally constant objects have the (unique) path-lifting property.

Theorem 5.4 *Let $\mathscr{E}$ be a locally connected topos bounded over $\mathscr{S}$. Then there exists a geometric morphism*

$$\kappa_{\mathscr{E}} : \Pi_1^p(\mathscr{E}) \to \Pi_1^c(\mathscr{E})$$

for which the triangle

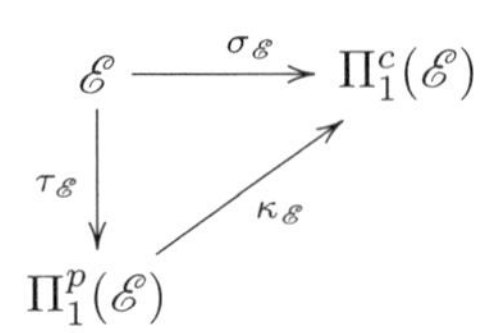

commutes.

Proof Let U be a cover in $\mathscr{E}$ and let $X \in \mathrm{Spl}_U(\mathscr{E})$. We wish to show that X has a (canonical) action by paths. We resort to Galois closure in $\mathscr{G}_U(\mathscr{E}) \simeq \mathrm{Spl}_U(\mathscr{E})$. Let $\zeta : A \to e^* e_! U$ be a generating $\mathscr{S}$-Galois family for $\mathscr{G}_U(\mathscr{E})$ as obtained in Theorem 4.2. Recall from Theorem 4.2 that the diagram

$$\begin{array}{ccc} \mathscr{E}/A & \xrightarrow{\varphi_A} & \mathscr{E} \\ \downarrow & & \downarrow \\ \mathscr{S}/e_!U & \xrightarrow[p_U]{} & \mathscr{G}_U \end{array}$$

is a pullback (in particular, it commutes). Consider now the diagram

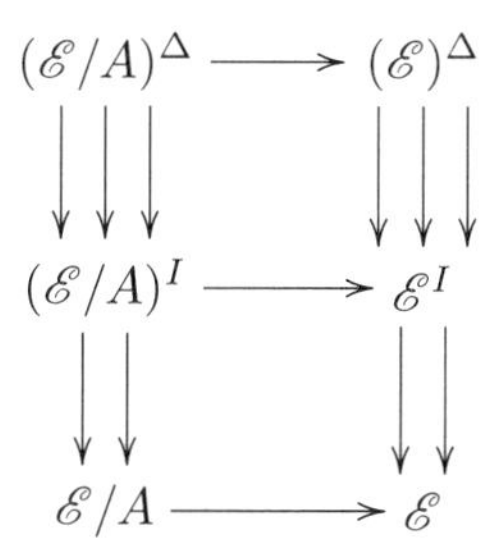

It follows now from Proposition 5.3 and Proposition 5.2 that the three horizontal arrows in the above descent diagram are open surjections, hence of effective descent. Therefore, it is sufficient to prove that $(\varphi_A)^* X$, as an object of $\mathscr{E}/A$ which arises as $\rho^*(\zeta)$ for some $\zeta : S \to e_! U$, has an action by paths. In turn, we claim that is enough to prove that $(\varphi_A)^* X$, as an object of $\mathscr{E}/A$ regarded as a topos over $\mathscr{S}/e_! U$, has an action by paths. This is because, since $\mathscr{S}/e_! U \to \mathscr{S}$ is a surjective local homeomorphism, the top arrow in the pullback square below is also one:

$$\begin{array}{ccc} [(\mathscr{E}/A)^I]_U & \longrightarrow & \mathscr{E}/A^I \\ \downarrow & & \downarrow \varepsilon_0 \\ \mathscr{S}/e_!U & \xrightarrow[\varphi_{e_!U}]{} & \mathscr{S} \end{array}$$

and similarly for $\Delta = \Delta_1$ and the other simplices locales.

The next observation is that for $i = 0, 1$ the triangle

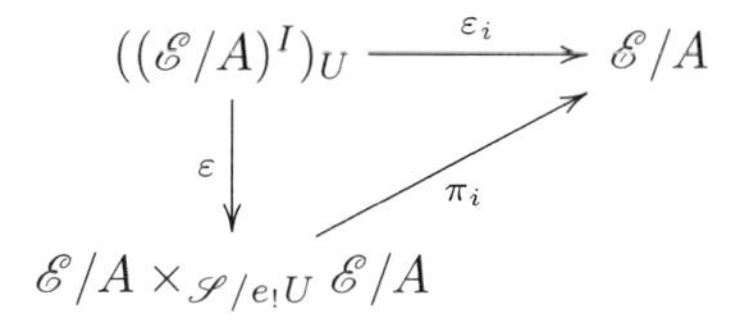

commutes up to iso 2-cell. By the theorem of Moerdijk and Wraith [34], ε is an open surjection since $\mathscr{E}/A \to \mathscr{S}/e_! U$ is connected locally connected. Therefore, since (for the same reason) the latter is the *coequalizer* of the pair π_0, π_1, it coequalizes the pair $\varepsilon_0, \varepsilon_1$ considered above. It follows easily from this that the object $(\varphi_U)^* X = \rho^* \zeta$ has an action by paths.

The preceding argument shows the existence of a canonical geometric morphism $\kappa_U : \Pi_1^p(\mathscr{E}) \to \mathscr{G}_U$, for each cover U, from which the existence of the desired morphism $\kappa_{\mathscr{E}} : \Pi_1^p(\mathscr{E}) \to \mathscr{G}(\mathscr{E})$ follows. $\square$

An obvious problem with the definition of $\Pi_1^c(\mathscr{E})$ for a locally connected topos $\mathscr{E}$ is the lack of pseudofunctoriality, by contrast with the pseudofunctoriality of $\Pi_1^p(\mathscr{E})$, as noted in [15]. As in topology, it is primarily the *paths* fundamental groupoid that is of interest, with the *coverings* version as a means for calculating the latter for spaces where the two are equivalent.

We have seen in the previous section that there is always a comparison map

$$\kappa_{\mathscr{E}} : \Pi_1^p(\mathscr{E}) \to \Pi_1^c(\mathscr{E})$$

and that this is a connected locally connected geometric morphism. For $\mathscr{E}$ a Galois topos, $\kappa_{\mathscr{E}}$ is an equivalence, in which case (that is defined on $\mathscr{S}$-Galois toposes) the assignment of $\Pi_1^c(\mathscr{E})$ to $\mathscr{E}$ is pseudofunctorial. But this case is not too interesting if what we are seeking is some sort of reflection of a *comprehensive* 2-category of toposes including the $\mathscr{S}$-Galois toposes into the 2-category of $\mathscr{S}$-Galois toposes, both as full sub 2-categories of $\mathbf{LTop}_{\mathscr{S}}$. Since the assignment of $\mathscr{E}_{lc}$ (the full subcategory of $\mathscr{E}$ consisting of its locally constant objets) to $\mathscr{E}$ is an idempotent operation, as is more generally the assignment of the full subcategory of $\mathscr{E}$ *generated* by the locally constant objects, the reflection would immediately result from the pseudofunctoriality and from it, in turn, a *van Kampen* theorem (fundamental group topos version) in the form of the preservation of pushouts could be directly inferred.

Let $\mathbf{GTop}_{\mathscr{S}}$ the full sub 2-category of $\mathbf{LTop}_{\mathscr{S}}$ with objects the $\mathscr{S}$-Galois toposes and let $I : \mathbf{GTop}_{\mathscr{S}} \to \mathbf{LTop}_{\mathscr{S}}$ be the inclusion 2-functor.

Remark 5.5 Denote by $\mathrm{lsc}(\mathbf{LTop}_{\mathscr{S}})$ (respectively $\mathrm{lsc}(\mathbf{GTop}_{\mathscr{S}})$) the full sub 2-category of $\mathbf{LTop}_{\mathscr{S}}$ (respectively of $\mathbf{GTop}_{\mathscr{S}}$) whose objects are locally simply connected toposes. The inclusion I above restricts to an inclusion $I : \mathrm{lsc}\mathbf{GTop}_{\mathscr{S}} \to \mathrm{lsc}\mathbf{LTop}_{\mathscr{S}}$. The only apparent way to render pseudofunctorial this assignment is (as done in [14]) to replace $\mathrm{lsc}(\mathbf{LTop}_{\mathscr{S}})$ by the non-full sub 2-category of $\mathbf{LTop}_{\mathscr{S}}$ whose objects are pairs $(\mathscr{E}\ ,\ A)$ with $\mathscr{E}$ locally connected and A a *universal cover* in $\mathscr{E}$ (by which we mean a locally constant object A of $\mathscr{E}$ with global support which splits all of the locally constant objects in $\mathscr{E}$, as implicitly in Proposition 4.4), and where a morphism $(\mathscr{E}, A) \to (\mathscr{F}, B)$ is given by a pair (φ, s) consisting of a geometric morphism $\varphi : \mathscr{E} \to \mathscr{F}$ and a morphism $s : A \to \varphi^*(B)$ in $\mathscr{E}$. However, this results in a loss of the reflection property, so that a simple-minded van Kampen theorem is not a consequence. In this non-full setting, however, a van Kampen theorem can be derived by imposing certain restrictions on the pushouts, as in [14] Theorem 6.5; the same restrictions reappear naturally in the coverings version of a van Kampen theorem, as in [14] or, more generally as in the next section.

Notice now that if $\mathscr{E}$ is a locally simply connected (and locally connected) topos with universal cover A, then $\mathscr{E}/A$ has the property that $\Pi_1^c(\mathscr{E}/A) \simeq \mathscr{S}/e_!A$, in such a way that the equivalence is compatible with the given morphisms with domain $\mathscr{E}/A$. From the existence of the comparison map follows then that to require that $\Pi_1^p(\mathscr{E}/A) \simeq \mathscr{S}/e_!A$ is a stronger condition. We thus make the following definition (see [15] for a related notion in the case of Grothendieck toposes).

Definition 5.6 A locally connected topos $\mathscr{E}$ is said to be *locally paths simply connected* if $\mathscr{E}$ is locally simply connected and the universal cover A has the property that $\Pi_1^p(\mathscr{E}/A) \simeq \mathscr{S}/e_!A$ by an equivalence which commutes with $\tau_{\mathscr{E}/A} : \mathscr{E}/A \to \Pi_1^p(\mathscr{E}/A)$ and $\rho_{\eta_A} : \mathscr{E}/A \to \mathscr{S}/e_!A$.

Theorem 5.7 *For $\mathscr{E}$ a locally connected and locally paths simply connected topos, the comparison map $\kappa_{\mathscr{E}} : \Pi_1^p(\mathscr{E}) \to \Pi_1^c(\mathscr{E})$ is an equivalence.*

Proof It is enough to prove that there exists a unique geometric morphism $\lambda_{\mathscr{E}} : \Pi_1^c(\mathscr{E}) \to \Pi_1^p(\mathscr{E})$ for which the triangle

$$\begin{array}{ccc} \mathscr{E} & \xrightarrow{\tau_{\mathscr{E}}} & \Pi_1^p(\mathscr{E}) \\ {\scriptstyle \sigma_{\mathscr{E}}}\downarrow & \nearrow_{\lambda_{\mathscr{E}}} & \\ \Pi_1^c(\mathscr{E}) & & \end{array}$$

commutes and is such hat $\lambda_{\mathscr{E}}\kappa_{\mathscr{E}} = \mathrm{id}_{\Pi_1^p(\mathscr{E})}$.

To this end, consider the diagram

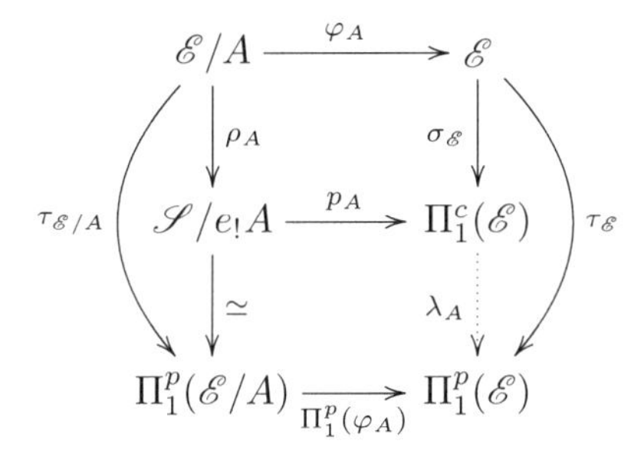

where A is a universal cover of $\mathscr{E}$. Since A splits every locally constant object in $\mathscr{E}$, the top square is a pushout. The desired morphism $\lambda_{\mathscr{E}}$ arises from the pseudofunctoriality of Π_1^p and pseudonaturality of τ, using now that A is a universal cover and $\mathscr{E}$ is locally paths simply connected, so that $\Pi_1^p(\mathscr{E}/A) \simeq \mathscr{S}/e_!A$ by an equivalence which commutes with $\tau_{\mathscr{E}/A} : \mathscr{E}/A \to \Pi_1^p(\mathscr{E}/A)$. □

With obvious notations, we now have the following.

Proposition 5.8 *The inclusion I : lpsc**GTop**$_{\mathscr{S}}$ $\to$ lpsc**LTop**$_{\mathscr{S}}$ has a left biadjoint given by the 2-functor Π_1^c : lpsc**LTop**$_{\mathscr{S}}$ $\to$ lpsc**GTop**$_{\mathscr{S}}$. The unit of adjointness has as components the (connected locally connected) geometric morphisms $\sigma_{\mathscr{E}} : \mathscr{E} \to \Pi_1^c(\mathscr{E})$.*

Proof The pseudofunctoriality of the assigment of the topos $\Pi_1^c(\mathscr{E})$ to a locally connected and locally paths simply connected topos $\mathscr{E}$ in the sense of Definition 5.6 is easily deduced from Theorem 5.7 and the stability of locally constant objects under pullback along arbitrary geometric morphisms. That $\Pi_1^c(\mathscr{E})$ is a (locally paths simply connected) $\mathscr{S}$-Galois topos for each locally connected (and locally paths simply connected) topos $\mathscr{E}$ follows in turn (since $\Pi_1^c(\mathscr{E})$ is again locally connected) from the claim that the induced geometric morphism

$$\Pi_1^c(\sigma_{\mathscr{E}}) : \Pi_1^c(\mathscr{E}) \to \Pi_1^c(\Pi_1^c(\mathscr{E}))$$

is an equivalence. To see this it is sufficient to observe that the operation of carving out $\mathscr{E}_{lc}$ from $\mathscr{E}$ is an idempotent one. It is now obvious that Π_1^c is a reflection onto lpsc**GTop**$_{\mathscr{S}}$ with σ the unit of the adjointness. This proves the first assertion; the second assertion is a consequence of the first and Corollary 4.11. □

It follows from Proposition 5.8 that pushouts in lpsc**LTop**$_{\mathscr{S}}$ are taken by $\Pi_1^c(-)$ into pushouts in **Top**$_{\mathscr{S}}$ in which all toposes in the pushout are locally paths simply connected Galois toposes, a statement which may be interpreted as a van Kampen theorem for (locally paths simply connected) toposes in terms of fundamental groupoid toposes [14]. We now derive a version of the van Kampen theorem in terms of fundamental groupoids.

Proposition 5.9 *Let*

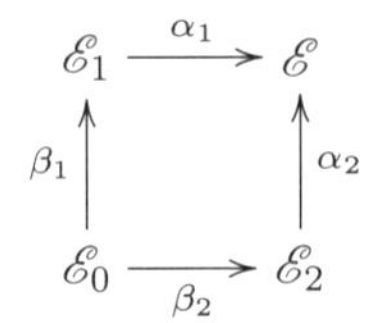

be a pullback in $\mathbf{Top}_{\mathscr{S}}$ *in which all four geometric morphisms are inclusions, with the induced* $\alpha : \mathscr{E}_1 + \mathscr{E}_2 \to \mathscr{E}$ *a locally connected surjection, and where all four toposes are locally connected and locally paths simply connected. Then, the induced diagram*

$$\begin{array}{ccc} \mathscr{B}(G_1) & \xrightarrow{\alpha_1} & \mathscr{B}(G) \\ {\scriptstyle \beta_1}\uparrow & & \uparrow{\scriptstyle \alpha_2} \\ \mathscr{B}(G_0) & \xrightarrow[\beta_2]{} & \mathscr{B}(G_2) \end{array}$$

is a pushout of classifying toposes of discrete groupoids, where all four morphisms commute with the canonical localic points.

Proof The first remark to make is that, under the assumptions made (enough for this that α be of effective descent, for instance, an open surpjection), the given pullback is also a pushout ([14]). In particular, the pseudofunctor Π_1^p carries it into a pushout which, under the further assumption that all four toposes are locally simply connected, translates into the pullback of classifying toposes of discrete groupoids as in the statement of the theorem.

It remains to prove that the morphisms in that pushout commute with the base points. Notice first that all four morphisms in the original pullback diagram are locally connected. Indeed, since α is locally connected, also α_1 and α_2 (composites of α with locally connected coproduct injections) are locally connected and, in turn (by the pullback condition), also β_1 and β_2 are locally connected. Thus, is enough to prove that, if $\varphi : \mathscr{E} \to \mathscr{F}$ is locally connected, with $\mathscr{E}$ and $\mathscr{F}$ locally paths simply connected, then the induced $\Pi_1^c(\varphi) : \Pi_1^c(\mathscr{E}) \to \Pi_1^c(\mathscr{F})$, alternatively viewed as a geometric morphism $\mathscr{B}(\varphi) : \mathscr{B}(G(\mathscr{E})) \to \mathscr{B}(G(\mathscr{F}))$, is a +-geometric morphism of classifying toposes in the sense that it commutes with the canonical localic points.

To see this, observe that since φ is locally connected, it follows that if we let U be a cover in$\mathscr{E}$ splitting all locally constant objects in $\mathscr{E}$, then $W = \varphi_!(U)$ splits all locally constant objects in $\mathscr{F}$. However, it does not follow from U a cover in $\mathscr{E}$ that also W is a cover in $\mathscr{F}$. Without loss of generality, however, we may replace W by $V = \varphi_!(U) + K$, where K is any cover splitting all locally constant objects in $\mathscr{F}$; V is now both a cover and splits all locally constant objects in $\mathscr{F}$. Moreover, there is a morphism $e_!U \to f_!V$, determined by the first coproduct injection and the isomorphism $f_!(\varphi_!(U) + K) \cong e_!U + f_!K$.

Consider now the universal cover $\tilde{V}$ associated with V, taken together with the canonical morphisms $\delta : \tilde{V} \to f^* f_! V$ which represents (the inverse image part of) the canonical localic point p_V of $\mathscr{F}$.

We have a cube

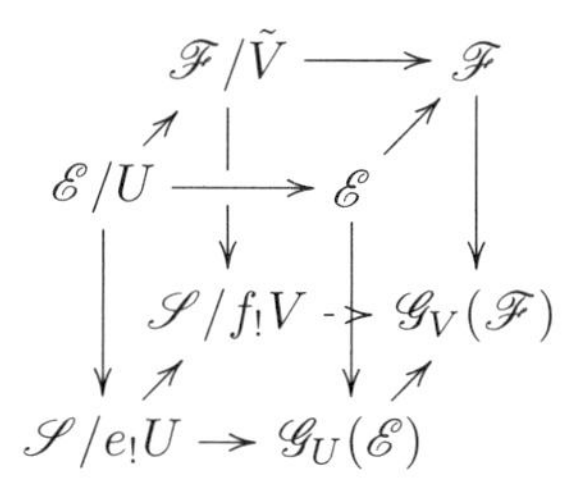

where the front square is a pushout and the back square is (a pushout and) a pullback.

Since there is given a morphism $e_!U \to f_!V$ in $\mathscr{S}$, as pointed out above, there is first of all a unique geometric morphism $\mathscr{E}/U \to \mathscr{F}/\tilde{V}$ making the left side and the top squares commute, using for this that the back square is a pullback (universal cover $\tilde{V}$ in $\mathscr{F}$). Since the front square is a pushout, we now obtain a unique geometric morphism $\mathscr{G}_U(\mathscr{E}) \to \mathscr{G}_V(\mathscr{F})$ which makes the right side and the bottom squares commute. In particular, this is a morphism $\mathscr{B}(G_U(\mathscr{E})) \to \mathscr{B}(G_V(\mathscr{F}))$ which commutes with the canonical localic points p_U and p_V as claimed.

It remains to see that the above argument can applied consistently to the four geometric morphisms in the pushout square, but this can be done just as in [14] (section 6). □

Denote by **Gpd**($\mathscr{S}$) the 2-category of (discrete localic) groupoids in $\mathscr{S}$.

Corollary 5.10 *Let*

$$\begin{array}{ccc} \mathscr{E}_1 & \xrightarrow{\alpha_1} & \mathscr{E} \\ {\scriptstyle \beta_1}\uparrow & & \uparrow{\scriptstyle \alpha_2} \\ \mathscr{E}_0 & \xrightarrow[\beta_2]{} & \mathscr{E}_2 \end{array}$$

be a diagram in $\mathbf{Top}_{\mathscr{S}}$ *satisfying the conditions of Proposition 5.9. Then the diagram*

$$\begin{array}{ccc} \pi_1^c(\mathscr{E}_1) & \xrightarrow{\pi_1^c(\alpha_1)} & \pi_1^c(\mathscr{E}) \\ {\scriptstyle \Pi_1^c(\beta_1)}\uparrow & & \uparrow{\scriptstyle \pi_1^c(\alpha_2)} \\ \pi_1^c(\mathscr{E}_0) & \xrightarrow[\pi_1^c(\beta_2)]{} & \pi_1^c(\mathscr{E}_2) \end{array}$$

is a pushout in **Gpd**.

Proof This is an immediate consequence of Proposition 5.9 and the Morita theorem for localic groupoids from [8]. □

6 Generalized covering morphisms and a van Kampen theorem

In topology, covering morphisms (in a more general sense than that of covering projections) are usually required to be local homeomorphisms which furthermore satisfy a path-lifting property. The latter should be thought of as topological coverings. Since morphisms not satisfying either condition, such as certain complete spreads in topos theory [11, 12], have also been thought of as covering morphisms "with singularities", we want to also make precise in what (generalized) sense are these still to be considered as coverings.

We shall restrict, as we have done so far in this paper, to locally connected toposes over $\mathscr{S}$, even if this assumption is not always strictly required. Denote by $\mathbf{LTop}_{\mathscr{S}}$ the full sub 2-category of $\mathbf{Top}_{\mathscr{S}}$ whose objects are the locally connected toposes.

A *notion* $\mathscr{C}$ *of covering morphism* on $\mathbf{LTop}_{\mathscr{S}}$ will be given by an assignment, for each $\mathscr{E} \in \mathbf{LTop}_{\mathscr{S}}$, of a category $\mathscr{C}(\mathscr{E}) \subseteq \mathbf{LTop}_{\mathscr{S}}/\mathscr{E}$, where the slice category $\mathbf{LTop}_{\mathscr{S}}/\mathscr{E}$ has as objects, geometric morphisms with codomain $\mathscr{E}$, with a morphism from $\varphi : \mathscr{F} \to \mathscr{E}$ to $\psi : \mathscr{G} \to \mathscr{E}$ given by a geometric morphism $\chi : \mathscr{F} \to \mathscr{G}$ for which there is a natural isomorphism $\alpha : \varphi \to \psi\chi$.

Denote by $\mathscr{L} : \mathscr{K}^{\text{op}} \to \mathbf{CAT}$ the pseudofunctor which assigns to a topos $\mathscr{E}$ the slice category $\mathbf{LTop}_{\mathscr{S}}/\mathscr{E}$ and to a geometric morphism $\varphi : \mathscr{F} \to \mathscr{E}$ the functor given by pulling back along φ.

We will say (following [16]) that a subpseudofunctor $\mathscr{C}$ of $\mathscr{L} \colon \mathscr{K}^{\text{op}} \to \mathbf{CAT}$ is a *stack* for a class Φ of morphisms of effective descent in $\mathbf{Top}_{\mathscr{S}}$, assumed to be closed under composition and pullbacks, if for any object $\mathscr{E}$ of $\mathscr{K}$, given $\alpha \in \mathscr{L}(\mathscr{E})$ such that for some $\varphi : \mathscr{F} \to \mathscr{E}$ with $\varphi \in \Phi$, it is the case that $\varphi^*(\alpha) \in \mathscr{C}(\mathscr{F})$, then it follows that $\alpha \in \mathscr{C}(\mathscr{E})$.

Definition 6.1 Let $\mathscr{K}$ be a sub 2-category of $\mathbf{LTop}_{\mathscr{S}}$. A notion $\mathscr{C}$ of covering morphisms is said to define *a fibration of covering morphisms for* $\mathscr{K}$ with respect to a class Φ as above if,

1. $\mathscr{C}$ extends to a pseudofunctor $\mathscr{C} : \mathscr{K}^{\text{op}} \to \mathbf{CAT}$,
2. there is a fully faithful pseudonatural transformation $H : \mathscr{C} \to \mathscr{L}$, and
3. $\mathscr{C}$ is a stack for the class Φ.

We say that a pseudofunctor $\mathscr{C} : \mathscr{K}^{\text{op}} \to \mathbf{CAT}$ *preserves binary products* if for each pair of objects E and F in $\mathscr{K}$, the canonical functors

$$\mathscr{C}(E + F) \to \mathscr{C}(E) \times \mathscr{C}(F)$$

induced by the injections $i : E \to E + F$ and $j : F \to E + F$ is an equivalence. In the terminology of [30], $\mathscr{C}$ is an *intensive quantity* on $\mathscr{K}$.

The assignment to a topos $\mathscr{E}$ of the class of *local homeomorphisms* over it constitutes a pseudofunctor $\mathscr{A} \colon \mathbf{LTop}_{\mathscr{S}}^{\text{op}} \to \mathbf{CAT}$; given any geometric morphism $\varphi : \mathscr{F} \to \mathscr{E}$ in $\mathbf{LTop}_{\mathscr{S}}$, pulling back along φ restricts to a functor $A(\varphi) : \mathscr{A}(\mathscr{E}) \to \mathscr{A}(\mathscr{F})$. The pseudofunctor $\mathscr{A}$ is trivially a stack for any (morphism of effective descent) class Φ. That $\mathscr{A}$ preserves binary products follows from the fact that colimits in $\mathbf{Top}_{\mathscr{S}}$ (or $\mathbf{LTop}_{\mathscr{S}}$) are calculated as limits in $\mathbf{CAT}$ of the corresponding diagrams of inverse image parts [33].

Notice that the 2-extensivity of $\mathscr{K}$ (for $\mathscr{K} = \mathbf{Top}_{\mathscr{S}}$, or $\mathscr{K} = \mathbf{LTop}_{\mathscr{S}}$) says that $\mathscr{L}$ preserves binary products. Notice also that there is a fully faithful pseudonatural transformation $J : \mathscr{A} \to \mathscr{L}$. Indeed, given any composite $\zeta = \psi\varphi$, where both ζ and ψ are local homeomorphisms, so is φ.

The following is a version of the *van Kampen theorem* in terms of covering morphisms. It is an abstraction of Theorem 5.7 of [14], given therein for covering projections on toposes; the proof is analogous.

Theorem 6.2 *Let* $\mathscr{C}$ *be a fibration of covering morphisms on* $\mathscr{K}$ *with respect to a class* Φ *of morphisms of effective descent in* $\mathscr{K}$. *Assume furthermore that* $\mathscr{C}$ *preserves binary products and that* $H : \mathscr{C} \to \mathscr{L}$ *factors through the canonical* *inclusion* $J : \mathscr{A} \to \mathscr{L}$. *Let*

$$\begin{array}{ccc} \mathscr{E}_1 & \xrightarrow{\alpha_1} & \mathscr{E} \\ {\scriptstyle\beta_1}\uparrow & & \uparrow{\scriptstyle\alpha_2} \\ \mathscr{E}_0 & \xrightarrow[\beta_2]{} & \mathscr{E}_2 \end{array}$$

be a pushout diagram in $\boldsymbol{LTop}_{\mathscr{S}}$ *in which the induced map* $\alpha : \mathscr{E}_1 + \mathscr{E}_2 \to \mathscr{E}$ *is a morphism in* Φ. *Then*

$$\begin{array}{ccc} \mathscr{C}(\mathscr{E}_1) & \xrightarrow{\beta_1^*} & \mathscr{C}(\mathscr{E}_0) \\ {\scriptstyle\alpha_1^*}\uparrow & & \uparrow{\scriptstyle\beta_2^*} \\ \mathscr{C}(\mathscr{E}) & \xrightarrow[\alpha_2^*]{} & \mathscr{C}(\mathscr{E}_2) \end{array}$$

is a pullback in **CAT**.

Another example of a fibration of covering morphisms is given by letting $\mathscr{C}(\mathscr{E}) = \mathscr{E}_{lc}$ be the full subcategory of $\mathscr{E}$ determined by the (locally trivial) covering morphisms in terms of the adjunction $e_! \dashv e^*$ in the sense of [24]; equivalently $\mathscr{C}(\mathscr{E})$ is the category of local homeomorphisms over $\mathscr{E}$ defined by a locally constant object in the sense of Definition 2.1. The pseudofunctoriality of $\mathscr{C}$ on $\mathbf{LTop}_{\mathscr{S}}$ follows from the stability of locally constant objects under pullbacks along arbitrary geometric morphisms, as remarked in Section 2. There is a pseudonatural transformation $H : \mathscr{C} \to \mathscr{A}$ (with fully faithful components) by the very definition of $\mathscr{C}$. It is shown in [14] that H is φ-cartesian for every surjective locally connected φ. This means that the pseudonaturality square

$$\begin{array}{ccc} \mathscr{A}(X) & \xrightarrow{\mathscr{A}(\varphi)} & \mathscr{A}(Y) \\ {\scriptstyle HX}\uparrow & & \uparrow{\scriptstyle HY} \\ \mathscr{C}(X) & \xrightarrow[\mathscr{C}(\varphi)]{} & \mathscr{C}(Y) \end{array}$$

is a pullback in $\mathscr{K}$. This readily implies that $\mathscr{C}$ is a stack for the class Φ of locally connected surjections. It is also shown in [14] that $\mathscr{C}$ preserves binary products.

In particular, there is a *van Kampen theorem* for $\mathscr{C}$ as an application of Theorem 6.2, already shown in [14].

We shall now consider a different fibration of covering morphisms for which wc need to recall some basic facts about complete spreads from [11, 12]. Let $\mathscr{E}$ be a (locally connected) topos presented over $\mathscr{S}$ by a site $\mathbf{C}$. A geometric morphism $\varphi : \mathscr{F} \to \mathscr{E}$ in $\mathbf{LTop}_{\mathscr{S}}$ is a *complete spread* if, for the discrete opfibration $D : \mathbf{D} \to \mathbf{C}$ which is associated with the functor $D : \mathbf{C} \to \mathscr{S}$ which is the composite

$$\mathbf{C} \xrightarrow{\varepsilon} \mathscr{E} \xrightarrow{\varphi^*} \mathscr{F} \xrightarrow{f_!} \mathscr{S} \; ,$$

the square

$$\begin{array}{ccc} \mathscr{F} & \longrightarrow & \mathscr{S}^{\mathbf{D}^{\mathrm{op}}} \\ {\scriptstyle\varphi}\downarrow & & \downarrow{\scriptstyle u} \\ \mathscr{E} & \longrightarrow & \mathscr{S}^{\mathbf{C}^{\mathrm{op}}} \end{array}$$

where u is induced by D, is a pullback.

Complete spreads are related to distributions in the sense of Lawvere [30]. A distribution on a (locally connected) topos $\mathscr{E}$ over $\mathscr{S}$ is any $\mathscr{S}$-cocontinuous functor $\mu : \mathscr{E} \to \mathscr{S}$. Denote by $\mathbf{Dist}(\mathscr{E})$ the category of distributions on $\mathscr{E}$. Explicitly, $\mathbf{Dist}(\mathscr{E}) = \mathrm{Coc}_{\mathscr{S}}(\mathscr{A}(\mathscr{E})\ ,\ \mathscr{S})$ and each geometric morphism $\varphi : \mathscr{F} \to \mathscr{E}$ induces, by composition with the $\mathscr{S}$-cocontinuous functor $\mathscr{A}(\varphi) = \varphi^* : \mathscr{E} \to \mathscr{F}$, a functor $\mathbf{Dist}(\varphi) : \mathbf{Dist}(\mathscr{F}) \to \mathbf{Dist}(\mathscr{E})$.

For each $\mathscr{E} \in \mathbf{Top}_{\mathscr{S}}$ there is a topos $\mathscr{M}(\mathscr{E})$ [10] (the "symmetric topos") and an $\mathscr{S}$-essential geometric morphism $\delta : \mathscr{E} \to \mathscr{M}(\mathscr{E})$ such that the pair $\langle \mathscr{M}(\mathscr{E})\ ,\ \delta \rangle$ classifies distributions on $\mathscr{E}$.

If $\varphi : \mathscr{F} \to \mathscr{E}$ is $\mathscr{S}$-essential, then composition with the $\mathscr{S}$-cocontinuous functor $\varphi_! : \mathscr{F} \to \mathscr{E}$ is a functor $\mathscr{R}_\varphi : \mathbf{Dist}(\mathscr{E}) \to \mathbf{Dist}(\mathscr{F})$ which is right adjoint to the functor $\mathbf{Dist}(\varphi) : \mathbf{Dist}(\mathscr{F}) \to \mathbf{Dist}(\mathscr{E})$, as the latter is given by composition with φ^*. Under the equivalence between distributions and complete spreads, it is the right adjoint $\mathscr{R}_\varphi$ which corresponds to pulling back complete spreads over $\mathscr{E}$ along the $\mathscr{S}$-essential morphism $\varphi : \mathscr{F} \to \mathscr{E}$ to give a complete spread over $\mathscr{F}$.

For objects $\mathscr{E}$ and $\mathscr{F}$ in $\mathbf{Top}_{\mathscr{S}}$, their coproduct $\mathscr{E} + \mathscr{F}$ exists in $\mathbf{Top}_{\mathscr{S}}$ and is given by the product category $\mathscr{E} \times \mathscr{F}$; the injections $\mathrm{i} : \mathscr{E} \to \mathscr{E} + \mathscr{F}$ and $\mathrm{j} : \mathscr{F} \to \mathscr{E} + \mathscr{F}$ have as inverse image parts the left adjoints to the projections. As geometric morphisms over $\mathscr{S}$, the injections into the coproduct are locally connected, since the projections have an $\mathscr{S}$-indexed left adjoint as well as a right adjoint. It is shown in [14] that $\mathbf{Top}_{\mathscr{S}}$ is an extensive 2-category; this says that for the 2-categorical versions of the slices, the coproduct pseudofunctor $F : \mathbf{Top}_{\mathscr{S}} \times \mathbf{Top}_{\mathscr{S}} \to \mathbf{Top}_{\mathscr{S}}$ induces biequivalences

$$F_{\langle \mathscr{E}, \mathscr{F} \rangle} : \mathbf{Top}_{\mathscr{S}}/\mathscr{E} \times \mathbf{Top}_{\mathscr{S}}/\mathscr{F} \to \mathbf{Top}_{\mathscr{S}}/(\mathscr{E} + \mathscr{F})$$

for all $\mathscr{E}$ and $\mathscr{F}$ in $\mathbf{Top}_{\mathscr{S}}$. Similar statements hold for $\mathbf{LTop}_{\mathscr{S}}$ [14].

Denote by $\mathscr{D}(\mathscr{E})$ the category of complete spreads (with locally connected domain) over $\mathscr{E}$. There is an functor $B : \mathscr{D}(\mathscr{E}) \to \mathbf{Dist}(\mathscr{E})$ which assigns, to a complete spread (in fact to any geometric morphism) $\varphi : \mathscr{F} \to \mathscr{E}$ in $\mathbf{LTop}_{\mathscr{S}}$ the distribution $\mu : \mathscr{E} \to \mathscr{S}$ given by the composite $\mu = f_! \varphi^*$. This functor is an equivalence. This gives rise to a (unique) factorization of any given geometric morphism $\varphi : \mathscr{F} \to \mathscr{E}$ with locally connected domain into a composite of a *pure* morphism $\pi : \mathscr{F} \to \mathscr{D}$ followed by a complete spread $\psi : \mathscr{D} \to \mathscr{E}$ with locally connected domain.

All of the above considerations remain true if $\mathbf{Top}_{\mathscr{S}}$ is replaced by $\mathbf{LTop}_{\mathscr{S}}$ [14].

Proposition 6.3 *Let $\mathscr{K}$ be $\mathbf{LTop}_{\mathscr{S}}$. The assignment of $\mathscr{D}(\mathscr{E})$ to $\mathscr{E}$ preserves binary products, in the sense that for any locally connnected toposes $\mathscr{E}$ and $\mathscr{F}$, there is a (canonical) equivalence $\mathscr{D}(\mathscr{E} + \mathscr{F}) \simeq \mathscr{D}(E) \times \mathscr{D}(F)$ of categories.*

Proof Consider the functor

$$F : \mathscr{D}(\mathscr{E} + \mathscr{F}) \to \mathscr{D}(\mathscr{E}) \times \mathscr{D}(\mathscr{F})$$

induced by pulling back along the coproduct injections. Since the latter are $\mathscr{S}$-essential geometric morphisms (in fact, locally connected), this is well defined [13].

Consider now the coproduct pullback diagram

$$\begin{array}{ccccc} \mathscr{A} & \xrightarrow{u} & \mathscr{A}+\mathscr{B} & \xleftarrow{v} & \mathscr{B} \\ \downarrow{\scriptstyle\psi} & & \downarrow{\scriptstyle\psi+\zeta} & & \downarrow{\scriptstyle\zeta} \\ \mathscr{E} & \xrightarrow[i]{} & \mathscr{E}+\mathscr{F} & \xleftarrow[j]{} & \mathscr{F} \end{array}$$

with ψ and ζ both complete spreads with locally connected domain.

We claim that the assignment of $\psi + \zeta$ to $\langle \psi, \zeta \rangle$ defines a functor

$$G : \mathscr{D}(\mathscr{E}) \times \mathscr{D}(\mathscr{F}) \to \mathscr{D}(\mathscr{E} + \mathscr{F})$$

To see that G is well defined we must prove that $\psi + \zeta$ is a complete spread (with locally connected domain). Consider the factorization $\psi + \zeta \simeq \varphi\rho$, with ρ pure and φ a complete spread with locally connected domain $\mathscr{C}$. By extensivity of $\mathscr{K}$ it follows first that $\mathscr{C} \simeq \mathscr{C}_0 + \mathscr{C}_1$ and $\varphi \simeq \varphi_0 + \varphi_1$, and second, that $\rho = \rho_0 + \rho_1$ From the stability of the pure/complete spread factorization under pullbacks along locally connected geometric morphisms, and the fact that ψ and ζ are complete spreads, it follows that ρ_0 and ρ_1 are (pure) equivalences and hence that $\varphi \simeq \psi + \zeta$, so that $\psi + \zeta$ is a complete spread as claimed.

The pair $\langle F, G \rangle$ consitutes an equivalence by the extensivity of $\mathbf{LTop}_{\mathscr{S}}$.

□

Let $\mathscr{U}(\mathscr{E})$ be the full subcategory of $\mathscr{E}$ determined by its complete spread objects [12, 19], alternatively viewed as the "unramified morphisms" over $\mathscr{E}$, consisting of those complete spreads over $\mathscr{E}$ which are also local homeomorphisms. Denote by $\mathscr{U}$ the intersection of $\mathscr{A}$ and $\mathscr{D}$.

Proposition 6.4 *$\mathscr{U}$ is a fibration of covering morphisms on $\mathbf{LTop}_{\mathscr{S}}$ for the class Φ of surjective local homeomorphisms. Furthermore, $\mathscr{U}$ preserves binary products.*

Proof The stability of complete spread objects (or of the unramified morphisms they determine) under pullback in $\mathbf{LTop}_{\mathscr{S}}$ has been shown in [19]. Hence, $\mathscr{U} : \mathbf{LTop}_{\mathscr{S}}^{\mathrm{op}} \to \mathbf{CAT}$ may be regarded as a sub pseudofunctor of $\mathscr{A}$ (not just of $\mathscr{L}$) . The remaining properties are a consequence Proposition 6.3. □

Corollary 6.5 *Let*

$$\begin{array}{ccc} \mathscr{E}_1 & \xrightarrow{\alpha_1} & \mathscr{E} \\ \uparrow{\scriptstyle\beta_1} & & \uparrow{\scriptstyle\alpha_2} \\ \mathscr{E}_0 & \xrightarrow[\beta_2]{} & \mathscr{E}_2 \end{array}$$

be a pushout diagram (or a pullpack diagram with α_1 and α_2 are pseudomonic) in $\mathbf{LTop}_{\mathscr{S}}$ in which the induced $\alpha : \mathscr{E}_1 + \mathscr{E}_2 \to \mathscr{E}$ is a surjective local homeomorphism. Then

$$\begin{array}{ccc} \mathscr{U}(\mathscr{E}_1) & \xrightarrow{\beta_1^*} & \mathscr{U}(\mathscr{E}_0) \\ \uparrow{\scriptstyle\alpha_1^*} & & \uparrow{\scriptstyle\beta_2^*} \\ \mathscr{U}(\mathscr{E}) & \xrightarrow[\alpha_2^*]{} & \mathscr{U}(\mathscr{E}_2) \end{array}$$

is a pullback in **CAT**.

Proof It follows directly from Theorem 6.2 and Proposition 6.4. □

A topological notion of covering morphism on $\mathbf{LTop}_{\mathscr{S}}$ ought to restrict at least to all local homeomorphisms satisfying a path-lifting property.

Definition 6.6 A fibration $\mathscr{C}$ of covering morphisms in the sense of Definition 6.1 is said to be *topological* if:

1. $\mathscr{C}$ is a subpseudofunctor of $\mathscr{A} : \mathbf{LTop}^{\mathrm{op}}_{\mathscr{S}} \to \mathbf{CAT}$, and
2. for every (locally connected) topos $\mathscr{E}$, every cover $X \in \mathscr{C}(\mathscr{E})$ has the unique path-lifting property in the sense of Definition 5.1.

Let us examine how the three examples of fibrations of covering morphisms (given so far) fare from the topological viewpoint expressed in Definition 6.6. The *second condition* in the above definition excludes $\mathscr{A}$ as a topological fibration of covering morphisms since not every cover X in an arbitrary (locally connected) topos $\mathscr{E}$ has the unique path-lifting property, or not every local homeomorphism is a (discrete) fibration. Thus, arbitrary local homeomorphisms, though covering morphisms, are not in general topological covering morphisms. The *first condition* in the above definition excludes also $\mathscr{D}$ as a topological fibration of covering morphisms since $\mathscr{D}$ is not a pseudofuctor on $\mathbf{LTop}^{\mathrm{op}}_{\mathscr{S}}$ but also it is not the case that for every $\mathscr{E}$, $\mathscr{D}(\mathscr{E})$ is included in $\mathscr{A}(\mathscr{E})$, as not all complete spreads are local homeomorphisms. By contrast, the fibration $\mathscr{C}$ of locally trivial covering morphisms satisfies the required conditions.

Proposition 6.7 *Let $\mathscr{C}(\mathscr{E}) = \mathscr{E}_{lc}$ be the full subcategory of $\mathscr{E}$ determined by the locally constant objects in the sense of Definition 2.1. Then $\mathscr{C}$ defines a topological fibration of covering morphisms in the sense of Definition 6.6.*

Proof That $\mathscr{C}$ is a subpseudofunctor of $\mathscr{A}$ is part of the definition of $\mathscr{C}$. That every object of every $\mathscr{C}(\mathscr{E})$ satisfies the unique path-lifting property was established in Proposition 5.3. □

It is natural to ask whether the topological notion expressed in Definition 6.6 includes examples other than the locally trivial covering morphisms.

The following result was shown in [12] for the locally constant objects in the sense of Barr and Diaconescu [3]; we need to establish it here for the locally constant objects in the sense of Definition 2.1.

Proposition 6.8 *Let Y be a locally constant object in a locally connected topos $\mathscr{E}$. Then, Y is a complete spread object in $\mathscr{E}$.*

Proof Let Y be a locally constant object in $\mathscr{E}$ in the sense of Definition 2.1. By Proposition 2.2 an equivalent condition (expressed therein as condition (3)) is that there is a cover U in $\mathscr{E}$, a morphism $\alpha : J \to I$ in $\mathscr{S}$, a morphism $\eta : U \to e^*I$, and a morphism $\zeta : Y \times U \to e^*J$, for which the square

$$\begin{array}{ccc} Y \times U & \xrightarrow{\pi_2} & U \\ \downarrow{\scriptstyle \zeta} & & \downarrow{\scriptstyle \eta} \\ e^*J & \xrightarrow[e^*\alpha]{} & e^*I \end{array}$$

is a pullback. By the above and the local character of complete spreads along surjective local homeomorphisms [12](Proposition 7.1), it is sufficient to prove that any geometric morphism $\mathscr{E}/e^*J \to \mathscr{E}/e^*I$ (induced by a morphism $\alpha : J \to I$ in $\mathscr{S}$) is a complete spread. We know [12](proof of Theorem 7.3) that constant objects are complete spreads. Therefore the composite $\mathscr{E}/e^*J \to \mathscr{E}/e^*I \to \mathscr{E}$ as well as the second morphisms in the composite are both complete spreads. By [11] (Proposition 1.2.3) this implies that $\mathscr{E}/e^*J \to \mathscr{E}/e^*I$ is a complete spread. □

Theorem 6.9 *Let X be a complete spread cover (i.e., a complete spread object with full supoort) in a locally connected topos $\mathscr{E}$. Then X has the unique path-lifting property in the sense of Definition 5.1.*

Proof The following is a characterization of complete spread covers [12] Proposition 7.5. Let $\mathscr{E}$ be a locally connected topos with pure inclusion $\mathscr{E} \hookrightarrow \mathscr{S}^{\mathbf{C}^{\mathrm{op}}}$. Assume that X is a cover in $\mathscr{E}$ with corresponding (surjective) discrete opfibration $X : \mathbf{X} \to \mathbf{C}$ and associated discrete fibration $D : \mathbf{D} \to \mathbf{C}$. Then X is a complete spread object if and only if the square

$$\begin{array}{ccc} \mathscr{E}/X & \longrightarrow & \mathscr{S}^{\mathbf{D}^{\mathrm{op}}} \\ \downarrow & & \downarrow \\ \mathscr{E} & \longrightarrow & \mathscr{S}^{\mathbf{C}^{\mathrm{op}}} \end{array}$$

is a pullback, where the right vertical arrow is induced by D and therefore is a surjective discrete fibration (and opfibration); the geometric morphism that it induces has the unique paths-lifting property.

We now show that the local homeomorphism $\mathscr{E}/X \to \mathscr{E}$ has the unique path-lifting property. Consider the cube:

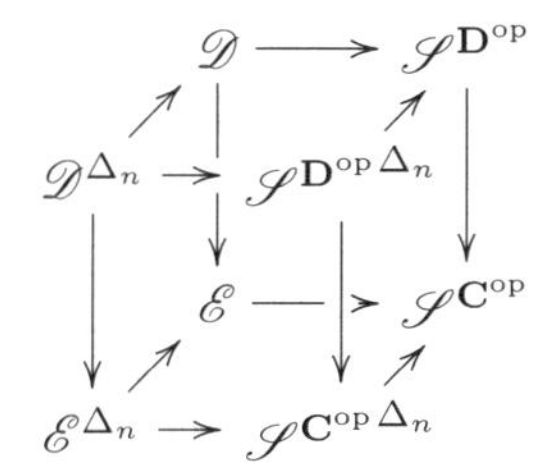

In it, the front, back, and right side faces are pullbacks, hence so is the left side square. This gives the unique paths-lifting property for the geometric morphism $\mathscr{D} \to \mathscr{E}$, as claimed. □

Proposition 6.10 *$\mathscr{U}$ is a topological fibration of covering morphisms.*

Proof By Proposition 6.4 $\mathscr{U}$ is a fibration of covering morphisms. By definition, $\mathscr{U}$ is a subfibration of $\mathscr{A}$. The path-lifting property is shown in Theorem 6.9. □

We note that, as shown explicitly in [19], the inclusion $\mathscr{C} \hookrightarrow \mathscr{U}$ is proper. Hence this is a genuine example of a notion of covering morphism having analogous topological properties as that of the locally trivial covering morphisms, yet it is not (seemingly) an instance of the Janelidze notion. Although $\mathscr{U}$ is a stack and contains all trivial morphisms, it is not the stack completion of it; indeed the stack completion of the trivial morphisms is $\mathscr{C}$. Although not discussed in this paper, it can be shown, just as in topology, that in the case of locally paths simply connected toposes, the two classes $\mathscr{C}$ and $\mathscr{U}$ agree.

Acknowledgements. I am very grateful to George Janelidze for directing my attention to an interesting connection between his work and mine, and to Eduardo Dubuc for valuable comments on Galois toposes and on my construction of the fundamental groupoid of a topos. Thanks are due to Steve Vickers, Steve Lack and George Janelidze for useful concrete remarks on a preliminary version of this paper, and to Bill Lawvere, Ronnie Brown and Claudio Hermida for motivating general discussions.

References

[1] M. Artin, A. Grothendieck, and J.-L. Verdier, *Théorie des Topos et Cohomologie Etale des Schemas (SGA4), Lecture Notes in Mathematics 269, 270, 305*, Springer-Verlag, Berlin-Heidelberg-New York, 1972.

[2] M. Barr and R. Diaconescu, Atomic Toposes, *J. Pure Appl. Alg.* 17: 1–24, 1980.

[3] M. Barr and R. Diaconescu, On locally simply connected toposes and their fundamental groups, *Cahiers Top. Geo. Diff. Cat.* 22:301–314, 1981.

[4] M. Barr and R. Paré, Molecular Toposes, *J. Pure Appl. alg.* 17: 265–280, 1980.

[5] F. Borceux and G. Janelidze, *Galois theories*, Cambridge Studies in Advanced Mathematics 72, Cambridge Univ. Press, 2001.

[6] R. Brown, Groupoids and van Kampen theorems, *Proc. London Math. Soc.* 3-17:385–401, 1967.

[7] M. Bunge, Internal Presheaf Toposes, *Cahiers de Top. et Géom. Diff.* 18-3:291–330, 1977.

[8] M. Bunge, An application of descent to a classification theorem for toposes, *Math. Proc. Camb. Phil. Soc.* 107:59–79, 1990.

[9] M. Bunge, Classifying toposes and fundamental localic groupoids, in R.A.G.Seely, ed., *Category Theory '91, CMS Conference Proceedings* 13:73–96, 1992.

[10] M. Bunge and A. Carboni, The symmetric topos, *J. Pure Appl. Alg.* **105** (1995) 233-249.

[11] M. Bunge and J. Funk, Spreads and the symmetric topos, *J. Pure Appl. Alg.* 113:1–38, 1996.

[12] M. Bunge and J. Funk, Spreads and the symmetric topos II, *J. Pure Appl. Alg.* 130:49–84, 1998.

[13] M. Bunge and J. Funk, On a bicomma object condition for KZ-doctrines, *J. Pure Appl. Alg.* **143** (1999) 69–105.

[14] M. Bunge and S. Lack, Van Kampen theorems for toposes, *Advances in Mathematics* 179-2:291–317, 2003.

[15] M. Bunge and I. Moerdijk, On the construction of the Grothendieck fundamental group of a topos by paths, *J. Pure Appl. Alg.* 116:99–113, 1997.

[16] M. Bunge and R. Paré, Stacks and equivalences of categories, *Cah. de Top. et Géom. Diff.* 20-4:373–399, 1979.

[17] M. Bunge, Stack completions and Morita equivalence for categories in a topos, *Cah. de Top. et Géom. Diff.* 20-4:401–436, 1979.

[18] J. Duskin, An outline of non-abelian cohomology in a topos: (1) The theory of bouquets and gerbes, *Cahiers de Top. et Géo. Diff. Catégoriques* 23:165–191, 1982.

[19] J. Funk and E .D. Tymchatyn, Unramified maps, *JP Jour. Geom. Topol.* 1:249–280, 2001.

[20] A. Grothendieck, *Revetements étales et groupe fondamental (SGA1), Lecture Notes in Mathematics* 224, Springer-Verlag, Berlin-Heidelberg-New York, 1971.

[21] A. Grothendieck, *La longue marche a travers la théorie de Galois, Tome I*, Transcription d'un manuscrit inédit, par J. Malgoire, Université Montpellier II, 7–48, 1995.

[22] A. Grothendieck, Esquisse d'un programme, in: L. Scnepps and P. Lochak, *Geometric Galois Actions. 1. Around Grothendieck's Esquisse d'un Programme London Mathematical Society Lecture Notes Series 242*, Cambridge University Press, 1997.

[23] G. Janelidze, Precategories and Galois theory, in: A. Carboni et al., *Category Theory. proceedings Como 1990* Lecture Notes in Mathematics 1488, 157–173, 1991.

[24] G. Janelidze, Pure Galois theory in categories, *J. Alg.* 132:270–286, 1990.

[25] G. Janelidze, A note on Barr-Diaconescu covering theory, *Contemp. Math* 131:121–124, 1992.

[26] G. Janelidze, D. Schumacher, and R. Street, Galois theory in variable categories, *Applied Categorical Structures* 1:103–110, 1993.

[27] P. T. Johnstone, *Topos Theory*, Academic Press, 1977.

[28] A. Joyal and M. Tierney, An extension of the Galois theory of Grothendieck, *Mem. Amer. Math. Soc.* 309, 1984.

[29] J. Kennison, The fundamental localic groupoid of a topos, *J. Pure Appl. Algebra* 77:67–86, 1992.

[30] F. W. Lawvere, *Extensive and intensive quantities*, Lectures given at the Workshop on Categorical Methods in Geometry, Aarhus, 1983.

[31] I. Moerdijk, Prodiscrete groups and Galois toposes, *Proc. Koninklijke Ned. Akad. van Wetenschappen* 51:219–234, 1989.

[32] I. Moerdijk, Continuous fibrations and inverse limits of toposes, *Compositio Math.* 58:45–72, 1986.

[33] I. Moerdijk, The classifying topos of a continuous groupoid I, *Trans. Amer. Math. Soc.* 310:629-668, 1988.

[34] I. Moerdijk and G. C. Wraith, Connected locally connected toposes are path-connected, *Trans. Amer. Math. Soc.* 295:849–859, 1986.

Received August 28, 2002; in revised form October 21,2003

Fields Institute Communications
Volume **43**, 2004

Galois Corings From the Descent Theory Point of View

S. Caenepeel
Faculty of Applied Sciences
Vrije Universiteit Brussel, VUB
Pleinlaan 2
B-1050 Brussels, Belgium
scaenepe@vub.ac.be

Abstract. We introduce Galois corings, and give a survey of properties that have been obtained so far. The Definition is motivated using descent theory, and we show that classical Galois theory, Hopf-Galois theory and coalgebra Galois theory can be obtained as a special case.

Introduction

Galois descent theory [36] has many applications in several branches of mathematics, such as number theory, commutative algebra and algebraic geometry; to name such one example, it is an essential tool in computing the Brauer group of a field. In the literature, several generalizations have appeared. Galois theory of commutative rings has been studied by Auslander and Goldman [3] and by Chase, Harrison and Rosenberg [16], see also [21]. The group action can be replaced by a Hopf algebra (co)action, leading to Hopf-Galois theory, see [17] (in the case where the Hopf algebra is finitely generated projective), and [24], [35] in the general case. More recently, coalgebra Galois extensions were introduced by Brzeziński and Hajac [10]. It became clear recently that a nice unification of all these theories can be formulated using the language of corings. Let us briefly sketch the history.

During the nineties, several unifications of the various kinds of Hopf modules that had appeared in the literature have been proposed. Doi [22] and Koppinen [30] introduced Doi-Hopf modules. A more general concept, entwined modules, was proposed by Brzeziński and Majid [11]. Böhm introduced Doi-Hopf modules over a weak bialgebra ([6]), and the author and De Groot proposed weak entwined modules [12]. Takeuchi [39] observed that all types of modules can be viewed as comodules over a coring, a concept that was already introduced by Sweedler [38], but then more or less forgotten, at least by Hopf algebra theorists; the idea was further investigated by Brzeziński [9]. He generalized several properties that had been studied in special cases to the situation where one works over a general coring, such

2000 *Mathematics Subject Classification.* Primary 16W30; Secondary 13B05.
Key words and phrases. Coring, Galois extension, descent theory.

as separability and Frobenius type properties, and it turned out that computations sometimes become amazingly simple if one uses the language of corings, indicating that this is really the right way to look at the problem. Brzeziński also introduces the notion of Galois coring: to a ring extension $i: B \to A$, one can associate the so-called canonical coring; a morphism from the canonical coring to another coring $\mathcal{C}$ is determined completely by a grouplike element x; if this morphism is an isomorphism, then we say that $(\mathcal{C}, x)$ is a Galois coring.

The canonical coring leads to an elegant formulation of descent theory: the category of descent data associated to the extension $i: B \to A$ is nothing else then the category of comodules over the canonical coring. This is no surprise: to an A-coring, we can associate a comonad on the category of A-modules, and the canonical coring is exactly the comonad associated to the adjoint pair of functors, given by induction and restriction of scalars. Thus, if a coring is isomorphic to the canonical coring, and if the induction functor is comonadic, then the category of descent data is isomorphic to the category of comodules over this coring, and equivalent to the category of B-modules. This unifies all the versions of descent theory that we mentioned at the beginning of this introduction.

In this paper, we present a survey of properties of Galois corings that have been obtained so far. We have organized it as follows: in Section 1, we recall definition, basic properties and examples of comodules over corings; in Section 2, we explain how descent theory can be formulated using the canonical coring. We included a full proof of Proposition 2.3, which is the noncommutative version of the fact that the induction functor is comonadic if and only if the ring morphism is pure as a map of modules. In Section 3, we introduce Galois corings, and discuss some properties, taken from [9] and [41]. In Section 4, Morita theory is applied to find some equivalent properties for a progenerator coring to be Galois; in fact, Theorem 4.7 is a new result, and generalizes results of Chase and Sweedler [17]. In Section 5, we look at special cases, and we show how to recover the "old" Galois theories. In Section 6, we present a recent generalization, due to El Kaoutit and Gómez Torrecillas [25].

1 Corings

Let A be a ring (with unit). The category of A-bimodules is a braided monoidal category, and an A-coring is by definition a coalgebra in the category of A-bimodules. Thus an A-coring is a triple $\mathcal{C} = (\mathcal{C}, \Delta_{\mathcal{C}}, \varepsilon_{\mathcal{C}})$, where

- $\mathcal{C}$ is an A-bimodule;
- $\Delta_{\mathcal{C}}: \mathcal{C} \to \mathcal{C} \otimes_A \mathcal{C}$ is an A-bimodule map;
- $\varepsilon_{\mathcal{C}}: \mathcal{C} \to A$ is an A-bimodule map

such that

$$(\Delta_{\mathcal{C}} \otimes_A I_{\mathcal{C}}) \circ \Delta_{\mathcal{C}} = (I_{\mathcal{C}} \otimes_A \Delta_{\mathcal{C}}) \circ \Delta_{\mathcal{C}}, \tag{1.1}$$

and

$$(I_{\mathcal{C}} \otimes_A \varepsilon_{\mathcal{C}}) \circ \Delta_{\mathcal{C}} = (\varepsilon_{\mathcal{C}} \otimes_A I_{\mathcal{C}}) \circ \Delta_{\mathcal{C}} = I_{\mathcal{C}}. \tag{1.2}$$

Sometimes corings are considered as coalgebras over noncommutative rings. This point of view is not entirely correct: a coalgebra over a commutative ring k is a k-coring, but not conversely: it could be that the left and and right action of k on the coring are different.

The Sweedler-Heyneman notation is also used for a coring $\mathcal{C}$, namely

$$\Delta_{\mathcal{C}}(c) = c_{(1)} \otimes_A c_{(2)},$$

where the summation is implicitely understood. (1.2) can then be written as

$$\varepsilon_{\mathcal{C}}(c_{(1)})c_{(2)} = c_{(1)}\varepsilon_{\mathcal{C}}(c_{(2)}) = c.$$

This formula looks like the corresponding formula for usual coalgebras. Notice however that the order matters in the above formula, since $\varepsilon_{\mathcal{C}}$ now takes values in A which is noncommutative in general. Even worse, the expression $c_{(2)}\varepsilon_{\mathcal{C}}(c_{(1)})$ makes no sense at all, since we have no well-defined switch map $\mathcal{C} \otimes_A \mathcal{C} \to \mathcal{C} \otimes_A \mathcal{C}$. A morphism between two corings $\mathcal{C}$ and $\mathcal{D}$ is an A-bimodule map $f: \mathcal{C} \to \mathcal{D}$ such that

$$\Delta_{\mathcal{D}}(f(c)) = f(c_{(1)}) \otimes_A f(c_{(2)}) \text{ and } \varepsilon_{\mathcal{D}}(f(c)) = \varepsilon_{\mathcal{C}}(c),$$

for all $c \in \mathcal{C}$. A right $\mathcal{C}$-comodule $M = (M, \rho)$ consists of a right A-module M together with a right A-linear map $\rho: M \to M \otimes_A \mathcal{C}$ such that:

$$(\rho \otimes_A I_{\mathcal{C}}) \circ \rho = (I_M \otimes_A \Delta_{\mathcal{C}}) \circ \rho, \tag{1.3}$$

and

$$(I_M \otimes_A \varepsilon_{\mathcal{C}}) \circ \rho = I_M. \tag{1.4}$$

We then say that $\mathcal{C}$ coacts from the right on M. Left $\mathcal{C}$-comodules and $\mathcal{C}$-bicomodules can be defined in a similar way. We use the Sweedler-Heyneman notation also for comodules:

$$\rho(m) = m_{[0]} \otimes_A m_{[1]}.$$

(1.4) then takes the form $m_{[0]}\varepsilon_{\mathcal{C}}(m_{[1]}) = m$. A right A-linear map $f: M \to N$ between two right $\mathcal{C}$-comodules M and N is called right $\mathcal{C}$-colinear if $\rho(f(m)) = f(m_{[0]}) \otimes m_{[1]}$, for all $m \in M$.

Corings were already considered by Sweedler in [38]. The interest in corings was revived after a mathematical review written by Takeuchi [39], in which he observed that entwined modules can be considered as comodules over a coring. This will be discussed in the examples below.

Example 1.1 As we already mentioned, if A is a commutative ring, then an A-coalgebra is also an A-coring.

Example 1.2 Let $i: B \to A$ be a ring morphism; then $\mathcal{D} = A \otimes_B A$ is an A-coring. We define

$$\Delta_{\mathcal{D}}: \mathcal{D} \to \mathcal{D} \otimes_A \mathcal{D} \cong A \otimes_B A \otimes_B A$$

and

$$\varepsilon_{\mathcal{D}}: \mathcal{D} = A \otimes_B A \to A$$

by

$$\Delta_{\mathcal{D}}(a \otimes_B b) = (a \otimes_B 1_A) \otimes_A (1_A \otimes_B b) \cong a \otimes_B 1_A \otimes_B b$$

and

$$\varepsilon_{\mathcal{D}}(a \otimes_B b) = ab.$$

Then $\mathcal{D} = (\mathcal{D}, \Delta_{\mathcal{D}}, \varepsilon_{\mathcal{D}})$ is an A-coring. It is called the canonical coring associated to the ring morphism i. We will see in the next section that this coring is crucial in descent theory.

Example 1.3 Let k be a commutative ring, G a finite group, and A a G-module algebra. Let $\mathcal{C} = \oplus_{\sigma\in G} Av_\sigma$ be the left free A-module with basis indexed by G, and let $p_\sigma : \mathcal{C} \to A$ be the projection onto the free component Av_σ. We make $\mathcal{C}$ into a right A-module by putting

$$v_\sigma a = \sigma(a)v_\sigma.$$

A comultiplication and counit on $\mathcal{C}$ are defined by putting

$$\Delta_{\mathcal{C}}(av_\sigma) = \sum_{\tau\in G} av_\tau \otimes_A v_{\tau^{-1}\sigma} \text{ and } \varepsilon_{\mathcal{C}} = p_e,$$

where e is the unit element of G. It is straightforward to verify that $\mathcal{C}$ is an A-coring. Notice that, in the case where A is commutative, we have an example of an A-coring, which is not an A-coalgebra, since the left and right A-action on $\mathcal{C}$ do not coincide.

Let us give a description of the right $\mathcal{C}$-comodules. Assume that $M = (M, \rho)$ is a right $\mathcal{C}$-comodule. For every $m \in M$ and $\sigma \in G$, let $\overline{\sigma}(m) = m_\sigma = (I_M \otimes_A p_\sigma)(\rho(m))$. Then we have

$$\rho(m) = \sum_{\sigma\in G} m_\sigma \otimes_A v_\sigma.$$

$\overline{e}$ is the identity, since $m = (I_M \otimes_A \varepsilon_{\mathcal{C}}) \circ \rho(m) = m_e$. Using the coassociativity of the comultiplication, we find

$$\sum_{\sigma\in G} \rho(m_\sigma) \otimes v_\sigma = \sum_{\sigma,\tau\in G} m_\sigma \otimes_A v_\tau \otimes_A v_{\tau^{-1}\sigma} = \sum_{\rho,\tau\in G} m_{\tau\rho} \otimes_A v_\tau \otimes_A v_\rho,$$

hence $\rho(m_\sigma) = \sum_{\tau\in G} m_{\tau\sigma} \otimes_A v_\tau$, and $\overline{\tau}(\overline{\sigma}(m)) = m_{\tau\sigma} = \overline{\tau\sigma}(m)$, so G acts as a group of k-automorphisms on M. Moreover, since ρ is right A-linear, we have that

$$\rho(ma) = \sum_{\sigma\in G} \overline{\sigma}(ma) \otimes_A v_\sigma = \sum_{\sigma\in G} \overline{\sigma}(m) \otimes_A v_\sigma a = \sum_{\sigma\in G} \overline{\sigma}(m)\sigma(a) \otimes_A v_\sigma$$

so $\overline{\sigma}$ is A-semilinear (cf. [29, p. 55]): $\overline{\sigma}(ma) = \overline{\sigma}(m)\sigma(a)$, for all $m \in M$ and $a \in A$. Conversely, if G acts as a group of right A-semilinear automorphims on M, then the formula

$$\rho(m) = \sum_{\sigma\in G} \overline{\sigma}(m) \otimes_A v_\sigma$$

defines a right $\mathcal{C}$-comodule structure on $\mathcal{M}$.

Example 1.4 Now let k be a commutative ring, G an arbitrary group, and A a G-graded k-algebra. Again let $\mathcal{C}$ be the free left A-module with basis indexed by G:

$$\mathcal{C} = \oplus_{\sigma\in G} Au_\sigma$$

Right A-action, comultiplication and counit are now defined by

$$u_\sigma a = \sum_{\tau\in G} a_\tau u_{\sigma\tau} \ ; \ \Delta_{\mathcal{C}}(u_\sigma) = u_\sigma \otimes_A u_\sigma \ ; \ \varepsilon_{\mathcal{C}}(u_\sigma) = 1.$$

$\mathcal{C}$ is an A-coring; let $M = (M, \rho)$ be a right $\mathcal{C}$-comodule, and let $M_\sigma = \{m \in M \mid \rho(m) = m \otimes_A u_\sigma\}$. It is then clear that $M_\sigma \cap M_\tau = \{0\}$ if $\sigma \neq \tau$. For any $m \in M$, we can write in a unique way:

$$\rho(m) = \sum_{\sigma\in G} m_\sigma \otimes_A u_\sigma.$$

Using the coassociativity, we find that $m_\sigma \in M_\sigma$, and using the counit property, we find that $m = \sum_\sigma m_\sigma$. So $M = \oplus_{\sigma\in G} M_\sigma$. Finally, if $m \in M_\sigma$ and $a \in A_\tau$, then it follows from the right A-linearity of ρ that

$$\rho(ma) = (m \otimes_A u_\sigma)a = ma \otimes_A u_{\sigma\tau},$$

so $ma \in M_{\sigma\tau}$, and $M_\sigma A_\tau \subset M_{\sigma\tau}$, and M is a right G-graded A-module. Conversely, every right G-graded A-module can be made into a right $\mathcal{C}$-comodule.

Example 1.5 Let H be a bialgebra over a commutative ring k, and A a right H-comodule algebra. Now take $\mathcal{C} = A \otimes H$, with A-bimodule structure

$$a'(b \otimes h)a = a'ba_{[0]} \otimes ha_{[1]}.$$

Now identify $(A\otimes H)\otimes_A(A\otimes H) \cong A\otimes H\otimes H$, and define the comultiplication and counit on $\mathcal{C}$, by putting $\Delta_{\mathcal{C}} = I_A \otimes \Delta_H$ and $\varepsilon_{\mathcal{C}} = I_A \otimes \varepsilon_H$. Then $\mathcal{C}$ is an A-coring. The category $\mathcal{M}^{\mathcal{C}}$ is isomorphic to the category of relative Hopf modules. These are k-modules M with a right A-action and a right H-coaction ρ, such that

$$\rho(ma) = m_{[0]}a_{[0]} \otimes_A m_{[1]}a_{[1]}$$

for all $m \in M$ and $a \in A$.

Example 1.6 Let k be a commutative ring, A a k-algebra, and C a k-coalgebra, and consider a k-linear map $\psi: C \otimes A \to A \otimes C$. We use the following Sweedler type notation, where the summation is implicitely understood:

$$\psi(c \otimes a) = a_\psi \otimes c^\psi = a_\Psi \otimes c^\Psi.$$

(A, C, ψ) is called a (right-right) entwining structure if the four following conditions are satisfied:

$$(ab)_\psi \otimes c^\psi = a_\psi b_\Psi \otimes c^{\psi\Psi}; \tag{1.5}$$

$$(1_A)_\psi \otimes c^\psi = 1_A \otimes c; \tag{1.6}$$

$$a_\psi \otimes \Delta_C(c^\psi) = a_{\psi\Psi} \otimes c_{(1)}^\Psi \otimes c_{(2)}^\psi; \tag{1.7}$$

$$\varepsilon_C(c^\psi)a_\psi = \varepsilon_C(c)a. \tag{1.8}$$

Let $\mathcal{C} = A \otimes C$ as a k-module, with A-bimodule structure

$$a'(b \otimes c)a = a'ba_\psi \otimes c^\psi.$$

Comultiplication and counit on $A \otimes C$ are defined as in Example 1.5. $\mathcal{C}$ is a coring, and the category $\mathcal{M}^{\mathcal{C}}$ is isomorphic to the category $\mathcal{M}(\psi)_A^C$ of entwined modules. These are k-modules M with a right A-action and a right C-coaction ρ such that

$$\rho(ma) = m_{[0]}a_\psi \otimes_A m_{[1]}^\psi,$$

for all $m \in M$ and $a \in A$.

Actually Examples 1.3, 1.4 and 1.5 are special cases of Example 1.6

- Example 1.3: take $C = (kG)^* = \oplus_{g\in G} kv_\sigma$, the dual of the group ring kG, and $\psi(v_\sigma \otimes a) = \sigma(a) \otimes v_\sigma$;
- Example 1.4: take $C = kG = \oplus_{g\in G} ku_\sigma$, the group ring, and $\psi(u_\sigma \otimes a) = \sum_{\tau\in G} a_\tau \otimes u_{\sigma\tau}$;
- Example 1.5: take $C = H$, and $\psi(h \otimes a) = a_{[0]} \otimes ha_{[1]}$.

If $\mathcal{C}$ is an A-coring, then its left dual ${}^*\mathcal{C} = {}_A\mathrm{Hom}(\mathcal{C}, A)$ is a ring, with (associative) multiplication given by the formula

$$f\#g = g \circ (I_\mathcal{C} \otimes_A f) \circ \Delta_\mathcal{C} \text{ or } (f\#g)(c) = g(c_{(1)}f(c_{(2)})), \tag{1.9}$$

for all left A-linear $f, g : \mathcal{C} \to A$ and $c \in \mathcal{C}$. The unit is $\varepsilon_\mathcal{C}$. We have a ring homomorphism $i : A \to {}^*\mathcal{C}$, $i(a)(c) = \varepsilon_\mathcal{C}(c)a$. We easily compute that

$$(i(a)\#f)(c) = f(ca) \text{ and } (f\#i(a))(c) = f(c)a, \tag{1.10}$$

for all $f \in {}^*\mathcal{C}$, $a \in A$ and $c \in \mathcal{C}$. We have a functor $F : \mathcal{M}^\mathcal{C} \to \mathcal{M}_{{}^*\mathcal{C}}$, where $F(M) = M$ as a right A-module, with right ${}^*\mathcal{C}$-action given by $m \cdot f = m_{[0]}f(m_{[1]})$, for all $m \in M$, $f \in {}^*\mathcal{C}$. If $\mathcal{C}$ is finitely generated and projective as a left A-module, then F is an isomorphism of categories: given a right ${}^*\mathcal{C}$-action on M, we recover the right $\mathcal{C}$-coaction by putting $\rho(m) = \sum_j (m \cdot f_j) \otimes_A c_j$, where $\{(c_j, f_j) \mid j = 1, \cdots, n\}$ is a finite dual basis of $\mathcal{C}$ as a left A-module. ${}^*\mathcal{C}$ is a right A-module, by (1.10): $(f \cdot a)(c) = f(c)a$, and we can consider the double dual $({}^*\mathcal{C})^* = \mathrm{Hom}_A({}^*\mathcal{C}, A)$. We have a canonical morphism $i : \mathcal{C} \to ({}^*\mathcal{C})^*$, $i(c)(f) = f(c)$, and we call $\mathcal{C}$ reflexive (as a left A-module) if i is an isomorphism. If $\mathcal{C}$ is finitely generated projective as a left A-module, then $\mathcal{C}$ is reflexive. For any $\varphi \in ({}^*\mathcal{C})^*$, we then have that $\varphi = i(\sum_j \varphi(f_j)c_j)$.

Corings with a grouplike element. Let $\mathcal{C}$ be an A-coring, and suppose that $\mathcal{C}$ coacts on A. Then we have a map $\rho : A \to A \otimes_A \mathcal{C} \cong \mathcal{C}$. The fact that ρ is right A-linear implies that ρ is completely determined by $\rho(1_A) = x$: $\rho(a) = xa$. The coassociativity of the coaction yields that $\Delta_\mathcal{C}(x) = x \otimes_A x$ and the counit property gives us that $\varepsilon_\mathcal{C}(x) = 1_A$. We say that x is a *grouplike element* of $\mathcal{C}$ and we denote $G(\mathcal{C})$ for the set of all grouplike elements of $\mathcal{C}$. If $x \in G(\mathcal{C})$ is grouplike, then the associated $\mathcal{C}$-coaction on A is given by $\rho(a) = xa$.

If $x \in G(\mathcal{C})$, then we call $(\mathcal{C}, x)$ a *coring with a fixed grouplike element.* For $M \in \mathcal{M}^\mathcal{C}$, we call

$$M^{\mathrm{co}\mathcal{C}} = \{m \in M \mid \rho(m) = m \otimes_A x\}$$

the submodule of coinvariants of M; note that this definition depends on the choice of the grouplike element. Also observe that

$$A^{\mathrm{co}\mathcal{C}} = \{b \in A \mid bx = xb\}$$

is a subring of A. Let $i : B \to A$ be a ring morphism. i factorizes through $A^{\mathrm{co}\mathcal{C}}$ if and only if

$$x \in G(\mathcal{C})^B = \{x \in G(\mathcal{C}) \mid xb = bx, \text{ for all } b \in B\}.$$

We then have two pairs of adjoint functors (F, G) and (F', G'), respectively between the categories $\mathcal{M}_B$ and $\mathcal{M}^\mathcal{C}$ and the categories ${}_B\mathcal{M}$ and ${}^\mathcal{C}\mathcal{M}$. For $N \in \mathcal{M}_B$ and $M \in \mathcal{M}^\mathcal{C}$,

$$F(N) = N \otimes_B A \text{ and } G(M) = M^{\mathrm{co}\mathcal{C}}.$$

The unit and counit of the adjunction are

$$\nu_N : N \to (N \otimes_B A)^{\mathrm{co}\mathcal{C}}, \ \nu_N(n) = n \otimes_B 1;$$

$$\zeta_M : M^{\mathrm{co}\mathcal{C}} \otimes_B A \to M, \ \zeta_M(m \otimes_B a) = ma.$$

The other adjunction is defined in a similar way. We want to discuss when (F, G) and (F', G') are category equivalences. In Section 2, we will do this for the canonical coring associated to a ring morphism; we will study the general case in Section 3.

2 The canonical coring and descent theory

Let $i: B \to A$ be a ring morphism. The problem of descent theory is the following: suppose that we have a right A-module M. When do we have a right B-module N such that $M = N \otimes_B A$? The same problem can be stated for modules with an additional structure, for example algebras. In the case where A and B are commutative, this problem has been discussed in a purely algebraic context in [29]. In fact the results in [29] are the affine versions of Grothendieck's descent theory for schemes, see [28]. In the situation where A and B are arbitrary, descent theory has been discussed by Cipolla [18], and, more recently, by Nuss [33]. For a purely categorical treatment of the problem, making use of monads, we refer to [7]. Here we will show that the results in [29] and [18] can be restated elegantly in terms of comodules over the canonical coring.

Let $\mathcal{D} = A \otimes_B A$ be the canonical coring associated to the ring morphism $i: B \to A$, and let $M = (M, \rho)$ be a right $\mathcal{D}$-comodule. We will identify $M \otimes_A \mathcal{D} \cong M \otimes_B A$ using the natural isomorphism. The coassociativity and the counit property then take the form

$$\rho(m_{[0]}) \otimes m_{[1]} = m_{[0]} \otimes_B 1_A \otimes_B m_{[1]} \text{ and } m_{[0]} m_{[1]} = m.$$

$1_A \otimes_B 1_A$ is a grouplike element of $\mathcal{D}$. As we have seen at the end of Section 1, we have two pairs of adjoint functors, respectively between $\mathcal{M}_B$ and $\mathcal{M}^{\mathcal{D}}$, and ${}_B\mathcal{M}$ and ${}^{\mathcal{D}}\mathcal{M}$, which we will denote by (K, R) and (K', R'). The unit and counit of the adjunction will be denoted by η and ε. K is called the comparison functor. If (K, R) is an equivalence of categories, then the "descent problem" is solved: $M \in \mathcal{M}_A$ is isomorphic to some $N \otimes_B A$ if and only if we can define a right $\mathcal{D}$-coaction on M.

Recall that a morphism of left B-modules $f: M \to M'$ is called pure if and only if $f_N = I_N \otimes_B f: N \otimes_B M \to N \otimes_B M'$ is monic, for every $N \in \mathcal{M}_B$. $i: B \to A$ is pure as a morphism of left B-modules if and only if η_N being injective, for all $N \in \mathcal{M}_B$, since η_N factorizes through i_N.

Proposition 2.1 *The comparison functor K is fully faithful if and only if $i: B \to A$ is pure as a morphism of left B-modules.*

Proof The comparison functor K is fully faithful if and only if η_N is bijective, for all $N \in \mathcal{M}_B$. From the above observation, it follows that it suffices to show that left purity of i implies that η_N is surjective. Since η_N is injective, we have that $N \subset (N \otimes_B A)^{\mathrm{co}\mathcal{D}} \subset N \otimes_B A$. Take $q = \sum_i n_i \otimes_B a_i \in (N \otimes_B A)^{\mathrm{co}\mathcal{D}}$. Then

$$\rho(\sum_i n_i \otimes_B a_i) = \sum_i n_i \otimes_B a_i \otimes_B 1 = \sum_i n_i \otimes_B 1 \otimes_B a_i. \tag{2.1}$$

Consider the right B-module $P = (P \otimes_B A)/N$, and let $\pi: N \otimes_B A \to P$ be the canonical projection. Applying $\pi \otimes_B I_A$ to (2.1), we obtain

$$\pi(q) \otimes_B 1 = \sum_i \pi(n_i \otimes_B 1) \otimes_B a_i = 0 \text{ in } P \otimes_B A,$$

hence $\pi(q) = 0$, since i_P is an injection. This means that $q \in N$, which is exactly what we needed. □

We also have an easy characterization of the fact that R is fully faithful.

Proposition 2.2 *The right adjoint R of the comparison functor K is fully faithful if and only if $\bullet \otimes_B A$ preserves the equalizer of ρ and i_M, for every $(M,\rho) \in \mathcal{M}^{\mathcal{D}}$. In particular, if A is flat as a left B-module, then R is fully faithful.*

Proof $M^{\mathrm{co}\mathcal{D}}$ is the equalizer of the maps

$$0 \longrightarrow M^{\mathrm{co}\mathcal{D}} \xrightarrow{\ j\ } M \underset{i_M}{\overset{\rho}{\rightrightarrows}} M \otimes_B A.$$

First assume that $M^{\mathrm{co}\mathcal{D}} \otimes_B A$ is the equalizer

$$0 \longrightarrow M^{\mathrm{co}\mathcal{D}} \otimes_B A \xrightarrow{j \otimes_B I_A} M \otimes_B A \underset{i_M \otimes_B I_A}{\overset{\rho \otimes_B I_A}{\rightrightarrows}} M \otimes_B A \otimes_B A. \tag{2.2}$$

From the coassociativity of ρ, it now follows that $\rho(m) \in M^{\mathrm{co}\mathcal{D}} \otimes_B A \subset M \otimes_B A \cong M \otimes_A (A \otimes_B A)$, for all $m \in M$, and we have a map $\rho: M \to M^{\mathrm{co}\mathcal{D}} \otimes_B A$. From the counit property, it follows that $\varepsilon_M \circ \rho = I_M$. For $m \in M^{\mathrm{co}\mathcal{D}}$ and $a \in A$, we have

$$\rho(\varepsilon_M(m \otimes_B a)) = \rho(ma) = \rho(m)a = (m \otimes_B 1)a = m \otimes_B a.$$

Thus the counit ε_M has an inverse, for all M, and R is fully faithful.

Conversely, assume that ε_M is bijective. Take $\sum_i m_i \otimes_B a_i \in M^{\mathrm{co}\mathcal{D}} \otimes_B A$, and put $m = \sum_i m_i a_i \in M$. Then $\rho(m) = m_{[0]} \otimes_B m_{[1]} = \sum_i m_i \otimes_B a_i \in M \otimes_B A$. Consequently, if $\sum_i m_i \otimes_B a_i = 0 \in M \otimes_B A$, then $m = m_{[0]}m_{[1]} = 0$, so $\sum_i m_i \otimes_B a_i = 0 \in M^{\mathrm{co}\mathcal{D}} \otimes_B A$, and we have shown that the canonical map $M^{\mathrm{co}\mathcal{D}} \otimes_B A \to M \otimes_B A$ is injective. □

If A and B are commutative, and $i: B \to A$ is pure as a morphism of B-modules, then $\bullet \otimes_B A$ preserves the equalizer of ρ and i_M, for every $(M,\rho) \in \mathcal{M}^{\mathcal{D}}$, and therefore R is fully faithful. This result is due to Joyal and Tierney (unpublished); an elementary proof was given recently by Mesablishvili [32]. We will now adapt Mesablishvili's proof to the noncommutative situation. In view of Proposition 2.1, one would expect that a sufficient condition for the fully faithfulness of R is the fact that i is pure as a morphism of left B-modules. It came as a surprise to the author that we need right purity instead of left purity.

We consider the contravariant functor $C = \mathrm{Hom}_{\mathbb{Z}}(\bullet, \mathbb{Q}/\mathbb{Z}) : \underline{\underline{\mathrm{Ab}}} \to \underline{\underline{\mathrm{Ab}}}$. $\mathbb{Q}/\mathbb{Z}$ is an injective cogenerator of $\underline{\underline{\mathrm{Ab}}}$, and therefore C is exact and reflects isomorphisms. If B is a ring, then C induces functors

$$C: \mathcal{M}_B \to {}_B\mathcal{M} \text{ and } {}_B\mathcal{M} \to \mathcal{M}_B.$$

For example, if $M \in \mathcal{M}_B$, then $C(M)$ is a left B-module, by putting $(b \cdot f)(m) = f(mb)$. For $M \in \mathcal{M}_B$ and $P \in {}_BM$, we have the following isomorphisms, natural in M and P:

$$\mathrm{Hom}_B(M, C(P)) \cong {}_B\mathrm{Hom}(P, C(M)) \cong C(M \otimes_B P) \tag{2.3}$$

If $P \in {}_B\mathcal{M}_B$, then $C(P) \in {}_B\mathcal{M}_B$, and the above isomorphisms are isomorphisms of left B-modules.

Proposition 2.3 *Let $i: B \to A$ be a ring morphism, and assume that i is pure as a morphism of right B-modules. Then the adjoint R of the comparison functor is fully faithful.*

Proof We have to show that (2.2) is exact, for all $(M,\rho) \in \mathcal{M}^{\mathcal{D}}$. If i is pure in $\mathcal{M}_B$, then $i_{C(B)} : B \otimes_B C(B) \to A \otimes_B C(B)$ is a monomorphism in $\mathcal{M}_B$, hence

$$C(i_{C(B)}) : C(A \otimes_B C(B)) \to C(B \otimes_B C(B))$$

is an epimorphism in ${}_B\mathcal{M}$. Applying (2.3), we find that

$$C(i) \circ \bullet : {}_B\mathrm{Hom}(C(B), C(A)) \to {}_B\mathrm{Hom}(C(B), C(B))$$

is also an epimorphism. This implies that $C(i) : C(A) \to C(B)$ is a split epimorphism in ${}_B\mathcal{M}$, and then it follows that for every $M \in \mathcal{M}_B$,

$$C(i) \circ \bullet : \mathrm{Hom}_B(M, C(A)) \to \mathrm{Hom}_B(M, C(B))$$

is a split epimorphism in ${}_B\mathcal{M}$. Applying (2.3) again, we find that

$$C(i_M) : C(M \otimes_B A) \to C(M)$$

is a split epimorphism in ${}_B\mathcal{M}$.
In $\mathcal{M}_B$, we have the following commutative diagram with exact rows:

$$\begin{array}{ccccccc}
0 & \longrightarrow & M^{\mathrm{co}\mathcal{D}} & \xrightarrow{j} & M & \overset{\rho}{\underset{i_M}{\rightrightarrows}} & M \otimes_B A \\
 & & \downarrow{\scriptstyle j} & & \downarrow{\scriptstyle i_M} & & \downarrow{\scriptstyle i_{M\otimes_B A}} \\
0 & \longrightarrow & M & \xrightarrow{\rho} & M \otimes_B A & \overset{\rho\otimes I_A}{\underset{i_M \otimes I_A}{\rightrightarrows}} & M \otimes_B A \otimes_B A
\end{array}$$

Applying the functor C to this diagram, we obtain the following commutative diagram with exact rows. We also know that $C(i_M)$ and $C(i_{M\otimes_B A})$ have right inverses h and h'.

$$\begin{array}{ccccccc}
C(M \otimes_B A \otimes_B A) & \overset{C(\rho\otimes I_A)}{\underset{C(i_M\otimes I_A)}{\rightrightarrows}} & C(M \otimes_B A) & \xrightarrow{C(\rho)} & C(M) & \longrightarrow & 0 \\
{\scriptstyle h'}\uparrow\ \downarrow{\scriptstyle C(i_{M\otimes_B A})} & & {\scriptstyle h}\uparrow\ \downarrow{\scriptstyle C(i_M)} & & {\scriptstyle k}\uparrow\ \downarrow{\scriptstyle C(j)} & & \\
C(M \otimes_B A) & \overset{C(\rho)}{\underset{C(i_M)}{\rightrightarrows}} & C(M) & \xrightarrow{C(j)} & C(M^{\mathrm{co}\mathcal{D}}) & \longrightarrow & 0
\end{array}$$

Diagram chasing leads to the existence of a right inverse k of $C(j)$, such that $k \circ C(j) = C(\rho) \circ h$. But this means that the bottom row of the above diagram is a split fork in ${}_B\mathcal{M}$, split by the morphisms

$$C(M \otimes_B A) \xleftarrow{h} C(M) \xleftarrow{k} C(M^{\mathrm{co}\mathcal{D}})$$

(see [31, p.149] for the definition of a split fork). Split forks are preserved by arbitrary functors, so applying ${}_B\mathrm{Hom}(A, \bullet)$, we obtain a split fork in ${}_B\mathcal{M}$; using (2.3), we find that this split fork is isomorphic to

$$C(M \otimes_B A \otimes_B A) \overset{C(\rho\otimes I_A)}{\underset{C(i_M\otimes I_A)}{\rightrightarrows}} C(M \otimes_B A) \xrightarrow{C(j\otimes I_A)} C(M^{\mathrm{co}\mathcal{D}} \otimes_B A)$$

The functor C is exact and reflects isomorphisms, hence it also reflects coequalizers. It then follows that (2.2) is an equalizer in $\mathcal{M}_B$, as needed. □

The converse of Proposition 2.3 is not true in general: the natural inclusion $i: \mathbb{Z} \to \mathbb{Q}$ is not pure in $\mathcal{M}_{\mathbb{Z}}$, but the functor R is fully faithful. Indeed, if $(M,\rho) \in \mathcal{M}^{\mathcal{D}}$, then M is a $\mathbb{Q}$-vector space, and $\rho: M \to M \otimes_{\mathbb{Z}} \mathbb{Q} \cong M$ is the identity, $M^{\mathrm{co}\mathcal{D}} = M$, and $\varepsilon_M: M^{\mathrm{co}\mathcal{D}} \otimes_{\mathbb{Z}} \mathbb{Q} \cong M \to M$ is also the identity.

It would be interesting to know if there exists a ring morphism $i: B \to A$ which is pure in ${}_B\mathcal{M}$, but not in $\mathcal{M}_B$, and such that (K,R) is an equivalence of categories.

Consider $K' = A \otimes_B \bullet : {}_B\mathcal{M} \to {}^{\mathcal{D}}\mathcal{M}$ and $R' = {}^{\mathrm{co}\mathcal{D}}(\bullet): {}^{\mathcal{D}}\mathcal{M} \to {}_B\mathcal{M}$. The next result is an immediate consequence of Propositions 2.2 and 2.3 and their left handed versions, and can be viewed as the noncommutative version of the Joyal-Tierney Theorem.

Theorem 2.4 *Let $i: B \to A$ be a morphism of rings. Then the following assertions are equivalent.*

1. *(K,R) and (K',R') are equivalences of categories;*
2. *K and K' are fully faithful;*
3. *i is pure in $\mathcal{M}_B$ and ${}_B\mathcal{M}$.*

We have seen in Proposition 2.2 that R is fully faithful if A is flat as a left B-module.

Proposition 2.5 *Let $i: B \to A$ be a morphism of rings, and assume that A is flat as a left B-module. Then (K,R) is an equivalence of categories if and only if A is faithfully flat as a left B-module.*

Proof First assume that A is faithfully flat as a left B-module. It follows from Proposition 2.1 that it suffices to show that A is pure as a left B-module. For $N \in \mathcal{M}_B$, the map

$$f = i_N \otimes_B I_A : N \otimes_B A \to N \otimes_B A \otimes_B A$$

is injective: if $f(\sum_i n_i \otimes_B a_i) = \sum_i n_i \otimes_B 1 \otimes_B a_i = 0$, then, multiplying the second and third tensor factor, we find that $\sum_i n_i \otimes_B a_i = 0$. Since A is faithfully flat as a left B-module, it follows that i_N is injective.

Conversely assume that (K,R) is an equivalence of categories. Then the functor R is exact. Let

$$0 \to N' \to N \to N'' \to 0 \tag{2.4}$$

be a sequence of right B-modules such that the sequence

$$0 \to N' \otimes_B A \to N \otimes_B A \to N'' \otimes_B A \to 0$$

is exact. Applying the functor R to the sequence, and using the fact that η is an isomorphism, we find that (2.4) is exact, so it follows that A is faithfully flat as a left B-module. □

The descent data that are considered in [18] are nothing else then comodules over the canonical coring (although the author of [18] was not aware of this). The descent data in [29] are different, so let us indicate how to go from descent data to comodules over the canonical coring.

Let $i: B \to A$ be a morphism of commutative rings. A *descent datum* consists of a pair (M,g), with $M \in \mathcal{M}_A$, and $g: A \otimes_B M \to M \otimes_B A$ an $A \otimes_B A$-module homorphism such that

$$g_2 = g_3 \circ g_1 : A \otimes_B A \otimes_B M \to A \otimes_B M \otimes_B A \tag{2.5}$$

and

$$\mu_M(g(1 \otimes_B m)) = m, \tag{2.6}$$

for all $m \in M$. Here g_i is obtained by applying I_A to the i-th tensor position, and g to the two other ones. It can be shown that (2.6) can be replaced by the condition that g is a bijection. A morphism of two descent data (M, g) and (M', g') consists of an A-module homomorphism $f: M \to M'$ such that

$$(f \otimes_B I_A) \circ g = g' \circ (I_A \otimes_B f).$$

$\underline{\underline{\mathrm{Desc}}}(A/B)$ will be the category of descent data.

Proposition 2.6 *Let $i: B \to A$ be a morphism of commutative rings. We have an isomorphism of categories*

$$\underline{\underline{\mathrm{Desc}}}(A/B) \cong \mathcal{M}^{A\otimes_B A}$$

Proof (sketch) For a right $\mathcal{D}$-comodule (M, ρ), we define $g: A \otimes_B M \to M \otimes_B A$ by $g(a \otimes_B m) = m_{[0]} a \otimes_B m_{[1]}$. Then (M, g) is a descent datum. Conversely, given a descent datum (M, g), the map $\rho: M \to M \otimes_B A$, $\rho(m) = g(1 \otimes_B m)$ makes M into a right $\mathcal{D}$-comodule. □

3 Galois corings

Let A be a ring, $(\mathcal{C}, x)$ a coring with a fixed grouplike element, and $i: B \to A^{\mathrm{co}\mathcal{C}}$ a ring morphism. We have seen at the end of Section 1 that we have two pairs of adjoint functors (F, G) and (F', G'). We also have a morphism of corings

$$\mathrm{can}: \mathcal{D} = A \otimes_B A \to \mathcal{C}, \ \mathrm{can}(a \otimes_B a') = axa'.$$

Proposition 3.1 *With notation as above, we have the following results.*

1. *If F is fully faithful, then $i: B \to A^{\mathrm{co}\mathcal{C}}$ is an isomorphism;*
2. *if G is fully faithful, then* can $: \mathcal{D} = A \otimes_B A \to \mathcal{C}$ *is an isomorphism.*

Proof 1. F is fully faithful if and only if ν is an isomorphism; it then suffices to observe that $i = \nu_B$.
2. G is fully faithful if and only if ζ is an isomorphism. $\mathcal{C} \in \mathcal{M}^{\mathcal{C}}$, the right coaction being induced by the comultiplication. The map $f: A \to \mathcal{C}^{\mathrm{co}\mathcal{C}}$, $f(a) = ax$, is an isomorphism of (A, B)-bimodules; the inverse of f is the restriction of $\varepsilon_{\mathcal{C}}$ to $\mathcal{C}^{\mathrm{co}\mathcal{C}}$. Indeed, if $c \in \mathcal{C}^{\mathrm{co}\mathcal{C}}$, then $\Delta_{\mathcal{C}}(c) = c \otimes_A x$, hence $c = \varepsilon(c)x = f(\varepsilon(c))$. It follows that $\mathrm{can} = \zeta_{\mathcal{C}} \circ (f \otimes_B I_A)$ is an isomorphism. □

Proposition 3.1 leads us to the following Definition.

Definition 3.2 Let $(\mathcal{C}, x)$ be an A-coring with a fixed grouplike, and let $B = A^{\mathrm{co}\mathcal{C}}$. We call $(\mathcal{C}, x)$ a Galois coring if the canonical coring morphism can : $\mathcal{D} = A \otimes_B A \to \mathcal{C}$, $\mathrm{can}(a \otimes_B b) = axb$ is an isomorphism.

Let $i: B \to A$ be a ring morphism. If $x \in G(\mathcal{C})^B$, then we can define a functor

$$\Gamma: \mathcal{M}^{\mathcal{D}} \to \mathcal{M}^{\mathcal{C}}, \ \Gamma(M, \rho) = (M, \widetilde{\rho})$$

with $\widetilde{\rho}(m) = m_{[0]} \otimes_A x m_{[1]} \in M \otimes_A \mathcal{C}$ if $\rho(m) = m_{[0]} \otimes_B m_{[1]} \in M \otimes_B A$. It is easy to see that $\Gamma \circ K = F$, and therefore we have the following result.

Proposition 3.3 *Let $(\mathcal{C}, x)$ be a Galois A-coring. Then Γ is an isomorphism of categories. Consequently R (resp. K) is fully faithful if and only if G (resp. F) is fully faithful.*

Let us now give some alternative characterizations of Galois corings; for the proof, we refer to [41, 3.6].

Proposition 3.4 *Let $(\mathcal{C}, x)$ be an A-coring with fixed grouplike element, and $B = A^{\mathrm{co}\mathcal{C}}$. The following assertions are equivalent.*

1. *$(\mathcal{C}, x)$ is Galois;*
2. *if $(M, \rho) \in \mathcal{M}^{\mathcal{C}}$ is such that $\rho: M \to M \otimes_A \mathcal{C}$ is a coretraction, then the evaluation map*
$$\varphi_M : \mathrm{Hom}^{\mathcal{C}}(A, M) \otimes_B A \to M, \ \varphi_M(f \otimes_B m) = f(m)$$
is an isomorphism;
3. *$\varphi_{\mathcal{C}}$ is an isomorphism.*

From Theorem 2.4 and Proposition 3.3, we immediately obtain the following result.

Theorem 3.5 *Let $(\mathcal{C}, x)$ be a Galois A-coring, and put $B = A^{\mathrm{co}\mathcal{C}}$. Then the following assertions are equivalent.*

1. *(F, G) and (F', G') are equivalences of categories;*
2. *the functors F and F' are fully faithful;*
3. *$i: B \to A$ is pure in ${}_B\mathcal{M}$ and $\mathcal{M}_B$.*

Remark 3.6 Let us make some remarks about terminology. In the literature, there is an inconsistency in the use of the term "Galois". An alternative definition is to require that $(\mathcal{C}, x)$ satisfies the equivalent definitions of Theorem 3.5, so that (F, G) and (F', G') are category equivalences. In Section 5, we will discuss special cases that have appeared in the literature before. In some cases, there is an agreement with Definition 3.2 (see e.g. [8], [35]), while in other cases, category equivalence is required (see e.g. [17], [21]).

In the particular situation where $\mathcal{C} = A \otimes H$, as in Example 1.5, the property that (F, G) is an equivalence (resp. G is fully faithful) is called the Strong (resp. Weak) Structure Theorem (see [24]). Let $i: B \to A$ be a ring morphism, and $\mathcal{D}$ the canonical coring. In the situation where A and B are commutative, i is called a descent morphism (resp. an effective descent morphism) if K is fully faithful (resp. (K, R) is an equivalence). In the general situation, (K, R) is an equivalence if and only if the functor $\bullet \otimes_B A: \mathcal{M}_B \to \mathcal{M}_A$ is comonadic (see e.g. [7, Ch. 4]).

Let us next look at the case where A is flat as a left B-module. Wisbauer [41] calls the following two results the *Galois coring Structure Theorem*.

Proposition 3.7 *Let $(\mathcal{C}, x)$ be an A-coring with fixed grouplike element, and $B = A^{\mathrm{co}\mathcal{C}}$. Then the following statements are equivalent.*

1. *$(\mathcal{C}, x)$ is Galois and A is flat as a left B-module;*
2. *G is fully faithful and A is flat as a left B-module;*
3. *$\mathcal{C}$ is flat as a left A-module, and A is a generator in $\mathcal{M}^{\mathcal{C}}$.*

Proof 1) $\Rightarrow$ 2) follows from Propositions 2.2 and 3.3. 2) $\Rightarrow$ 1) follows from Proposition 3.1. For the proof of 1) $\Leftrightarrow$ 3), we refer to [41, 3.8]. □

Proposition 3.8 *Let $(\mathcal{C}, x)$ be an A-coring with fixed grouplike element, and $B = A^{\mathrm{co}\mathcal{C}}$. Then the following statements are equivalent.*

1. *$(\mathcal{C}, x)$ is Galois and A is faithfully flat as a left B-module;*
2. *(F, G) is an equivalence and A is flat as a left B-module;*

3. *$\mathcal{C}$ is flat as a left A-module, and A is a projective generator in $\mathcal{M}^{\mathcal{C}}$.*

Proof The equivalence of 1) and 2) follows from Propositions 2.5 and 3.3. For the remaining equivalence, we refer to [41]. □

A right $\mathcal{C}$-comodule N is called semisimple (resp. simple) in $\mathcal{M}^{\mathcal{C}}$ if every $\mathcal{C}$-monomorphism $U \to N$ is a coretraction (resp. an isomorphism). Similar definitions apply to left $\mathcal{C}$-comodules and $(\mathcal{C},\mathcal{C})$-bicomodules. $\mathcal{C}$ is said to be right (left) semisimple if it is semisimple as a right (left) $\mathcal{C}$-comodule. $\mathcal{C}$ is called a simple coring if it is simple as a $(\mathcal{C},\mathcal{C})$-bicomodule. For the proof of the following result, we refer to [26].

Proposition 3.9 *For an A-coring $\mathcal{C}$, the following assertions are equivalent:*

1. *$\mathcal{C}$ is right semisimple;*
2. *$\mathcal{C}$ is projective as a left A-module and $\mathcal{C}$ is semisimple as a left ${}^*\mathcal{C}$-module;*
3. *$\mathcal{C}$ is projective as a right A-module and $\mathcal{C}$ is semisimple as a right $\mathcal{C}^*$-module;*
4. *$\mathcal{C}$ is left semisimple.*

The connection to Galois corings is the following, due to Wisbauer [41, 3.12], and to El Kaoutit, Goméz Torrecillas and Lobillo [26, Theorem 4.4]:

Proposition 3.10 *For an A-coring with a fixed grouplike element $(\mathcal{C}, x)$, the following assertions are equivalent:*

1. *$\mathcal{C}$ is a simple and left (or right) semisimple coring;*
2. *$(\mathcal{C}, x)$ is Galois and $\mathrm{End}^{\mathcal{C}}(A)$ is simple and left semisimple;*
3. *$(\mathcal{C}, x)$ is Galois and B is a simple left semisimple subring of A;*
4. *$\mathcal{C}$ is flat as a right A-module, A is a projective generator in ${}^{\mathcal{C}}\mathcal{M}$, and $\mathrm{End}^{\mathcal{C}}(A)$ is simple and left semisimple (the left $\mathcal{C}$-coaction on A being given by $\rho^l(a) = ax$).*

4 Galois corings and Morita theory

Let $(\mathcal{C}, x)$ be a coring with a fixed grouplike element, $B = A^{\mathrm{co}\mathcal{C}}$, and $\mathcal{D} = A \otimes_B A$. We can consider the left dual of the map can:

$$^*\mathrm{can}:\ {}^*\mathcal{C} \to {}^*\mathcal{D} \cong {}_B\mathrm{End}(A)^{\mathrm{op}},\ \ {}^*\mathrm{can}(f)(a) = f(xa).$$

The following result is obvious.

Proposition 4.1 *If $(\mathcal{C}, x)$ is Galois, then ${}^*\mathrm{can}$ is an isomorphism. The converse property holds if $\mathcal{C}$ and A are finitely generated projective, respectively as a left A-module, and a left B-module.*

Let $Q = \{q \in {}^*\mathcal{C} \mid c_{(1)}q(c_{(2)}) = q(c)x,\ \text{for all}\ c \in \mathcal{C}\}$. A straightforward computation shows that Q is a $({}^*\mathcal{C}, B)$-bimodule. Also A is a left $(B, {}^*\mathcal{C})$-bimodule; the right ${}^*\mathcal{C}$-action is induced by the right $\mathcal{C}$-coaction: $a \cdot f = f(xa)$. Now consider the maps

$$\tau:\ A \otimes_{^*\mathcal{C}} Q \to B,\ \ \tau(a \otimes_{^*\mathcal{C}} q) = q(xa);$$

$$\mu:\ Q \otimes_B A \to {}^*\mathcal{C},\ \ \mu(q \otimes_B a) = q\#i(a).$$

With this notation, we have the following property (see [14]).

Proposition 4.2 *$(B, {}^*\mathcal{C}, A, Q, \tau, \mu)$ is a Morita context.*

Properties of this Morita context are studied in [1], [2], [14] and [15]. It generalizes (and unifies) Morita contexts discussed in [5], [17], [19], [20] and [23]. We recall the following properties from [14] and [15].

Proposition 4.3 [14, Th. 3.3 and Cor. 3.4] *If τ is surjective, then $M^{\text{co}\mathcal{C}} = M^{^*\mathcal{C}} = \{m \in M \mid m \cdot f = mf(x),$ for all $f \in {}^*\mathcal{C}\}$, for all $M \in \mathcal{M}^{\mathcal{C}}$.*
The following assertions are equivalent:

1. *τ is surjective;*
2. *there exists $q \in Q$ such that $q(x) = 1$;*
3. *for every $M \in \mathcal{M}^{^*\mathcal{C}}$, the map*

$$\omega_M : M \otimes_{^*\mathcal{C}} Q \to M^{^*\mathcal{C}}, \ \omega_M(m \otimes_{^*\mathcal{C}} q) = m \cdot q$$

   is bijective.

Proposition 4.4 [15], [14, Th. 3.5] *The following assertions are equivalent:*

1. *μ is surjective;*
2. *$\mathcal{C}$ is finitely generated and projective as a left A-module and G is fully faithful.*

As an application of Proposition 4.3, we have the following result.

Corollary 4.5 *Assume that $\mathcal{C}$ is finitely generated projective as a left A-module. Consider the adjoint pair $(F = \bullet \otimes_B A, G = (\bullet)^{\text{co}\mathcal{C}})$, and the functors $\widetilde{F} = \bullet \otimes_B A$ and $\widetilde{G} = \bullet \otimes_{^*\mathcal{C}} Q$ coming from the Morita context of Proposition 4.2. Then $F \cong \widetilde{F}$ and $G \cong \widetilde{G}$ if τ is surjective.*

Proof Take $N \in \mathcal{M}_B$. $F(N)$ corresponds to $\widetilde{F}$ under the isomorphism $\mathcal{M}^{\mathcal{C}} \cong \mathcal{M}_{^*\mathcal{C}}$. If τ is surjective, then it follows from Proposition 4.3 that $\omega : \widetilde{G} \to G$ is bijective. □

Let us now compute the Morita context associated to the canonical coring.

Proposition 4.6 *Let $i : B \to A$ be a ring morphism, and assume that i is pure as a morphism of left B-modules. Then the Morita context associated to the canonical coring $(\mathcal{D} = A \otimes_B A, 1 \otimes_B 1)$ is the Morita context $(B, {}_B\text{End}(A)^{\text{op}}, A, {}_B\text{Hom}(A, B), \varphi, \psi)$ associated to A as a left B-module (see [4, II.4]).*

Proof From Proposition 2.1, it follows that

$$A^{\text{co}\mathcal{D}} = \{b \in A \mid b \otimes_B 1 = 1 \otimes_B b\} = B.$$

Take $q \in Q \subset {}_A\text{Hom}(A \otimes_B A, A)$ and the corresponding $\widetilde{q} \in {}_B\text{Hom}(A, A)$, given by $\tilde{q}(a) = q(1 \otimes_B a)$. Then

$$q(a' \otimes_B a)(1 \otimes_B 1) = (a' \otimes_B 1)q(1 \otimes_B a).$$

Taking $a' = 1$, we find

$$\tilde{q}(a) \otimes_B 1 = 1 \otimes_B \tilde{q}(a)$$

hence $\tilde{q}(a) \in B$, and it follows that $Q \subset {}_B\text{Hom}(A, B)$. The converse inclusion is proved in a similar way. A straightforward verification shows that $\varphi = \tau$ and $\psi = \mu$. □

Recall that the context associated to the left A-module B is strict if and only if A is a left B-progenerator. We are now ready to prove the following result. In Section 5, we will see that it is a generalization of [17, Th. 9.3 and 9.6].

Theorem 4.7 *Let $(\mathcal{C}, x)$ be a coring with fixed grouplike element, and assume that $\mathcal{C}$ is a left A-progenerator. We take a subring B' of $B = A^{\mathrm{co}\mathcal{C}}$, and consider the map*

$$\mathrm{can}': \mathcal{D}' = A \otimes_{B'} A \to \mathcal{C}, \ \mathrm{can}'(a \otimes_{B'} a') = axa'$$

Then the following statements are equivalent:

1. • can' *is an isomorphism;*
   • A *is faithfully flat as a left B'-module.*
2. • ${}^*\mathrm{can}'$ *is an isomorphism;*
   • A *is a left B'-progenerator.*
3. • $B = B'$;
   • *the Morita context $(B, {}^*\mathcal{C}, A, Q, \tau, \mu)$ is strict.*
4. • $B = B'$;
   • (F, G) *is an equivalence of categories.*

Proof 1) $\Leftrightarrow$ 2). Obviously ${}^*\mathrm{can}'$ is an isomorphism if can' is an isomorphism, and the converse holds if $\mathcal{C}$ is a left A-progenerator and A is a left B'-progenerator. If can' is an isomorphism, then $A \otimes_{B'} A = \mathcal{D}' \cong \mathcal{C}$ is a left A-progenerator, hence A is a left B'-progenerator.
1) $\Rightarrow$ 3). Since A is faithfully flat as a left B'-module, $A^{\mathrm{co}\mathcal{D}'} = B'$. Since can' is an isomorphism, it follows that $B = A^{\mathrm{co}\mathcal{C}} = A^{\mathrm{co}\mathcal{D}'} = B'$. Then $\mathrm{can} = \mathrm{can}'$ is an isomorphism, hence the Morita contexts associated to $(\mathcal{C}, x)$ and $(\mathcal{D}, 1 \otimes_B 1)$ are isomorphic. From the equivalence of 1) and 2), we know that A is a left B-progenerator, so the context associated to $(\mathcal{D}, 1 \otimes_B 1)$ is strict, see the remark preceding Theorem 4.7. Therefore the Morita context $(B, {}^*\mathcal{C}, A, Q, \tau, \mu)$ associated to $(\mathcal{C}, x)$ is also strict.
3) $\Rightarrow$ 1). and 3) $\Rightarrow$ 4). If $(B, {}^*\mathcal{C}, A, Q, \tau, \mu)$ is strict, then A is a left B-progenerator, and a fortiori faithfully flat as a left B-module. τ is surjective, so it follows from Corollary 4.5 that $F \cong \widetilde{F}$ and $G \cong \widetilde{G}$. $(\widetilde{F}, \widetilde{G})$ is an equivalence, so (F, G) is also an equivalence. Then $(\mathcal{C}, x)$ is Galois by Proposition 3.1.
4) $\Rightarrow$ 1). can is an isomorphism, by Proposition 3.1, and we have already seen that this implies that A is a left B-progenerator, so A is faithfully flat as a left B-module. □

5 Application to particular cases

5.1 Coalgebra Galois extensions. From [10], we recall the following Definition.

Definition 5.1 Let $i: B \to A$ be a morphism of k-algebras, and C a k-coalgebra. A is called a C-Galois extension of B if the following conditions hold:

1. A is a right C-comodule;
2. $\mathrm{can}: A \otimes_B A \to A \otimes C$, $\mathrm{can}(a \otimes_B a') = aa'_{[0]} \otimes_B a'_{[1]}$ is an isomorphism;
3. $B = \{a \in A \mid \rho(a) = a\rho(1)\}$.

Proposition 5.2 *Let $i: B \to A$ be a morphism of k-algebras, and C a k-coalgebra. A is called a C-Galois extension of B if and only if there exists a right-right entwining structure (A, C, ψ) and $x \in G(A \otimes C)$ such that $A^{\mathrm{co}A\otimes C} = B$ and $(A \otimes C, x)$ is a Galois coring.*

Proof Let (A, C, ψ) be an entwining structure. We have seen in Example 1.6 that $\mathcal{C} = A \otimes C$ is a coring. Given a grouplike element $x = \sum_i a_i \otimes c_i$, we have a right $\mathcal{C}$-coaction on A, hence A is an entwined module (see Example 1.6), and therefore a C-comodule. The C-coaction is given by the formula

$$\rho(a) = a_{[0]} \otimes a_{[1]} = \sum_i a_i a_\psi \otimes c_i^\psi.$$

Then the conditions of Definition 5.1 are satisfied, and A is C-Galois extension of B.

Conversely, let A be a C-Galois extension of B. can is bijective, so the coring structure on $A \otimes_B A$ induces a coring structure on $A \otimes C$. We will show that this coring structure comes from an entwining structure (A, C, ψ).

It is clear that the natural left A-module structure on $A \otimes C$ makes can into a left A-linear map. The right A-module structure on $A \otimes C$ induced by can is given by

$$(b \otimes c)a = \mathrm{can}(\mathrm{can}^{-1}(b \otimes c)a).$$

Since $\mathrm{can}^{-1}(1_{[0]} \otimes 1_{[1]}) = 1 \otimes_B 1$, we have

$$(1_{[0]} \otimes 1_{[1]})a = \mathrm{can}(1 \otimes a) = a_{[0]} \otimes a_{[1]}. \tag{5.1}$$

The comultiplication Δ on $A \otimes C$ is given by

$$\Delta(a \otimes c) = (\mathrm{can} \otimes_A \mathrm{can})\Delta_{\mathcal{C}}(\mathrm{can}^{-1}(a \otimes c)) \in (A \otimes C) \otimes_A (A \otimes C),$$

for all $a \in A$ and $c \in C$. can is bijective, so we can find $a_i, b_i \in A$ such that

$$\mathrm{can}(\sum_i a_i \otimes_B b_i) = \sum_i a_i b_{i[0]} \otimes_B b_{i[1]} = a \otimes c,$$

and we compute that

$$\begin{aligned}
\Delta(a \otimes c) &= (\mathrm{can} \otimes_A \mathrm{can})\Delta_{\mathcal{C}}(\sum_i a_i \otimes_B b_i) \\
&= \sum_i \mathrm{can}(a_i \otimes_B 1) \otimes_A \mathrm{can}(1 \otimes_B b_i) \\
&= \sum_i (a_i 1_{[0]} \otimes 1_{[1]}) \otimes_A (b_{i[0]} \otimes b_{i[1]}) \\
&= \sum_i (a_i 1_{[0]} \otimes 1_{[1]}) b_{i[0]} \otimes_A (1 \otimes b_{i[1]}) \\
\stackrel{(5.1)}{=}& \sum_i (a_i b_{i[0]} \otimes b_{i[1]}) \otimes_A (1 \otimes b_{i[2]}) \\
&= (a \otimes c_{(1)}) \otimes_A (1 \otimes c_{(2)}).
\end{aligned}$$

Finally

$$\begin{aligned}
\varepsilon_{\mathcal{C}}(a \otimes c) &= \varepsilon_{\mathcal{C}}(\sum_i a_i \otimes b_i) = \sum_i a_i b_i \\
&= \sum_i a_i b_{i[0]} \varepsilon_C(b_{i[1]}) = a\varepsilon_C(c).
\end{aligned}$$

Now define $\psi : C \otimes A \to A \otimes C$ by $\psi(c \otimes a) = (1_A \otimes c)a$. It follows from [9] that (A, C, ψ) is an entwining structure. $\square$

Let (A, C, ψ) be an entwining structure, and consider $g \in C$ grouplike. Then $x = 1_A \otimes g$ is a grouplike element of $A \otimes C$. Let us first describe the Morita context from the previous Section.

Observe that ${}^*\mathcal{C} = {}_A\mathrm{Hom}(A \otimes C, A) \cong \mathrm{Hom}(C, A)$ as a k-module. The ring structure on ${}^*\mathcal{C}$ induces a k-algebra structure on $\mathrm{Hom}(C, A)$, and this k-algebra is denoted $\#(C, A)$. The product is given by the formula

$$(f\#g)(c) = f(c_{(2)})_\psi g(c_{(1)}^\psi). \tag{5.2}$$

We have a natural algebra homomorphism $i: A \to \#(C, A)$, $i(a)(c) = \varepsilon_C(c)a$, and we have, for all $a \in A$ and $f: C \to A$:

$$(i(a)\#f)(c) = a_\psi f(c^\psi) \text{ and } (f\#i(a))(c) = f(c)a. \tag{5.3}$$

$\mathrm{Hom}(C, A)$ will denote the k-algebra with the usual convolution product, that is

$$(f * g)(c) = f(c_{(1)})g(c_{(2)}). \tag{5.4}$$

The ring of coinvariants is

$$B = A^{\mathrm{co}C} = \{b \in A \mid b_\psi \otimes g^\psi = b \otimes g\}, \tag{5.5}$$

and the bimodule Q is naturally isomorphic to

$$Q = \{q \in \#(C, A) \mid q(c_{(2)})_\psi \otimes c_{(1)}^\psi = q(c) \otimes g\}.$$

We have maps

$$\mu: Q \otimes_{B'} A \to \#(C, A),\ \mu(q \otimes_B a)(c) = q(c)a,$$

$$\tau: A \otimes_{\#(C,A)} Q \to B,\ \tau(a \otimes q) = a_\psi q(x^\psi),$$

and $(B, \#(C, A), A, Q, \tau, \mu)$ is a Morita context.

Proposition 5.3 [14, Prop. 4.3] *Assume that $\lambda: C \to A$ is convolution invertible, with convolution inverse λ^{-1}. Then the following assertions are equivalent:*

1. *$\lambda \in Q$;*
2. *for all $c \in C$, we have*

$$\lambda^{-1}(c_{(1)})\lambda(c_{(3)})_\psi \otimes c_{(2)}^\psi = \varepsilon(c)1_A \otimes g; \tag{5.6}$$

3. *for all $c \in C$, we have*

$$\lambda^{-1}(c_{(1)}) \otimes c_{(2)} = \lambda^{-1}(c)_\psi \otimes g^\psi. \tag{5.7}$$

Notice that condition 3) means that λ^{-1} is right C-colinear. If such a $\lambda \in Q$ exists, then we call (A, C, ψ, g) cleft.

Proposition 5.4 [14, Prop. 4.4] *Assume that (A, C, ψ, g) is a cleft entwining structure. Then the map τ in the associated Morita context is surjective.*

We say that the entwining structure (A, C, ψ, g) satisfies the *right normal basis property* if there exists a left B-linear and right C-colinear isomorphism $B \otimes C \to A$. The following is one of the main results in [14]. As before, we consider the functor $F = \bullet \otimes_B A: \mathcal{M}_B \to \mathcal{M}(\psi)_A^C$ and its right adjoint $G = (\bullet)^{\mathrm{co}C}$.

Theorem 5.5 [14, Theorem 4.5] *Let (A, C, ψ, g) be an entwining structure with a fixed grouplike element. The following assertions are equivalent:*

1. *(A, C, ψ, g) is cleft;*
2. *(F, G) is a category equivalence and (A, C, ψ, g) satisfies the right normal basis property;*

3. *(A, C, ψ, g) is Galois, and satisfies the right normal basis property;*
4. *the map ${}^*\mathrm{can}:\ \#(C,A) \to \mathrm{End}_B(A)^{\mathrm{op}}$ is bijective and (A, C, ψ, g) satisfies the right normal basis property.*

5.2 Hopf-Galois extensions. Let H be a Hopf algebra over a commutative ring k with bijective antipode, and A a right H-comodule algebra (cf. Example 1.5). Then $\mathcal{C} = A \otimes H$ is an A-coring, and $1_A \otimes 1_H \in G(\mathcal{C})$. Let $B = A^{\mathrm{co}H}$. The canonical map is now the following:

$$\mathrm{can}:\ A \otimes_B A \to A \otimes H,\ \mathrm{can}(a' \otimes_B a) = a' a_{[0]} \otimes a_{[1]}$$

Definition 5.6 (see e.g. [24, Def. 1.1]) A is a Hopf-Galois extension of B if and only if can is an isomorphism.

Obviously A is a Hopf-Galois extension of B if and only if $(A \otimes H, 1_A \otimes 1_H)$ is a Galois coring.

Assume now that H is a progenerator as a k-module, i.e. H is finitely generated, faithful, and projective as a k-module. Then $A \otimes H$ is a left A-progenerator, so we can apply the results of Section 4. We will show that we recover results from [17]. To this end, we will describe the Morita context associated to $(A \otimes H, 1_A \otimes 1_H)$. First we compute ${}^*\mathcal{C}$. We have already seen in Section 5.1 that ${}^*\mathcal{C} \cong \#(H, A)$. As a module, $\mathrm{Hom}(H, A) \cong H^* \otimes A$, since H is finitely generated and projective. The multiplication on $\#(H, A)$ can be transported into a multiplication on $H^* \otimes A$, giving us a k-algebra denoted by $H^* \# A$. A straightforward computation shows that this multiplication is given by the following formula. H^* is a coalgebra, since H is finitely generated projective, and H^* acts on A from the left: $h^* \rightharpoonup a = \langle h^*, a_{[1]} \rangle a_{[0]}$. Then we can compute that

$$(h^* \# a)(k^* \# b) = (k^*_{(1)} * h^*) \# (k^*_{(2)} \rightharpoonup a_{[0]}) b. \tag{5.8}$$

Consider the map $\mathrm{can}':\ A \otimes A \to A \otimes H$; its dual ${}^*\mathrm{can}':\ H^* \# A \to \mathrm{End}(A)^{\mathrm{op}}$ is given by

$$^*\mathrm{can}'(h^* \# a)(b) = (h^* \rightharpoonup b) a. \tag{5.9}$$

Take $y = \sum_i h_i^* \# a_i \in H^* \# A$. $y \in Q$ if and only if

$$\sum_i \langle h_i^*, h_{(2)} \rangle a_{i[0]} \otimes h_{(1)} a_{i[1]} = \sum_i \langle h_i^*, h \rangle a_i \otimes 1,$$

for $h \in H$. Since H is finitely generated and projective, this is also equivalent to

$$\sum_i \langle h_i^*, h_{(2)} \rangle a_{i[0]} \langle h^*, h_{(1)} a_{i[1]} \rangle = \sum_i \langle h_i^*, h \rangle a_i \langle h^*, 1 \rangle,$$

for all $h \in H$ and $h^* \in H^*$. The left hand side equals

$$\sum_i \langle h^*_{(1)} * h_i^*, h \rangle \langle h^*_{(2)}, a_{i[1]} \rangle a_{i[0]},$$

so we find that $y \in Q$ if and only if

$$\sum_i (h^*_{(1)} * h_i^*) \# (h^*_{(2)} \rightharpoonup a_i) = \langle h^*, 1 \rangle \sum_i h_i^* \# a_i,$$

or

$$y(h^* \# 1) = \langle h^*, 1 \rangle y.$$

for all $h^* \in H^*$. Thus

$$Q = \{y \in H^*\#A \mid y(h^*\#1) = \langle h^*, 1\rangle y, \text{ for all } h^* \in H^*\}. \tag{5.10}$$

Elementary computations show that the maps μ and τ from the Morita context are the following:

$$\tau: A \otimes_{H^*\#A} Q \to B, \ \tau(a \otimes y) = {}^*\mathrm{can}(y)(a);$$

$$\mu: Q \otimes_B A \to H^*\#A, \ \mu(y \otimes a) = y(\varepsilon_C\#a),$$

where $B = A^{\mathrm{co}H}$, as usual. Theorem 4.7 now takes the following form.

Proposition 5.7 *Let H be a k-progenerator Hopf algebra over a commutative ring k, and A a right H-comodule algebra. Then the following statements are equivalent (with notation as above):*

1. • $\mathrm{can}': A \otimes A \to A \otimes H$, $\mathrm{can}'(a' \otimes a) = a'a_{[0]} \otimes a_{[1]}$ *is bijective;*
   • *A is faithfully flat as a k-module.*
2. • ${}^*\mathrm{can}': H^*\#A \to \mathrm{End}(A)^{\mathrm{op}}$, ${}^*\mathrm{can}(h^*\#a)(b) = (h^*\rightharpoonup b)a$ *is an isomorphism;*
   • *A is a k-progenerator.*
3. • $A^{\mathrm{co}H} = k$;
   • *the Morita context $(k, H^*\#A, A, Q, \tau, \mu)$ is strict.*
4. • $A^{\mathrm{co}H} = k$;
   • *the adjoint pair of functors $(F = \bullet \otimes A, G = (\bullet)^{\mathrm{co}H})$ is an equivalence between the categories $\mathcal{M}_k$ and $\mathcal{M}_A^H$.*

[17, Theorems 9.3 and 9.6] follow from Proposition 5.7.

5.3 Classical Galois Theory. As in Example 1.3, let G be a finite group, and A a G-module algebra. We have seen that $\mathcal{C} = A \otimes (kG)^* = \oplus_{\sigma\in G} Av_\sigma$ is an A-coring. $\sum_\sigma v_\sigma$ is a grouplike element. Since $(kG)^*$ is finitely generated and projective, we can apply Proposition 5.7. We have

$$\mathrm{can}': A \otimes A \to \oplus_{\sigma\in G} Av_\sigma, \ \mathrm{can}'(a \otimes b) = \sum_\sigma a\sigma(b)v_\sigma.$$

$${}^*\mathcal{C} = \oplus_\sigma u_\sigma A,$$

with multiplication

$$(u_\sigma a)(u_\tau b) = u_{\tau\sigma}\tau(a)b,$$

and

$${}^*\mathrm{can}': \oplus_\sigma u_\sigma A \to \mathrm{End}(A)^{\mathrm{op}}, \ {}^*\mathrm{can}'(u_\sigma a)(b) = \sigma(b)a.$$

We also have

$$Q = \{\sum_\sigma u_\sigma \sigma(a) \mid a \in A\} \cong A,$$

which is not surprising since $(kG)^*$ is a Frobenius Hopf algebra (see [14] and [20]). If $A^G = k$, then we have a Morita context $(k, {}^*\mathcal{C}, A, A, \tau, \mu)$, where the connecting maps are the following:

$$\tau: A \otimes_{{}^*\mathcal{C}} A \to k, \ \tau(a \otimes b) = \sum_\sigma \sigma(ab);$$

$$\mu: A \otimes A \to {}^*\mathcal{C}, \ \mu(a \otimes b) = \sum_\sigma u_\sigma \sigma(b)a.$$

Proposition 5.7 now takes the following form (compare to [21, Prop. III.1.2]).

Proposition 5.8 *Let G be a finite group, k a commutative ring and A a G-module algebra. Then the following statements are equivalent:*

1. • can' *is an isomorphism;*
   • *A is faithfully flat as a k-module.*
2. • $^*\mathrm{can}'$ *is an isomorphism;*
   • *A is a k-progenerator.*
3. • $A^G = k$;
   • *the Morita context $(k, {}^*\mathcal{C}, A, A, \tau, \mu)$ is strict.*
4. • $A^G = k$;
   • *the adjoint pair of functors $(F = \bullet \otimes A, G = (\bullet)^G)$ is an equivalence between the categories of k-modules and right A-modules on which G acts as a group of right A-semilinear automorphisms.*

In the case where A is a commutative G-module algebra, we have some more equivalent conditions.

Proposition 5.9 *Let G be a finite group, k a commutative ring and A a commutative G-module algebra. Then the statements of Proposition 5.8 are equivalent to*

5. • $A^G = k$;
   • *for each non-zero idempotent $e \in A$ and $\sigma \neq \tau \in G$, there exists $a \in A$ such that $\sigma(a)e \neq \tau(a)e$;*
   • *A is a separable k-algebra (i.e. A is projective as an A-bimodule).*
6. • $A^G = k$;
   • *there exist $x_1, \cdots, x_n, y_1, \cdots y_n \in A$ with*
   $$\sum_{j=1}^{n} x_j \sigma(y_j) = \delta_{\sigma,e}$$
   for all $\sigma \in G$.
7. • $A^G = k$;
   • *for each maximal ideal m of A, and for each $\sigma \neq e \in G$, there exists $x \in A$ such that $\sigma(x) - x \notin m$.*

Proof We refer to [16, Th 1.3] and [21, Prop. III.1.2]. □

If $A = l$ is a field, then the second part of condition 7. is satisfied. Let l be a finite field extension of a field k, and G the group of k-automorphisms of l. Then $l^G = k$ if and only if l is a normal and separable (in the classical sense) extension of k (see e.g. [37, Th. 10.8 and 10.10]). Thus we recover the classical definition of a Galois field extension.

5.4 Strongly graded rings. As in Example 1.4, let G be a group, and A a G-graded ring, and $\mathcal{C} = \oplus_{\sigma \in G} A u_\sigma$. Fix $\lambda \in G$, and take the grouplike element $u_\lambda \in G(\mathcal{C})$. Then $M^{\mathrm{co}\mathcal{C}} = M_\lambda$, for any right G-graded A-module, and $B = A^{\mathrm{co}\mathcal{C}} = A_e$. Since B is a direct factor of A, A is flat as a left and right B-module, and $i:\ B \to A$ is pure in $\mathcal{M}_B$ and ${}_B\mathcal{M}$. Also

$$\mathrm{can}:\ A \otimes_B A \to \oplus_{\sigma \in G} A u_\sigma,\ \mathrm{can}(a' \otimes a) = \sum_{\sigma \in G} a' a_\sigma u_{\lambda\sigma}.$$

Proposition 5.10 *With notation as above, the following assertions are equivalent.*

1. *A is strongly G-graded, that is, $A_\sigma A_\tau = A_{\sigma\tau}$, for all $\sigma, \tau \in G$;*
2. *the pair of adjoint functors ($F = \bullet \otimes_B A, G = (\bullet)_\lambda$) is an equivalence between $\mathcal{M}_B$ and $\mathcal{M}_A^G$, the category of G-graded right A-modules;*
3. *$(\mathcal{C}, u_\lambda)$ is a Galois coring.*

In this case A is faithfully flat as a left (or right) B-module.

Proof 1) $\Rightarrow$ 2) is a well-known fact from graded ring theory. We sketch a proof for completeness sake. The unit of the adjunction between $\mathcal{M}_B$ and $\mathcal{M}_A^G$ is given by

$$\eta_N: \ N \to (N \otimes_B A)_\lambda, \ \eta_N(n) = n \otimes_B 1_A.$$

η_N is always bijective, even if A is not strongly graded. Let us show that the counit maps $\zeta_M: \ M_\lambda \otimes_B A \to M$, $\zeta_M(m \otimes_B a) = ma$ are surjective. For each $\sigma \in G$, we can find $a_i \in A_{\sigma^{-1}}$ and $a'_i \in A_\sigma$ such that $\sum_i a_i a'_i = 1$. Take $m \in M_\tau$ and put $\sigma = \lambda_{-1}\tau$. Then $m = \zeta_M(\sum_i ma_i \otimes_B a'_i)$, and ζ_M is surjective.
If $m_j \in M_\lambda$ and $c_j \in A$ are such that $\sum_j m_j c_j = 0$, then for each $\sigma \in G$, we have

$$\sum_j m_j \otimes c_{j\sigma} = \sum_{i,j} m_j \otimes c_{j\sigma} a_i a'_i = \sum_{i,j} m_j c_{j\sigma} a_i \otimes a'_i = 0.$$

hence $\sum_j m_j \otimes c_j = \sum_{\sigma \in G} = \sum_j m_j \otimes c_{j\sigma} = 0$, so ζ_M is also injective.
2) $\Rightarrow$ 3) follows from Proposition 3.1.
3) $\Rightarrow$ 1) follows from Theorem 3.5 and the fact that $i: \ B \to A$ is pure in $\mathcal{M}_B$ and ${}_B\mathcal{M}$.

The final statement follows from Proposition 2.5 and the fact that A is flat as a B-module. □

Notice that, in this situation, the fact that $(\mathcal{C}, u_\lambda)$ is Galois is independent of the choice of λ.

6 A more general approach: comatrix corings

Let $\mathcal{C}$ be an A-coring. We needed a grouplike $x \in \mathcal{C}$ in order to make A into a right $\mathcal{C}$-comodule. In [25], the following idea is investigated. A couple $(\mathcal{C}, \Sigma)$, consisting of a coring $\mathcal{C}$ and a right $\mathcal{C}$-comodule Σ which is finitely generated and projective as a right A-module, will be called a coring with a fixed finite comodule. Let $T = \mathrm{End}^{\mathcal{C}}(\Sigma)$. Then we have a pair of adjoint functors

$$F = \bullet \otimes_T \Sigma: \ \mathcal{M}_T \to \mathcal{M}^{\mathcal{C}} \ ; \ G = \mathrm{Hom}^{\mathcal{C}}(\Sigma, \bullet): \ \mathcal{M}^{\mathcal{C}} \to \mathcal{M}_T,$$

with unit ν and counit ζ given by

$$\nu_N: \ N \to \mathrm{Hom}^{\mathcal{C}}(\Sigma, N \otimes_T \Sigma), \ \nu_N(n)(u) = n \otimes_T u;$$

$$\zeta_M: \ \mathrm{Hom}^{\mathcal{C}}(\Sigma, M) \otimes_T \Sigma \to M, \ \zeta_M(f \otimes_T u) = f(u).$$

In the situation where $\Sigma = A$, we recover the adjoint pair discussed at the end of Section 1. A particular example is the *comatrix coring*, generalizing the *canonical coring*. Let A and B be rings, and $\Sigma \in {}_B\mathcal{M}_A$, with Σ finitely generated and projective as a right A-module. Let

$$\{(e_i, e_i^*) \mid i = 1, \cdots, n\} \subset \Sigma \times \Sigma^*$$

be a finite dual basis of Σ as a right A-module. $\mathcal{D} = \Sigma^* \otimes_B \Sigma$ is an (A, A)-bimodule, and an A-coring, via

$$\Delta_\mathcal{D}(\varphi \otimes_B u) = \sum_i \varphi \otimes_B e_i \otimes_A e_i^* \otimes_B u \text{ and } \varepsilon_\mathcal{D}(\varphi \otimes_B u) = \varphi(u).$$

Furthermore $\Sigma \in \mathcal{M}^\mathcal{D}$ and $\Sigma^* \in {}^\mathcal{D}\mathcal{M}$. The coactions are given by

$$\rho^r(u) = \sum_i e_i \otimes_A e_i^* \otimes_B u \ ; \ \rho^l(\varphi) = \sum_i \varphi \otimes_B e_i \otimes_A e_i^*.$$

We also have that ${}^*\mathcal{D} \cong {}_B\mathrm{End}(\Sigma)^{\mathrm{op}}$. El Kaoutit and Gómez Torrecillas proved the following generalization of the Faithfully Flat Descent Theorem.

Theorem 6.1 *Let $\Sigma \in {}_B\mathcal{M}_A$ be finitely generated and projective as a right A-module, and $\mathcal{D} = \Sigma^* \otimes \Sigma$. Then the following are equivalent*

1. *Σ is faithfully flat as a left B-module;*
2. *Σ is flat as a left B-module and $(\bullet \otimes_B \Sigma, \mathrm{Hom}^\mathcal{D}(\Sigma, \bullet))$ is a category equivalence between $\mathcal{M}_B$ and $\mathcal{M}^\mathcal{D}$.*

Let $(\mathcal{C}, \Sigma)$ be a coring with a fixed finite comodule, and $T = \mathrm{End}^\mathcal{C}(\Sigma)$. We have an isomorphism $f : \Sigma^* \to \mathrm{Hom}^\mathcal{C}(\Sigma, \mathcal{C})$ given by

$$f(\varphi) = (\varphi \otimes_A I_\mathcal{C}) \circ \rho \text{ and } f^{-1}(\phi) = \varepsilon_\mathcal{C} \circ \phi,$$

for all $\varphi \in \Sigma^*$ and $\phi \in \mathrm{Hom}^\mathcal{C}(\Sigma, \mathcal{C})$. Consider the map

$$\mathrm{can} = \zeta_\mathcal{C} \circ (f \otimes_B I_\Sigma) : \ \mathcal{D} = \Sigma^* \otimes_T \Sigma \to \mathrm{Hom}^\mathcal{C}(\Sigma, \mathcal{C}) \otimes_T \Sigma \to \mathcal{C}.$$

We compute easily that $\mathrm{can}(\varphi \otimes_B u) = \varphi(u_{[0]})u_{[1]}$. can is a morphism of corings, and can is an isomorphism if and only if $\zeta_\mathcal{C}$ is an isomorphism.

Definition 6.2 [25, 3.4] Let $(\mathcal{C}, \Sigma)$ be a coring with a fixed finite comodule, and $T = \mathrm{End}^\mathcal{C}(\Sigma)$. $(\mathcal{C}, \Sigma)$ is termed Galois if can : $\Sigma^* \otimes_T \Sigma \to \mathcal{C}$ is an isomorphism.

Theorem 6.3 [25, 3.5] *If $(\mathcal{C}, \Sigma)$ is Galois, and Σ is faithfully flat as a left T-module, then (F, G) is an equivalence of categories.*

For further results, we refer to [25].

ACKNOWLEDGEMENTS

The author thanks George Janelidze for stimulating discussions about descent theory, and the referee for pointing out that the proper assumption in Proposition 2.3 is right purity of i, instead of left purity.

References

[1] Abuhlail J. *Morita contexts for corings and equivalences*, in "Hopf algebras in non-commutative geometry and physics", Caenepeel S. and Van Oystaeyen, F. (eds.), Lecture Notes Pure Appl. Math., Dekker, New York, to appear.
[2] Abuhlail, J. *Rational modules for corings*, Comm. Algebra **31** (2003), 5793–5840.
[3] Auslander, M. and Goldman, O. *The Brauer group of a commutative ring.* Trans. Amer. Math. Soc. **97** (1960), 367–409.
[4] Bass, H. "Algebraic K-theory", Benjamin, New York, 1968.
[5] Beattie, M., Dăscălescu, S. and Raianu, Ş. *Galois extensions for co-Frobenius Hopf algebras*, J. Algebra **198** (1997), 164–183.
[6] Böhm, G. *Doi-Hopf modules over weak Hopf algebras*, Comm. Algebra **28** (2000), 4687–4698.
[7] Borceux, F. Handbook of categorical algebra 2, Encyclopedia Math. Appl. **51**, Cambridge University Press, Cambridge, 1994.

[8] Brzeziński, T. *Coalgebra-Galois extensions from the extension point of view*, in "Hopf algebras and quantum groups", Caenepeel S. and Van Oystaeyen, F. (eds.), Lecture Notes in Pure and Appl. Math. **209**, Marcel Dekker, New York, 2000.

[9] T. Brzeziński, *The structure of corings. Induction functors, Maschke-type theorem, and Frobenius and Galois properties*, Algebr. Representat. Theory **5** (2002), 389–410.

[10] Brzeziński, T. and Hajac, P.M. *Coalgebra extensions and algebra coextensions of Galois type*, Comm. Algebra **27** (1999), 1347-1367.

[11] Brzeziński, T. and Majid, S. *Coalgebra bundles*, Comm. Math. Phys. **191** (1998), 467–492.

[12] Caenepeel, S. and De Groot, E. *Modules over weak entwining structures*, Contemp. Math. **267** (2000), 31–54.

[13] Caenepeel, S., Militaru, G. and Zhu, Shenglin "Frobenius and separable functors for generalized module categories and nonlinear equations", Lecture Notes in Math. **1787**, Springer Verlag, Berlin, 2002.

[14] Caenepeel, S., Vercruysse, J. and Wang, Shuanhong *Morita Theory for corings and cleft entwining structures*, J. Algebra, to appear.

[15] Caenepeel, S., Vercruysse, J. and Wang, Shuanhong *Rationality properties for Morita contexts associated to corings*, preprint.

[16] Chase, S., Harrison, D. and Rosenberg, A. *Galois theory and Galois cohomology of commutative rings*, Mem. Amer. Math. Soc. **52** (1965), 1–19.

[17] Chase, S. and Sweedler, M. E. "Hopf algebras and Galois theory", Lect. Notes in Math. **97**, Springer Verlag, Berlin, 1969.

[18] Cipolla, M. *Discesa fedelmente piatta dei moduli*, Rendiconti del Circolo Matematico di Palermo, Serie II **25** (1976).

[19] Cohen, M. and Fischman, D. *Semisimple extensions and elements of trace 1*, J. Algebra **149** (1992), 419–437.

[20] Cohen, M., Fischman, D. and Montgomery, S. *Hopf Galois extensions, smash products, and Morita equivalence*, J. Algebra **133** (1990), 351–372.

[21] DeMeyer F. and Ingraham, E. "Separable algebras over commutative rings", Lecture Notes in Math. **181**, Springer Verlag, Berlin, 1971.

[22] Doi, Y. *Unifying Hopf modules*, J. Algebra **153** (1992), 373–385.

[23] Doi, Y. *Generalized smash products and Morita contexts for arbitrary Hopf algebras*, in "Advances in Hopf algebras", Bergen, J. and Montgomery, S. (eds.), Lect. Notes Pure Appl. Math. **158**, Dekker, New York, 1994.

[24] Doi, Y. and Takeuchi, M. *Hopf-Galois extensions of algebras, the Miyashita-Ulbrich action and Azumaya algebras*, J. Algebra **121** (1989), 488–516.

[25] El Kaoutit, L. and Gómez Torrecillas, J. *Comatrix corings: Galois corings, descent theory, and a structure Theorem for cosemisimple corings*, Math. Z., **244** (2003), 887–906.

[26] El Kaoutit, L., Gómez Torrecillas, J. and Lobillo, F. J. *Semisimple corings*, preprint 2001.

[27] Gómez Torrecillas, J. *Separable functors in corings*, Int. J. Math. Math. Sci. **30** (2002), 203–225.

[28] Grothendieck, A. *Technique de Descente I*, Sém. Bourbaki, exp. **190** (1959-1960).

[29] Knus, M. A. and Ojanguren, M. "Théorie de la descente et algébres d'Azumaya", Lecture Notes in Math. **389**, Springer Verlag, Berlin, 1974.

[30] Koppincn, M. *Variations on thc smash product with applications to group-gradcd rings*, J. Pure Appl. Algebra **104** (1995), 61–80.

[31] Mac Lane, S. "Categories for the working mathematician", second edition, Graduate Texts in Mathematics **5**, Springer Verlag, Berlin, 1997.

[32] Mesablishvili, B. *Pure morphisms of commutative rings are effective descent morphisms for modules - a new proof*, Theory Appl. Categories **7** (2000), 38–42.

[33] Nuss, P. *Noncommutative descent and nonabelian cohomology*, K-Theory, **12** (1997), 23–74.

[34] Schauenburg, P. *Doi-Koppinen modules versus entwined modules*, New York J. Math., **6** (2000), 325–329.

[35] Schneider, H.J. *Principal homogeneous spaces for arbitrary Hopf algebras*, Israel J. Math. **70** (1990), 167–195.

[36] Serre, J.P. "Cohomologie Galoisienne", Lect. Notes in Math. **5**, Springer Verlag, Berlin, 1965.

[37] Stewart, I. "Galois theory", second edition, Chapman and Hall, London, 1989.

[38] Sweedler, M. E. *The predual Theorem to the Jacobson-Bourbaki Theorem*, Trans. Amer. Math. Soc. **213** (1975), 391–406.

[39] Takeuchi, M., as referred to in MR 2000c 16047, by A. Masuoka.

[40] Wisbauer, R. *On the category of comodules over corings*, in "Mathematics and mathematics education (Bethlehem, 2000)", World Sci. Publishing, River Edge, NJ, 2002, 325–336.

[41] Wisbauer, R. *On Galois corings*, in "Hopf algebras in non-commutative geometry and physics", Caenepeel S. and Van Oystaeyen, F. (eds.), Lecture Notes Pure Appl. Math., Dekker, New York, to appear.

Received January 27, 2003; in revised form June 10, 2003

Fields Institute Communications
Volume **43**, 2004

Quantum Categories, Star Autonomy, and Quantum Groupoids

Brian Day
Centre of Australian Category Theory
Macquarie University
N.S.W. 2109, Australia

Ross Street
Centre of Australian Category Theory
Macquarie University
N.S.W. 2109, Australia
street@math.mq.edu.au

Abstract. A useful general concept of bialgebroid seems to be resolving itself in recent publications; we give a treatment in terms of modules and enriched categories. Generalizing this concept, we define the term "quantum category" in a braided monoidal category with equalizers distributed over by tensoring with an object. The definition of antipode for a bialgebroid is less resolved in the literature. Our suggestion is that the kind of dualization occurring in Barr's star-autonomous categories is more suitable than autonomy (= compactness = rigidity). This leads to our definition of quantum groupoid intended as a "Hopf algebra with several objects".

1 Introduction

This paper has several purposes. We wish to introduce the concept of quantum category. We also wish to generalize the theory of $*$-autonomous categories in the sense of [1]. The connection between these two concepts is that they lead to our notion of quantum groupoid.

It was shown in [32] that $*$-autonomous categories provide models of the linear logic described in [19]. This suggests an interesting possibility of interactions between computer science and quantum group theory. Perhaps it will be possible, in future papers, to exploit the dichotomy between categories as structures and categories of structures. For example, what is the quantum category of finite sets, or the quantum category of finite dimensional vector spaces?

It is well known that ordinary categories are not models of an ordinary algebraic (Lawvere) theory; rather, they are models of a finite-limit theory, requiring

---

2000 *Mathematics Subject Classification.* Primary 81R50, 18D10; Secondary 03F52, 20L05.

operations to be defined in stages since some of them are defined on finite limits of earlier operations. Quantum categories, in a braided monoidal category with equalizers distributed over by tensoring, similarly involve operations defined on objects created by tensoring and taking equalizers of previously defined objects and operations.

The section headings are as follows:

1. Introduction
2. Ordinary categories revisited
3. Takeuchi bialgebroids
4. The lax monoidal operation $\times_R$
5. Monoidal star autonomy
6. Modules and promonoidal enriched categories
7. Forms and promonoidal star autonomy
8. The star and Chu constructions
9. Star autonomy in monoidal bicategories
10. Ordinary groupoids revisited
11. Hopf bialgebroids
12. Quantum categories and quantum groupoids

Before looking at quantum categories we will develop, in this introduction, a definition of "category" which suggests the definition of "quantum category". We will then relate this definition to the literature.

We use the terminology of Eilenberg-Kelly [18] for monoidal categories and monoidal functors; so we use the adjective "*strong* monoidal" for a functor which preserves tensor and unit up to coherent natural isomorphisms. A comonoidal category would have, instead of a tensor product, a tensor coproduct $\mathcal{A} \to \mathcal{A} \times \mathcal{A}$ and a counit with appropriately coherent constraints; this concept is not so interesting for ordinary categories but becomes more so for enriched categories. Comonoidal functors would go between comonoidal categories. So, for monoidal categories $\mathcal{A}$ and $\mathcal{X}$, like [27], we use the term *opmonoidal functor* for a functor

$$F : \mathcal{A} \to \mathcal{X}$$

equipped with a natural family of morphisms $\delta_{A,B} : F(A \otimes B) \to FA \otimes FB$ and a morphism $\varepsilon : FI \to I$ that are coherent.

For any set X, consider the monoidal category $\mathrm{Set}/X \times X$ of sets over $X \times X$ with the tensor product defined by

$$\left(A \xrightarrow{(s,t)} X \times X \right) \otimes \left(B \xrightarrow{(u,v)} X \times X \right) = \left(P \xrightarrow{(sop,voq)} X \times X \right)$$

where P is the pullback of $t : A \to X$ and $u : B \to X$ with projections $p : P \to A$ and $q : P \to B$. The objects of $\mathrm{Set}/X \times X$ are directed graphs with vertex-set X and the monoids are the categories with object-set X; this is well known (see [25]) and easy. Less well known, but also easy, is the fact that category structures on the graph $A \xrightarrow{(s,t)} X \times X$ amount to monoidal structures on the category Set/A of sets over A together with a strong monoidal structure on the functor

$$\Sigma_{(s,t)} : \ \mathrm{Set}/A \to \ \mathrm{Set}/X \times X$$

defined on objects by composing the function into A with (s, t).

To see this, notice that every object of a slice category Set$/C$ is a coproduct of elements $c : 1 \to C$ of C (here 1 is a chosen set with precisely one element) so that any tensor product on Set$/C$, which preserves coproducts in each variable, will be determined by its value on elements (which may not be another element in general). The tensor product on Set$/X \times X$ is such, and its value on elements is given by $(x, y) \otimes (u, v) = (x, v)$ when $y = u$ (which is in fact another element) but is the unique function $\emptyset \to X \times X$ when $y \neq u$. Since $\Sigma_{(x,t)}$ is conservative and coproduct preserving, and is to be strong monoidal, the tensor product on Set$/A$ preserves coproducts in each variable. An object of Set$/A$ has the same source set as its value under $\Sigma_{(x,t)}$. So, for elements a and b of A, the tensor product $a \otimes b$ is an element of A if and only if $t(a) = s(b)$; in this case, $s(a \otimes b) = s(a)$ and $t(a \otimes b) = t(b)$; otherwise, $a \otimes b$ is the unique function $\emptyset \to A$. The unit for the monoidal category Set$/X \times X$ is the diagonal $X \to X \times X$, so the unit for Set$/A$ has the form $i : X \to A$ with $s(i(x)) = t(i(x)) = x$ for all $x \in X$. Thus we have a reflexive graph X, A, s, t, i with a composition operation. We leave the reader to check that the associativity and unity constraints of the monoidal category Set$/A$ give associativity for the composition and that each $i(x)$ is an identity.

Conversely, suppose we have a category $\mathbf{A}$ with underlying graph $A \xrightarrow{(s,t)} X \times X$. Notice that Set$/A$ is canonically equivalent to the category $[A,\text{Set}]$ of functors from the discrete category A to Set. We can define a promonoidal structure (in the sense of [9]) on A by

$$P(a, b; c) = \begin{cases} 1 & \text{when } c \text{ is the composite of } a \text{ and } b; \\ \emptyset & \text{otherwise;} \end{cases} \quad \text{and}$$

$$J(a) = \begin{cases} 1 & \text{when } a \text{ is an identity;} \\ \emptyset & \text{otherwise.} \end{cases}$$

Then $[A,\text{Set}]$ becomes a monoidal category under **convolution**; this transports to a monoidal structure on Set$/A$ for which $\Sigma_{(s,t)}$ is strong monoidal.

When our category $\mathbf{A}$ is actually a groupoid (that is, every arrow is invertible), there is a bijection $S : A \to A$ defined by $Sa = a^{-1}$. We draw attention to the isomorphisms

$$P(a, b; Sc) \cong P(b, c; Sa)$$

noting here that S is its own inverse and that the diagram

$$\begin{array}{ccc} A & \xrightarrow{\;S\;} & A \\ {\scriptstyle (s,t)}\downarrow & & \downarrow{\scriptstyle (s,t)} \\ X \times X & \xrightarrow[\;S\;]{} & X \times X \end{array}$$

commutes, where the lower S is the switch map — which is inversion for X as a chaotic category (meaning, the category whose object set is X and each homset has exactly one element). We will relate this kind of "antipode" structure to $*$-autonomy.

Now suppose we have a category $\mathbf{A} : A \xrightarrow{(s,t)} X \times X$ and suppose we regard Set$/A$ as monoidal in the manner described above. The functor $\Sigma_{(s,t)}$ has a right adjoint $(s, t)^*$ defined by pulling back along (s, t). The strong monoidal structure on $\Sigma_{(s,t)}$ is obviously both monoidal and opmonoidal; the opmonoidal structure transforms to a monoidal structure on the right adjoint $(s, t)^*$ in such a way that the unit and counit for the adjunction are monoidal natural transformations. The composite of monoidal functors is monoidal; so the endofunctor $G_A = \Sigma_{(s,t)}(s,t)^*$ is also monoidal. The adjunction also generates a comonad structure on G_A in such a way that the counit and comultiplication are monoidal natural transformations; we have a monoidal comonad G_A on $\mathrm{Set}/X \times X$. Remember the term "monoidal comonad"!

It is also important to notice that $(s,t)^*$ has a right adjoint $\Pi_{(s,t)}$; so the endofunctor G_A has a right adjoint $(s,t)^*\Pi_{(s,t)}$. By Beck's Theorem (see [25] for example), $\Sigma_{(s,t)}$ is comonadic since it is obviously conservative (that is, reflects isomorphisms) and preserves equalizers. On the other hand, any monoidal comonad on a monoidal category leads to a monoidal structure on the category of Eilenberg-Moore coalgebras in such a way that the forgetful functor is strong monoidal (see [28] or [27] for example). Any cocontinuous endofunctor of $\mathrm{Set}/X \times X$ has the form $\Sigma_{(s,t)}(s,t)^*$ for some graph $A \xrightarrow{(s,t)} X \times X$. Assembling all this, we obtain:

Proposition 1.1 *Categories with underlying graph $A \xrightarrow{(s,t)} X \times X$ are in bijection with monoidal comonad structures on the endofunctor $\Sigma_{(s,t)}(s,t)^*$ of $\mathrm{Set}/X \times X$.*

Let us compare the combinatorial context of Proposition 1.1 with the linear algebra context. Szlachányi [37] has shown that, for a k-algebra R, the $\times_R$-bialgebras of Takeuchi [38] are opmonoidal monads on the monoidal category $\mathrm{Vect}_k^{R\otimes R^\circ}$ of left R-, right R-bimodules over R where the underlying endofunctor of the monad is a left adjoint. These $\times_R$-bialgebras of Takeuchi have been convincingly proposed (see [41, 24, 31]) as the good concept of "bialgebroid" based on R (that is, with "object of objects R").

Here we face the usual dilemma. Given a k-bialgebra H, is it better to consider the category of modules for the underlying algebra with the monoidal structure coming from the comultiplication, or, the category of comodules for the underlying coalgebra with the monoidal structure coming from the multiplication? Our preference is definitely the latter since the obvious linearization of the group case leads to this decision; also see [21]. When H is finite dimensional (as a vector space over a field k) there is essentially no difference. We feel that the functor from the category of sets to the category of k-vector spaces should provide the mechanism for regarding classical categories as quantum categories. For this we need to dualize the $\times_R$-bialgebras of Takeuchi to be based on a k-coalgebra C rather than a k-algebra R; indeed, Brzezinski-Militaru [8] have already made this dualization of the $\times_R$-bialgebras of Takeuchi based on a k-coalgebra C rather than a k-algebra R. We take this as our concept of quantum category; it involves a monoidal comonad. Actually, our general setting of a monoidal bicategory formalizes this duality.

The basic examples of quantum groups are Hopf algebras with braidings (also called quasitriangular elements or R-matrices) or cobraidings, depending how the dilemma is resolved. Indeed, these basic quantum groups are cotortile bialgebras (see [21]). We leave it to a future paper to define and discuss braidings and twists on quantum categories.

So what is a quantum groupoid? It should be a quantum category with an "antipode". We first develop a notion of antipode for the $\times_R$-bialgebras of Takeuchi. We are influenced by the chaotic example $R^\circ \otimes R$ itself where we believe the antipode should be the switch isomorphism $(R^\circ \otimes R)^\circ \to R^\circ \otimes R$. This is not a dualization in the sense of [14] but a dualization of the kind that arises in Barr's $*$-autonomous monoidal categories [1].

Consequently we are led to study $*$-autonomy for enriched categories. In fact, we define $*$-autonomous promonoidal $\mathcal{V}$-categories and show this notion is preserved under convolution. There is always the canonical promonoidal structure on $\mathcal{A}^{\mathrm{op}} \otimes \mathcal{A}$. (see the concluding remarks of [9]) which is $*$-autonomous (as remarked by Luigi Santocanale after the talk [12]) and leads under convolution to the tensor product of bimodules. The Chu construction as described in [3] and [36] is purely for ordinary categories: it needs the repetition and deletion of variables that are available in a cartesian closed base category such as Set. We vastly extend the notion of $*$-autonomy to include enriched categories and other contexts. We provide a general star-construction which leads to the Chu construction as a special case.

Equipped with this we can define when a Takeuchi $\times_R$-bialgebra is "Hopf". Then, by dualizing from k-algebras to k-coalgebras, we define quantum groupoids to be $*$-autonomous quantum categories.

2 Ordinary categories revisited

Let us consider Proposition 1.1 from a slightly different viewpoint. A left adjoint (or cocontinuous) functor F: $\mathrm{Set}/X \to \mathrm{Set}/Y$ between slice categories is determined by its restriction to the elements $x : 1 \to X$ of X, and so, by a functor

$$X \to \mathrm{Set}/Y \xrightarrow{\sim} [Y, \mathrm{Set}],$$

where we regard the sets X and Y as discrete categories and write $[\mathcal{A}, \mathcal{B}]$ for the category of functors and natural transformations from $\mathcal{A}$ to $\mathcal{B}$. However, the functors $X \to [Y, \mathrm{Set}]$ are in bijection with functors $S : X \times Y \to \mathrm{Set}$ which we think of as matrices

$$S = (S(x;y))_{(x,y)\in X\times Y} \,.$$

This gives us an equivalent (actually "biequivalent") way of looking at the 2-category whose objects are (small) sets, whose morphisms $F : X \to Y$ are cocontinuous functors $\mathrm{Set}/X \to \mathrm{Set}/Y$, and whose 2-cells are natural transformations; however, rather than a 2-category we only have a bicategory which we call Mat(Set) (compare [6] for example). Again, the objects are sets, the morphisms $S : X \to Y$ are matrices, and the 2-cells $\theta : S \Rightarrow T$ are matrices of functions

$$\theta = (\theta(x;y) : S(x;y) \to T(x;y))_{(x,y)\in X\times Y} \,;$$

vertical composition of 2-cells is defined by entrywise composition of functions, horizontal composition of morphisms $S : X \to Y$ and $T : Y \to Z$ is defined by matrix multiplication

$$(T \circ S)(x;z) = \sum_{y\in Y} S(x;y) \times T(y;z),$$

and horizontal composition is extended in the obvious way to 2-cells. We write X: $X \to X$ for the identity matrix (or Kronecker delta):

$$X(x;y) = \begin{cases} 1 & \text{for } x = y, \\ \emptyset & \text{otherwise.} \end{cases}$$

Of course Mat(Set) is also biequivalent to the bicategory Span(Set) of spans (in the sense of Bénabou [5]) in the category Set of sets.

In fact, Mat(Set) is an autonomous monoidal bicategory in the sense of the authors [15]. That is, there is a reasonably well behaved tensor product pseudo-functor

$$\mathrm{Mat}(\mathrm{Set}) \times \mathrm{Mat}(\mathrm{Set}) \to \mathrm{Mat}(\mathrm{Set})$$

which is simply defined on objects by cartesian product of sets and likewise, by cartesian product entrywise, on morphisms and 2-cells. Each object Y is actually self-dual since a matrix $X \times Y \to Z$ can be identified with a matrix $X \to Y \times Z$. This means that $Y \times Z$ is the internal hom in Mat(Set) of Y and Z (mimicking the fact that in finite-dimensional vector spaces the vector space of linear functions from V to W is isomorphic to $V^* \otimes W$). In particular, $X \times X$ is the internal endohom of X; and so we expect it to be a pseudomonoid in Mat(Set) (mimicking the fact that the internal endohom of an object in a monoidal category is an internal monoid).

Let us be more specific about this pseudomonoid structure on $X \times X$ in Mat(Set). The multiplication

$$P : (X \times X) \times (X \times X) \to X \times X$$

is defined by $P(y_2, x_2, y_1, x_1; x, y) = X(y; x_1) \times X(y_1; x_2) \times X(y_2; x)$. The unit $J : 1 \to X \times X$ is defined by $J(\bullet; x, y) = X(x; y)$. One easily checks the canonical associativity and unit isomorphisms

$$\begin{aligned} P \circ (P \times (X \times X)) &\cong P \circ ((X \times X) \times P) \\ P \circ (J \times (X \times X)) &\cong X \times X \cong P \circ ((X \times X) \times J). \end{aligned}$$

Thinking of the set $X \times X$ as a discrete category, we see that P, J and these isomorphisms form a promonoidal structure on $X \times X$. Noting that, under the equivalence of categories

$$[X \times X, \mathrm{Set}] \xrightarrow{\sim} \mathrm{Set}/X \times X,$$

the convolution monoidal structure for $X \times X$ transports across the equivalence to the monoidal structure on $\mathrm{Set}/X \times X$ described in the Introduction, the following result becomes a corollary of Proposition 1.1.

Proposition 2.1 *Categories with object set X are equivalent to monoidal comonads on the internal endohom pseudomonoid $X \times X$ in the monoidal bicategory Mat(Set).*

It may be instructive to sketch a direct proof of this result. A monoidal comonad G on $X \times X$ comes equipped with 2-cells

$$\delta : G \to G \circ G,\ \varepsilon : G \to X \times X,\ \mu : P \circ (G \times G) \to G \circ P \ \text{ and } \ \eta : J \to G \circ J,$$

subject to appropriate axioms. The mere existence of ε is quite a strong condition since $X(x; u) \times X(y; v)$ is empty unless $x = u$ and $y = v$; so $G(x, y; u, v)$ is empty unless $x = u$ and $y = v$. This leads us to put

$$A(x, y) = G(x, y; x, y)$$

which defines the homsets of our category A. It is then easy to check that μ defines composition and η provides the identities for the category A. We note finally that δ is forced to be a genuine diagonal morphism: we are dealing here with the categories of "commutative geometry".

3 Takeuchi bialgebroids

We are now ready to move from set theory to linear algebra. Let k be any commutative ring and write $\mathcal{V}$ for the monoidal category of k-modules; we write $\otimes$ for the tensor product of k-modules. Monoids R in $\mathcal{V}$ will be called k-algebras and we write $\mathcal{V}^R$ for the category of left R-modules; we can think of R as a one-object $\mathcal{V}$-category [18] so that $\mathcal{V}^R$ is the category of $\mathcal{V}$-functors from R to $\mathcal{V}$. From this viewpoint the k-algebra R^0, which is just R with opposite multiplication, is the opposite $\mathcal{V}$-category of R.

We briefly recall the preliminaries of Morita theory starting with Watts' Theorem [39] characterizing cocontinuous functors between categories of modules. For k-algebras R and S, a left adjoint (or cocontinuous) functor $F : \mathcal{V}^R \to \mathcal{V}^S$ between module categories is, up to isomorphism, determined by its restriction to the $\mathcal{V}$-dense (see [13]) full subcategory of $\mathcal{V}^R$ consisting of R itself as a left R-module. This full subcategory is isomorphic to R^0. So the left S-module $F(R) = M$ is also a right R-, left S-bimodule which we call a *module* from R to S and use the arrow notation $M : R \to S$. (The fact that R is actually on the left of the arrow and S on the right, rather than the other way around, has to do with our convention to compose functions in the usual order.) We also identify M with an object of $\mathcal{V}^{R^0 \otimes S}$.

There is a 2-category whose objects are k-algebras, whose morphisms $R \to S$ are left adjoint functors $F : \mathcal{V}^R \to \mathcal{V}^S$, and whose 2-cells are natural transformations between such functors F; the compositions are the usual ones for functors and natural transformations. This 2-category is biequivalent to the bicategory $\mathrm{Mod}(\mathcal{V})$ whose objects are k-algebras, whose morphisms are modules $M : R \to S$, and whose 2-cells are 2-sided module morphisms; the horizontal composite $N \circ M : R \to T$ of $M : R \to S$ and $N : S \to T$ is the tensor product $N \otimes_S M$ of the modules M and N over S ; vertical composition of 2-cells is the usual composition of module morphisms.

Indeed, like Mat(Set), the bicategory $\mathrm{Mod}(\mathcal{V})$ is autonomous monoidal. The tensor product is that of $\mathcal{V}$: k-algebras R and S are taken to the k-algebra $R \otimes S$, modules $M : R \to S$ and $M' : R' \to S'$ are taken to the module $M \otimes M' : R \otimes R' \to S \otimes S'$, and module morphisms are tensored using the functoriality of $M \otimes M'$ in the two variables. The opposite k-algebra S^0 acts as a dual for S since the category of modules $R \otimes X \to T$ is equivalent to the category of modules $R \to S^0 \otimes T$.

It follows that $R^0 \otimes R$ is an internal endohom for R and, as such, is a pseudomonoid in $\mathrm{Mod}(\mathcal{V})$. The multiplication

$$P : (R^0 \otimes R) \otimes (R^0 \otimes R) \to R^0 \otimes R$$

is $P = R \otimes R \otimes R$ as a k-module, with the further actions defined by

$$(x \otimes y)(a \otimes b \otimes c)(y_1 \otimes x_1 \otimes y_2 \otimes x_2) = (yax_1) \otimes (y_1bx_2) \otimes (y_2cx)$$

for $a \otimes b \otimes c \in P$, $x \otimes y \in R^0 \otimes R$ and $x_1 \otimes y_1 \otimes x_2 \otimes y_2 \in R^0 \otimes R \otimes R^0 \otimes R$. The unit

$$J : k \to R^0 \otimes R$$

is just $J = R$ as a k-module, with the further action $(x \otimes y)a = yax$. One easily checks that there are canonical isomorphisms

$$\begin{aligned} P \otimes_{R^e \otimes R^e} (R^e \otimes P) &\cong P \otimes_{R^e \otimes R^e} (P \otimes R^e) \quad \text{and} \\ P \otimes_{R^e \otimes R^e} (R^e \otimes J) &\cong R^e \cong P \otimes_{R^e \otimes R^e} (J \otimes R^e) \end{aligned}$$

where we have used the traditional notation $R^e = R^0 \otimes R$ for this pseudomonoid; the "e" superscript could be thought to stand for "endo" as well as the usual "envelope".

Definition 3.1 A *Takeuchi bialgebroid* is a k-module R together with an opmonoidal monad on R^e in the monoidal bicategory $\mathrm{Mod}(\mathcal{V})$.

To see that this definition agrees with that of $\times_R$-bialgebra as defined by Takeuchi [38] (and developed by [24, 41, 31, 8, 37]) we shall be more explicit about what an opmonoidal monad A on any pseudomonoid E involves.

In any monoidal bicategory $\mathcal{B}$ (with tensor product $\otimes$ and unit k) we use the term *pseudomonoid* (or "monoidal object") for an object E equipped with a binary multiplication $P : E \otimes E \to E$ and a unit $J : k \to E$ which are associative and unital up to coherent invertible 2-cells. A *monoidal morphism* $f : E \to E'$ is a morphism equipped with coherent 2-cells $P \circ (f \otimes f) \Rightarrow f \circ P$ and $J \Rightarrow f \circ J$. A *monoidal 2-cell* is a 2-cell compatible with these last coherent 2-cells. With the obvious compositions, this defines a bicategory $\mathrm{Mon}\mathcal{B}$ of pseudomonoids in $\mathcal{B}$. For example, if $\mathcal{B}$ is the cartesian-monoidal 2-category Cat of categories, functors and natural transformations then $\mathrm{Mon}\mathcal{B}$ is the 2-category MonCat of monoidal categories, monoidal functors and monoidal natural transformations as in [18].

We write $\mathcal{B}^{\mathrm{co}}$ for the bicategory obtained from $\mathcal{B}$ on reversing 2-cells. We put

$$\mathrm{Opmon}\mathcal{B} = (\mathrm{Mon}\mathcal{B}^{\mathrm{co}})^{\mathrm{co}};$$

the objects are again pseudomonoids, the morphisms are *opmonoidal morphisms*, and the 2-cells are *opmonoidal 2-cells.* An *opmonoidal monad* in $\mathcal{B}$ is a monad in $\mathrm{Opmon}\mathcal{B}$.

A monoidal morphism $f : E \to E'$ is called strong when the 2-cells $J \Rightarrow f \circ J$ and $P \circ (f \otimes f) \Rightarrow f \circ P$ are invertible. The inverses for these 2-cells equip such a strong f with the structure of opmonoidal morphism.

Now we return to the case of opmonoidal monads in $\mathcal{B} = \mathrm{Mod}(\mathcal{V})$. First of all, we have a module $A : E \to E$. The monad structure consists of module morphisms

$$\mu : A \otimes_E A \to A \qquad \text{and} \qquad \eta : E \to A$$

satisfying the usual conditions of associativity and unitality:

$$\mu \circ (\mu \otimes_E 1_A) = \mu \circ (1_A \otimes_E \mu), \qquad \mu \circ (\eta \otimes_E 1_A) = 1_A = \mu \circ (1_A \otimes_E \eta).$$

The opmonoidal structure consists of module morphisms

$$\delta : A \otimes_E P \to P \otimes_{E\otimes E} (A \otimes A) \quad \text{and} \quad \varepsilon : A \otimes_E J \to J$$

satisfying the following conditions:

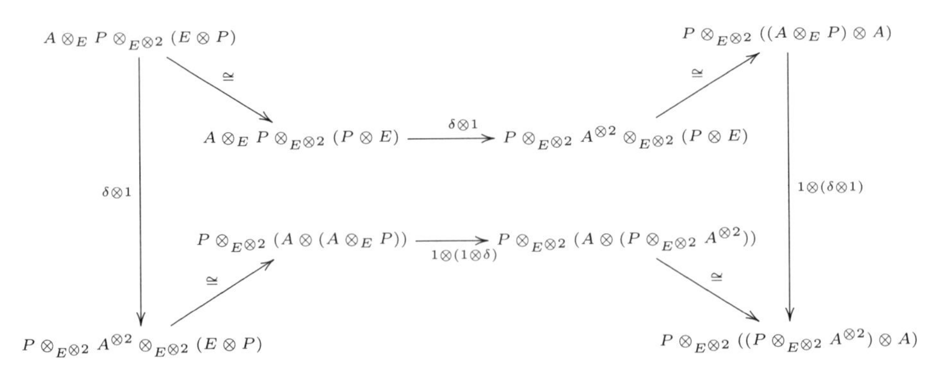

$$\begin{array}{ccc} A \otimes_E P \otimes_{E^{\otimes 2}} (E \otimes J) & \xrightarrow{\delta \otimes 1} & P \otimes_{E^{\otimes 2}} A^{\otimes 2} \otimes_{E^{\otimes 2}} (E \otimes J) \cong P \otimes_{E^{\otimes 2}} (A \otimes (A \otimes_E J)) \\ & \searrow^{\cong} & \downarrow{\scriptstyle 1\otimes(1\otimes\varepsilon)} \\ & & A \cong P \otimes_{E^{\otimes 2}} (A \otimes J) \end{array}$$

$$\begin{array}{ccc} A \otimes_E P \otimes_{E^{\otimes 2}} (J \otimes E) & \xrightarrow{\delta \otimes 1} & P \otimes_{E^{\otimes 2}} A^{\otimes 2} \otimes_{E^{\otimes 2}} (J \otimes E) \cong P \otimes_{E^{\otimes 2}} ((A \otimes_E J) \otimes A) \\ & \searrow^{\cong} & \downarrow{\scriptstyle 1\otimes(1\otimes\varepsilon)} \\ & & A \cong P \otimes_{E^{\otimes 2}} (J \otimes A) \end{array}$$

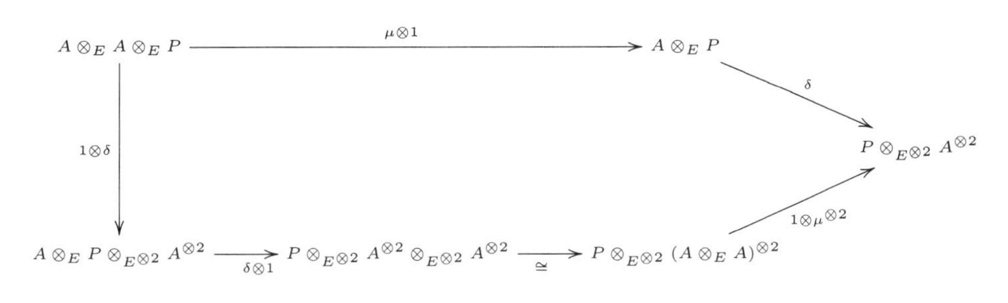

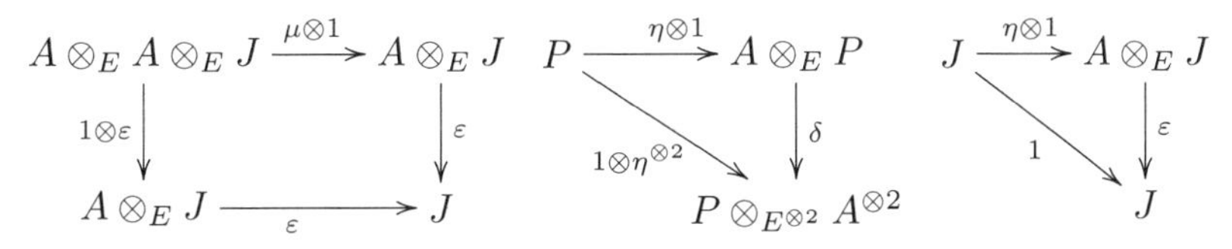

Notice in particular that A becomes a k-algebra with multiplication defined by composing μ with the quotient morphism $A \otimes A \to A \otimes_E A$ and with unit $\eta(1)$. Indeed, $\eta : E \to A$ becomes a k-algebra morphism. Moreover, the structure on A as a module $A : E \to E$ is induced by $\eta : E \to A$ via $eae' = \eta(e)a\eta(e')$.

From time to time we will require special properties of bicategories such as $\mathrm{Mod}(\mathcal{V})$. In particular, at this moment, we need to point out that $\mathrm{Mod}(\mathcal{V})$ admits both the Kleisli and Eilenberg-Moore constructions for monads. For monads in 2-categories rather than bicategories, the universal nature of these constructions was defined in [33]; however, for the kind of phenomenon for modules we are about to explain, a better reference is [34]. To be explicit, a *monad* in a bicategory $\mathcal{B}$ is an object A of $\mathcal{B}$ together with a monoid t in the monoidal category $\mathcal{B}(A, A)$ in which the tensor product is horizontal composition in $\mathcal{B}$. An *Eilenberg-Moore* object for (A, t) is an object denoted A^t for which there is an equivalence of categories

$$\mathcal{B}(X, A^t) \simeq \mathcal{B}(X, A)^{\mathcal{B}(X,t)}$$

pseudonatural in objects X of $\mathcal{B}$, where the right-hand side is the category of Eilenberg- Moore algebras for the monad $\mathcal{B}(X, t)$ on the category $\mathcal{B}(X, A)$ in the familiar sense of say [25]. The existence of Eilenberg-Moore objects is a completeness condition on $\mathcal{B}$; that condition on $\mathcal{B}^{\mathrm{op}}$ is the Kleisli construction, the notion of monad being invariant under this kind of duality. That is, a *Kleisli object* for (A, t) is an object denoted A_t for which there is an equivalence of categories

$$\mathcal{B}(A_t, X) \simeq \mathcal{B}(A, X)^{\mathcal{B}(t,X)}$$

pseudonatural in objects X of $\mathcal{B}$.

Now we move more explicitly to the bicategory $\mathrm{Mod}(\mathcal{V})$. Notice that each k-algebra morphism $f : R \to S$ leads to two modules $f_* : R \to S$ and $f^* : S \to R$ which are both equal to S as k-modules but with the module actions defined by

$$sxr = sxf(r) \qquad \text{and} \qquad rys = f(r)ys$$

for $x \in f_*$, $y \in f^*$, $r \in R$ and $s \in S$. What is more, there are module morphisms

$$R \to f^* \otimes_S f_* \qquad \text{and } f_* \otimes_R f^* \to S,$$

the former defined by f and the latter defined by multiplication in S, forming the unit and counit of an adjunction in which f^* is right adjoint to f_*.

Suppose $A : E \to E$ is a monad on the k-algebra E in the bicategory $\mathrm{Mod}(\mathcal{V})$. The multiplication $\mu : A \otimes_E A \to A$ and unit $\eta : E \to A$ morphisms compose with the quotient morphism $A \otimes A \to A \otimes_E A$ and the unit $k \to E$, respectively, to provide the k-module A with a k-algebra structure with $\eta : E \to A$ becoming a morphism of k-algebras. Then μ can be regarded as a 2-cell

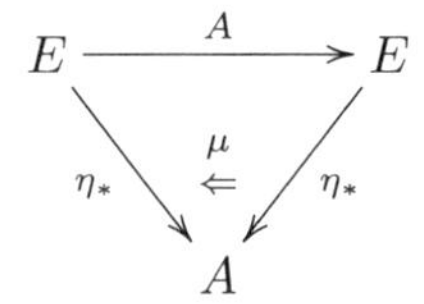

in $\mathrm{Mod}(\mathcal{V})$; it is a right action of the monoid A on η_*. Indeed, this is the universal right action of A on modules out of E; that is, the above triangle exhibits A as the Kleisli construction for the monad A on E. Since the homcategories of $\mathrm{Mod}(\mathcal{V})$ are cocomplete and composition with a given module preserves these colimits, the triangle

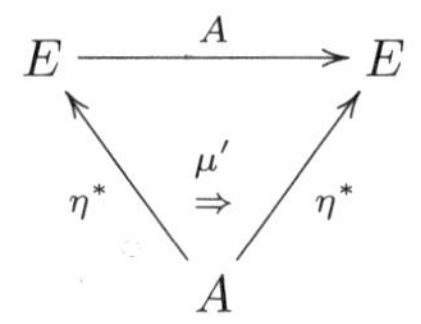

in which μ' is the mate of μ under the adjunction $\eta_* \dashv \eta^*$, exhibits A as the Eilenberg-Moore construction for the monad A on E. That is, μ' is the universal left action of A on modules into E.

The following result abstracts Proposition 2.16 of [27].

Lemma 3.2 *If the monoidal bicategory $\mathcal{B}$ admits the Eilenberg-Moore construction for monads then so does* $\mathrm{Opmon}\mathcal{B}$. *Furthermore, the forgetful morphism*

$$\mathrm{Opmon}\mathcal{B} \to \mathcal{B}$$

preserves the Eilenberg-Moore construction.

In particular, this means that $\mathrm{Opmon}\mathrm{Mod}(\mathcal{V})$ admits the Eilenberg-Moore construction. (That the Kleisli construction exists for promonoidal monads was remarked in Section 3 of [10].)

Proposition 3.3 *Suppose E is a pseudomonoid in Mod($\mathcal{V}$) and $\eta : E \to A$ is a k-algebra morphism. There is an equivalence between the category of opmonoidal monad structures μ, δ, ε on $A : E \to E$ inducing η and the category of pseudomonoid structures on A for which $\eta^* : A \to E$ is a strong monoidal morphism.*

Proof In one direction, given the opmonoidal monad A on E inducing the given η, Lemma 3.2 lifts the triangle involving μ' to a triangle in OpmonMod($\mathcal{V}$) where it is again the Eilenberg-Moore construction. In particular, the adjunction $\eta_* \dashv \eta^*$ lifts to OpmonMod($\mathcal{V}$) and so, for general reasons explained in [22], $\eta^* : A \to E$ is strong monoidal. In the other direction, any k-algebra morphism $\eta : E \to A$ always has the property that η_* is opmonadic in Mod($\mathcal{V}$); that is, it supplies the Kleisli construction for the opmonoidal monad $\eta^* \otimes_A \eta_*$ on E generated by the adjunction $\eta_* \dashv \eta^*$. This opmonoidal monad has the form $A, \mu, \delta, \varepsilon, \eta$ as required. These two directions are the object functions for an obvious equivalence of categories. □

It follows that a Takeuchi bialgebroid can equally be defined as consisting of a k-algebra R, a k-algebra morphism $\eta : R^e \to A$, and a pseudomonoid structure on A for which η^* is strong monoidal.

In preparation for interpreting Takeuchi bialgebroids in terms of module categories, we need to clarify further some monoidal terminology. The concepts are not new but the terminology is inconsistent in the literature.

We say that a monoidal $\mathcal{V}$-category $\mathcal{A}$ is *left closed* when, for all pairs of objects B, C, there is an object $[B, C]_\ell$, called the *left internal hom of* B and C, for which there are isomorphisms

$$\mathcal{A}(A, [B, C]_\ell) \cong \mathcal{A}(A \otimes B, C),$$

$\mathcal{V}$-natural in A. A *right internal hom* $[B, C]$ satisfies

$$\mathcal{A}(A, [B, C]_r) \cong \mathcal{A}(B \otimes A, C).$$

We call a monoidal $\mathcal{V}$-category *closed* when it is both left and right closed. (This differs from Eilenberg-Kelly [18] who use "closed" for left closed. However, they were mainly interested in the symmetric case where left closed implies right closed.)

As pointed out in [18], if $\mathcal{A}$ and $\mathcal{X}$ are closed monoidal, a monoidal $\mathcal{V}$-functor $F : \mathcal{A} \to \mathcal{X}$, with its (lax) constraints

$$\phi_0 : I \to FI \qquad \text{and} \qquad \phi_{2;A,B} : FA \otimes FB \to F(A \otimes B)$$

subject to axioms, could equally be called a *left closed $\mathcal{V}$-functor* since these constraints are in bijection with pairs

$$\phi_0 : I \to FI \qquad \text{and} \qquad \phi^\ell_{2;B;C} : F[B, C]_\ell \to [FB, FC]_\ell$$

satisfying corresponding axioms. Equally F could be called a *right closed $\mathcal{V}$-functor* since the constraints are in bijection with pairs

$$\phi_0 : I \to FI \qquad \text{and} \qquad \phi^r_{2;A;;C} : F[A, C]_r \to [FA, FC]_r$$

satisfying corresponding axioms. We call a monoidal $\mathcal{V}$-functor F *normal* when ϕ_0 is invertible. As usual we call F *strong monoidal* when it is normal and each $\phi_{2;A,B}$ is invertible. We define F to be *strong left closed* when it is normal and each $\phi^\ell_{2;B;C}$ is invertible; it is *strong right closed* when it is normal and each $\phi^r_{2;A;C}$ is invertible; and it is *strong closed* when it both strong left and strong right closed.

Pseudomonoid structures on A in Mod($\mathcal{V}$) are equivalent to closed monoidal structures on the $\mathcal{V}$-category $\mathcal{V}^A = \text{Mod}(\mathcal{V})(k, A)$ of left A-modules; this is a special case of convolution in the sense of [9]. In fact, since k is a comonoid in $\mathrm{Mod}(\mathcal{V})$, we have a monoidal pseudofunctor

$$\mathrm{Mod}(\mathcal{V})(k,-):\ \mathrm{Mod}(\mathcal{V}) \to \mathcal{V}\text{-Cat},$$

which, as such, takes pseudomonoids to pseudomonoids. Since it is representable by k, it also preserves Eilenberg-Moore constructions (and all weighted limits for that matter). This means that when we apply $\mathrm{Mod}(\mathcal{V})(k,-)$ to a Takeuchi bialgebroid $\eta : R^e \to A$, we obtain a strong monoidal monadic functor

$$\mathcal{V}^A \to \mathcal{V}^{R^e}.$$

Conversely, given a k-algebra morphism $\eta : R^e \to A$, a $\mathcal{V}$-monoidal structure on $\mathcal{V}^A$, and a strong monoidal structure on the functor $\mathcal{V}^A \to \mathcal{V}^{R^e}$, we obtain a Takeuchi bialgebroid structure on $\eta : R^e \to A$. This is because $\mathcal{V}^A \to \mathcal{V}^{R^e}$ has both adjoints and is conservative (= reflects isomorphisms), so is monadic; but being strong monoidal and colimit preserving, any monoidal structure on $\mathcal{V}^A$ will be automatically closed, reflecting the fact that the monoidal $\mathcal{V}$-category $\mathcal{V}^{R^e}$ is closed. Consequently, by [9], the monoidal structure on $\mathcal{V}^A$ is obtained by convolution of a pseudomonoid structure on A.

By Theorem 5.1 of [30] (also see Theorem 3.1 of [8]) characterizing the $\times_R$-bialgebras of Takeuchi as monoidal structures on $\mathcal{V}^A$ for which $\mathcal{V}^A \to \mathcal{V}^{R^e}$ is strong monoidal, we have shown that our Takeuchi bialgebroids are the $\times_R$-bialgebras. We will see this in another way in the next section.

4 The lax monoidal operation $\times_R$

In order to define a bimonoid (or bialgebra) in a monoidal category, the monoidal category requires some kind of commutativity of the tensor product such as a braiding. A braiding can be regarded as a second monoidal structure on the category for which the new tensor is strongly monoidal with respect to the old. The so-called Eckmann-Hilton argument forces the new tensor to be isomorphic to the old and forces a braiding to appear (see [20]).

Ah, but what if the second tensor is only a lax multitensor and is only monoidal with respect to the old monoidal structure? Then there is certainly no need for the two structures to coincide. However, it is still possible to speak of a bimonoid: there is sufficient structure to express compatibility between a monoid structure for one tensor and a comonoid structure (on the same object) for the other tensor. After some preliminaries about right extensions in bicategories, we shall describe in detail just such a situation.

On top of the already discussed diverse properties and rich structure enjoyed by $\mathrm{Mod}(\mathcal{V})$, we also have the property that all right liftings and right extensions exist. Despite the terminology (from [33] for example), these concepts are very familiar in the usual theory of modules.

Suppose M and M' are modules $R \to S$. We put

$$\mathrm{Hom}_R^S(M,M') = \mathrm{Mod}(\mathcal{V})(R,S)(M,M');$$

that is, traditionally, it is the k-module of left S, right R-bimodule morphisms from M to M' . Now consider three modules as in the triangle

$$\begin{array}{ccc} R & \xrightarrow{M} & S \\ & {\scriptstyle N}\searrow \quad \swarrow{\scriptstyle L} & \\ & T & \end{array}$$

Let $\mathrm{Hom}_R(M,N) : S \to T$ denote the k-module of right R-module morphisms with right S- and left T-actions defined by $(tfs)(m) = tf(sm)$ for

$$s \in S,\ t \in T,\ f \in \ \mathrm{Hom}_R(M,N) \quad \text{and} \quad m \in M.$$

Let $\mathrm{Hom}^T(L,N) : R \to S$ denote the k-module of left T-module morphisms with right R- and left S-actions defined by $(sgr)(\ell) = g(\ell s)r$ for

$$r \in R,\ s \in S,\ g \in \ \mathrm{Hom}^T(L,N) \quad \text{and } \ell \in L.$$

There are natural isomorphisms

$$\mathrm{Hom}_S^T(L, \mathrm{Hom}_R(M,N)) \cong \ \mathrm{Hom}_R^T(L \otimes_S M, N) \cong \ \mathrm{Hom}_R^S(M, \mathrm{Hom}^T(L,N)).$$

induced by evaluation morphisms

$$ev_N^M : \ \mathrm{Hom}_R(M,N) \otimes_S M \to N \quad \text{and} \quad ev_N^L : L \otimes_S \ \mathrm{Hom}^T(L,N)) \to N.$$

In bicategorical terms, $\mathrm{Hom}_R(M,N)$ is the right extension of N along M, while $\mathrm{Hom}^T(L,N)$ is the right lifting of N through L.

We require *normal lax monoidal categories* in the sense of [16] and [17]. These structures have been considered by Michael Batanin; they are the algebras for the categorical operad defined on page 88 of [4]. A lax monoidal structure on a category $\mathcal{E}$ amounts to a sequence of functors

$$\underset{n}{\bullet} : \underbrace{\mathcal{E} \times \cdots \times \mathcal{E}}_{n} \to \mathcal{E}$$

(thought of as multiple tensor products) together with substitution operations μ_ξ in the direction we will give below in our main example, and a unit $\eta : X \to \bullet_1 X$, satisfying three axioms. This is called *normal* when η is invertible (and so can be replaced by an identity).

Consider any pseudomonoid E, with multiplication P and unit J, in a monoidal bicategory $\mathcal{B}$ which admits all right extensions (where we have in mind $\mathcal{B} = \mathrm{Mod}(\mathcal{V})$). Then the endohom category $\mathrm{End}(E) = \mathcal{B}(E,E)$ becomes a lax monoidal category as follows. We define

$$P_n : E^{\otimes n} \to E$$

to be the composite

$$E^{\otimes n} \xrightarrow{P \otimes E^{\otimes(n-2)}} E^{\otimes(n-1)} \xrightarrow{P \otimes E^{\otimes(n-3)}} \cdots \xrightarrow{P \otimes E} E^{\otimes 2} \xrightarrow{P} E$$

for $n \geq 2$, to be the identity of $n = 1$, and to be J when $n = 0$. The coherence conditions for a pseudomonoid ensure that $P_m \cong P_n \circ (P_{m_1} \otimes \cdots \otimes P_{m_n})$ for each partition $\xi : m_1 + \cdots m_n = m$.

We define the multiple tensor $\bullet_n(M_1, \ldots, M_n)$ of objects $M_1, \ldots, M_n$ of $\mathrm{End}(E)$ to be the right extension of $P_n \circ (M_1 \otimes \cdots \otimes M_n)$ along P_n; that is,

$$\underset{n}{\bullet}(M_1, \ldots, M_n) = \ \mathrm{Hom}_{E^{\otimes n}}(P_n, P_n \otimes_{E^{\otimes n}} (M_1 \otimes \cdots \otimes M_n)).$$

The lax associativity constraint

$$\mu_\xi : \underset{n}{\bullet}(\underset{m_1}{\bullet}(M_{11},\ldots,M_{1m_1}),\ldots,\underset{m_n}{\bullet}(M_{n1},\ldots,M_{nm_n})) \to \underset{m}{\bullet}(M_{11},\ldots,M_{nm_n})$$

for each partition $\xi : m_1 + \cdots m_n = m$ is, by using the right extension property of the target, induced by the morphism

$$\underset{n}{\bullet}(\underset{m_1}{\bullet}(M_{11},\ldots,M_{1m_1}),\ldots,\underset{m_n}{\bullet}(M_{n1},\ldots,M_{nm_n})) \circ P_m \to P_m \circ (M_{11},\ldots,M_{nm_n})$$

which, after "conjugation" with $P_m \cong P_n \circ (P_{m_1} \otimes \cdots \otimes P_{m_n})$, is the composite

$$\begin{aligned}
&\underset{n}{\bullet}(\underset{m_1}{\bullet}(M_{11},\ldots,M_{1m_1}),\ldots,\underset{m_n}{\bullet}(M_{n1},\ldots,M_{nm_n})) \circ P_n \\
&\qquad \circ\ (P_{m_1} \otimes \cdots \otimes P_{m_n}) \xrightarrow{ev\circ 1} \\
&P_n \circ (\underset{m_1}{\bullet}(M_{11},\ldots,M_{1m_1}),\ldots,\underset{m_n}{\bullet}(M_{n1},\ldots,M_{nm_n})) \\
&\qquad \circ\ (P_{m_1} \otimes \cdots \otimes P_{m_n}) \xrightarrow{\cong} \\
&P_n \circ ((\underset{m_1}{\bullet}(M_{11},\ldots,M_{1m_1}) \circ P_{m_1}) \otimes \cdots \otimes \\
&\qquad (\bullet_{m_n}(M_{n1},\ldots,M_{nm_n})) \circ P_{m_n}) \xrightarrow{1\circ(ev\otimes\cdots\otimes ev)} \\
&P_n \circ ((P_{m_1} \circ \underset{m_1}{\bullet}(M_{11},\ldots,M_{1m_1})) \otimes \cdots \otimes \\
&\qquad P_{m_n} \circ \underset{m_n}{\bullet}(M_{n1},\ldots,M_{nm_n}))) \xrightarrow{\cong} \\
&P_n \circ (P_{m_1} \otimes \cdots \otimes P_{m_n}) \circ (M_{11} \otimes \cdots \otimes M_{nm_n}).
\end{aligned}$$

The three axioms for a lax monoidal category can be verified. Since $P_1 : E \to E$ is the identity, we see that $\bullet_1 M = M$; so the lax monoidal structure on $\mathrm{End}(E)$ is normal.

As an endomorphism category $\mathrm{End}(E)$ is also a monoidal category for which the tensor product is composition. So $\mathrm{End}(E)$ is an object of the 2-category MonCat. Now MonCat is a monoidal 2-category with cartesian product as tensor. We will now see that $\mathrm{End}(E)$ is a lax monoid in MonCat.

Proposition 4.1 *Regard End(E) as a monoidal category under composition. The functors $\bullet_n$: End(E)^n $\to$ End(E) are equipped with canonical monoidal structures such that the substitutions μ_ξ are monoidal natural transformations.*

Proof The structure in question is the family of morphisms

$$\underset{n}{\bullet}(N_1,\ldots,N_n) \circ \underset{n}{\bullet}(M_1,\ldots,M_n) \to \underset{n}{\bullet}(N_1 \circ M_1,\ldots,N_n \circ M_n)$$

which, using the right extension property of the target, are induced by the composites

$$\begin{aligned}
&\bullet_n(N_1,\ldots,N_n) \circ \bullet_n(M_1,\ldots,M_n) \circ P_n \xrightarrow{1\circ ev} \\
&\bullet_n(N_1,\ldots,N_n) \circ P_n \circ (M_1,\ldots,M_n) \xrightarrow{ev\circ 1}
\end{aligned}$$

$$P_n \circ (N_1 \otimes \cdots \otimes N_n) \circ (M_1 \otimes \cdots \otimes M_n) \xrightarrow{\cong} P_n \circ ((N_1 \circ M_1) \otimes \cdots \otimes (N_n \circ M_n)).$$

The compatibility of these morphisms with the lax associativity morphisms is readily verified. □

A monoid for composition in $\mathrm{End}(E)$ is a monad on E in $\mathcal{B}$. We write $\mathrm{MonEnd}(E)$ for the category of monads on E; the morphisms are 2-cells between the endofunctors of the monads that are compatible with the units and multiplications. It follows from Proposition 4.1 that the lax monoidal structure on $\mathrm{End}(E)$ lifts to the category $\mathrm{MonEnd}(E)$.

The concept of comonoid makes sense in any lax monoidal category.

Proposition 4.2 *A Takeuchi bialgebroid can equally be defined as a k-algebra R together with a comonoid in the lax monoidal category $\mathrm{MonEnd}(R^e)$.*

Proof Both a Takeuchi bialgebroid $A : R^e \to R^e$ and a comonoid in $\mathrm{MonEnd}(R^e)$ start with a monad $A : R^e \to R^e$ on R^e in $\mathrm{Mod}(\mathcal{V})$. To make this a comonoid in $\mathrm{MonEnd}(R^e)$ we need a comultiplication $\delta' : A \to \underset{2}{\bullet}(A, A)$ and a counit $\varepsilon' : A \to \underset{0}{\bullet}$ satisfying axioms. By the right extension properties of their targets, these morphisms are determined by morphisms $\delta : A \circ P_2 \to P_2 \circ (A \otimes A)$ and $\varepsilon : A \circ P_0 \to P_0$, exactly as for a Takeuchi bialgebroid. The condition that δ' and ε' should form a comonoid translates to the first three diagrams on δ and ε describing an opmonoidal monad (as in Section 3) while the conditions that δ' and ε' should respect the monad structure translate to the last four diagrams on δ and ε. So the comonoid is equivalently a Takeuchi bialgebroid. □

The operation $\underset{2}{\bullet}$ on $\mathrm{MonEnd}(R^e)$ is precisely the operation $\times_R$ of Takeuchi [38]; also compare Section 2 of [31][1] whose $\alpha : (M \times_R P) \times_R N \to M \times_R P \times_R N$, for example, is our substitution $\mu_\xi : \underset{2}{\bullet}(\underset{2}{\bullet}(M, P), N) \to \underset{3}{\bullet}(M, P, N)$ for $\xi : 2 + 1 = 3$. To help the reader make these identifications explicit, let $E = R^e = R^0 \otimes R$ take left-E, right E-bimodules M and N, and recall that $P_2 = R \otimes R \otimes R$ with the actions explained in Section 3. There is a canonical isomorphism

$$P_2 \otimes_{E^{\otimes 2}} (M \otimes N) \cong M \otimes_R N$$

where $M \otimes_R N = M \otimes N / \langle ((x \otimes 1)m) \otimes n \sim m \otimes ((1 \otimes x)n) \rangle$. Then we have the following calculation where the third isomorphism is obtained by evaluating the homomorphisms at $1 \otimes 1 \otimes 1 \in R \otimes R \otimes R$.

$$\begin{aligned} \underset{2}{\bullet}(M, N) &\cong \mathrm{Hom}_{E^{\otimes 2}}(P_2, P_2 \otimes_{E^{\otimes 2}} (M \otimes N)) \cong \mathrm{Hom}_{E^{\otimes 2}}(P_2, M \otimes_R N) \\ &\cong \left\{ \sum_i m_i \otimes_R n_i \in M \otimes_R N \,\middle|\, \sum_i m_i(x \otimes 1) \otimes_R n_i \right. \\ &\qquad \left. = \sum_i m_i \otimes_R n_i(1 \otimes x) \ \forall\, x \in R \right\} \\ &= M \times_R N. \end{aligned}$$

5 Monoidal star autonomy

In this section we extend the theory of $*$-autonomous categories in the sense of Barr (see [1], and, for the non-symmetric case, see [3]) to enriched categories in the sense of Eilenberg-Kelly [18]. The kind of duality present in a $*$-autonomous

[1] In its basic form the integral notation attributed to Mac Lane in [31] is originally due to Yoneda; see page 546 of [42]. It was adopted by [13] for their concept of "end" and "coend" in the general enriched context; however, their use of subscripts and superscripts on the integral (adopted by [25]) is the reverse of [31]. This reversal is reproduced in [8].

category is closer than compactness (also called rigidity or autonomy) to what is needed for an antipode in a bialgebroid or quantum category, and so for a concept of Hopf bialgebroid or quantum groupoid (see Example 7.4).

A $\mathcal{V}$-functor $F : \mathcal{A} \to \mathcal{B}$ is called *eso* (for "essentially surjective on objects") when every object of $\mathcal{B}$ is isomorphic to one of the form FA for some object A of $\mathcal{A}$.

A *left star operation* for a monoidal $\mathcal{V}$-category $\mathcal{A}$ is an eso $\mathcal{V}$-functor

$$S_\ell : \mathcal{A} \to \mathcal{A}^{\mathrm{op}}$$

together with a $\mathcal{V}$-natural family of isomorphisms (called the *left star constraint*)

$$\mathcal{A}(A \otimes B, S_\ell C) \cong \mathcal{A}(A, S_\ell(B \otimes C)).$$

It follows that $\mathcal{A}$ is left closed with $[B, C]_\ell \cong S_\ell(B \otimes D)$ where $S_\ell D \cong C$.

A *right star operation* for a monoidal $\mathcal{V}$-category $\mathcal{A}$ is an eso $\mathcal{V}$-functor

$$S_r : \mathcal{A}^{\mathrm{op}} \to \mathcal{A}$$

together with a $\mathcal{V}$-natural family of isomorphisms (called the *right star constraint*)

$$\mathcal{A}(A \otimes B, S_r C) \cong \mathcal{A}(B, S_r(C \otimes A)).$$

It follows that $\mathcal{A}$ is then right closed with $[A, C]_r \cong S_r(E \otimes A)$ where $S_r E \cong C$.

A monoidal $\mathcal{V}$-category $\mathcal{A}$ is called $*$-*autonomous* when it is equipped with a left star operation which is fully faithful. Since it follows that S_ℓ is then an equivalence of $\mathcal{V}$-categories, we write S_r for its adjoint equivalence so that the left star constraint can be written as

$$\mathcal{A}(A \otimes B, S_\ell C) \cong \mathcal{A}(B \otimes C, S_r A).$$

We see from this that S_r is a right star operation and $*$-autonomy can equally be defined in terms of a fully faithful right star operation. It follows that $*$-autonomous monoidal $\mathcal{V}$-categories are closed, with internal homs given by the formulas

$$[B, C]_\ell \cong S_\ell(B \otimes S_r C) \text{ and } [A, C]_r \cong S_r(S_\ell C \otimes A).$$

Notice that

$$\mathcal{A}(A, S_\ell I) \cong \mathcal{A}(I \otimes A, S_\ell I) \cong \mathcal{A}(A \otimes I, S_r I) \cong \mathcal{A}(A, S_r I),$$

so that $S_\ell I \cong S_r I$ (by the Yoneda Lemma). The object $S_\ell I$ is called the *dualizing object* and determines the left star operation via $[B, S_\ell I]_\ell \cong S_\ell B$.

For the reader interested in checking that our $*$-autonomous monoidal categories agree with Michael Barr's $*$-autonomous categories, we recommend Definition 2.3 of [2] as the appropriate one for comparison. Also see [35].

A monoidal $\mathcal{V}$-category is autonomous if and only if there exists a left star operation S_ℓ and a family of $\mathcal{V}$-natural isomorphisms

$$S_\ell(A \otimes B) \cong S_\ell B \otimes S_\ell A.$$

If $\mathcal{A}$ is autonomous then taking the left dual provides a left star operation with isomorphisms as required which *a fortiori* satisfy the conditions for a strong monoidal $\mathcal{V}$-functor. To see the less obvious implication, suppose we have an S_ℓ and the isomorphisms. Then $[B, C]_\ell \cong S_\ell(B \otimes D) \cong S_\ell D \otimes S_\ell B \cong C \otimes S_\ell B$ where $S_\ell D \cong C$, so $S_\ell B$ is a left dual for B. So every object B has a left dual $S_\ell B$. However, every object B is isomorphic to $S_\ell D$ for some D. This implies that D is a right dual for B.

6 Modules and promonoidal enriched categories

An important part of our goal is to extend star autonomy from monoidal categories to promonoidal categories. In preparation, in this section we shall discuss some basic facts about enriched categories and modules between them. Then we will review promonoidal categories and promonoidal functors in the enriched context [9]. We obtain a result about restriction along a promonoidal functor.

Let $\mathcal{V}$ denotes any complete and cocomplete symmetric monoidal closed category. We write $\mathcal{V}$-Mod for the symmetric monoidal bicategory (in the sense of [15]) whose objects are $\mathcal{V}$-categories and whose hom-categories are defined by

$$\mathcal{V}\text{-Mod}(\mathcal{A},\mathcal{B}) = [\mathcal{A}^{\mathrm{op}} \otimes \mathcal{B}, \mathcal{V}].$$

The objects $M : \mathcal{A} \to \mathcal{B}$ of $\mathcal{V}$-Mod$(\mathcal{A},\mathcal{B})$ are called *modules from $\mathcal{A}$ to $\mathcal{B}$*. The composite of modules $M : \mathcal{A} \to \mathcal{B}$ and $N : \mathcal{B} \to \mathcal{C}$ is defined by the equation

$$(N \circ M)(A, C) = \int^{B} N(B, C) \otimes M(A, B);$$

the integral here is the "coend" in the sense of [13] (also see [23]). The tensor product for $\mathcal{V}$-Mod is the usual tensor product of $\mathcal{V}$-categories in the sense of [18] (also see [23]); explicitly, an object of $\mathcal{A} \otimes \mathcal{B}$ is a pair (A, B) where A is an object of $\mathcal{A}$ and B is an object of $\mathcal{B}$, and the homs are defined by

$$(\mathcal{A} \otimes \mathcal{B})((A, B), (A', B')) = \mathcal{A}(A, A') \otimes \mathcal{B}(B, B').$$

Actually $\mathcal{V}$-Mod is autonomous since we have

$$\mathcal{V}\text{-Mod}(\mathcal{A} \otimes \mathcal{B}, \mathcal{C}) \cong \mathcal{V}\text{-Mod}(\mathcal{B}, \mathcal{A}^{\mathrm{op}} \otimes \mathcal{C})$$

since both sides are isomorphic to $[\mathcal{B}^{\mathrm{op}} \otimes \mathcal{A}^{\mathrm{op}} \otimes \mathcal{C}, \mathcal{V}]$.

We have reversed the direction of modules from that in [?] so that a promonoidal $\mathcal{V}$-category $\mathcal{A}$ is precisely a pseudomonoid (monoidal object) of $\mathcal{V}$-Mod (rather than $\mathcal{A}^{\mathrm{op}}$ being such). The multiplication module $P : \mathcal{A} \otimes \mathcal{A} \to \mathcal{A}$ and the unit module $J : \mathcal{I} \to \mathcal{A}$ are equally $\mathcal{V}$-functors

$$P : \mathcal{A}^{\mathrm{op}} \otimes \mathcal{A}^{\mathrm{op}} \otimes \mathcal{A} \to \mathcal{V} \quad \text{and} \quad J : \mathcal{A} \to \mathcal{V},$$

and we have associativity constraints

$$\int^{X} P(X, C; D) \otimes P(A, B; X) \cong \int^{Y} P(A, Y; D) \otimes P(B, C; Y)$$

and unital constraints

$$\int^{X} P(X, A; B) \otimes JX \cong \mathcal{A}(A, B) \cong \int^{Y} P(A, Y; B) \otimes JY,$$

satisfying the usual two axioms (see [9]) which yield coherence. It is convenient to introduce the $\mathcal{V}$-functors

$$P_n : \underbrace{\mathcal{A}^{\mathrm{op}} \otimes \cdots \otimes \mathcal{A}^{\mathrm{op}} \otimes \mathcal{A}}_{n} \to \mathcal{V},$$

for all natural numbers n, which we define as follows:

$$P_0 A = JA,\ P_1(A_1; A) = \mathcal{A}(A_1, A),\ P_2(A_1, A_2; A) = P(A_1, A_2; A)$$

and

$$P_{n+1}(A_1, \ldots, A_{n+1}; A) = \int^{X} P(X, A_{n+1}; A) \otimes P(A_1, \ldots, A_n; X).$$

We think of $P_n(A_1, \ldots, A_n; A)$ as the object of multimorphisms from $A_1, \ldots, A_n$ to A in $\mathcal{A}$. For example, when $\mathcal{A}$ is a monoidal $\mathcal{V}$-category, we have a promonoidal structure on $\mathcal{A}$ with

$$P_n(A_1, \ldots, A_n; A) \cong \mathcal{A}(A_1 \otimes \cdots \otimes A_n, A),$$

where the multitensor product is, say, bracketed from the left.

It will also be convenient to define a *multimorphism structure* on a $\mathcal{V}$-category $\mathcal{A}$ to be a sequence of $\mathcal{V}$-functors

$$P_n : \underbrace{\mathcal{A}^{\mathrm{op}} \otimes \cdots \otimes \mathcal{A}^{\mathrm{op}}}_{n} \otimes \mathcal{A} \to \mathcal{V}$$

subject to no constraints. So a promonoidal structure is an example where all the P_n are obtained from the particular ones for $n = 0, 1, 2$. A *multitensor structure* on $\mathcal{A}$ is a multimorphism structure for which each $P_n(A_1, \ldots, A_n; -)$ is representable; so we have objects $\underset{n}{\otimes}(A_1, \ldots, A_n)$ of $\mathcal{A}$ and a $\mathcal{V}$-natural family of isomorphisms

$$P_n(A_1, \ldots, A_n; A) \cong \mathcal{A}(\underset{n}{\otimes}(A_1, \ldots, A_n), A).$$

For example, when $\mathcal{A}$ is monoidal, we obtain $\underset{n}{\otimes}(A_1, \ldots, A_n)$ inductively from the cases $n = 0, 1$, and 2 where it is the unit, the identity functor, and the binary tensor product, respectively.

Suppose $\mathcal{A}$ and $\mathcal{E}$ are promonoidal $\mathcal{V}$-categories. A $\mathcal{V}$-functor $H : \mathcal{E} \to \mathcal{A}$ is called promonoidal when it is equipped with $\mathcal{V}$-natural families of morphisms

$$\phi_{2;U,V;W} : P(U, V; W) \to P(HU, HV; HW) \quad \text{and} \quad \phi_{0;U} : JU \to JHU$$

that are compatible in the obvious way with the associativity and unital constraints [10]. For any such promonoidal H, we can inductively define $\mathcal{V}$-natural families of morphisms

$$\phi_{n;U_1,\ldots,U_n;U} : P_n(U_1, \ldots, U_n; U) \to P_n(HU_1, \ldots, HU_n; HU)$$

using the inductive definition of P_n. In particular, $\phi_{1;U;V} : \mathcal{E}(U, V) \to \mathcal{A}(HU, HV)$ is the effect of H on homs. We say that H is *promonoidally fully faithful* when each $\phi_{n;U_1,\ldots,U_n;U}$ is invertible. We say F is *normal* when each $\phi_{0;U}$ is invertible.

A promonoidal $\mathcal{V}$-functor $H : \mathcal{E} \to \mathcal{A}$ also gives rise in the obvious way to $\mathcal{V}$-natural families of morphisms

$$\bar{\phi}_{2;A,B;W} : \int^{U,V} P(U, V; W) \otimes \mathcal{A}(A, HU) \otimes \mathcal{A}(B, HV) \to P(A, B; HW),$$

$$\phi^{\ell}_{2;U,B;C} : \int^{B,C} P(U, V; W) \otimes \mathcal{A}(B, HV) \otimes \mathcal{A}(HW, C) \to P(HU, B; C),$$

and

$$\phi^{r}_{2;A,V;C} : \int^{U,W} P(U, V; W) \otimes \mathcal{A}(A, HU) \otimes \mathcal{A}(HW, C) \to P(A, HV; C).$$

We need to say a little bit about convolution (see [9, 11, 17]). For $\mathcal{V}$-categories $\mathcal{A}$ and $\mathcal{X}$ equipped with multimorphism structures, the *convolution multimorphism structure* on the $\mathcal{V}$-functor $\mathcal{V}$-category $[\mathcal{A}, \mathcal{X}]$ is defined by

$$P_n(M_1, \ldots, M_n; M) = \int_{A_1,\ldots,A_n} [P_n(A_1, \ldots, A_n; A), P_n(M_1A_1, \ldots, M_nA_n; MA)]$$

whenever these ends all exist (for example, when $\mathcal{A}$ is small). In the case where $\mathcal{X}$ is multitensored, the convolution is also multitensored by the formula

$$\underset{n}{\star}(M_1, \ldots, M_n)(A) = \int^{A_1,\ldots,A_n} P_n(A_1, \ldots, A_n; A) \otimes \underset{n}{\otimes}(M_1A_1, \ldots, M_nA_n),$$

provided the appropriate weighted colimits (expressed here by coends and tensors) exist in $\mathcal{X}$. In the case where $\mathcal{A}$ is promonoidal, if $\mathcal{X}$ is cocomplete closed monoidal then so is $[\mathcal{A}, \mathcal{X}]$ (see [9]).

Proposition 6.1 *Suppose $H : \mathcal{E} \to \mathcal{A}$ is a normal promonoidal $\mathcal{V}$-functor. The restriction $\mathcal{V}$-functor*

$$[H, 1] : [\mathcal{A}, \mathcal{V}] \to [\mathcal{E}, \mathcal{V}]$$

is a normal monoidal $\mathcal{V}$-functor. It is strong monoidal if and only if each $\bar{\phi}_{2;A,B;W}$ is invertible. It is strong left (respectively, strong right) closed if and only if each $\phi^{\ell}_{2;U,B;C}$ (respectively, $\phi^{r}_{2;A,V;C}$) is invertible.

Proof The monoidal unital constraint for $[H, 1]$ is $\phi_{0;U} : JU \to JHU$. To obtain the associativity constraint, we use the Yoneda Lemma to replace

$$(MH \star NH)W = \int^{U,V} P(U, V; W) \otimes MHU \otimes NHV$$

by the isomorphic expression

$$\int^{U,V,A,B} P(U, V; W) \otimes \mathcal{A}(A, HU) \otimes \mathcal{A}(B, HV) \otimes MHU \otimes NHV$$

and take the morphism into

$$(M \star N)HW = \int^{A,B} P(A, B; HW) \otimes MA \otimes NB$$

of the form $\int^{A,B} \bar{\phi}_{2;A,B;W} \otimes 1 \otimes 1$ which is clearly invertible if $\bar{\phi}_{2;A,B;W}$ is. The converse comes by taking M and N to be representable and using Yoneda.

Similarly, the left closed constraint for $[H, 1]$ is obtained by composing the morphism $\int_{B,C}[\phi^{\ell}_{2;U,B;C} \otimes 1, 1]$ from

$$[N, L]_{\ell}HU = \int_{B,C} [P(HU, B; C) \otimes NB, LC]$$

to

$$\int_{V,W,B,C} [P(U, V; W) \otimes \mathcal{A}(B, HV) \otimes \mathcal{A}(HW, C) \otimes NB, LC]$$

with the Yoneda isomorphism between this last expression and

$$[NH, LH]_{\ell}U = \int_{V,W} [P(U, V; W) \otimes NHV, LHW];$$

this constraint is clearly invertible if $\phi^{\ell}_{2;U,V;C}$ is, and the converse comes by taking N and L to be representable. The right closed case is dual. □

7 Forms and promonoidal star autonomy

A problem with $*$-autonomy is that the common base categorics (likc thc catcgory of sets and the category of vector spaces) are not themselves $*$-autonomous. So we do not expect the convolution monoidal structure on $[\mathcal{A}, \mathcal{V}]$ to be $*$-autonomous even when $\mathcal{A}$ is. We introduce the notion of *form* to address this problem: forms do exist on base categories and carry over to convolutions, while $*$-autonomy is to be equipped with a special kind of form. The definition of a $*$-autonomous promonoidal $\mathcal{V}$-category will be expressed in terms of forms.

A *form* for a promonoidal $\mathcal{V}$-category $\mathcal{A}$ is a module $\sigma : \mathcal{A} \otimes \mathcal{A} \to \mathcal{I}$ (where $\mathcal{I}$ is the usual one-object $\mathcal{V}$-category) together with an isomorphism $\sigma \circ (P \otimes 1) \cong \sigma \circ (1 \otimes P)$. In other words, a form is a $\mathcal{V}$-functor

$$\sigma : \mathcal{A}^{\mathrm{op}} \otimes \mathcal{A}^{\mathrm{op}} \to \mathcal{V}$$

together with a $\mathcal{V}$-natural family of isomorphisms

$$\int^X \sigma(X, C) \otimes P(A, B; X) \cong \int^Y \sigma(A, Y) \otimes P(B, C; Y)$$

called *form constraints.* Indeed, we can inductively obtain isomorphisms

$$\int^X \sigma(X, A_{n+1}) \otimes P_n(A_1, \ldots, A_n; X) \cong \int^Y \sigma(A_1, Y) \otimes P_n(A_2, \ldots, A_{n+1}; Y)$$

called the *generalized form constraints.* A promonoidal $\mathcal{V}$-category with a chosen form is called *formal.*

For example, every object K of any promonoidal $\mathcal{V}$-category $\mathcal{A}$ defines a form $\sigma(A, B) = P(A, B; K)$; the form constraints are provided by the promonoidal associativity and unit constraints. Other examples are $*$-autonomous monoidal categories, as we shall soon discover. Moreover, we will also see that forms carry over to various constructions such as tensor products and general convolutions of $\mathcal{V}$-categories.

If $\mathcal{A}$ is monoidal, using Yoneda, the form constraints become

$$\sigma(A \otimes B, C) \cong \sigma(A, B \otimes C).$$

A form is called *continuous* when $\sigma(A, -)$ and $\sigma(-, B) : \mathcal{A}^{\mathrm{op}} \to \mathcal{V}$ are small (weighted) limit preserving for all objects A and B of $\mathcal{A}$.

Proposition 7.1 *Let $\mathcal{A}$ and $\mathcal{X}$ be formal promonoidal $\mathcal{V}$-categories.*

(a) If $\mathcal{A}$ and $\mathcal{X}$ are formal then the tensor product $\mathcal{A} \otimes \mathcal{X}$ with promonoidal structure

$$P_n((A_1, X_1), \ldots, (A_n, X_n); (A, X)) = P_n(A_1, \ldots, A_n; A) \otimes P_n(X_1, \ldots, X_n; X)$$

admits the form $\sigma((A, X), (B, Y)) = \sigma(A, B) \otimes \sigma(X, Y)$.

(b) If $\mathcal{A}$ is small and $\mathcal{X}$ is cocomplete closed monoidal with a continuous form then the convolution monoidal $\mathcal{V}$-category $[\mathcal{A}, \mathcal{X}]$ admits the continuous form

$$\sigma(M, N) = \int_{A,B} [\sigma(A, B), \sigma(MA, NB)].$$

Proof (a) This is trivial.

(b) We have the calculation

$$
\begin{aligned}
\sigma(M \star N, L) &= \int_{U,C} [(M \star N)U \otimes LC, \sigma(U,C)] \\
&\cong \int_{U,C} \left[\sigma(U,C), \sigma \left(\int^{A,B} P(A,B;U) \otimes MA \otimes NB, LC \right) \right] \\
&\cong \int_{U,A,B,C} [\sigma(U,C) \otimes P(A,B;U), \sigma(MA \otimes NB, LC)] \\
&\cong \int_{(,A,B,C} [\sigma(A,U) \otimes P(B,C;U), \sigma(MA, NB \otimes LC)] \\
&\cong \int_{U,A} \left[\sigma(A,U), \sigma \left(MA, \int^{B,C} P(B,C;U) \otimes NB \otimes LC \right) \right] \\
&\cong \int_{U,A} [\sigma(A,U), \sigma(MA, (N \star L)U)] \cong \sigma(M, N \star L).
\end{aligned}
$$

□

A form $\sigma : \mathcal{A} \otimes \mathcal{A} \to \mathcal{I}$ transforms under the duality of $\mathcal{V}$-modules to a $\mathcal{V}$-module $\hat{\sigma} : \mathcal{A} \to \mathcal{A}^{\mathrm{op}}$. We say the form σ is *non-degenerate* when $\hat{\sigma}$ is an equivalence as a $\mathcal{V}$-module (a Morita equivalence if you prefer). A form σ is said to be representable when there exists a $\mathcal{V}$-functor $S_\ell : A \to \mathcal{A}^{\mathrm{op}}$ and a $\mathcal{V}$-natural isomorphism

$$\sigma(A,B) \cong \mathcal{A}(A, S_\ell B).$$

A promonoidal $\mathcal{V}$-category is $*$-*autonomous* when it is equipped with a representable non-degenerate form. In fact, if $\mathcal{A}$ satisfies a minimal completeness condition ("Cauchy completeness") then "representable" is redundant. Notice that S_ℓ is necessarily an equivalence, with adjoint inverse S_r, say, and the form constraints have the cyclic appearance

$$P(A,B;S_\ell C) \cong P(B,C;S_r A).$$

More generally, using Yoneda, the generalized form constraints become

$$
\begin{aligned}
&P_n(A_1, \ldots, A_n; S_\ell A_{n+1}) \\
&\cong \int^X \mathcal{A}(X, S_\ell A_{n+1}) \otimes P_n(A_1, \ldots, A_n; X) \\
&\cong \int^X \sigma(X, A_{n+1}) \otimes P_n(A_1, \ldots, A_n; X) \simeq \int^Y \sigma(A_1, Y) \otimes P_n(A_2, \ldots, A_{n+1}; Y) \\
&\cong \int^Y \mathcal{A}(Y, S_r A_1) \otimes P_n(A_2, \ldots, A_{n+1}; Y) \cong P_n(A_2, \ldots, A_{n+1}; S_r A_1).
\end{aligned}
$$

A monoidal category is $*$-autonomous in the monoidal sense if and only if it is $*$-autonomous in the promonoidal sense.

Corollary 7.2 *In Proposition 7.1, if $\mathcal{A}$ and $\mathcal{X}$ are $*$-autonomous then so are*

(a) $\mathcal{A} \otimes \mathcal{X}$ *and* *(b)* $[\mathcal{A}, \mathcal{X}]$.

Proof (a)

$$
\begin{aligned}
\sigma((A,X),(B,Y)) &= \sigma(A,B) \otimes \sigma(X,Y) \cong \mathcal{A}(A, S_\ell B) \otimes \mathcal{X}(X, S_\ell Y) \\
&\cong (\mathcal{A} \otimes \mathcal{X})((A,X),(S_\ell B, S_\ell Y)).
\end{aligned}
$$

(b)

$$\begin{aligned}\sigma(M,N) &= \int_{A,B}[\sigma(A,B),\sigma(MA,NB)] \cong \int_{A,B}[\mathcal{A}(A,S_\ell B),\mathcal{X}(MA,S_\ell NB)] \\ &\cong \int_B \mathcal{X}(MS_\ell B,S_\ell NB) \cong [\mathcal{A},\mathcal{X}](MS_\ell,S_\ell N) \cong [\mathcal{A},\mathcal{X}](M,S_\ell NS_r).\end{aligned}$$

□

Example 7.3 As noted in the final remarks of [9], for any $\mathcal{V}$-category $\mathcal{C}$, there is a canonical promonoidal structure on $\mathcal{C}^{\mathrm{op}} \otimes \mathcal{C}$. It is explicitly defined by

$$P_0(C,D) = J(C,D) = \mathcal{C}(C,D)$$

and

$$P_2((D_1,C_1),(D_2,C_2);(C_3,D_3)) = \mathcal{C}(C_3,D_1)\otimes\mathcal{C}(C_1,D_2)\otimes\mathcal{C}(C_2,D_3).$$

More generally,

$$\begin{aligned}&P_n((D_1,C_1),\ldots,(D_n,C_n);(C_{n+1},D_{n+1})) \\ &\quad = \mathcal{C}(C_{n+1},D_1)\otimes\mathcal{C}(C_1,D_2)\otimes\cdots\otimes\mathcal{C}(C_n,D_{n+1}).\end{aligned}$$

After the lecture [12], Luigi Santocanale observed that $\mathcal{C}^{\mathrm{op}}\otimes\mathcal{C}$ is $*$-autonomous. To be precise, define $S : (\mathcal{C}^{\mathrm{op}}\otimes\mathcal{C})^{\mathrm{op}} \to \mathcal{C}^{\mathrm{op}}\otimes\mathcal{C}$ by $S(D,C) = (C,D)$. Clearly

$$\begin{aligned}&P_n((D_1,C_1),\ldots,(D_n,C_n);(C_{n+1},D_{n+1})) \\ &\quad = P_n((D_2,C_2),\ldots,(D_{n+1},C_{n+1});(C_1,D_1)),\end{aligned}$$

so that $S_r = S_\ell = S$ for $*$-autonomy. To relate this to our discussion of bialgebroids (Section 3), note that a k-algebra $\mathcal{C} = R$ is a one-object $\mathcal{V}$-category (for $\mathcal{V}$ the category of k-modules) and so the "chaotic bialgebroid" $\mathcal{C}^{\mathrm{op}}\otimes\mathcal{C} = R^e$ is $*$-autonomous.

Example 7.4 The notion of Hopf $\mathcal{V}$-algebroid appearing in Definition 21 of [15] is an example of a $*$-autonomous promonoidal $\mathcal{V}$-category. Suppose that the $\mathcal{V}$-category $\mathcal{C}$ is comonoidal [9]; that is, $\mathcal{C}$ is a pseudomonoid (or monoidal object) in $(\mathcal{V}\text{-Cat})^{\mathrm{op}}$: this means we have $\mathcal{V}$-functors $\Delta : \mathcal{C} \to \mathcal{C}\otimes\mathcal{C}$ and $E : \mathcal{C} \to \mathcal{I}$, coassociative and counital up to coherent $\mathcal{V}$-natural isomorphisms. It is easy to see that Δ must be given by the diagonal $\Delta C = (C,C)$ on objects. A multimorphism structure Q on $\mathcal{C}^{\mathrm{op}}\otimes\mathcal{C}$ is then defined by

$$Q_n(C;C_1,\ldots,C_n) = \mathcal{C}(C,C_1)\otimes\cdots\otimes\mathcal{C}(C,C_n);$$

the actions on hom-objects require the $\mathcal{V}$-functors Δ and E. Indeed, Q defines a promonoidal structure (compare Section 5 of [9]). If this promonoidal $\mathcal{V}$-category is $*$-autonomous then the condition $Q(A,B;S_\ell C) \cong Q(B,C;S_r A)$ becomes

$$\mathcal{C}(A,S_\ell C)\otimes\mathcal{C}(B,S_\ell C) \cong \mathcal{C}(B,S_r A)\otimes\mathcal{C}(C,S_r A) \cong \mathcal{C}(B,S_r A)\otimes\mathcal{C}(A,S_\ell C),$$

which precisely gives the condition

$$\mathcal{C}(A,C)\otimes\mathcal{C}(B,C) \cong \mathcal{C}(B,S_r,A)\otimes\mathcal{C}(A,C)$$

for the authors' concept of Hopf $\mathcal{V}$-algebroid [15].

A promonoidal functor $H : \mathcal{E} \to \mathcal{A}$ between $*$-autonomous promonoidal $\mathcal{V}$-categories is called *$*$-autonomous* when it is equipped with a $\mathcal{V}$-natural transformation

$$\tau^{\ell} : HS_{\ell} \to S_{\ell}H$$

such that the following diagram commutes

$$\begin{array}{ccccc} P(U,V;S_{\ell}W) & \xrightarrow{\phi_{2;U,V;S_{\ell}W}} & P(HU,HV;SH_{\ell}W) & \xrightarrow{P(1,1;\tau^{\ell})} & P(HU,HV;S_{\ell}HW) \\ \downarrow\cong & & & & \downarrow\cong \\ P(V,W;S_rU) & \xrightarrow[\phi_{2;V,W;S_rU}]{} & P(HV,HW;HS_r,U) & \xrightarrow[P(1,1;\tau^{r})]{} & P(HV,HW;S_rHU) \end{array}$$

where $\tau^r : HS_r \to S_rH$ is the mate of τ^{ℓ} under the adjunction between S_{ℓ} and S_r. We call H *strong $*$-autonomous* when τ^{ℓ} is invertible; it follows that τ^r is invertible.

Proposition 7.5 *Suppose $H : \mathcal{E} \to \mathcal{A}$ is a strong $*$-autonomous promonoidal $\mathcal{V}$-functor. If the restriction $\mathcal{V}$-functor $[H,1] : [\mathcal{A},\mathcal{V}] \to [\mathcal{E},\mathcal{V}]$ is strong monoidal then it is strong closed.*

Proof The idea of the proof is to use $*$-autonomy to cycle the criterion of Proposition 6.1 for $[H,1]$ to be strong monoidal into the criteria for it to be strong closed. The precise calculation for strong left closed is as follows:

$$\begin{aligned} &\int^{V,W} P(U,V;W) \otimes \mathcal{A}(B,HV) \otimes \mathcal{A}(HW,C) \\ \cong\ &\int^{V,W} P(U,V;W) \otimes \mathcal{A}(B,HV) \otimes \mathcal{A}(HW,S_{\ell}S_rC) \\ \cong\ &\int^{V,W} P(U,V;W) \otimes \mathcal{A}(B,HV) \otimes \mathcal{A}(S_rC,S_rHW) \\ \cong\ &\int^{V,W} P(U,V;W) \otimes \mathcal{A}(B,HV) \otimes \mathcal{A}(S_rC,HS_rW) \\ \cong\ &\int^{V,W} P(U,V;S_{\ell}W) \otimes \mathcal{A}(B,HV) \otimes \mathcal{A}(S_rC,HW) \\ \cong\ &\int^{V,W} P(U,V;S_rU) \otimes \mathcal{A}(B,HV) \otimes \mathcal{A}(S_rC,HW) \\ \cong\ &P(B,S_rC;HS_rU) \cong P(P,S_rC;S_rHU) \cong P(HU,B;S_{\ell}S_rC) \cong P(HU,B;C). \end{aligned}$$

□

The next simple observation can be useful in this context.

Proposition 7.6 *Suppose $U : \mathcal{A} \to \mathcal{X}$ is any $\mathcal{V}$-functor with a left adjoint F, and suppose there are equivalences $S : \mathcal{A} \to \mathcal{A}^{\mathrm{op}}$ and $S : \mathcal{X} \to \mathcal{X}^{\mathrm{op}}$ such that $S \circ U \cong U \circ S$. Then U has a right adjoint $S^{-1} \circ F \circ S$ and the monad $T = U \circ F$ generated by the original adjunction has a right adjoint comonad $G = U \circ S^{-1} \circ F \circ S$. Dually, F has a left adjoint $S^{-1} \circ U \circ S$. A doubly infinite string of adjunctions is thereby created.*

Proof Clearly $U : \mathcal{A}^{\mathrm{op}} \to \mathcal{X}^{\mathrm{op}}$ has F as right adjoint whereas the mutually inverse equivalences S and S^{-1} are adjoint to each other on both sides. The results now follow by composing adjunctions. □

An *opform* for a promonoidal $\mathcal{V}$-category $\mathcal{A}$ is a $\mathcal{V}$-functor

$$\sigma : \mathcal{A} \otimes \mathcal{A} \to \mathcal{V}$$

and $\mathcal{V}$-natural isomorphisms

$$\int_X [P(A, B; X), \sigma(X, C)] \cong \int_Y [P(B, C; Y), \sigma(A, Y)],$$

called *opform constraints.* For a monoidal $\mathcal{V}$-category, we see by Yoneda's Lemma that an opform on $\mathcal{A}$ is the same as a form on $\mathcal{A}^{\mathrm{op}}$. Moreover, in general, if σ is a form on $\mathcal{A}$ and K is any object of $\mathcal{V}$ then an opform σ_K on $\mathcal{A}$ is defined by the equation

$$\sigma_K(A, B) = [\sigma(A, B), K].$$

Proposition 7.7 *Let $\mathcal{A}$ be a small promonoidal $\mathcal{V}$-category. Each opform σ for $\mathcal{A}$ determines a continuous form for the convolution monoidal $\mathcal{V}$-category $[\mathcal{A}, \mathcal{V}]$ via the formula*

$$\sigma(M, N) = \int_{A,B} [MA \otimes NB, \sigma(A, B)].$$

Furthermore, every continuous form on $[\mathcal{A}, \mathcal{V}]$ arises thus from an opform on $\mathcal{A}$.

Proof We have the calculation

$$\begin{aligned}
\sigma(M \star N, L) &= \int_{U,C} [(M \star N)U \otimes LC, \sigma(U, C)] \\
&\cong \int_{U,C} \left[\int^{A,B} P(A, B; U) \otimes MA \otimes NB \otimes LC, \sigma(U, C) \right] \\
&\cong \int_{U,A,B,C} [MA \otimes NB \otimes LC, [P(A, B; U), \sigma(U, C)]] \\
&\cong \int_{U,A,B,C} [MA \otimes NB \otimes LC, [P(B, C; U), \sigma(A, U)]] \\
&\cong \int_{U,A,B,C} [MA, [P(B, C; U) \otimes NB \otimes LC, \sigma(A, U)]] \\
&\cong \int_{U,A} \left[MA, \left[\int^{B,C} P(B, C; U) \otimes NB \otimes LC, \sigma(A, U) \right] \right] \\
&\cong \sigma(M, N \star L).
\end{aligned}$$

Conversely, any continuous form σ on $[\mathcal{A}, \mathcal{V}]$ will have

$$\begin{aligned}
\sigma(M, N) &\cong \sigma\left(\int^A MA \otimes \mathcal{A}(A, -), \int^B NB \otimes \mathcal{A}(B, -) \right) \\
&\cong \int_{A,B} [MA \otimes MB, \sigma(\mathcal{A}(A, -), \mathcal{A}(B, -))].
\end{aligned}$$

so that σ will be determined by its value on representables. We define σ for $\mathcal{A}$ by

$$\sigma(A, B) = \sigma(\mathcal{A}(A, -), \mathcal{A}(B, -)).$$

We have the calculation

$$
\begin{aligned}
&\int_U [P(A,B;U), \sigma(\mathcal{A}(U,-), \mathcal{A}(C,-))] \\
&\cong \ \sigma\left(\left(\int^U P(A,B;U) \otimes \mathcal{A}(U,-), \mathcal{A}(C,-)\right)\right) \\
&\cong \ \sigma(P(A,B;-), \mathcal{A}(C,-)) \cong \sigma(\mathcal{A}(A,-) \star \mathcal{A}(B,-), \mathcal{A}(C,-)) \\
&\cong \ \sigma(\mathcal{A}(A,-), \mathcal{A}(B,-) \star \mathcal{A}(C,-)) \\
&\cong \ \sigma(\mathcal{A}(A,-), P(B,C;-)) \cong \int_V [P(B,C;V), \sigma(\mathcal{A}(A,-), \mathcal{A}(V,-))].
\end{aligned}
$$

□

8 The star and Chu constructions

We adhere to the spirit of the review [36] where the Chu construction is defined at the multimorphism level. The star construction on a multimorphism structure yields one that is $*$-autonomous. When applied to a promonoidal $\mathcal{V}$-category, the result may not be promonoidal — hence the need to work at the more general level.

For that, we define a general multimorphism structure to be *$*$-autonomous* when there exists an equivalence $S_\ell : \mathcal{A} \to \mathcal{A}^{\text{op}}$ of $\mathcal{V}$-categories and a sequence of $\mathcal{V}$-natural isomorphisms

$$P_n(A_1, \ldots, A_n; S_\ell A_{n+1}) \cong P_n(A_2, \ldots, A_{n+1}; S_r A_1)$$

where S_r is an adjoint inverse for S_ℓ.

In this section we will show how to modify a multimorphism structure, with a prescribed S_ℓ, to obtain a $*$-autonomous one with the same S_ℓ. We first need a natural definition: an *equivalence* $F : \mathcal{A} \to \mathcal{B}$ of multimorphism structures is an equivalence F of $\mathcal{V}$-categories together with natural isomorphisms

$$P_n(A_1, \ldots, A_n; A) \cong P_n(FA_1, \ldots, FA_n; FA);$$

the inverse equivalence of F is obviously also a multimorphism equivalence.

Notice that, for any $*$-autonomous multimorphism structure, $S_\ell \circ S_\ell : \mathcal{A} \to \mathcal{A}$ is a multimorphism equivalence: for we have the calculation

$$
\begin{aligned}
&P_n(A_1, \ldots, A_n; A) \\
&\cong \ P_n(A_1, \ldots, A_n; S_r S_\ell A) \\
&\cong \ P_n(S_\ell A, A_1, \ldots, A_{n-1}; S_\ell A_n) \cong P_n(S_\ell A, A_1, \ldots, A_{n-1}; S_r S_\ell S_\ell A_n) \\
&\cong \ \cdots \cong P_n(S_\ell S_\ell A_2, \ldots, S_\ell S_\ell A_n, S_\ell A; S_\ell A_1) \cong P_n(S_\ell S_\ell A_1, \ldots, S_\ell S_\ell A_n; S_\ell S_\ell A).
\end{aligned}
$$

Now to our construction. Suppose we have a multimorphism structure P on any $\mathcal{V}$-category $\mathcal{A}$ equipped with a contravariant $\mathcal{V}$-functor $S_\ell : \mathcal{A}^{\text{op}} \to \mathcal{A}$ such that $S_\ell \circ S_\ell : \mathcal{A} \to \mathcal{A}$ is an equivalence of multimorphism structures. It follows that S_ℓ is an equivalence; we write S_r for the adjoint equivalence. The *starring* of this situation is the multimorphism structure P^* on $\mathcal{A}$ defined by the formula

$$
\begin{aligned}
&P_n^*(X_1, \ldots, X_n; S_\ell X_{n+1}) \\
&= \ \int^{U_{ij} (1 \le i < j \le n+1)} \bigotimes_{m=1}^{n+1} P_n(U_{m\,m+1}, \ldots U_{m\,n+1}, S_r U_{1\,m}, \ldots, S_r U_{m-1\,m}; S_r X_m).
\end{aligned}
$$

Proposition 8.1 *The starring P^* produces a $*$-autonomous multimorphism structure on $\mathcal{A}$ with the given S_ℓ.*

Proof Extend the definition of the U_{ij} and X_i by putting $U_{ji} = S_r U_{ij}$ and $X_{n+i+1} = S_r S_r X_i$. From the definition, we have

$$P_n^*(X_2, \ldots, X_{n+1}; S_r X_1)$$
$$= \int^{V_{ij}(1 \le i < j \le n+1)} \bigotimes_{m=1}^{n+1} P_n(V_{m\,m+1}, \ldots V_{m\,n+1}, S_r V_{1\,m}, \ldots, S_r V_{m-1\,m}; S_r X_{m+1}),$$

which we notice is isomorphic to the formula for $P_n^*(X_1, \ldots, X_n; S_\ell X_{n+1})$ on making the change of variables $V_{ij} = U_{i+1\,j+1}$ and using the isomorphisms

$$P_n(U_{12}, \ldots, U_{1\,n+1}; S_r X_1) \cong P_n(S_r S_r U_{12}, \ldots, S_r S_r U_{1\,n+1}; S_r S_r S_r X_1).$$

□

Let $\mathcal{C}$ be a $\mathcal{V}$-category with a multimorphism structure P and a multimorphism equivalence $T : \mathcal{C} \to \mathcal{C}$. We suppose furthermore that $\mathcal{C}$ is a comonoidal $\mathcal{V}$-category with derived promonoidal structure Q as made explicit in Example 7.4. We require that $T : \mathcal{C}^{\mathrm{op}} \to \mathcal{C}^{\mathrm{op}}$ is an equivalence for the multimorphism structure Q (that is, that $T : \mathcal{C} \to \mathcal{C}$ is a comonoidal equivalence).

We want to apply the star construction to $\mathcal{A} = \mathcal{C}^{\mathrm{op}} \otimes \mathcal{C}$ with $S_\ell(C, D) = (D, T^{-1}C)$, so that $S_r(C, D) = (TD, C)$, and with the tensor product multimorphism structure $Q \otimes P$ for the P and Q as described in the last paragraph. Notice $S_\ell S_\ell(C, D) = (T^{-1}C, T^{-1}D)$ so that $S_\ell \circ S_\ell : \mathcal{C}^{\mathrm{op}} \otimes \mathcal{C} \to \mathcal{C}^{\mathrm{op}} \otimes \mathcal{C}$ is indeed a multimorphism equivalence.

Let us calculate the star R^* of $R = Q \otimes P$:

$$R_n^*\left((X_1, Y_1), \ldots, (X_n, Y_n); (Y_{n+1}, T^{-1}X_{n+1})\right)$$
$$= \int^{(U_{ij}, V_{ij})} \bigotimes_{m=1}^{n+1} \begin{pmatrix} Q_n(TY_m; U_{m\,m+1}, \ldots, U_{m\,n+1}, TV_{1\,m}, \ldots, TV_{m-1\,m}) \\ \otimes P_n(V_{m\,m+1}, \ldots, V_{m\,n+1}, U_{1\,m}, \ldots, U_{m-1\,m}; X_m) \end{pmatrix}$$
$$\cong \int^{(U_{ij}, V_{ij})} \bigotimes_{m=1}^{n+1} \begin{pmatrix} \mathcal{C}(TY_m, U_{m\,m+1}) \otimes \cdots \otimes \mathcal{C}(TY_m, U_{m\,n+1}) \\ \otimes \mathcal{C}(Y_m, V_{1\,m}) \otimes \cdots \otimes \mathcal{C}(Y_m, V_{m-1\,m}) \\ \otimes P_n(V_{m\,m+1}, \ldots, V_{m\,n+1}, U_{1\,m}, \ldots, U_{m-1\,m}; X_m) \end{pmatrix}$$
$$\cong \int^{(U_{ij}, V_{ij})} \bigotimes_{r<s} (\mathcal{C}(TY_r, U_{rs}) \otimes \mathcal{C}(Y_s, V_{rs}))$$
$$\otimes \bigotimes_{m=1}^{n+1} P_n(V_{m\,m+1}, \ldots, V_{m\,n+1}, U_{1\,m}, \ldots, U_{m-1\,m}; X_m)$$
$$\cong \bigotimes_{m=1}^{n+1} P_n(Y_{m+1}, \ldots, Y_{n+1}, TY_1, \ldots, TY_{m-1}; X_m)$$

which has the same shape as the multimorphism structure described in [36].

Proposition 8.2 *In the situation just described, if P is actually a monoidal structure on $\mathcal{C}$, then R^* is a $*$-autonomous promonoidal structure on $\mathcal{C}^{\mathrm{op}} \otimes \mathcal{C}$.*

Proof After Proposition 8.1, it suffices to show that R^* is promonoidal. We need to see that each R_n^* is determined by the $n = 0$ and $n = 2$ cases. The general calculation is by induction so we trust that the following exemplary step will be sufficient indication for the reader:

$$
\begin{aligned}
&\int^{A_1,B_1} R_2^*\left((X_1,Y_1),(X_2,Y_2);(B_1,T^{-1}A_1)\right) \\
&\qquad\qquad \otimes R_2^*\left((TB_1,A_1),(X_3,Y_3);(B_4,T^{-1}A_4)\right) \\
&\cong \int^{A_1,B_1} \begin{pmatrix} P_2(Y_2,B_1;X_1)\otimes P_2(B_1,TY_1;X_2)\otimes P_2(TY_1,TY_2;A_1) \\ \otimes P_2(Y_3,Y_4;TB_1)\otimes P_2(Y_4,A_1;X_3)\otimes P_2(A_1,TY_3;X_4) \end{pmatrix} \\
&\cong \int^{A_1,B_1} \begin{pmatrix} \mathcal{C}(Y_2\otimes B_1,X_1)\otimes \mathcal{C}(B_1\otimes TY_1,X_2)\otimes \mathcal{C}(TY_1\otimes TY_2,A_1) \\ \otimes \mathcal{C}(Y_3\otimes Y_4,TB_1)\otimes \mathcal{C}(Y_4\otimes A_1,X_3)\otimes \mathcal{C}(A_1\otimes TY_3,X_4) \end{pmatrix} \\
&\cong \begin{pmatrix} \mathcal{C}(Y_2\otimes T(Y_3\otimes Y_4),X_1)\otimes \mathcal{C}(T(Y_3\otimes Y_4)\otimes TY_1,X_2) \\ \otimes \mathcal{C}(Y_4\otimes TY_1\otimes TY_2,X_3)\otimes \mathcal{C}(TY_1\otimes TY_2\otimes TY_3,X_4) \end{pmatrix} \\
&\cong P_3(Y_2,TY_3,TY_4;X_1)\otimes P_3(TY_3,TY_4,TY_1;X_2)\otimes P_3(Y_4,TY_1,TY_2;X_3) \\
&\qquad\qquad \otimes P_3(TY_1,TY_2,TY_3;X_4) \\
&\cong R_3^*\left((X_1,Y_1),(X_2,Y_2),(X_3,Y_3);(Y_4,T^{-1}X_4)\right).
\end{aligned}
$$

□

Proposition 8.3 *In the situation of the Proposition 8.2, further suppose that P is closed monoidal and that the comonoidal structure on $\mathcal{C}$ is representable by an object K, an operation $B \bullet C$, and $\mathcal{V}$-natural isomorphisms*

$$
\mathcal{C}(A,K)\cong I \qquad \text{and} \qquad \mathcal{C}(A,B\bullet C)\cong (\mathcal{C}(A,B)\otimes \mathcal{C}(A,C)
$$

where the right-hand sides require the counit and comultiplication for their effects on homs. Then R^* is a $*$-autonomous monoidal structure on $\mathcal{C}^{\mathrm{op}}\otimes\mathcal{C}$.

Proof We have the calculations:

$$
\begin{aligned}
&R_2^*(X_1,Y_1),(X_2,Y_2);(Y_3,T^{-1}X_3)) \\
&\quad\cong\ P_2(Y_2,Y_3;X_1)\otimes P_2(Y_3,TY_1;X_2)\otimes P_2(TY_1,TY_2;X_3) \\
&\quad\cong\ \mathcal{C}(Y_2\otimes Y_3,X_1)\otimes \mathcal{C}(Y_3\otimes TY_1,X_2)\otimes \mathcal{C}(TY_1\otimes TY_2,X_3) \\
&\quad\cong\ \mathcal{C}\left(Y_3,[Y_2,X_1]_r\right)\otimes \mathcal{C}(Y_3,[TY_1,X_2]_\ell)\otimes \mathcal{C}(TY_1\otimes TY_2,X_3) \\
&\quad\cong\ \mathcal{C}\left(Y_3,[Y_2,X_1]_r\bullet[TY_1,X_2]_\ell\right)\otimes \mathcal{C}(Y_1\otimes Y_2,T^{-1}X_3) \\
&\quad\cong\ (\mathcal{C}^{\mathrm{op}}\otimes\mathcal{C})\left(([Y_2,X_1]_r\bullet[TY_1,X_2]_\ell,Y_1\otimes Y_2),(Y_3,T^{-1}X_3)\right)
\end{aligned}
$$

and

$$
\begin{aligned}
R_0^*(Y,T^{-1}X) \ &\cong\ P_0(X)\cong \mathcal{C}(I,X)\cong \mathcal{C}(Y,K)\otimes \mathcal{C}(T^{-1}I,T^{-1}X) \\
&\cong\ (\mathcal{C}^{\mathrm{op}}\otimes\mathcal{C})((K,I),(Y,T^{-1}X)).
\end{aligned}
$$

so that $\mathcal{C}^{\mathrm{op}}\otimes\mathcal{C}$ is monoidal with unit (K,I) and tensor product

$$
(X_1,Y_1)\otimes(X_2,Y_2)=([Y_2,X_1]_r\bullet[TY_1,X_2]_\ell,Y_1\otimes Y_2).
$$

□

A particular case of Proposition 8.3 is the Chu construction of [3]. Here $\mathcal{V}$ is the category of sets with cartesian monoidal structure (although any cartesian closed base would do). Then every $\mathcal{V}$-category $\mathcal{C}$ is comonoidal via the diagonal functor Δ. The representability of this structure as required in Proposition 8.3 amounts to $\mathcal{C}$ having finite limits; so K is the terminal object and $B \bullet C = B \times C$ is the product of B and C. Then R^* is the $\ast$-autonomous monoidal structure on $\mathcal{C}^{\mathrm{op}} \otimes \mathcal{C}$ arising from any monoidal closed category $\mathcal{C}$ with finite products and a monoidal endoequivalence T.

However, the case of finite products for ordinary categories is not the only example where the representable comonoidal structure can be found. For any $\mathcal{V}$, such structure exists for example on any $\mathcal{C}$ which is a free $\mathcal{V}$-category on an ordinary category with finite products.

9 Star autonomy in monoidal bicategories

In order to exploit duality, we need to generalise the notion of star autonomy to pseudomonoids in a monoidal bicategory $\mathcal{B}$. The work of Sections 5 to 8 is a special case taking place in the autonomous monoidal bicategory $\mathcal{V}$-Mod of $\mathcal{V}$-categories and $\mathcal{V}$-modules as defined in Section 6.

As mentioned in Section 3, for pseudomonoids A and E in $\mathcal{B}$, where we write p and j for the multiplications and units, *a monoidal morphism* $g : A \to E$ is a morphism equipped with coherent 2-cells

$$\phi_2 : p \circ (g \otimes g) \Rightarrow g \circ p \text{ and } \phi_0 : j \Rightarrow g \circ j.$$

The morphism is called *strong monoidal* when ϕ_2 and ϕ_0 are invertible. When g has a left adjoint h, there are 2-cells

$$\phi_2^\ell : h \circ p \circ (1 \otimes g) \Rightarrow p \circ (h \otimes 1) \text{ and } \phi_2^r : h \circ p \circ (g \otimes 1) \Rightarrow p \circ (1 \otimes h)$$

obtained from ϕ_2 as mates under adjunction. We say g is *strong left [right] closed* when ϕ_2^ℓ [respectively, ϕ_2^r] is invertible; it is strong closed when it is both.

For a pseudomonoid A in $\mathcal{B}$, the category $\mathcal{B}(I, A)$ is monoidal with tensor product defined by

$$m \star n = p \circ (m \otimes n).$$

The internal homs, provided $\mathcal{B}$ has the relevant right liftings, are defined as follows: $[n, r]_\ell$ is the right lifting of r through $p \circ (1_A \otimes n)$ while $[m, r]_r$ is the right lifting of r through $p \circ (m \otimes 1_A)$.

Proposition 9.1 *If $g : A \to E$ is a strong monoidal morphism between pseudomonoids then $\mathcal{B}(I, g) : \mathcal{B}(I, A) \to B(I, E)$ is a strong monoidal functor. If g has a left adjoint h and is strong closed then the functor $B(I, g)$ is strong closed.*

Proof For the first sentence we have

$$\begin{aligned}\mathcal{B}(I, g)(m \star n) &= g \circ p \circ (m \otimes n) \cong p \circ (g \otimes g) \circ (m \otimes n) \\ &\cong p \circ (g \circ m) \otimes (g \circ n) \cong (g \circ m) \star (g \circ n) \\ &\cong \mathcal{B}(I, g)(m) \star \mathcal{B}(I, g)(n).\end{aligned}$$

For the second sentence consider the diagram

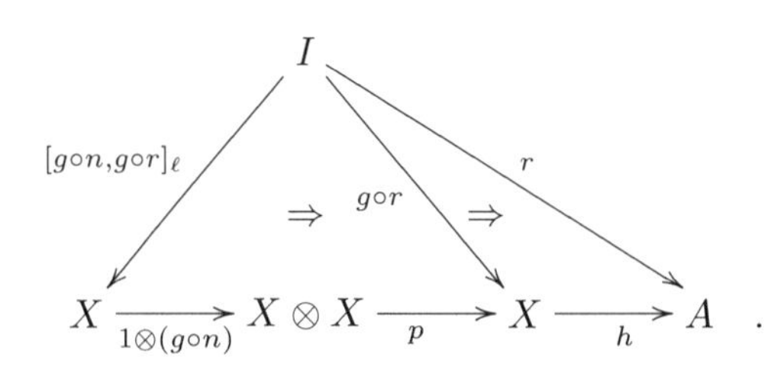

The right-hand triangle is a right lifting since h is left adjoint to g. The left-hand triangle is a right lifting by definition of the left internal hom. So the outside triangle exhibits $[g \circ n, g \circ r]_\ell$ as a right lifting of r along the bottom composite. However, if g is strong left closed, the bottom composite is isomorphic to

$$h \circ p \circ (1 \otimes g) \circ (1 \otimes n) \cong p \circ (h \otimes 1) \circ (1 \otimes n) \cong p \circ (1 \otimes n) \circ h.$$

However, the right lifting of r through $p \circ (1 \otimes n)$ is $[n, r]_\ell$, and the right lifting of $[n, r]_\ell$ through h is $g \circ [n, r]_\ell$. So we have $g \circ [n, r]_\ell \cong [g \circ n, g \circ r]_\ell$ proving $\mathcal{B}(I, g)$ strong left closed. Right closedness is dual. □

A *form* for a pseudomonoid A in $\mathcal{B}$ is a morphism $\sigma : A \otimes A \to I$ together with an isomorphism

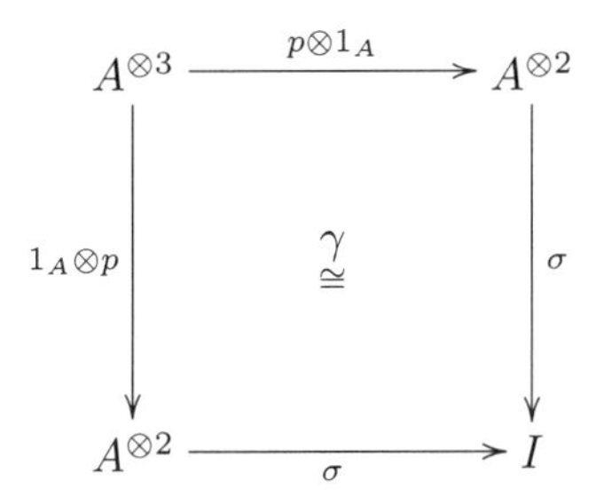

called the *form constraint*.

In the bicategories $\mathcal{B}$ that we have in mind there are special morphisms (as abstracted by Wood [40]). The special morphisms h have right adjoints h^* and, in some cases, are precisely the morphisms with right adjoints, sometimes called *maps* in $\mathcal{B}$. For example, in Mat(Set) the maps are precisely the matrices arising from functions, and these are the special morphisms we want. For Mod($\mathcal{V}$), the special morphisms are those modules isomorphic to h_* for some algebra morphism h. In the bicategory of $\mathcal{V}$-categories and $\mathcal{V}$-modules the special modules are those arising from $\mathcal{V}$-functors.

Suppose $\mathcal{B}$ has selected special maps and that $\mathcal{B}$ is autonomous. Each form $\sigma : A \otimes A \to I$ corresponds to a morphism $\hat{\sigma} : A \to A^0$. We say that the form σ is *representable* when $\hat{\sigma}$ is isomorphic to a special map. We say that σ is *non-degenerate* when $\hat{\sigma}$ is an equivalence.

A pseudomonoid in $\mathcal{B}$ is defined to be $*$-*autonomous* when it is equipped with a non-degenerate representable form. For example, for any object R of $\mathcal{B}$ and any equivalence $\upsilon : R \to R^{00}$ the canonical endohom pseudomonoid $R^e = R^0 \otimes R$ becomes $*$-autonomous when equipped with the form $\sigma : R^e \otimes R^e \to I$ defined by

$$\hat{\sigma} = 1_{R^0} \otimes \upsilon : R^e = R^0 \otimes R \to R^0 \otimes R^{00} = R^{e0}.$$

An opmorphism $h : E \to A$ between $*$-autonomous pseudomonoids is called $*$-autonomous when there is an isomorphism

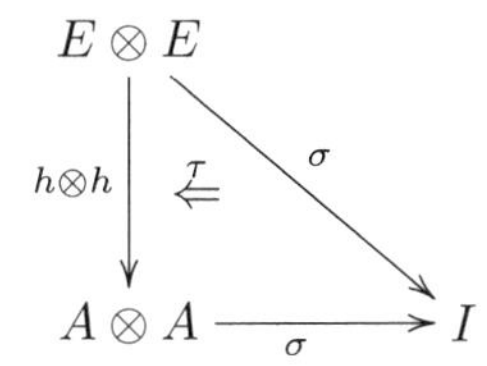

such that the following equation holds:

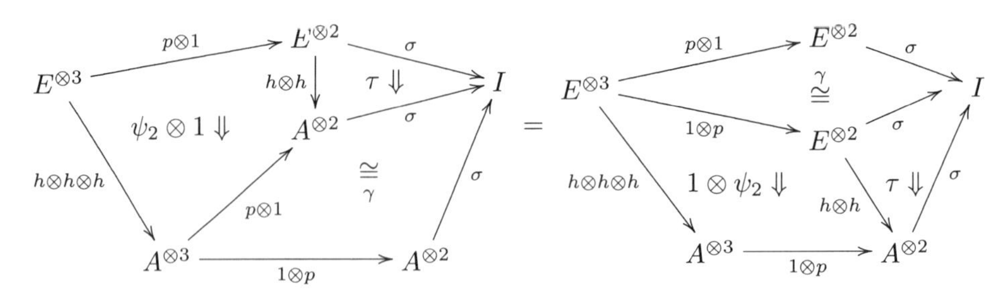

We are particularly interested in opmorphisms h that are maps. Then the right adjoint h^* is a morphism of pseudomonoids. Under these circumstances we define h to be *strong $*$-autonomous* when the mate

$$\tau^\ell : \sigma \circ (h^* \otimes 1) \Rightarrow \sigma \circ (1 \otimes h)$$

of τ is invertible. It follows that $\tau^r : \sigma \circ (1 \otimes h^*) \Rightarrow \sigma \circ (h \otimes 1)$ is also invertible.

Proposition 9.2 *Suppose $h : E \to A$ is a strong $*$-autonomous special opmorphism between $*$-autonomous pseudomonoids in $\mathcal{B}$. If h^* is strong monoidal then h^* is strong closed.*

Proof We have the calculation

$$\begin{aligned}
&\sigma \circ (h \otimes 1) \circ (p \otimes 1) \circ (1 \otimes h^* \otimes 1)\\
&\cong \quad \sigma \circ (1 \otimes h^*) \circ (p \otimes 1) \circ (1 \otimes h^* \otimes 1) \cong \sigma \circ (p \otimes 1) \circ (1 \otimes h^*) \circ (1 \otimes h^* \otimes 1)\\
&\cong \quad \sigma \circ (1 \otimes p) \circ (1 \otimes h^* \otimes h^*) \cong \sigma \circ (1 \otimes h^*) \circ (1 \otimes p)\\
&\cong \quad \sigma \circ (h \otimes 1) \circ (1 \otimes p) \cong \sigma \circ (1 \otimes p) \circ (h \otimes 1 \otimes 1)\\
&\cong \quad \sigma \circ (1 \otimes p) \circ (h \otimes 1 \otimes 1).
\end{aligned}$$

It follows that $\hat{\sigma} \circ h \circ p \circ (1 \otimes h^*) \cong \hat{\sigma} \circ p \circ (h \otimes 1)$. Left strong closedness follows since σ is non-degenerate. Right closedness is dual. □

Motivated by Proposition 3.3, we define *basic data* in an autonomous monoidal bicategory $\mathcal{B}$ to consist of an object R equipped with a special opmorphism $h : R^0 \otimes R \to A$ into a pseudomonoid A such that h^* is strong monoidal. Here $R^0 \otimes R$ has the canonical endohom pseudomonoid structure. Suppose further that $R^e = R^0 \otimes R$ is $*$-autonomous via a form arising as above from an equivalence $\upsilon : R \to R^{00}$. The basic data is called *Hopf* when A is equipped with a $*$-autonomous structure and h is strong $*$-autonomous.

From basic data, by applying the pseudofunctor $\mathcal{B}(I, -) : \mathcal{B} \to \mathrm{Cat}$, we obtain an adjunction

$$\mathcal{B}(I, h) \dashv \mathcal{B}(I, h^*) : \mathcal{B}(I, A) \to \mathcal{B}(I, R^0 \otimes R)$$

which transports via the equivalence $\mathcal{B}(I, R^0 \otimes R) \xrightarrow{\sim} \mathcal{B}(R, R)$ to an adjunction between $\mathcal{B}(I, A)$ and $\mathcal{B}(R, R)$. The pseudomonoid structure on A induces a monoidal structure on $\mathcal{B}(I, A)$ and the canonical endohom pseudomonoidal structure on $R^0 \otimes R$ induces the monoidal structure on $\mathcal{B}(R, R)$ whose tensor product is composition in $\mathcal{B}$. Since h^* is strong monoidal, the right adjoint $\mathcal{B}(I, A) \to \mathcal{B}(R, R)$ is strong monoidal. By Propositions 9.1 and 9.2, this right adjoint is also strong closed in the Hopf case.

Since basic and Hopf basic data are expressible purely in terms of the monoidal bicategory structure and the special maps of $\mathcal{B}$, the next result is clear.

Proposition 9.3 *Strong monoidal pseudofunctors that preserve special maps also preserve basic and Hopf basic data.*

Remark 9.4 The day after we submitted this paper to the Fields Workshop organizers, the preprint [7] appeared on math.arXiv. We contacted Dr. Gabriella Böhm who pointed out that, in our original preprint, we had not been specific about the $*$-autonomous structure on $R^e = R^0 \otimes R$ in our definition of Hopf basic data. This was indeed an omission and we had in mind the symmetric case where we had the opportunity to take $R^{00} = R$ and $\hat{\sigma} : R^0 \otimes R \to R^0 \otimes R^{00}$ the identity.

10 Ordinary groupoids revisited

Let us return to the definition of ordinary category as formulated in Propositions 1.1 and 2.1. Let G be a monoidal comonad on the internal endohom pseudomonoid $X \times X$ in the monoidal bicategory Mat(Set). Recall that $G(x, y; u, v)$ is empty unless $x = u$ and $y = v$, and we put

$$\mathbf{A}(x, y) = G(x, y; x, y)$$

which defines the homsets of our category $\mathbf{A}$. Let A denote the set of arrows of the category $\mathbf{A}$; we have the triangle

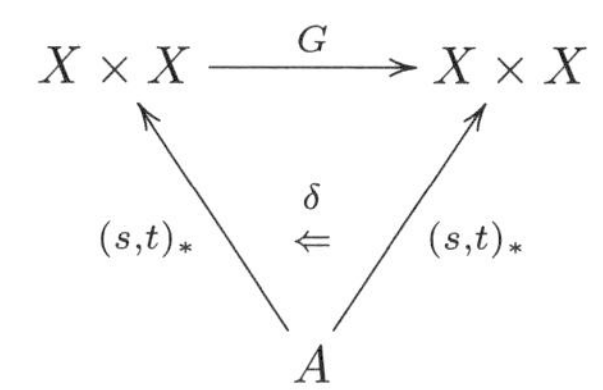

which is the universal coaction of G on a morphism into $X \times X$; it is the Eilenberg-Moore construction for the comonad G. By a dual of Lemma 3.2, there is a pseudomonoid structure on A such that the whole triangle lifts to the Eilenberg-Moore construction in the bicategory MonMat(Set).

We already pointed out in the Introduction what the pseudomonoidal structure on A is; that on $X \times X$ is the special case of a chaotic category. Referring to the definition of basic data at the end of Section 9, we have:

Proposition 10.1 *An equivalent definition of ordinary small categories is that they are basic data in the autonomous monoidal bicategory Mat(Set)^{co} where the special morphisms are all the maps.*

Proof Reversing 2-cells interchanges left and right adjunctions. So for a morphism to have a right adjoint in Mat(Set)^{co} is to be a right adjoint in Mat(Set); that is, to be the reverse of a function. Basic data in Mat(Set)^{co} therefore consists of a set X , a pseudomonoid A in Mat(Set), and a function $(s, t) : A \to X \times X$ that is strong monoidal. The functor $\mathcal{B}(I, h^*)$ as at the end of Section 9 transports to the left-adjoint functor

$$\Sigma_{(s,t)} : \text{Set}/A \to \text{Set}/X \times X$$

of the Introduction, which by Section 9 is strong monoidal. So we have a category $\mathbf{A}$. Conversely, if $\mathbf{A}$ is a category, clearly (s, t) is strong monoidal. $\square$

The discussion of the Introduction already shows that, if $\mathbf{A}$ is a groupoid, then it is $*$-autonomous in Mat(Set) with $Sa = a^{-1}$. In particular (the chaotic case), the endohom $X \times X$ is $*$-autonomous with $S(x, y) = (y, x)$. For $\mathbf{A}$ a groupoid,

$(s,t)^* : X \times X \longrightarrow A$ is a strong $*$-autonomous map in $\mathrm{Mat(Set)^{co}}$. So we have Hopf basic data in $\mathrm{Mat(Set)^{co}}$. The converse almost holds.

Proposition 10.2 *Consider a category as basic data in Mat(Set)^{co}. The category is a groupoid iff the basic data are Hopf.*

Proof The characterizing property of $S = S_\ell$ is that

$$b \circ a = Sc \qquad \text{iff} \qquad c \circ b = S^{-1}a.$$

For each object x, put $e_x = S1_x$. Taking $c = 1_x$ and $b = S^{-1}a$ to ensure $c \circ b = S^{-1}a$, we deduce that $S^{-1}a \circ a = e_x$ for all $a : x \longrightarrow y$. Taking $a = 1_x$ we see that $e_x = S^{-1}1_x$ so $Se_x = 1_x$. Now go back to the characterizing property with $c = e_x$, b arbitrary, and $a = S(e_x \circ b)$ to ensure $c \circ b = S^{-1}a$: so we deduce that $b \circ S(e_x \circ b) = Se_x = 1_x$. It follows that every morphism b has a right inverse. So the category is a groupoid. □

Remark 10.3 (This arose in lunchtime conversation with John Baez and Isar Stubbe.) The operation S_ℓ of $*$-autonomy is not unique. For a groupoid $\mathbf{A}$ as we have been considering, we can choose any endomorphism e_x of each object x and define $S_\ell a = a^{-1} \circ e_x$ so that $S_r a = e_x \circ a^{-1}$. This defines another $*$-autonomous structure on our pseudomonoid A.

Remark 10.4 The argument of this section can be internalized to any finitely complete category $\mathcal{E}$. In particular, groupoids internal to $\mathcal{E}$ can be identified with Hopf basic data in the monoidal bicategory $\mathrm{Span}(\mathcal{E})^{co}$. More details will be provided in Example 12.3.

11 Hopf bialgebroids

A bialgebroid A based on a k-algebra R is an opmonoidal monad on R^e in $\mathrm{Mod}(\mathcal{V})$ (see Section 3). We have already seen that A becomes a k-algebra and that $\eta^* : A \longrightarrow R^e$ provides the Eilenberg-Moore object for the monad, thereby lifting to the bicategory of pseudomonoids in $\mathrm{Mod}(\mathcal{V})$.

In the terminology of Section 9, a bialgebroid is precisely basic data $\eta : R^e \longrightarrow A$ in $\mathcal{B} = \mathrm{Mod}(\mathcal{V})$. We define a bialgebroid $\eta : R^e \longrightarrow A$ to be *Hopf* when this basic data in $\mathrm{Mod}(\mathcal{V})$ is Hopf; that is, A should be $*$-autonomous and $\eta^* : A \longrightarrow R^e$ should be strong $*$-autonomous. It follows from Section 9 that $\mathrm{Mod}(\mathcal{V})(k, \eta^*)$ is strong monoidal and strong closed; this is none other than the functor

$$\mathcal{V}^A \longrightarrow \mathcal{V}^{R^e}$$

defined by restriction along $\eta : R^e \longrightarrow A$; compare Proposition 7.5 in the case of one-object $\mathcal{V}$-categories.

Preservation of internal homs was taken as paramount in the Hopf algebroid notions of [15] and [31]. Example 7.4 explains the connection between our work here and that of [15] while we see from the last paragraph that our Hopf bialgebroids are more restrictive than the Hopf algebroids of [31].

Remark 11.1 In the correspondence mentioned in Remark 9.4 Dr. Böhm advised us that her notion of Hopf bialgebroid in [7] fits our setting, where $\mathcal{V}$ is the category of vector spaces, and that she has examples where the $*$-autonomous structure S_ℓ on $R^e = R^0 \otimes R$ is defined by

$$S_\ell(x \otimes y) = x \otimes u(y)$$

with u a non-identity k-algebra automorphism of R. This kind of perturbation fits well with our treatment of the Chu construction in Section 8.

Example 11.2 Let $\mathcal{V}$ continue to be the category of k-vector spaces and let $\mathcal{A}$ denote the category of commutative k-algebras. The category $\mathcal{A}$ is finitely cocomplete; the pushout of two morphisms out of an object A is given by tensoring over A the codomains of the two morphisms. Definition B.3.7 of [29] labels groupoids internal to $\mathcal{A}^{\mathrm{op}}$ as "Hopf algebroids" (generalizing the idea that a commutative Hopf algebra is exactly a group in $\mathcal{A}^{\mathrm{op}}$). In fact, these are examples of Hopf bialgebroids in our sense. To see this we make use of the strong monoidal pseudofunctor

$$\mathrm{Span}(\mathcal{A}^{\mathrm{op}})^{\mathrm{co}} \to \mathrm{Mod}(\mathcal{V})$$

which takes each commutative algebra A to itself as an algebra and each cospan C from A to B in $\mathcal{A}$ to C with actions of A and B coming from the morphisms into C. By Remark 10.4, each Ravenel "Hopf algebroid" is Hopf basic data in $\mathrm{Span}(\mathcal{A}^{\mathrm{op}})^{\mathrm{co}}$. Then, by Proposition 9.3 our pseudofunctor applies to give Hopf basic data in $\mathrm{Mod}(\mathcal{V})$; that is, to give a Hopf bialgebroid. We are grateful to Terry Bisson for pointing out the book [29] which features good examples of groupoids internal to $\mathcal{A}^{\mathrm{op}}$ occurring in algebraic topology.

12 Quantum categories and quantum groupoids

It remains to state the main definitions of the paper. We now have the motivation and concepts readily at hand.

Let $\mathcal{V}$ be a braided monoidal category with coreflexive equalizers (that is, equalizers of pairs of morphisms with a common left inverse). We begin by recalling the definition of the right autonomous monoidal bicategory $\mathrm{Comod}(\mathcal{V})$ as appearing in [14]. We assume the condition:

each of the functors $X \otimes - : \mathcal{V} \to \mathcal{V}$ preserves coreflexive equalizers.

Briefly, $\mathrm{Comod}(\mathcal{V}) = \mathrm{Mod}(\mathcal{V}^{\mathrm{op}})^{\mathrm{coop}}$. To make calculations we will need to make the definition more explicit.

The objects of $\mathrm{Comod}(\mathcal{V})$ are comonoids C in $\mathcal{V}$; the comultiplication and counit are denoted by $\delta : C \to C \otimes C$ and $\varepsilon : C \to I$. The hom-category $\mathrm{Comod}(\mathcal{V})(C, D)$ is the category of Eilenberg-Moore coalgebras for the comonad $C \otimes - \otimes D$ on the category $\mathcal{V}$. This implies that the morphisms $M : C \to D$ in $\mathrm{Comod}(\mathcal{V})$ are comodules from C to D; that is, left C-, right D-comodules. So M is an object of $\mathcal{V}$ together with a coaction $\delta : M \to C \otimes M \otimes D$ satisfying the expected equations. It is sometimes useful to deal with the left and right actions $\delta_\ell : M \to C \otimes M$ and $\delta_r : M \to M \otimes D$ which are obtained from δ using the counit. The 2-cells $f : M \Rightarrow M' : C \to D$ in $\mathrm{Comod}(\mathcal{V})$ are morphisms $f : M \to M'$ in $\mathcal{V}$ respecting the coactions.

Composition of comodules $M : C \to D$ and $N : D \to E$ is given by the equalizer

$$N \circ M = M \underset{D}{\otimes} N \to M \otimes N \underset{1 \otimes \delta_\ell}{\overset{\delta_r \otimes 1}{\rightrightarrows}} M \otimes D \otimes N.$$

The identity comodule $C \to C$ is C with the obvious coaction. We point out that the pair of morphisms being equalized here have a common left inverse $1 \otimes \varepsilon \otimes 1$; so the equalizer is coreflexive.

The remaining details describing $\mathrm{Comod}(\mathcal{V})$ as a bicategory should now be clear.

Remark 12.1 (a) When $\mathcal{V} = \mathrm{Set}$, it is readily checked that $\mathrm{Comod}(\mathcal{V})$ is biequivalent to $\mathrm{Mat}(\mathrm{Set})$.

(b) The main case that should be kept in mind is when $\mathcal{V}$ is the category of vector spaces over a field k; then the objects of $\mathrm{Comod}(\mathcal{V})$ are precisely k-coalgebras.

(c) If $\mathcal{V}$ itself is a $*$-autonomous monoidal category then the distinction between $\mathrm{Mod}(\mathcal{V})$ and $\mathrm{Comod}(\mathcal{V})$ evaporates.

(d) By the Chu construction, any complete cocomplete closed monoidal $\mathcal{V}$ can be embedded into a complete cocomplete $*$-autonomous monoidal $\mathcal{E} = \mathcal{V}^{\mathrm{op}} \otimes \mathcal{V}$ taking V to $(1, V)$ where 1 is the terminal object of $\mathcal{V}$. The embedding is strong monoidal and preserves colimits and connected limits. So we can take full advantage of remark (c) by working in $\mathcal{E}$-Mod and deducing results for both $\mathrm{Mod}(\mathcal{V})$ and $\mathrm{Comod}(\mathcal{V})$.

Returning to general $\mathcal{V}$, we note that each comonoid morphism $f : C \to D$ determines a comodule $f_* : C \to D$ defined to be C together with the coaction

$$C \xrightarrow{\delta} C \otimes C \xrightarrow{\delta \otimes f} C \otimes C \otimes D,$$

and a comodule $f^* : D \to C$ defined to be C together with the coaction

$$C \xrightarrow{\delta} C \otimes C \xrightarrow{f \otimes \delta} D \otimes C \otimes C.$$

Notice that we have $\gamma_f : f_* \circ f^* \Rightarrow 1_D$ which is defined to be $f : C \to D$ since $f_* \circ f^* = f^* \underset{C}{\otimes} f_* = C$ with coaction $C \xrightarrow{\delta} C \otimes C \xrightarrow{1 \otimes \delta} C \otimes C \otimes C \xrightarrow{f \otimes 1 \otimes f} D \otimes C \otimes D$. Also, $f_* \underset{D}{\otimes} f^* = f^* \circ f_*$ is the equalizer

$$f_* \underset{D}{\otimes} f^* \to C \otimes C \underset{(C \otimes f \otimes C) \circ (C \otimes \delta)}{\overset{(C \otimes f \otimes C) \circ (\delta \otimes C)}{\rightrightarrows}} C \otimes D \otimes C \quad ;$$

and, since

$$C \longrightarrow C \otimes C \underset{C \otimes \delta}{\overset{\delta \otimes C}{\rightrightarrows}} C \otimes C \otimes C$$

is an (absolute) equalizer, we have a unique morphism $C \to f_* \underset{D}{\otimes} f^*$ commuting with the morphisms into $C \otimes C$; this gives us $\omega_f : 1_C \Rightarrow f^* \circ f_*$. Indeed, γ_f, ω_f are the counit and unit for an adjunction $f_* \dashv f^*$ in the bicategory $\mathrm{Comod}(\mathcal{V})$.

The comodules f^* provide the special maps for the bicategory $\mathrm{Comod}(\mathcal{V})^{\mathrm{co}}$. Suppose C, D are comonoids. Then $C \otimes D$ becomes a comonoid with coaction

$$C \otimes D \xrightarrow{\delta \otimes \delta} C \otimes C \otimes D \otimes D \xrightarrow{C \otimes c_{C,D} \otimes D} C \otimes D \otimes C \otimes D$$

where c is the braiding and, as justified by coherence theorems, we ignore associativity in $\mathcal{V}$. For comodules $M : C \to C'$ and $N : D \to D'$, we obtain a comodule $M \otimes N : C \otimes D \to C' \otimes D'$ where the coaction is given in the obvious way using the braiding. This extends to a pseudofunctor $\otimes :\ \mathrm{Comod}(\mathcal{V}) \times \mathrm{Comod}(\mathcal{V}) \to \mathrm{Comod}(\mathcal{V})$. The remaining structure required to obtain $\mathrm{Comod}(\mathcal{V})$ as a monoidal bicategory should be obvious.

Write C^0 for C with the comultiplication

$$C \xrightarrow{\delta} C \otimes C \xrightarrow{c_{C,C}} C \otimes C.$$

There is a pseudonatural equivalence between the category of comodules $M : C \otimes D \to E$. and the category of comodules $\hat{M} : D \to C^0 \otimes E$, where $M = \hat{M}$ as objects. It follows that C^0 is a right bidual for C. This defines the structure of right autonomous monoidal bicategory on Comod($\mathcal{V}$).

Each $C^0 \otimes C$ has the canonical structure of a pseudomonoid in Comod($\mathcal{V}$) because it is an endohom in the autonomous monoidal bicategory.

A *quantum category* in $\mathcal{V}$ is basic data $C, h : C^0 \otimes C \to A$ in Comod$(\mathcal{V})^{\mathrm{co}}$.

A *quantum groupoid* in $\mathcal{V}$ is Hopf basic data in Comod$(\mathcal{V})^{\mathrm{co}}$.

Our referee has sensibly recommended that we unpackage these definitions for the utility of the reader and for comparison with the definition of "bicoalgebroid" in [8].

A *quantum graph* $\mathbf{A}$ in $\mathcal{V}$ consists of

- a comonoid C, called the *object object* of $\mathbf{A}$,
- a comonoid A, called the *arrow object* of $\mathbf{A}$, and
- comonoid morphisms $s : A \to C^0$ and $t : A \to C$, called *source* and *target morphisms* of $\mathbf{A}$,

such that the following diagram commutes.

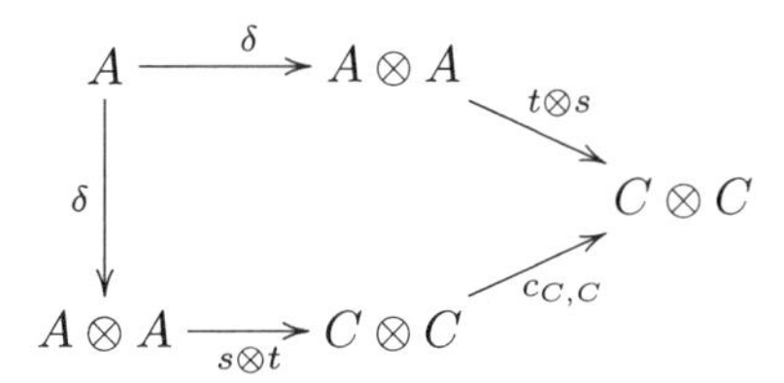

It follows that $r : A \xrightarrow{\delta} A \otimes A \xrightarrow{s\otimes t} C^0 \otimes C$ is a comonoid morphism. Therefore we have a comodule $I \xrightarrow{\varepsilon^*} A \xrightarrow{r_*} C^0 \otimes C$ which corresponds, under $C \dashv C^0$, to a comodule $C \to C$; explicitly, it is $A : C \to C$ with coactions

$$\delta_\ell : A \xrightarrow{\delta} A \otimes A \xrightarrow{1\otimes s} A \otimes C \xrightarrow{c_{C,A}^{-1}} C \otimes A$$

$$\delta_r : A \xrightarrow{\delta} A \otimes A \xrightarrow{1\otimes t} A \otimes C.$$

Then we can define the *composable pairs object* $P = A \underset{C}{\otimes} A$ as the composite comodule $C \xrightarrow{A} C \xrightarrow{A} C$; explicitly, it is the equalizer

$$P \xrightarrow{\iota} A \otimes A \overset{\delta_r\otimes 1}{\underset{1\otimes\delta_\ell}{\rightrightarrows}} A \otimes C \otimes A$$

which becomes a comodule $P : C \to C$ via right and left coactions induced by

$$A \otimes A \xrightarrow{1\otimes\delta_r} A \otimes A \otimes C \quad \text{and} \quad A \otimes A \xrightarrow{\delta_\ell\otimes 1} C \otimes A \otimes A.$$

Although in general P is not a comonoid with ι a comonoid morphism, there is a unique morphism $\delta_\ell : P \to A \otimes A \otimes P$ such that the following diagram commutes.

$$\begin{array}{ccccc} P & \xrightarrow{\iota} & A \otimes A & \xrightarrow{\delta\otimes\delta} & A \otimes A \otimes A \otimes A \\ {\scriptstyle\delta_\ell}\downarrow & & & & \downarrow{\scriptstyle 1\otimes c_{A,A}\otimes 1} \\ A \otimes A \otimes P & & \xrightarrow{\quad 1\otimes 1\otimes\iota \quad} & & A \otimes A \otimes A \end{array}$$

This is because the diagram

$$P \xrightarrow{\iota} A \otimes A \xrightarrow{\delta \otimes \delta} A \otimes A \otimes A \otimes A \xrightarrow{1 \otimes c_{A,A} \otimes 1} A \otimes A \otimes A \otimes A \underset{1 \otimes 1 \otimes 1 \otimes \delta_\ell}{\overset{1 \otimes 1 \otimes \delta_r \otimes 1}{\rightrightarrows}} A \otimes A \otimes A \otimes C \otimes A$$

commutes, and $1 \otimes 1 \otimes \iota$ is the equalizer of $1 \otimes 1 \otimes \delta_r \otimes 1$ and $1 \otimes 1 \otimes 1 \otimes \delta_\ell$. A small calculation (four steps using string diagrams) proves that $\delta_\ell : P \to A \otimes A \otimes P$ is a left coaction of the comonoid $A \otimes A$ on P.

A *composition morphism* for a quantum graph $\mathbf{A}$ is a comodule morphism

$$\mu : P \to A : C \to C$$

that satisfies the axioms CM0, CM1 and CM2 below.

CM0. $\mu : A \underset{C}{\otimes} A \to A$ is associative in the monoidal category $\mathrm{Comod}(\mathcal{V})(C, C)$.

CM1. The following diagram commutes:

$$P \xrightarrow{\delta_\ell} A \otimes A \otimes P \underset{\varepsilon \otimes s \otimes 1}{\overset{t \otimes \varepsilon \otimes 1}{\rightrightarrows}} C \otimes P \xrightarrow{1 \otimes \mu} C \otimes A.$$

Before stating CM2 we need to notice, using CM1, that there exists a unique morphism $\delta_r : P \to P \otimes A$ such that the following diagram commutes.

$$\begin{array}{ccc} P & \xrightarrow{\delta_\ell} & A \otimes A \otimes P \\ {\scriptstyle \delta_r}\downarrow & & \downarrow{\scriptstyle 1 \otimes 1 \otimes \mu} \\ P \otimes A & \xrightarrow[\iota \otimes 1]{} & A \otimes A \otimes A \end{array}$$

This is because the diagram

$$P \xrightarrow{\delta_\ell} A \otimes A \otimes P \xrightarrow{1 \otimes 1 \otimes \mu} A \otimes A \otimes A \underset{1 \otimes \delta_\ell \otimes 1}{\overset{\delta_r \otimes 1 \otimes 1}{\rightrightarrows}} A \otimes C \otimes A \otimes A$$

commutes, and $\iota \otimes 1$ is the equalizer of $\delta_r \otimes 1 \otimes 1$ and $1 \otimes \delta_\ell \otimes 1$. Now we can state:

CM2. The following diagram commutes.

$$\begin{array}{ccc} P & \xrightarrow{\mu} & A \\ {\scriptstyle \delta_r}\downarrow & & \downarrow{\scriptstyle \delta} \\ P \otimes A & \xrightarrow[\mu \otimes 1]{} & A \otimes A \end{array}$$

It can now be shown that $P : A \otimes A \to A$ is a comodule with coactions δ_ℓ and δ_r as above.

An *identities morphism* for $\mathbf{A}$ is a comodule morphism $\eta : C \to A : C \to C$ satisfying the axioms

IM0. η is a unit for μ in $\mathrm{Comod}(\mathcal{V})(C, C)$.

IM1. The following diagram commutes.

$$\begin{array}{ccc} C & \xrightarrow{\eta} & A \\ & {\scriptstyle \varepsilon}\searrow \quad \swarrow{\scriptstyle \varepsilon} & \\ & I & \end{array}$$

IM2. The following diagram commutes.

$$\begin{array}{c} \xrightarrow{t\otimes 1} C\otimes A \xrightarrow{\eta\otimes 1} \\ C \xrightarrow{\eta} A \xrightarrow{\delta} A\otimes A \xrightarrow{\qquad 1_{A\otimes A} \qquad} A\otimes A \\ \xrightarrow{s\otimes 1} C\otimes A \xrightarrow{\eta\otimes 1} \end{array}$$

It follows that A becomes a pseudomonoid in Comod$(\mathcal{V})$ when equipped with the multiplication P, the unit $J = \eta_*$, and the canonical associativity and unit constraints. Furthermore, $r_* : A \to C^0 \otimes C$ becomes strong monoidal.

Notice that we obtain a morphism $\varsigma : P \to C \otimes C \otimes C$ by taking either of the equal routes in the diagram

$$P \xrightarrow{\iota} A\otimes A \underset{1\otimes\delta_\ell}{\overset{\delta_r\otimes 1}{\rightrightarrows}} A\otimes C\otimes A \xrightarrow{s\otimes 1\otimes t} C\otimes C\otimes C.$$

A quantum category is the same as a quantum graph equipped with a composition morphism and an identities morphism. The basic data in Comod$(\mathcal{V})^{\mathrm{co}}$ is the comodule $r_* : A \to C^0 \otimes C$.

When $\mathcal{V}$ is the monoidal category of vector spaces over a field k, our quantum graph corresponds to BC1 of [8] while our axioms CM0-CM2 amount to BC2 of [8] and our axioms IM0-IM2 amount to BC3 of [8].

The *chaotic quantum category* $\mathbf{A} = \mathbf{C}_{\mathrm{ch}}$ *on* C is defined by $A = C^0 \otimes C$, $s = 1_{C^0} \otimes \varepsilon$ and $t = \varepsilon \otimes 1_C$. Thus $P = C \otimes C \otimes C$ with $\iota = 1_C \otimes \delta \otimes 1_C$, $\delta_r = 1_C \otimes 1_C \otimes \delta$ and $\delta_\ell = \delta \otimes 1_C \otimes 1_C$. Finally, $\mu = 1_C \otimes \varepsilon \otimes 1_C$, $\eta = \delta$ and $\varsigma = 1_{C\otimes C\otimes C}$.

A quantum groupoid is a quantum category $\mathbf{A}$ equipped with comonoid equivalences

$$v : C \to C^{00} \qquad \text{and} \qquad \nu : A \to A^0$$

such that $s\nu \cong t$ and $t\nu \cong vs$, and for which there is a left $A \otimes A \otimes A$-comodule isomorphism $\gamma : P_\ell \cong P_r$, where P_ℓ is P with the left coaction

$$P \xrightarrow{\delta} A\otimes A\otimes P\otimes A \xrightarrow{1\otimes 1\otimes 1\otimes \nu} A\otimes A\otimes P\otimes A \xrightarrow{1\otimes 1\otimes c_{P,A}} A\otimes A\otimes P$$

and P_r is P with the left coaction

$$P \xrightarrow{\delta} A\otimes A\otimes P\otimes A \xrightarrow{1\otimes 1\otimes 1\otimes \nu'} A\otimes A\otimes P\otimes A \xrightarrow{c_{A\otimes A\otimes P,A}} A\otimes A\otimes P$$

in which ν' is an inverse equivalence for ν and δ is the coaction associated with the comodule $P : A \otimes A \to A$, such that the following square commutes.

$$\begin{array}{ccc} P & \xrightarrow{\varsigma} & C\otimes C\otimes C \\ {\scriptstyle\gamma}\downarrow & & \downarrow{\scriptstyle c_{C,C\otimes C}} \\ P & \xrightarrow[\varsigma]{} & C\otimes C\otimes C \end{array}$$

Example 12.2 Let $\mathcal{V}$ be the symmetric monoidal category of vector spaces over a field k. For any set X, let FX be the vector space with X as basis. This F is the object function for a strong monoidal functor $F : \mathrm{Set} \to \mathcal{V}$ that preserves coreflexive equalizers (exercise!). It therefore induces a strong monoidal pseudofunctor

$$\bar{F} : \ \mathrm{Comod(Set)}^{\mathrm{co}} \to \ \mathrm{Comod}(\mathcal{V})^{\mathrm{co}}.$$

Special maps are preserved by $\bar{F}$. It follows from Proposition 9.3 that $\bar{F}$ takes each category to a quantum category and each groupoid to a quantum groupoid.

Example 12.3 Following up on Remark 10.4 where $\mathcal{E}$ is a category with finite limits, we shall lead the reader into showing how quantum categories and quantum groupoids in $\mathcal{V} = \mathcal{E}$ (where the tensor product is cartesian product) are precisely categories and groupoids in $\mathcal{E}$. Every object of $\mathcal{E}$ has a unique comonoid structure defined by the diagonal morphism, every morphism of $\mathcal{E}$ is a comonoid morphism, and the only 2-cells between morphisms are equalities. Also each object C has $C^0 = C$. So a quantum graph $\mathbf{A}$ in $\mathcal{E}$ is just a pair of morphisms

$$s, t : A \to C;$$

that is, $\mathbf{A}$ is a (directed) graph in $\mathcal{E}$. The equalizer $P = A \underset{C}{\otimes} A$ is now easily seen to be the pullback of s and t; that is, P is the usual object of composable pairs in the graph. A composition morphism μ and an identities morphism η are precisely what is required to make $\mathbf{A}$ a category in $\mathcal{E}$. If $\mathbf{A}$ is a quantum groupoid then, because of the absence of 2-cells, $\upsilon : C \to C$ and $\nu : A \to A$ are isomorphisms while $s\nu = t$ and $t\nu = \upsilon s$. Arguing as for Proposition 10.2, we see that υ is actually the identity and ν makes $\mathbf{A}$ a groupoid in $\mathcal{E}$.

References

[1] M. Barr, $*$-Autonomous categories, with an appendix by Po Hsiang Chu. Lecture Notes in Mathematics 752 (Springer, Berlin, 1979).

[2] M. Barr, Nonsymmetric $*$-autonomous categories, Theoretical Computer Science 139 (1995) 115-130.

[3] M. Barr, The Chu construction, Theory Appl. Categories 2 (1996) 17-35.

[4] M. A. Batanin, Homotopy coherent category theory and A_∞-structures in monoidal categories, J. Pure Appl. Algebra 123 (1998) 67-103.

[5] J. Bénabou, Introduction to bicategories, Lecture Notes in Math. 47 (Springer, Berlin, 1967) 1- 77.

[6] R. Betti, A. Carboni, R. Street and R. Walters, Variation through enrichment, J. Pure Appl. Algebra 29 (1983) 109-127.

[7] G. Böhm, An alternative notion of Hopf algebroid, arXiv:math.QA/0301169 (16 Jan 2003).

[8] T. Brzezinski and G. Militaru, Bialgebroids, $\times_R$-bialgebras and duality, arXiv:math.QA/0012164 v2 (7 Nov 2001).

[9] B.J. Day, On closed categories of functors, Lecture Notes in Math. 137 (Springer- Verlag, Berlin 1970) 1-38.

[10] B.J. Day, Note on monoidal monads, J. Austral. Math. Soc. Ser. A 23 (1977) 292-311.

[11] B.J. Day, Promonoidal functor categories, J. Austral. Math. Soc. Ser. A 23 (1977) 312-328.

[12] B.J. Day, $*$-autonomous convolution (Talk to the Australian Category Seminar on 5 March 1999).

[13] B.J. Day and G.M. Kelly, Enriched functor categories, Lecture Notes in Math. 106 (Springer, Berlin, 1969) 178-191.

[14] B.J. Day, P. McCrudden and R. Street, Dualizations and antipodes, Appl. Categorical Structures 11 (2003) 229-260.

[15] B.J. Day and R. Street, Monoidal bicategories and Hopf algebroids, Advances in Math. 129 (1997) 99-157.

[16] B.J. Day and R. Street, Lax monoids, pseudo-operads, and convolution, in: "Diagrammatic Morphisms and Applications", Contemporary Mathematics 318 (AMS; April 2003) 75-96.

[17] B.J. Day and R. Street, Abstract substitution in enriched categories, J. Pure Appl. Algebra 179 (2003) 49-63.

[18] S. Eilenberg and G.M. Kelly, Closed categories, Proceedings of the Conference on Categorical Algebra at La Jolla (Springer, 1966) 421-562.

[19] J.-Y. Girard, Linear logic, Theoretical Computer Science 50 (1987) 1-102.

[20] A. Joyal and R. Street, Braided tensor categories, Advances in Math. 102 (1993) 20-78.

[21] A. Joyal and R. Street, An introduction to Tannaka duality and quantum groups, Lecture Notes in Math. 1488 (Springer-Verlag Berlin, Heidelberg 1991) 411-492.
[22] G.M. Kelly, Doctrinal adjunction, Lecture Notes in Math. 420 (Springer, Berlin, 1974) 257-280.
[23] G.M. Kelly, Basic Concepts of Enriched Category Theory, London Math. Soc. Lecture Notes Series 64 (Cambridge University Press 1982).
[24] J.-H. Lu, Hopf algebroids and quantum groupoids, Int. J. Math. 7 (1996) 47-70.
[25] S. Mac Lane, Categories for the Working Mathematician, Graduate Texts in Math. 5 (Springer-Verlag, Berlin 1971).
[26] G. Maltsiniotis, Groupoï des quantiques, Comptes Rendus Acad. Sci. Paris 314, Série I (1992) 249-252.
[27] P. McCrudden, Opmonoidal monads, Theory Appl. Categories 10 (2002) 469-485.
[28] I. Moerdijk, Monads on tensor categories, J. Pure Appl. Algebra 168 (2002) 189-208.
[29] D.C. Ravenal, Nilpotence and periodicity in stable homotopy theory, Annals of Mathematics Studies 128 (Princeton University Press 1992).
[30] P. Schauenburg, Bialgebras over non-commutative rings and a structure theorem for Hopf bimodules, Applied Categorical Structures 6 (1998) 193-222.
[31] P. Schauenburg, Duals and doubles of quantum groupoids ($\times_R$-Hopf algebras), Contemporary Math. 267 (2000) 273-299.
[32] R.A.G. Seely, Linear logic, $*$-autonomous categories and cofree coalgebras, Contemporary Math. 92 (Amer. Math. Soc., Providence, RI, 1989) 371-382.
[33] R. Street, The formal theory of monads, J. Pure Appl. Algebra 2 (1972) 149-168.
[34] R. Street, Cauchy characterization of enriched categories, Rendiconti del Seminario Matematico e Fisico di Milano 51 (1981) 217-233.
[35] R. Street, Quantum Groups: an entrée to modern algebra, <http://www-texdev.mpce.mq.edu.au/Quantum/Quantum.html>
[36] R. Street, Review of [3], Math. Rev. 97f:18004 (1997).
[37] K. Szlachányi, The monoidal Eilenberg-Moore construction and bialgebroids, J. Pure Appl. Algebra 182 (2003) 287-315.
[38] M. Takeuchi, Groups of algebras over $A \otimes \bar{A}$, J. Math. Soc. Japan 29 (1977) 459-492.
[39] C. Watts, Intrinsic characterizations of some additive functors, Proc. American Math. Soc. 11 (1960) 5-8.
[40] R.J. Wood, Abstract proarrows I, Cahiers Topologie Géom. Différentielle 23 (1982) 279-290.
[41] P. Xu, Quantum groupoids, Commun. Math. Phys. 216 (2001) 539-581.
[42] N. Yoneda, On Ext and exact sequences, J. Fac. Sci. Univ. Tokyo Sect. I 8 (1960) 507-576.

Received January 16, 2003; in revised form September 2,2003

Fields Institute Communications
Volume **43**, 2004

Morphisms of 2-groupoids and Low-dimensional Cohomology of Crossed Modules

John W. Duskin
Department of Mathematics
244 Mathematics Building
State University of New York at Buffalo
Buffalo, NY 14260-2900 U.S.A.
duskin@math.buffalo.edu

Rudger W. Kieboom
Department of Mathematics
Vrije Universiteit Brussel
Pleinlaan 2 F 10
1050 Brussels, Belgium
rkieboom@vub.ac.be

Enrico M. Vitale
Département de Mathématique
Université catholique de Louvain
Chemin du Cyclotron 2
1348 Louvain-la-Neuve, Belgium
vitale@math.ucl.ac.be

Abstract. Given a morphism $P\colon \mathcal{G} \to \mathcal{H}$ of 2-groupoids, we construct a 6-term 2-exact sequence of cat-groups and pointed groupoids. We use this sequence to obtain an analogue for cat-groups (and, in particular, for crossed modules) of the fundamental exact sequence of non-abelian group cohomology. The link with simplicial topology is also explained.

Introduction

The aim of this paper is to obtain a basic result in low-dimensional cohomology of crossed modules. Homology and cohomology of crossed modules have been studied extensively, and a satisfactory theory has been developed (see [7] and the references therein, [14, 19, 20, 26]). The existing literature on this subject considers crossed modules and their morphisms as a category. Our point of view is that crossed modules are in a natural way the objects of a 2-category, and therefore they should be studied in a 2-dimensional context. This different point of view leads to

2000 *Mathematics Subject Classification.* Primary 18G50; Secondary 18B40, 18D05, 18D35, 20L05.

Third author supported by FNRS grant 1.5.116.01.

a choice of limits and colimits which are the natural ones in our 2-categorical setting, that is bilimits, but which do not have a universal property in the underlying category of crossed modules. Accordingly, the notions of exactness and extension we consider are not equivalent to those studied in the previous papers devoted to this subject.

The result we look for to test our theory is a generalization to crossed modules of the fundamental exact sequence in non-abelian cohomology of groups [25]. To get this result, we adapt to crossed modules the method developed by Brown in [5], where the fundamental sequence is obtained as a special case of an exact sequence associated to a fibration of groupoids. In fact, to follow in a more transparent way the analogy with groups, we work with cat-groups instead of crossed modules, since the 2-category of (strict and small) cat-groups is biequivalent to the 2-category of crossed modules [6, 23].

The paper is organized as follows. In the first section we recall, for the reader's convenience and in a way convenient to be generalized to 2-groupoids, the result due to Brown. Section 2 is devoted to the construction of a 6-term 2-exact sequence of strict cat-groups and pointed groupoids from any morphism of 2-groupoids. For basic facts on cat-groups and 2-exact sequences we refer to [18, 27] and the bibliography therein; we recall in Section 2 the definitions we need. The idea of an higher-dimensional version of Brown's exact sequence comes from the paper [17] by Hardie, Kamps and the second author. The precise link between the main result of [17] and our 2-exact sequence is explained in Remark 2.7. In the third section we fix a cat-group $\mathbb{G}$ and an extension

$$\mathbb{A} \xrightarrow{i} \mathbb{B} \xrightarrow{j} \mathbb{C}$$

of $\mathbb{G}$-cat-groups. From such an extension we obtain, as a particular case of the sequence in Section 2, a 6-term 2-exact sequence of cat-groups and pointed groupoids

$$\mathbb{A}^{\mathbb{G}} \to \mathbb{B}^{\mathbb{G}} \to \mathbb{C}^{\mathbb{G}} \to H^1(\mathbb{G}, \mathbb{A}) \to H^1(\mathbb{G}, \mathbb{B}) \to H^1(\mathbb{G}, \mathbb{C})$$

which is the 2-dimensional generalization of the fundamental sequence in non-abelian group cohomology. As a corollary of the main result of Section 2, we also get a 9-term exact sequence of groups and pointed sets. Instead of exploiting the homological notion of 2-exactness, this 9-term sequence can also be obtained using classical results from simplicial topology. This is the content of Section 4.

1 Brown's exact sequence

As in the topological case, it is better to work with the homotopy fibre instead of the "set-theoretical" fibre. In this way, we can obtain an exact sequence from any functor between groupoids (and not only from a fibration). Moreover, we avoid some choices which would be quite hard to handle in the higher dimensional analogue developed in Section 2.

Recall that a groupoid $\mathbb{G}$ is a category in which each arrow is an isomorphism. Consider now a functor between groupoids

$$P \colon \mathbb{G} \to \mathbb{H}$$

and fix an object H in $\mathbb{H}$. The homotopy fibre $\mathbb{F}_H$ of P at the point H is the following comma groupoid:

- objects of $\mathbb{F}_H$ are pairs $(Y, y \colon P(Y) \to H)$, with Y an object of $\mathbb{G}$ and y an arrow in $\mathbb{H}$;

- an arrow $f\colon (Y_1, y_1) \to (Y_2, y_2)$ in $\mathbb{F}_H$ is an arrow $f\colon Y_1 \to Y_2$ in $\mathbb{G}$ such that $P(f) \cdot y_2 = y_1$ (composition denoted from left to right).

There is an obvious faithful functor $j\colon \mathbb{F}_H \to \mathbb{G}$. If X is an object of $\mathbb{G}$, we write $\mathbb{F}_X$ for $\mathbb{F}_{P(X)}$.

Now fix an object X in $\mathbb{G}$ and consider the following groups and pointed sets:

- $\pi_0(\mathbb{G})$, the set of isomorphism classes of objects of $\mathbb{G}$, pointed by the class of X; $\pi_0(\mathbb{H})$, pointed by the class of $P(X)$; $\pi_0(\mathbb{F}_X)$, pointed by the class of $(X, 1_{P(X)})$;
- $\mathbb{G}(X) = \mathbb{G}(X, X)$, the group of automorphisms of the object X in $\mathbb{G}$; $\mathbb{H}(X) = \mathbb{H}(P(X), P(X))$, the group of automorphisms of the object $P(X)$ in $\mathbb{H}$; $\mathbb{F}_X(X) = \mathbb{F}_X((X, 1_{P(X)}), (X, 1_{P(X)}))$, the group of automorphisms of $(X, 1_{P(X)})$ in $\mathbb{F}_X$.

They can be connected by the following morphisms (square brackets are isomorphism classes):

- $j_X\colon \mathbb{F}_X(X) \to \mathbb{G}(X) \quad j_X(f\colon X \to X) = f$;
- $P_X\colon \mathbb{G}(X) \to \mathbb{H}(X) \quad P_X(f\colon X \to X) = P(f)$;
- $\pi_0(P)\colon \pi_0(\mathbb{G}) \to \pi_0(\mathbb{H}) \quad [X] \mapsto [P(X)]$;
- $\pi_0(j)\colon \pi_0(\mathbb{F}_X) \to \pi_0(\mathbb{G}) \quad [Y, y\colon P(Y) \to P(X)] \mapsto [Y]$;
- $\delta\colon \mathbb{H}(X) \to \pi_0(\mathbb{F}_X) \quad \delta(x\colon P(X) \to P(X)) = [X, x\colon P(X) \to P(X)]$.

Proposition 1.1 *With the previous notations, the sequence*

$$0 \to \mathbb{F}_X(X) \xrightarrow{j_X} \mathbb{G}(X) \xrightarrow{P_X} \mathbb{H}(X) \xrightarrow{\delta} \pi_0(\mathbb{F}_X) \xrightarrow{\pi_0(j)} \pi_0(\mathbb{G}) \xrightarrow{\pi_0(P)} \pi_0(\mathbb{H})$$

is exact.

Proof Consider an element $[Y, y]$ in $\pi_0(\mathbb{F}_X)$ and assume that $[Y] = [X]$ in $\pi_0(\mathbb{G})$. Then there is an arrow $y'\colon Y \to X$ in $\mathbb{G}$ and therefore $[Y, y] = [X, P(y')^{-1} \cdot y] = \delta(P(y')^{-1} \cdot y)$ because $y'\colon (Y, y) \to (X, P(y')^{-1} \cdot y)$ is an arrow in $\mathbb{F}_X$. The rest of the proof is straightforward. □

Now consider the strict fibre $\mathbb{S}_H$ of P at the point H:

- objects of $\mathbb{S}_H$ are the objects Y of $\mathbb{G}$ such that $P(Y) = H$;
- an arrow $f\colon Y_1 \to Y_2$ of $\mathbb{G}$ is in $\mathbb{S}_H$ if $P(f) = 1_H$.

There is, for each object H in $\mathbb{S}_H$, a full and faithful functor $i_H\colon \mathbb{S}_H \to \mathbb{F}_H$. Clearly, P is a fibration of groupoids [5] if and only if for each H the functor i_H is essentially surjective on objects. Therefore, if P is a fibration, we can replace $\mathbb{F}_X(X)$ and $\pi_0(\mathbb{F}_X)$ by $\mathbb{S}_X(X)$ and $\pi_0(\mathbb{S}_X)$ and we obtain Brown's exact sequence associated to a fibration of groupoids (Theorem 4.3 in [5]).

2 The 2-exact sequence

In this section we fix a morphism of 2-groupoids

$$P\colon \mathcal{G} \to \mathcal{H}$$

that is a 2-functor between 2-categories in which each arrow is an equivalence and each 2-cell is an isomorphism.

Fix an object H in $\mathcal{H}$; the homotopy fibre $\mathcal{F}_H$ of P at the point H is the following 2-groupoid:

- objects are pairs $(Y, y\colon P(Y) \to H)$, with Y an object in $\mathcal{G}$ and y an arrow in $\mathcal{H}$;

- arrows $(f, \varphi)\colon (Y_1, y_1) \to (Y_2, y_2)$ are pairs with $f\colon Y_1 \to Y_2$ an arrow in $\mathcal{G}$ and $\varphi\colon y_1 \Rightarrow P(f) \cdot y_2\colon P(Y_1) \to H$ a 2-cell in $\mathcal{H}$;
- a 2-cell $\alpha\colon (f, \varphi) \Rightarrow (g, \psi)\colon (Y_1, y_1) \to (Y_2, y_2)$ is a 2-cell $\alpha\colon f \Rightarrow g$ in $\mathcal{G}$ such that the following diagram commutes

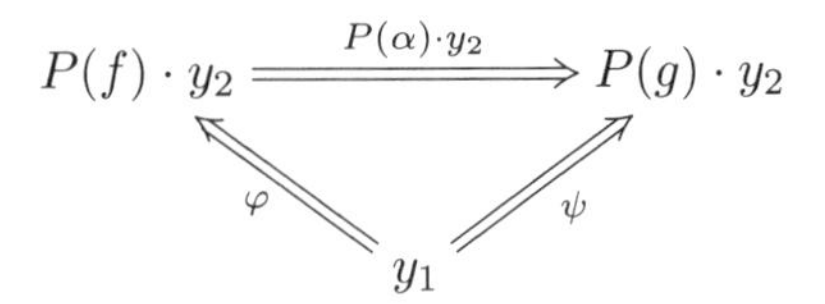

There is a morphism $j\colon \mathcal{F}_H \to \mathcal{G}$ which sends $\alpha\colon (f, \varphi) \Rightarrow (g, \psi)\colon (Y_1, y_1) \to (Y_2, y_2)$ to $\alpha\colon f \Rightarrow g\colon Y_1 \to Y_2$. The morphism j is faithful on arrows and on 2-cells. If X is an object of $\mathcal{G}$, we write $\mathcal{F}_X$ for $\mathcal{F}_{P(X)}$.

We recall now the notion of 2-exact sequence for pointed groupoids (and, in particular, for cat-groups, i.e. monoidal groupoids in which each object is invertible, up to isomorphisms, w.r.t. the tensor product). Morphisms of pointed groupoids (cat-groups) are pointed functors (monoidal functors). A natural transformation between pointed (monoidal) functors is always assumed to be pointed (monoidal). Let $F\colon \mathbb{G} \to \mathbb{H}$ be a morphism of pointed groupoids; its homotopy kernel $kF\colon KerF \to \mathbb{G}$ is the homotopy fibre (in the sense of Section 1) of F on the base object I of $\mathbb{H}$. There is a natural transformation $\kappa F\colon kF \cdot F \Rightarrow 0$ (0 is the morphism which sends each arrow to the identity of I) given, for each object (Y, y) of $KerF$, by $y\colon F(Y) \to I$.

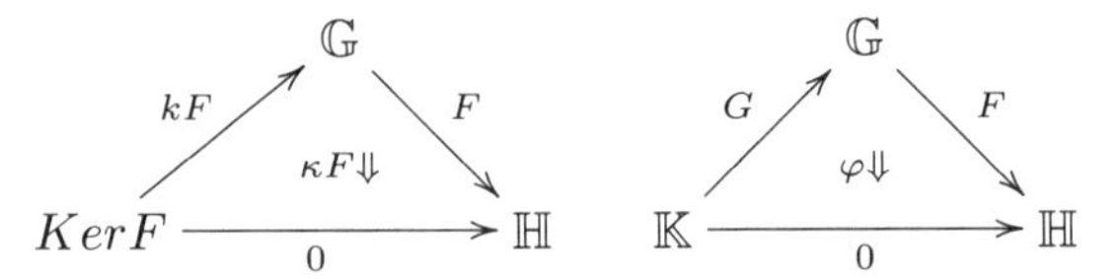

Moreover, given a pointed groupoid $\mathbb{K}$, a morphism G and a natural transformation φ as in the previous diagram, there is a unique comparison morphism $G'\colon \mathbb{K} \to KerF$, $G'(g\colon A_1 \to A_2) = G(g)\colon (G(A_1), \varphi_{A_1}) \to (G(A_2), \varphi_{A_2})$, such that $G' \cdot kF = G$ and $G' \cdot \kappa F = \varphi$ (compare with [15]). The universal property of $(KerF, kF, \kappa F)$ as a bilimit, discussed in [18, 27], determines it uniquely, up to equivalence.

Definition 2.1 Consider two morphisms G, F and a natural transformation φ of pointed groupoids as in the previous diagram; we say that the triple (G, φ, F) is 2-exact if the comparison $G'\colon \mathbb{K} \to KerF$ is full and essentially surjective on objects.

Now come back to the 2-functor between 2-groupoids $P\colon \mathcal{G} \to \mathcal{H}$ and fix an object X of $\mathcal{G}$. We can consider the following three hom-categories, which are in fact strict cat-groups: $\mathcal{G}(X) = \mathcal{G}(X, X)$, $\mathcal{H}(X) = \mathcal{H}(P(X), P(X))$ and $\mathcal{F}_X(X) = \mathcal{F}_X((X, 1_{P(X)}), (X, 1_{P(X)}))$. Moreover, we can consider the classifying groupoid $cl(\mathcal{G})$ of the 2-groupoid $\mathcal{G}$: $cl(\mathcal{G})$ has the same objects as $\mathcal{G}$ and 2-isomorphism classes of arrows of $\mathcal{G}$ as arrows. The groupoid $cl(\mathcal{G})$ is pointed by the object X. Similarly, we have the groupoid $cl(\mathcal{H})$ pointed by $P(X)$ and the groupoid $cl(\mathcal{F}_X)$ pointed by $(X, 1_{P(X)})$. These cat-groups and pointed groupoids can be connected by the following morphisms (square brackets are 2-isomorphism classes of arrows):

- $j_X \colon \mathcal{F}_X(X) \to \mathcal{G}(X)$
  $\alpha \colon (f,\varphi) \Rightarrow (g,\psi) \colon (X, 1_{P(X)}) \to (X, 1_{P(X)}) \ \mapsto\ \alpha \colon f \Rightarrow g \colon X \to X$
- $P_X \colon \mathcal{G}(X) \to \mathcal{H}(X)$
  $\alpha \colon f \Rightarrow g \colon X \to X \ \mapsto\ P(\alpha) \colon P(f) \Rightarrow P(g) \colon P(X) \to P(X)$
- $cl(j) \colon cl(\mathcal{F}_X) \to cl(\mathcal{H})$
  $[f,\varphi] \colon (Y_1, y_1) \to (Y_2, y_2) \ \mapsto\ [f] \colon Y_1 \to Y_2$
- $cl(P) \colon cl(\mathcal{G}) \to cl(\mathcal{H})$
  $[f] \colon Y_1 \to Y_2 \ \mapsto\ [P(f)] \colon P(Y_1) \to P(Y_2)$
- $\delta \colon \mathcal{H}(X) \to cl(\mathcal{F}_X)$
  $\beta \colon h \Rightarrow k \colon P(X) \to P(X) \ \mapsto\ [1_X, \beta] \colon (X,h) \to (X,k)$.

Proposition 2.2 *With the previous notations, the sequence*

$$\mathcal{F}_X(X) \xrightarrow{j_X} \mathcal{G}(X) \xrightarrow{P_X} \mathcal{H}(X) \xrightarrow{\delta} cl(\mathcal{F}_X) \xrightarrow{cl(j)} cl(\mathcal{G}) \xrightarrow{cl(P)} cl(\mathcal{H})$$

with the obvious natural transformations $j_X \cdot P_X \Rightarrow 0$, $P_X \cdot \delta \Rightarrow 0$, $\delta \cdot cl(j) \Rightarrow 0$, $cl(j) \cdot cl(P) \Rightarrow 0$, *is 2-exact.*

Proof 1) 2-exactness in $\mathcal{G}(X)$: it is straightforward to verify that the functor $j_X \colon \mathcal{F}_X(X) \to \mathcal{G}(X)$ is exactly the kernel of $P_X \colon \mathcal{G}(X) \to \mathcal{H}(X)$.

2) 2-exactness in $\mathcal{H}(X)$: consider the comparison $P'_X \colon \mathcal{G}(X) \to Ker\delta$
  $\alpha \colon f \Rightarrow g \colon X \to X \ \mapsto\ P(\alpha) \colon (P(f), [f, 1_{P(f)}]) \Rightarrow (P(g), [g, 1_{P(g)}])$
  - P'_X is essentially surjective: given an object $(h, [f,\varphi])$ in $Ker\delta$, we obtain an arrow $\varphi \colon (h, [f,\varphi]) \Rightarrow \delta(f)$ in $Ker\delta$;
  - P'_X is full: given an arrow $\beta \colon \delta(f) \Rightarrow \delta(g)$ in $Ker\delta$, then $\delta(\beta) \cdot [g, 1_{P(g)}] = [f, 1_{P(f)}]$, but this means that there exists a 2-cell $\alpha \colon f \Rightarrow g$ such that $P(\alpha) = \beta$.

3) 2-exactness in $cl(\mathcal{G})$: consider the comparison $j' \colon cl(\mathcal{F}_X) \to Ker(cl(P))$
  $[f,\varphi] \colon (Y_1, y_1) \to (Y_2, y_2) \ \mapsto\ [f] \colon (Y_1, [y_1]) \to (Y_2, [y_2])$
  - j' is essentially surjective: obvious;
  - j' is full: let $[f] \colon j'(Y_1, y_1) \to j'(Y_2, y_2)$ be an arrow in $Ker(cl(P))$, this means that there exists a 2-cell $\varphi \colon y_1 \Rightarrow P(f) \cdot y_2$ and then $[f,\varphi] \colon (Y_1, y_1) \to (Y_2, y_2)$ is an arrow in $cl(\mathcal{F}_X)$.

4) 2-exactness in $cl(\mathcal{F}_X)$: consider the comparison $\delta' \colon \mathcal{H}(X) \to Ker(cl(j))$
  $\beta \colon h \Rightarrow k \colon P(X) \to P(X) \ \mapsto\ [1_X, \beta] \colon (X, h, [1_X]) \to (X, k, [1_X])$
  - δ' is full: let $[f,\varphi] \colon \delta'(h) \to \delta'(k)$ be an arrow in $Ker(cl(j))$, then $[f] \cdot [1_X] = [1_X]$, that is there exists a 2-cell $\alpha \colon 1_X \Rightarrow f$. We obtain $\beta \colon h \Rightarrow k$ in $\mathcal{H}(X)$ in the following way :

$$\beta = (h \xRightarrow{\varphi} P(f) \cdot k \xRightarrow{P(\alpha)^{-1} \cdot k} k) \ ;$$

  - δ' is essentially surjective: consider an object

$$(Y, y \colon P(Y) \to P(X), [x] \colon Y \to X)$$

  in $Ker(cl(j))$, then $P(x)^{-1} \cdot y \colon P(X) \to P(X)$ is an object in $\mathcal{H}(X)$ and $[x, c] \colon (Y, y, [x]) \to \delta'(P(x)^{-1} \cdot y)$ is an arrow in $Ker(cl(j))$, where c is the canonical 2-cell $c \colon y \Rightarrow P(x) \cdot P(x)^{-1} \cdot y$.

□

As in Section 1, if $(\mathbb{G}, I)$ is a pointed groupoid (a cat-group), we write $\pi_0(\mathbb{G})$ for the pointed set (the group) of isomorphism classes of objects and $\pi_1(\mathbb{G})$ for the (abelian) group of automorphisms $\mathbb{G}(I, I)$. π_0 and π_1 extend to morphisms and carry

2-exact sequences on exact sequences of pointed sets or groups. Finally, observe that if $\mathcal{G}$ is a 2-groupoid and X is a chosen object in $\mathcal{G}$, then $\pi_1(cl(\mathcal{G})) = \pi_0(\mathcal{G}(X))$. In a similar way, if $P\colon \mathcal{G} \to \mathcal{H}$ is a 2-functor, then $\pi_1(cl(P)) = \pi_0(P_X)$.

Corollary 2.3 *Let $P\colon \mathcal{G} \to \mathcal{H}$ be a 2-functor between 2-groupoids and fix an object X in $\mathcal{G}$; the following is an exact sequence of groups and pointed sets (the last three terms)*

$$0 \to \pi_1(\mathcal{F}_X(X)) \to \pi_1(\mathcal{G}(X)) \to \pi_1(\mathcal{H}(X)) \to$$

$$\pi_1(cl(\mathcal{F}_X)) = \pi_0(\mathcal{F}_X(X)) \to \pi_1(cl(\mathcal{G})) = \pi_0(\mathcal{G}(X)) \to \pi_1(cl(\mathcal{H})) = \pi_0(\mathcal{H}(X))$$

$$\to \pi_0(cl(\mathcal{F}_X)) \to \pi_0(cl(\mathcal{G})) \to \pi_0(cl(\mathcal{H}))\,.$$

Proof As far as exactness in $\pi_1(\mathcal{F}_X(X))$ is concerned, observe that j_X is the kernel of P_X, so it is faithful and then $\pi_1(j_X)$ is injective. The rest follows from the 2-exactness of the sequence in Proposition 2.2 and the previous remarks on π_0 and π_1. □

Remark 2.4 If $P\colon \mathbb{G} \to \mathbb{H}$ is a functor between groupoids, we can look at it as a 2-functor between discrete 2-groupoids (2-groupoids with no non-trivial 2-cells). The exact sequence of Corollary 2.3 reduces then to the exact sequence of Proposition 1.1, because the first non-trivial term is $\pi_0(\mathcal{F}_X(X))$.

Remark 2.5 Brown's exact sequence of Proposition 1.1 satisfies a strong exactness condition in $\pi_0(\mathbb{F}_X)$, which is the transition point between groups and pointed groupoids. The 2-dimensional analogue of strong exactness has been formulated in [16]. It is not difficult to prove that the sequence of Proposition 2.2 is strongly 2-exact in $cl(\mathcal{F}_X)$, let us just observe that the needed action $\mathcal{H}(X) \times cl(\mathcal{F}_X) \to cl(\mathcal{F}_X)$ sends $(f\colon P(X) \to P(X), (Y, y\colon P(Y) \to P(X)))$ into $(Y, y \cdot f\colon P(Y) \to P(X) \to P(X))$.

Remark 2.6 Proposition 2.2 and Corollary 2.3 hold also for a morphism $P\colon \mathcal{G} \to \mathcal{H}$ of bigroupoids, that is a pseudo-functor between bicategories [2, 3] in which each arrow is an equivalence and each 2-cell is an isomorphism. The generalization is strightforward: just observe that the homotopy fibre $\mathcal{F}_H$ inherits a structure of bicategory from that of $\mathcal{G}$. Clearly, if $\mathcal{G}$ and $\mathcal{H}$ are bigroupoids, the cat-groups $\mathcal{F}_X(X), \mathcal{G}(X)$ and $\mathcal{H}(X)$ of Proposition 2.2 are no longer strict. In Section 3 we will use this more general version of Proposition 2.2.

Remark 2.7 Recall that a morphism of bigroupoids $P\colon \mathcal{G} \to \mathcal{H}$ is a fibration if the functor $cl(P)\colon cl(\mathcal{G}) \to cl(\mathcal{H})$ is a fibration of groupoids and for each Y_1, Y_2 in $\mathcal{G}$ the functor $P_{Y_1,Y_2}\colon \mathcal{G}(Y_1, Y_2) \to \mathcal{H}(P(Y_1), P(Y_2))$ is a fibration of groupoids [17, 22]. This is equivalent to ask that for each object X in $\mathcal{G}$, the induced functor $St_P(X)\colon St_{\mathcal{G}}(X) \to St_{\mathcal{H}}(P(X))$ (where the "star-groupoid" $St_{\mathcal{G}}(X)$ is the groupoid having morphisms $y\colon X \to Y$ as objects and 2-cells $\alpha\colon y_1 \Rightarrow y_2\colon X \to Y$ as arrows) is an essentially surjective fibration. When P is a fibration, one easily checks that for each object H of $\mathcal{H}$, the homotopy fibre $\mathcal{F}_H$ is biequivalent to the strict fibre $\mathcal{S}_H$ (i.e. the sub-bigroupoid of $\mathcal{G}$ having as 2-cells the 2-cells $\alpha\colon f \Rightarrow g\colon Y_1 \to Y_2$ such that $P(\alpha)$ is the identity 2-cell of 1_H). Therefore, if P is a fibration, $\mathcal{F}_X(X)$ and $cl(\mathcal{F}_X)$ are equivalent to $\mathcal{S}_X(X)$ and $cl(\mathcal{S}_X)$ and the sequence of Corollary 2.3 is exactly the Hardie-Kamps-Kieboom 9-term exact sequence associated to a fibration of bigroupoids (Theorem 2.4 in [17]).

Remark 2.8 Proposition 2.2 can be also used to construct a Picard-Brauer 2-exact sequence from a homomorphism of unital commutative rings. In fact such a morphism induces a pseudo-functor between the bigroupoids having Azumaya algebras as objects, invertible bimodules as arrows and bimodule isomorphisms as 2-cells. Compare with [27], where a similar 2-exact sequence is obtained using homotopy cokernels instead of homotopy fibres.

3 The cohomology sequence

Let us fix a cat-group $\mathbb{G}$. A $\mathbb{G}$-cat-group is a pair $(\mathbb{C}, \gamma)$ where $\mathbb{C}$ is a cat-group and $\gamma\colon \mathbb{G} \to Aut\mathbb{C}$ is a monoidal functor with codomain the cat-group of monoidal auto-equivalences of $\mathbb{C}$. $\mathbb{G}$-cat-groups are the objects of a 2-category, having equivariant monoidal functors as arrows and compatible monoidal transformations as 2-cells (see [9, 12] for more details and for an equivalent definition of $\mathbb{G}$-cat-group in terms of an action $\mathbb{G} \times \mathbb{C} \to \mathbb{C}$). Observe that homotopy kernels in the 2-category of $\mathbb{G}$-cat-groups are computed as in the 2-category of cat-groups (in other words, if $j\colon (\mathbb{B}, \beta) \to (\mathbb{C}, \gamma)$ is a morphism of $\mathbb{G}$-cat-groups and $i\colon \mathbb{A} \to \mathbb{B}$ is its kernel as a morphism of cat-groups, then $\mathbb{A}$ inherits from $\mathbb{B}$ a structure $\alpha\colon \mathbb{G} \to Aut\mathbb{A}$ of $\mathbb{G}$-cat-group such that $i\colon \mathbb{A} \to \mathbb{B}$ is a morphism of $\mathbb{G}$-cat-groups).

If $(\mathbb{C}, \gamma)$ is a $\mathbb{G}$-cat-group, a *derivation* is a pair $\langle M\colon \mathbb{G} \to \mathbb{C}, \mathbf{m}\rangle$ where M is a functor and

$$\mathbf{m} = \{m_{X,Y}\colon M(X) \otimes \gamma(X)(M(Y)) \to M(X \otimes Y)\}_{X,Y \in \mathbb{G}}$$

is a natural family of coherent isomorphisms (for more details, see [13], where $\mathbb{C}$ is assumed to be braided, or [11], where $\mathbb{G}$ is discrete). Derivations are the objects of a bigroupoid $\mathcal{Z}^1(\mathbb{G}, \mathbb{C})$:

- an arrow is a pair $\langle C, \mathbf{c}\rangle\colon \langle M, \mathbf{m}\rangle \to \langle N, \mathbf{n}\rangle$ with $C \in \mathbb{C}$ and

$$\mathbf{c} = \{c_X\colon M(X) \otimes \gamma(X)(C) \to C \otimes N(X)\}_{X \in \mathbb{G}}$$

  is a natural family of isomorphisms, compatible with $\mathbf{m}$ and $\mathbf{n}$;
- a 2-cell $f\colon \langle C, \mathbf{c}\rangle \Rightarrow \langle C', \mathbf{c}'\rangle$ is an arrow $f\colon C \to C'$ in $\mathbb{C}$ compatible with $\mathbf{c}$ and $\mathbf{c}'$.

In $\mathcal{Z}^1(\mathbb{G}, \mathbb{C})$ there is a trivial derivation

$$\theta_{\mathbb{C}} = \langle 0\colon \mathbb{G} \to \mathbb{C}, \{I \otimes \gamma(X)(I) \simeq I\}\rangle$$

and the cat-group $\mathcal{Z}^1(\mathbb{G}, \mathbb{C})(\theta_{\mathbb{C}}, \theta_{\mathbb{C}})$ is the cat-group $\mathbb{C}^{\mathbb{G}}$ of $\mathbb{G}$-*invariant objects*. Explicitly:

- an object of $\mathbb{C}^{\mathbb{G}}$ is a pair $\langle C, \mathbf{c}\rangle$ with $C \in \mathbb{C}$ and

$$\mathbf{c} = \{c_X\colon \gamma(X)(C) \to C\}_{X \in \mathbb{G}}$$

  a natural family of isomorphisms compatible with the monoidal structure of $\mathbb{G}$;
- an arrow $f\colon \langle C, \mathbf{c}\rangle \to \langle D, \mathbf{d}\rangle$ in $\mathbb{C}^{\mathbb{G}}$ is an arrow $f\colon C \to D$ in $\mathbb{C}$ such that the following diagram commutes for each $X \in \mathbb{G}$

$$\begin{array}{ccc} \gamma(X)(C) & \xrightarrow{c_X} & C \\ {\scriptstyle \gamma(X)(f)}\downarrow & & \downarrow{\scriptstyle f} \\ \gamma(X)(D) & \xrightarrow[d_X]{} & D \end{array}$$

A morphism $j\colon (\mathbb{B},\beta) \to (\mathbb{C},\gamma)$ of $\mathbb{G}$-cat-groups induces a pseudo-functor

$$j_*\colon \mathcal{Z}^1(\mathbb{G},\mathbb{B}) \to \mathcal{Z}^1(\mathbb{C},\mathbb{C}).$$

This pseudo-functor j_* sends a derivation $\langle H\colon \mathbb{G} \to \mathbb{B}, \mathbf{h}\rangle$ into the derivation $\langle H \cdot j\colon \mathbb{G} \to \mathbb{B} \to \mathbb{C}, j(\mathbf{h})\rangle$, where $j(\mathbf{h})$ is defined by the following composition

$$\begin{array}{c} j(H(X)) \otimes \gamma(X)(j(H(Y))) \\ \downarrow \simeq \\ j(H(X)) \otimes j(\beta(X)(H(Y))) \\ \downarrow \simeq \\ j(H(X) \otimes \beta(X)(H(Y))) \\ \downarrow j(h_{X,Y}) \\ j(H(X \otimes Y)) \end{array}$$

(the first isomorphism is the equivariant structure of j, the second one is its monoidal structure) and is defined in an obvious way on arrows and 2-cells.

In the next lemma we need the homotopy fibre $\mathbb{F}$ of j_* at the point $\theta_{\mathbb{B}}$. Let us describe explicitly the objects of $\mathbb{F}$ (without loosing in generality, we can assume that $j(I) = I$, so that $j_*(\theta_{\mathbb{B}}) = \theta_{\mathbb{C}}$). An object of $\mathbb{F}$ is a 4-tuple

$$\langle\, \mathcal{D} = \langle H\colon \mathbb{G} \to \mathbb{B}, \mathbf{h}\rangle \in \mathcal{Z}^1(\mathbb{G},\mathbb{B}),\ \langle \overline{H} \in \mathbb{C}, \overline{\mathbf{h}}\rangle \in \mathcal{Z}^1(\mathbb{G},\mathbb{C})(j_*(\mathcal{D}),\theta_{\mathbb{C}})\,\rangle$$

with

$$\mathbf{h} = \{h_{X,Y}\colon H(X) \otimes \beta(X)(H(Y)) \to H(X \otimes Y)\}_{X,Y \in \mathbb{G}}$$
$$\overline{\mathbf{h}} = \{\overline{h}_X\colon j(H(X)) \otimes \gamma(X)(\overline{H}) \to \overline{H}\}_{X \in \mathbb{G}}$$

Lemma 3.1 *Consider an essentially surjective morphism $j\colon \mathbb{B} \to \mathbb{C}$ of $\mathbb{G}$-cat-groups and its homotopy kernel*

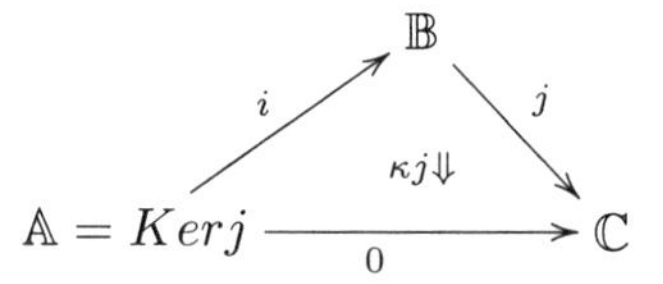

The homotopy fibre $\mathbb{F}$ of $j_*\colon \mathcal{Z}^1(\mathbb{G},\mathbb{B}) \to \mathcal{Z}^1(\mathbb{G},\mathbb{C})$ at the point $\theta_{\mathbb{B}}$ is biequivalent to the bigroupoid of derivations $\mathcal{Z}^1(\mathbb{G},\mathbb{A})$.

Proof Given an object in $\mathcal{Z}^1(\mathbb{G},\mathbb{A})$

$$\langle F\colon \mathbb{G} \to \mathbb{A}, \mathbf{f} = \{f_{X,Y}\colon F(X) \otimes \alpha(X)(F(Y)) \to F(X \otimes Y)\}_{X,Y \in \mathbb{G}}\rangle$$

we get an object in $\mathbb{F}$

$$\langle\, \langle F \cdot i\colon \mathbb{G} \to \mathbb{A} \to \mathbb{B}, i(\mathbf{f})\rangle,\ \langle I \in \mathbb{C}, \{\kappa j_{F(X)}\colon j(i(F(X))) \to I\}_{X \in \mathbb{G}}\rangle\,\rangle$$

This construction extends to a 2-functor $\epsilon\colon \mathcal{Z}^1(\mathbb{G},\mathbb{A}) \to \mathbb{F}$ which is always locally an equivalence (even if $j\colon \mathbb{B} \to \mathbb{C}$ is not essentially surjective). Let us check that ϵ is surjective on objects up to equivalence. Let $\langle\, \langle H\colon \mathbb{G} \to \mathbb{B}, \mathbf{h}\rangle, \langle \overline{H} \in \mathbb{C}, \overline{\mathbf{h}}\rangle\,\rangle$ be an object of $\mathbb{F}$. Since $j\colon \mathbb{B} \to \mathbb{C}$ is essentially surjective, there is an object $Z \in \mathbb{B}$ and an arrow $z\colon \overline{H} \to j(Z)$. Now we can construct a functor

$$D\colon \mathbb{G} \to \mathbb{B} \qquad X \mapsto Z^* \otimes H(X) \otimes \beta(X)(Z)$$

(Z^* is a dual of Z in the cat-group $\mathbb{B}$) which has a structure of derivation $\mathbf{d} = \{d_{X,Y} \colon D(X) \otimes \beta(X)(D(Y)) \to D(X \otimes Y)\}$ obtained from that of H in the following way

$$
\begin{array}{c}
D(X) \otimes \beta(X)(D(Y)) \\
\Big\downarrow{=} \\
Z^* \otimes H(X) \otimes \beta(X)(Z) \otimes \beta(X)(Z^* \otimes H(Y) \otimes \beta(Y)(Z)) \\
\Big\downarrow{\simeq} \\
Z^* \otimes H(X) \otimes \beta(X)(Z) \otimes \beta(X)(Z^*) \otimes \beta(X)(H(Y) \otimes \beta(Y)(Z)) \\
\Big\downarrow{\simeq} \\
Z^* \otimes H(X) \otimes \beta(X)(H(Y)) \otimes \beta(X)(\beta(Y)(Z)) \\
\Big\downarrow{\simeq} \\
Z^* \otimes H(X) \otimes \beta(X)(H(Y)) \otimes \beta(X \otimes Y)(Z) \\
\Big\downarrow{1 \otimes h_{X,Y} \otimes 1} \\
Z^* \otimes H(X \otimes Y) \otimes \beta(X \otimes Y)(Z) = D(X \otimes Y)
\end{array}
$$

Observe now that the functor $D \colon \mathbb{G} \to \mathbb{B}$ factors through the kernel of j. Indeed, if $X \in \mathbb{G}$, we have

$$
\begin{array}{c}
j(D(X)) = j(Z^* \otimes H(X) \otimes \beta(X)(Z)) \\
\Big\downarrow{\simeq} \\
j(Z^*) \otimes j(H(X)) \otimes j(\beta(X)(Z)) \\
\Big\downarrow{\simeq} \\
j(Z^*) \otimes j(H(X)) \otimes \gamma(X)(j(Z)) \\
\Big\downarrow{z^* \otimes 1 \otimes z^{-1}} \\
\overline{H}^* \otimes j(H(X)) \otimes \gamma(X)(\overline{H}) \\
\Big\downarrow{1 \otimes \overline{h}_X} \\
\overline{H}^* \otimes \overline{H} \simeq I
\end{array}
$$

Let us call $\widetilde{D} \colon \mathbb{G} \to \mathbb{A}$ the factorization of $D \colon \mathbb{G} \to \mathbb{B}$ through the kernel $\mathbb{A}$. The structure $\mathbf{d}$ of the derivation D pass to $\widetilde{D}$ because $\overline{\mathbf{h}}$ is compatible with $j(\mathbf{h})$ and with the structure of the trivial derivation $\theta_{\mathbb{C}}$. In this way, we have built up an object $\langle \widetilde{D} \colon \mathbb{G} \to \mathbb{A}, \tilde{\mathbf{d}} \rangle$ of $\mathcal{Z}^1(\mathbb{G}, \mathbb{A})$. Finally, an arrow

$$\langle\, \langle H \colon \mathbb{G} \to \mathbb{B}, \mathbf{h} \rangle,\ \langle \overline{H} \in \mathbb{C}, \overline{\mathbf{h}} \rangle\, \rangle \to \epsilon \langle \widetilde{D} \colon \mathbb{G} \to \mathbb{A}, \tilde{\mathbf{d}} \rangle$$

in $\mathbb{F}$ is provided by $Z \in \mathbb{B}$, $z \colon \overline{H} \to j(Z)$ and by the family of canonical isomorphisms

$$\{H(X) \otimes \beta(X)(Z) \simeq Z \otimes Z^* \otimes H(X) \otimes \beta(X)(Z) = Z \otimes D(X)\}_{X \in \mathbb{G}}$$

□

An essentially surjective morphism with its homotopy kernel

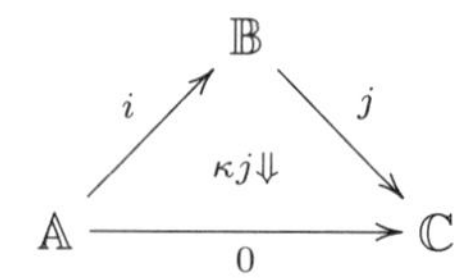

is called in [4, 8, 24] an *extension*. Putting together Proposition 2.2 and the previous lemma, we get our generalization of the fundamental sequence in non-abelian group cohomology. We write $H^1(\mathbb{G},\mathbb{C})$ for $cl(\mathcal{Z}^1(\mathbb{G},\mathbb{C}))$.

Corollary 3.2 *Consider an extension of* $\mathbb{G}$*-cat-groups*

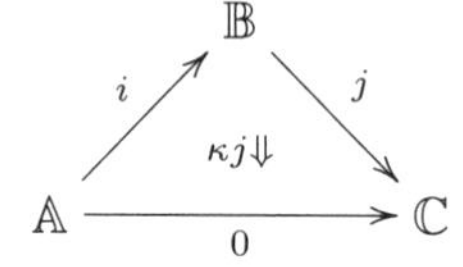

There is a 2-exact sequence of cat-groups and pointed groupoids

$$\mathbb{A}^{\mathbb{G}} \to \mathbb{B}^{\mathbb{G}} \to \mathbb{C}^{\mathbb{G}} \to H^1(\mathbb{G},\mathbb{A}) \to H^1(\mathbb{G},\mathbb{B}) \to H^1(\mathbb{G},\mathbb{C})\,.$$

Remark 3.3 Observe that if the cat-groups $\mathbb{G},\mathbb{B},\mathbb{C}$ and the monoidal functor $j\colon \mathbb{B} \to \mathbb{C}$ are strict, then $j_*\colon \mathcal{Z}^1(\mathbb{G},\mathbb{B}) \to \mathcal{Z}^1(\mathbb{G},\mathbb{C})$ is a 2-functor between 2-groupoids, and the cat-groups involved in the previous corollary are strict, that is they are crossed modules.

To end, we sketch an equivalent description of the bigroupoid $\mathcal{Z}^1(\mathbb{G},\mathbb{C})$ of derivations using the semi-direct product. Let us start with a general construction: if $\mathcal{G}$ and $\mathcal{H}$ are bicategories, $[\mathcal{G},\mathcal{H}]$ is the bicategory of pseudo-functors $\mathcal{G} \to \mathcal{H}$, pseudo-natural transformations and modifications [2, 3]. If $\mathbb{G}$ and $\mathbb{H}$ are cat-groups, we can see them as bicategories with only one object, and $[\mathbb{G},\mathbb{H}]$ is now a bigroupoid. Explicitly:

- an object of $[\mathbb{G},\mathbb{H}]$ is a monoidal functor $F\colon \mathbb{G} \to \mathbb{H}$;
- an arrow $(H,\varphi)\colon F \to G\colon \mathbb{G} \to \mathbb{H}$ is an object H of $\mathbb{H}$ and a natural transformation

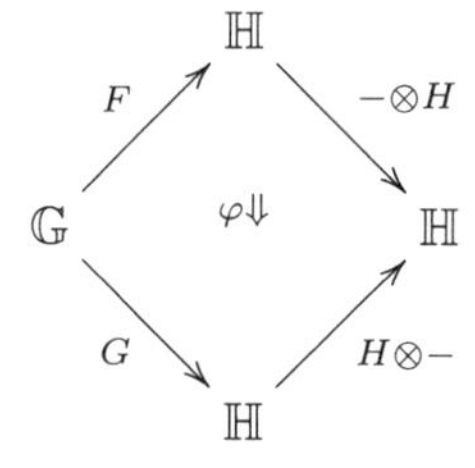

making commutative the following diagrams

$$\begin{array}{ccccc} F(X)\otimes F(Y)\otimes H & \xrightarrow{1\otimes\varphi_Y} & F(X)\otimes H\otimes G(Y) & \xrightarrow{\varphi_X\otimes 1} & H\otimes G(X)\otimes G(Y) \\ \simeq\downarrow & & & & \downarrow\simeq \\ F(X\otimes Y)\otimes H & & \xrightarrow[\varphi_{X\otimes Y}]{} & & H\otimes G(X\otimes Y) \end{array}$$

$$\begin{array}{ccc} F(I)\otimes H & \xleftarrow{\simeq} & I\otimes H \simeq H \\ {\scriptstyle \varphi_I}\downarrow & & \downarrow{\scriptstyle 1} \\ H\otimes G(I) & \xleftarrow[\simeq]{} & H\otimes I \simeq H \end{array}$$

Observe that composition of arrows and parallel composition of 2-cells are defined using the tensor product in $\mathbb{H}$. Finally, a morphism of cat-groups $P\colon \mathbb{H}\to\mathbb{K}$ induces a pseudo-functor $P_*\colon [\mathbb{G},\mathbb{H}]\to[\mathbb{G},\mathbb{K}]$.

Now we apply the previous construction to a particular case. Consider a cat-group $\mathbb{G}$ and a $\mathbb{G}$-cat-group $(\mathbb{C},\gamma\colon \mathbb{G}\to Aut\mathbb{C})$. Following [12], we can construct the semi-direct product $\mathbb{G}\times_\gamma\mathbb{C}$ together with the projection $P\colon \mathbb{G}\times_\gamma\mathbb{C}\to\mathbb{G}$, which is a monoidal functor. Therefore, we have a pseudo-functor

$$P_*\colon [\mathbb{G},\mathbb{G}\times_\gamma\mathbb{C}]\to[\mathbb{G},\mathbb{G}]$$

and it is possible to construct a biequivalence from $\mathcal{Z}^1(\mathbb{G},\mathbb{C})$ to the homotopy fibre of P_* at the point $Id_\mathbb{G}\in[\mathbb{G},\mathbb{G}]$. This biequivalence is another way to formulate the universal property of the semi-direct product studied in [13].

4 The simplicial topological point of view

As with the original paper [5] of Brown, we are motivated by homotopy theory and the classical exact sequences of homotopy groups and pointed sets which occur there, although in the previous sections the linkage to 2-types of topological spaces is far in the background. Indeed, because our bicategorical notion of 2-exact sequence, the presentation has been more homological/algebraic in feeling.

Nevertheless, the link to simplicial homotopy theory as pioneered by Daniel Kan, John Moore, and John Milnor in the fifties is quite direct. To briefly review the relevant parts of this theory [1], recall that as observed by Kan, the property of the singular complex of a topological space which permits one to combinatorially define *all* of the homotopy groups at any base point is that the singular complex has a simple simplicial horn-lifting property which makes it a "Kan complex", and that a corresponding "Kan fibration" property (that corresponds essentially to the lifting properties of a fibration of spaces) is all that is needed to associate to a pointed simplicial Kan fibration f with fiber F :

$$F\subset E\longrightarrow B$$

[1] For more detail see [21], [10], or [1]

a long exact sequence of pointed sets, groups and abelian groups:

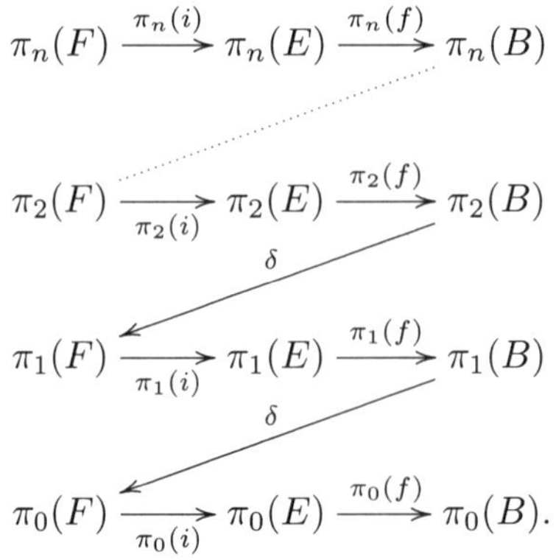

$$\begin{array}{ccccc}
\pi_n(F) & \xrightarrow{\pi_n(i)} & \pi_n(E) & \xrightarrow{\pi_n(f)} & \pi_n(B) \\
\vdots & & & & \\
\pi_2(F) & \xrightarrow[\pi_2(i)]{} & \pi_2(E) & \xrightarrow{\pi_2(f)} & \pi_2(B) \\
 & & \delta & & \\
\pi_1(F) & \xrightarrow[\pi_1(i)]{} & \pi_1(E) & \xrightarrow{\pi_1(f)} & \pi_1(B) \\
 & & \delta & & \\
\pi_0(F) & \xrightarrow[\pi_0(i)]{} & \pi_0(E) & \xrightarrow{\pi_0(f)} & \pi_0(B).
\end{array}$$

Now associated with any pointed simplicial complex X is the contractible complex $\mathcal{P}(X)$ of "based paths" of X which has as its 0-simplices the 1-simplices of X of the form $x \longrightarrow 0$, where $0 \in X_0$ is the base point. It is supplied with a canonical pointed simplicial map (last face) $d_n(X)\colon \mathcal{P}(X) \longrightarrow X$ which is a Kan fibration provided that X is a Kan complex. The fiber of this simplicial map is then also a Kan complex and is, of course, the complex $\Omega(X)$ of loops of X at the base point of X.

$$\Omega(X) \subset \mathcal{P}(X) \longrightarrow X$$

The long exact sequence associated with this fibration (since $\pi_i(\mathcal{P}(X)) = \{0\}$) just recapitulates the familiar sequence of isomorphisms

$$\pi_i(\Omega(X)) \simeq \pi_{i+1}(X) \quad i \geq 0.$$

Thus if $f\colon X \longrightarrow Y$ is a pointed simplicial map of Kan complexes, one can form the pullback along f of the fibration $d_n(Y)\colon \mathcal{P}(Y) \longrightarrow Y$

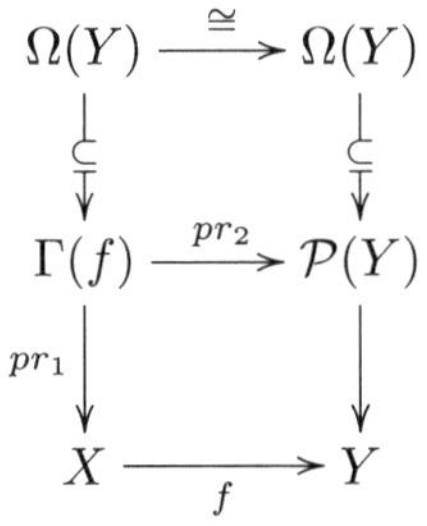

$$\begin{array}{ccc}
\Omega(Y) & \xrightarrow{\cong} & \Omega(Y) \\
\big\downarrow{\scriptstyle\subset} & & \big\downarrow{\scriptstyle\subset} \\
\Gamma(f) & \xrightarrow{pr_2} & \mathcal{P}(Y) \\
{\scriptstyle pr_1}\big\downarrow & & \big\downarrow \\
X & \xrightarrow[f]{} & Y
\end{array}$$

The fibers at the base point are then isomorphic and pr_1, as a pullback of a fibration, is itself a fibration. We then obtain a long exact sequence associated with the fibration pr_1,

$$\begin{array}{ccccc}
\pi_n(\Omega(Y)) & \xrightarrow{\pi_n(i)} & \pi_n(\Gamma(f)) & \xrightarrow{\pi_n(f)} & \pi_n(X) \\
 & & \cdots & & \\
\pi_2(\Omega(Y)) & \xrightarrow[\pi_2(i)]{} & \pi_2(\Gamma(f)) & \xrightarrow{\pi_2(pr_1)} & \pi_2(X) \\
 & & \delta & & \\
\pi_1(\Omega(Y)) & \xrightarrow[\pi_1(i)]{} & \pi_1(\Gamma(f)) & \xrightarrow{\pi_1(pr_1)} & \pi_1(X) \\
 & & \delta & & \\
\pi_0(\Omega(Y)) & \xrightarrow[\pi_0(i)]{} & \pi_0(\Gamma(f)) & \xrightarrow{\pi_0(pr_1)} & \pi_0(X)
\end{array}$$

which then becomes the *long exact sequence of the pointed simplicial mapping* $f\colon X \longrightarrow Y$:

$$\begin{array}{ccccc}
\pi_{n+1}(Y) & \xrightarrow{\pi_n(i)} & \pi_n(\Gamma(f)) & \xrightarrow{\pi_n(f)} & \pi_n(X) \\
 & & \cdots & & \\
\pi_3(Y) & \xrightarrow[\pi_2(i)]{} & \pi_2(\Gamma(f)) & \xrightarrow{\pi_2(pr_1)} & \pi_2(X) \\
 & & \delta & & \\
\pi_2(Y) & \xrightarrow[\pi_1(i)]{} & \pi_1(\Gamma(f)) & \xrightarrow{\pi_1(pr_1)} & \pi_1(X) \\
 & & \delta & & \\
\pi_1(Y) & \xrightarrow[\pi_0(i)]{} & \pi_0(\Gamma(f)) & \xrightarrow{\pi_0(pr_1)} & \pi_0(X) \\
 & & \pi_0(f) & & \\
\pi_0(Y) & & & &
\end{array}$$

or, equivalently,

$$\begin{array}{ccccc}
\pi_n(\Gamma(f)) & \xrightarrow{\pi_n(pr_1)} & \pi_n(X) & \xrightarrow{\pi_n(f)} & \pi_n(Y) \\
 & & \cdots & & \\
\pi_2(\Gamma(f)) & \xrightarrow[\pi_2(pr_1)]{} & \pi_2(X) & \xrightarrow{\pi_2(f)} & \pi_2(Y) \\
 & & \pi_1(i) & & \\
\pi_1(\Gamma(f)) & \xrightarrow[\pi_1(pr_1)]{} & \pi_1(X) & \xrightarrow{\pi_1(f)} & \pi_1(Y) \\
 & & \pi_0(i) & & \\
\pi_0(\Gamma(f)) & \xrightarrow[\pi_0(pr_1)]{} & \pi_0(X) & \xrightarrow{\pi_0(f)} & \pi_0(Y)
\end{array}$$

and thus another justification for calling $\Gamma(f)$ the *homotopy fiber* of $f\colon X \longrightarrow Y$. If f is already a fibration with fiber F, then $pr_2\colon \Gamma(f) \longrightarrow \mathcal{P}(Y)$ is a fibration and its long exact sequence combined with the contractibility of $\mathcal{P}(Y)$ then gives that $\pi_n(F) \simeq \pi_n(\Gamma(f))$, as expected. If f is an inclusion $f\colon X \subseteq Y$ then $\Gamma(f)$ defines the *homotopy groups of* Y *relative to* X with $\pi_{n-1}(\Gamma(f)) = \pi_n(Y;X)$.

In the previous sections, we have used bigroupoids, that is bicategories in which every 2-cell is an isomorphism and every 1-cell is an equivalence (*i.e.*, invertible up to isomorphism with respect to the (horizontal)"tensor product " composition). The link to simplicial topology now comes from the fact that every bicategory $\mathcal{G}$ has a simplicial set *nerve*, $\mathbf{Ner}(\mathcal{G})$, and *this simplicial set is is a Kan complex, precisely when the bicategory is a bigroupoid, i.e.*, has exactly the same invertibility requirements as those required for the bicategory to be a bigroupoid [10]. This nerve is minimal in dimensions ≥ 2 and if one chooses a base point $0 \in \mathbf{Ner}(\mathcal{G})_0$, which is the set of objects or 0-cells of $\mathcal{G}$, then $\mathbf{Ner}(\mathcal{G})$ has at that basepoint:

- $\pi_0(\mathbf{Ner}(\mathcal{G}))$ = the pointed set of categorical equivalence classes of the objects of $\mathcal{G}$, pointed by the class of 0.
- $\pi_1(\mathbf{Ner}(\mathcal{G}))$ = the group of homotopy classes of 1-cells of the form $f : 0 \longrightarrow 0$ under (horizontal) tensor composition of 1-cells
  = $\pi_0(\mathcal{G}(0,0))$, the set of connected components of the groupoid $\mathcal{G}(0,0) =_{\mathrm{DEF}} \mathcal{G}(0)$, whose objects are 1-cells $f \colon 0 \longrightarrow 0$ and whose arrows are the 2-cell isomorphisms $\alpha \colon f \Longrightarrow g$ and whose nerve is $\Omega(\mathbf{Ner}(\mathcal{G}))$ at the basepoint 0.
  = $\pi_1(cl(\mathcal{G}))$, where $cl(\mathcal{G})$ denotes, as in Section 2, the groupoid which has the same objects as $\mathcal{G}$ but has 2-cell isomorphism classes of 1-cells for arrows and is the *fundamental groupoid* $\Pi_1(\mathbf{Ner}(\mathcal{G}))$ of the Kan complex $\mathbf{Ner}(\mathcal{G})$.
- $\pi_2(\mathbf{Ner}(\mathcal{G}))$ = the abelian group of 2-simplices all of whose 1-simplex faces are at the base point $s_0(0) \colon 0 \longrightarrow 0$[2]
  = $\mathrm{Aut}(1_0)$ in the groupoid $\mathcal{G}(0,0))$, where $1_0 = s_0(0) \colon 0 \longrightarrow 0$ is the pseudo-identity 1-cell for 0 under tensor composition
  = $\pi_1(\mathcal{G}(0))$ in the notation of Section 2, and equivalently, $\pi_1(\Omega(\mathbf{Ner}(\mathcal{G}))$ in conventional simplicial notation.
- For $i \geq 3$, $\pi_i(\mathbf{Ner}(\mathcal{G})) = \mathbf{0}$, since, by definition, the canonical map

$$\mathbf{Ner}(\mathcal{G}) \longrightarrow \mathrm{Cosk}^3(\mathbf{Ner}(\mathcal{G}))$$

  is an isomorphism and this forces all higher dimensional homotopy groups of the pointed Kan complex $\mathbf{Ner}(\mathcal{G})$ to be trivial.

Now it is easy to verify that simplicial maps between nerves of bigroupoids correspond exactly to strictly unitary homomorphisms $P \colon \mathcal{G} \longrightarrow \mathcal{H}$ of bigroupoids. With this in mind and choosing $P(0) = 0$ as the base point of $\mathcal{H}$, we obtain a pointed simplicial mapping of Kan complexes $\mathbf{Ner}(P) \colon \mathbf{Ner}(\mathcal{G}) \longrightarrow \mathbf{Ner}(\mathcal{H})$. Thus all one need note is that the nerve of the "homotopy fiber bigroupoid" $\mathcal{F}_0$ of Section 2 is just $\Gamma(\mathbf{Ner}(P))$, $\mathcal{F}_0(0) \simeq \Omega(\Gamma(\mathbf{Ner}(P)))$, and that the long exact sequence of the pointed simplicial mapping $\mathbf{Ner}(P)$ is precisely the nine term sequence of Corollary 2.3.

Similar remarks apply to Brown's original paper: every category $\mathcal{G}$ has canonically associated to a simplicial set, its Grothendieck nerve, whose n-simplices can be identified with "composable sequences of length n of arrows of the category". The resulting simplicial set is a Kan complex if, and only if, every arrow of $\mathcal{G}$ is invertible, *i.e.*, $\mathcal{G}$ is a groupoid. For any object 0 in $\mathbf{Ner}(\mathcal{G})$ as a base point, $\mathbf{Ner}(\mathcal{G})$ has only the pointed set of isomorphism classes of its objects as π_0 and $Aut(0)$ as π_1. The long exact sequence above then reduces to a six term one of exactly the same form.

[2]The complex is minimal in this dimension (and higher), so homotopic 2-simplices are equal.

Note that in both cases, the weakest tenable notion of $f\colon X \longrightarrow Y$ is a *fibration* is that which guarantees that for any choice of basepoint, the canonical simplicial mapping from the true fiber $Fib(f)$ of the mapping f to the "homotopy fiber" $\Gamma(f)$ of the same mapping be a weak equivalence.

References

[1] André, M. *Generalized Homotopy Theory*, Batell Memorial Institute International Division, Special Technical and Scientific Report **2** (1963).

[2] Bénabou, J. *Introduction to bicategories*, Lecture Notes in Mathematics vol. 40, Springer, 1967, 1-77.

[3] Borceux, F. *Handbook of categorical algebra*, Cambridge University Press, 1994.

[4] Bourn, D.; Vitale, E. M. *Extensions of symmetric categorical groups*, Homology, Homotopy and Applications **4** (2002), 103–162 (http://www.rmi.acnet.ge/hha/).

[5] Brown, R. *Fibrations of groupoids*, Journal of Algebra **15** (1970), 103–132.

[6] Brown, R.; Spencer, C. B. *G-groupoids, crossed modules and the fundamental group of a topological group*, Proc. Kon. Ned. Akad. v. Wetensch. **79** (1976), 296–302.

[7] Carrasco, P.; Cegarra, A. M.; Grandjeán, A. R. *(Co)Homology of crossed modules*, Journal of Pure and Applied Algebra **168** (2002), 147–176.

[8] Carrasco, P.; Garzón, A. R. *Obstruction theory for extensions of categorical groups*, Applied Categorical Structures (to appear).

[9] Carrasco, P.; Garzón, A. R.; Miranda, J. G. *Schreier theory for singular extensions of categorical groups and homotopy classification*, Communications in Algebra **28** (2000), 2585–2613.

[10] Duskin, J. W. *Simplicial Matrices and the Nerves of Weak n-Categories I : Nerves of Bicategories*, Theory and Applications of Categories **9** (2002), 198–308 (http://www.tac.mta.ca/tac/).

[11] Garzón, A. R.; del Río, A. *The Whitehead categorical group of derivations*, Georgian Mathematical Journal **9** (2002), 709–721.

[12] Garzón, A. R.; Inassaridze, H. *Semidirect products of categorical groups. Obstruction theory*, Homology, Homotopy and Applications **3** (2001), 111–138 (http://www.rmi.acnet.ge/hha/).

[13] Garzón, A. R.; Inassaridze, H.; del Río, A. *Derivations of categorical groups*, preprint (2002).

[14] Gilbert, N. D. *The low-dimensional homology of crossed modules*, Homology, Homotopy and Applications **2** (2000), 41–50 (http://www.rmi.acnet.ge/hha/).

[15] Grandis, M. *Homotopical algebra in homotopical categories*, Applied Categorical Structures **2** (1994), 351–406.

[16] Grandis, M.; Vitale, E. M. *A higher dimensional homotopy sequence*, Homology, Homotopy and Applications **4** (2002), 59–69 (http://www.rmi.acnet.ge/hha/).

[17] Hardie, K. A.; Kamps, K. H.; Kieboom, R. W. *Fibrations of bigroupoids*, Journal of Pure and Applied Algebra **168** (2002), 35–43.

[18] Kasangian, S.; Vitale, E. M. *Factorization systems for symmetric cat-groups*, Theory and Applications of Categories **7** (2000), 47–70 (http://www.tac.mta.ca/tac/).

[19] Ladra, M.; Grandjeán, A. R. *Crossed modules and homology*, Journal of Pure and Applied Algebra **95** (1994), 41–55.

[20] Lue, A. S.-T. *Cohomology of groups relative to a variety*, Journal of Algebra **69** (1981), 155–174.

[21] May, J. P. *Simplicial Objects in Algebraic topology*, University of Chicago Press, 1992 (First published 1967).

[22] Moerdijk, I. *Lectures on 2-dimensional groupoids*, Rapport Inst. Math. Pure Appl. Univ. catholique Louvain **175** (1990).

[23] Porter, T. *Extensions, crossed modules and internal categories in categories of groups with operations*, Proceedings of the Edimburg Mathematical Society **30** (1987), 373–381.

[24] Rousseau, A. *Extensions de Gr-catégories*, Thèse de Doctorat, Université de Paris 13, 2000.

[25] Serre, J. P. *Cohomologie Galoisienne*, Lecture Notes in Mathematics vol. 5, Springer, 1964.

[26] Vieites, A. M.; Casas, J. M. *Some results on central extensions of crossed modules*, Homology, Homotopy and Applications **4** (2002), 29–42 (http://www.rmi.acnet.ge/hha/).

[27] Vitale, E. M. *A Picard-Brauer exact sequence of categorical groups*, Journal of Pure and Applied Algebra **175** (2002), 383–408.

Received November 13, 2002; in revised form June 24, 2003

Fields Institute Communications
Volume **43**, 2004

Applications of Categorical Galois Theory in Universal Algebra

Marino Gran
Université du Littoral Côte d'Opale
Laboratoire de Mathématiques Pures et Appliquées
50, Rue F. Buisson BP 699
62228 Calais, France
gran@lmpa.univ-littoral.fr

Abstract. We introduce factor permutable categories, which are a common generalization of Maltsev categories, of congruence modular varieties and of strongly unital varieties. We prove some basic centrality properties in these categories, which are useful to investigate several new aspects of the categorical theory of central extensions. We extend the equivalence between two natural and independent notions of central extension to exact factor permutable categories. These results are applied to semi-abelian categories, where several characterizations of central extensions are given, on the model of the category of groups. We finally show that the so-called Galois pregroupoid associated with an extension is an internal groupoid in any semi-abelian category.

Introduction

The categorical Galois theory developed by Janelidze [29] has contributed to clarify the deep relationship between several independent investigations in universal algebra and in homological algebra.

More specifically, the categorical theory of central extensions [31], which is a special case of it, provides a complete classification of the category $Centr(B)$ of central extensions of an object B in any exact category $\mathcal{C}$, where the property of centrality depends on the choice of an admissible subcategory $\mathcal{X}$ of $\mathcal{C}$. On the one hand, this theory includes several classical homological descriptions of $Centr(B)$ in the varieties of groups, rings, associative algebras and Lie algebras. More generally, it extends the theory developed by Frölich [22], Lue [39] and Furtado-Coelho [23] in the context of Ω-groups. On the other hand, the categorical notion of central extension includes the one naturally arising from the theory of commutators in

2000 *Mathematics Subject Classification.* Primary 18E10, 18G50, 08B10; Secondary 18B40, 08C15 .

Key words and phrases. Galois theory, commutator theory, abelian objects, modular and Maltsev varieties, factor permutable and semi-abelian categories.

universal algebra [32], [25]. More precisely, when $\mathcal{C}$ is a congruence modular variety and $\mathcal{X}$ is its admissible subvariety of abelian algebras, an extension $f\colon A \to B$ is *categorically central* (i.e. central in the sense of the categorical Galois theory) if and only if it is *algebraically central*, by which we mean that its kernel congruence $R[f]$ is contained in the centre of A, or, equivalently, $[R[f], \nabla_A] = \Delta_A$ (where ∇_A and Δ_A are the largest and the smallest congruences on A).

The notion of *factor permutable* category introduced in the present paper provides an appropriate framework to develop several new aspects of the categorical theory of central extensions. This notion is a categorical formulation of the axiom defining factor permutable varieties [26]: among the examples of factor permutable categories there are regular Maltsev categories, modular varieties and strongly unital varieties. We develop some basic aspects of a theory of centrality in these categories and establish some stability properties of the abelian objects. We prove that there is a direct product decomposition of central equivalence relations. This important property allows us to establish the equivalence between the notions of categorically central extension and of algebraically central extension in exact factor permutable categories. This theorem extends various previous results in this direction [32], [33], [25], and confirms the importance of the structural approach to commutator theory, which is at the same time simple and general.

We then consider some further aspects of the theory in the richer context of semi-abelian categories [34]. This part has been developed in collaboration with D. Bourn in [14]. We provide several characterizations of central extensions in semi-abelian categories on the model of the category of groups. We then show that the notion of central extension becomes intrinsic in this context: an extension $f\colon A \to B$ is central if and only if the diagonal $s\colon A \to R[f]$ is a kernel. We finally consider some applications of the categorical Galois theory, which provides in particular a complete classification of the category of central extensions. In the context of semi-abelian varieties, the category $Centr(B)$ can be decribed as a category of actions of the internal Galois groupoid of an appropriate extension.

The paper is structured as follows:

1. Internal structures
2. Connectors and commutators in algebra
3. Abelian objects in factor permutable categories
4. Algebraically central extensions
5. Categorically central extensions
6. Equivalence of the two notions
7. Central extensions in semi-abelian categories
8. Galois groupoids

In the first two sections we revise some basic categorical structures and we explain their relationship with the commutator of congruence relations as defined in universal algebra. In sections 3, 4, 5 and 6 we study the properties of abelian objects, of algebraically central extensions and of categorically central extensions in exact factor permutable categories. We include a brief and self-contained account of the categorical theory of central extensions in the form needed for our purposes. In the last two sections we work in the richer context of semi-abelian categories, where many stronger results can be established. For instance, the so-called Galois pregroupoid associated with an extension is proved to be always an internal groupoid.

1 Internal structures

In this section we briefly introduce some internal categorical structures, which will play a central role in the following. We refer to [2] or [40] for more details. In this article $\mathcal{C}$ will always denote a finitely complete category.

Internal categories and groupoids. We begin with some basic definitions:

1.1. Definition. An *internal category* X in $\mathcal{C}$ is a diagram in $\mathcal{C}$ of the form

$$X_1 \times_{X_0} X_1 \underset{p_2}{\overset{p_1}{\underset{m}{\rightrightarrows}}} X_1 \underset{d_2}{\overset{d_1}{\underset{\overset{s}{\leftarrow}}{\rightrightarrows}}} X_0,$$

where X_0 can be thought as the "object of objects", X_1 as the "object of arrows", $X_1 \times_{X_0} X_1$ as the "object of composable pairs of arrows" given by the following pullback

$$\begin{array}{ccc} X_1 \times_{X_0} X_1 & \xrightarrow{p_2} & X_1 \\ {\scriptstyle p_1}\downarrow & & \downarrow{\scriptstyle d_1} \\ X_1 & \xrightarrow[d_2]{} & X_0. \end{array}$$

The morphisms d_1, d_2 are called "domain" and "codomain" respectively, s is the "identity", p_1, p_2 are the projections, m is the "composition". These data must satisfy the usual axioms:

1. $d_1 \circ s = 1_{X_0} = d_2 \circ s$
2. $m \circ (1_{X_1}, s \circ d_2) = 1_{X_1} = m \circ (s \circ d_1, 1_{X_1})$
3. $d_1 \circ p_1 = d_1 \circ m, \quad d_2 \circ p_2 = d_2 \circ m$
4. $m \circ (1_{X_1} \times_{X_0} m) = m \circ (m \times_{X_0} 1_{X_1})$

(where the domain of the two composites in 4. is the "object of triples of composable arrows" $X_1 \times_{X_0} X_1 \times_{X_0} X_1$: it can be obtained as the pullback of $d_2 \circ p_2$ along d_1 or, equivalently, as the pullback of $d_1 \circ p_1$ along d_2).

1.2. Definition. An *internal groupoid* in $\mathcal{C}$ is an internal category in $\mathcal{C}$ equipped with a morphism $\sigma \colon X_1 \to X_1$, called the "inversion", such that

1. $d_1 \circ \sigma = d_2, \qquad d_2 \circ \sigma = d_1$
2. $m \circ (1_{X_1}, \sigma) = s \circ d_1, \qquad m \circ (\sigma, 1_{X_1}) = s \circ d_2$

Of course, when $\mathcal{C}$ is the category *Sets* of sets, an internal category (groupoid) is just an ordinary small category (groupoid). We denote by $Cat(\mathcal{C})$ and by $Grpd(\mathcal{C})$ the categories of internal categories and of internal groupoids in $\mathcal{C}$, respectively. The arrows in these categories are called *internal functors*: they are pairs (f_0, f_1) of arrows in $\mathcal{C}$ as in the diagram

$$\begin{array}{ccccc} X_1 \times_{X_0} X_1 & \xrightarrow{m_X} & X_1 & \underset{d_1}{\overset{d_2}{\underset{\overset{s}{\leftarrow}}{\rightrightarrows}}} & X_0 \\ \vdots\, {\scriptstyle f_2} & & \downarrow {\scriptstyle f_1} & & \downarrow {\scriptstyle f_0} \\ Y_1 \times_{Y_0} Y_1 & \xrightarrow{m_Y} & Y_1 & \underset{d_1}{\overset{d_2}{\underset{\overset{s}{\leftarrow}}{\rightrightarrows}}} & Y_0 \end{array}$$

such that

1. $d_1 \circ f_1 = f_0 \circ d_1, \qquad d_2 \circ f_1 = f_0 \circ d_2$

2. $f_1 \circ s = s \circ f_0$
3. $f_1 \circ m_X = m_Y \circ f_2$

(where f_2 is the arrow induced by the universal property of the pullback).
The following kind of internal functors will be useful:

1.3. Definition. If X and Y are internal categories in $\mathcal{C}$, an internal functor $(f_0, f_1)\colon X \to Y$ is a *discrete fibration* if the commutative square

$$\begin{array}{ccc} X_1 & \xrightarrow{f_1} & Y_1 \\ {\scriptstyle d_2}\downarrow & & \downarrow{\scriptstyle d_2} \\ X_0 & \xrightarrow[f_0]{} & Y_0 \end{array}$$

is a pullback.

When X and Y are groupoids, an internal functor is a discrete fibration if and only if the commutative square $f_0 \circ d_1 = d_1 \circ f_1$ is a pullback.

Internal equivalence relations. An *internal relation* R from X to Y in a category $\mathcal{C}$ is a pair of arrows in $\mathcal{C}$

$$X \xleftarrow{d_1} R \xrightarrow{d_2} Y,$$

which is jointly monic, i.e. the factorization $(d_1, d_2)\colon R \to X \times Y$ is a monomorphism. Again, in the category *Sets* of sets, this is equivalent to saying that $(d_1, d_2)\colon R \to X \times Y$ determines a relation in the usual sense on the set X. When $X = Y$ a relation $(d_1, d_2)\colon R \to X \times X$ is called a relation on X, and is also denoted by (R, X), or by R.

A relation R on X in $\mathcal{C}$ is *reflexive* if there is an arrow $s\colon X \to R$ with $d_1 \circ s = 1_X = d_2 \circ s$.

A relation R on X in $\mathcal{C}$ is *symmetric* if there is an arrow $\sigma\colon R \to R$ with $d_1 \circ \sigma = d_2$ and $d_2 \circ \sigma = d_1$.
Let

$$\begin{array}{ccc} R \times_X R & \xrightarrow{p_2} & R \\ {\scriptstyle p_1}\downarrow & & \downarrow{\scriptstyle d_1} \\ R & \xrightarrow[d_2]{} & X \end{array}$$

denote the pullback of d_1 along d_2. Then the relation R in $\mathcal{C}$ is *transitive* if there is an arrow $m\colon R \times_X R \to R$ such that $d_1 \circ m = d_1 \circ p_1$ and $d_2 \circ m = d_2 \circ p_2$.

1.4. Definition. A relation R on X is an *equivalence relation* if it is reflexive, symmetric and transitive.

1.5. Remark. Again, an internal equivalence relation in the category of sets is an equivalence relation in the usual sense. When $\mathcal{C}$ is a variety of universal algebras, an internal equivalence relation R on an algebra X in $\mathcal{C}$ is simply a *congruence*, i.e. an equivalence relation (on the underlying set of X) which is at the same time a subalgebra of $X \times X$.

Any internal equivalence relation (R, X) as just defined is in particular an internal groupoid with m and σ defined as above. We shall write $Eq(\mathcal{C})$ for the full subcategory of the category $Grpd(\mathcal{C})$ whose objects are the internal equivalence relations in $\mathcal{C}$. The category whose objects are the equivalence relations on a fixed object X in $\mathcal{C}$, with obvious arrows, will be denoted by $Eq_X(\mathcal{C})$.

The *kernel pair* $(p_1, p_2)\colon R[f] \to X \times X$ of any arrow $f\colon X \to Y$ is an equivalence relation. When an equivalence relation R on X is a kernel pair, R is said to be *effective*. The largest and the smallest equivalence relations ∇_X and Δ_X on a given object X are always effective: indeed, if $\tau_X\colon X \to 1$ is the unique arrow from an object X to the terminal object 1, then $\nabla_X = R[\tau_X]$, while $\Delta_X = R[1_X]$.

If $f\colon X \to Y$ is an arrow in $\mathcal{C}$ and R is an equivalence relation on Y, the inverse image of R along f is the equivalence relation $f^{-1}(R)$ defined by the following pullback:

$$\begin{array}{ccc} f^{-1}(R) & \xrightarrow{\tilde{f}} & R \\ {\scriptstyle (d_1,d_2)}\downarrow & & \downarrow{\scriptstyle (d_1,d_2)} \\ X \times X & \xrightarrow[f\times f]{} & Y \times Y \end{array}$$

Internal connectors. The notion of internal connector, introduced in [13], slightly modifies the notion of internal pregroupoid, which was discovered by Pedicchio to play a very important role in commutator theory [42], [43]. Unlike pregroupoids, connectors are also suitable to deal with non-effective equivalence relations [15]. The notion of internal pregroupoid goes back to the work of Kock [38].

If R and S are two equivalence relations on X, we denote by $R \times_X S$ the pullback

$$\begin{array}{ccc} R \times_X S & \xrightarrow{p_2} & S \\ {\scriptstyle p_1}\downarrow & & \downarrow{\scriptstyle d_1} \\ R & \xrightarrow[d_2]{} & X. \end{array}$$

1.6. Definition. An (internal) *connector* on R and S is an arrow $p\colon R \times_X S \to X$ in $\mathcal{C}$ such that

1. $p(x, x, y) = y$
2. $xSp(x, y, z)$
3. $p(x, y, p(y, u, v)) = p(x, u, v)$

1*. $p(x, y, y) = x$

2*. $zRp(x, y, z)$

3*. $p(p(x, y, u), u, v) = p(x, y, v)$.

Let us point out that when an arrow $p\colon R \times_X S \to X$ satisfies the conditions 1, 1*, 2 and 2*, then it satisfies also 3 and 3* if and only if it satisfies the classical associativity $p(x, y, p(z, u, v)) = p(p(x, y, z), u, v)$.

The relationship between connectors and internal groupoids is clarified in the first of the following examples:

1.7. Example. Let X be a graph in a category $\mathcal{C}$

$$X_1 \underset{d_2}{\overset{d_1}{\rightrightarrows}} X_0 \quad (s\colon X_0 \to X_1)$$

with $d_1 \circ s = 1_{X_0} = d_2 \circ s$, namely a *reflexive graph*. The connectors on $R[d_1]$ and $R[d_2]$ are in bijection with the groupoid structures on this reflexive graph.

Indeed, on the one hand, if p is a connector on $R[d_1]$ and $R[d_2]$ then the arrow $m\colon X_1 \times_{X_0} X_1 \to X_1$ defined by $m(x,y) = p(y,(s \circ d_2)(x),x)$ (for any (x,y) in $X_1 \times_{X_0} X_1$) gives the composition of a groupoid structure. On the other hand, if X is equipped with a groupoid composition m and an inversion σ, a connector is obtained by setting $p(x,y,z) = m(m(z,\sigma(y)),x)$ (for any (x,y,z) in $R[d_1] \times_{X_1} R[d_2]$).

1.8. Example. A ternary operation $p\colon X \times X \times X \to X$ is called an *associative Maltsev operation* when it satisfies the properties $p(x,x,y) = y$, $p(x,y,y) = x$ and $p(x,y,p(z,u,v)) = p(p(x,y,z),u,v)$. An (internal) associative Maltsev operation is therefore the same thing as a connector between ∇_X and ∇_X.

1.9. Example. Given two objects X and Y, there is always a canonical connector arising from the product $X \times Y$. By using the suitable product projections let us form the pullback

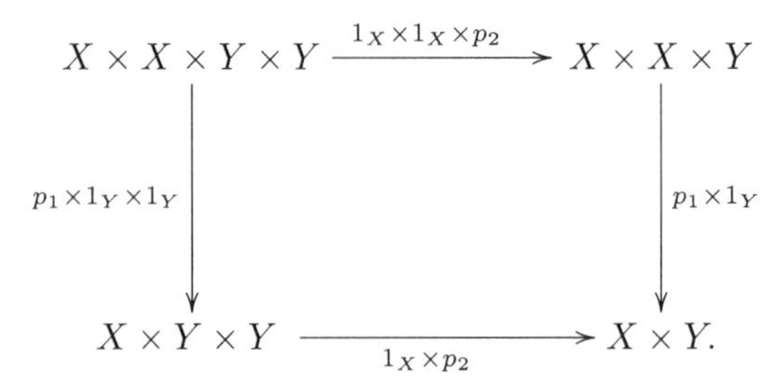

Then the canonical connector $p\colon X \times X \times Y \times Y \to X \times Y$ on $R[p_X]$ and $R[p_Y]$ (where $p_X\colon X \times Y \to X$ and $p_Y\colon X \times Y \to Y$ are the projections) is defined by

$$p(x,x',y,y') = (x',y).$$

Internal double equivalence relations. If $\mathcal{C}$ is a finitely complete category, the category $Eq(\mathcal{C})$ is itself finitely complete. Therefore, one can consider the category $Eq(Eq(\mathcal{C}))$ of internal equivalence relations in $Eq(\mathcal{C})$, which is called the category of internal double equivalence relations in $\mathcal{C}$.

A double equivalence relation C in $\mathcal{C}$ can be represented by a diagram of the form

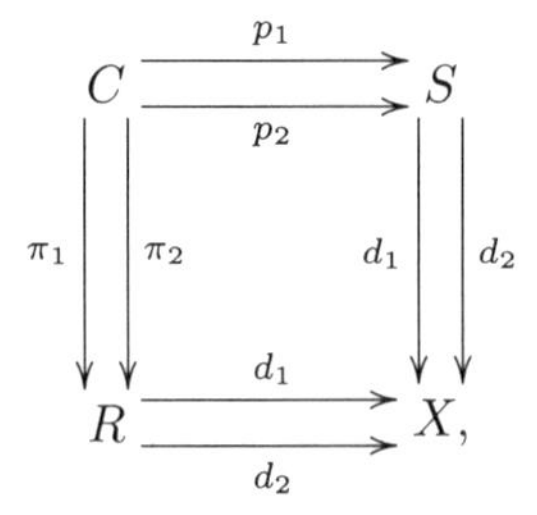

where each pair of parallel arrows represents an internal equivalence relation in $\mathcal{C}$, and

$$d_1 \circ \pi_1 = d_1 \circ p_1, d_1 \circ \pi_2 = d_2 \circ p_1, d_2 \circ \pi_1 = d_1 \circ p_2, d_2 \circ \pi_2 = d_2 \circ p_2.$$

C is called a *double equivalence relation on R and S*. Given two equivalence relations R and S on X there is a canonical largest double equivalence relation on R and S, which is denoted by $R\Box S$ (see [20],[36]):

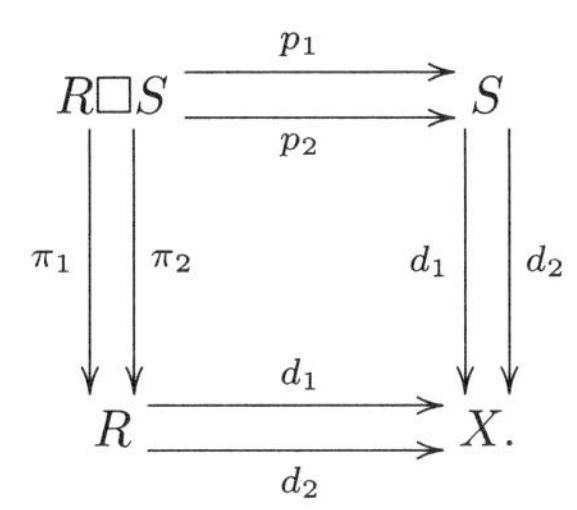

The double relation $R\Box S$ is defined by the following pullback:

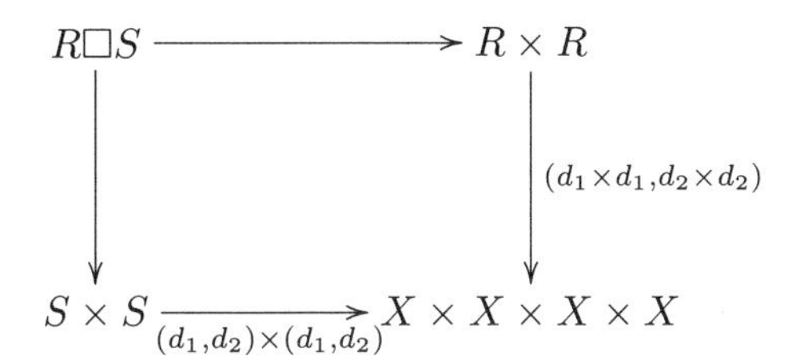

$R\Box S$ is the subobject of X^4 consisting, in the set-theoretical context, of the quadruples (x, y, t, z) with xRy, tRz, xSt and ySz. Any double equivalence relation C on R and S is contained in $R\Box S$: an element in C will be denoted also by a matrix

$$\begin{pmatrix} x & t \\ y & z \end{pmatrix}$$

The following kind of double equivalence relations will be useful:

1.10. Definition. [20], [44] A double equivalence relation C on R and S in $\mathcal{C}$ as above is called a *centralizing relation* when the following square is a pullback:

$$\begin{array}{ccc} C & \xrightarrow{p_2} & S \\ {\scriptstyle \pi_1}\downarrow & & \downarrow{\scriptstyle d_1} \\ R & \xrightarrow[d_2]{} & X. \end{array}$$

Connectors and centralizing relations actually determine the same structure:

1.11. Lemma. [20], [15] *If R and S are two equivalence relations on the same object X, then the following conditions are equivalent:*

1. *there is a connector on R and S*
2. *there is a centralizing relation on R and S*

Proof 1. $\Rightarrow$ 2. If $p\colon R \times_X S \to X$ is a connector between R and S, then by defining $\pi_1(x, y, z) = (x, p(x, y, z))$ and $\pi_2(x, y, z) = (p(x, y, z), z)$ one gets a centralizing relation on R and S:

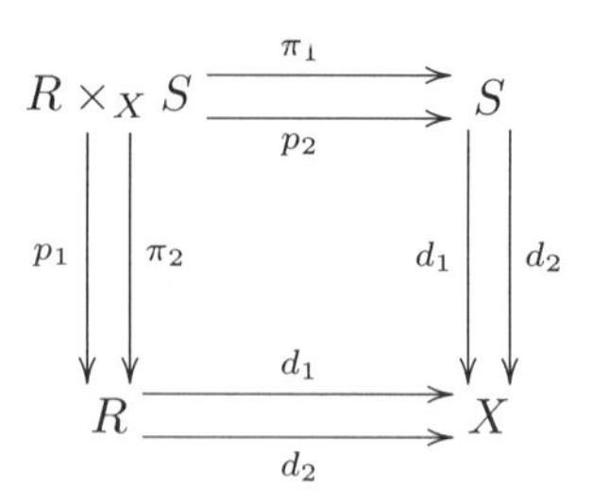

2. $\Rightarrow$ 1. If

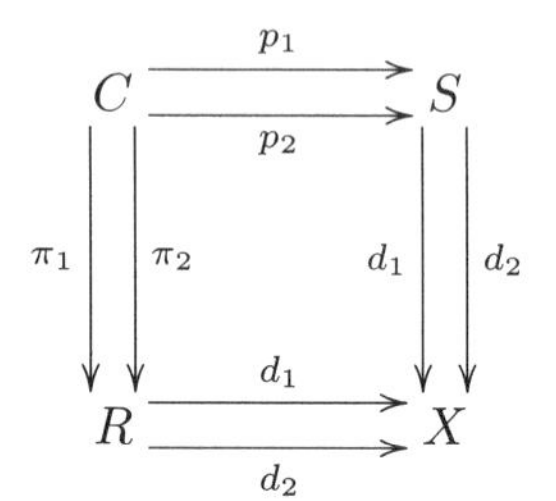

is a centralizing relation on R and S, then the arrow $d_2 \circ p_1 \colon C = R \times_X S \to S \to X$ defines a connector between R and S. $\square$

2 Connectors and commutators in algebra

In this section we recall some basic properties of regular, exact and Maltsev categories that will be useful in the following. We show how the notion of connector is deeply related to the notion of commutator of congruences in universal algebra. The discovery of this relationship is due to the pioneering work by Pedicchio [42], [43], and by Pedicchio and Janelidze [35], [36]. Recently, some new results in this direction have been obtained by Bourn and Gran [14], [15], [16], [12].

For the sake of simplicity we shall explain the relationship between connectors and commutators in the context of exact Maltsev categories. For more general varieties of universal algebras or categories the reader may consult [36], [15], [16].

Connectors in Maltsev categories. First of all, let us consider the case of the category *Grp* of groups. It is well known that in the category of groups there is a bijection between the congruences (=internal equivalence relations) and the normal subgroups. With a normal subgroup H of a group G one associates the congruence R_H defined by xR_Hy if and only if $x \cdot y^{-1}$ is in H. Conversely, with a congruence R on G one associates the normal subgroup $H = \{g \in G \mid gR1\}$ of G given by the elements in G which are in relation R with the unit element 1 of the group.

In the category of groups the existence of a connector between the congruences R_H and R_K on a group G is equivalent to the fact that the corresponding normal subgroups H and K commute in the usual sense, namely $[H, K] = \{1\}$. Let us recall the argument. If there is a connector $p \colon R_H \times_G R_K \to G$ on R_H and R_K, then we can construct an arrow $\alpha \colon H \times K \to G$ defined by $\alpha(h, k) = p(h, 1, k)$, which is a group homomorphism. Then, for all $h \in H$ and $k \in K$, we have: $h \cdot k = \alpha(h, 1) \cdot \alpha(1, k) = \alpha((h, 1) \cdot (1, k)) = \alpha(h, k) = \alpha((1, k) \cdot (h, 1)) = \alpha(1, k) \cdot \alpha(h, 1) = k \cdot h$. Conversely, suppose that $H \cdot K = K \cdot H$. Then the map $p \colon R_H \times_G R_K \to G$ defined by $p(h, k, l) = h \cdot k^{-1} \cdot l$ is a group homomorphism: indeed, for any $(a, b), (d, e) \in R_H$

and $(b,c),(e,f) \in R_K$ we have $(b^{-1}\cdot c)\cdot(d\cdot e^{-1}) = (d\cdot e^{-1})\cdot(b^{-1}\cdot c)$, since $b^{-1}\cdot c \in K$ and $d\cdot e^{-1} \in H$. This implies that $(a\cdot b^{-1}\cdot c)\cdot(d\cdot e^{-1}\cdot f) = (a\cdot d)\cdot(b\cdot e)^{-1}(c\cdot f)$, as desired. The axioms of connector are clearly satisfied by p, and it follows that p is a connector on R_H and R_K.

Similarly, in the category *Rng* of rings, two ideals I and J centralize each other in the usual sense, namely $IJ + JI = 0$, if and only if there is a connector on the corresponding congruences R_I and R_J.

It is useful to mention that a connector on two congruences in the category of groups or rings is necessarily unique, when it exists (a proof of this fact will be given in Theorem 2.6 in a more general context).

The fact that the existence of a connector on two congruences corresponds to the triviality of the commutator holds in very general categories. In order to explain this fact let us recall the following important notion due to Carboni, Lambek and Pedicchio:

2.1. Definition. [19] A finitely complete category $\mathcal{C}$ is a *Maltsev* category if any internal reflexive relation in $\mathcal{C}$ is an equivalence relation.

The terminology is motivated by the classical Maltsev Theorem:

2.2. Theorem. [41] *A variety $\mathcal{V}$ of universal algebras is Maltsev if and only if its theory has a ternary term $p(x,y,z)$ satisfying the axioms $p(x,x,y) = y$ and $p(x,y,y) = x$.*

Many varieties are Maltsev: groups, rings, Heyting algebras, Lie algebras, associative algebras, quasigroups and crossed modules. Among the examples of Maltsev categories there are also any abelian category, the dual of an elementary topos, the category of torsion-free abelian groups, the categories of Hausdorff groups and of topological groups [18].

As we shall see here below, the Maltsev property is equivalent to the fact that any internal relation is difunctional. Let us recall this latter notion:

2.3. Definition. A relation $R \rightarrowtail X \times Y$ from X to Y is *difunctional* if, whenever xRy, zRy and zRt, one then has that xRt.

Remark that the difunctionality property can be expressed in any finitely complete category (actually, pullbacks suffice). Indeed, let T be the object

$$T = \{(x,y,z,t) \in X \times Y \times X \times Y \mid xRy,\ zRy \,\text{and}\, zRt\},$$

which can be constructed as a limit. Then the relation R is difunctional if the canonical projections $p_1 \colon T \to X$ and $p_4 \colon T \to Y$ both factorize through R, i.e. there is an arrow $n \colon T \to R$ with $d_1 \circ n = p_1$ and $d_2 \circ n = p_4$. It is well known that a reflexive relation is an equivalence relation if and only if it is an internal groupoid: in the same way a relation $(d_1,d_2) \colon R \rightarrowtail X \times Y$ is difunctional if and only if it is a pregroupoid [36] (i.e. if and only if there is a connector on $R[d_1]$ and $R[d_2]$).

We are now ready to prove the following theorem:

2.4. Theorem. [20] *Let $\mathcal{C}$ be a finitely complete category. The following conditions are equivalent:*

1. *$\mathcal{C}$ is a Maltsev category*
2. *any reflexive relation in $\mathcal{C}$ is symmetric*
3. *any reflexive relation in $\mathcal{C}$ is transitive*
4. *for any $X, Y \in \mathcal{C}$ any relation $R \rightarrowtail X \times Y$ is difunctional*

5. *for any $X \in \mathcal{C}$ any relation $R \rightarrowtail X \times X$ is difunctional*

Proof For any relation $R \rightarrowtail X \times Y$ one defines a new relation $S \rightarrowtail R \times R$ as follows:

$$(a,b)S(c,d) \quad \Leftrightarrow \quad aRd.$$

This relation is reflexive by definition.
1. $\Rightarrow$ 2. Trivial.
2. $\Rightarrow$ 3. If aRb and bRc, then $(b,c)S(a,b)$ and by symmetry $(a,b)S(b,c)$. It follows that aRc.
3. $\Rightarrow$ 4. If aRb, cRb and cRd, then $(a,b)S(c,b)$ and $(c,b)S(c,d)$, so by transitivity $(a,b)S(c,d)$ and aRd.
4. $\Rightarrow$ 5. Trivial.
5. $\Rightarrow$ 1. Let R be a reflexive relation on X and let us prove that it is symmetric: if aRb, then from bRb aRb and aRa it follows that bRa. R is also transitive: if aRb and bRc, then aRb, bRb and bRc imply that aRc. □

In a Maltsev category the notion of internal connector becomes very simple:

2.5. Lemma. [42], [13] *Let $\mathcal{C}$ be a Maltsev category and let R and S be two equivalence relations on X. For an arrow $p\colon R \times_X S \to X$ the following conditions are equivalent:*

1. *p has the property that $p(x,y,y) = x$ and $p(x,x,y) = y$*
2. *p is a connector between R and S*

Proof 1. $\Rightarrow$ 2. Let us check the condition $zRp(x,y,z)$. For this, define a relation $T \rightarrowtail X^2 \times X$ as follows: $(x,y)Tz$ if and only if $(x,y,z) \in R \times_X S$ and $(z, p(x,y,z)) \in R$. Clearly, for any $(x,y,z) \in R \times_X S$, one has that $(x,y)Ty$, $(y,y)Ty$ and $(y,y)Tz$: by difunctionality it follows that $(x,y)Tz$. The condition $xSp(x,y,z)$ is proved similarly. In order to check that $p(p(x,y,z),u,v) = p(x,y,p(z,u,v))$, one defines a relation $W \rightarrowtail X^2 \times X^3$ by setting $(x,y)W(z,u,v)$ if and only if $(x,y,z) \in R \times_X S$, $(z,u,v) \in R \times_X S$ and $p(p(x,y,z),u,v) = p(x,y,p(z,u,v))$. From $(x,y)W(z,u,u)$,$(z,z)W(z,u,u)$ and $(z,z)W(z,u,v)$ it follows that $(x,y)W(z,u,v)$ for any x,y,z,u,v with the property that $(x,y,z) \in R \times_X S$ and $(z,u,v) \in R \times_X S$. □

Moreover, having a connector for two equivalence relations R and S on X becomes a property:

2.6. Theorem. [42], [13] *When there is a connector $p\colon R \times_X S \to X$ on R and S in a Maltsev category, it is necessarily unique.*

Proof Given two connectors p and p' on R and S, one defines a relation $D \rightarrowtail X^2 \times X$ as follows: $(x,y)Dz$ if and only if $(x,y,z) \in R \times_X S$ and $p(x,y,z) = p'(x,y,z)$. Then, for any $(x,y,z) \in R \times_X S$ one clearly has that $(x,y)Dy$, $(y,y)Dy$ and $(y,y)Dz$. By difunctionality of the relation D it follows that $(x,y)Dz$, as desired. □

Regular and exact Maltsev categories. We now revise some properties of the permutability of the composition of equivalence relations in regular and Barr-exact categories. We show that the Maltsev property is equivalent to the permutability of the composition of equivalence relations.

2.7. Definition. A finitely complete category $\mathcal{C}$ is *regular* when kernel pairs have coequalizers and regular epimorphisms are stable under pullbacks.

In a regular category any map $f\colon X \to Y$ factorizes through a unique smallest subobject $i\colon I \rightarrowtail Y$ of Y called its "regular image". Given two relations $(d_1, d_2)\colon R \rightarrowtail X \times Y$ and $(d_1, d_2)\colon S \rightarrowtail Y \times Z$ in a regular category, one can define the composite $S \circ R$ as the regular image of the arrow $(d_1 \circ p_1, d_2 \circ p_2)\colon R \times_Y S \to X \times Z$

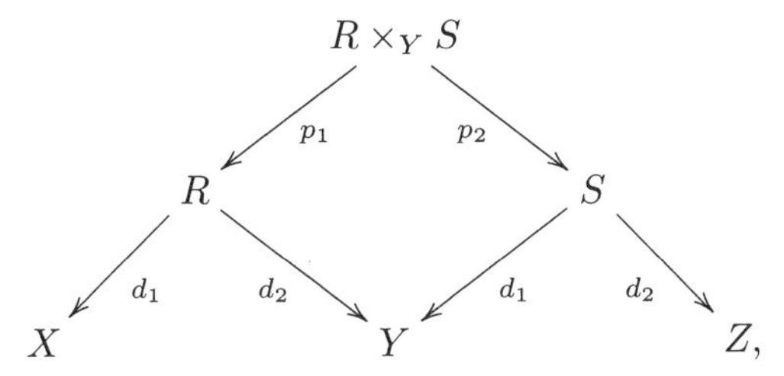

where $(R \times_Y S, p_1, p_2)$ is the pullback of $d_2\colon R \to Y$ along $d_1 : S_2 \to Y$. In a regular category the composition of relations is associative. A map $f\colon X \to Y$ can be considered as a relation by identifying it with its graph $(1_X, f)\colon X \rightarrowtail X \times Y$. In particular the identity map $1_X\colon X \to X$ gives the equivalence relation Δ_X.

For any relation $(d_1, d_2)\colon R \rightarrowtail X \times Y$ one can consider the opposite relation R^o given by $(d_2, d_1)\colon R \rightarrowtail Y \times X$. The fact that a relation $(d_1, d_2)\colon R \rightarrowtail X \times X$ is reflexive can be expressed by saying that $\Delta_X \leq R$; similarly $R = R^o$ says that the relation R is symmetric, and $R \circ R \leq R$ that R is transitive.

Given two maps $f\colon X \to Y$ and $g\colon Z \to Y$ one can consider their composite as relations $g^o \circ f$ which is simply the pullback of f and g. In particular, when $f = g$, the relation $f^o \circ f\colon X \times_Y X \rightarrowtail X \times X$ is the kernel pair of f. One can check also the following useful facts: 1) for any relation $(d_1, d_2)\colon R \rightarrowtail X \times Y$ one has that $R = d_2 \circ d_1^o$, 2) for any regular epimorphism $f\colon X \to Y$ one has that $f \circ f^o = \Delta_Y$.

2.8. Remark. In a regular category a relation $R \rightarrowtail X \times Y$ is *difunctional* exactly when $R \circ R^o \circ R = R$. It is clear that any relation f that is a map is difunctional, so that $f \circ f^o \circ f = f$.

In a regular category the Maltsev property corresponds to the permutability of the equivalence relations:

2.9. Theorem. [18], [19] *Let $\mathcal{C}$ be a regular category. The following statements are equivalent:*

1. *$\mathcal{C}$ is a Maltsev category*
2. *for any $X \in \mathcal{C}$, $\forall R, S \in Eq_X(\mathcal{C})$ the composite $R \circ S$ of two equivalence relations on X is an equivalence relation*
3. *the composition of equivalence relations is permutable:*
   for any $X \in \mathcal{C}$, $\forall R, S \in Eq_X(\mathcal{C})$, one has $R \circ S = S \circ R$

Proof 1. $\Rightarrow$ 2. It follows from the fact that $\Delta_X \leq R \circ S$.

2. $\Rightarrow$ 3. By assumption $R \circ S$ and $S \circ R$ are equivalence relations; this implies that $(R \circ S) \circ (R \circ S) \leq R \circ S$ and $(S \circ R) \circ (S \circ R) \leq S \circ R$. Then

$$R \circ S = \Delta_X \circ R \circ S \circ \Delta_X \leq (S \circ R) \circ (S \circ R) \leq S \circ R$$

and similarly $S \circ R \leq R \circ S$.
3. $\Rightarrow$ 1. We are going to prove that any relation $R \rightarrowtail X \times Y$ is difunctional (the result will then follow by Theorem 2.4). If $R = d_2 \circ d_1^o$, one has

$$R \circ R^o \circ R = d_2 \circ d_1^o \circ d_1 \circ d_2^o \circ d_2 \circ d_1^o = d_2 \circ d_2^o \circ d_2 \circ d_1^o \circ d_1 \circ d_1^o = d_2 \circ d_1^o = R,$$

since by assumption the equivalence relations $d_1^o \circ d_1$ and $d_2^o \circ d_2$ permute, and the difunctionality always holds for relations that are maps. □

Later on we shall be interested in regular categories which are also Barr-exact:

2.10. Definition. [1] A regular category $\mathcal{C}$ is *Barr-exact* if any equivalence relation in $\mathcal{C}$ is effective.

In an exact category, when two equivalence relations R and S permute, the following result holds:

2.11. Theorem. [18] *Let $\mathcal{C}$ be an exact category. Let r and s be two regular epimorphisms with the same domain X and kernel pairs R and S, respectively. Assume that $R \circ S = S \circ R$ and that the exterior part of the diagram*

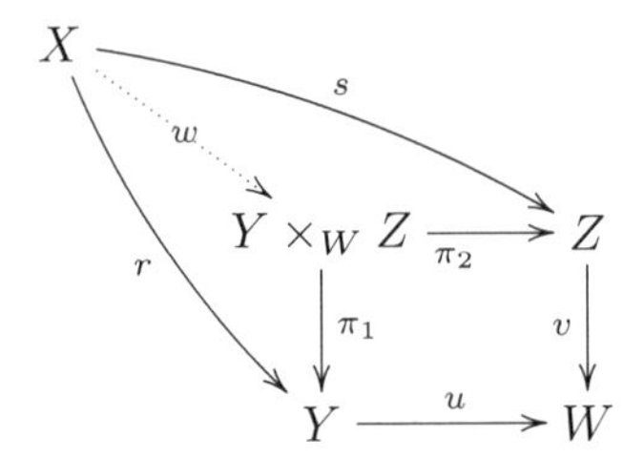

is a pushout. Then the comparison map w to the pullback is a regular epimorphism.

Proof Let T be the kernel pair of $t = v \circ s = u \circ r$. Since the exterior of the diagram is a pushout, $t = r \wedge s$ as quotients of X, hence T is the join $R \vee S$ of R and S in $Eq_X(\mathcal{C})$. By assumption we have that

$$R \circ S = S \circ R = R \vee S = T.$$

Since $R = r^o \circ r$, $S = s^o \circ s$, $\Delta_Y = r \circ r^o$ and $\Delta_Z = s \circ s^o$, it follows that

$$s^o \circ s \circ r^o \circ r = S \circ R = T = t^o \circ t,$$

from which, by multiplying on the left by s and on the right by r^o,

$$s \circ s^o \circ s \circ r^o \circ r \circ r^o = s \circ (v \circ s)^o \circ (u \circ r) \circ r^o;$$

hence, $s \circ r^o = s \circ s^o \circ v^o \circ u \circ r \circ r^o = v^o \circ u$. Since $v^o \circ u$ is the pullback of v and u and $s \circ r^o$ is the image of $(r, s) \colon X \to Y \times Z$, one concludes that w is a regular epimorphism. □

In particular, let us notice that any pushout of regular epimorphisms in an exact Maltsev category satisfies the assumptions of the previous Theorem.

Connectors and commutators. In any exact Maltsev category with coequalizers one can define the commutator of two equivalence relations R and S [42], [43], which extends the one originally defined by Smith in Maltsev varieties [44].

We assume that $\mathcal{C}$ is an exact Maltsev category with coequalizers. As in the case of varieties of universal algebra, one first defines an equivalence relation Δ_R^S on S: Δ_R^S is the kernel pair of the coequalizer $q\colon S \to Q$ of the morphisms $s \circ d_1 \colon R \to X \to S$ and $s \circ d_2 \colon R \to X \to S$. When $\mathcal{C}$ is a Maltsev variety, Δ_R^S is the smallest congruence on S containing all pairs of the form $((x, x), (y, y))$, in which (x, y) is in R.

By definition Δ_R^S is an equivalence relation on S. It is also an equivalence relation on R, because it is easily seen to be a reflexive relation, and $\mathcal{C}$ is a Maltsev category:

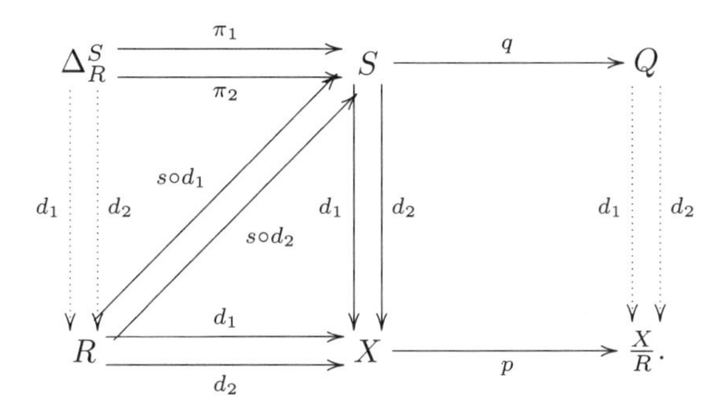

Accordingly, Δ_R^S is a double relation on R and S. The commutator of R and S is then defined (see [42], [43]) as

$$[R, S] = \{(x, y) \in X \times X \mid \exists z \in X \quad \textit{with} \quad \begin{pmatrix} z & x \\ z & y \end{pmatrix} \in \Delta_R^S\}.$$

Categorically, the relation $[R, S]$ on X can be constructed as follows: let $R[d_1]$ denote the kernel pair of $d_1 \colon S \to X$ and let $R[d_1] \cap \Delta_R^S$ denote the intersection of $R[d_1]$ and Δ_R^S as relations on S. Then, if i is the inclusion of $R[d_1] \cap \Delta_R^S$ in Δ_R^S and $\nu = (d_2 \circ \pi_1, d_2 \circ \pi_2) \colon \Delta_R^S \to X \times X$, then the commutator $[R, S]$ is the regular image of the arrow $\nu \circ i$.

The fact that Δ_R^S is a double equivalence relation on R and S implies that $[R, S]$ is a reflexive relation on X, hence an equivalence relation, because $\mathcal{C}$ is Maltsev.

Now, the crucial observation, due to Pedicchio, relating commutators and connectors is the following

2.12. Theorem. [42] *Let R and S be two equivalence relations on X in an exact Maltsev category with coequalizers. Then the following conditions are equivalent:*

1. $[R, S] = \Delta_X$
2. *there is a connector on R and S*

Proof Let us first prove that $[R,S]=\Delta_X$ if and only if the canonical factorization $\alpha\colon \Delta_R^S \to R\times_X S$ through the usual pullback is a monomorphism:

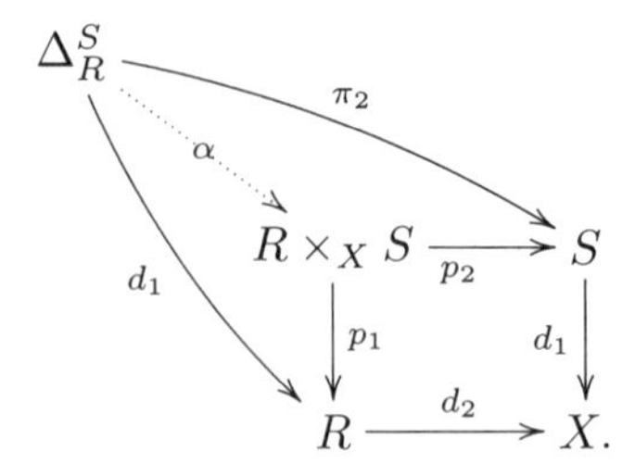

For this, let us assume that $[R,S]=\Delta_X$ and that

$$\alpha\begin{pmatrix} x & t \\ y & z\end{pmatrix} = \alpha\begin{pmatrix} x & t' \\ y & z\end{pmatrix}.$$

By the symmetry and the transitivity of Δ_R^S on S, one knows that $\begin{pmatrix} x & t \\ x & t'\end{pmatrix}$ is in Δ_R^S, hence (t,t') is in $[R,S]$. It follows that $t=t'$. Conversely, let α be a mono and let (t,t') be in $[R,S]$. Then, on the one hand, $\begin{pmatrix} z & t \\ z & t'\end{pmatrix}$ is in Δ_R^S for some z in X; on the other hand $\begin{pmatrix} z & t' \\ z & t'\end{pmatrix}$ is in Δ_R^S because Δ_R^S is a double equivalence relation. Since α is a monomorphism, it follows that $t=t'$.

To complete the proof one needs to observe that this factorization α is always a regular epi thanks to Theorem 2.11. It follows that Δ_R^S is a centralizing relation on R and S if and only if $[R,S]=\Delta_X$. By Lemma 1.11 this fact is equivalent to the fact that there is a connector on R and S. □

The discovery of the relationship between connectors and commutators opened the way to the categorical approach to centrality. Indeed, the main stability properties of the commutators correspond to the properties of some internal categorical structures [36], [15], [16], [12].

3 Abelian objects in factor permutable categories

In this section we define *factor permutable categories* and prove some basic centrality properties in these categories. Our definition is a categorical version of Gumm's definition of a factor permutable variety [26]. Abelian objects and central relations still behave well in these categories, and this fact makes them suitable to extend several results of the categorical theory of central extensions.

Let $p_A\colon A\times B\to A$ and $p_B\colon A\times B\to B$ denote the product projections:

3.1. Definition. A regular category $\mathcal{C}$ is *factor permutable* if any equivalence relation R on a direct product $A\times B$ permutes with $R[p_A]$ and with $R[p_B]$.

3.2. Examples. 1. By Theorem 2.9, any regular Maltsev category is factor permutable.
2. Any congruence modular variety is factor permutable. This important property was proved by Gumm (Corollary 4.5 in [26]). Hence in particular distributive varieties are factor permutable, as for instance the variety of lattices and the variety of implication algebras.

3. Any strongly unital variety is factor permutable. In particular, the variety of left closed magmas is strongly unital [10]. Strongly unital varieties have been characterized as follows [5]: a variety $\mathcal{V}$ is strongly unital if and only if its theory has exactly one constant 0 and a ternary operation $p(x, y, z)$ satisfying the axioms $p(x, x, y) = y$ and $p(x, 0, 0) = x$. To see that these varieties are factor permutable [26], consider a product $A \times B$ in a strongly unital variety $\mathcal{V}$. Let R be a congruence on $A \times B$, and let $(a, b)R[p_A](a, c)R(d, e)$. Then

$$p((a, c), (0, c), (0, b))\, R\, p((d, e), (0, c), (0, b))$$

gives $(a, b)R(d, p(e, c, b))$. This implies that $(a, b)R(d, p(e, c, b))R[p_A](d, e)$, as desired.

The following result will be very useful in the following:

3.3. Lemma. *If $\mathcal{C}$ is a factor permutable category, then the* weak shifting property *holds: for any equivalence relation R and S on $A\times B$ such that $R\cap R[p_A] \leq S$, given $(a, b), (a, c), (d, e)$ and (d, f) related as in the diagram*

$$\begin{array}{ccc} (a,b) & \overset{R[p_A]}{\text{———}} & (a,c) \\ R\,\big| & & \big|\,R \\ (d,e) & \underset{R[p_A]}{\text{———}} & (d,f), \end{array} \quad \text{with } (a,b)\ S\ (d,e)$$

one always has that $(a, c)S(d, f)$.

Proof By the factor permutability of $\mathcal{C}$, the assumption $(d, e)(R\cap S)\circ R[p_A](a, c)$ implies that there is an element (d, g) such that $(d, e)R[p_A](d, g)R\cap S(a, c)$, so that $(d, f)R(a, c)R \cap S(d, g)$ and then $(d, f)R \cap R[p_A](d, g)$. By the assumption that $R\cap R[p_A] \leq S$ it follows that $(d, f)S(d, g)$, and then $(d, f)S(a, c)$. □

Of course, the same kind of property holds when we consider equivalence relations R and S on $A\times B$ such that $R\cap R[p_B] \leq S$. The weak shifting property is equivalent to the following categorical property:

3.4. Lemma. *Given a product $A\times B$ in a factor permutable category, given an equivalence relation R on $A\times B$ the following property holds: for any $U \in Eq_{A\times B}(\mathcal{C})$ with $R\cap R[p_A] \leq U \leq R$ the canonical inclusion of equivalence relations*

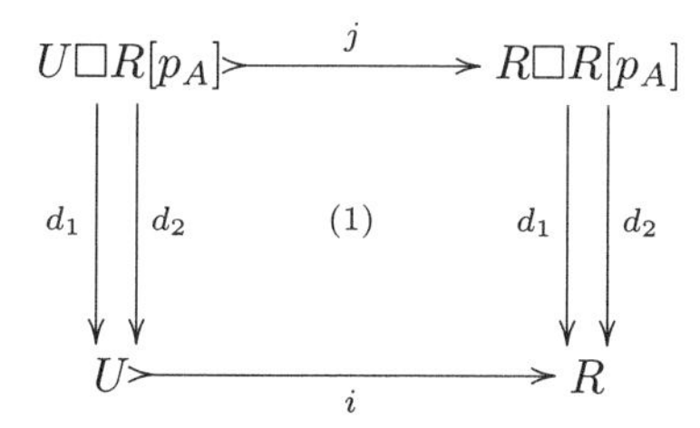

is a discrete fibration.

Proof Observe that in the formulation of the weak shifting property we could assume that S is less or equal to R: indeed, if not, then we could replace S by its intersection $R\cap S = U$ with R. After making this assumption the discrete fibration condition becomes nothing but the direct translation of the weak shifting property into the language of pullbacks. □

3.5. Remark. From now on we shall adopt a simplification in the notations: in the diagrams we shall often write p_A also for the equivalence relation $R[p_A]$. Furthermore, if $f\colon A \times B \to C$ is an arrow, then the diagram on a product $A \times B$

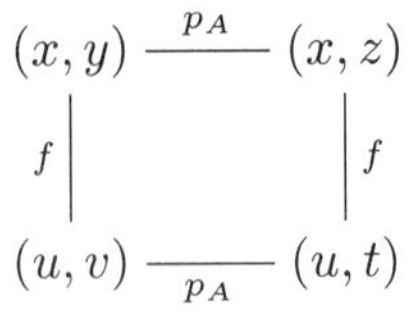

$$\begin{array}{ccc} (x,y) & \overset{p_A}{\text{———}} & (x,z) \\ f \Big| & & \Big| f \\ (u,v) & \underset{p_A}{\text{———}} & (u,t) \end{array}$$

indicates that $(x,y)R[p_A](x,z)$, $(u,v)R[p_A](u,t)$, $(x,y)R[f](u,v)$ and $(x,z)R[f](u,t)$.

In the rest of this section $\mathcal{C}$ will denote a factor permutable category. First of all, we remark that the weak shifting property has some interesting consequences:

3.6. Lemma. *Given an equivalence relation R on an object X, there is at most one internal partial Maltsev operation $p\colon R \times X \to X$. In particular there is at most one connector on R and ∇_X.*

Proof Let $p\colon R \times X \to X$ be an internal partial Maltsev operation. We first prove that $R[p] \cap R[p_R] = \Delta_{R\times X}$. Consider two elements (x,y,z) and (x,y,u) in $R \times X$ with the property that $p(x,y,z) = p(x,y,u)$. Then the weak shifting property applied to

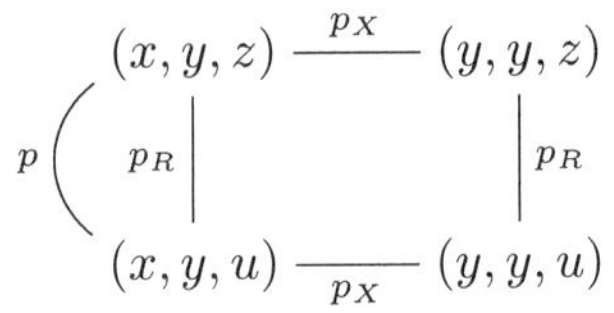

$$p\left(\begin{array}{ccc} (x,y,z) & \overset{p_X}{\text{———}} & (y,y,z) \\ p_R \Big| & & \Big| p_R \\ (x,y,u) & \underset{p_X}{\text{———}} & (y,y,u) \end{array}\right.$$

gives $z = p(y,y,z) = p(y,y,u) = u$. Let then $p\colon R \times X \to X$ and $p'\colon R \times X \to X$ be two internal Maltsev operations. The fact that $R[p] \cap R[p_R] = \Delta_{R\times X}$ allows one to apply the weak shifting property to

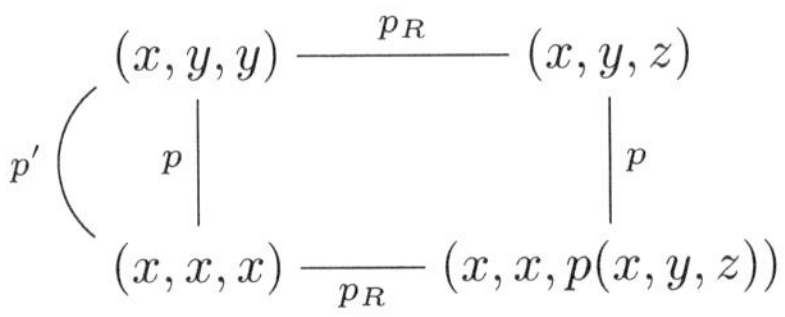

$$p'\left(\begin{array}{ccc} (x,y,y) & \overset{p_R}{\text{————}} & (x,y,z) \\ p \Big| & & \Big| p \\ (x,x,x) & \underset{p_R}{\text{———}} & (x,x,p(x,y,z)) \end{array}\right.$$

for any (x,y,z) in $R \times X$. Accordingly, $p'(x,y,z) = p'(x,x,p(x,y,z)) = p(x,y,z)$. □

If there is a connector on R and ∇_X, we say that R is *(algebraically) central.*

3.7. Corollary. *Given an object X in a factor permutable category $\mathcal{C}$, there is at most one internal Maltsev operation $p\colon X \times X \times X \to X$ on X. This operation is always associative.*

Proof Thanks to the previous lemma, only the associativity of p needs to be checked. Let us think of $X \times X \times X$ as being the product $(X \times (X \times X), p_1, p_{2,3})$.

Then the weak shifting property applied to the diagram

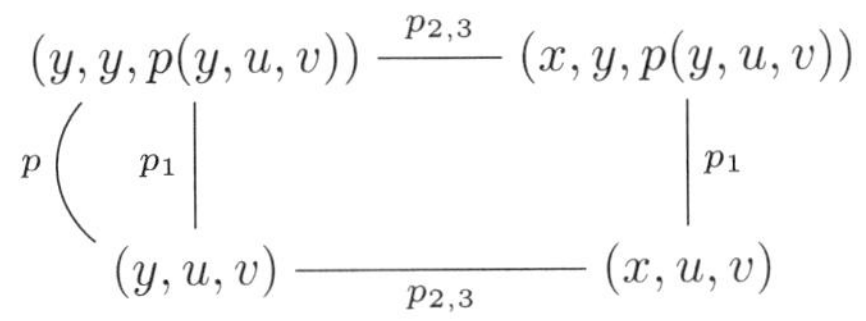

$$\begin{array}{ccc} (y,y,p(y,u,v)) & \overset{p_{2,3}}{\text{———}} & (x,y,p(y,u,v)) \\ p \Big(\;\; p_1 \Big| & & \Big| p_1 \\ (y,u,v) & \underset{p_{2,3}}{\text{———}} & (x,u,v) \end{array}$$

gives $p(x,y,p(y,u,v)) = p(x,u,v)$. Similarly one checks that $p(p(x,y,u),u,v) = p(x,y,v)$. Consequently,

$$p(x,y,p(z,u,v)) = p(p(x,y,z),z,p(z,u,v)) = p(p(x,y,z),u,v).$$

□

According to the previous corollary, being abelian for an object X in a factor permutable category is a property.

3.8. Definition. An object X in a factor permutable category is *abelian* if there is a (unique) Maltsev operation $p\colon X \times X \times X \to X$ on X.

Let $\mathcal{C}_{Ab}$ be the *category of abelian objects* in $\mathcal{C}$. Objects in $\mathcal{C}_{Ab}$ are the abelian objects in $\mathcal{C}$, and arrows in $\mathcal{C}_{Ab}$ are the arrows $f\colon X \to Y$ in $\mathcal{C}$ that respect the Maltsev operation, i.e. such that $p_Y \circ f^3 = f \circ p_X$.

3.9. Lemma. *$\mathcal{C}_{Ab}$ is full in $\mathcal{C}$.*

Proof Let $f\colon X \to Y$ be an arrow in $\mathcal{C}$, let $p_X\colon X^3 \to X$ and $p_Y\colon Y^3 \to Y$ be Maltsev operations on X and Y, respectively. Since $R[p_X] \cap R[p_{1,2}] = \Delta_{X^3}$ (as shown in the proof of Lemma 3.6) we can apply the weak shifting property to the diagram

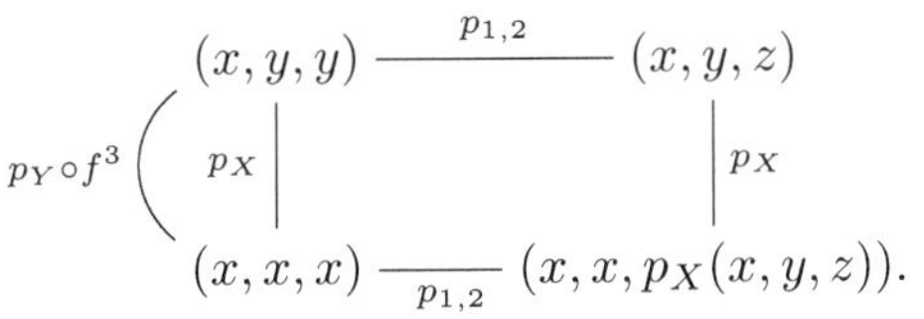

$$\begin{array}{ccc} (x,y,y) & \overset{p_{1,2}}{\text{———}} & (x,y,z) \\ p_Y \circ f^3 \Big(\;\; p_X \Big| & & \Big| p_X \\ (x,x,x) & \underset{p_{1,2}}{\text{———}} & (x,x,p_X(x,y,z)). \end{array}$$

This gives $(p_Y \circ f^3)(x,y,z) = (p_Y \circ f^3)(x,x,p_X(x,y,z)) = (f \circ p_X)(x,y,z)$, for any $(x,y,z) \in X^3$. □

Johnstone introduced in [37] the notion of a naturally Maltsev category: a finitely complete category is *naturally Maltsev* if there is a natural transformation p from $1_{\mathcal{C}} \times 1_{\mathcal{C}} \times 1_{\mathcal{C}}$ to $1_{\mathcal{C}}$ with the following property: for any A in $\mathcal{C}$, the A-component p_A of p is an internal Maltsev operation. It is easy to see that any naturally Maltsev category is Maltsev.

3.10. Corollary. *$\mathcal{C}_{Ab}$ is a naturally Maltsev category.*

Proof By definition of $\mathcal{C}_{Ab}$ any object in it is equipped with an internal Maltsev operation. Moreover, the fact that $\mathcal{C}_{Ab}$ is full in $\mathcal{C}$ means that the various Maltsev operations organize themselves in a natural transformation from $1_{\mathcal{C}} \times 1_{\mathcal{C}} \times 1_{\mathcal{C}}$ to $1_{\mathcal{C}}$. □

The following property of factor permutable categories will have some very strong consequences:

3.11. Proposition. *Any double equivalence relation D on ∇_X and R has the property that the canonical arrow $\alpha\colon D \to R \times X$ defined by*

$$\alpha\begin{pmatrix} x & t \\ y & z \end{pmatrix} = (x, y, z)$$

is a regular epi.

Proof Thanks to the Barr embedding for regular categories, it is sufficient to prove this fact in the category of sets [1] (for an explanation of this fact see also Metatheorem 1.8 in [3]). Now, let D be any double equivalence relation on R and ∇_X, let (x, y, z) be an element in $R \times X$, and let $p_1\colon X \times X \to X$ be the first projection. Since the category is factor permutable, the equivalence relations $R[p_1]$ and D on $X \times X$ permute. Accordingly, from

$$(x, y)R[p_1](x, x)D(y, y)R[p_1](y, z)$$

it follows that there is an element (x, t) in $X \times X$ such that

$$(x, y)R[p_1](x, x)R[p_1](x, t)D(y, z).$$

Consequently, there is an element $\begin{pmatrix} x & t \\ y & z \end{pmatrix}$ in D, and this shows that the arrow α is a regular epi. $\square$

3.12. Proposition. *Let R be an equivalence relation on Y. If R is central and $i\colon X \to Y$ is a monomorphism, then $i^{-1}(R)$ is central.*

Proof We denote by D the centralizing double relation on R and ∇_Y. Then consider the inverse image of the equivalence relation R along i:

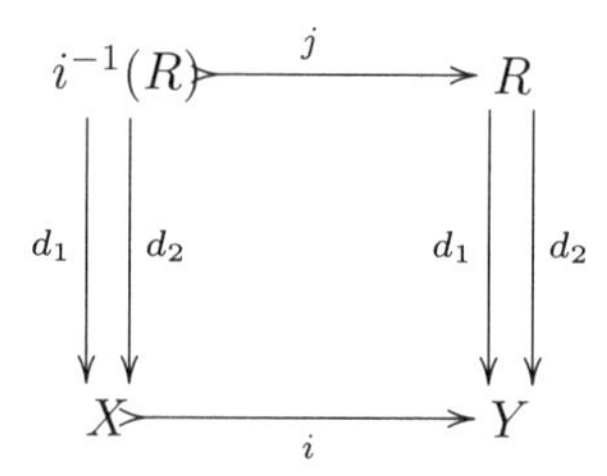

The inverse image $j^{-1}(D)$ of D along j determines a double equivalence relation on $i^{-1}(R)$ and ∇_X:

$$\begin{array}{ccc} j^{-1}(D) & \overset{p_1}{\underset{p_2}{\rightrightarrows}} & X \times X \\ \pi_1 \downarrow\downarrow \pi_2 & & d_1 \downarrow\downarrow d_2 \\ i^{-1}(R) & \overset{d_1}{\underset{d_2}{\rightrightarrows}} & X. \end{array}$$

There is an induced arrow α from $j^{-1}(D)$ to the pullback $i^{-1}(R) \times X$ of $d_2\colon i^{-1}(R) \to X$ along $d_1\colon X \times X \to X$. The arrow α is a regular epi thanks to the previous Proposition. On the other hand, the fact that $i \times i$ and j are monos implies that α is a mono, and then an iso. This means that $j^{-1}(D)$ is a centralizing relation on $i^{-1}(R)$ and ∇_X. $\square$

3.13. Proposition. *Let R and S be two equivalence relations on X, with $R \leq S$. If S is central, then R is central.*

Proof Let $j\colon R \to S$ denote the inclusion of R in S. If $p\colon S \times X \to X$ is the connector on S and ∇_X, then the arrow $p \circ (j \times 1_X)\colon R \times X \to X$ is the connector between R and ∇_X. The only axiom in the definition of a connector that is not trivial is the one asserting that $p(x, y, z)Rz$ for any (x, y, z) in $R \times X$. In order to check this axiom, consider the relation D on $X \times X$ defined by

$$(x, t)D(y, z) \quad \Leftrightarrow \quad xRy \text{ and } t = p(x, y, z).$$

One can see that D is an equivalence relation on $X \times X$ with the property that $D \cap R[p_1] = \Delta_{X \times X}$. Consequently, for any $(x, y, z) \in R \times X$, we can apply the weak shifting property to the situation

$$R\square\nabla_X \left(\begin{array}{ccc} (x,x) & \overset{p_1}{\text{———}} & (x, p(x,y,z)) \\ D \Big| & & \Big| D \\ (y,y) & \underset{p_1}{\text{———}} & (y,z), \end{array} \right.$$

It follows that $(x, p(x, y, z))R\square\nabla_X(y, z)$, thus $p(x, y, z)Rz$, as desired. □

3.14. Corollary. *If $i\colon R \to S$ is a monomorphism in the category of equivalence relations in $\mathcal{C}$ and S is central, then R is central.*

Proof It follows by the two previous propositions. □

In particular we obtain a stability property of abelian objects:

3.15. Corollary. *$\mathcal{C}_{Ab}$ is closed under subobjects in $\mathcal{C}$.*

Proof If Y is an abelian object and $i\colon X \to Y$ is a mono, then $(i, i \times i)\colon \nabla_X \to \nabla_Y$ is trivially an arrow in the category of equivalence relations, and the result follows by the previous corollary. □

3.16. Lemma. *Let R be a central equivalence relation on X, and let S be an equivalence relation on Y. Then, if $\overline{f}$ in the commutative diagram*

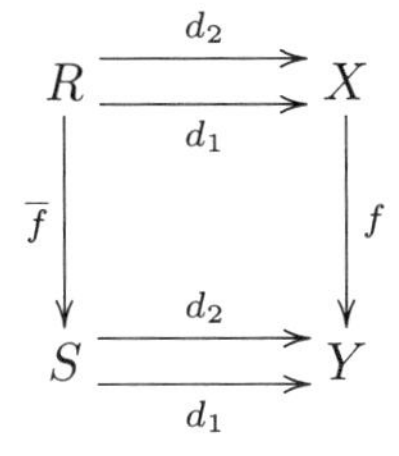

is a regular epimorphism, then S is central.

Proof Let $p\colon R \times X \to X$ be the connector on R and ∇_X. Since f and $\overline{f}$ are regular epimorphisms, and the category is regular, both $f \times f\colon X \times X \to Y \times Y$ and $\overline{f} \times f\colon R \times X \to S \times Y$ are regular epimorphisms as well. Let $p_R\colon R \times X \to R$ and $p_X\colon R \times X \to X$ be the projections, let us first prove that $R[\overline{f} \times f] \cap R[p_R] \leq R[f \circ p]$.

Indeed, if $[(x,y,z),(x,y,u)]$ is in $R[\overline{f} \times f] \cap R[p_R]$, one has that $f(z) = f(u)$. By the weak shifting property applied to the situation

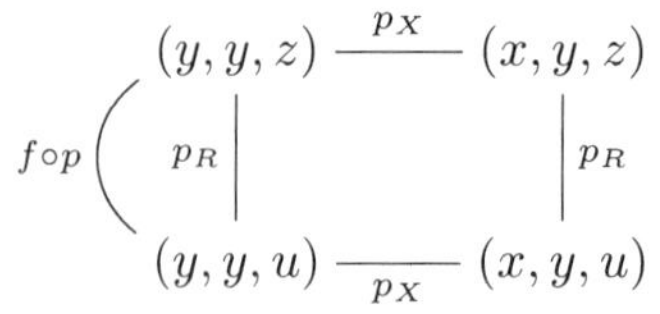

$$\begin{array}{ccc} (y,y,z) & \overset{p_X}{\text{———}} & (x,y,z) \\ {\scriptstyle f\circ p}\Big(\; {\scriptstyle p_R}\Big| & & \Big|{\scriptstyle p_R} \\ (y,y,u) & \underset{p_X}{\text{———}} & (x,y,u) \end{array}$$

it follows that $[(x,y,z),(x,y,u)]$ is in $R[f \circ p]$, as desired. Let us then show that the kernel pair $R[\overline{f} \times f]$ of $\overline{f} \times f$ is contained in $R[f \circ p]$. If $[(x,y,z),(u,v,w)]$ is an element in $R[\overline{f} \times f]$, then clearly

$$(f \circ p)(x,y,y) = f(x) = f(u) = (f \circ p)(u,v,v)$$

so that we can form the diagram

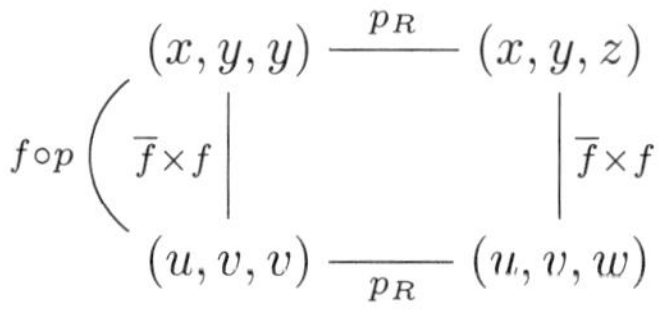

$$\begin{array}{ccc} (x,y,y) & \overset{p_R}{\text{———}} & (x,y,z) \\ {\scriptstyle f\circ p}\Big(\; {\scriptstyle \overline{f}\times f}\Big| & & \Big|{\scriptstyle \overline{f}\times f} \\ (u,v,v) & \underset{p_R}{\text{———}} & (u,v,w) \end{array}$$

We have just proved that $R[\overline{f} \times f] \cap R[p_R] \leq R[f \circ p]$, so that the weak shifting property implies that $[(x,y,z),(u,v,w)]$ is also in $R[f \circ p]$. Consequently, the universal property of the coequalizer $\overline{f} \times f$ yields a unique arrow $\pi \colon S \times Y \to Y$ such that $\pi \circ (\overline{f} \times f) = f \circ p$. This arrow π is the connector on S and ∇_Y (we leave the verification to the reader). $\square$

3.17. Corollary. *$\mathcal{C}_{Ab}$ is closed in $\mathcal{C}$ under regular quotients.*

Proof Given a regular epi $f \colon X \to Y$ with X an abelian object, the commutative diagram

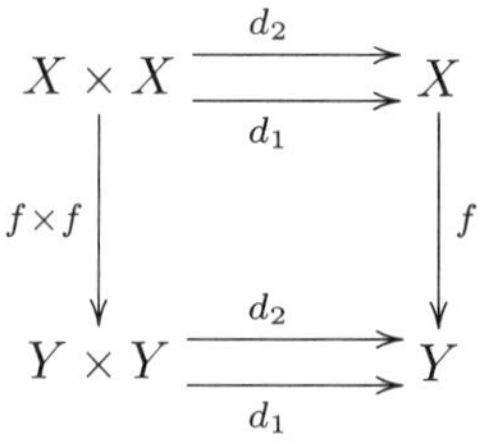

$$\begin{array}{ccc} X \times X & \underset{d_1}{\overset{d_2}{\rightrightarrows}} & X \\ {\scriptstyle f\times f}\downarrow & & \downarrow{\scriptstyle f} \\ Y \times Y & \underset{d_1}{\overset{d_2}{\rightrightarrows}} & Y \end{array}$$

is clearly of the type considered in the previous lemma. Accordingly, Y is abelian. $\square$

4 Algebraically central extensions

In this section we prove a useful direct product decomposition of central relations, and we introduce the so-called algebraically central extensions. $\mathcal{C}$ will always denote an exact factor permutable category.

4.1. Proposition. *An algebraically central equivalence relation R on A in $\mathcal{C}$ is canonically isomorphic to a product $A \times Q$, with Q an abelian object.*

Proof Let C be the centralizing relation on R and ∇_A:

$$\begin{array}{ccc} C & \overset{p_1}{\underset{p_2}{\rightrightarrows}} & A \times A \\ \pi_1 \downarrow\downarrow \pi_2 & & a_1 \downarrow\downarrow a_2 \\ R & \overset{r_1}{\underset{r_2}{\rightrightarrows}} & A \end{array}$$

Now, by taking the coequalizer q of π_1 and π_2 we obtain the commutative diagram

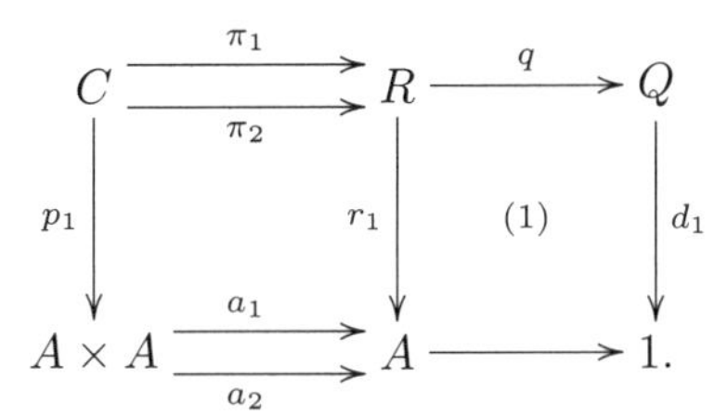

Since the category $\mathcal{C}$ is exact, the equivalence relation

$$C \overset{\pi_1}{\underset{\pi_2}{\rightrightarrows}} R$$

is the kernel pair of its coequalizer q, and this latter is a pullback stable regular epi. By assumption the arrow (p_1, r_1) from the equivalence relation (C, R) to (∇_A, A) is a discrete fibration of internal equivalence relations, so that the square (1) is a pullback (see Corollary 2 in [8]). Accordingly, the equivalence relation R is isomorphic to

$$A \times Q \overset{\pi_A}{\underset{\delta_2}{\rightrightarrows}} A \ .$$

The object Q is abelian: indeed, the fact that there is a connector on the equivalence relations $R[r_1]$ and $R[r_2]$ (Example 1.7) implies that Q is abelian. This fact essentially follows from the fact that the induced arrows $\tilde{q_1}$ and $\tilde{q_2}$ in the diagram

$$\begin{array}{ccccc} R[r_1] & \rightrightarrows & R & \leftleftarrows & R[r_2] \\ \tilde{q_1} \downarrow & & q \downarrow & & \downarrow \tilde{q_2} \\ Q \times Q & \rightrightarrows & Q & \leftleftarrows & Q \times Q. \end{array}$$

are regular epimorphisms, as is the arrow $q_3 \colon R[r_1] \times_R R[r_2] \to Q \times Q \times Q$ induced by the universal property of pullbacks. $\square$

In the special case of exact Maltsev categories the previous proposition was proved in [15].

By an *extension of* B we mean a regular epi $f \colon A \to B$ whose codomain is B.

4.2. Definition. An extension $f \colon A \to B$ is *algebraically central* if its kernel pair $R[f]$ is algebraically central.

We remark that this definition has its plain meaning in factor permutable categories: indeed, for an extension being algebraically central becomes a property (thanks to Lemma 3.6). A consequence of the previous product decomposition is that any algebraically central extension $f\colon A \to B$ which is a split epi is isomorphic to the projection $\pi_B\colon B \times Q \to B$, where Q is the abelian object canonically associated with $R[f]$:

4.3. Corollary. *Let $f\colon A \to B$ be an algebraically central extension split by an arrow $i\colon B \to A$. Then $A \simeq B \times Q$, with Q an abelian object.*

Proof Let $A \times Q$ be the canonical product decomposition of $R[f]$, as obtained in the previous Proposition:

$$A \times Q \underset{\delta_2}{\overset{\pi_A}{\rightrightarrows}} A \xrightarrow{f} B.$$

Then both squares in the following commutative diagram are pullbacks:

$$\begin{array}{ccccc} B \times Q & \xrightarrow{i \times 1} & A \times Q & \xrightarrow{\delta_2} & A \\ \downarrow{\scriptstyle \pi_B} & & \downarrow{\scriptstyle \pi_A} & & \downarrow{\scriptstyle f} \\ B & \xrightarrow[i]{} & A & \xrightarrow[f]{} & B. \end{array}$$

Since $f \circ i = 1_B$, it follows that $\delta_2 \circ (i \times 1)$ is an isomorphism, and then f is isomorphic to the (split) extension $\pi_B\colon B \times Q \to B$. $\square$

The following result, asserting that algebraically central extensions are pullback stable, will be very useful:

4.4. Lemma. *If the extension $f\colon A \to B$ is algebraically central and the square*

$$\begin{array}{ccc} C & \xrightarrow{\pi_A} & A \\ \downarrow{\scriptstyle \pi_E} & & \downarrow{\scriptstyle f} \\ E & \xrightarrow[g]{} & B \end{array}$$

is a pullback, then π_E is algebraically central.

Proof Let R and $\overline{R}$ be the kernel pairs of f and π_E, respectively. Let D be the centralizing relation on R and ∇_A. Let $p\colon \overline{R} \to R$ denote the arrow induced by the universal property of kernel pairs. The relation $p^{-1}(D)$, defined by the pullback

$$\begin{array}{ccc} p^{-1}(D) & \xrightarrow{\overline{q}} & D \\ \downarrow{\scriptstyle (\overline{\pi}_1,\overline{\pi}_2)} & & \downarrow{\scriptstyle (\pi_1,\pi_2)} \\ \overline{R} \times \overline{R} & \xrightarrow[p \times p]{} & R \times R \end{array}$$

determines a double equivalence relation on $\overline{R}$ and ∇_C:

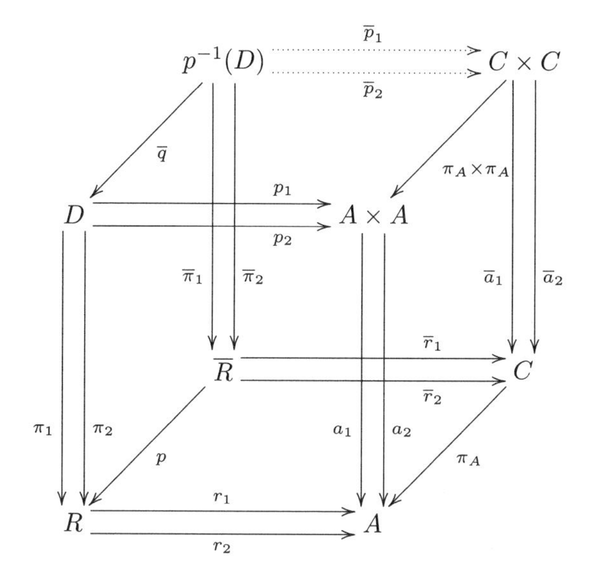

It is an easy exercise on pullbacks to check that $p^{-1}(D)$ is a centralizing relation on $\overline{R}$ and ∇_C. $\square$

5 Categorically central extensions

We are now going to prove that whenever the subcategory $\mathcal{C}_{Ab}$ is reflective in a Barr-exact factor permutable category $\mathcal{C}$, it is necessarily "admissible" in the sense of the categorical theory of central extensions [31]. This will allow us to define the so-called categorically central extensions.

Let us first recall some terminology: for an object B in an exact category $\mathcal{C}$, the category $Ext(B)$ of "extensions" of B is the full subcategory of the comma category $\mathcal{C} \downarrow B$ whose objects are the regular epimorphisms with codomain B. We also write $\mathcal{C} \Downarrow B$ for $Ext(B)$.

When $\mathcal{X}$ is a full replete reflective subcategory of an exact category $\mathcal{C}$, closed in $\mathcal{C}$ under subobjects and regular quotients, one says that $\mathcal{X}$ is a *Birkhoff subcategory* of $\mathcal{C}$. This terminology is motivated by the classical Birkhoff's theorem which, in our terminology, says that a subcategory $\mathcal{X}$ of a variety $\mathcal{C}$ is a subvariety if and only if it is a Birkhoff subcategory.

We write $\eta_B \colon B \to HIB$ for the B-component of the unit of the adjunction, and we often drop H from the notations and write $\eta_B \colon B \to IB$. The left adjoint I to the inclusion functor induces, for all B in $\mathcal{C}$, a functor $I^B \colon \mathcal{C} \Downarrow B \to \mathcal{X} \Downarrow IB$ sending the extension $f \colon A \to B$ to the extension $If \colon IA \to IB$. This functor I^B has a right adjoint $H^B \colon \mathcal{X} \Downarrow IB \to \mathcal{C} \Downarrow B$ defined as follows: with an extension $\phi \colon X \to IB$ in $\mathcal{X}$ it associates the extension $s \colon C \to B$ given by the pullback

$$
\begin{array}{ccc}
C & \xrightarrow{\;t\;} & HX \\
{\scriptstyle s}\downarrow & (1) & \downarrow{\scriptstyle H\phi} \\
B & \xrightarrow[\eta_B]{} & HIB.
\end{array}
$$

The categorical theory of central extensions can be developed when $\mathcal{X}$ is a Birkhoff subcategory of the exact category $\mathcal{C}$ satisfying the following additional property:

5.1. Definition. A Birkhoff subcategory $\mathcal{X}$ of the exact category $\mathcal{C}$ is *admissible* when, for any B in $\mathcal{C}$, the functor $H^B : \mathcal{X} \Downarrow IB \to \mathcal{C} \Downarrow B$ is fully faithful.

Remark that for every extension ϕ lying in $\mathcal{X}$, the pullback (1) can be seen as the exterior rectangle of the following diagram:

$$\begin{array}{ccccc}
C & \xrightarrow{\eta_C} & IC & \xrightarrow{\beta} & X \\
{\scriptstyle s}\downarrow & (2) & \downarrow{\scriptstyle Is} & (3) & \downarrow{\scriptstyle \phi} \\
B & \xrightarrow[\eta_B]{} & IB & \xrightarrow[1_{IB}]{} & IB
\end{array}$$

where $\beta \colon IC \to X$ is the unique arrow such that $\beta \circ \eta_C = t$. The ϕ-component of the counit $\varepsilon^B : I^B H^B \to 1$ of the induced adjunction is this arrow $\beta \colon Is \to \phi$. The functor H^B is fully faithful, and $\mathcal{X}$ is admissible, exactly when the counit is an isomorphism [40] or, equivalently, when each β as above is an iso.

We are now going to show that, in our context, $\mathcal{C}_{Ab}$ is always admissible in $\mathcal{C}$, provided it is reflective in $\mathcal{C}$.

From now on we shall assume that $\mathcal{C}$ is an exact factor permutable category, and that the inclusion functor $H \colon \mathcal{C}_{Ab} \to \mathcal{C}$ (which is full by Lemma 3.9) has a left adjoint, that will be denoted by $I \colon \mathcal{C} \to \mathcal{C}_{Ab}$. Such a left adjoint does exist when $\mathcal{C}$ is a congruence modular variety [32], and also when $\mathcal{C}$ is an exact Maltsev category with coequalizers [24] (see also [12]). In both cases the functor $I \colon \mathcal{C} \to \mathcal{C}_{Ab}$ sends an object A to the quotient $\frac{A}{[\nabla_A, \nabla_A]}$ of A by the largest commutator $[\nabla_A, \nabla_A]$, this latter being constructed as explained at the end of section 2. Thanks to our previous results, we already know that the category $\mathcal{C}_{Ab}$ is then a Birkhoff subcategory in $\mathcal{C}$, being a full subcategory of $\mathcal{C}$ closed in $\mathcal{C}$ under subobjects and regular quotients (by Corollary 3.15 and 3.17). Let us now prove that $\mathcal{C}_{Ab}$ is also admissible:

5.2. Theorem. *Let $\mathcal{C}$ be an exact factor permutable category such that $\mathcal{C}_{Ab}$ is reflective in $\mathcal{C}$. Then $\mathcal{C}_{Ab}$ is admissible.*

Proof Let us consider the canonical decomposition (2) + (3) of any pullback (1) as indicated here above, and we are going to show that the arrow $\beta \colon IC \to X$ is always an isomorphism. By taking the kernel pairs of the arrows s and Is in the commutative square (2) we obtain the diagram

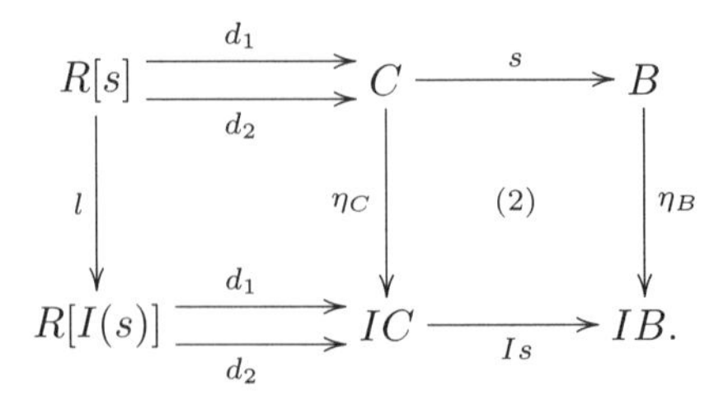

We shall first show that the induced arrow l is a regular epi: to check this, let us consider the regular epi-mono decomposition $i \circ p$ of l, which yields the following

diagram

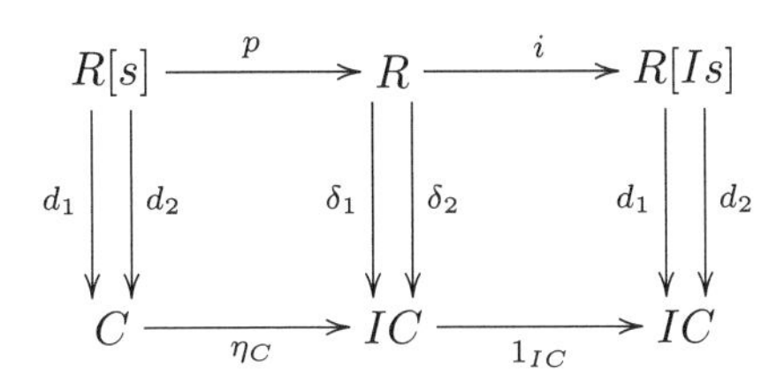

Now, the middle vertical arrows represent a reflexive relation R in the category $\mathcal{C}_{Ab}$, since $\mathcal{C}_{Ab}$ is closed in $\mathcal{C}$ under subobjects and products. Since $\mathcal{C}_{Ab}$ is naturally Maltsev (Corollary 3.10), R actually is an equivalence relation. It is easy to see that δ_1 and δ_2 have Is as coequalizer. The equivalence relations are effective in $\mathcal{C}$, then $R \simeq R[Is]$, i is an iso and l a regular epi.

Let us then recall that any extension f in $\mathcal{C}_{Ab}$ is algebraically central: in a naturally Maltsev category the connector $p\colon R[f] \times A \to A$ is the restriction of the A-component of the natural transformation from $1_{\mathcal{C}} \times 1_{\mathcal{C}} \times 1_{\mathcal{C}}$ to $1_{\mathcal{C}}$. Algebraically central extensions are stable under pulling back (by Lemma 4.4), hence the extension s is algebraically central because ϕ is so, and $R[s]$ is then isomorphic to $C \times Q$, where Q is an abelian object. Consequently in the following commutative diagram

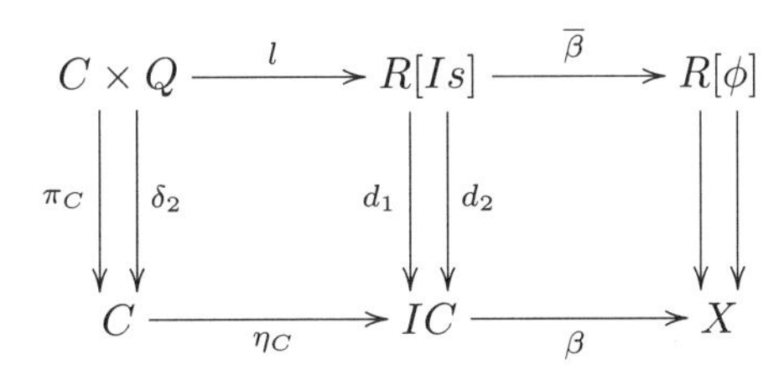

any of the left-hand commutative squares is a pushout of regular epimorphisms such that, by factor permutability, $R[\pi_C] \circ R[l] = R[l] \circ R[\pi_C]$. This implies that the canonical arrow $\alpha\colon C \times Q \to C \times_{IC} R[Is]$ to the corresponding pullback of η_C along $d_1\colon R[Is] \to IC$ is a regular epi (by Theorem 2.11). The arrow α is also a mono, since π_C and $\overline{\beta} \circ l$ are jointly monic (since the square (1) is a pullback), so that π_C and l are jointly monic as well. Accordingly, the arrow $(\eta_C, l)\colon (C \times Q, C) \to (R[Is], IC)$ is a discrete fibration, then (2) is a pullback, and this implies that (3) is a pullback and β is an iso, as desired. $\square$

As it was observed in [31], admissibility can be seen as an exactness condition on the left adjoint I. In our context, it can be expressed as follows:

5.3. Corollary. *Let $\mathcal{C}$ be an exact factor permutable category such that $\mathcal{C}_{Ab}$ is reflective in $\mathcal{C}$. Then the left adjoint $I\colon \mathcal{C} \to \mathcal{C}_{Ab}$ preserves all pullbacks of the form*

$$\begin{array}{ccc} C & \xrightarrow{u} & Z \\ {\scriptstyle s}\big\downarrow & & \big\downarrow{\scriptstyle \psi} \\ B & \xrightarrow[v]{} & Y, \end{array}$$

where Y, Z are in $\mathcal{C}_{Ab}$ and ψ is a regular epi.

Proof Let us write v as the composite $w \circ \eta_B$ for a unique $w\colon IB \to Y$, and let us consider the following composite of two pullbacks:

$$\begin{array}{ccccc} C & \xrightarrow{t} & X & \xrightarrow{x} & Z \\ {\scriptstyle s}\downarrow & & \downarrow{\scriptstyle \phi} & & \downarrow{\scriptstyle \psi} \\ B & \xrightarrow[\eta_B]{} & IB & \xrightarrow[w]{} & Y. \end{array}$$

The object X is in $\mathcal{C}_{Ab}$, and the arrow ϕ is a regular epi, being the pullback of ψ along w. Consequently the functor I preserves the left-hand pullback thanks to the previous theorem, and the right-hand pullback because it lies in $\mathcal{C}_{Ab}$. □

Let us now introduce some more terminology.

5.4. Definition. An extension $f\colon A \to B$ is *trivial* if it lies in the image of the functor H^B. This precisely means that the following square is a pullback:

$$\begin{array}{ccc} A & \xrightarrow{\eta_A} & HIA \\ {\scriptstyle f}\downarrow & & \downarrow{\scriptstyle HIf} \\ B & \xrightarrow[\eta_B]{} & HIB \end{array}$$

5.5. Definition. An extension $f\colon A \to B$ is (E,p)-split, where $p\colon E \to B$ is itself an extension of B, when $s\colon E \times_B A \to E$ in the following pullback is a trivial extension of E

$$\begin{array}{ccc} E \times_B A & \xrightarrow{t} & A \\ {\scriptstyle s}\downarrow & & \downarrow{\scriptstyle f} \\ E & \xrightarrow[p]{} & B. \end{array}$$

We shall write $Spl(E,p)$ for the full subcategory of $Ext(B)$ whose objects are the extensions which are split by p.

5.6. Definition. An extension $f\colon A \to B$ is *categorically central* when there exists an extension $p\colon E \to B$ such that f is (E,p)-split.

If one denotes by $Triv(B)$ and by $Centr(B)$ the full subcategories of the category $Ext(B)$ whose objects are the trivial and the categorically central extensions of B, one has

$$Triv(B) \subseteq Centr(B) \subseteq Ext(B),$$

where the inclusions are proper in general. Trivial and categorically central extensions are pullback stable:

5.7. Proposition. [31] *For any extension $g\colon E \to B$ the pullback functor*

$$g^*\colon Ext(B) \to Ext(E)$$

takes trivial (categorically central) extensions of B to trivial (categorically central) extensions of E.

Proof Let the left-hand square below be the pullback along $g\colon E \to B$ of a trivial extension $f\colon A \to B$:

$$\begin{array}{ccccc} C & \xrightarrow{h} & A & \xrightarrow{\eta_A} & HIA \\ {\scriptstyle k}\downarrow & & \downarrow{\scriptstyle f} & & \downarrow{\scriptstyle HIf} \\ E & \xrightarrow[g]{} & B & \xrightarrow[\eta_B]{} & HIB \end{array} \tag{4}$$

Since both the squares are pullbacks, so is the exterior rectangle. By the naturality of η this last pullback is equal to the exterior rectangle in

$$\begin{array}{ccccc} C & \xrightarrow{\eta_C} & HIC & \xrightarrow{HIh} & HIA \\ {\scriptstyle k}\downarrow & & \downarrow{\scriptstyle HIk} & & \downarrow{\scriptstyle HIf} \\ E & \xrightarrow[\eta_E]{} & HIE & \xrightarrow[HIg]{} & HIB \end{array} \tag{5}$$

which is accordingly a pullback. Now applying HI to the exterior rectangle of (4) gives the right-hand square of (5), so this last too is a pullback by Corollary 5.3. This implies that the left-hand square of (5) is a pullback and $k\colon C \to E$ is a trivial extension of E.

Let us then prove that categorically central extensions are pullback stable. Let $f\colon A \to B$ be a categorically central extension, and let us assume that f belongs to $Spl(E,p)$ for an extension $p\colon E \to B$. For any arrow $g\colon D \to B$ we can then form the pullback

$$\begin{array}{ccc} E \times_B D & \xrightarrow{q} & D \\ {\scriptstyle h}\downarrow & & \downarrow{\scriptstyle g} \\ E & \xrightarrow[p]{} & B. \end{array}$$

Then the extension which is obtained by pulling back f along $g \circ q = p \circ h$ is trivial by the first part of the proof. Accordingly, the pullback of f along g belongs to $Spl(E \times_B D, q)$ and it is then (categorically) central. □

5.8. Remark. From the previous proposition it follows that if $p\colon E \to B$ and $p'\colon E' \to B$ are two extensions of B, then $Spl(E',p') \subset Spl(E,p)$ whenever there is a map $g\colon E \to E'$ with $p' \circ g = p$.

6 Equivalence of the two notions

In this section we prove the equivalence between the two notions of categorically central and of algebraically central extension. A consequence of this result is that any central extension is normal, which means that it is split by itself.

In this section $\mathcal{C}$ will denote an exact factor permutable category such that $\mathcal{C}_{Ab}$ is a reflective (and hence admissible) subcategory of $\mathcal{C}$.

6.1. Theorem. *An extension $f\colon A \to B$ is algebraically central if and only if it is categorically central.*

Proof The category $\mathcal{C}_{Ab}$ is naturally Maltsev and consequently any extension in $\mathcal{C}_{Ab}$ is algebraically central. For this reason, in order to prove that any categorically central extension is algebraically central it is sufficient to prove that in any

pullback

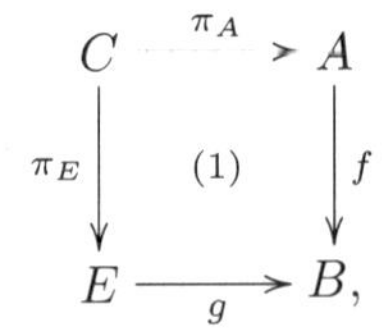

$$\begin{array}{ccc} C & \xrightarrow{\pi_A} & A \\ {\scriptstyle \pi_E}\downarrow & (1) & \downarrow{\scriptstyle f} \\ E & \xrightarrow[g]{} & B, \end{array}$$

where g is a regular epi, the extension f is algebraically central if and only if π_E is algebraically central. In Lemma 4.4 it was proved that algebraically central extensions are stable by pullbacks, so that we only need to prove that when π_E is an algebraically central extension then f is algebraically central. For this, let us consider the diagram

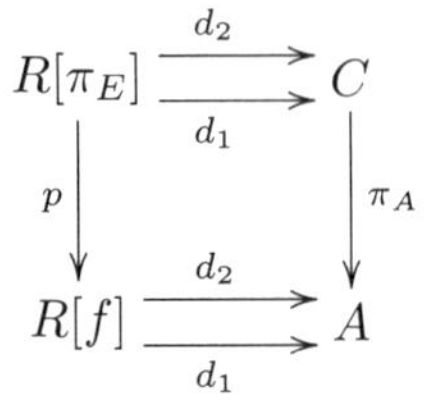

$$\begin{array}{ccc} R[\pi_E] & \overset{d_2}{\underset{d_1}{\rightrightarrows}} & C \\ {\scriptstyle p}\downarrow & & \downarrow{\scriptstyle \pi_A} \\ R[f] & \overset{d_2}{\underset{d_1}{\rightrightarrows}} & A \end{array}$$

where p is the arrow induced by the universal property of the kernel pairs. This arrow p is a regular epi because the square (1) is a pullback, while π_A is a regular epi because g is a regular epi, so that also $\pi_A \times \pi_A$ is a regular epi. By Lemma 3.16 it follows that $R[f]$ is central, and f is algebraically central.

Conversely, let us assume that $f: A \to B$ is algebraically central. By Proposition 4.1 we know that its kernel pair is given (up to isomorphism) by

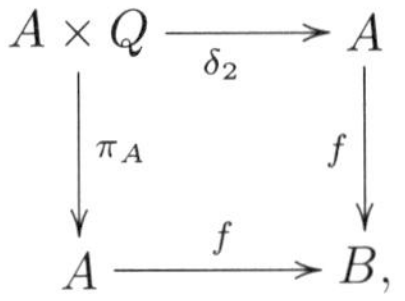

$$\begin{array}{ccc} A \times Q & \xrightarrow[\delta_2]{} & A \\ {\scriptstyle \pi_A}\downarrow & & \downarrow{\scriptstyle f} \\ A & \xrightarrow{f} & B, \end{array}$$

where Q is an object in $\mathcal{C}_{Ab}$. Let $t \circ m$ be the regular epi-mono factorization of the unique arrow from Q to 1, and let I be the regular image of this arrow. The arrow $m: Q \to I$ belongs to $\mathcal{C}_{Ab}$, and there is an induced arrow $i: A \to I$ with $i \circ \pi_A = m \circ \pi_Q$. The extension π_A appears then as the pullback of the trivial extension m along i, hence π_A is a trivial extension, and f is categorically central, as desired. □

Let us recall from [31] that an extension $f: A \to B$ is said to be *normal* when it is split by itself, i.e. when f is in $Spl(A, f)$. By definition any normal extension is central, and there are examples of central extensions which are not normal [31]. The proof of the previous theorem shows that, under our assumptions, we have:

6.2. Corollary. *Any central extension is normal.*

7 Central extensions in semi-abelian categories

In this section we are going to give several characterizations of central extensions in semi-abelian categories. The relationship with (the natural generalization of) the definition given by Frölich for varieties of Ω-groups is also clarified. The main results of this section were established in [14].

Let us first briefly recall the notions of protomodular category (see Bourn [6]) and of semi-abelian category (see Janelidze, Marki, Tholen [34]). We shall often refer to the survey by Borceux [3] in this volume, where many important properties of semi-abelian categories are proved in detail.

When $\mathcal{C}$ is a finitely complete category, let $Pt(\mathcal{C})$ denote the category whose objects are the split epimorphisms with a given splitting, and arrows pairs (f_0, f_1) of arrows f_0 and f_1 in $\mathcal{C}$ making commutative the two squares between these data:

$$\begin{array}{ccc} A & \xrightarrow{f_1} & A' \\ {\scriptstyle i}\uparrow\downarrow{\scriptstyle p} & & {\scriptstyle i'}\uparrow\downarrow{\scriptstyle p'} \\ B & \xrightarrow[f_0]{} & B'. \end{array}$$

Let $\pi \colon Pt(\mathcal{C}) \to \mathcal{C}$ be the functor sending any split epimorphism to its codomain. The functor π is a fibration, called the fibration of pointed objects. A *protomodular category* is a finitely complete category $\mathcal{C}$ such that every change of base functor with respect to the fibration π is conservative, i.e. it reflects isomorphisms [6]. When $\mathcal{C}$ has a zero object, this condition is equivalent to the split short five lemma.

7.1. Definition. [34] A *semi-abelian category* is an exact protomodular category with a zero object and finite coproducts.

An exact category is protomodular if and only if the following property holds: for any commutative diagram

$$\begin{array}{ccccc} A & \longrightarrow & B & \longrightarrow & C \\ \downarrow & (1) & \downarrow{\scriptstyle f} & (2) & \downarrow \\ D & \longrightarrow & E & \longrightarrow & F \end{array}$$

whenever (1) and (1) + (2) are pullbacks and f is a regular epi, then the square (2) is a pullback as well. This equivalent formulation of the protomodularity property will be useful in our study of central extensions.

Any semi-abelian category $\mathcal{C}$ is in particular an exact Maltsev category, (see, for instance, Theorem 3.7 in [3] for a proof of this fact, which is a special case of the more general results in [7]). In particular $\mathcal{C}$ is then an exact factor permutable category, so that all the results of the previous sections hold in any semi-abelian category.

7.2. Examples. Semi-abelian varieties of universal algebras have been recently characterized by Bourn and Janelidze [17] as follows: the theory has a unique constant 0, a $(n+1)$-ary operation β and n binary operations α_i satisfying the conditions $\alpha_i(x, x) = 0$ et $\beta(\alpha_1(x, y), , \alpha_n(x, y), y) = x$. The categories of groups, rings, Lie algebras, crossed modules and, more generally, any variety of Ω-groups [28] are semi-abelian. Further important examples of semi-abelian categories are given by the category of Heyting algebras, by any abelian category and by the dual category of the category of pointed sets.

If $f\colon A \to B$ is an arrow in a finitely complete pointed category $\mathcal{C}$, we denote by $Ker(f)\colon K[f] \to A$ the kernel of f, which is given by the pullback

$$\begin{array}{ccc} K[f] & \xrightarrow{Ker(f)} & A \\ \downarrow & & \downarrow f \\ 0 & \longrightarrow & B. \end{array}$$

7.3. Remark. The kernel of f can be built up from the kernel pair $R[f]$ in a very simple way. Consider the diagram

$$\begin{array}{ccccc} K & \xrightarrow{k} & R[f] & \xrightarrow{d_2} & A \\ \downarrow & & \downarrow d_1 & & \downarrow f \\ 0 & \longrightarrow & A & \xrightarrow[f]{} & B \end{array}$$

where the left-hand square represents the kernel k of d_1: it is then clear that $d_2 \circ k = Ker(f)$.

In order to prove the main results of this section, a few preliminary lemmas will be needed. We begin with the following characterization of pushouts of regular epimorphisms, which holds more generally in any exact Maltsev category [18], [11]:

7.4. Lemma. *Consider a commutative diagram of regular epimorphisms*

$$\begin{array}{ccc} A & \xrightarrow{h} & C \\ f \downarrow & (1) & \downarrow g \\ B & \xrightarrow[l]{} & D, \end{array}$$

in an exact Maltsev category $\mathcal{C}$. This square is a pushout if and only if the arrow $\tilde{h}\colon R[f] \to R[g]$ induced by the kernel pairs is a regular epimorphism.

Proof Let us assume that (1) is a pushout of regular epimorphisms in an exact Maltsev category. Consider the diagram

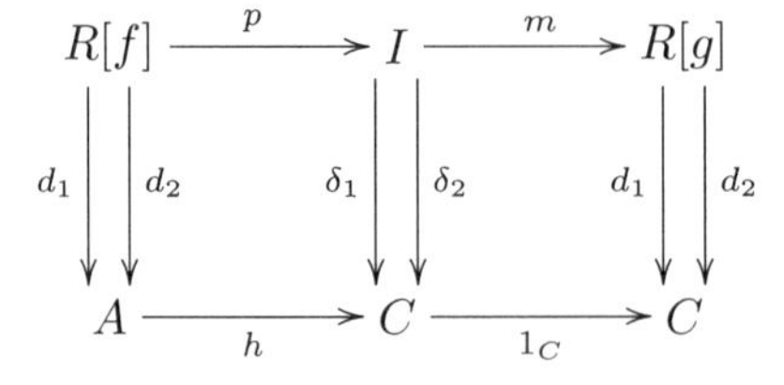

where $m \circ p = \tilde{h}$ is the regular epi-mono factorization of the arrow $\tilde{h}$ and the central part of the diagram represents the reflexive relation determined by this factorization. The relation I is then an equivalence relation on C because $\mathcal{C}$ is a Maltsev category, and the fact that the square (1) is a pushout implies that g is the coequalizer of δ_1 and δ_2. Since $\mathcal{C}$ is exact, I is then the kernel pair of g, m is an isomorphism and $\tilde{h}$ a regular epi, as desired. Conversely, if $\tilde{h}$ is a regular epi, then the square (1) is a pushout in any category. $\square$

The following result due to Bourn [9] will be needed. Its proof can be also found in the survey [3].

7.5. Lemma. *Consider a commutative diagram in a semi-abelian category*

$$\begin{array}{ccccc} A & \xrightarrow{k} & B & \xrightarrow{p} & C \\ {\scriptstyle a}\downarrow & (1) & \downarrow{\scriptstyle b} & (2) & \downarrow{\scriptstyle c} \\ A' & \xrightarrow[k']{} & B' & \xrightarrow[p']{} & C' \end{array}$$

where $k = Ker(p)$, $k' = Ker(p')$ and p, p' are regular epis. Then, when a and c are regular epis, b is a regular epi as well.

We are now in a position to prove the following

7.6. Proposition. [14] *Consider a commutative diagram of regular epimorphisms*

$$\begin{array}{ccc} A & \xrightarrow{h} & C \\ {\scriptstyle f}\downarrow & (1) & \downarrow{\scriptstyle g} \\ B & \xrightarrow[l]{} & D, \end{array}$$

in a semi-abelian category. This square is a pushout if and only if the restriction $\overline{h}\colon K[f] \to K[g]$ of h to the kernels is a regular epimorphism.

Proof Thanks to Lemma 7.5 it is sufficient to prove that the restriction $\overline{h}\colon K[f] \to K[g]$ of h to the kernel is a regular epi if and only if the induced arrow $\tilde{h}\colon R[f] \to R[l]$ is a regular epi. Let us recall that, by Remark 7.3, the objects $K[f]$ and $K[d_1]$ are isomorphic (with $d_1\colon R[f] \to A$) and, for the same reason, $K[g] \simeq K[d_1]$ (where this time $d_1\colon R[g] \to C$). Consequently, whenever $\overline{h}$ is a regular epi, the previous lemma applied to the diagram

$$\begin{array}{ccccc} K[f] & \xrightarrow{Ker(d_1)} & R[f] & \xrightarrow{d_1} & A \\ {\scriptstyle \overline{h}}\downarrow & & \downarrow{\scriptstyle \tilde{h}} & & \downarrow{\scriptstyle h} \\ K[g] & \xrightarrow[Ker(d_1)]{} & R[g] & \xrightarrow[d_1]{} & C \end{array}$$

implies that $\tilde{h}$ is a regular epi.

Conversely, when $\tilde{h}$ is a regular epi, then the square

$$\begin{array}{ccc} R[f] & \xrightarrow{\tilde{h}} & R[g] \\ {\scriptstyle d_1}\downarrow & & \downarrow{\scriptstyle d_1} \\ A & \xrightarrow[h]{} & C \end{array}$$

is a pushout (because the vertical arrows are split epis). Since the category is exact Maltsev, it follows that the factorization $\gamma\colon R[f] \to A \times_C R[g]$ is a regular epi:

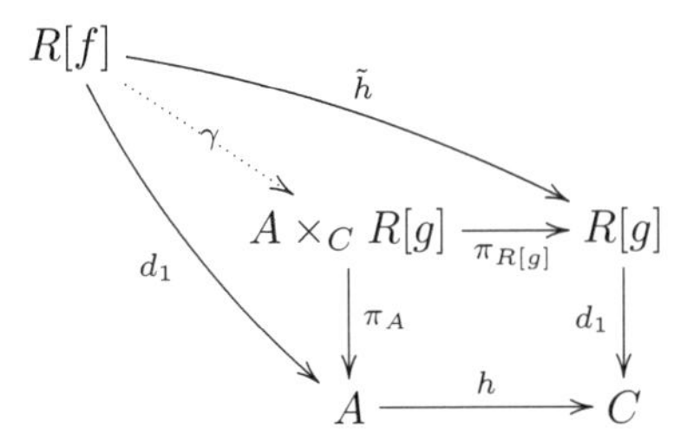

There exists a unique arrow $\alpha\colon K[g] \to A \times_C R[g]$ which is the kernel of π_A and has the property that $\pi_{R[g]} \circ \alpha = Ker(d_1)$. If $k\colon K[f] \to R[f]$ is the kernel of $d_1\colon R[f] \to A$, then the square

$$\begin{array}{ccc} K[f] & \xrightarrow{\overline{h}} & K[g] \\ {\scriptstyle k}\downarrow & & \downarrow{\scriptstyle \alpha} \\ R[f] & \xrightarrow[\gamma]{} & A \times_C R[g] \end{array}$$

is a pullback. It follows that the arrow $\overline{h}$ is a regular epi because γ is a regular epi. □

When $\mathcal{C}$ is semi-abelian and $\mathcal{C}_{Ab}$ is its full subcategory of abelian objects in $\mathcal{C}$, there exists a reflection functor $I\colon \mathcal{C} \to \mathcal{C}_{Ab}$ which sends an object A to its quotient by the largest commutator $[\nabla_A, \nabla_A]$ on A. The reflective subcategory $\mathcal{C}_{Ab}$ is then always admissible in the sense of Definition 5.1. We denote by KA the kernel of the A-component of the unit η_A of the adjunction, by $K(g)\colon KA \to KB$ the restriction to KA of an arrow $g\colon A \to B$ in $\mathcal{C}$.

When $\mathcal{C}$ is semi-abelian there is a simple characterization of trivial extensions:

7.7. Lemma. *An extension $f\colon A \to B$ is trivial if and only if $K(f)\colon KA \to KB$ is a monomorphism or, equivalently, if and only if $K(f)$ is an isomorphism.*

Proof Let us consider the canonical commutative square

$$\begin{array}{ccc} A & \xrightarrow{\eta_A} & HIA \\ {\scriptstyle f}\downarrow & (1) & \downarrow{\scriptstyle HIf} \\ B & \xrightarrow[\eta_B]{} & HIB. \end{array}$$

In any pointed category when (1) is a pullback, then the restriction $K(f)\colon KA \to KB$ to the kernel of η_A is an iso. Conversely, when $\mathcal{C}$ is semi-abelian and $K(f)$ is an iso, the exterior rectangle and the left-hand square in the diagram

$$\begin{array}{ccccc} KA & \xrightarrow{k} & A & \xrightarrow{f} & B \\ \downarrow & & \downarrow{\scriptstyle \eta_A} & & \downarrow{\scriptstyle \eta_B} \\ 0 & \longrightarrow & HIA & \xrightarrow[HIf]{} & HIB \end{array}$$

are pullbacks. By protomodularity it follows that the right-hand square is a pullback as well, because η_A is a regular epi. Thanks to the previous Proposition the proof is complete, since the square (1) is always a pushout of regular epimorphisms (this latter fact follows from the fact that $\mathcal{C}_{Ab}$ is closed under quotients in $\mathcal{C}$). □

If $f: A \to B$ is an extension, we denote by d_1, d_2 the projections of the kernel pair $R[f]$ and by $s: A \to R[f]$ the diagonal.

7.8. Theorem. [14] *For an extension $f: A \to B$ in a semi-abelian category $\mathcal{C}$ the following conditions are equivalent:*

1. *f is central*
2. *f is normal*
3. *$K(d_1)$ is a mono*
4. *$K(d_1)$ is an iso*
5. *$K(s)$ is a regular epi*
6. *$K(s)$ is an iso*
7. *$K(d_1) = K(d_2)$*
8. *For any $x, y: D \to A$ such that $f \circ x = f \circ y$, one has $K(x) = K(y)$.*

Proof 1. and 2. are equivalent by Corollary 6.2, 2. and 3. are equivalent by Lemma 7.7. The conditions 3., 4., 5. and 6. are trivially equivalent and 6. implies 7. To prove that 7. implies 3. consider two arrows γ and δ from any object D to $K(R[f])$ with the property that $K(d_1) \circ \gamma = K(d_1) \circ \delta$. One has $Ker(\eta_A) \circ K(d_1) \circ \gamma = Ker(\eta_A) \circ K(d_1) \circ \delta$, then $d_1 \circ Ker(\eta_{R[f]}) \circ \gamma = d_1 \circ Ker(\eta_{R[f]}) \circ \delta$. By assumption it follows that $Ker(\eta_A) \circ K(d_2) \circ \gamma = Ker(\eta_A) \circ K(d_2) \circ \delta$, so that $d_2 \circ Ker(\eta_{R[f]}) \circ \gamma = d_2 \circ Ker(\eta_{R[f]}) \circ \delta$ and then $\gamma = \delta$.

Since 8. clearly implies 7., the proof will be complete if we show that 7. implies 8. Let x and y be two arrows from an object D to A such that $f \circ x = f \circ y$. By the universal property of the kernel pair $R[f]$ there is an arrow $\sigma: D \to R[f]$ such that $d_1 \circ \sigma = x$ and $d_2 \circ \sigma = y$. It follows $K(x) = K(d_1) \circ K(\sigma) = K(d_2) \circ K(\sigma) = K(y)$. □

7.9. Remark. The condition 8. above literally extends the definition of central extension given by Frölich [22] and by Lue [39] for varieties of Ω-groups with respect to the subvariety of abelian algebras (see also [33]). We can then conclude that, when one considers a semi-abelian category $\mathcal{C}$ and its admissible subcategory $\mathcal{C}_{Ab}$, the three possible definitions of central extensions (categorically central, algebraically central and central in the sense of Frölich) are all equivalent.

It is also possible to give a characterization of central extensions which is intrinsic in $\mathcal{C}$, namely without referring to the subcategory $\mathcal{C}_{Ab}$:

7.10. Proposition. [14] *An extension $f: A \to B$ in a semi-abelian category is central if and only if the diagonal $s: A \to R[f]$ is a kernel.*

Proof Let us first assume that the extension f is central. Let D be the centralizing relation on $R[f]$ and ∇_A:

$$\begin{array}{ccc} D & \overset{p_1}{\underset{p_2}{\rightrightarrows}} & A \times A \\ {\scriptstyle \pi_1}\downarrow\downarrow{\scriptstyle \pi_2} & & {\scriptstyle a_1}\downarrow\downarrow{\scriptstyle a_2} \\ R[f] & \overset{r_1}{\underset{r_2}{\rightrightarrows}} & A \end{array}$$

If $\sigma\colon A \times A \to D$ is the arrow giving the reflexivity of the relation D on $A \times A$, then the internal functor

$$\begin{array}{ccc} A \times A & \xrightarrow{\sigma} & D \\ {\scriptstyle a_1}\downarrow\downarrow{\scriptstyle a_2} & & {\scriptstyle \pi_1}\downarrow\downarrow{\scriptstyle \pi_2} \\ A & \xrightarrow[s]{} & R[f] \end{array}$$

is a discrete fibration. It follows that s is the kernel of the quotient $q\colon R[f] \to \frac{R[f]}{D}$ of $R[f]$ by the equivalence relation D.

Conversely, let us assume that s is a kernel of a map g, and let D be the kernel pair of g. There is a commutative diagram

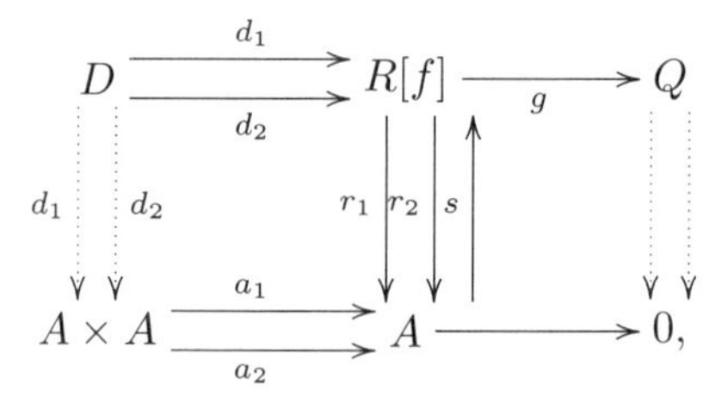

where the left vertical dotted arrows are induced by the universal property of kernel pairs, and determine a reflexive relation on $A \times A$, hence an equivalence relation, since $\mathcal{C}$ is a Maltsev category. By using the cancellation property for pullbacks in a semi-abelian category (recalled after Definition 7.1), one can check that D is a centralizing relation on $R[f]$ and ∇_A, and the extension f is then central. $\square$

8 Galois groupoids

The categorical theory of central extensions is a special instance of the general categorical Galois theory developed by Janelidze (see [29],[30], [4]). In this section we show that the so-called Galois pregroupoid associated with an extension actually is an internal groupoid in the semi-abelian context. We conclude by briefly recalling how a description of the central extensions of an object B can be obtained by applying the results of the categorical Galois theory (the reader may find a thorough presentation on this subject in [4]).

Let us first recall that an internal *precategory* in a category $\mathcal{C}$ [30] is a diagram of the form

$$P_2 \; \overset{p_1,\ m,\ p_2}{\Rrightarrow} \; P_1 \; \overset{d_1,\ s,\ d_2}{\rightleftarrows} \; P_0$$

with

1. $d_1 \circ s = 1_{P_0} = d_2 \circ s$

2. $d_2 \circ p_1 = d_1 \circ p_2$
3. $d_1 \circ p_1 = d_1 \circ m$, $d_2 \circ p_2 = d_2 \circ m$

Roughly speaking, a precategory is what remains of the definition of an internal category when one cancels all references to pullbacks. An arrow in the category $Precat(\mathcal{C})$ of internal precategories is simply a natural transformation between two such diagrams.

8.1. Definition. Let X be an internal groupoid in $\mathcal{C}$:

$$X_1 \times_{X_0} X_1 \overset{p_1}{\underset{p_2}{\overset{\longrightarrow}{\underset{\longrightarrow}{\xrightarrow{m}}}}} X_1 \overset{d_1}{\underset{d_2}{\overset{\longrightarrow}{\underset{\longrightarrow}{\xleftarrow{s}}}}} X_0.$$

An internal *covariant presheaf* P on the groupoid X (also called an *internal action* of X)

$$\begin{array}{ccccc} P_2 & \overset{p_1}{\underset{p_2}{\overset{\longrightarrow}{\underset{\longrightarrow}{\xrightarrow{m}}}}} & P_1 & \overset{d_1}{\underset{d_2}{\overset{\longrightarrow}{\underset{\longrightarrow}{\xleftarrow{s}}}}} & P_0 \\ \downarrow{\scriptstyle f_2} & & \downarrow{\scriptstyle f_1} & & \downarrow{\scriptstyle f_0} \\ X_1 \times_{X_0} X_1 & \overset{p_1}{\underset{p_2}{\overset{\longrightarrow}{\underset{\longrightarrow}{\xrightarrow{m}}}}} & X_1 & \overset{d_1}{\underset{d_2}{\overset{\longrightarrow}{\underset{\longrightarrow}{\xleftarrow{s}}}}} & X_0. \end{array}$$

consists in a precategory in $\mathcal{C}$ (the upper line), a natural transformation (f_0, f_1, f_2) from the precategory to X with the property that the squares

$$f_0 \circ d_1 = d_1 \circ f_1 \quad \text{and} \quad p_1 \circ f_2 = f_1 \circ p_1$$

are pullbacks.

Given two covariant presheaves P and P' on X, an internal *natural transformation* α from P to P' is given by three arrows $(\alpha_0, \alpha_1, \alpha_2)$ such that ${f'}_i \circ \alpha_i = f_i$ for $i = 0, 1, 2$ and all the squares of "corresponding arrows" in the diagram

$$\begin{array}{ccccc} P_2 & \overset{p_1}{\underset{p_2}{\overset{\longrightarrow}{\underset{\longrightarrow}{\xrightarrow{m}}}}} & P_1 & \overset{d_1}{\underset{d_2}{\overset{\longrightarrow}{\underset{\longrightarrow}{\xleftarrow{s}}}}} & P_0 \\ \downarrow{\scriptstyle \alpha_2} & & \downarrow{\scriptstyle \alpha_1} & & \downarrow{\scriptstyle \alpha_0} \\ {P_2}' & \overset{{p_1}'}{\underset{{p_2}'}{\overset{\longrightarrow}{\underset{\longrightarrow}{\xrightarrow{m'}}}}} & {P_1}' & \overset{{d_1}'}{\underset{{d_1}'}{\overset{\longrightarrow}{\underset{\longrightarrow}{\xleftarrow{s'}}}}} & {P_0}' \end{array}$$

are commutative.

Let now $\mathcal{C}$ be a semi-abelian category and let X be an internal groupoid in $\mathcal{C}_{Ab}$. We denote by $\{X, \mathcal{C}_{Ab}\}$ the category whose objects are the internal covariant presheaves in $\mathcal{C}_{Ab}$ on the internal groupoid X with the property that each f_i is a regular epi for $i = 0, 1, 2$ (or, equivalently, just for $i = 0$). The arrows in $\{X, \mathcal{C}_{Ab}\}$ are the internal natural transformations.

Now, each extension $f\colon A \to B$ in $\mathcal{C}$ determines the internal groupoid

$$R[f] \times_A R[f] \overset{p_1}{\underset{p_2}{\overset{\longrightarrow}{\underset{\longrightarrow}{\xrightarrow{m}}}}} R[f] \overset{d_1}{\underset{d_2}{\overset{\longrightarrow}{\underset{\longrightarrow}{\xleftarrow{s}}}}} A$$

in $\mathcal{C}$ which is its kernel pair. By applying the left adjoint $I: \mathcal{C} \to \mathcal{C}_{Ab}$ of the inclusion functor $H: \mathcal{C}_{Ab} \to \mathcal{C}$ to this internal groupoid one obtains an internal precategory

$$I(R[f] \times_A R[f]) \overset{I(p_1)}{\underset{I(p_2)}{\xrightarrow{I(m)}}} I(R[f]) \overset{I(d_1)}{\underset{I(d_2)}{\xleftarrow{I(s)}}} I(A)$$

in $\mathcal{C}$. This special kind of precategory, which is called the *internal Galois pregroupoid of f*, turns out to be always an internal groupoid in our context. This fact will be a consequence of the following lemma:

8.2. Lemma. *Let $\mathcal{C}$ be a semi-abelian category. Then the functor $I: \mathcal{C} \to \mathcal{C}_{Ab}$ preserves pullbacks of split epimorphisms along regular epimorphisms.*

Proof Let

$$\begin{array}{ccc} P & \xrightarrow{p_B} & B \\ j \uparrow \downarrow p_A & & i \uparrow \downarrow g \\ A & \xrightarrow[f]{} & C \end{array}$$

be a pullback of a split epi g along a regular epi f. By applying I to it one obtains a pushout of regular epis in $\mathcal{C}_{Ab}$

$$\begin{array}{ccc} IP & \xrightarrow{Ip_B} & IB \\ Ij \uparrow \downarrow Ip_A & & Ii \uparrow \downarrow Ig \\ IA & \xrightarrow[If]{} & IC. \end{array}$$

By Theorem 2.11 we know that the factorization α from IP through the pullback $IA \times_{IC} IB$ of If and Ig is a regular epi. Let $\beta: P \to IA \times_{IC} IB$ be the unique arrow induced by the universal property of the pullback, so that $\beta = \alpha \circ \eta_P$. In order to prove that α is an isomorphism, it will be sufficient to check that $Ker(\eta_P) \simeq Ker(\beta)$. Indeed, in a semi-abelian category it is easy to see that two regular epis determine the same quotient if and only if their kernels are isomorphic. Now, by using the same notation as in the previous section, it is clear that the canonical factorization $\gamma: KP \to KA \times_{KC} KB \simeq Ker(\beta)$ is a monomorphism in any pointed category. It is also a regular epi, as one can see by applying Theorem 2.11 this time to the square

$$\begin{array}{ccc} KP & \xrightarrow{K(p_B)} & KB \\ K(j) \uparrow \downarrow K(p_A) & & K(i) \uparrow \downarrow K(g) \\ KA & \xrightarrow[K(f)]{} & KC. \end{array}$$

Indeed, $K(p_B)$ and $K(f)$ are regular epis by Proposition 7.6, therefore the square above is a pushout of regular epimorphisms. $\square$

8.3. Theorem. *If $\mathcal{C}$ is a semi-abelian category, any Galois pregroupoid in $\mathcal{C}_{Ab}$ is an internal groupoid.*

Proof It follows from the previous lemma, since all the pullbacks in the definition of an internal groupoid are then preserved by the functor I. □

The Galois groupoid associated with the extension $f\colon A \to B$ is denoted by $Gal(f)$. The results in [30] then give, in our terminology, an equivalence of categories between the category of extensions which are split by an extension $f\colon A \to B$ and the category $\{Gal(f), \mathcal{C}_{Ab}\}$:

8.4. Theorem. *For any extension $f\colon A \to B$ in $\mathcal{C}$, the categories $Spl(A, f)$ and $\{Gal(f), \mathcal{C}_{Ab}\}$ are equivalent.*

Thanks to Remark 5.8, if there exists a regular epi $\overline{p}\colon \overline{E} \to B$ with $\overline{E}$ projective with respect to regular epimorphisms, one has $Spl(E,p) \subset Spl(\overline{E}, \overline{p})$ for all extensions $p\colon E \to B$. When $\mathcal{C}$ is a semi-abelian variety of universal algebras such a projective extension $\overline{p}\colon \overline{E} \to B$ always exists. Consequently, in this case, the category $Centr(B)$ is simply $Spl(\overline{E}, \overline{p})$, since this last category is the union of all $Spl(E,p)$.

8.5. Theorem. [31] *Let $\mathcal{C}$ be a semi-abelian variety. If $\overline{p}\colon \overline{E} \to B$ is an extension of B projective with respect to surjective homomorphisms, then*

$$\{Gal(\overline{p}), \mathcal{C}_{Ab}\} \simeq Centr(B).$$

References

[1] M. Barr, *Exact Categories*, Springer Lect. Notes in Math. 236, 1971, 1-120.

[2] F. Borceux, *Handbook of categorical algebra*, Encycl. of Math. and its Applications 50, Vol. 1, Cambridge Univ. Press, 1994.

[3] F. Borceux, *A survey of semi-abelian categories*, to appear in this volume.

[4] F. Borceux - G. Janelidze, *Galois Theories*, Cambr. Studies in Adv. Math. 72, Cambr. Univ. Press, 2001.

[5] F. Borceux - D. Bourn, *Mal'cev, Protomodular, Homological and Semi-Abelian Categories*, Kluwer, 2004.

[6] D. Bourn, *Normalization, equivalence, kernel equivalence and affine categories*, Springer Lect. Notes in Math. 1488, 1991, 43-62.

[7] D. Bourn, *Mal'cev categories and fibration of pointed objects*, Appl. Categ. Struct., 4, 1996, 307-327.

[8] D. Bourn, *Normal subobjects and abelian objects in protomodular categories*, J. Algebra, 228, 2000, 143-164.

[9] D. Bourn, 3×3 *lemma and Protomodularity*, J. Algebra, 236, 2001, 778-795.

[10] D. Bourn, *Intrinsic centrality and associated classifying properties*, J. Algebra, 256, 2002, 126-145.

[11] D. Bourn, *Denormalized* 3×3 *lemma*, J. Pure Appl. Algebra, 177, 2003, 113-129.

[12] D. Bourn, *Commutator theory in regular Mal'cev categories*, to appear in this volume.

[13] D. Bourn - M. Gran, *Centrality and normality in protomodular categories*, Theory Appl. Categ., Vol. 9, No 8, 2002, 151-165.

[14] D. Bourn - M. Gran, *Central extensions in semi-abelian categories*, J. Pure Appl. Algebra, 175, 2002, 31-44.

[15] D. Bourn - M. Gran, *Centrality and connectors in Maltsev categories*, Algebra Univers., 48, 2002, 309-331.

[16] D. Bourn - M. Gran, *Categorical aspects of modularity*, to appear in this volume.

[17] D. Bourn - G. Janelidze, *Characterization of protomodular varieties of universal algebras*, Theory Appl. Categ., Vol.11, No.6, 2003, 143-147.

[18] A. Carboni - G.M. Kelly - M.C. Pedicchio, *Some remarks on Maltsev and Goursat categories* Appl. Categ. Struct., 1, 1993, 385-421.
[19] A. Carboni - J. Lambek - M.C. Pedicchio, *Diagram chasing in Mal'cev categories,* J. Pure Appl. Algebra, 69, 1990, 271-284.
[20] A. Carboni - M.C. Pedicchio - N. Pirovano, *Internal graphs and internal groupoids in Mal'cev categories,* Proc. Conference Montreal 1991, 1992, 97-109.
[21] R. Freese - R. McKenzie, *Commutator theory for congruence modular varieties*, Lond. Math. Soc. Lect. Notes Series, 125, Cambr. Univ. Press, 1987.
[22] A. Frölich, *Baer-invariants of algebras,* Trans. Amer. Math. Soc., 109, 1963, 221-244.
[23] J. Furtado-Coelho, *Homology and Generalized Baer Invariants,* J. Algebra, 40, 1976, 596-600.
[24] M. Gran, *Central extensions and internal groupoids in Maltsev categories,* J. Pure Appl. Algebra, 155, 2001, 139-166.
[25] M. Gran, *Commutators and central extensions in universal algebra,* J. Pure Appl. Algebra, 174, 2002, 249-261.
[26] H. P. Gumm, *Geometrical Methods in Congruence Modular Algebras*, Mem. Amer. Math. Soc., 45, 286, 1983.
[27] J. Hagemann - C. Herrmann, *A concrete ideal multiplication for algebraic systems and its relation to congruence distributivity*, Arch. Math. (Basel), 32, 1979, 234-245.
[28] P.J. Higgins, *Groups with multiple operators*, Proc. London Math. Soc., (3), 6, 1956, 366-416.
[29] G. Janelidze, *Pure Galois theory in categories*, J. Algebra, 132, 1990, 270-286.
[30] G. Janelidze, *Precategories and Galois theory*, Springer Lect. Notes in Math. 1488, 1991, 157-173.
[31] G. Janelidze - G.M. Kelly, *Galois theory and a general notion of central extension*, J. Pure Appl. Algebra, 97, 1994, 135-161.
[32] G. Janelidze - G.M. Kelly, *Central extensions in universal algebra: a unification of three notions*, Algebra Univers., 44, 2000, 123-128.
[33] G. Janelidze - G.M. Kelly, *Central extensions in Mal'tsev varieties*, Theory Appl. Categ., Vol. 7, No. 10, 2000, 219-226.
[34] G. Janelidze - L. Marki - W. Tholen, *Semi-abelian categories*, J. Pure Appl. Algebra, 168, 2002, 367-386.
[35] G. Janelidze - M.C. Pedicchio, *Internal categories and groupoids in congruence modular varieties*, J. Algebra 193, 1997, 552-570.
[36] G. Janelidze - M.C. Pedicchio, *Pseudogroupoids and commutators*, Theory Appl. Categ., Vol. 8, No. 15, 2001, 405-456.
[37] P.T. Johnstone, *Affine categories and naturally Mal'cev categories,* J. Pure Appl. Algebra, 61, 1989, 251-256.
[38] A. Kock, *Fibre bundles in general categories,* J. Pure Appl. Algebra, 56, 1989, 233-245.
[39] A.S.-T. Lue, *Baer-invariants and extensions relative to a variety*, Proc. Cambridge Philos. Soc., 63, 1967, 569-578.
[40] S. Mac Lane *Categories for the working mathematician*, Second Edition, Grad. Texts in Math., Springer, 1998.
[41] A.I. Mal'cev, *On the general theory of algebraic systems,* Mat. Sbornik N. S., 35, 1954, 3-20.
[42] M.C. Pedicchio, *A categorical approach to commutator theory*, J. Algebra, 177, 1995, 647-657.
[43] M.C. Pedicchio, *Arithmetical categories and commutator theory*, Appl. Categ. Struct., 4, 1996, 297-305.
[44] J.D.H. Smith, *Mal'cev Varieties*, Springer Lect. Notes in Math. 554, 1976.

Received January 27, 2003; in revised form November 5, 2003

Fields Institute Communications
Volume **43**, 2004

Fibrations for Abstract Multicategories

Claudio Hermida
School of Computing
Queen's University
Kingston, ON, Canada K7L 3N6
chermida@cs.queensu.ca

Abstract. Building upon the theory of 2-dimensional fibrations and that of (abstract) multicategories, we present the basics of a theory of *fibred multicategories*. We show their intrinsic role in the general theory: a multicategory is representable precisely when it is covariantly fibrant over the terminal one. Futhermore, such fibred structures allow for a treatment of *algebras for operads* in the internal category setting. We obtain thus a conceptual proof of the 'slices of categories of algebras are categories of algebras' property, which is instrumental in setting up Baez-Dolan's opetopes.

1 Introduction

We introduce the notion of *fibration for multicategories*, the latter understood in their most general sense of (normal) lax algebras on bimodules, as we recall below. Given the space constraints, we limit ourselves to a brief introduction of the attendant theory of fibred multicategories, taking it as an opportunity to review some aspects of our work on 2-fibrations [12] and the theory of representable multicategories [13, 14]. We omit most proofs, occasionally outlining interesting arguments.

In [13] we introduced the notion of **representable multicategory** as an alternative axiomatisation of the notion of **monoidal category**, *representability* being a *universal property* of a multicategory; it demands the existence of universal 'multilinear' morphisms, $\pi_{\vec{x}} : \vec{x} \to \otimes\, \vec{x}$ for every tuple of objects $\vec{x}$, whose codomain endows the underlying the underlying category of 'linear' morphisms with a 'tensor product'. The basics of this theory (axiomatics of universal morphisms, strictness, and coherence) were developed upon the heuristic

$$\textit{universal morphism} \sim \textit{cocartesian morphism} \tag{1}$$

2000 *Mathematics Subject Classification.* Primary 18D10, 18D30; Secondary 18D50.

so that the theory of representable multicategories should parallel that of (co)fibred categories *cf.*[13, Table I]. Subsequently, in [14] we gave a general treatment of the above transformation

monoidal category $\mapsto$ representable multicategory

in the setting of pseudo-algebras for a cartesian monad M on a '2-regular 2-category' $\mathcal{K}$, *i.e.* a 2-category admitting a 'calculus of bimodules' (the 2-dimensional analogue of the calculus of relations available in a regular category), so that M induces a pseudo-monad $\mathsf{Bimod}(\mathsf{M}) : \mathsf{Bimod}(\mathcal{K}) \to \mathsf{Bimod}(\mathcal{K})$ on bimodules. Given these data, we constructed a 2-category $\mathcal{K}_\dashv$ (consisting of normal lax algebras for $\mathsf{Bimod}(\mathsf{M})$) equipped with a 2-monad $\mathsf{T}_\dashv$ such that

1. $\mathsf{T}_\dashv$ has the adjoint-pseudo-algebra property, *i.e.* a psuedo-algebra structure $x : \mathsf{T}_\dashv X \to X$ on an object X is a left-adjoint to the unit $\eta_X : X \to \mathsf{T}_\dashv X$.
2. The 2-categories of pseudo-algebras, strong morphisms and transformations of M and $\mathsf{T}_\dashv$ are equivalent.

This construction achieves the transformation

coherent structure $\mapsto$ universally characterised structure

which subsumes the case of monoidal categories above, as well as the classical Grothendieck transformation of psuedo-functors into (co)fibred categories (see Remark 1.1 below).

In [13, footnote p.169] we argued that in the analogy of representable multicategories with cofibred categories, the former lack a *base*. Here we rectify this statement, showing that representable multicategories are precisely those (covariantly) fibred over the *terminal* one (Theorem 4.1), thereby formalising the heuristic (1) above. Hence, when reasoning about certain categorical structures characterised by universal properties, we can soundly consider them as (covariantly) fibred structures. This correspondence provides yet another argument for the importance of fibred category theory in analysing categorical structure, complementary to the 'philosophical arguments' in [3].

Note on Terminology: Fibrations in categories involve the 'lifting' of morphisms from codomain to domain (or target to source) and thus give rise to contravariant pseudo-functors. For the dual notion, working in the locally dual 2-category $\mathcal{C}at^{co}$, there are two conflicting terminologies in the literature: Gray [11] advocates the use of the term *opfibration* while the Grothendieck school would use the term *cofibration*. Unfortunately, this latter clashes with the notion of cofibration of Quillen in the context of model structures on categories. Given that both notions would come to be used simultaneously in our subsequent work on coherence, it seems sensible to adopt a dissambiguating terminology right here: we shall refer to those fibred structures which give rise to covariant psuedo-functors as **covariant fibrations**, reserving the term **cofibration** for the algebraic topologists' stablished use.

In our foray into the theory of fibrations for multicategories, we would like the reader to keep in mind the following three levels of (increasing) generality and abstraction:

1. The ordinary $\mathcal{S}et$-based notion of multicategory introduced by Lambek [17].

2. Multicategories as monads in a 'Kleisli bicategory of spans' $\mathbf{Spn}_{\mathsf{M}}(\mathbb{B})$, where $\mathsf{M} : \mathbb{B} \to \mathbb{B}$ is a cartesian monad on a category with pullbacks, as in [13](*cf.* [4, 18]).
3. Multicategories as normal lax algebras in $\mathsf{Bimod}(\mathcal{K})$, the bicategory of bimodules in a 2-regular 2-category $\mathcal{K}$ ([15]), with respect to the pseudo-monad $\mathsf{Bimod}(\mathsf{M})$ induced by a 2-monad $\mathsf{M} : \mathcal{K} \to \mathcal{K}$, compatible with the calculus of bimodules, as in [14].

1.1. Remark. Since the Grothendieck correspondence between covariant psuedo-functors and cofibrations (covariant fibrations)

$$\mathsf{Ps}\text{-}[\mathbb{C}, \mathcal{C}at] \simeq \mathcal{C}o\mathcal{F}ib/\mathbb{C}$$

on a category $\mathbb{C}$ is not exhibited explicitly in [14, §11] as a consequence of the general theory, we outline here how it can be achieved. This example shows that our 'abstract multicategories' may not look like multicategories at first sight.

Recall that a bimodule $\alpha \colon X \not\to Y$ is called *reprepresentable* when it is of the form $x_{\#} = x \downarrow id$ (the span given by the two projections out of the comma-object) for a functor $x : X \to Y$. We observe that pseudo-functors out of $\mathbb{C}$ correspond to pseudo-algebras in the 2-category $[C_0, \mathcal{C}at]$, where C_0 is the object-of-objects of $\mathbb{C}$ (we are in an internal category setting $[C_0, \mathcal{C}at] \cong \mathcal{C}at[C_0, \mathcal{S}et]$). The appropriate 2-monad is $\mathsf{M} = \mathcal{C}at(\mathbb{C} \star \_)$, whose (1-dimensional) cartesian monad $\mathbb{C} \star \_ : [C_0, \mathcal{S}et] \to [C_0, \mathcal{S}et]$ expresses the free action of the (morphisms of) $\mathbb{C}$ on a C_0-family of sets. Remark 11.3 of *ibid.* shows that

$$\mathsf{Ps}\text{-}[\mathbb{C}, \mathcal{C}at] \simeq \textit{Representable}-\mathsf{Lax}_{rep}[\mathbb{C}, \mathsf{Bimod}(\mathcal{C}at)]$$

where $\mathsf{Lax}_{rep}[\mathbb{C}, \mathsf{Bimod}(\mathcal{C}at)]$ is the 2-category of lax functors into bimodules and representable transformations between them (that is, induced by functors), obtained by our transformation process out of the 2-monad M and the 2-category $[C_0, \mathcal{S}et]$. The *Representable*$-$ qualificative means the adjoint pseudo-algebras over such with respect to the 2-monad induced by the 'envelope' adjunction, which we recall in §2.1. See also Remark 2.2.(1) below for a reminder of the characterisation of such pseudo-algebras.

We want to show that the new basis of aximotisation is equivalent to the 2-category $\mathcal{C}at/\mathbb{C}$. Theorem 8.2 of *ibid.* entails :

$$\mathsf{Lax}_{rep}[\mathbb{C}, \mathsf{Bimod}(\mathcal{C}at)] \simeq \mathcal{M}ulticat_{\mathbb{C}\star\_}([C_0, \mathcal{S}et])$$

where C_0 is the object-of-objects of $\mathbb{C}$. This equivalence means that since we are working in an internal category setting we can simplify from bimodules to spans (and their easy composition via pullbacks). We now appeal to the well-known (and easy) equivalence $[C_0, \mathcal{S}et] \simeq Set/C_0$, which expresses the two canonical ways of viewing a family of sets. We obtain the following equivalences of 2-categories:

$$\mathcal{M}ulticat_{\mathbb{C}\star\_}([C_0, \mathcal{S}et]) \simeq \mathcal{M}ulticat_{\mathbb{C}\star\_}(\mathcal{S}et/C_0) \simeq \mathcal{C}at/\mathbb{C}$$

the last equivalence resulting from a mere inspection of the diagrams involved. Hence, 'multicategories' in this situation amount to functors into the category $\mathbb{C}$. Remark 11.2 of *ibid.* completes the argument, in the sense that the resulting adjoint pseudo-algebras over $\mathcal{C}at/\mathbb{C}$ obatined by our transformation are indeed cofibrations (covariant fibrations):

$$\textit{Representable}-\mathcal{C}at/\mathbb{C} \simeq \mathcal{C}o\mathcal{F}ib/\mathbb{C} \quad \square$$

In [20], Street develops some basic aspects of the theory of fibrations internally in a 2-category $\mathcal{K}$ in a representable fashion, *i.e.* a morphism $p : E \to B$ is a fibration iff $\mathcal{K}(X,p) : \mathcal{K}(X,E) \to \mathcal{K}(X,B)$ is a fibration in $\mathcal{C}at$, for every object X. Starting with the concrete setting 1) above, we expect a fibration of multicategories to involve a lifting of (multi)morphisms. Here we ran into the problem that in the 2-category $\mathcal{M}ulticat$, the 2-cells refer only to *linear* morphisms, *i.e.* those morphisms whose domain lies in the image of the unit of the monad M (a 'singleton sequence'). Hence the representable definition is unsuitably weak for multicategories.

The expected lifting of (multi)morphisms is obtained, abstractly, by means of the 'fundamental' monadic adjunction

$$\mathsf{Lax\text{-}Bimod_T}(\mathcal{K})\text{-}\mathsf{alg} \underset{U}{\overset{\mathsf{Env}}{\rightleftarrows}} \mathsf{T\text{-}alg} \quad (\bot)$$

so that the left adjoint 'envelope' 2-functor *reflects (covariant) fibrations* (Theorem 2.4). This point of view allows us to reduce the situation to the ordinary case of (representable) fibrations in a 2-category mentioned above. We could draw an analogy with modules for a Lie algebra $\mathfrak{G}$, which correspond to ordinary modules for its universal envelope $U\mathfrak{G}$.

This incipient theory of fibred structures in a multicategory scenario has several foreseeable applications besides our motivational correspondence with representability above. We illustrate this by elucidating some aspects of the theory of *algebras for operads* (in the non-Σ case), giving a conceptual proof of the fact that the 'slices' of such a category of algebras are categories of algebras for a multicategory (Theorem 5.2). Among the topics we have left out for lack of space are the pseudo-functorial (or 'indexed') version of the fibred structures, the 'comprehensive' factorisation system associated to (discrete) covariant fibrations and a related Yoneda structure, which we will present elsewhere.

As for related work, we should mention that quite independently of our developments, Clementino, Hofmann and Tholen have used V-enriched multicategories (relative to a monad) as an abstract setting for categorical topology, with emphasis on the theory of descent [9, 5, 7]. In particular, their analysis of exponentiability involves liftings of factorisations of 'multimorphisms', analogous to Giraud's characterisation of exponentiability in $\mathcal{C}at$ [10] (so that covariant fibrations in our sense are exponentiable, just like in $\mathcal{C}at$). While their setting of V-enriched modules cannot deal with internal structures (which has been our emphasis), their developments and ours can be put into a common framework of 'abstract proarrows' in the sense of [21]. In fact, as we pointed out in [14, §2], the theory developed therein is essentially based on such an axiomatic of 'bicategories of bimodules'.

2 Fibrations for multicategories

We start concretely with $\mathcal{S}et$-based multicategories and introduce the elementary definitions of (covariant) fibrations for them. The covariant situation features more prominently in our applications than the contravariant one.

2.1. Definition. Let $p : \mathbb{T} \to \mathbb{B}$ be a morphism of multicategories.

- A morphism $f : \langle x_1 \dots x_n \rangle \to y$ in $\mathbb{T}$ is (p-)**cocartesian** iff every morphism $g : \langle x_1 \dots x_n \rangle \to z$, with the same image on the base $pg = pf$, admits a *unique* factorisation $g = \hat{g} \circ f$, with $\hat{g} : y \to z$ a *vertical* morphism ($p\hat{g} =$

$id_{py} = id_{pz}$). Diagrammatically,

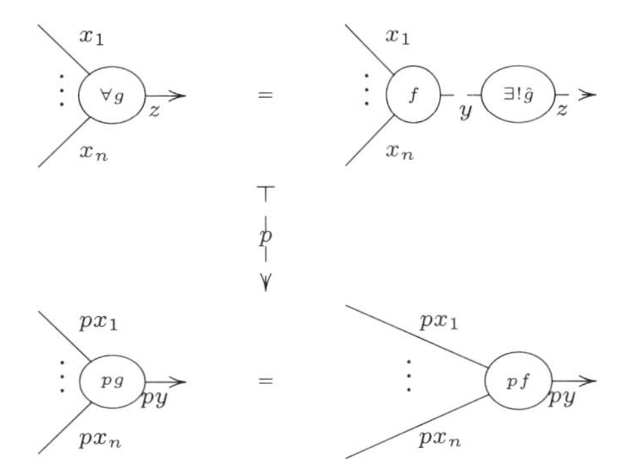

- The morphism $p : \mathbb{T} \to \mathbb{B}$ is a **covariant fibration** if the following hold:
  1. for every list of objects $\vec{x} = \langle x_1, \ldots, x_n \rangle$ in $\mathbb{T}$ and every morphism $u : p\vec{x} \to j$ in $\mathbb{B}$, there is a cocartesian morphism $\underline{u} : \vec{x} \to u_! \vec{x}$ over u ($p\underline{u} = u$).
  2. Cocartesian morphisms are closed under composition.

2.2. Remarks.

1. Just like in the ordinary categorical situation, we could have phrased the definition of covariant fibration appealing to a stronger notion of cocartesian morphism (so that its universal property holds with respect to morphisms which factorise through its projection) and thereby dispense with the composition requirement above. But one of our basic results [14, Theorem 5.4] shows that the given formulation is more fundamental: a lax algebra $\alpha \colon MA \not\to A$ admits an adjoint pseudo-$\mathsf{T}_\dashv$-algebra structure iff
   (a) the bimodule α is representable (which in our context amounts to the existence of cocartesian morphisms), and
   (b) the structural 2-cell $\mu : \alpha \bullet \alpha \Rightarrow \alpha$ is an isomorphism (cocartesian morphisms are closed under multicategory composition).
2. We shall distinguish between fibrations of multicategories, as defined above, and *fibrations in Multicat* (in the representable sense), which have cartesian liftings of *linear* morphisms only.

Dually, we have a notion of (p-)cartesian morphism and of fibration (contravariant lifting).

2.3. Examples.

1. Given a functor $q : \mathbb{E} \to \mathbb{C}$ in *Cat*, consider the induced morphism of multicategories $q_\blacktriangleright : \mathbb{E}_\blacktriangleright \to \mathbb{C}_\blacktriangleright$ (the multicategories of discrete cocones [13, Example 2.2(2)]):
   - If q is a fibration in *Cat*, $q_\blacktriangleright$ is a fibration of multicategories (cartesian cocones consist of cartesian morphisms in $\mathbb{E}$). In particular, the multicategory $\mathbb{C}_\blacktriangleright$ is fibred over the terminal multicategory.
   - If q is a covariant fibration, and $\mathbb{E}$ has cofibred coproducts (coproducts in the fibres preserved by direct images), $q_\blacktriangleright$ is a covariant fibration of multicategories: given a list of objects $\langle x_1, \ldots, x_n \rangle$ of $\mathbb{E}_\blacktriangleright$ and a morphism $\langle u^1 : qx_1 \to j, \ldots, u^n : qx_n \to j \rangle$ we obtain a cocartesian

lifting by considering individual cocartesian liftings $\underline{u^i} : qx_i \to u^i_!(x_i)$ and forming the coproduct $\vec{u}_!(\vec{x}) = \coprod_i u^i_!(x_i)$ in $\mathbb{E}_j$ with coproduct injections $\kappa^i : u^i_!(x_i) \to \vec{u}_!(\vec{x})$. The composite cocone $\langle \kappa^i \circ \underline{u^i} \rangle_i$ is a cocartesian lifting of $\vec{u}$.

2. Let Rng be the category of commuatitive rings with unit and Rng_m the corresponding multicategory of multilinear maps: $\mathsf{Rng}_m(\langle R_1, \ldots, R_n \rangle, S) = \mathsf{Rng}(R_1 \otimes \ldots \otimes R_n, S)$. Let Mod_m be the multicategory whose objects are pairs (R, M), with R a ring and M an R-module. A morphism in $\mathsf{Mod}_m(\langle (R_1, M_1) \ldots (R_n, M_n) \rangle, (S, N))$ consists of a pair of morphisms (h, a), with $h : R_1 \otimes \ldots \otimes R_n \to S$ in Rng and a $R_1 \otimes \ldots \otimes R_n$-module morphism $a : M_1 \otimes \ldots \otimes M_n \to h^*(N)$ (the tensor product of the abelian groups M_is has componentwise action by the tensor product of the rings). The evident forgetful functor $U : \mathsf{Mod}_m \to \mathsf{Rng}_m$ is a covariant fibration of multicategories: a cocartesian lifting of $\langle (R_1, M_1) \ldots (R_n, M_n) \rangle$ at $h : R_1 \otimes \ldots \otimes R_n \to S$ is the direct image $(M_1 \otimes \ldots \otimes M_n) \otimes_{R_1 \otimes \ldots \otimes R_n} h^*(S)$, where $h^*(S)$ is S regarded as a $(R_1 \otimes \ldots \otimes R_n)$-module via h.

   A sophisticated variation of this example is explored in [19], where the total multicategory has 'multilinear maps with singularities' and the (implicit) base category consists of the full subcategory of Rng on the tensor powers of a Hopf algebra H. It provides a framework for *vertex algebras*.

2.1 Fibrations and the enveloping adjunction. We recall that the monadic adjunction $F \dashv U : \mathcal{MonCat} \to \mathcal{Multicat}$ (for our second view of multicategories (2) as monads in in the Kleisli bicategory of spans $\mathbf{Spn}_{\mathsf{M}}(\mathbb{B})$) acts as follows: given a monoidal category $\mathbb{C}$ with objects C_0 and arrows C_1, $U\mathbb{C}$ is

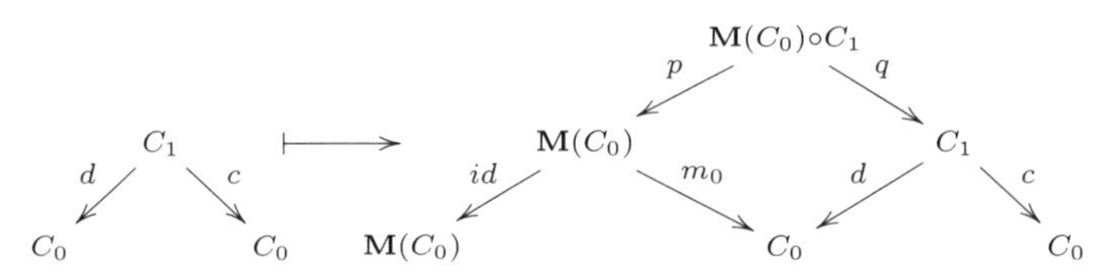

where $\mathbf{M}$ is the free-monoid monad in the ambient category (in $\mathcal{S}et$, $\mathbf{M}X = X^*$ the monoid of sequences under concatenation). Given a multicategory $\mathbb{D}$ with objects D_0 and arrows D_1, the free monoidal category $F\mathbb{D}$ is

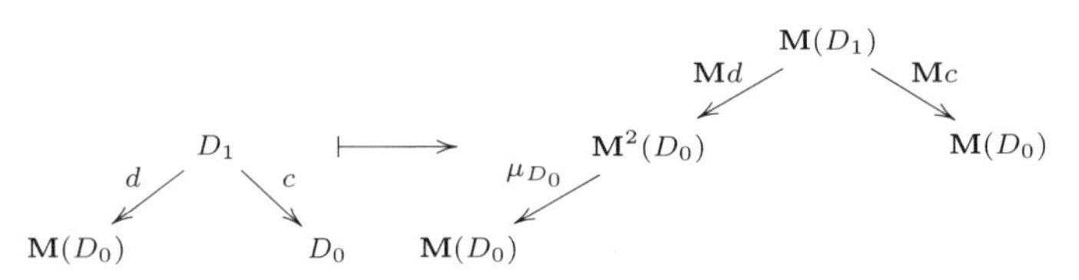

For more details of how this construction works, see [13, §8.3]. A more involved construction (via a lax colimit for a monad in $\mathsf{Bimod}(\mathcal{K})$ [14, §2.2]) yields

$$F \dashv U : \mathbf{M}\text{-alg} \to \text{Lax-Bimod}_{\mathbf{M}}(\mathcal{K})\text{-alg},$$

with the same intuitive content: a generalised 2-cell (or 'morphism') of Fx is a 'tuple' of 'morphims' of x (generalised 2-cells of the top object of the bimodule x) whose domain is the 'concatenation' of the domains of its components. The adjunction induces a cartesian 2-monad $\mathsf{T}_{\dashv} : \text{Lax-Bimod}_{\mathbf{M}}(\mathcal{K})\text{-alg} \to \text{Lax-Bimod}_{\mathbf{M}}(\mathcal{K})\text{-alg}$.

The basic relationship between the notions of fibrations for multicategories and categories is the following:

2.4. Theorem. *Let $p : \mathbb{T} \to \mathbb{B}$ be a morphism of multicategories. Tfae:*

1. *p is a (covariant) fibration of multicategories.*
2. *$Fp : F\mathbb{T} \to F\mathbb{B}$ is a (covariant) fibration of categories.*
3. *$Fp : F\mathbb{T} \to F\mathbb{B}$ is a (covariant) fibration of monoidal categories.*
4. *$\mathsf{T}_{\dashv}p : \mathsf{T}_{\dashv}\mathbb{T} \to \mathsf{T}_{\dashv}\mathbb{B}$ is a (covariant) fibration in $\mathcal{M}ulticat$ (in the sense of Remark 2.2.(2)).*

A fibration of monoidal categories in (3) above is the one-object case of a 2-fibration in the sense of [12]. We remind the reader that a 2-functor $P : \mathscr{E} \to \mathscr{B}$ is a *2-fibration* if it is a fibration at the 1-dimensional level and every 'local hom' functor $P_{X,Y} : \mathscr{E}(X,Y) \to \mathscr{B}(PX,PY)$ is a fibration, whose cartesian (2-)cells are preserved by precomposition with 1-cells (they have a pointwise nature). Hence, a fibration of strict monoidal categories amounts to a fibration of categories whose cartesian morphisms are closed under tensoring.

Notice that both (2) and (4) in the above characterisation make sense for our abstract multicategories as lax algebras in $\mathsf{Bimod}(\mathcal{K})$ and either of them can be adopted as a *definition* of (covariant) fibration in this setting.

From the above characterisation theorem, since both F and $\mathsf{T}_{\dashv}$ preserve pullbacks (because $\mathbf{M}$ is a cartesian monad), we deduce a change-of-base result for fibrations of multicategories, *i.e.* they are stable under pullback. Let $\mathcal{F}ib(\mathcal{M}ulticat)$ denote the 2-category whose objects are fibrations of multicategories, morphisms are commuting squares where the top morphism preserves cartesian morphisms of the total multicategories, and the evident 2-cells (*cf.* [12]).

2.5. Proposition. *The forgetful 2-functor base : $\mathcal{F}ib(\mathcal{M}ulticat) \to \mathcal{M}ulticat$ taking a fibration to its base multicategory, is a 2-fibration.*

Hence all the basic results of 2-fibrations of [12] (factorisation of adjunctions, construction of Kleisli objects, *etc.*) carry through to the setting of multicategories.

2.2 Adjoint characterisation. In [20] Street gave a characterisation of fibrations internal to a 2-category admitting comma-objects; the existence of cartesian liftings amounts to the existence of a right adjoint to the unit mapping a morphism (object over B) to its free fibration. We reformulate this characterisation using $\mathsf{Hom}_{\_}$ (cotensors with the $\to$ category) and pullbacks, which we can fruitfully reinstantiate in the setting of lax algebras on bimodules.

2.6. Lemma. *A functor $p : \mathbb{E} \to \mathbb{B}$ in $\mathcal{C}at$ is fibration iff the functor η canonically induced into the pullback*

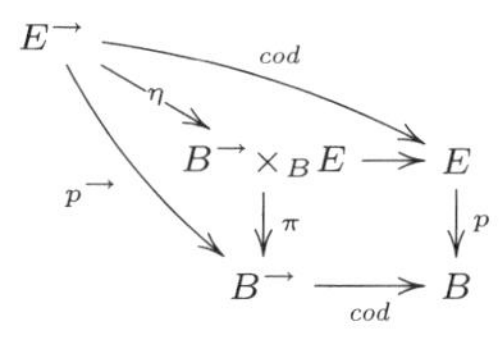

admits a right adjoint in $\mathcal{C}at/(\mathbb{B}^{\to})$.

While in $\mathcal{C}at$ the situation for covariant fibrations is entirely dual (simply replacing *cod* by *dom* and *right* by *left* above), the asymmetry between the domain and codomain of (multi)morphisms means that we have to state the characterisations of covariant and contravariant fibrations of multicategories separately:

2.7. Proposition. (Adjoint characterisation of fibrations of multicategories)
Consider a morphism $p : E \to B$ in $\mathsf{Lax}\text{-}\mathsf{Bimod}_{\mathbf{M}}(\mathcal{K})\text{-}\mathsf{alg}$, with lax algebras on bimodules $E = \mathbf{M}E_0 \overset{d}{\leftarrow} E_1 \overset{c}{\to} E_0$ and $B = \mathbf{M}B_0 \overset{d}{\leftarrow} B_1 \overset{c}{\to} B_0$

1. *p is a fibration iff $p_0 : E_0 \to B_0$ is a fibration (in $\mathcal{K}$) and the canonical morphism η into the pullback*

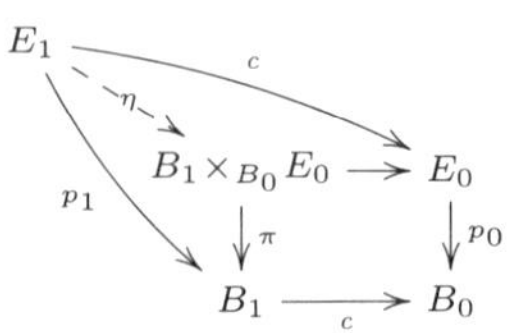

   admits a right adjoint in $\mathcal{K}/B_1$.

2. *p is a covariant fibration iff $p_0 : E_0 \to B_0$ is a covariant fibration (in $\mathcal{K}$) and the canonical morphism η into the pullback*

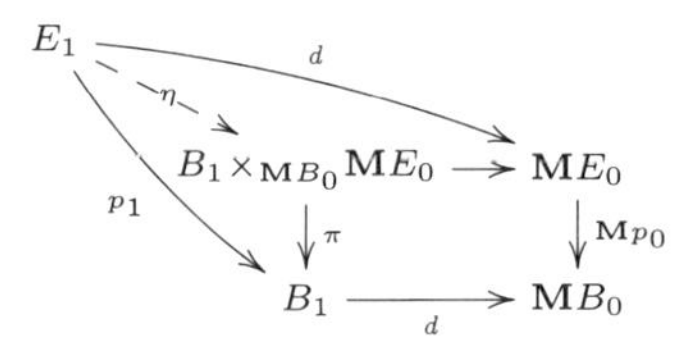

   admits a left adjoint in $\mathcal{K}/B_1$.

2.8. Corollary. *Given algebras $x : \mathbf{M}X \to X$ and $y : \mathbf{M}Y \to Y$, and a morphism $f : x \to y$, such that the underlying morphism $f : X \to Y$ is a covariant fibration in $\mathcal{K}$, the induced morphism $Uf : x_\# \to y_\#$ between lax algebras is a covariant fibration.*

Proof Applying Proposition 2.7.(2), the corresponding left adjoint is obtained from the given one characterising f as a covariant fibration in $\mathcal{K}$, by pulling this latter back along the algebra structure $x : \mathbf{M}X \to X$.

□

3 Coherence for fibrations of multicategories

Fixing a base multicategory $\mathbb{B}$, let $\mathcal{F}ib/\mathbb{B}$ denote the fibre over $\mathbb{B}$ of the 2-fibration *base* : $\mathcal{F}ib(\mathcal{M}ulticat) \to \mathcal{M}ulticat$ and similarly let $Split(\mathcal{F}ib/\mathbb{B})$ the corresponding sub-2-category of split fibrations (*i.e.* those with a choice of cartesian liftings closed under composition and identities) and morphisms between such preserving the splittings. Using the usual coherence theorem for fibrations of categories, the characterisation Theorem 2.4, and the fact that the unit of the monadic adjunction $F \dashv U$ is cartesian, we deduce the following coherence result:

3.1. Theorem. *The inclusion $Split(\mathcal{F}ib/\mathbb{B}) \hookrightarrow \mathcal{F}ib/\mathbb{B}$ has a left biadjoint whose unit is a pseudo-natural equivalence (with a section). Thus, every fibration is equivalent to a split one.*

The dual statement for covariant fibrations also holds.

3.2. Remark. An equivalence with a section is a split-epi at the 'object' level and thus both a covariant and (contravariant) fibration in any 2-category (in the representable sense). By Theorem 2.4, the same holds for equivalences with sections between abstract multicategories.

4 Covariant fibrations and representability

When the 2-category $\mathcal{K}$ has a terminal object $\mathbf{1}$, it bears a unique M-algebra structure, which makes it the terminal object in M-alg. Consequently, $U\mathbf{1}$ is the terminal object in $\mathsf{Lax\text{-}Bimod_M}(\mathcal{K})$-$\mathsf{alg}$. For any M-algebra $x : \mathsf{M}X \to X$, the unique morphism $! : X \to \mathbf{1}$ is (rather trivially) a (covariant) fibration, since the terminal object is discrete. For multicategories, the situation is interestingly different (we work in the framework of a strong 2-regular 2-category in the sense of [15]):

4.1. Theorem. *A multicategory $\mathbb{B}$ (qua lax algebra) is representable iff the unique morphism $! : \mathbb{B} \to \mathbf{1}$ is a covariant fibration of multicategories.*

Proof

($\Leftarrow$) Since the unit $\eta : id \Rightarrow \mathsf{T}_\dashv$ of the adjunction

$$F \dashv U : \mathsf{M\text{-}alg} \to \mathsf{Lax\text{-}Bimod_M}(\mathcal{K})\text{-}\mathsf{alg}$$

is cartesian with respect to (representable) covariant fibrations (this is where the axioms of 2-regularity come into play), the following square is a pullback

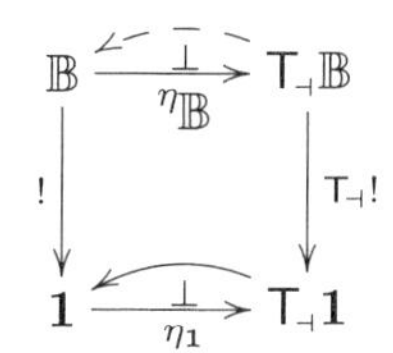

The existence of a left adjoint to the bottom morphism ($\mathbf{1}$ is clearly representable, since it comes from a M-algebra), the lifting of adjoints in a 2-fibration ([12, Lemma 4.1]) implies the existence of the 'dashed' left adjoint on top, which shows that $\mathbb{B}$ is representable.

($\Rightarrow$) For an M-algebra X, the unique $! : X \to \mathbf{1}$ is a covariant fibration, and so is $U! : UX \to U\mathbf{1}$ (Corollary 2.8). By coherence [14, Thm.7.4], any representable $\mathbb{B}$ is equivalent (via a covariant fibration, see Remark 2.8) to some UX (a strict M-algebra). The composite $\mathbb{B} \to UX \to \mathbf{1}$ is then a covariant fibration, as required. □

4.2. Remark. In the setting of $\mathcal{S}et$-based multicategories, the above theorem has the following concrete interpretation:

- The terminal multicategory has underlying multigraph

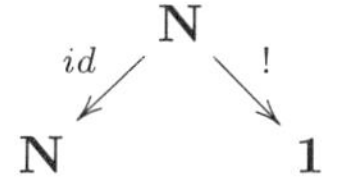

  where $\mathbf{N}$ is the set of natural numbers. Thus we have a unique arrow $n \succ \bullet$ for every n. Notice that this object is 'discrete' with respect to 2-cells into it, but it has non-trivial (multi)morphisms.
- Universal morphisms in a multicategory $\mathbb{C}$, $\pi_{\vec{x}} : \vec{x} \to \otimes\, \vec{x}$, are precisely the cocartesian morphisms for $! : \mathbb{C} \to \mathbf{1}$ over $|\vec{x}| \succ \bullet$ (see our heuristic (1) in §1).

Of course, in this simple setting, the above correspondence can be seen by mere inspection of the definitions involved. The general proof however requires vastly different methods. It is worth emphasizing that the ($\Rightarrow$) argument is quintessentially 2-fibrational. In the opposite direction, we have used coherence for adjoint pseudo-algebras. Although this use of coherence is not strictly necessary, the given argument does show several pieces of the theory at work.

4.3. Corollary. *The assignment* $\mathbb{M} \mapsto (! : \mathbb{M} \to \mathbf{1})$ *yields an equivalence of the 2-categories of representable multicategories and covariant fibrations over the terminal:*

$$\mathcal{RepMulticat} \cong \mathcal{CoFib}/\mathbf{1}$$

Notice that the coherence theorem 3.1 and the above corollary do allow us to recover the coherence theorem for (abstract) representable multicategories.

The relationship between representability and covariant fibrations is completed by the following two results:

4.4. Proposition. *Let* $p : \mathbb{T} \to \mathbb{B}$ *be a morphism of multicategories.*

- *If both* $\mathbb{T}$ *and* $\mathbb{B}$ *are representable and* p *preserves universals, then* p *is a covariant fibration of multicategories iff* p *is a covariant fibration in* $\mathcal{Multicat}$.
- *If* p *is a covariant fibration of multicategories and* $\mathbb{B}$ *is representable, then* $\mathbb{E}$ *is representable and* p *preserves universals.*

The first item above means that a covariant fibration between representable multicategories is the same thing as a *fibration of categories* (assuming that the induced functor is strong monoidal). The second result has the following logical interpretation: as we have argued in [12] and the references therein, the notion of *logical relation* between models of (various kinds of) type-theories can be fruitfully understood in terms of categorical structure in the total category of a fibration (over the base 'models'). The above result shows how to obtain a *logical tensor* in a *multicagory of predicates*, which is one covariantly fibred over a representable multicategory, that is, a base which admits a 'tensor'. For instance, in Example 2.3.(2), since the base multicategory Rng_m is representable, so is the multicategory Mod_m.

5 Operads and algebras

As an application of the theory of covariant fibrations, we show their role in the theory of *operads* and their *algebras*. From their origin in algebraic topology [16], these tools have found their way into various approaches to higher-dimensional category theory ([1, 2, 18]).

The basic setting of [16] is a (symmetric) monoidal category $\mathbb{C}$. In order to treat these notions with our multicategorical formulation, we would assume that $\mathbb{C}$ has finite limits and admits a free-monoid cartesian monad M, so that we consider multicategories as monads in $\mathbf{Spn}_{\mathsf{M}}(\mathbb{C})$. We have the following identification:

one-object multicategory $\equiv$ (non-permutative) operad

Indeed, a one-object multicategory amounts to an $\mathbf{N}$-indexed family $O = \{O_n\}$ (elements of O_n should be thought of as n-ary operations) closed under composition and identities. Thus, an operad is a structure which groups together the *operations* of a (restricted) algebraic theory. On the other hand, a *monad* describes the result of applying such operations to some generators, thereby describing the *free algebras* of the theory. With these identifications in mind, it is easy to see that an operad O gives rise to a monad $(\_) \otimes_{\mathsf{M}} O$: the category $\mathbb{C}$ embeds in $\mathbf{Spn}_{\mathsf{M}}(\mathbb{C})(\mathbf{1},\mathbf{1})$ ($J : \mathbb{C} \to \mathbf{Spn}_{\mathsf{M}}(\mathbb{C})(\mathbf{1},\mathbf{1})$ regards an object X as a span $\mathsf{M}\mathbf{1} \overset{\eta}{\leftarrow} \mathbf{1} \leftarrow X \to \mathbf{1}$), while taking the top object of the span gives a functor $D : \mathbf{Spn}_{\mathsf{M}}(\mathbb{C})(\mathbf{1},\mathbf{1}) \to \mathbb{C}$. Given an operad O as an object in $\mathbf{Spn}_{\mathsf{M}}(\mathbb{C})(\mathbf{1},\mathbf{1})$, we set $(\_) \otimes_{\mathsf{M}} O = D \circ (\_) \bullet O \circ J$, which inherits the monoid structure from O, thereby yielding a monad $(\_) \otimes_{\mathsf{M}} O : \mathbb{C} \to \mathbb{C}$. Notice that the composite $(\_)\bullet O$ is the application of the operations to generators, the latter suitably reinterpreted as a family of operations. We can identify algebras for the operad O with those of its associated monad:

$$\boxed{O\text{-algebras} \equiv (\_ \otimes_{\mathbf{M}} O)\text{-}\mathsf{alg}}$$

In concrete terms, an O-algebra amounts to an object A of $\mathbb{C}$ endowed with actions $a(o) : A^n \to A$ for every n-ary operation o, associative and unitary (with respect to the monoid structure on O). By inspection of the resulting diagrams, we notice that such actions can be equivalently phrased in terms of discrete covariant fibrations:

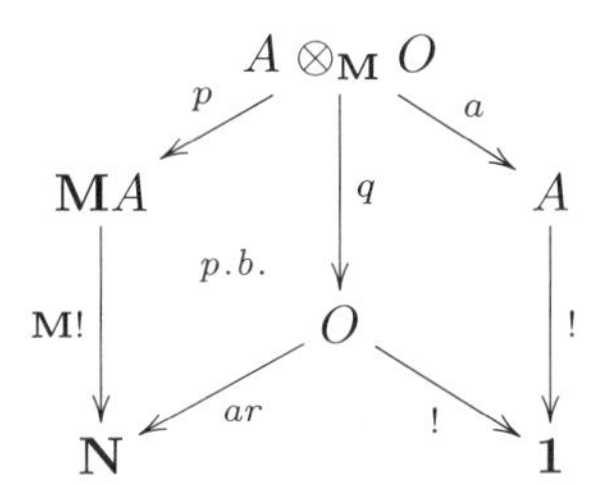

where $\mathbf{N} = \mathbf{M1}$ and the left-hand square is a pullback. The top span is then a multicategory (because the actions are associative and unitary), which we write $(A, a)^+$, and $! : A \to \mathbf{1}$ is a covariant fibration. The pullback square means that the fibres are discrete, so that a (multi)morphism of $(A, a)^+$ is uniquely determined by its source and its image in O. We arrive to the following:

5.1. Proposition.

$$\boxed{O\text{-}algebras \equiv \textit{discrete covariant fibrations over the multicategory } O}$$

This identification indicates that we can consider more generally a notion of **algebra for a multicategory** as a discrete covariant fibration over it. From the algebraic-theory perspective above, operads correspond to single-sorted restricted theories (equations must involve the same variables in the same order on both sides [8], in the non-permutative case), while multicategories correspond to the many-sorted version.

We now consider slice categories of algebras. In order to see that such slice categories are themselves categories of algebras, we appeal to the following two 'fibrational facts':

- Given covariant fibrations of multicategories $p : \mathbb{A} \to \mathbb{B}$ and $q : \mathbb{C} \to \mathbb{B}$, a morphism of covariant fibrations $h : p \to q$ is a covariant fibration in $\mathcal{C}o\mathcal{F}ib/\mathbb{B}$ iff it is a covariant fibration of multicategories $h : \mathbb{A} \to \mathbb{C}$. More concisely

  $$\{\mathcal{C}o\mathcal{F}ib - in(\mathcal{C}o\mathcal{F}ib/\mathbb{C})\}/q \equiv \mathcal{C}o\mathcal{F}ib/\mathbb{C}$$

  This property can be deduced from the corresponding one for ordinary fibrations in a 2-category, via the 2-fibrational argument in [12, §4.3] and the adjoint characterisation of covariant fibrations in Proposition 2.7.
- In the same situation, if the base covariant fibration q has discrete fibres, any morphism into it is a covariant fibration:

  $$\{\mathcal{C}o\mathcal{F}ib - in(\mathcal{C}o\mathcal{F}ib/\mathbb{C})\}/q \cong (\mathcal{C}o\mathcal{F}ib/\mathbb{C})/q$$

Combining these two facts we obtain the following slicing result:

5.2. Theorem. *For a multicatgory O, any slice of the category of O-algebras is again a category of algebras:*

$$\boxed{O\text{−algebras}/(A, a) \equiv (A, a)^{+}\text{−algebras}}$$

5.3. Remark. For brevity we do not deal here with permutative operads (those whose operations have symmetric-group actions), which we will take us into the more involved setting of lax algebras on bimodules rather than spans (we would work with $\mathsf{Lax\text{-}Bimod_S}(\mathcal{K})\text{-}\mathsf{alg}$, where S is the free-symmetric-monoidal-category monad). This is the set-up of [2]: an operad in their sense is a lax algebra. One technical subtlety of this extension is that S is not quite compatible with the calculus of bimodules and the resulting gadget $\mathsf{Bimod_S}(\mathcal{K})$ is only a *lax* bicategory. Nevertheless, the notion of monad applies equally to this setting and the conceptual identifications above regarding algebras carry through. In particular Theorem 5.2 gives an alternative (fibrational) view of the slicing result claimed in *ibid.* See [6] for a more detailed account of the slicing process.

We conclude pointing out that the consideration of algebras for an operad O via the *endomorphism operad* $\mathsf{End}(A, A)$ of an object A (O-algebra structure on $A \equiv$ operad morphism $O \to \mathsf{End}(A, A)$) is available in our setting if the ambient category $\mathbb{C}$ is locally cartesian closed, but we forego the details for lack of space.

Acknowledgments: The author thanks the organizers of the Workshop, especially George Janelidze, for the opportunity to present and publish the above material.

References

[1] M. Batanin. Monoidal globular categories as a natural environment for the theory of weak n-categories. *Advances in Mathematics*, 136:39–103, 1998.

[2] J. Baez and J. Dolan. Higher-dimensional algebra III: n-categories and the algebra of opetopes. *Advances in Mathematics*, 135:145–206, 1998.

[3] J. Bénabou. Fibred categories and the foundation of naive category theory. *Journal of Symbolic Logic*, 50, 1985.

[4] A. Burroni. T-categories. *Cahiers Topologie Géom. Differentielle Catégoriques*, 12:215–321, 1971.

[5] M.M. Clementino and D. Hofmann. Effective descent morphisms in categories of lax algebras. preprint 02-20, CMUC, Universidade de Coimbra, 2002.

[6] E. Cheng. Equivalence between approaches to the theory of opetopes. 2000. talk delivered at CT2000, Como.

[7] M.M. Clementino, D. Hofmann, and W. Tholen. The convergence approach to exponentiable maps. preprint, 2003.

[8] A. Carboni and P. Johnstone. Connected limits, familial representability and Artin glueing. *Mathematical Structures in Computer Science*, 5(4):441–459, 1995.

[9] M.M. Clementino and W. Tholen. Metric, topology and multicategory - A common approach. to appear in *Journal of Pure and Applied Algebra*, 2001.

[10] J. Giraud. Méthode de la descente. *Bull. Soc. Math. Fr., Suppl., Mm. 2*, 1964.

[11] J. W. Gray. Fibred and cofibred categories. In S. Eilenberg, editor, *Proceedings of the Conference on Categorical Algebra*. Springer Verlag, 1966.

[12] C. Hermida. Some properties of **fib** as a fibred 2-category. *Journal of Pure and Applied Algebra*, 134(1):83–109, 1999. Presented at ECCT'94, Tours, France.

[13] C. Hermida. Representable multicategories. *Advances in Mathematics*, 151:164–225, 2000. Available at `http://www.cs.math.ist.utl.pt/s84.www/cs/claudio.html`.

[14] C. Hermida. From coherent structures to universal properties. *Journal of Pure and Applied Algebra*, 165(1):7–61, 2001. preprint available math.CT/0006161.

[15] C. Hermida. Descent on 2-fibrations and 2-regular 2-categories. to appear in special issue of *Applied Categorical Structures* (coproceedings of *Workshop on Categorical Structures for Descent and Galois Theory, Hopf Algebras and Semiabelian Categories*, Fields Institute, Toronto, September 23-28), 2002.

[16] I. Kriz and P. May. Operads, algebras, modules and motives. *Asterisque*, 1995.

[17] J. Lambek. Deductive systems and categories (II). In *Category Theory, Homology Theory and their applications I*, volume 86 of *Lecture Notes in Mathematics*, pages 76–122. Springer Verlag, 1969. Battelle Institute Conference 1968, vol. I.

[18] T. Leinster. *Operads in higher-dimensional category theory.* PhD thesis, DPMMS, University of Cambridge, 2000. preprint at `math.CT/0011106`.

[19] Craig T. Snydal. Relaxed multicategory structure of a global category of rings and modules. *J. Pure Appl. Algebra*, 168(2-3):407–423, 2002.

[20] R. Street. Fibrations and Yoneda's lemma in a 2-category. In *Category Seminar*, volume 420 of *Lecture Notes in Mathematics*. Springer Verlag, 1973.

[21] R. Wood. Proarrows II. *Cahiers Topologie Géom. Differentielle Catégoriques*, XXVI:135–168, 1985.

Received March 6, 2003; in revised form October 23, 2003

Fields Institute Communications
Volume **43**, 2004

Lie-Rinehart Algebras, Descent, and Quantization

Johannes Huebschmann
Université des Sciences et Technologies de Lille
U. F. R. de Mathématiques
CNRS-UMR 8524
F-59 655 VILLENEUVE D'ASCQ, France
Johannes.Huebschmann@AGAT.UNIV-LILLE1.FR

Abstract. A *Lie-Rinehart algebra* (A, L) consists of a commutative algebra A and a Lie algebra L with additional structure which generalizes the mutual structure of interaction between the algebra of smooth functions and the Lie algebra of smooth vector fields on a smooth manifold. Lie-Rinehart algebras provide the correct categorical language to solve the problem whether Kähler quantization commutes with reduction which, in turn, may be seen as a descent problem.

Introduction

The algebra of smooth functions $C^\infty(N)$ on a smooth manifold N and its Lie algebra of smooth vector fields $\mathrm{Vect}(N)$ have an interesting structure of interaction. For reasons which will become apparent below, we will refer to a pair (A, L) which consists of a commutative algebra A and a Lie algebra L with additional structure modeled on a pair of the kind $(C^\infty(N), \mathrm{Vect}(N))$ as a *Lie-Rinehart algebra*. In this article we will show that the notion of Lie-Rinehart algebra provides the correct categorical language to solve a problem which we will describe shortly. Lie-Rinehart algebras occur in other areas of mathematics as well; an overview will be given in Section 1 below.

According to a philosophy going back to Dirac, the correspondence between a classical theory and its quantum counterpart should be based on an analogy between their mathematical structures. In one direction, this correspondence, albeit not well defined, is referred to as *quantization*. Given a classical system with constraints which, in turn, determine what is called the *reduced system*, the question arises whether quantization *descends* to the reduced system in such a way that, once the unconstrained system has been successfully quantized, imposing the symmetries on the quantized unconstrained system is equivalent to quantizing the reduced system. This question goes back to the early days of quantum mechanics and appears already in Dirac's work on the electron and positron [15], [16].

2000 *Mathematics Subject Classification.* Primary 17B63 17B65 17B66 32Q15 53D17 53D20 58F06 81S10; Secondary 14L24 14L30 17B81 32C17 32C20 32S05 32S60 53D50 58F05.

In the framework of Kähler quantization, the problem may be phrased as a *descent problem* and, indeed, under favorable circumstances which, essentially come down to requiring that the unreduced and reduced spaces be both ordinary quantizable Kähler manifolds, the problem has been known for long to have a solution [22] which, among other things, involves a version of what is referred to as KEMPF'S *descent lemma* [46] in geometric invariant theory; see e. g. [17] and Remark 8.6 below. In the present article we will advertise the idea that the concept of *Lie-Rinehart algebra* provides the appropriate categorical language to solve the problem, spelled out as a *descent problem*, under suitable more general circumstances so that, in a sense, reduction after quantization is then equivalent to quantization after reduction; here the term "descent" should, perhaps, not be taken in too narrow a sense.

Given a classical system, its dynamical behaviour being encapsulated in a Poisson bracket among the classical observables, according to DIRAC's idea of correspondence between the classical and quantum system, the Poisson bracket should then be the classical analogue of the quantum mechanical commutator. Thus, on the physics side, the Poisson bracket is a crucial piece of structure. Mathematically, it is a crucial piece of structure as well; in particular, when the classical phase space involves singularities, these may be understood in terms of the Poisson structure. More precisely, when the classical phase space carries a stratified symplectic structure, the Poisson structure encapsulates the mutual positions of the symplectic structures on the strata. See e. g. [31], [32], [37], [39].

Up to now, the available methods have been insufficient to attack the problem of quantization of reduced observables, once the reduced phase space is no longer a smooth manifold; we will refer to this situation as the *singular case*. The singular case is the rule rather than the exception. For example, simple classical mechanical systems and the solution spaces of classical field theories involve singularities; see e. g. [3] and the references there. In the presence of singularities, restricting quantization to a smooth open dense part, the "top stratum", leads to a loss of information and in fact to inconsistent results, cf. Section 4 of [41]. To overcome these difficulties on the classical level, in [39], we isolated a certain class of "Kähler spaces with singularities", which we call *stratified Kähler spaces*. On such a space, the complex analytic structure alone is unsatisfactory for issues related with quantization because it overlooks the requisite Poisson structures. In [41] we developed the Kähler quantization scheme over (complex analytic) stratified Kähler spaces. A suitable notion of *prequantization*, phrased in terms of *prequantum modules* introduced in [30], yields the requisite representation of the Poisson algebra; in particular, this representation satisfies the Dirac condition. A suitably defined concept of stratified Kähler polarization then takes care of the irreducibility problem, as does an ordinary polarization in the smooth case. Over a stratified space, the appropriate quantum phase space is what we call a *costratified* Hilbert space; this is a system of Hilbert spaces, one for each stratum, which arises from quantization on the closure of that stratum, the stratification provides linear maps between these Hilbert spaces reversing the partial ordering among the strata, and these linear maps are compatible with the quantizations. The main result obtained in [41] says that, for a positive Kähler manifold with a hamiltonian action of a compact Lie group, when suitable additional conditions are imposed, reduction after quantization coincides with quantization after reduction in the sense that not only the reduced and unreduced quantum phase spaces correspond but the *invariant unreduced and reduced quantum observables as well.* Examples abound; one such class of examples, involving holomorphic nilpotent orbits and in particular angular momentum zero spaces, has been treated in [41]. A particular case thereof will be reproduced in Section 6 below, for the sake of illustration.

A stratified polarization, see Section 5 below for details, is defined in terms of an appropriate Lie-Rinehart algebra which, for any Poisson algebra, serves as a replacement for the tangent bundle of a smooth symplectic manifold. The question whether quantization commutes with reduction includes the question whether what is behind the phrase "in terms of an appropriate Lie-Rinehart algebra" descends to the reduced level. This hints at interpreting this question as a descent problem.

To our knowledge, the idea of Lie-Rinehart algebra was first used by JACOBSON in [43] (without being explicitly identified as a structure in its own) to study certain field extensions. Thereafter this idea occurred in other areas including differential geometry and differential Galois theory. More details will be given below.

I am indebted to the organizers of the meeting for having given me the chance to illustrate an application of Lie-Rinehart algebras to a problem phrased in a language entirely different from that of Lie-Rinehart algebras. Perhaps one can build a general Galois theory including ordinary Galois theory, differential Galois theory, and ordinary principal bundles, in which Lie-Rinehart algebras appear as certain objects which capture infinitesimal symmetries.

1. Lie-Rinehart algebras

Let R be a commutative ring with 1 taken as ground ring which, for the moment, may be arbitrary. For a commutative R-algebra A, we denote by $\mathrm{Der}(A)$ the R-Lie algebra of derivations of A, with its standard Lie algebra structure. An (R,A)-*Lie algebra* [66] is a Lie algebra L over R which acts on (the left of) A (by derivations) and is also an A-module satisfying suitable compatibility conditions which generalize the usual properties of the Lie algebra of vector fields on a smooth manifold viewed as a module over its ring of functions; these conditions read

$$\begin{aligned}[]
[\alpha, a\beta] &= \alpha(a)\beta + a[\alpha,\beta],\\
(a\alpha)(b) &= a(\alpha(b)),
\end{aligned}$$

where $a,b \in A$ and $\alpha,\beta \in L$. When the emphasis is on the pair (A,L) with the mutual structure of interaction between A and L, we refer to the pair (A,L) as a *Lie-Rinehart* algebra. Given an arbitrary commutative algebra A over R, an obvious example of a Lie-Rinehart algebra is the pair $(A,\mathrm{Der}(A))$, with the obvious action of $\mathrm{Der}(A)$ on A and obvious A-module structure on $\mathrm{Der}(A)$. There is an obvious notion of morphism of Lie-Rinehart algebras and, with this notion of morphism, Lie-Rinehart algebras constitute a category. More details may be found in RINEHART [66] and in our papers [29] and [30].

We will now briefly spell out some of the salient features of Lie-Rinehart algebras. Given an (R,A)-Lie algebra L, its *universal algebra* $(U(A,L),\iota_L,\iota_A)$ is an R-algebra $U(A,L)$ together with a morphism $\iota_A\colon A \longrightarrow U(A,L)$ of R-algebras and a morphism $\iota_L\colon L \longrightarrow U(A,L)$ of Lie algebras over R having the properties

$$\iota_A(a)\iota_L(\alpha) = \iota_L(a\,\alpha), \quad \iota_L(\alpha)\iota_A(a) - \iota_A(a)\iota_L(\alpha) = \iota_A(\alpha(a)),$$

and $(U(A,L),\iota_L,\iota_A)$ is *universal* among triples (B,ϕ_L,ϕ_A) having these properties. For example, when A is the algebra of smooth functions on a smooth manifold N

and L the Lie algebra of smooth vector fields on N, then $U(A,L)$ is the *algebra of (globally defined) differential operators on* N. An explicit construction for the R-algebra $U(A,L)$ is given in RINEHART [66]. See our paper [29] for an alternate construction which employs the MASSEY-PETERSON [59] algebra.

The universal algebra $U(A,L)$ admits an obvious filtered algebra structure $U_{-1} \subseteq U_0 \subseteq U_1 \subseteq \ldots$, cf. [66], where $U_{-1}(A,L) = 0$ and where, for $p \geq 0$, $U_p(A,L)$ is the left A-submodule of $U(A,L)$ generated by products of at most p elements of the image $\overline{L}$ of L in $U(A,L)$, and the associated graded object $E^0(U(A,L))$ inherits a commutative graded A-algebra structure. The Poincaré-Birkhoff-Witt Theorem for $U(A,L)$ then takes the following form where $S_A[L]$ denotes the symmetric A-algebra on L, cf. (3.1) of [66].

Theorem 1.1. (Rinehart) *For an (R,A)-Lie algebra L which is projective as an A-module, the canonical A-epimorphism $S_A[L] \longrightarrow E^0(U(A,L))$ is an isomorphism of A-algebras.*

Consequently, for an (R,A)-Lie algebra L which is projective as an A-module, the morphism $\iota_L \colon L \longrightarrow U(A,L)$ is injective.

The construction of the ordinary Koszul complex computing Lie algebra cohomology carries over as well: Let $\Lambda_A(sL)$ be the exterior Hopf algebra over A on the suspension sL of L, where "suspension" means that sL is L except that its elements are regraded up by 1. RINEHART [66] has proved that the ordinary Chevalley-Eilenberg operator induces an $U(A,L)$-linear operator d on $U(A,L) \otimes_A \Lambda_A(sL)$ (this is not obvious since L is not an ordinary A-Lie algebra unless L acts trivially on A) having square zero. We will refer to

$$K(A,L) = (U(A,L) \otimes_A \Lambda_A(sL), d) \tag{1.2}$$

as the *Rinehart complex* for (A,L). It is manifest that the Rinehart complex is functorial in (A,L). Moreover, as a graded A-module, the resulting chain complex $\mathrm{Hom}_{U(A,L)}(K(A,L),A)$ underlies the A-algebra $\mathrm{Alt}_A(L,A)$ of A-multilinear functions on L but, beware, the differential is linear only over the ground ring R and turns $\mathrm{Alt}_A(L,A)$ into a differential graded cocommutative algebra over R; we will refer to this differential graded R-algebra as the *Rinehart algebra* of (A,L). Rinehart also noticed that, when L is projective or free as a left A-module, $K(A,L)$ is a projective or free resolution of A in the category of left $U(A,L)$-modules according as L is a projective or free left A-module; details may be found in [66]. In particular, the Rinehart algebra $(\mathrm{Alt}_A(L,A),d)$ then computes the Ext-algebra $\mathrm{Ext}^*_{U(A,L)}(A,A)$.

Rinehart also noticed that, when A is the algebra of smooth functions on a smooth manifold N and L the Lie algebra of smooth vector fields on N, then $(\mathrm{Alt}_A(L,A),d)(=\mathrm{Hom}_{U(A,L)}(K(A,L),A))$ is the ordinary *de Rham complex of* N whence, as an algebra, the de Rham cohomology of N amounts to the Ext-algebra $\mathrm{Ext}^*_{U(A,L)}(A,A)$ over the algebra $U(A,L)$ of differential operators on N. Likewise, for a Lie algebra L over R acting trivially on R, $K(R,L)$ is the ordinary Koszul complex; in particular, when L is projective as an R-module, $K(R,L)$ is the ordinary Koszul resolution of the ground ring R. *Thus the cohomology of Lie-Rinehart algebras comprises de Rham- as well as Lie algebra cohomology.* In particular, this offers a possible explanation why CHEVALLEY and EILENBERG [12], when they first isolated Lie algebra cohomology, derived their formulas by abstracting from the de Rham operator of a smooth manifold. Suitable graded versions of the cohomology of Lie-Rinehart algebras comprise as well Hodge cohomology and coherent sheaf cohomology of complex manifolds [36, 38].

The classical differential geometry notions of connection, curvature, characteristic classes, etc. may be developed for arbitrary Lie-Rinehart algebras [29], [30], [34], and there are notions of duality for Lie-Rinehart algebras generalizing Poincaré duality [35]; the idea of duality has been shown in [33] to cast new light on Gerstenhaber- and Batalin-Vilkovisky algebras. In a sense these homological algebra interpretations of Batalin-Vilkovisky algebras push further Rinehart's observations related with the interpretation of de Rham cohomology as certain Ext-groups. Graded versions of duality for Lie-Rinehart algebras [36], [38] may be used to study e. g. complex manifolds, CR-structures, and the mirror conjecture.

Lie-Rinehart algebras were implicitly used already by JACOBSON [43] and later by HOCHSCHILD [27]. The idea of Lie-Rinehart algebra has been introduced by a very large number of authors, most of whom independently proposed their own terminology. I am indebted to K. Mackenzie for his help with compiling the following list in chronological order: Pseudo-algèbre de Lie: Herz, 1953 [24]—actually, Herz seems to be the first to describe the structure in a form which makes its generality clear—; Lie d-ring: Palais, 1961 [62]; (R,C)-Lie algebra: Rinehart, 1963 [66]; (R,C)-éspace d'Elie Cartan régulier et sans courbure: de Barros, 1964 [14]; (R,C)-algèbre de Lie: Bkouche, 1966 [7]; Lie algebra with an associated module structure: Hermann, 1967 [23]; Lie module: Nelson, 1967 [61]; Pseudo-algèbre de Lie: Pradines, 1967 [64]; (A, C) system: Kostant and Sternberg, 1971 [52]; Sheaf of twisted Lie algebras: Kamber and Tondeur, 1971 [44]; Algèbre de Lie sur C/R: Illusie, 1972 [42]; Lie algebra extension: N. Teleman, 1972 [76]; Lie-Cartan pair: Kastler and Stora, 1985 [45]; Atiyah algebra: Beilinson and Schechtmann, 1988 [6]; Lie-Rinehart algebra: Huebschmann, 1990 [29]; Differential Lie algebra: Kosmann–Schwarzbach and Magri, 1990 [50]. Hinich and Schechtman (1993) [26] have used the term Lie algebroid for the general algebraic concept. In differential Galois theory, Lie-Rinehart algebras occur under the name "algèbre différentielle" in a paper by FAHIM [18]. Lie-Rinehart algebras occur as well in CHASE [10] and STASHEFF [75]. We have chosen to use the terminology *Lie-Rinehart algebra* since, as already pointed out, Rinehart subsumed the cohomology of these objects under standard homological algebra and established a Poincaré-Birkhoff-Witt theorem for them. In differential geometry, (R, A)-Lie algebras arise as spaces of sections of Lie algebroids. These, in turn, were introduced in 1966 by PRADINES [63] and, in that paper, Pradines raised the issue whether Lie's third theorem holds for Lie algebroids in the sense that any Lie algebroid integrates to a Lie groupoid. CRAINIC and FERNANDES [13] have recently given a solution of this problem in terms of suitably defined obstructions. See MACKENZIE [56] for a complete account of Lie algebroids and Lie groupoids, as well as CANAS DA SILVA-WEINSTEIN [9] and MACKENZIE [57] for more recent surveys on particular aspects. The idea of Lie algebroid is lurking behind a construction in FUCHSSTEINER [19] (see Remark 2.6 below). A descent construction for Lie algebroids may be found in HIGGINS-MACKENZIE [25]. A general notion of morphism of Lie algebroids has been introduced by ALMEIDA-KUMPERA [2]. This notion has been used by S. CHEMLA [11] to study a version of duality for Lie algebroids in complex algebraic geometry generalizing Serre duality. Lie-Rinehart algebras are lurking as well behind the nowadays very active research area of D-modules.

Remark (out of context). At the end of his paper [66], RINEHART introduced an operator on the Hochschild complex of a commutative algebra which, some 20 years later, was reinvented by A. CONNES in order to define cyclic cohomology.

2. Poisson algebras

For intelligibility, we recall briefly how for an arbitrary Poisson algebra an appropriate Lie-Rinehart algebra serves as a replacement for the tangent bundle of a smooth symplectic manifold.

Let $(A, \{\cdot,\cdot\})$ be a Poisson algebra, and let D_A be the A-module of formal differentials of A the elements of which we write as du, for $u \in A$. For $u, v \in A$, the assignment to (du, dv) of $\pi(du, dv) = \{u, v\}$ yields an A-valued A-bilinear skew-symmetric 2-form $\pi = \pi_{\{\cdot,\cdot\}}$ on D_A, the *Poisson 2-form* for $(A, \{\cdot,\cdot\})$. Its adjoint

$$\pi^\sharp \colon D_A \longrightarrow \mathrm{Der}(A) = \mathrm{Hom}_A(D_A, A) \tag{2.1}$$

is a morphism of A-modules, and the formula

$$[adu, bdv] = a\{u, b\}dv + b\{a, v\}du + abd\{u, v\} \tag{2.2}$$

yields a Lie bracket $[\cdot,\cdot]$ on D_A, viewed as an R-module. More details may be found in [29]. For the record we recall the following, established in [29] (3.8).

Proposition 2.3. *The A-module structure on D_A, the bracket $[\cdot,\cdot]$, and the morphism $\pi^\sharp$ of A-modules turn the pair (A, D_A) into a Lie-Rinehart algebra in such a way that $\pi^\sharp$ is a morphism of Lie-Rinehart algebras.*

We write $D_{\{\cdot,\cdot\}} = (D_A, [\cdot,\cdot], \pi^\sharp)$. The 2-form $\pi_{\{\cdot,\cdot\}}$, which is defined for *every* Poisson algebra, is plainly a 2-cocycle in the Rinehart algebra $(\mathrm{Alt}_A(D_{\{\cdot,\cdot\}}, A), d)$. In [29], we defined the *Poisson cohomology* $\mathrm{H}^*_{\mathrm{Poisson}}(A, A)$ of the Poisson algebra $(A, \{\cdot,\cdot\})$ to be the cohomology of this Rinehart algebra, that is,

$$\mathrm{H}^*_{\mathrm{Poisson}}(A, A) = \mathrm{H}^*\left(\mathrm{Alt}_A(D_{\{\cdot,\cdot\}}, A), d\right).$$

Henceforth we shall take as ground ring that of the reals $\mathbb{R}$ or that of the complex numbers $\mathbb{C}$. We shall consider spaces N with an algebra of continuous $\mathbb{R}$-valued or $\mathbb{C}$-valued functions, deliberately denoted by $C^\infty(N, \mathbb{R})$ or $C^\infty(N, \mathbb{C})$ as appropriate, or by just $C^\infty(N)$, for example ordinary smooth manifolds and ordinary smooth functions; such an algebra $C^\infty(N)$ will then be referred to as a *smooth* structure on N, and $C^\infty(N)$ will be viewed as part of the structure of N. A space may support *different* smooth structures, though. Given a space N with a smooth structure $C^\infty(N)$, we shall write $\Omega^1(N)$ for the space of formal differentials with those differentials divided out that are zero at each point, cf. [53]; for example, over the real line with its ordinary smooth structure, the formal differentials $d\sin x$ and $\cos x dx$ do not coincide but the formal differential $d\sin x - \cos x dx$ is zero at each point. At a point of N, the object $\Omega^1(N)$ comes down to the ordinary space of differentials for the smooth structure on N; see Section 1.3 of our paper [41] for details. When N is an ordinary smooth manifold, $\Omega^1(N)$ amounts to the space of smooth sections of the cotangent bundle. For a general smooth space N over the reals, when $A = C^\infty(N, \mathbb{R})$ is endowed with a Poisson structure, the formula (1.2) yields a Lie-bracket $[\cdot,\cdot]$ on the A-module $\Omega^1(N)$ and the 2-form $\pi_{\{\cdot,\cdot\}}$ is still defined on $\Omega^1(N)$; its adjoint then yields an A-linear map $\pi^\sharp$ from $\Omega^1(N)$ to $\mathrm{Der}(A)$ in such a way that $([\cdot,\cdot], \pi^\sharp)$ is an $(\mathbb{R}, A)$-Lie algebra structure on $\Omega^1(N)$

and that the adjoint $\pi^\sharp$ is a morphism of $(\mathbb{R}, A)$-Lie algebras. Here the notation $[\cdot,\cdot]$, $\pi^\sharp$, $\pi_{\{\cdot,\cdot\}}$ is abused somewhat. The obvious projection map from $D_{C^\infty(N)}$ to $\Omega^1(N)$ is plainly compatible with the Lie-Rinehart structures. The fact that $D_{C^\infty(N)}$ is "bigger" than $\Omega^1(N)$ in the sense that the surjection from the former to the latter has a non-trivial kernel causes no problem here since the A-dual of this surjection, that is, the induced map from $\mathrm{Hom}(\Omega^1(N), A)$ to $\mathrm{Hom}(D_{C^\infty(N)}, A)$, is an isomorphism. We shall write $\Omega^1(N)_{\{\cdot,\cdot\}} = (\Omega^1(N), [\cdot,\cdot], \pi^\sharp)$. When N is a smooth manifold in the usual sense, the range $\mathrm{Der}(A)$ of the adjoint map $\pi^\sharp$ from $\Omega^1(N)$ to $\mathrm{Der}(A)$ boils down to the space $\mathrm{Vect}(N)$ of smooth vector fields on N. In this case, the Poisson structure on N is symplectic, that is, arises from a (uniquely determined) symplectic structure on N, if and only if $\pi^\sharp$, which may now be written as a morphism of smooth vector bundles from T^*N to $\mathrm{T}N$, is an isomorphism.

The 2-form $\pi_{\{\cdot,\cdot\}}$ is a 2-cocycle in the Rinehart algebra $(\mathrm{Alt}_A(\Omega^1(N)_{\{\cdot,\cdot\}}, A), d)$ and the canonical map of differential graded algebras from $(\mathrm{Alt}_A(\Omega^1(N)_{\{\cdot,\cdot\}}, A), d)$ to $(\mathrm{Alt}_A(D_{\{\cdot,\cdot\}}, A), d)$ is an isomorphism. In particular, we may take the cohomology of the Rinehart algebra $(\mathrm{Alt}_A(\Omega^1(N)_{\{\cdot,\cdot\}}, A), d)$ as the definition of the Poisson cohomology $\mathrm{H}^*_{\mathrm{Poisson}}(A, A)$ of the Poisson algebra $(A, \{\cdot,\cdot\})$ as well. When N is an ordinary smooth manifold, its algebra of ordinary smooth functions being endowed with a Poisson structure, this notion of Poisson cohomology comes down to that introduced by LICHNEROWICZ [55]. For a general Poisson algebra A, the 2-form $\pi_{\{\cdot,\cdot\}}$, be it defined on $\Omega^1(N)_{\{\cdot,\cdot\}}$ for the case where A is the structure algebra $C^\infty(N)$ of a space N or on $D_{\{\cdot,\cdot\}}$ for an arbitrary Poisson algebra, generalizes the symplectic form of a symplectic manifold; see Section 3 of [29] for details. Suffice it to make the following observation, relevant for quantization: Consider a space N with a smooth structure $C^\infty(N)$ which, in turn, is endowed with a Poisson bracket $\{\cdot,\cdot\}$. The Poisson 2-form $\pi_{\{\cdot,\cdot\}}$ determines an *extension* of Lie-Rinehart algebras which is central as a Lie algebra extension. For technical reasons it is more convenient to take here the extension

$$0 \longrightarrow A \longrightarrow \overline{L}_{\{\cdot,\cdot\}} \longrightarrow \Omega^1(N)_{\{\cdot,\cdot\}} \longrightarrow 0 \tag{2.4}$$

which corresponds to the negative of the Poisson 2-form. Here, as A-modules, $\overline{L}_{\{\cdot,\cdot\}} = A \oplus \Omega^1(N)_{\{\cdot,\cdot\}}$, and the Lie bracket on $\overline{L}_{\{\cdot,\cdot\}}$ is given by

$$[(a, du), (b, dv)] = (\{u, b\} + \{a, v\} - \{u, v\}, d\{u, v\}), \quad a, b, u, v \in A. \tag{2.5}$$

Here we have written "$\overline{L}$" rather than simply L to indicate that the extension (2.4) represents the negative of the class of $\pi_{\{\cdot,\cdot\}}$ in the second cohomology group $\mathrm{H}^2(\mathrm{Alt}_A(D_{\{\cdot,\cdot\}}, A), d)$ of the corresponding Rinehart algebra, cf. [29], and the notation du, dv etc. is abused somewhat. Now, any principal circle bundle admits as its infinitesimal object an *Atiyah*-sequence [5] whose spaces of sections constitute a central extension of Lie-Rinehart algebras; see [56] for a complete account of Atiyah-sequences and [34] for a theory of characteristic classes for extensions of general Lie-Rinehart algebras. When the Poisson structure is an ordinary smooth symplectic Poisson structure whose symplectic form represents an integral cohomology class, the Lie-Rinehart algebra extension (2.4) comes down to the space of sections of the Atiyah-sequence of the principal circle bundle classified by that cohomology class.

Remark 2.6. For the special case where N is an ordinary smooth manifold, $C^\infty(N)$ its algebra of ordinary smooth functions, and where $\{\cdot,\cdot\}$ is a Poisson structure on $C^\infty(N)$, the Lie-Rinehart structure on the pair $(C^\infty(N), \Omega^1(N))$ (where $\Omega^1(N)$ amounts to the space of ordinary smooth 1-forms on N) was discovered by a number of authors during the 80's most of whom phrased the structure in terms of the corresponding Lie algebroid structure on the cotangent bundle of N; some historical comments may be found in [29]. The first reference I am aware of where versions of the Lie algebroid bracket and of the anchor map may be found is [19]; in that paper, the notion of "implectic operator" is introduced—this is the operator nowadays referred to as *Poisson tensor*—and the Lie bracket and anchor map are the formula (2) and morphism written as Ω_ϕ, respectively, in that paper. *The construction in terms of formal differentials carried out in* [29] (and reproduced above)—as opposed to the Lie algebroid construction—is *more general, though, since it refers to an arbitrary Poisson structure, not necessarily one which is defined on an algebra of smooth functions on an ordinary smooth manifold.* In fact, the aim of the present article is to demonstrate the significance of this more general construction which works as well for Poisson algebras of continuous functions defined on spaces with *singularities* where among other things it yields a tool to relate the Poisson structures on the strata of a stratified symplectic space; suitably translated into the language of sheaves, it also works over not necessarily non-singular varieties.

3. Quantization

According to DIRAC [15], [16], a *quantization* of a classical system described by a real Poisson algebra $(A, \{\cdot,\cdot\})$ is a representation $a \mapsto \hat{a}$ of a certain Lie subalgebra B of A, A and B being viewed merely as Lie algebras, by symmetric or, whenever possible, self-adjoint, operators $\hat{a}$ on a Hilbert space $\mathcal{H}$ such that (i) the Dirac condition

$$i\,[\hat{a}, \hat{b}] = \widehat{\{a, b\}}$$

holds; that (ii) for a constant c, the operator $\hat{c}$ is given by $\hat{c} = c\,\mathrm{Id}$; and that (iii) the representation is irreducible. Here the factor i in the Dirac condition is forced by the interpretation of quantum mechanics: Observables are to be represented by symmetric (or self-adjoint) operators but the ordinary commutator of two symmetric operators is not symmetric. The second requirement rules out the adjoint representation, and the irreducibility condition is forced by the requiremend that phase transitions be possible between two different states. See e. g. SNIATYCKI [71] or WOODHOUSE [79]. Also it is known that for $B = A$ the problem has no solution whence the requirement that only a sub Lie algebra of A be represented. The physical constant $\hbar$ is here absorbed in the Poisson structure. It has become common to refer to a procedure furnishing a representation that satisfies only (i) and (ii) above as *prequantization*.

Under suitable circumstances, over a smooth symplectic manifold, the *geometric quantization scheme* developed by KIRILLIOV [48], KOSTANT [51], SOURIAU [74], and I. SEGAL [67], furnishes a quantization; see [71] or [79] for complete accounts. We confine ourselves with the remark that geometric quantization proceeds in two steps. The first step, prequantization, yields a representation of the Lie algebra underlying the whole Poisson algebra which satisfies (i) and (ii) but such a representation is not irreducible; the second step involves a choice of *polarization* to force the irreducibility condition. In particular, a *Kähler polarization* leads to what is called *Kähler quantization*. The existence of a Kähler polarization entails that the underlying manifold carries an ordinary Kähler structure. In the singular case, the ordinary geometric quantization scheme is no longer available, though. In the rest of the paper we shall describe how, under certain favorable circumstances, the difficulties in the singular case can be overcome in the framework of Kähler quantization. An observation crucial in the singular case is that the notion of polarization can be given a meaning by means of appropriately defined Lie-Rinehart algebras. Before going into details, we will briefly explain one of the origins of singularities.

4. Symmetries

Recall that a *symplectic* manifold is a smooth manifold N together with a closed non-degenerate 2-form σ. Given a function f, the identity $\sigma(X_f,\cdot) = df$ then associates a uniquely determined vector field X_f to f, the *Hamiltonian vector field* of f and, given two functions f and h, their *Poisson bracket* $\{f,h\}$ is defined by $\{f,h\} = X_f h$. This yields a Poisson bracket $\{\cdot,\cdot\}$ on the algebra $C^\infty(N)$ of ordinary smooth functions on N, referred to as a *symplectic* Poisson bracket.

Given a Lie group G, a *hamiltonian* G-space is a smooth symplectic G-manifold (N,σ) together with a smooth G-equivariant map μ from N to the dual $\mathfrak g^*$ of the Lie algebra $\mathfrak g$ of G satisfying the formula

$$\sigma(X_N,\cdot) = X \circ d\mu \tag{4.1}$$

for every $X \in \mathfrak g$; here X_N denotes the vector field on N induced by $X \in \mathfrak g$ via the G-action, and X is viewed as a linear form on $\mathfrak g^*$. The map μ is called a *momentum mapping* (or *moment map*). We recall that (4.1) says that, given $X \in \mathfrak g$, the vector field X_N is the hamiltonian vector field for the smooth function $X\circ\mu$ on N. See e. g. [1] for details. Given a hamiltonian G-space (N,σ,μ), the space $N_{\mathrm{red}} = \mu^{-1}(0)/G$ is called the *reduced space*. When G is not compact, this space may have bad properties; for example, it is not even a Hausdorff space when there are non-closed G-orbits in N.

Let $C^\infty(N_{\mathrm{red}}) = (C^\infty(N))^G)/I^G$, where I^G refers to the ideal (in the algebra $(C^\infty(N))^G$) of smooth G-invariant functions on N which vanish on the zero locus $\mu^{-1}(0)$; this is a smooth structure on the reduced space N_{red}. As observed by Arms-Cushman-Gotay [3], the Noether Theorem implies that the symplectic Poisson structure on $C^\infty(N)$ descends to a Poisson structure $\{\ ,\ \}_{\mathrm{red}}$ on $C^\infty(N_{\mathrm{red}})$, and Sjamaar-Lerman [70] have shown that, when G is compact and when the momentum mapping is proper, the orbit type decomposition of N_{red} is a stratification in the sense of Goresky-MacPherson [20]. The idea that the orbit type decomposition is a stratification (in a somewhat weaker sense) may be found already in [4]. For intelligibility, we recall some of the requisite technical details.

A decomposition of a space Y into pieces which are smooth manifolds such that these pieces fit together in a certain precise way is called a *stratification* [20]. More precisely: Let Y be a Hausdorff paracompact topological space and let $\mathcal I$ be a partially ordered set with order relation denoted by $\leq$. An *$\mathcal I$-decomposition* of Y is a locally finite collection of disjoint locally closed manifolds $S_i \subseteq Y$ called *pieces* (recall that a collection $\mathcal A$ of subsets of Y is said to be *locally finite* provided every $x \in Y$ has a neighborhood U_x in Y such that $U_x \cap A \neq \emptyset$ for at most finitely many A in $\mathcal A$) such that the following hold:

$Y = \cup S_i \ (i \in \mathcal{I})$,
$S_i \cap \overline{S}_j \neq \emptyset \Longleftrightarrow S_i \subseteq \overline{S}_j \Longleftrightarrow i \leq j \ (i, j \in \mathcal{I})$.
The space Y is then called a *decomposed* space. A decomposed space Y is said to be a *stratified space* if the pieces of Y, called *strata*, satisfy the following condition: Given a point x in a piece S there is an open neighborhood U of x in Y, an open ball B around x in S, a stratified space Λ, called the *link* of x, and a decomposition preserving homeomorphism from $B \times C^\circ(\Lambda)$ onto U. Here $C^\circ(\Lambda)$ refers to the open cone on Λ and, as a stratified space, Λ is less complicated than $C^\circ(\Lambda)$ whence the definition is not circular; the idea of complication is here made precise by means of the notion of *depth*.

A *stratified symplectic space* [70] is a stratified space Y together with a Poisson algebra $(C^\infty(Y), \{\ ,\ \})$ of continuous functions on Y which, on each piece of the decomposition, restricts to an ordinary smooth symplectic Poisson structure; in particular, $C^\infty(Y)$ is a smooth structure on Y.

Example 4.2. On the ordinary plane, with coordinates x_1, x_2, consider the algebra A of smooth functions in the coordinate functions x_1, x_2 together with, which is *crucial* here, an additional function r which is the radius function, subject to the relation $x_1^2 + x_2^2 = r^2$. Notice that r is *not* a smooth function in the usual sense whence the algebra A is strictly larger than that of ordinary smooth functions on the plane. The Poisson structure $\{\cdot, \cdot\}$ on A given by the formulas

$$\{x_1, x_2\} = 2r, \quad \{x_1, r\} = 2x_2, \quad \{x_2, r\} = -2x_1 \tag{4.2.1}$$

turns the plane into a stratified symplectic space. Geometrically, the plane is taken here as a half cone, the algebra A being that of *Whitney*-smooth functions on the half cone, with reference to the embedding into 3-space; there are two strata, the vertex of the half cone and the complement thereof. On the complement of the vertex, which is a punctured plane, the Poisson structure is symplectic. In physics, the Poisson algebra $(A, \{\cdot, \cdot\})$ arises, for $n \geq 2$, as the reduced Poisson algebra of a single particle in $\mathbb{R}^n$ with $\mathrm{O}(n, \mathbb{R})$-symmetry and angular momentum zero. For $n = 1$, the example still makes sense: the symmetry group is then just a copy of $\mathbb{Z}/2$, and the angular momentum is zero.

Given a hamiltonian G-space (N, σ, μ) with G compact, in view of an observation in [70], the Arms-Cushman-Gotay construction turns $(N_{\mathrm{red}}, C^\infty(N_{\mathrm{red}}), \{\ ,\ \}_{\mathrm{red}})$ (more precisely: each connected component of N_{red} in case the momentum mapping is not proper) into a stratified symplectic space. When N_{red} is smooth, i. e. has a single stratum, this space is just a smooth symplectic manifold, the ordinary MARSDEN-WEINSTEIN reduced space [58].

5. Stratified complex polarizations

Within the ordinary geometric quantization scheme, the irreducibility requirement is taken care of by means of a *polarization*. In particular, a *complex polarization* for an ordinary symplectic manifold N is an integrable Lagrangian distribution $F \subseteq \mathrm{T}^{\mathbb{C}}N$ of the complexified tangent bundle $\mathrm{T}^{\mathbb{C}}N$ [79]; under the identification of $\mathrm{T}^{\mathbb{C}}N$ with its (complex) dual coming from the symplectic structure, a complex polarization F then corresponds to a certain uniquely defined $(\mathbb{C}, C^\infty(N, \mathbb{C}))$-Lie subalgebra P of $\Omega^1(N, \mathbb{C})_{\{\cdot,\cdot\}}$.

Given a stratified symplectic space N, we refer to a $(\mathbb{C}, C^\infty(X, \mathbb{C}))$-Lie subalgebra P of $\Omega^1(X, \mathbb{C})_{\{\cdot,\cdot\}}$ as a *stratified complex polarization* for N if, for every stratum Y, under the identification of $\mathrm{T}^{\mathbb{C}}Y$ with its (complex) dual coming from the symplectic structure on that stratum, the $(\mathbb{C}, C^\infty(Y,\mathbb{C}))$-Lie subalgebra P_Y of $\Omega^1(Y,\mathbb{C})_{\{\cdot,\cdot\}}$ generated by the restriction of P to Y is identified with the space of sections of an ordinary complex polarization. A stratified complex polarization is, then, a *Kähler* polarization provided on any stratum it comes from an ordinary (not necessarily positive) Kähler polarization. We say that a stratified Kähler polarization is *complex analytic* provided it is induced from a complex analytic structure on N, and we define a complex analytic stratified Kähler structure to be a *normal Kähler structure* provided the complex analytic structure is normal. A normal Kähler structure is *positive* provided it is positive on each stratum. See Section 2 of [39] for more details.

Let G be a compact Lie group, denote its complex form by $G^{\mathbb{C}}$, and recall the following, cf. Proposition 4.2 of [39].

Proposition 5.1. *Given a positive Kähler manifold N with a holomorphic $G^{\mathbb{C}}$-action whose restriction to G preserves the Kähler structure and is hamiltonian, the Kähler polarization F induces a positive normal (complex analytic stratified) Kähler polarization P^{red} on the reduced space N^{red}, the latter being endowed with its stratified symplectic Poisson algebra $(C^\infty(N^{\mathrm{red}}), \{\cdot,\cdot\}^{\mathrm{red}})$.*

Under these circumstances, the underlying complex analytic structure of N^{red} is that of a geometric invariant theory quotient; the existence thereof may be found in [47] and [49]. The existence problem of this complex analytic structure may be seen as one of descent.

6. Examples

We will now illustrate the notions introduced so far by means of a number of examples. The interested reader will find more details in [39].

Example 6.1. For $\ell \geq 1$, consider the constrained system of ℓ particles in $\mathbb{R}^s$ with total angular momentum zero. Its unreduced phase space N is a product $(\mathrm{T}^*\mathbb{R}^s)^\ell$ of ℓ copies of $\mathrm{T}^*\mathbb{R}^s$, and we write the points of N in the form

$$(\mathbf{q}, \mathbf{p}) = (\mathbf{q}_1, \mathbf{p}_1, \ldots, \mathbf{q}_\ell, \mathbf{p}_\ell).$$

Let $H = \mathrm{O}(s,\mathbb{R})$. With reference to the obvious H-symmetry, the momentum mapping of this system has the form

$$\mu\colon N \longrightarrow \mathfrak{h}^*, \quad \mu(\mathbf{q}_1, \mathbf{p}_1, \ldots, \mathbf{q}_\ell, \mathbf{p}_\ell) = \mathbf{q}_1 \wedge \mathbf{p}_1 + \cdots + \mathbf{q}_\ell \wedge \mathbf{p}_\ell,$$

where the Lie algebra $\mathfrak{h} = \mathfrak{so}(s,\mathbb{R})$ is identified with its dual in the standard fashion. To elucidate the reduced space, observe that the assignment to $(\mathbf{q}, \mathbf{p}) = (\mathbf{q}_1, \mathbf{p}_1, \ldots, \mathbf{q}_\ell, \mathbf{p}_\ell)$ of the real symmetric $(2\ell \times 2\ell)$-matrix

$$\xi(\mathbf{q}, \mathbf{p}) = \begin{bmatrix} \mathbf{q}_j\mathbf{q}_k & \mathbf{q}_j\mathbf{p}_k \\ \mathbf{p}_j\mathbf{q}_k & \mathbf{p}_j\mathbf{p}_k \end{bmatrix}_{1\leq j,k\leq \ell}$$

yields a real algebraic map $\xi\colon N \longrightarrow \mathrm{S}^2_{\mathbb{R}}[\mathbb{R}^{2\ell}]$ from N to the real vector space $\mathrm{S}^2_{\mathbb{R}}[\mathbb{R}^{2\ell}]$ of real symmetric $(2\ell \times 2\ell)$-matrices which passes to an embedding of the reduced space $N^{\mathrm{red}} = \mu^{-1}(0)/H$ into $\mathrm{S}^2_{\mathbb{R}}[\mathbb{R}^{2\ell}]$, in fact, realizes N^{red} as a semi-algebraic set in $\mathrm{S}^2_{\mathbb{R}}[\mathbb{R}^{2\ell}]$. Let J be the standard complex structure on $\mathbb{R}^{2\ell}$. Now, on the one hand, the association $S \mapsto JS$ identifies $\mathrm{S}^2_{\mathbb{R}}[\mathbb{R}^{2\ell}]$ with $\mathfrak{sp}(\ell,\mathbb{R})$ in an $\mathrm{Sp}(\ell,\mathbb{R})$-equivariant fashion (with reference to the obvious actions) and hence identifies N^{red} with a subset of $\mathfrak{sp}(\ell, \mathbb{R})$ which, as observed in [54], is the closure of a certain nilpotent orbit which has been identified as a *holomorphic nilpotent orbit* in [39]. The Killing form transforms the Lie-Poisson structure on $\mathfrak{sp}(\ell, \mathbb{R})^*$ to a Poisson structure on $\mathfrak{sp}(\ell, \mathbb{R})$ which, restricted to N^{red}, yields a stratified symplectic structure. Another observation in [54] entails that this stratified symplectic structure coincides with the Sjamaar-Lerman stratified symplectic structure [70] mentioned earlier arising by symplectic reduction from $N = (\mathrm{T}^*\mathbb{R}^s)^\ell$. We mention in passing that, $\mathfrak{sp}(\ell, \mathbb{R})$ being identified with its dual by means of an appropriate positive multiple of the Killing form, as well as with $\mathrm{S}^2_{\mathbb{R}}[\mathbb{R}^{2\ell}]$, the map ξ is essentially the momentum mapping for the obvious $\mathrm{Sp}(\ell, \mathbb{R})$-action on N.

On the other hand, the choice of J determines a maximal compact subalgebra of $\mathfrak{sp}(\ell, \mathbb{R})$ which is just a copy of $\mathfrak{u}(\ell)$ and, furthermore, a Cartan decomposition $\mathfrak{sp}(\ell, \mathbb{R}) = \mathfrak{u}(\ell) \oplus \mathfrak{p}$. Now matrix multiplication by J from the left induces a complex structure on $\mathfrak{p}$ and, with this structure, as a complex vector space, $\mathfrak{p}$ amounts to the complex symmetric square $\mathrm{S}^2_{\mathbb{C}}[\mathbb{C}^\ell]$ on $\mathbb{C}^\ell$. In particular, orthogonal projection to $\mathfrak{p}$ induces a linear surjection of real vector spaces from $\mathrm{S}^2_{\mathbb{R}}[\mathbb{R}^{2\ell}]$ to $\mathrm{S}^2_{\mathbb{C}}[\mathbb{C}^\ell]$, uniquely determined by J; it is given by the assignment to a real symmetric $(2\ell \times 2\ell)$-matrix of the corresponding complex symmetric $(\ell \times \ell)$-matrix with respect to the standard complex structure J on $\mathbb{R}^{2\ell}$. This projection, restricted to N^{red}, is injective and yields a complex analytic structure on N^{red}. The two structures are compatible and yield a normal (complex analytic stratified) Kähler structure on N^{red}; see [39] for details. We will describe the requisite (complex analytic) stratified Kähler polarization P shortly. For $\ell \geq s$, as a complex analytic space, N^{red} comes down to $\mathrm{S}^2_{\mathbb{C}}[\mathbb{C}^\ell]$ whereas, for $\ell < s$, as a complex analytic space, N^{red} may be described as a complex determinantal variety in $\mathrm{S}^2_{\mathbb{C}}[\mathbb{C}^\ell]$, that is, as an affine variety given by determinantal equations; see e. g. [8] for determinantal varieties. This may be deduced from standard geometric invariant theory results combined with the standard description of the invariants of the classical groups which, in turn, may be found e. g. in [78]. As a stratified symplectic space, the singularity structure of N^{red} is finer than that of the complex analytic structure, though: Once ℓ is fixed, for $s = \ell$, the smooth structure $C^\infty(N_\ell)$ and hence the Poisson structure on $N_\ell = N^{\mathrm{red}} \cong \mathbb{C}^d, d = \frac{\ell(\ell+1)}{2}$, is not standard and, as a stratified symplectic space, N_ℓ has $\ell + 1$ strata. For $s < \ell$, the space $N^{\mathrm{red}} = N_s$ (say) may be described as the closure of a stratum in N_ℓ; moreover, a system of ℓ particles in $\mathbb{R}^{s-1}$ being viewed as a system of ℓ particles in $\mathbb{R}^s$ via the standard inclusion of $\mathbb{R}^{s-1}$ into $\mathbb{R}^s$ yields an injection of N_{s-1} into N_s. Thus we get a sequence

$$\{o\} = N_0 \subseteq N_1 \subseteq \ldots N_{s-1} \subseteq N_s \subseteq \cdots \subseteq N_\ell$$

of injections of normal (complex analytic stratified) Kähler spaces in such a way that, for $1 \leq s \leq \ell$, $N_{s-1} \subseteq N_s$ is the singular locus of N_s in the sense of stratified symplectic spaces, and the stratified Kähler structure on N_s, in particular the requisite Poisson structure, is then simply obtained by restriction from N_ℓ. For example, for $\ell = 1$, $(N_1, C^\infty(N_1), \{\cdot,\cdot\})$ is just the reduced space and reduced Poisson algebra of a system of a single particle in $\mathbb{R}^n$ $(n \geq 2)$ with angular momentum zero explained in the Example 4.2 above. For $\ell = s = 2$, the space $N_2 = N^{\mathrm{red}}$ is complex analytically a copy of $\mathbb{C}^3$ which, as a stratified symplectic space, sits inside $\mathfrak{sp}(2, \mathbb{R})$, and we need ten generators to describe the Poisson structure on N_2. The reduced space N_1 for $\ell = 2$ and $s = 1$ is here complex analytically realized inside $N_2 \cong \mathbb{C}^3$ as the quadric $Y^2 = XZ$.

To introduce coordinates, and to spell out a description of the complex analytic stratified Kähler polarizations, consider the complexification $\mathfrak{sp}(\ell, \mathbb{C})$ of $\mathfrak{sp}(\ell, \mathbb{R})$; this complexification sits inside the complex polynomial algebra

$$\mathbb{C}[z_1, \ldots, z_\ell, \overline{z}_1, \ldots, \overline{z}_\ell]$$

as its homogeneous quadratic constituent. The complexification $\mathfrak{k}^{\mathbb{C}} \cong \mathfrak{gl}(\ell, \mathbb{C})$ of the maximal compact subalgebra $\mathfrak{k} = \mathfrak{u}(\ell)$ of $\mathfrak{sp}(\ell, \mathbb{R})$ is the span of the $z_j \overline{z}_k$'s and, with reference to the Cartan decomposition $\mathfrak{sp}(\ell, \mathbb{R}) = \mathfrak{u}(\ell) \oplus \mathfrak{p}$ of $\mathfrak{sp}(\ell, \mathbb{R})$, the constituents $\mathfrak{p}^+$ and $\mathfrak{p}^-$ of the decomposition $\mathfrak{p}^{\mathbb{C}} = \mathfrak{p}^+ \oplus \mathfrak{p}^-$ are the spans of the $z_j z_k$'s and the $\overline{z}_j \overline{z}_k$'s, respectively; this gives an explicit description of $\mathfrak{p}^+$ and $\mathfrak{p}^-$ as $\mathrm{S}^2_{\mathbb{C}}[\mathbb{C}^\ell]$ and $\overline{\mathrm{S}^2_{\mathbb{C}}[\mathbb{C}^\ell]}$, respectively. Furthermore, $\mathfrak{k} = \mathfrak{u}(\ell)$ sits inside $\mathfrak{sp}(\ell, \mathbb{C})$ as the real span of the $z_j \overline{z}_k + \overline{z}_j z_k$'s and $i(z_j \overline{z}_k - \overline{z}_j z_k)$'s, and $\mathfrak{p}$ sits inside $\mathfrak{sp}(\ell, \mathbb{C})$ as the real span of the $z_j z_k + \overline{z}_j \overline{z}_k$'s and $i(z_j z_k - \overline{z}_j \overline{z}_k)$'s; the assignment to a real symmetric $(2\ell \times 2\ell)$-matrix of the corresponding complex symmetric $(\ell \times \ell)$-matrix is given by the association

$$z_j z_k + \overline{z}_j \overline{z}_k \longmapsto z_j z_k, \quad i(z_j z_k - \overline{z}_j \overline{z}_k) \longmapsto i z_j z_k.$$

The summands $\mathfrak{p}^+$ and $\mathfrak{p}^-$ are the irreducible $\mathfrak{k}^{\mathbb{C}}$-representations in $\mathfrak{sp}(\ell, \mathbb{C})$ complementary to $\mathfrak{k}^{\mathbb{C}}$.

The homogeneous quadratic polynomials in the variables $z_1, \ldots, z_\ell, \overline{z}_1, \ldots, \overline{z}_\ell$ yield coordinates on $\mathfrak{sp}(\ell, \mathbb{R})$ and hence, via restriction, on N^{red}, that is, the smooth structure $C^\infty(N^{\mathrm{red}}, \mathbb{C})$ may be described as the algebra of smooth functions in these variables, subject to the relations coming from the embedding of N^{red} into $\mathfrak{sp}(\ell, \mathbb{R})$. Now, the differentials $d(z_j z_k)$ of the coordinate functions $z_j z_k$ $(1 \leq j, k \leq \ell)$ (that is, of those coordinate functions which do not involve any of the $\overline{z}_j$'s) generate the corresponding complex analytic stratified Kähler polarization $P \subseteq \Omega^1(N^{\mathrm{red}}, \mathbb{C})$ as an $(\mathbb{C}, C^\infty(N^{\mathrm{red}}, \mathbb{C}))$-Lie subalgebra of $\Omega^1(N^{\mathrm{red}}, \mathbb{C})_{\{\cdot,\cdot\}}$.

In [39], we developed a theory of holomorphic nilpotent orbits of hermitian Lie algebras and established the fact that the (topological) closure of any holomorphic nilpotent orbit inherits a normal (complex analytic stratified) Kähler structure. The space N^{red}, realized as the closure of a holomorphic nilpotent orbit in $\mathfrak{sp}(\ell, \mathbb{R})$, is a special case thereof.

Example 6.2. A variant of the above example arises from the constrained system of ℓ harmonic oscillators in $\mathbb{R}^s$ with total angular momentum zero and constant energy. Its unreduced phase space Q is a copy of complex projective space $\mathbb{P}^{s\ell-1}\mathbb{C}$ of complex dimension $s\ell - 1$. For $\ell \geq s$, as a complex analytic space, Q^{red} coincides with the (complex) projectivization $\mathbb{P}\mathrm{S}^2_{\mathbb{C}}[\mathbb{C}^\ell]$ of $\mathrm{S}^2_{\mathbb{C}}[\mathbb{C}^\ell]$ whereas for $\ell < s$, as a complex analytic space, Q^{red} may be described as a complex projective determinantal variety in $\mathbb{P}\mathrm{S}^2_{\mathbb{C}}[\mathbb{C}^\ell]$. In fact, the determinantal equations mentioned in Example 6.1 above are homogeneous and yield the requisite homogeneous equations for the present case. In the same vein as before, we get a sequence

$$Q_1 \subseteq \ldots Q_{s-1} \subseteq Q_s \subseteq \cdots \subseteq Q_\ell \cong \mathbb{P}^d\mathbb{C}, \; d = \frac{\ell(\ell+1)}{2} - 1,$$

of injections of compact normal (complex analytic stratified) Kähler spaces in such a way that, for $2 \leq s \leq \ell$, $Q_{s-1} \subseteq Q_s$ is the singular locus of Q_s in the sense of stratified symplectic spaces, each Q_s being the closure of a stratum in Q_ℓ, and the stratified Kähler structure on Q_s, in particular the requisite Poisson structure, is simply obtained by restriction from Q_ℓ. Complex analytically, each Q_s is a projective variety. Again, the smooth structure $C^\infty(Q_\ell)$ and hence the Poisson structure on $Q_\ell \cong \mathbb{P}^d\mathbb{C}$ $(s = \ell)$ is not the standard one (which arises from the Fubini-Study metric on complex projective space) and, as a stratified symplectic space, Q_ℓ has ℓ strata. For example, for $\ell = s = 2$, the space Q_2 is complex analytically a copy of $\mathbb{P}^2\mathbb{C}$, and the corresponding reduced space Q_1 (for $\ell = 2, s = 1$), which is abstractly just complex projective 1-space, sits complex analytically inside $Q_2 \cong \mathbb{P}^2\mathbb{C}$ as the projective conic $Y^2 = XZ$. These spaces are particular cases of a systematic class of examples of *exotic projective varieties*, introduced and explored in our paper [39].

Remark 6.3. Given a Lie group G, a smooth hamiltonian G-space, and a real G-invariant polarization, the question arises whether the statement of Proposition 5.1 still holds for this real polarization. When we try to identify, on the reduced level, a stratified version of such a polarization, we may run into the following difficulty, though: Under the circumstances of the Example 6.1, let $\ell = 1$, and consider the vertical polarization on $N = \mathrm{T}^*\mathbb{R}^n$. This polarization integrates to the foliation—even fibration—defined by the projection map from $\mathrm{T}^*\mathbb{R}^n$ to $\mathbb{R}^n$. This foliation is clearly $\mathrm{O}(n,\mathbb{R})$-invariant and, in terms of the standard coordinates $\mathbf{q} = (q^1, \ldots, q^n)$ on $\mathbb{R}^n$, a leaf is given by the equation $\mathbf{q} = \mathbf{q}_0$ where $\mathbf{q}_0$ is a constant. We will now write the ordinary scalar product of two vectors $\mathbf{x}$ and $\mathbf{y}$ as $\mathbf{xy}$. With these preparations out of the way, under the present circumstances, the assignment to $(\mathbf{q}, \mathbf{p}) \in \mathrm{T}^*\mathbb{R}^n$ of $x_1 = \mathbf{qq} - \mathbf{pp}$ and $x_2 = 2\mathbf{qp}$ yields a map from $\mathrm{T}^*\mathbb{R}^n$ to the plane $\mathbb{R}^2$ which induces an isomorphism of stratified symplectic spaces from the reduced space N^{red} onto the exotic plane described in the Example 4.2. In particular, the radius function r is given by $r = \mathbf{qq} + \mathbf{pp}$. Now $2\mathbf{qq} = x_1 + r$ whence, under reduction, the leaf $\mathbf{q} = \mathbf{q}_0$ passes to the subspace of the plane given by the equation

$$x_1 + r = 2\mathbf{q}_0\mathbf{q}_0 = c \text{ (say)}.$$

For $\mathbf{q}_0 \neq 0$, in the (x_1, x_2)-plane, this is just the parabola $x_2^2 + 2cx_1 = c^2$ since $r^2 = x_1^2 + x_2^2$ while, for $\mathbf{q}_0 = 0$, it is the non-positive half x_1-axis. The reason for this degeneracy is that the leaf $\mathbf{q} = 0$ is not transverse to the momentum mapping μ in the sense that, whatever $\mathbf{p} \in \mathbb{R}^n$, $\mu(0, \mathbf{p}) = 0$ while $\ker(d\mu(0, \mathbf{p}))$ and the tangent space of the leaf at $(0, \mathbf{p})$ do *not* together span the tangent space of N at $(0, \mathbf{p})$. Thus the reduced space is still foliated, but one leaf is singular; however even the restriction of this foliation to the top stratum still has a singular leaf, the negative half x_1-axis. A little thought reveals that this implies that this foliation cannot result from a stratified real polarization, the notion of stratified real polarization being defined in the same fashion as a stratified complex polarization, except that, on each stratum, the polarization should come down to a real polarization. As a side remark we mention that the assignment to a leaf of its intersection point with the non-negative x_1-axis identifies the space of leaves with the non-negative x_1-axis, and the latter in fact coincides with the orbit space $\mathbb{R}^n/\mathrm{O}(n,\mathbb{R})$. This description visualizes the exceptional role played by the non-positive x_1-axis. The distribution parallel to this foliation, though, is given by the hamiltonian vector field of the function $\mathbf{qq}$ in $C^\infty(N^{\mathrm{red}})$; it has the form

$$\{\mathbf{qq}, -\} = \frac{1}{2}\{x_1 + r, -\} = -x_2\frac{\partial}{\partial x_1} + (x_1 + r)\frac{\partial}{\partial x_2}$$

and in particular vanishes on the non-positive half x_1-axis. The function $\mathbf{qq}$ generates a maximal abelian Poisson subalgebra of $\big(C^\infty(N^{\mathrm{red}}), \{\cdot,\cdot\}^{\mathrm{red}}\big)$. This phenomenon is typical for cotangent bundles with a hamiltonian action of a Lie group arising from an action of that group on the base with more than a single orbit type. *Thus we see that the question whether a polarization other than a Kähler polarization descends to a stratified polarization on the reduced level leads to certain delicacies, and we do not know to what extent we can interpret it merely as a descent problem.*

The question whether, under suitable circumstances so that in particular the reduced space is still a smooth manifold, a real polarization descends has been studied in [21].

7. Prequantization on spaces with singularities

To develop prequantization over stratified symplectic spaces and to describe the behaviour of prequantization under reduction, in our paper [41], we introduced *stratified prequantum modules* over stratified symplectic spaces. A stratified prequantum module is defined in terms of the appropriate Lie-Rinehart algebra and determines what we call a *costratified prequantum space* but the two notions, though closely related, should not be confused.

Let N be a stratified symplectic space, and let $(A, \{\cdot,\cdot\})$ be its stratified symplectic Poisson algebra; a special case would be the ordinary symplectic Poisson algebra of a smooth symplectic manifold. Consider the extension (2.4) of Lie-Rinehart algebras. Given an $(A \otimes \mathbb{C})$-module M, we refer to an $(A, \overline{L}_{\{\cdot,\cdot\}})$-module structure $\chi\colon \overline{L}_{\{\cdot,\cdot\}} \longrightarrow \mathrm{End}_{\mathbb{R}}(M)$ on M as a *prequantum module structure for* $(A, \{\cdot,\cdot\})$ provided (i) the values of χ lie in $\mathrm{End}_{\mathbb{C}}(M)$, that is to say, the operators $\chi(a, \alpha)$ are complex linear transformations, and (ii) for every $a \in A$, $\chi(a, 0) = i\, a \,\mathrm{Id}_M$ [30, 41].

We recall from [29] that the assignment to $a \in A$ of $(a, da) \in \overline{L}_{\{\cdot,\cdot\}}$ yields a morphism ι of real Lie algebras from A to $\overline{L}_{\{\cdot,\cdot\}}$; this reduces the construction of Lie algebra representations of the Lie algebra which underlies the Poisson algebra A to the construction of representations of $\overline{L}_{\{\cdot,\cdot\}}$. Thus, for any prequantum module (M, χ), the composite of ι with $-i\chi$ is a representation $a \mapsto \widehat{a}$ of the A underlying real Lie algebra on M, viewed as a complex vector space, by $\mathbb{C}$-linear operators so that the constants in A act by multiplication and so that the Dirac condition holds, even though M does not necessarily carry a Hilbert space structure. These operators are given by the formula

$$\widehat{a}(x) = \frac{1}{i}\chi(0, da)(x) + ax, \quad a \in A,\ x \in M. \tag{7.1}$$

For illustration, consider an ordinary quantizable symplectic manifold (N, σ), with ordinary *prequantum bundle* $\zeta\colon \Lambda \to N$, that is, ζ is a complex line bundle with a connection ∇ whose curvature equals $-i\sigma$; the assignments $\chi_\nabla(a, 0) = i\, a\, \mathrm{Id}_M$ $(a \in A)$ and $\chi_\nabla(0, \alpha) = \nabla_{\pi^\sharp(\alpha)}$ $(\alpha \in \Omega^1(N)_{\{\cdot,\cdot\}})$ then yield a prequantum module structure

$$\chi_\nabla\colon \overline{L}_{\{\cdot,\cdot\}} \to \mathrm{End}_{\mathbb{C}}(M) \subseteq \mathrm{End}_{\mathbb{R}}(M)$$

for $(A, \{\cdot,\cdot\})$. (Here $\pi^\sharp\colon \Omega^1(N) \to \mathrm{Vect}(N)$ refers to the adjoint of the 2-form π induced by the symplectic Poisson structure, cf. Section 2.) This is just the ordinary prequantization construction in another guise.

As before, consider a general stratified symplectic space N, with stratified symplectic Poisson algebra $(C^\infty(N), \{\cdot,\cdot\})$. For each stratum Y, let $(C^\infty(Y), \{\cdot,\cdot\}^Y)$ be its symplectic Poisson structure, and let

$$0 \to C^\infty(Y) \to \overline{L}_{\{\cdot,\cdot\}^Y} \to \Omega^1(Y)_{\{\cdot,\cdot\}^Y} \to 0$$

be the corresponding extension (2.4) of Lie-Rinehart algebras. As in (1.5) of [39], we define a *stratified prequantum module* for N to consist of
— a prequantum module (M, χ) for $(C^\infty(N), \{\cdot,\cdot\})$, together with,
— for each stratum Y, a prequantum module structure χ_Y for $(C^\infty(Y), \{\cdot,\cdot\}^Y)$ on $M_Y = C^\infty(Y) \otimes_{C^\infty(N)} M$ in such a way that the canonical linear map of complex vector spaces from M to M_Y is a morphism of prequantum modules from (M, χ) to (M_Y, χ_Y).

Given a stratified prequantum module (M, χ) for N, when Y runs through the strata of N, we refer to the system

$$\left(M_{\overline{Y}}, \chi_{\overline{Y}}\colon \overline{L}_{\{\cdot,\cdot\}^{\overline{Y}}} \to \mathrm{End}_{\mathbb{R}}(M_Y)\right)$$

of prequantum modules, together with, for every pair of strata Y, Y' such that $Y' \subseteq \overline{Y}$, the induced morphism

$$(M_{\overline{Y}}, \chi_{\overline{Y}}) \to (M_{\overline{Y}'}, \chi_{\overline{Y}'})$$

of prequantum modules, as a *costratified prequantum space*. More formally: Consider the category $\mathcal{C}_N$ whose objects are the strata of N and whose morphisms are the inclusions $Y' \subseteq \overline{Y}$. We define a *costratified complex vector space* on N to be a contravariant functor from $\mathcal{C}_N$ to the category of complex vector spaces, and a *costratified prequantum space* on N to be a costratified complex vector space together with a compatible system of prequantum module structures. Thus a stratified prequantum module (M, χ) for $(N, C^\infty(N), \{\cdot,\cdot\})$ determines a costratified prequantum space on N; see the (1.4) and (1.5) of [41] for details.

Theorem 7.2. *Given a symplectic manifold N with a hamiltonian action of a compact Lie group G, a G-equivariant prequantum bundle ζ descends to a stratified prequantum module $(\chi^{\mathrm{red}}, M^{\mathrm{red}})$ for the stratified symplectic space*

$$(N^{\mathrm{red}}, C^\infty(N^{\mathrm{red}}), \{\cdot,\cdot\}^{\mathrm{red}}).$$

Proof. See Theorem 2.1 of [41]. □

Thus, phrased in the language of prequantum modules, the relationship between the unreduced and reduced prequantum object may be interpreted as one of descent.

In particular, consider a complex analytic stratified Kähler space $(N, C^\infty(N), \{\cdot,\cdot\}, P)$ (cf. Section 5 above or Section 2 of [39]), and let (M, χ) be a stratified prequantum module for $(C^\infty(N), \{\cdot,\cdot\})$. We refer to (M, χ) as a *complex analytic* stratified prequantum module provided M is the space of ($C^\infty(N)$-) sections of a complex V-line bundle ζ on N in such a way that P endows ζ via χ with a complex analytic structure. If this happens to be the case, M^P necessarily amounts to the space of global sections of the sheaf of germs of holomorphic sections of ζ. See Section 3 of [39].

8. Kähler quantization and reduction

Let G be a compact Lie group, let (N, σ, μ) be a hamiltonian G-space of the kind as that in the circumstances of Proposition 5.1, and suppose that N is quantizable. Thus N is, in particular, a positive Kähler manifold with a holomorphic $G^{\mathbb{C}}$-action whose restriction to G preserves the Kähler structure and is hamiltonian. Write P for the corresponding Kähler polarization, necessarily G-invariant, viewed as a $(\mathbb{C}, C^\infty(N, \mathbb{C}))$-Lie subalgebra of the $(\mathbb{C}, C^\infty(N, \mathbb{C}))$-Lie algebra $\Omega^1(N, \mathbb{C})_{\{\cdot,\cdot\}}$, and let ζ be a prequantum bundle. Via its connection, it acquires a holomorphic structure, and the connection is the unique hermitian connection for a corresponding hermitian structure. The momentum mapping induces, in particular, an infinitesimal action of the Lie algebra $\mathfrak{g}$ of G on ζ preserving the connection and hermitian structure. Suppose that this action lifts to a G-action on ζ preserving the connection and lifting the G-action on N. For connected G, the assumption that the G-action lift to one on ζ is (well known to be) redundant (since the infinitesimal action is essentially given by the momentum mapping) and it will suffice to replace G by an appropriate covering group if need be. Prequantization turns the space of smooth sections of ζ into a prequantum module for the ordinary smooth symplectic Poisson algebra of N. We write this prequantum module as M; it inherits a G-action preserving the polarization P. Hence the quantum module M^P, that is, the space $\Gamma(\zeta)$ of global holomorphic sections of ζ, is a complex representation space for G. This quantum module is the corresponding *unreduced* quantum state space, except that there is no Hilbert space structure present yet, and *reduction after quantization*, for the *quantum state spaces*, amounts to taking the space $(M^P)^G$ of G-invariant holomorphic sections.

The projection map from the space of smooth G-invariant sections of ζ to M^{red} restricts to a linear map

$$\rho\colon \Gamma(\zeta)^G \longrightarrow (M^{\mathrm{red}})^{P^{\mathrm{red}}} \tag{8.1}$$

of complex vector spaces, defined on the space $(M^P)^G = \Gamma(\zeta)^G$ of G-invariant holomorphic sections of ζ. Here and below P^{red} refers to the stratified Kähler polarization the existence of which is asserted in Proposition 5.1 above, M^{red} to the prequantum module for the stratified Kähler space mentioned in Theorem 7.2 (without having been made explicit there), and $(M^{\mathrm{red}})^{P^{\mathrm{red}}}$ to the P^{red}-invariants; notice that P^{red} is, in particular, a Lie algebra whence it makes sense to talk about P^{red}-invariants. A module of the kind $(M^{\mathrm{red}})^{P^{\mathrm{red}}}$ is referred to as a *reduced quantum module* in [41]. It acquires a costratified Hilbert space structure, the requisite scalar products being induced from appropriate hermitian structures via integration.

As far as the comparison of G-invariant unreduced quantum observables and reduced quantum observables is concerned, the statement that *Kähler quantization commutes with reduction* amounts to the following, cf. Theorem 3.6 in [41].

Theorem 8.2. *The data (N, σ, μ, M, P) being fixed so that, in particular, (N, σ, μ) is a smooth hamiltonian G-space structure on a quantizable positive Kähler manifold N with a holomorphic $G^{\mathbb{C}}$-action, the Kähler polarization being written as P, let f be a smooth G-invariant function on N which is quantizable in the sense that it preserves P. Then its class*

$$[f] \in C^\infty(N^{\mathrm{red}})(= (C^\infty(N))^G/I^G)$$

is quantizable, i. e. preserves P^{red} *and, for every* $h \in (M^P)^G$,

$$\rho(\ddot{f}(h)) = \widehat{[f]}(\rho(h)). \tag{8.2.1}$$

So far, we did not make any claim to the effect that the reduced quantum module $(M^{\mathrm{red}})^{P^{\mathrm{red}}}$ amounts to a space of global holomorphic sections. We now recall that, under the circumstances of Theorem 8.2, the momentum mapping is said to be *admissible* provided, for every $m \in N$, the path of steepest descent through m is contained in a compact set [68], [49] (§9). For example, when the momentum mapping is proper it is admissible. Likewise, the momentum mapping for a unitary representation of a compact Lie group is admissible in this sense, see Example 2.1 in [68].

The statement "*Kähler quantization commutes with reduction*" is then completed by the following two observations, cf. [41] ((3.7) and (3.8)).

Proposition 8.3 *Under the circumstances of Theorem* 8.2, *when* μ *is admissible and when* N^{red} *has a top stratum (i. e. an open dense stratum), for example when* μ *is proper, the reduced stratified prequantum module* $(M^{\mathrm{red}}, \chi^{\mathrm{red}})$ *is complex analytic, that is, as a complex vector space,* M^{red} *amounts to the space of global holomorphic sections of a suitable holomorphic* V*-line bundle on* N^{red}.

The relevant V-line bundle on N^{red} may be found in [68] (Proposition 2.11).

Theorem 8.4 [68] (Theorem 2.15) *Under the circumstances of Theorem* 8.2, *when the momentum mapping* μ *is proper, in particular, when* N *is compact, the map* ρ *is an isomorphism of complex vector spaces.*

In this result, the properness condition, while sufficient, is not necessary, that is, the map ρ may be an isomorphism without the momentum mapping being proper.

A version of Theorem 8.4 has been established in (4.15) of [60]; cf. also [69] and the literature there, as well as [65] and [77] for generalizations to higher dimensional sheaf cohomology.

Remark 8.5. The statements of Theorems 8.2 and 8.4 are logically independent; in particular the statement of Theorem 8.2 makes sense whether or not ρ is an isomorphism, and its proof does not rely on ρ being an isomorphism.

Thus we have consistent Kähler quantizations on the unreduced and reduced spaces, including a satisfactory treatment of observables, as indicated by the formula (8.2.1). We have already pointed out in the introduction that examples in finite dimensions abound. We hope that this kind of approach, suitably adapted, will eventually yield the quantization of certain infinite dimensional systems arising from field theory.

Remark 8.6. KEMPF'S descent lemma [46] mentioned earlier characterizes, among the holomorphic V-line bundles which arise on a geometric invariant theory quotient by the standard geometric invariant theory construction, those which are ordinary (holomorphic) line bundles. In the circumstances of Theorem 8.4, complex analytically, the space N^{red} is a geometric invariant theory quotient, and the V-line bundle which underlies the reduced quantum module arises by the standard geometric invariant theory construction. Here the term "descent" is used in its strict sense.

Illustration 8.7. Under the circumstances of the Example 6.2, let $\mathcal{O}(1)$ be the ordinary hyperplane bundle on $Q = \mathbb{P}^{s\ell-1}\mathbb{C}$ and, as usual, for $k \geq 1$, write its k'th power as $\mathcal{O}(k)$. The unitary group $\mathrm{U}(s\ell)$ acts on $\mathbb{P}^{s\ell-1}\mathbb{C}$ in a hamiltonian fashion having as momentum mapping the familiar embedding of $\mathbb{P}^{s\ell-1}\mathbb{C}$ into $\mathfrak{u}(s\ell)^*$, and the adjoint thereof yields a morphism of Lie algebras from $\mathfrak{u}(s\ell)$ to $C^\infty(\mathbb{P}^{s\ell-1}\mathbb{C})$, the latter being endowed with its symplectic Poisson structure coming from the Fubini-Study metric. It is a standard fact that, for $k \geq 1$, Kähler quantization, with reference to $k\omega$ and $\mathcal{O}(k)$ (where ω is the Fubini-Study form), yields the k'th symmetric power of the standard representation defining the Lie algebra $\mathfrak{u}(s\ell)$, and this representation integrates to the k'th symmetric power E_s^k of the standard representation E_s defining the group $\mathrm{U}(s\ell)$. (We use the subscript $-_s$ since here and below ℓ is fixed while s varies.) The symmetry group $H = \mathrm{O}(s,\mathbb{R})$ of the constrained system in (6.1) above appears as a subgroup of $\mathrm{U}(s\ell)$ in an obvious fashion and, viewed as this subgroup, H centralizes the subgroup $\mathrm{U}(\ell) = \mathrm{Sp}(\ell,\mathbb{R}) \cap \mathrm{U}(s\ell)$ (the maximal compact subgroup $\mathrm{U}(\ell)$ of $\mathrm{Sp}(\ell,\mathbb{R})$); hence, for $k \geq 1$, the subspace $(E_s^k)^H$ of H-invariants is a $\mathrm{U}(\ell)$-representation. On the other hand, with an abuse of notation, let $\mathcal{O}(1)$ be the hyperplane bundle on the reduced space $Q_\ell = \mathbb{P}^d\mathbb{C}$, $d = \frac{\ell(\ell+1)}{2} - 1$ and, for $k \geq 1$, let $\mathcal{O}(k)$ be its k'th power. The space of holomorphic sections thereof, $\Gamma(\mathcal{O}(k))$, amounts to the k'th symmetric power $\mathrm{S}^k_{\mathbb{C}}[\mathfrak{p}^*]$ of the dual of $\mathfrak{p} = \mathrm{S}^2_{\mathbb{C}}[\mathbb{C}^\ell]$ (the space of homogeneous degree k polynomial functions on $\mathfrak{p}$). For $1 \leq s \leq \ell$ and $k \geq 1$, maintain the notation $\mathcal{O}(k)$ for the restriction of the k'th power of the hyperplane bundle to $Q_s \subseteq Q_\ell$; for $s < \ell$, the space of holomorphic sections $\widetilde{E}_s^k$ of $\mathcal{O}(k)$ is now a certain quotient of $\widetilde{E}_\ell^k = \mathrm{S}^k_{\mathbb{C}}[\mathfrak{p}^*]$ which will be made precise below.

For $1 \leq s \leq \ell$, the composite of the embedding of N_s into $\mathfrak{sp}(\ell,\mathbb{R})^*$ with the surjection from $\mathfrak{sp}(\ell,\mathbb{R})^*$ to $\mathfrak{u}(\ell)^*$ induced from the injection of $\mathfrak{u}(\ell)$ into $\mathfrak{sp}(\ell,\mathbb{R})$ yields a map from N_s to $\mathfrak{u}(\ell)^*$ which descends to a map from Q_s to $\mathfrak{u}(\ell)^*$, the adjoint of which induces a morphism of Lie algebras from $\mathfrak{u}(\ell)$ to $C^\infty(Q_s)$, the latter being endowed with its *stratified symplectic Poisson structure* explained earlier. For $k \geq 1$, the space of sections M^{red} (cf. Theorem 7.2) of $\mathcal{O}(k)$, with reference to a $C^\infty(Q_s)$-module structure constructed in [41] and not made precise here, inherits a stratified prequantum module structure; and stratified Kähler quantization yields a $\mathrm{U}(\ell)$-representation on the space $\widetilde{E}_s^k$, which amounts to that written earlier as $(M^{\mathrm{red}})^{P^{\mathrm{red}}}$, cf. (8.1), in such a way that the map ρ given as (8.1) above identifies the representation written above as $(E_s^{2k})^H$ with $\widetilde{E}_s^k$; moreover, the spaces $(E_s^{2k-1})^H$ are zero.

We conclude with an explicit description of the spaces $(E_s^{2k})^H$ or, equivalently, of the spaces $\widetilde{E}_s^k$: Introduce coordinates $x_1, \dots, x_\ell$ on $\mathbb{C}^\ell$. These give rise to coordinates $\{x_{i,j} = x_{j,i};\ 1 \leq i,j \leq \ell\}$ on $\mathfrak{p} = S^2_{\mathbb{C}}[\mathbb{C}^\ell]$, and the determinants

$$\delta_1 = x_{1,1},\ \delta_2 = \begin{vmatrix} x_{1,1} & x_{1,2} \\ x_{1,2} & x_{2,2} \end{vmatrix},\ \delta_3 = \begin{vmatrix} x_{1,1} & x_{1,2} & x_{1,3} \\ x_{1,2} & x_{2,2} & x_{2,3} \\ x_{1,3} & x_{2,3} & x_{3,3} \end{vmatrix},\ \text{etc.}$$

are highest weight vectors for certain $\mathrm{U}(\ell)$-representations. For $1 \leq s \leq r$ and $k \geq 1$, the $\mathrm{U}(\ell)$-representation $\widetilde{E}_s^k$ is the sum of the irreducible representations having as highest weight vectors the monomials

$$\delta_1^\alpha \delta_2^\beta \dots \delta_s^\gamma, \quad \alpha + 2\beta + \dots + s\gamma = k,$$

and the morphism from $\widetilde{E}^k_s$ to $\widetilde{E}^k_{s-1}$ coming from restriction from Q_s to Q_{s-1} is an isomorphism on the span of those irreducible representations which do not involve δ_s and has the span of the remaining ones as its kernel. In particular, this explains how $\widetilde{E}^k_s$ arises from $\widetilde{E}^k_\ell = \mathrm{S}^k_{\mathbb{C}}[\mathfrak{p}^*]$. For $1 \leq s \leq \ell$, the system $(\widetilde{E}^k_1, \widetilde{E}^k_2, \dots, \widetilde{E}^k_s)$ is an example of a costratified quantum space.

The alerted reader is invited to consult [41] for more details.

References

1. R. Abraham and J. E. Marsden [1978] *Foundations of Mechanics*, Benjamin/Cummings Publishing Company.
2. R. Almeida and A. Kumpera [1981] *Structure produit dans la catégorie des algébroïdes de Lie*, An. Acad. Brasil. Cienc. **53**, 247–250.
3. J. M. Arms, R. Cushman, and M. J. Gotay [1991] *A universal reduction procedure for Hamiltonian group actions*, The geometry of Hamiltonian systems, T. Ratiu, ed., MSRI Publ. vol. 20, Springer-Verlag, Berlin · Heidelberg · New York · Tokyo, pp. 33–51.
4. J. M. Arms, J. E. Marsden, and V. Moncrief [1981] *Symmetry and bifurcation of moment mappings*, Comm. Math. Phys. **78**, 455–478.
5. M. F. Atiyah [1957] *Complex analytic connections in fibre bundles*, Trans. Amer. Math. Soc. **85**, 181–207.
6. A. A. Beilinson and V. V. Schechtmann [1988] *Determinant bundles and Virasoro algebras*, Comm. Math. Physics **118**, 651–701.
7. R. Bkouche [1966] *Structures (K, A)-linéaires*, C. R. A. S. Paris Série A **262**, 373–376.
8. W. Bruns and U. Vetter [1988] *Determinantal Rings*, Lecture Notes in Mathematics, vol. 1327, Springer-Verlag, Berlin · Heidelberg · New York.
9. Ana Canas da Silva and Alan Weinstein [1999] *Geometric models for Noncommutative Algebras*, Berkeley Mathematical Lecture Notes, vol. 10, AMS, Boston Ma.
10. S. U. Chase [1976] *Group scheme actions by inner automorphisms*, Comm. Alg. **4**, 403–434.
11. S. Chemla [1999] *A duality property for complex Lie algebroids*, Math. Z. **232**, 367–388.
12. C. Chevalley and S. Eilenberg [1948] *Cohomology theory of Lie groups and Lie algebras*, Trans. Amer. Math. Soc. **63**, 85–124.
13. M. Crainic and R. L. Fernandes *Integrability of Lie brackets*, Ann. of Math. (to appear), `math.DG/0105033`.
14. C. M. de Barros [1964AA] *Espaces infinitésimaux*, Cahiers Topologie Géom. différentielle **7**.
15. P. A. M. Dirac [1964] *Lectures on Quantum Mechanics*, Belfer Graduate School of Science, Yeshiva University, New York.
16. P. A. M. Dirac [1950] *Generalized Hamiltonian systems*, Can. J. of Math. **12**, 129–148.
17. J.-M. Drezet and M.S. Narasimhan [1989] *Groupe de Picard des variétés de modules de fibrés semistables sur les courbes algébriques*, Invent. Math. **97**, 53–94.
18. A. Fahim [1997] *Extensions galoisiennes d'algèbres différentielles*, Pacific J. Math. **180**, 7-40.
19. B. Fuchssteiner [1982] *The Lie algebra structure of degenerate Hamiltonian and bi-hamiltonian systems*, Progr. Theor. Phys. **68**, 1082–1104.
20. M. Goresky and R. MacPherson [1980] *Intersection homology theory*, Topology **19**, 135–162.
21. M. J. Gotay [1986] *Constraints, reduction, and quantization*, J. of Math. Phys. **27**, 2051–2066.
22. V. W. Guillemin and S. Sternberg [1982] *Geometric quantization and multiplicities of group representations*, Invent. Math. **67**, 515–538.
23. R. Hermann [1967] *Analytic continuations of group representations. IV.*, Comm. Math. Phys. **5**, 131–156.
24. J. Herz [1953] *Pseudo-algèbres de Lie*, C. R. Acad. Sci. Paris **236**, 1935–1937.
25. P. J. Higgins and K. Mackenzie [1990] *Algebraic constructions in the category of Lie algebroids*, J. of Algebra **129**, 194–230.
26. V. Hinich and V. Schechtman [1997] *Deformation theory and Lie algebra homology.* I., `alg-geom/9405013`, Alg. Colloquium **4:2**, 213–240; II., 291–316.
27. G. Hochschild [1955] *Simple algebras with purely inseparable splitting field of exponent 1*, Trans. Amer. Math. Soc. **79**, 477–489.
28. G. Hochschild, B. Kostant, and A. Rosenberg [1962] *Differential forms on regular affine algebras*, Trans. Amer. Math. Soc. **102**, 383–408.

29. J. Huebschmann [1990] *Poisson cohomology and quantization*, J. für die reine und angewandte Mathematik **408**, 57–113.
30. J. Huebschmann [1991] *On the quantization of Poisson algebras*, Symplectic Geometry and Mathematical Physics, Actes du colloque en l'honneur de Jean-Marie Souriau, P. Donato, C. Duval, J. Elhadad, G.M. Tuynman, eds.; Progress in Mathematics, vol. 99, Birkhäuser-Verlag, Boston · Basel · Berlin, pp. 204–233.
31. J. Huebschmann [1996] *Poisson geometry of certain moduli spaces*, Lectures delivered at the "14th Winter School", Srni, Czeque Republic, January 1994, Rendiconti del Circolo Matematico di Palermo, Serie II **39**, 15–35.
32. J. Huebschmann [1996] *On the Poisson geometry of certain moduli spaces*, Proceedings of an international workshop on "Lie theory and its applications in physics", Clausthal, 1995, H. D. Doebner, V. K. Dobrev, J. Hilgert, eds., World Scientific, Singapore · New Jersey · London · Hong Kong, pp. 89–101.
33. J. Huebschmann [1998] *Lie-Rinehart algebras, Gerstenhaber algebras, and Batalin- Vilkovisky algebras*, Annales de l'Institut Fourier **48**, 425–440, `math.DG/9704005`.
34. J. Huebschmann [1999] *Extensions of Lie-Rinehart algebras and the Chern-Weil construction*, Festschrift in honour of Jim Stasheff's 60'th anniversary, Cont. Math. **227**, 145–176, `math.DG/9706002`.
35. J. Huebschmann [1999] *Duality for Lie-Rinehart algebras and the modular class*, Journal für die reine und angew. Math. **510**, 103–159, `math.DG/9702008`.
36. J. Huebschmann [2000] *Differential Batalin-Vilkovisky algebras arising from twilled Lie-Rinehart algebras*, Banach center publications **51**, 87–102.
37. J. Huebschmann [2001] *Singularities and Poisson geometry of certain representation spaces*, Quantization of Singular Symplectic Quotients, N. P. Landsman, M. Pflaum, M. Schlichenmaier, eds., Workshop, Oberwolfach, August 1999, Progress in Mathematics, vol. 198, Birkhäuser-Verlag, Boston · Basel · Berlin, pp. 119–135, `math.DG/0012184`.
38. J. Huebschmann *Twilled Lie-Rinehart algebras and differential Batalin-Vilkovisky algebras*, `math.DG/9811069`.
39. J. Huebschmann *Kähler spaces, nilpotent orbits, and singular reduction*, Memoirs AMS (to appear), `math.DG/0104213`.
40. J. Huebschmann *Severi varieties and holomorphic nilpotent orbits*, `math.DG/0206143`.
41. J. Huebschmann *Kähler reduction and quantization*, `math.SG/0207166`.
42. L. Illusie [1972] *Complexe cotangent et déformations*. II., Lecture Notes in Mathematics, vol. 238, Springer-Verlag, Berlin · Heidelberg · New York.
43. N. Jacobson [1944] *An extension of Galois theory to non-normal and non-separable fields*, Amer. J. Math. **66**, 1–29.
44. F. W. Kamber and Ph. Tondeur [1971] *Invariant differential operators and the cohomology of Lie algebra sheaves*, Memoirs of the Amer. Math. Soc. **113**, Amer. Math. Soc., Providence, R. I.
45. D. Kastler and R. Stora [1986] *A differential geometric setting for BRS transformations and anomalies*. I.II., J. Geom. Phys. **3**, 437–482; 483–506.
46. G. Kempf [1978] *Instability in invariant theory*, Ann. of Math. **108**, 299–316.
47. G. Kempf and L. Ness [1978] *The length of vectors in representation spaces*, Algebraic Geometry, Copenhagen, 1978, Lecture Notes in Mathematics, vol. 732, Springer-Verlag, Berlin · Heidelberg · New York, pp. 233–244.
48. A. A. Kirillov [1962] *Unitary representations of nilpotent Lie groups*, Uspehi Mat. Nauk. **17**, 57–101; Russ. Math. Surveys **17** (1962), 57–101.
49. F. Kirwan [1984] *Cohomology of quotients in symplectic and algebraic geometry*, Princeton University Press, Princeton, New Jersey.
50. Y. Kosmann-Schwarzbach and F. Magri [1989] *Poisson-Nijenhuis structures*, Annales Inst. H. Poincaré Série A (Physique théorique) **53**, 35–81.
51. B. Kostant [1970] *Quantization and unitary representations*, Lectures in Modern Analysis and Applications, III, ed. C. T. Taam, Lecture Notes in Math. vol. 170, Springer-Verlag, Berlin · Heidelberg · New York, pp. 87–207.
52. B. Kostant and S. Sternberg [1990] *Anti-Poisson algebras and current algebras*, unpublished manuscript.

53. I. S. Krasil'shchik, V. V. Lychagin, and A. M. Vinogradov [1986] *Geometry of Jet Spaces and Nonlinear Partial Differential Equations*, Advanced Studies in Contemporary Mathematics, vol. 1, Gordon and Breach Science Publishers, New York, London, Paris, Montreux, Tokyo.
54. E. Lerman, R. Montgomery and R. Sjamaar [1993] *Examples of singular reduction*, Symplectic Geometry, Warwick, 1990, D. A. Salamon, editor, London Math. Soc. Lecture Note Series, vol. 192, Cambridge University Press, Cambridge, UK, pp. 127–155.
55. A. Lichnerowicz [1977] *Les variétés de Poisson et leurs algèbres de Lie associées*, J. Diff. Geo. **12**, 253–300.
56. K. Mackenzie [1987] *Lie groupoids and Lie algebroids in differential geometry*, London Math. Soc. Lecture Note Series, vol. 124, Cambridge University Press, Cambridge, England.
57. K. Mackenzie [1995] *Lie algebroids and Lie pseudoalgebras*, Bull. London Math. Soc. **27 (2)**, 97 – 147.
58. J. Marsden and A. Weinstein [1974] *Reduction of symplectic manifolds with symmetries*, Rep. on Math. Phys. **5**, 121–130.
59. W. S. Massey and F. P. Petersen [1965] *The cohomology structure of certain fibre spaces.I*, Topology **4**, 47–65.
60. M. S. Narasimhan and T. R. Ramadas [1993] *Factorization of generalized theta functions*, Inventiones **114**, 565-623.
61. E. Nelson [1967] *Tensor Analysis*, Princeton University Press, Princeton, N. J..
62. R. S. Palais [1961] *The cohomology of Lie rings*, Proc. Symp. Pure Math. vol. III, Amer. Math. Soc., Providence, R. I., pp. 130–137.
63. J. Pradines [1966] *Théorie de Lie pour les groupoïdes différentiables. Relations entre propriétés locales et globales*, C. R. Acad. Sci. Paris Série A **263**, 907–910.
64. J. Pradines [1967] *Théorie de Lie pour les groupoïdes différentiables. Calcul différentiel dans la catégorie des groupoïdes infinitésimaux*, C. R. Acad. Sci. Paris Série A **264**, 245–248.
65. T. R. Ramadas [1996] *Factorization of generalised theta functions* II: *The Verlinde formula*, Topology **35**, 641–654.
66. G. Rinehart [1963] *Differential forms for general commutative algebras*, Trans. Amer. Math. Soc. **108**, 195–222.
67. I. E. Segal [1960] *Quantization of non-linear systems*, J. of Math. Phys. **1**, 468–488.
68. R. Sjamaar [1995] *Holomorphic slices, symplectic reduction, and multiplicities of representations*, Ann. of Math. **141**, 87–129.
69. R. Sjamaar [1996] *Symplectic reduction and Riemann-Roch formulas for multiplicities*, Bull. Amer. Math. Soc. **33**, 327–338.
70. R. Sjamaar and E. Lerman [1991] *Stratified symplectic spaces and reduction*, Ann. of Math. **134**, 375–422.
71. J. Śniatycki [1980] *Geometric quantization and quantum mechanics*, Applied Mathematical Sciences No. 30, Springer-Verlag, Berlin · Heidelberg · New York.
72. J. Śniatycki [1983] *Constraints and quantization*, Nonlinear partial differential operators and quantization procedures, Clausthal 1981, eds. S. I. Anderson and H. D. Doebner, Lecture Notes in Mathematics, vol. 1037, Springer-Verlag, Berlin · Heidelberg · New York, pp. 301–334.
73. J. Śniatycki and A. Weinstein [1983] *Reduction and quantization for singular moment mappings*, Lett. Math. Phys. **7**, 155–161.
74. J. M. Souriau [1966] *Quantification géométrique*, Comm. Math. Physics **1**, 374–398.
75. J. D. Stasheff [1997] *Homological reduction of constrained Poisson algebras*, J. of Diff. Geom. **45**, 221–240.
76. N. Teleman [1972] *A characteristic ring of a Lie algebra extension*, Accad. Naz. Lincei. Rend. Cl. Sci. Fis. Mat. Natur. (8) **52**, 498–506 and 708–711.
77. C. Teleman [2000] *The quantization conjecture revisited*, Ann. of Math. **152**, 1–43.
78. H. Weyl [1946] *The classical groups*, Princeton University Press, Princeton, New Jersey.
79. N. M. J. Woodhouse [1991] *Geometric quantization*, Second edition, Clarendon Press, Oxford.

Received December 20, 2002; in revised form February 28, 2003

Fields Institute Communications
Volume **43**, 2004

A Note on the Semiabelian Variety of Heyting Semilattices

Peter Johnstone
Department of Pure Mathematics
University of Cambridge
Cambridge CB3 0WB, UK
P.T.Johnstone@dpmms.cam.ac.uk

A *Heyting semilattice* is a partially ordered set which is cartesian closed when regarded as a category: equivalently, the variety **HSLat** of Heyting semilattices corresponds to the algebraic theory generated by a constant $\top$ and two binary operations $\wedge, \Rightarrow$ satisfying the equations

$$\top \wedge x = x \ , \quad x \wedge x = x \ , \quad x \wedge y = y \wedge x \ ,$$
$$x \wedge (y \wedge z) = (x \wedge y) \wedge z \ , \quad (x \Rightarrow x) = \top \ ,$$
$$x \wedge (x \Rightarrow y) = x \wedge y \ , \quad y \wedge (x \Rightarrow y) = y \ ,$$
$$\text{and } x \Rightarrow (y \wedge z) = (x \Rightarrow y) \wedge (x \Rightarrow z) \ .$$

In this note we shall also wish to consider the variety **HAlg** of *Heyting algebras*, obtained by adding a further constant $\bot$ and a binary operation $\vee$ satisfying

$$\bot \vee x = x \ , \quad x \vee x = x \ , \quad x \vee y = y \vee x \ ,$$
$$x \vee (y \vee z) = (x \vee y) \vee z \ , \quad x \wedge (x \vee y) = x \ ,$$
$$\text{and } x \vee (x \wedge y) = x \ ,$$

and the variety **BAlg** of Boolean algebras, obtained from **HAlg** by adding the further equation $x \vee (x \Rightarrow \bot) = \top$. (As usual, we shall normally abbreviate $(x \Rightarrow \bot)$ to $\neg x$, and we shall write $(x \Leftrightarrow y)$ for $((x \Rightarrow y) \wedge (y \Rightarrow x))$.)

The category of Heyting algebras is protomodular, by an argument due to the present author (but first published in [1]); the argument also works for Heyting semilattices, and since the latter variety is pointed (equivalently, the theory of Heyting semilattices has a unique constant) it is a semiabelian category. In [2], Bourn and Janelidze showed that a pointed variety is semiabelian iff its theory contains, for some positive integer n, a family of n binary operations α_i $(1 \leq i \leq n)$ and an $(n+1)$-ary operation β satisfying the equations $\alpha_i(x, x) = \top$ for all i (where $\top$ denotes the unique constant of the theory) and

$$\beta(\alpha_i(x, y), \alpha_2(x, y), \ldots, \alpha_n(x, y), y) = x \ .$$

(We shall refer to these equations as condition (BJ), although in fact they were first considered by Ursini [5].)

2000 *Mathematics Subject Classification.* Primary 08B05; Secondary 06D20, 18C05.

In his talk at the Fields Institute workshop, Francis Borceux noted that most familiar examples of semiabelian varieties satisfy condition (BJ) with $n = 1$, and asked whether there was a 'naturally occurring' variety where $n > 1$ was required. In this note we shall show that **HSLat** is such a variety: we shall exhibit a set of Heyting semilattice operations satisfying condition (BJ) for $n = 2$, but show that the condition cannot be satisfied for $n = 1$ even in the variety of Heyting algebras.

For the positive result, we set $\alpha_1(x, y) = (x \Rightarrow y)$, $\alpha_2(x, y) = (((x \Rightarrow y) \Rightarrow y) \Rightarrow x)$ and $\beta(u, v, w) = (u \Rightarrow w) \wedge v$. It is easily seen that we have $\alpha_1(x, x) = \top$ and

$$\alpha_2(x, x) = ((\top \Rightarrow x) \Rightarrow x) = (x \Rightarrow x) = \top \; ;$$

and

$$\beta(\alpha_1(x, y), \alpha_2(x, y), y) = ((x \Rightarrow y) \Rightarrow y) \wedge (((x \Rightarrow y) \Rightarrow y) \Rightarrow x) \leq x \; ,$$

but we also have $x \leq ((x \Rightarrow y) \Rightarrow y)$ since $x \wedge (x \Rightarrow y) \leq y$, and $x \leq (((x \Rightarrow y) \Rightarrow y) \Rightarrow x)$ since $x \wedge ((x \Rightarrow y) \Rightarrow y) \leq x$, so $\beta(\alpha_1(x, y), \alpha_2(x, y), y) = x$, as required.

For the negative result, note first that if we have binary operations α and β in any algebraic theory satisfying condition (BJ) for $n = 1$, then the implication '$x = y$ implies $\alpha(x, y) = \top$' must be reversible; for we must have $\beta(\top, y) = y$ for all y. Hence, if we have such operations in the theory of Heyting algebras, then $\alpha(x, y)$ must coincide in any Boolean algebra with the operation $(x \Leftrightarrow y)$, since bi-implication is the only Boolean binary operation with this property. In particular, the unary operation $\alpha(x, \bot)$ must coincide in any Boolean algebra with $(x \Leftrightarrow \bot) = \neg x$. However, we have an explicit description of the unary operations in the theory of Heyting algebras (equivalently, of the elements of the free Heyting algebra on one generator); they are pictured on p. 35 of [3], for example. From this description, it is easy to see that $\neg x$ is the unique unary operation in Heyting algebras which reduces to $\neg x$ in Boolean algebras; so we must actually have $\alpha(x, \bot) = \neg x$ in all Heyting algebras. But then, for any x, we have

$$x = \beta(\alpha(x, \bot), \bot) = \beta(\alpha(\neg\neg x, \bot), \bot) = \neg\neg x \; ;$$

so we have derived a contradiction.

In conclusion, we remark that if an algebraic theory contains a constant $\top$ and a Mal'cev operation μ which is (left) weakly associative in the sense of [4], then it satisfies condition (BJ) with $n = 1$: we simply set $\alpha(x, y) = \mu(x, y, \top)$ and $\beta(x, y) = \mu(x, \top, y)$. In [4], we showed that the theory of Heyting algebras does not contain a weakly associative Mal'cev operation by exhibiting a topological Heyting algebra whose underlying space is not homogeneous. The foregoing argument provides another proof of this fact.

References

[1] D. Bourn, Mal'cev categories and fibration of pointed objects, *Appl. Categ. Struct.* 4 (1996), 307–327.

[2] D. Bourn and G. Janelidze, Characterization of protomodular varieties of universal algebras, *Theory Appl. Categ.* 11 (2003), 143–147.

[3] P.T. Johnstone, *Stone Spaces*, Cambridge Studies in Advanced Math. no. 3 (Cambridge Univ. Press, 1982).

[4] P.T. Johnstone and M.C. Pedicchio, Remarks on continous Mal'cev algebras, *Rend. Ist. Mat. Univ. Trieste* 25 (1995), 277–297.

[5] A. Ursini, Osservazioni sulla varietà BIT, *Boll. Un. Mat. Ital.* (4) 7 (1973), 205–211.

Received November 15, 2002

Fields Institute Communications
Volume **43**, 2004

Monoidal Functors Generated by Adjunctions, with Applications to Transport of Structure

G.M. Kelly
School of Mathematics and Statistics
University of Sydney
NSW 2006, Australia
maxk@maths.usyd.edu.au

Stephen Lack
School of Mathematics and Statistics
University of Sydney NSW 2006
Australia
and
School of Quantitative Methods and Mathematical Sciences
University of Western Sydney
Locked Bag 1797 DC South Penrith 1797 NSW
Australia
s.lack@uws.edu.au

Abstract. Bénabou pointed out in 1963 that a pair $f \dashv u : A \to B$ of adjoint functors induces a monoidal functor $[f, u] : [A, A] \to [B, B]$ between the (strict) monoidal categories of endofunctors. We show that this result about adjunctions in the monoidal 2-category **Cat** extends to adjunctions in any right-closed monoidal 2-category $\mathcal{V}$, or more generally in any 2-category $\mathcal{A}$ with an action $*$ of a monoidal 2-category $\mathcal{V}$ admitting an adjunction $\mathcal{A}(T * A, B) \cong \mathcal{V}(T, \langle A, B\rangle)$; certainly such an adjunction exists when $*$ is the canonical action of $[\mathcal{A}, \mathcal{A}]$ on $\mathcal{A}$, provided that $\mathcal{A}$ is complete and locally small. This result allows a concise and general treatment of the transport of algebraic structure along an equivalence.

1 Introduction

We suppose given a *monoidal 2-category* $\mathcal{V}$: that is, a 2-category $\mathcal{V}$ along with a monoidal structure $(\otimes, I, a, l, r)$ for which $\otimes$ is a 2-functor and a, l, r are 2-natural. We further suppose given a 2-category $\mathcal{A}$ and an *action* of $\mathcal{V}$ on $\mathcal{A}$: that is, a 2-functor $* : \mathcal{V} \times \mathcal{A} \to \mathcal{A}$ together with 2-natural isomorphisms $\alpha : (X \otimes Y) * A \cong X * (Y * A)$ and $\lambda : I * A \cong A$ satisfying the usual two coherence axioms. Finally

2000 *Mathematics Subject Classification.* Primary 18D05; Secondary 18A40, 18D10.
The support of the Australian Research Council is gratefully acknowledged.

we suppose each 2-functor $- * A : \mathcal{V} \to \mathcal{A}$ to have a right adjoint $\langle\!\langle A, - \rangle\!\rangle$, so that we have a 2-natural isomorphism

$$\Phi : \mathcal{A}(X * A, B) \cong \mathcal{V}(X, \langle\!\langle A, B \rangle\!\rangle). \tag{1.1}$$

A first example is that where $\mathcal{A}$ is $\mathcal{V}$ itself, with $\otimes$ for $*$ and a, l for α, λ; then $\langle\!\langle A, B \rangle\!\rangle$ is an "internal hom", more commonly denoted by $[A, B]$, whose existence makes of $\mathcal{V}$ a *right-closed* monoidal 2-category. A second example is that where $\mathcal{A}$ is any 2-category which is locally small and complete, while $\mathcal{V}$ is the monoidal 2-category $[\mathcal{A}, \mathcal{A}]$ of endofunctors of $\mathcal{A}$ (meaning of course *endo-2-functors*, since $\mathcal{A}$ is a 2-category), with composition for its tensor product and the identity functor $1 = 1_{\mathcal{A}}$ for its unit object. The action $[\mathcal{A}, \mathcal{A}] \times \mathcal{A} \to \mathcal{A}$ we intend here is that given by evaluation, sending (T, A) to TA and similarly defined on morphisms. Now (1.1) takes the form

$$\Phi : \mathcal{A}(XA, B) \cong [\mathcal{A}, \mathcal{A}](X, \langle A, B \rangle), \tag{1.2}$$

where $\langle A, B \rangle$ is the right Kan extension of $B : 1 \to \mathcal{A}$ along $A : 1 \to \mathcal{A}$ given by $\langle A, B \rangle C = B^{\mathcal{A}(C,A)}$.

There is a sense in which the second example is "extremal". For in the context of a general example as in (1.1), we can still apply (1.2) (provided $\mathcal{A}$ is locally small and complete) to get

$$\mathcal{A}(X * A, B) \cong [\mathcal{A}, \mathcal{A}](X * -, \langle A, B \rangle),$$

so that we have a natural isomorphism

$$\mathcal{V}(X, \langle\!\langle A, B \rangle\!\rangle) \cong [\mathcal{A}, \mathcal{A}](X * -, \langle A, B \rangle). \tag{1.3}$$

Moreover, as was discussed in [6], it is common in examples of such actions for the 2-functor $\mathcal{V} \to [\mathcal{A}, \mathcal{A}]$ sending X to $X * -$ to have a right adjoint $\theta : [\mathcal{A}, \mathcal{A}] \to \mathcal{V}$; and when this is so, (1.3) gives a natural isomorphism $\langle\!\langle A, B \rangle\!\rangle \cong \theta \langle A, B \rangle$. In these circumstances, our main results below for the general case (1.1) are consequences of those for the extremal case (1.2).

However, it in fact costs nothing to consider the general case throughout, especially if we use the coherence to simplify the notation as follows. Forget for the moment that $\mathcal{V}$ and $\mathcal{A}$ are 2-categories. To give a monoidal category $\mathcal{V}$ and an action $* : \mathcal{V} \times \mathcal{A} \to \mathcal{A}$ of $\mathcal{V}$ on $\mathcal{A}$ is equally to give a bicategory $\mathbb{B}$ with just two object 0 and 1, having

$$\mathbb{B}(0,0) = \mathcal{V}, \ \mathbb{B}(1,0) = \mathcal{A} \ , \mathbb{B}(1,1) = 1, \ \mathbb{B}(0,1) = 0,$$

where the last 0 denotes the empty category. As shown by Mac Lane and Paré [15] — for a more elegant alternative proof attributed to Gordon and Power see also [7] — we can replace $\mathbb{B}$ by an equivalent bicategory $\mathbb{C}$, with the same objects 0 and 1, in which composition is strictly associative. (Recall that this is indeed an equivalence, and not merely a biequivalence: there are homomorphisms $\mathbb{B} \to \mathbb{C}$ and $\mathbb{C} \to \mathbb{B}$, each of whose composites is *isomorphic* to the identity via (invertible) strong transformations.) When $\mathcal{V}$ and $\mathcal{A}$ are in fact 2-categories as above, the 2-cells of $\mathcal{V}$ and of $\mathcal{A}$, which are 3-cells in $\mathbb{B}$, just go along for the ride in the equivalence. Accordingly, so long as we deal with properties stable under such an equivalence, we may simplify by supposing henceforth that both $\otimes$ and $*$ are strictly associative — which allows us to write XY for $X \otimes Y$ in $\mathcal{V}$ and XA for $X * A$ in $\mathcal{A}$, with 1 for I.

Moreover, because of the importance of the extremal case, we shall henceforth write $\langle A,B\rangle$ rather than $\langle\!\langle A,B\rangle\!\rangle$ in the general case, so that (1.1) becomes

$$\Phi : \mathcal{A}(XA,B) \cong \mathcal{V}(X,\langle A,B\rangle); \tag{1.4}$$

and we shall henceforth use Φ without further explanation to denote this isomorphism.

Of course $\langle -,-\rangle$ admits a unique structure of 2-functor $\mathcal{A}^{\mathrm{op}} \times \mathcal{A} \to \mathcal{V}$ for which Φ is 2-natural in each variable. Let us write

$$e = e_{A,B} : \langle A,B\rangle A \to B \tag{1.5}$$

for the 2-natural counit of the adjunction, and recall that we have a multiplication

$$M = M^B_{A,C} : \langle B,C\rangle\langle A,B\rangle \to \langle A,C\rangle \tag{1.6}$$

determined, using the adjunction, by the commutativity of

$$\begin{array}{ccc} \langle B,C\rangle\langle A,B\rangle A & \xrightarrow{MA} & \langle A,C\rangle A \\ {\scriptstyle \langle B,C\rangle e}\downarrow & & \downarrow{\scriptstyle e} \\ \langle B,C\rangle B & \xrightarrow[e]{} & C, \end{array} \tag{1.7}$$

as well as a "unit map" $J = J_A : 1 \to \langle A,A\rangle$ which is the mate under Φ of $\lambda : 1A \to A$ (here given by the identity), so that we have

$$1A \xrightarrow{JA} \langle A,A\rangle A \xrightarrow{e} A$$

equal to the identity. As is well known — see for example [6] — M and J provide the composition and the unit for a $\mathcal{V}$-category $\mathbb{A}$, whose underlying ordinary category is $\mathcal{A}$ and whose $\mathcal{V}$-valued hom $\mathbb{A}(A,B)$ is $\langle A,B\rangle$.

For each $A \in \mathcal{A}$ we have on $\langle A,A\rangle$ the structure of a monoid $(\langle A,A\rangle, i, m)$, where $m : \langle A,A\rangle\langle A,A\rangle \to \langle A,A\rangle$ is M^A_{AA} and $i : 1 \to \langle A,A\rangle$ is J_A. For a second object B of $\mathcal{A}$, let us write $(\langle B,B\rangle, j, n)$ for the monoid structure; in the extremal case where $\mathcal{V} = [\mathcal{A},\mathcal{A}]$, these monoids are of course *monads on* $\mathcal{A}$ (meaning 2-monads, since $\mathcal{A}$ is a 2-category).

Our central result concerns an adjunction

$$\eta, \varepsilon : f \dashv f^* : A \to B \tag{1.8}$$

in the 2-category $\mathcal{A}$. Write w for $\langle f, f^*\rangle : \langle A,A\rangle \to \langle B,B\rangle$, noting that it is the image under Φ of the composite

$$\langle A,A\rangle B \xrightarrow{\langle A,A\rangle f} \langle A,A\rangle A \xrightarrow{e} A \xrightarrow{f^*} B,$$

which we shall denote more briefly by $t : \langle A,A\rangle B \to B$.

In the very simple case where $\mathcal{A} = \mathcal{V} = \mathbf{Cat}$ with its cartesian monoidal structure, A is a category and $\langle A,A\rangle = [A,A]$ is the strict monoidal category of endofunctors of A. Now an adjunction $f \dashv f^* : A \to B$ in $\mathcal{A}$ is just an adjunction in the original sense of the word; and Bénabou [1] observed that here $w = \langle f, f^*\rangle$ is part of a *monoidal functor* $(w, w^\circ, \widetilde{w})$. Indeed w sends $u : A \to A$ to f^*uf, and we have only to take $w^\circ : 1 \to f^*1f$ to be $\eta : 1 \to f^*f$, and to take $\widetilde{w}_{u,v} : f^*uff^*vf \to f^*uvf$ to be $f^*u\varepsilon vf$. Our central aim is to prove a similar result in the general case, providing for w the structure of a *lax map of monoids in* $\mathcal{V}$. Doing so is equivalent to providing for $t : \langle A,A\rangle B \to B$ the structure of a *lax action on* B

of the monoid $\langle A, A\rangle$; and this observation allows us to enrich the central result as follows. The evaluation $e : \langle A, A\rangle A \to A$ is itself a strict action of $\langle A, A\rangle$ on A, and we show $f^* : A \to B$ to admit the structure of a lax map of lax $\langle A, A\rangle$-algebras, while $f : B \to A$ becomes a colax map of such algebras. Under further hypotheses on the adjunction $f \dashv f^*$, which are certainly satisfied when it is an equivalence (that is, when η and ε are invertible), the whole adjunction enriches to one in the 2-category Ps-$\langle A, A\rangle$-Alg of pseudo $\langle A, A\rangle$-algebras. When A has the structure of a T-algebra, the corresponding map $T \to \langle A, A\rangle$ of monoids provides a 2-functor from Ps-$\langle A, A\rangle$-Alg to Ps-T-Alg carrying the adjunction to one in Ps-T-Alg, which can be seen as a rule for transporting pseudo T-algebra structures along an equivalence. Finally, the 2-functor from Ps-T-Alg to T-Alg, which we have in the case of a *flexible* monoid T, carries the adjunction in Ps-T-Alg to one in T-Alg, giving a rule for transporting (strict) T-algebra structures along an equivalence when T is flexible.

We provide below the detailed statements of these and related results, along with their proofs. First, we recall in the next section the definitions of lax maps of monoids, of lax algebras, and of lax morphisms of lax algebras.

It is a pleasure to thank Ross Street for several helpful comments on the contents of this paper.

2 The definitions

The 2-category Colax$[\mathbf{2}, \mathcal{V}]$ has for objects the arrows $f : X \to Y$ of $\mathcal{V}$, for arrows $f \to f'$ the triples (u, ρ, v) of the form

$$\begin{array}{ccc} X & \xrightarrow{\;u\;} & X' \\ {\scriptstyle f}\downarrow & \overset{\rho}{\Rightarrow} & \downarrow{\scriptstyle f'} \\ Y & \xrightarrow[\;v\;]{} & Y', \end{array}$$

and for 2-cells $(u, \rho, v) \to (\bar{u}, \bar{\rho}, \bar{v})$ the pairs (α, β) where $\alpha : u \to \bar{u}$ and $\beta : v \to \bar{v}$ satisfy the obvious coherence condition [10, p.221]. This 2-category has an evident monoidal structure in which the tensor product of $f : X \to Y$ and $g : W \to Z$ is $fg : XW \to YZ$. For monoids $T = (T, i, m)$ and $S = (S, j, n)$ in $\mathcal{V}$ (recall that these are 2-monads in the case $\mathcal{V} = [\mathcal{A}, \mathcal{A}]$), a *lax map of monoids* or *lax monoid map* $w = (w, w^\circ, \widetilde{w}) : T \to S$ consists of a map $w : T \to S$ in $\mathcal{V}$ along with 2-cells

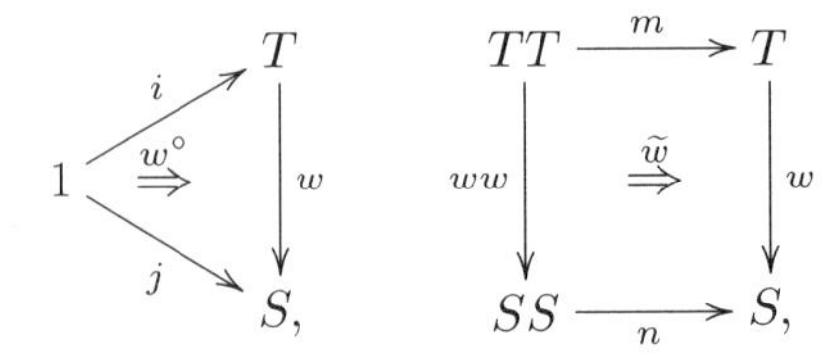

satisfying the three equations [9, (4.2-4.4)] which make of $w : T \to S$ a monoid in Colax$[\mathbf{2}, \mathcal{V}]$. If now $z = (z, z^\circ, \widetilde{z})$ is another lax monoid map, a 2-cell $\theta : w \to z$ is said to be a *monoid 2-cell* if it makes commutative the diagrams

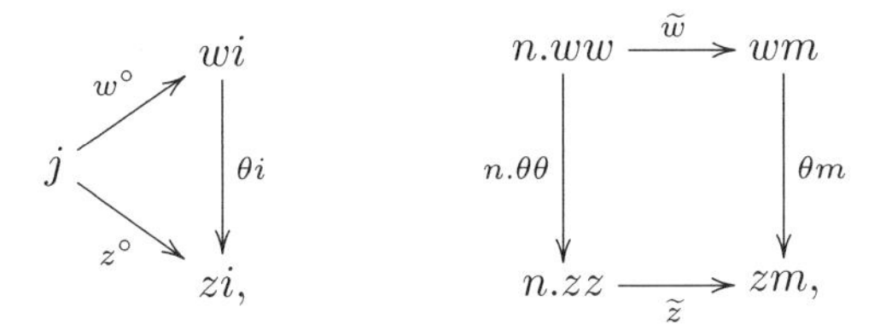

wherein $\theta\theta$ denotes the common value of

$$S\theta.\theta T : Sw.wT \to Sz.zT \quad \text{and} \quad \theta S.T\theta : wS.Tw \to zS.Tz.$$

Thus we have a 2-category $\mathrm{Mon}_l\mathcal{V}$ of monoids in $\mathcal{V}$, lax maps of these, and monoid 2-cells. A *pseudo map of monoids* is a lax one for which w° and $\widetilde{w}$ are invertible; with the same notion of 2-cell, these form a 2-category $\mathrm{Mon}_p\mathcal{V}$. And of course a *strict map of monoids*, or simply a *monoid map*, is just a lax one for which w° and $\widetilde{w}$ are identities; with the same notion of 2-cell once again, these form a 2-category $\mathrm{Mon}\mathcal{V}$, which may also be called $\mathrm{Mon}_s\mathcal{V}$ if we wish to emphasize the strictness of the maps.

When, in the definition above of lax monoid map, (S, j, n) is the monoid $(\langle B, B\rangle, j, n)$ described in Section 1, to give the arrow $w : T \to \langle B, B\rangle$ is equally, by (1.4), to give an arrow $t : TB \to B$ in $\mathcal{A}$; similarly to give the 2-cells w° and $\widetilde{w}$ is equally to give 2-cells

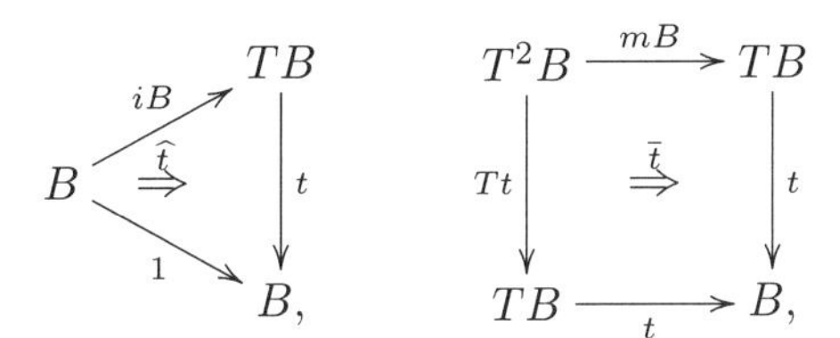

and the three axioms on $(w, w^\circ, \widetilde{w})$ are easily converted to the three axioms [9, (4.6-4.8)] for $(t, \widehat{t}, \overline{t})$ to be a *lax action* of the monoid T on the object B of $\mathcal{A}$. It is a *pseudo action* when $\widehat{t}$ and $\overline{t}$ are invertible, and a *strict action* — or merely an *action* — when $\widehat{t}$ and $\overline{t}$ are identities. Reversing the sense of $\widehat{t}$ and $\overline{t}$ produces the notion of a *colax action* of T on B. When $(t, \widehat{t}, \overline{t})$ is a lax action of T on B, we call the quadruple $(B, t, \widehat{t}, \overline{t})$ a *lax T-algebra*; similarly for the notions of *pseudo T-algebra*, of *strict T-algebra* (or merely *T-algebra*), and of *colax T-algebra*.

If $(B, b, \widehat{b}, \overline{b})$ and $(A, a, \widehat{a}, \overline{a})$ are lax T-algebras, a *lax morphism* (or *lax map*) from $(B, b, \widehat{b}, \overline{b})$ to $(A, a, \widehat{a}, \overline{a})$ is a pair $(f, \overline{f})$ where $f : B \to A$ is a morphism in $\mathcal{A}$ while $\overline{f}$ is a 2-cell $a.Tf \to fb$ satisfying the two axioms [9, (4.10) and (4.11)]. (Note that we have explained lax monoid maps as monoids in a suitable monoidal 2-category, and explained lax actions as lax monoid maps $T \to \langle B, B\rangle$; the corresponding rationale for the definition of lax morphisms of lax algebras will be given a little later.)

The lax morphism is said to be a *pseudo morphism*, or just a *morphism*, of the lax T-algebras when $\overline{f}$ is invertible, and to be a *strict morphism* when $\overline{f}$ is the identity; while reversing the sense of $\overline{f}$ gives the notion of a *colax morphism*. In the case where $\widehat{b}$, $\overline{b}$, $\widehat{a}$, and $\overline{a}$ are identities, we recover the usual notions of lax, pseudo, strict, or colax morphisms of (strict) T-algebras; even for these, following the lead of [4], we use "morphism" without a modifier to mean "pseudo morphism". In the same way, when $\widehat{b}$, $\bar{b}$, $\widehat{a}$, and $\bar{a}$ are invertible, we call $(f, \bar{f})$ a lax morphism of pseudo T-algebras, and so on.

The notion of *algebra 2-cell* $\varphi : f \to g : (B, b, \widehat{b}, \bar{b}) \to (A, a, \widehat{a}, \bar{a})$ is the same for lax algebras, pseudo ones, or strict ones: namely a 2-cell $\varphi : f \to g$ in $\mathcal{A}$ satisfying the single obvious equation. So we have 2-categories and inclusions

$$\text{Lax-}T\text{-Alg}_s \to \text{Lax-}T\text{-Alg}_p = \text{Lax-}T\text{-Alg} \to \text{Lax-}T\text{-Alg}_l$$

of lax T-algebras with, respectively, strict morphisms, pseudo morphisms (often just called morphisms), and lax morphisms; as well as the 2-category Lax-T-Alg_c whose morphisms are the colax ones. Similarly, there are strings of inclusions with Lax-T-Alg_l replaced by Ps-T-Alg_l (when we restrict to the pseudo algebras) or by T-Alg_l (when we restrict to the strict ones).

We promised to give a "rationale" for the definition of lax morphism of lax T-algebras; in fact our needs below in proving the central result make it more convenient to work with colax morphisms. We therefore define a 2-category $\text{Lax}[\mathbf{2}, \mathcal{A}]$, analogous to $\text{Colax}[\mathbf{2}, \mathcal{A}]$, in which an object is once again a morphism $f : B \to A$ in $\mathcal{A}$, while an arrow $(b, r, a) : f \to g$ consists of morphisms b and a together with a 2-cell r as in

$$\begin{array}{ccc} B & \xrightarrow{\;b\;} & D \\ {\scriptstyle f}\downarrow & \Downarrow r & \downarrow{\scriptstyle g} \\ A & \xrightarrow[\;a\;]{} & C, \end{array}$$

and we have the obvious definition of 2-cell. There is an evident action of the monoidal 2-category $\mathcal{V}$ on $\text{Lax}[\mathbf{2}, \mathcal{A}]$, sending $(T, f : B \to A)$ to $Tf : TB \to TA$, and defined in the obvious way on morphisms and 2-cells. To give a map $(b, r, a) : Tf \to g$ in $\text{Lax}[\mathbf{2}, \mathcal{A}]$ is equally to give β, ρ, and α as in

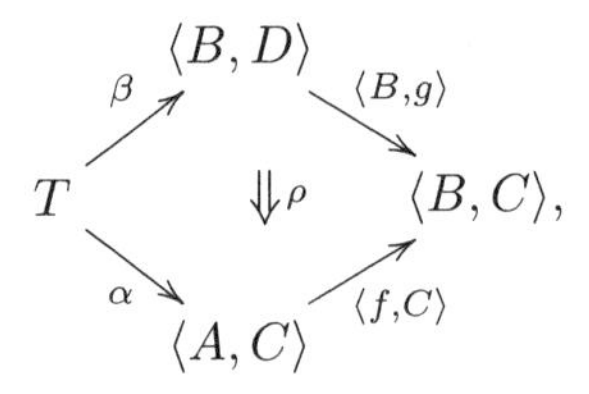

where $\beta = \Phi b$, $\alpha = \Phi a$, and $\rho = \Phi r$; so that b, r, and a are recovered, using the evaluation e, as $b = e_{B,D}.\beta B$, $r = e_{B,C}.\rho B$, and $a = e_{A,C}.\alpha A$. If we now write

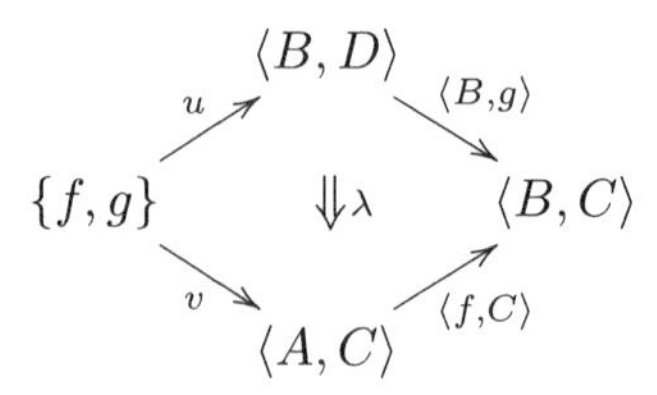

for the comma object, to give (β, ρ, α) is equally to give a map $\gamma : T \to \{f, g\}$ in $\mathcal{V}$: namely the unique map for which $u\gamma = \beta$, $\lambda\gamma = \rho$, and $v\gamma = \alpha$. These bijections sending γ to $(u, \lambda, v)\gamma = (\beta, \rho, \alpha)$ and sending (β, ρ, α) to $(e.\beta B, e.\rho B, e.\alpha A) = (b, r, a)$ clearly extend to 2-cells and become isomorphisms of categories, their composite being a natural isomorphism

$$\mathcal{V}(T, \{f, g\}) \cong \text{Lax}[\mathbf{2}, \mathcal{A}](Tf, g) \tag{2.1}$$

exhibiting $\{f, -\}$ as the right adjoint of the 2-functor $\mathcal{V} \to \mathrm{Lax}[\mathbf{2}, \mathcal{A}]$ sending T to Tf. The counit $E_{f,g} : \{f, g\}f \to g$ of the adjunction, which we may again call the *evaluation*, has the form

$$\begin{array}{ccc} \{f,g\}B & \xrightarrow{E^0_{f,g}} & D \\ {\scriptstyle \{f,g\}f}\downarrow & \Downarrow E^{\to}_{f,g} & \downarrow{\scriptstyle g} \\ \{f,g\}A & \xrightarrow[E^1_{f,g}]{} & C, \end{array}$$

and is obtained by setting $T = \{f, g\}$ and $\gamma = 1$, so that $E^0_{f,g} = e.uB$, $E^{\to}_{f,g} = e.\lambda B$, and $E^1_{f,g} = e.vA$. When $g = f$, the comma object $\{f, g\}$ becomes

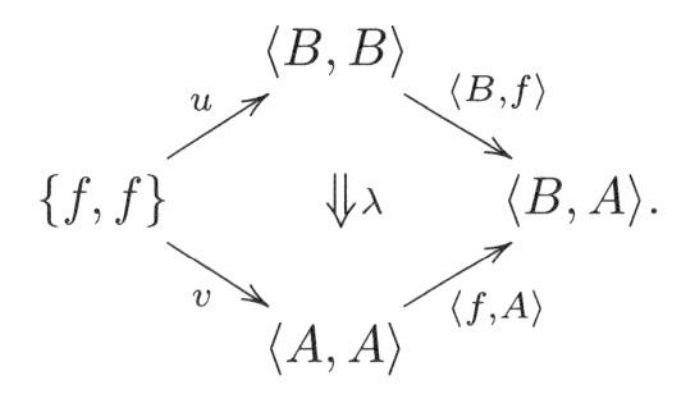

By the general results on actions in Section 1, for any $f : B \to A$ the object $\{f, f\}$ admits a canonical structure $(\{f, f\}, k, l)$ of monoid in $\mathcal{V}$, where $l : \{f, f\}\{f, f\} \to \{f, f\}$ is determined by the commutativity in $\mathrm{Lax}[\mathbf{2}, \mathcal{A}]$ of

$$\begin{array}{ccc} \{f,f\}\{f,f\}f & \xrightarrow{lf} & \{f,f\}f \\ {\scriptstyle \{f,f\}E_{f,f}}\downarrow & & \downarrow{\scriptstyle E_{f,f}} \\ \{f,f\}f & \xrightarrow[E_{f,f}]{} & f, \end{array} \tag{2.2}$$

and similarly $k : 1 \to \{f, f\}$ is determined by the equation $E_{f,f}.kf = 1_f$ in $\mathrm{Lax}[\mathbf{2}, \mathcal{A}]$. Moreover, for a monoid $T = (T, i, m)$ in $\mathcal{V}$, to give a lax monoid map $\gamma : T \to \{f, f\}$ is equivalently, by the earlier part of this section, to give (as its image under (2.1)) a lax action $(c, \widehat{c}, \overline{c}) : Tf \to f$. Here c consists of maps $b : TB \to B$ and $a : TA \to A$ in $\mathcal{A}$ and a 2-cell $\overline{f} : fb \to a.Tf$, while $\widehat{c}$ is a pair $(\widehat{b}, \widehat{a})$ of 2-cells $\widehat{b} : 1 \to b.iB$ and $\widehat{a} : 1 \to a.iA$, and $\overline{c}$ is a pair $(\overline{b}, \overline{a})$ of 2-cells $\overline{b} : b.Tb \to b.mB$ and $\overline{a} : a.Ta \to a.mA$; all these data satisfying equations which assert precisely that $(b, \widehat{b}, \overline{b})$ and $(a, \widehat{a}, \overline{a})$ are lax actions of T on B and A and that $(f, \overline{f})$ is a colax morphism $(B, b, \widehat{b}, \overline{b}) \to (A, a, \widehat{a}, \overline{a})$ of lax T-algebras.

In particular, a strict monoid map $\gamma : T \to \{f, f\}$ corresponds to strict actions $b : TB \to B$ and $a : TA \to A$, along with an $\overline{f}$ making $(f, \overline{f}) : (B, b) \to (A, a)$ a colax morphism of T-algebras. The strict monoid maps $\beta : T \to \langle B, B\rangle$ and $\alpha : T \to \langle A, A\rangle$ corresponding to the strict actions b and a are the composites of γ with $u : \{f, f\} \to \langle B, B\rangle$ and $v : \{f, f\} \to \langle A, A\rangle$; since γ here may be the identity map of $\{f, f\}$, we conclude that *u and v are themselves strict monoid maps.*

We can be more explicit about the value of $k : 1 \to \{f, f\}$: it corresponds of course under (2.1) to the identity $1f \to f$, and hence is the unique k for which uk and vk are the units $j : 1 \to \langle B, B\rangle$ and $i : 1 \to \langle A, A\rangle$ of the monoids $\langle B, B\rangle$ and $\langle A, A\rangle$ while $\lambda k = id$.

The explicit description of the multiplication $l : \{f, f\}\{f, f\} \to \{f, f\}$ is slightly more complicated. First we unravel (2.2) to obtain

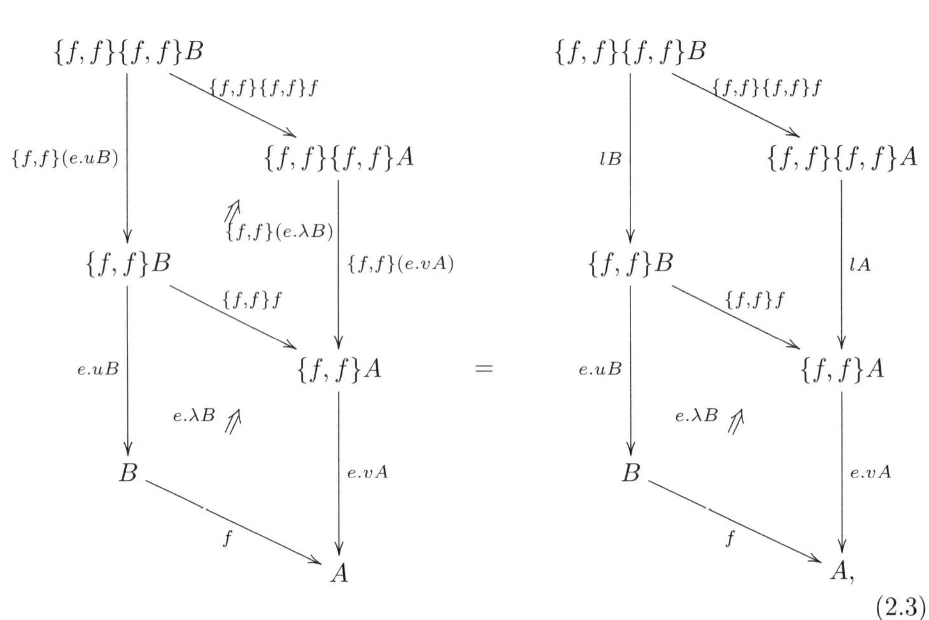

(2.3)

and then apply the isomorphism Φ to this equality. The resulting equality, at the level of 1-cells, asserts that ul and vl are the composites

$$\{f, f\}\{f, f\} \xrightarrow{uu} \langle B, B\rangle\langle B, B\rangle \xrightarrow{n} \langle B, B\rangle,$$

$$\{f, f\}\{f, f\} \xrightarrow{vv} \langle A, A\rangle\langle A, A\rangle \xrightarrow{m} \langle A, A\rangle,$$

(repeating our observation above that u and v are strict monoid maps); at the level of 2-cells, it reduces, as we indicate below, to the assertion that λl is given by

$$\begin{array}{ccccccc}
 & & \langle B,B\rangle\langle B,B\rangle & \xrightarrow{n} & & & \langle B,B\rangle \\
 & \nearrow^{u\langle B,B\rangle} & & \searrow^{\langle B,f\rangle\langle B,B\rangle} & & & \downarrow{\scriptstyle\langle B,f\rangle} \\
 & \{f,f\}\langle B,B\rangle & \Downarrow\lambda\langle B,B\rangle & & \langle B,A\rangle\langle B,B\rangle & \xrightarrow{M} & \\
\nearrow^{\{f,f\}u} & \searrow^{v\langle B,B\rangle} & & \nearrow^{\langle f,A\rangle\langle B,B\rangle} & & & \\
\{f,f\}\{f,f\} & & \langle A,A\rangle\langle B,B\rangle & & & & \langle B,A\rangle \\
\searrow^{v\{f,f\}} & \nearrow^{\langle A,A\rangle u} & & \searrow^{\langle A,A\rangle\langle B,f\rangle} & & & \\
 & \langle A,A\rangle\{f,f\} & \Downarrow\langle A,A\rangle\lambda & & \langle A,A\rangle\langle B,A\rangle & \xrightarrow{M} & \uparrow{\scriptstyle\langle f,A\rangle} \\
 & \searrow^{\langle A,A\rangle v} & & \nearrow^{\langle A,A\rangle\langle f,A\rangle} & & & \\
 & & \langle A,A\rangle\langle A,A\rangle & \xrightarrow[m]{} & & & \langle A,A\rangle.
\end{array}$$

(2.4)

Since the image under Φ of the right side of (2.3) is the composite λl, we must exhibit (2.4) as the image under Φ of the left side of (2.3); and this left side is the "vertical" composite of $e.\lambda B.\{f, f\}e.\{f, f\}uB$ with $e.vA.\{f, f\}e.\{f, f\}\lambda B$. Because the action $* : \mathcal{V} \times \mathcal{A} \to \mathcal{A}$, denoted by juxtaposition, is a 2-functor, we have $\lambda B.\{f,f\}e = \langle B,A\rangle e.\lambda\langle B,B\rangle B$, so that

$$\begin{aligned} e.\lambda B.\{f,f\}e.\{f,f\}uB &= e.\langle B,A\rangle e.\lambda\langle B,B\rangle B.\{f,f\}uB \\ &= e.MB.\lambda\langle B,B\rangle B.\{f,f\}uB, \end{aligned}$$

where the second step uses (1.7) to replace $e.\langle B,A\rangle e$ by $e.MB$; and the image $M.\lambda\langle B,B\rangle.\{f,f\}u$ of $e.MB.\lambda\langle B,B\rangle B.\{f,f\}uB$ under Φ is the top half of (2.4). Similar arguments justify the steps in

$$\begin{aligned} e.vA.\{f,f\}e.\{f,f\}\lambda B &= e.\langle A,A\rangle e.v\langle B,A\rangle B.\{f,f\}\lambda B \\ &= e.MB.\langle A,A\rangle\lambda B.v\{f,f\}B; \end{aligned}$$

and the image $M.\langle A,A\rangle\lambda.v\{f,f\}$ of this last under Φ is the bottom half of (2.4).

3 The central result

We begin with the following, due to Street [16, Proposition 5]:

Lemma 3.1 *Let*

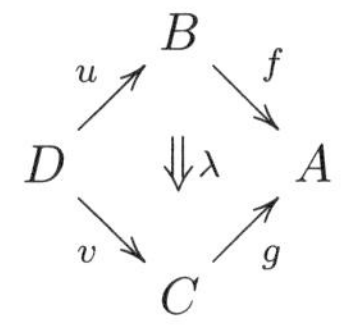

be a comma object in the 2-category $\mathcal{A}$. If f has a right adjoint given by $\eta,\varepsilon : f \dashv f^* : A \to B$, then v has a right adjoint given by $\zeta, id : v \dashv v^* : C \to D$, where v^* is the unique map satisfying $uv^* = f^*g$, $\lambda v^* = \varepsilon g$, and $vv^* = 1$; while $\zeta : 1 \to v^*v$ is the unique 2-cell for which $v\zeta : v \to vv^*v$ is the identity on $v(= vv^*v)$ and $u\zeta : u \to uv^*v$ is the composite

$$u \xrightarrow{\eta u} f^*fu \xrightarrow{f^*\lambda} f^*gv = uv^*v.$$

Remark 3.2 There is of course a similar result where we replace the comma object by an iso-comma object, and the adjunction $f \dashv f^*$ by an equivalence; but we shall not need to refer to this below.

Returning now to the general situation of a monoidal 2-category $\mathcal{V}$ acting on a 2-category $\mathcal{A}$ with a right adjoint expressed by the 2-natural isomorphism Φ of (1.4), consider an arbitrary adjunction $\eta,\varepsilon : f \dashv f^* : A \to B$ in $\mathcal{A}$, and note that the 2-functor $\langle B,-\rangle : \mathcal{A} \to \mathcal{V}$ takes this adjunction into an adjunction

$$\langle B,\eta\rangle, \langle B,\varepsilon\rangle : \langle B,f\rangle \dashv \langle B,f^*\rangle : \langle B,A\rangle \to \langle B,B\rangle$$

in $\mathcal{V}$. Supposing henceforth $\mathcal{V}$ to admit comma objects, we can apply Lemma 3.1 to this adjunction and to the comma object

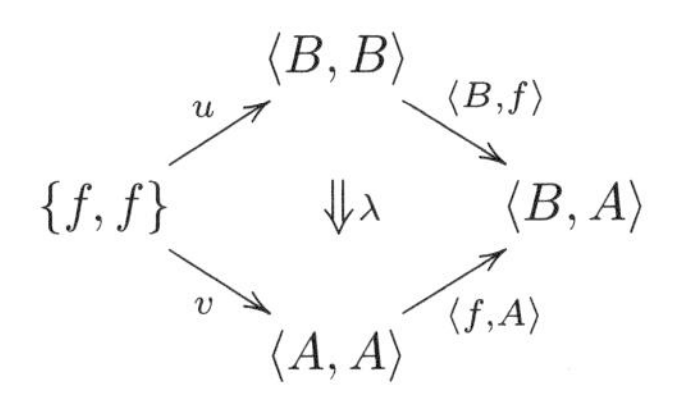

to get:

Proposition 3.3 *In the presence of the adjunction $\eta, \varepsilon : f \dashv f^* : A \to B$, the map $v : \{f,f\} \to \langle A, A\rangle$ has a right adjoint given by $\zeta, id : v \dashv z : \langle A, A\rangle \to \{f,f\}$, where z is the unique map satisfying $uz = \langle f, f^*\rangle$, $\lambda z = \langle f, \varepsilon\rangle : \langle f, ff^*\rangle \to \langle f, A\rangle$, and $vz = 1$; while $\zeta : 1 \to zv$ is the unique 2-cell for which $v\zeta = id$ and $u\zeta$ is the composite*

$$u \xrightarrow{\langle B,\eta\rangle u} \langle B, f^*f\rangle u \xrightarrow{\langle B,f^*\rangle\lambda} \langle f, f^*\rangle v\ .$$

The result of the following lemma is very like that of [8, Theorem 1.2], of which it is not, however, a consequence; the situation is rather that the proof-techniques of that paper adapt so readily to the present lemma that we can safely leave the details to the reader.

Lemma 3.4 *Let $\alpha, \beta : \rho \dashv \sigma : S \to T$ be an adjunction in the monoidal 2-category $\mathcal{V}$, where T and S have monoid structures (T, i, m) and (S, j, n). Then there is a bijection between enrichments of ρ to a colax monoid map $(\rho, \rho', \rho^\#)$ and enrichments of σ to a lax monoid map $(\sigma, \sigma^\circ, \widetilde{\sigma})$, where σ° and $\widetilde{\sigma}$ are given respectively by the pasting composites*

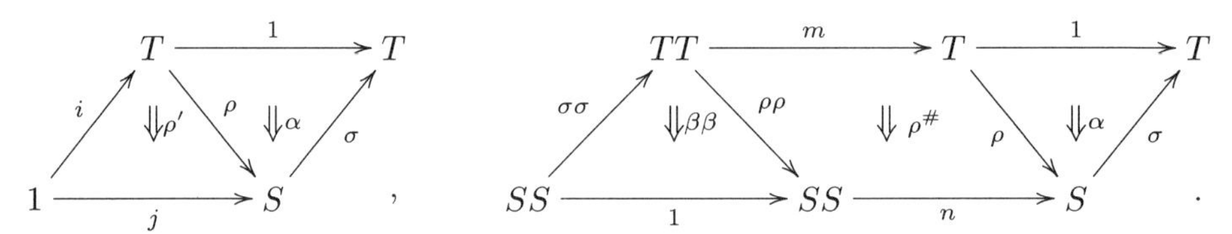

We now apply the lemma to the adjunction $\zeta, id : v \dashv z : \langle A, A\rangle \to \{f,f\}$. We saw in Section 2 that v is a strict monoid map, which we can see as a colax monoid map (v, id, id); it follows from the lemma that z admits an enrichment to a lax monoid map $(z, z^\circ, \widetilde{z})$ where z° is given by

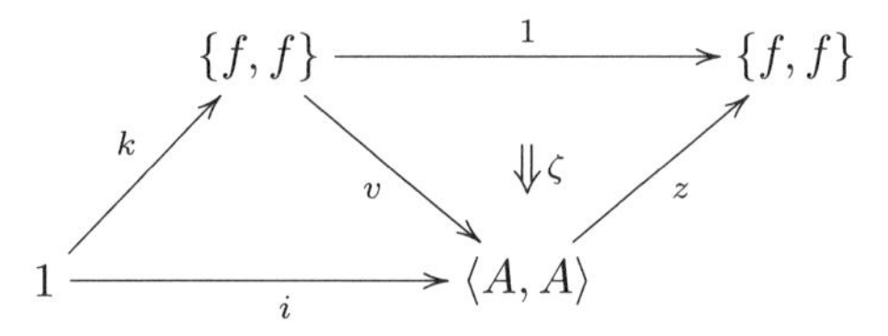

or simply ζk, while $\widetilde{z}$ is given by

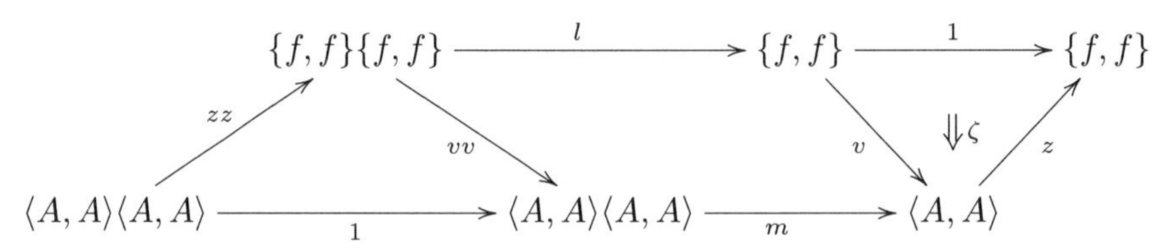

or simply $\zeta l.zz$. Note that, since $v\zeta = id$ by Proposition 3.3, we have $vz^\circ = id$ and $v\widetilde{z} = id$; thus the composite $(vz, vz^\circ, v\widetilde{z})$ of the strict monoid map v and the lax monoid map $z = (z, z^\circ, \widetilde{z})$ is the identity monoid map $1 = (1, id, id) : \langle A, A\rangle \to \langle A, A\rangle$.

Let us set $w = \langle f, f^*\rangle : \langle A, A\rangle \to \langle B, B\rangle$, as indicated in Section 1. Then $w = uz$ by the definition of z; and since u is a strict monoid map while $z = (z, z^\circ, \widetilde{z})$ is a lax monoid map, we have a lax monoid map

$$(w, w^\circ, \widetilde{w}) = (uz, uz^\circ, u\widetilde{z}) : \langle A, A\rangle \to \langle B, B\rangle.$$

To complete our central result, therefore, it remains to describe more explicitly w° and $\widetilde{w}$; or equivalently to describe the $\widehat{t}$ and the $\bar{t}$ which enrich the $t : \langle A, A\rangle B \to B$ of Section 1, given by

$$\langle A,A\rangle B \xrightarrow{\langle A,A\rangle f} \langle A,A\rangle A \xrightarrow{e} A \xrightarrow{f^*} B,$$

to the lax action $(t, \widehat{t}, \bar{t})$ of $\langle A, A\rangle$ on B corresponding to the lax monoid map $(w, w^\circ, \widetilde{w})$.

Now $w^\circ = uz^\circ = u\zeta k$, which, by Proposition 3.3 and the observation in Section 2 that $\lambda k = id$, is just the 2-cell $\langle B, \eta\rangle uk$ in

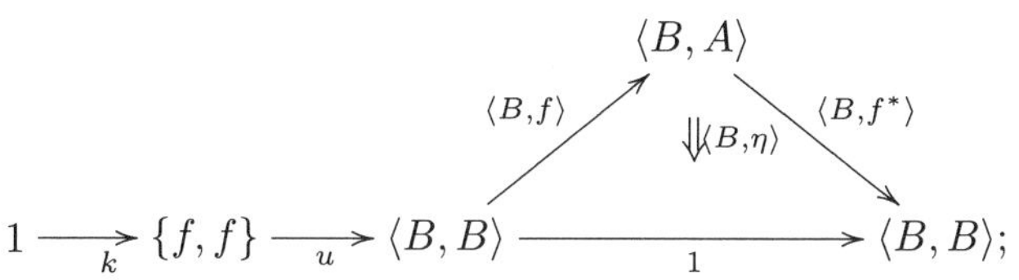

$$\begin{array}{ccccccc} & & & & \langle B,A\rangle & & \\ & & & \overset{\langle B,f\rangle}{\nearrow} & \Downarrow\langle B,\eta\rangle & \overset{\langle B,f^*\rangle}{\searrow} & \\ 1 & \xrightarrow[k]{} & \{f,f\} & \xrightarrow[u]{} \langle B,B\rangle & \xrightarrow[1]{\qquad\qquad} & & \langle B,B\rangle; \end{array}$$

and since $uk = j$, this is just $\langle B, \eta\rangle j$. Moreover, applying Φ^{-1} to each side of the equality $w^\circ = \langle B, \eta\rangle j$ shows that $\widehat{t} : 1 \to t.iB$ is given by $\eta : 1 \to f^*f$; observe here, using the naturality of i, that $f^*f = f^*e.iA.f = f^*e.\langle A, A\rangle f.iB = t.iB$.

It remains to describe the 2-cell

$$\begin{array}{ccc} \langle A,A\rangle\langle A,A\rangle & \xrightarrow{m} & \langle A,A\rangle \\ {\scriptstyle ww}\downarrow & \overset{\widetilde{w}}{\Rightarrow} & \downarrow{\scriptstyle w} \\ \langle B,B\rangle\langle B,B\rangle & \xrightarrow[n]{} & \langle B,B\rangle, \end{array}$$

or equivalently the component

$$\begin{array}{ccc} \langle A,A\rangle\langle A,A\rangle B & \xrightarrow{mB} & \langle A,A\rangle B \\ {\scriptstyle \langle A,A\rangle t}\downarrow & \overset{\bar{t}}{\Rightarrow} & \downarrow{\scriptstyle t} \\ \langle A,A\rangle B & \xrightarrow[t]{} & B \end{array}$$

of the lax action $(t, \widehat{t}, \bar{t})$ of $\langle A, A\rangle$ on B. Now $\widetilde{w} = u\widetilde{z} = u\zeta l.zz$, so from the definition of $u\zeta$ in Proposition 3.3 it follows that $\widetilde{w}$ is given by the pasting composite

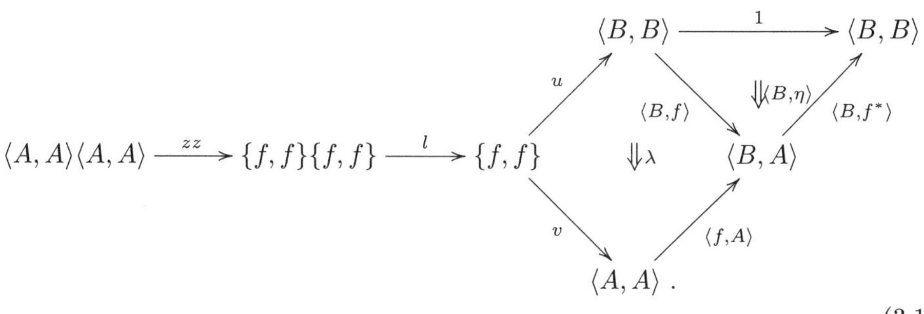

$$\begin{array}{ccccccccc} & & & & & \langle B,B\rangle & \xrightarrow{1} & & \langle B,B\rangle \\ & & & & \overset{u}{\nearrow} & & \overset{\langle B,f\rangle}{\searrow} & \Downarrow\langle B,\eta\rangle & \overset{\langle B,f^*\rangle}{\nearrow} \\ \langle A,A\rangle\langle A,A\rangle & \xrightarrow{zz} & \{f,f\}\{f,f\} & \xrightarrow{l} & \{f,f\} & & \Downarrow\lambda & \langle B,A\rangle & \\ & & & & \underset{v}{\searrow} & & \underset{\langle f,A\rangle}{\nearrow} & & \\ & & & & & \langle A,A\rangle\ . & & & \end{array} \qquad (3.1)$$

Using the description (2.4) of λl and the equations $vz = 1$ and $uz = \langle f, f^* \rangle$, we see that $\lambda l.zz$ may be written as

$$
\begin{array}{ccccccc}
 & & & \langle B,B\rangle\langle B,B\rangle & \xrightarrow{\quad n \quad} & & \langle B,B\rangle \\
 & & {\scriptstyle u\langle B,B\rangle}\nearrow & & \searrow{\scriptstyle \langle B,f\rangle\langle B,B\rangle} & & \downarrow{\scriptstyle \langle B,f\rangle} \\
 & \{f,f\}\langle B,B\rangle & & \Downarrow{\scriptstyle \lambda\langle B,B\rangle} & \langle B,A\rangle\langle B,B\rangle & & \\
{\scriptstyle z\langle f,f^*\rangle}\nearrow & & \searrow{\scriptstyle v\langle B,B\rangle} & & \nearrow{\scriptstyle \langle f,A\rangle\langle B,B\rangle} & \searrow{\scriptstyle M} & \\
\langle A,A\rangle\langle A,A\rangle & & & \langle A,A\rangle\langle B,B\rangle & & & \langle B,A\rangle \\
{\scriptstyle \langle A,A\rangle z}\searrow & & \nearrow{\scriptstyle \langle A,A\rangle u} & & \searrow{\scriptstyle \langle A,A\rangle\langle B,f\rangle} & \nearrow{\scriptstyle M} & \\
 & \langle A,A\rangle\{f,f\} & & \Downarrow{\scriptstyle \langle A,A\rangle\lambda} & \langle A,A\rangle\langle B,A\rangle & & \uparrow{\scriptstyle \langle f,A\rangle} \\
 & & \searrow{\scriptstyle \langle A,A\rangle v} & & \nearrow{\scriptstyle \langle A,A\rangle\langle f,A\rangle} & & \\
 & & & \langle A,A\rangle\langle A,A\rangle & \xrightarrow[\quad m \quad]{} & & \langle A,A\rangle.
\end{array}
\tag{3.2}
$$

Using the equation $\lambda z = \langle f, \varepsilon \rangle$ to simplify (3.2), and substituting the result into (3.1), we conclude that $\widetilde{w}$ is the composite:

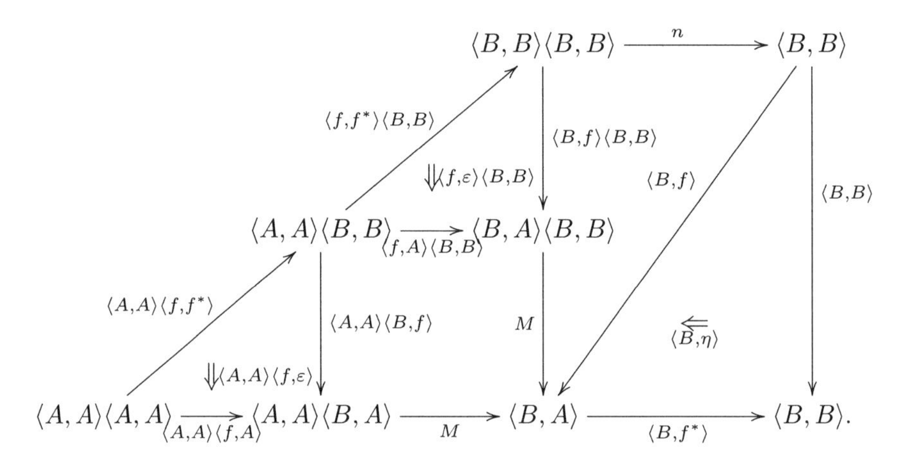

By the "extraordinary" naturality of the M's and one of the triangular equations, this reduces to

$$
\begin{array}{ccccc}
 & \langle A,A\rangle\langle B,B\rangle & & & \\
{\scriptstyle \langle A,A\rangle\langle f,f^*\rangle}\nearrow & \Downarrow{\scriptstyle \langle A,A\rangle\langle f,\varepsilon\rangle} & \searrow{\scriptstyle \langle A,A\rangle\langle B,f\rangle} & & \\
\langle A,A\rangle\langle A,A\rangle & \xrightarrow[\langle A,A\rangle\langle f,A\rangle]{\qquad\qquad} & \langle A,A\rangle\langle B,A\rangle & \xrightarrow[M]{} \langle B,A\rangle & \xrightarrow[\langle B,f^*\rangle]{} \langle B,B\rangle,
\end{array}
$$

and so, using extraordinary naturality once again, to

$$
\begin{array}{ccccc}
 & & & \langle B,B\rangle & \\
 & & {\scriptstyle \langle B,f^*\rangle}\nearrow & \Downarrow{\scriptstyle \langle B,\varepsilon\rangle} & \searrow{\scriptstyle \langle B,f\rangle} \\
\langle A,A\rangle\langle A,A\rangle \xrightarrow[m]{} \langle A,A\rangle & \xrightarrow[\langle f,A\rangle]{} & \langle B,A\rangle & \xrightarrow[\quad 1 \quad]{} & \langle B,A\rangle \xrightarrow[\langle B,f^*\rangle]{} \langle B,B\rangle.
\end{array}
\tag{3.3}
$$

Finally $\bar{t}$ is obtained from $\widetilde{w}$ by applying $(\ )B$ and composing with the evaluation $e : \langle B, B\rangle B \to B$; by the ordinary and the extraordinary 2-naturality of e this gives

$$\langle A,A\rangle\langle A,A\rangle B \xrightarrow[mB]{} \langle A,A\rangle B \xrightarrow[\langle A,A\rangle f]{} \langle A,A\rangle A \xrightarrow[e]{} A \xrightarrow[1]{} A \xrightarrow[f^*]{} B,$$

with the 2-cell $\Downarrow \varepsilon$ in the triangle $A \xrightarrow{f^*} B \xrightarrow{f} A$ over $1 : A \to A$,

which is perhaps more readily seen as a 2-cell $t.\langle A, A\rangle t \to t.mB$ by using the 2-naturality of e once more, to display it in the form

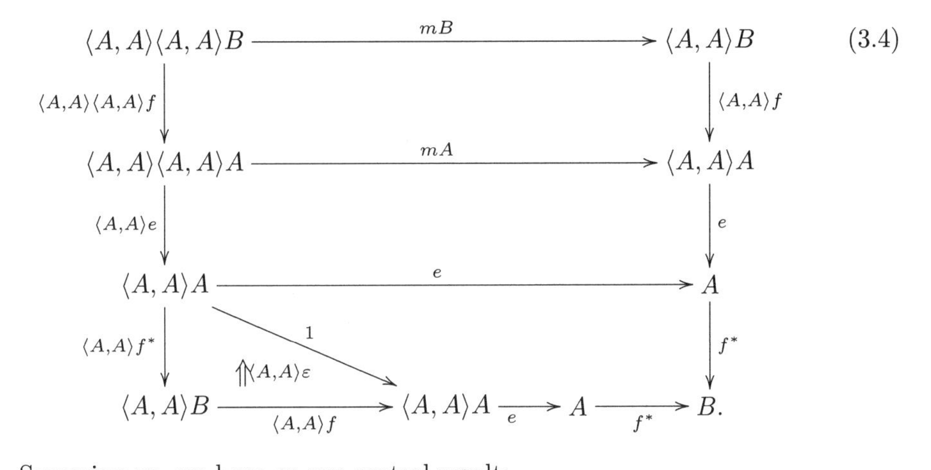

Summing up, we have as our central result:

Theorem 3.5 *Let the monoidal 2-category $\mathcal{V}$ admit comma objects, and let it so act on the 2-category $\mathcal{A}$ that we have the adjunction $\Phi : \mathcal{A}(XA, B) \cong \mathcal{V}(X, \langle A, B\rangle)$. Then each adjunction $\eta, \varepsilon : f \dashv f^* : A \to B$ in $\mathcal{A}$ gives rise to a lax map of monoids $(w, w^\circ, \widetilde{w}) : \langle A, A\rangle \to \langle B, B\rangle$ in $\mathcal{V}$, where $w = \langle f, f^*\rangle$ and w° is given by $\langle B, \eta\rangle j$, while $\widetilde{w}$ is given by (3.3). In fact to give a lax map $(w, w^\circ, \widetilde{w}) : \langle A, A\rangle \to \langle B, B\rangle$ of monoids is equally to give a lax action $(t, \widehat{t}, \bar{t})$ of the monoid $\langle A, A\rangle$ on B; and here $t : \langle A, A\rangle B \to B$ is the composite*

$$\langle A, A\rangle B \xrightarrow{\langle A,A\rangle f} \langle A, A\rangle A \xrightarrow{e} A \xrightarrow{f^*} B,$$

while $\widehat{t} = \Phi^{-1}(w^\circ)$ is given by η and $\bar{t} = \Phi^{-1}(\widetilde{w})$ is given by (3.4). When η and ε are invertible, so that the adjunction $f \dashv f^*$ is an equivalence, the 2-cells w°, $\widetilde{w}$, $\widehat{t}$, and $\bar{t}$ are invertible, so that $(w, w^\circ, \widetilde{w})$ is a pseudo map of monoids, while $(t, \widehat{t}, \bar{t})$ is a pseudo action of $\langle A, A\rangle$ on B.

4 The enrichments of f and f^*

We continue to suppose satisfied the hypotheses of Theorem 3.5. As we saw in Section 3, the lax monoid map $(z, z^\circ, \widetilde{z}) : \langle A, A\rangle \to \{f, f\}$ satisfies $u(z, z^\circ, \widetilde{z}) = (w, w^\circ, \widetilde{w})$ and $v(z, z^\circ, \widetilde{z}) = 1$ in $\mathrm{Mon}_l\,\mathcal{V}$. Since the lax monoid map $(w, w^\circ, \widetilde{w}) : \langle A, A\rangle \to \langle B, B\rangle$ corresponds to the lax action $(t, \widehat{t}, \bar{t}) : \langle A, A\rangle B \to B$ and the strict monoid map $1 : \langle A, A\rangle \to \langle A, A\rangle$ corresponds to the strict action $e : \langle A, A\rangle A \to A$, it follows from our observations in Section 2 that we have a colax map $(f, \bar{\bar{f}}) : (B, t, \hat{t}, \bar{t}) \to (A, e)$ of lax $\langle A, A\rangle$-algebras, where the diagram

$$\begin{array}{ccc} \langle A,A\rangle B & \xrightarrow{\ t\ } & B \\ {\scriptstyle \langle A,A\rangle f}\big\downarrow & \Downarrow \bar{\bar{f}} & \big\downarrow{\scriptstyle f} \\ \langle A,A\rangle A & \xrightarrow[\ e\ ]{} & A \end{array}$$

is the image under Φ^{-1} of λz. Since $\lambda z = \langle B, \varepsilon\rangle\langle f, A\rangle$ by Proposition 3.3, an easy calculation exhibits $\bar{\bar{f}}$ as the 2-cell

$$\begin{array}{ccccccc} \langle A,A\rangle B & \xrightarrow{\langle A,A\rangle f} & \langle A,A\rangle A & \xrightarrow{\ e\ } & A & \xrightarrow{\ f^*\ } & B \\ {\scriptstyle \langle A,A\rangle f}\big\downarrow & & & & & {\scriptstyle 1}\searrow \ \overset{\Leftarrow}{\scriptstyle \varepsilon} & \big\downarrow{\scriptstyle f} \\ \langle A,A\rangle A & & & \xrightarrow[\ e\ ]{} & & & A. \end{array}$$

It now follows from [8, Theorem 1.2] that we have a lax map $(f^*, \overline{f^*}) : (A, e) \to (B, t, \hat{t}, \bar{t})$ of lax $\langle A, A\rangle$-algebras, where the 2-cell $\overline{f^*}$ is the composite

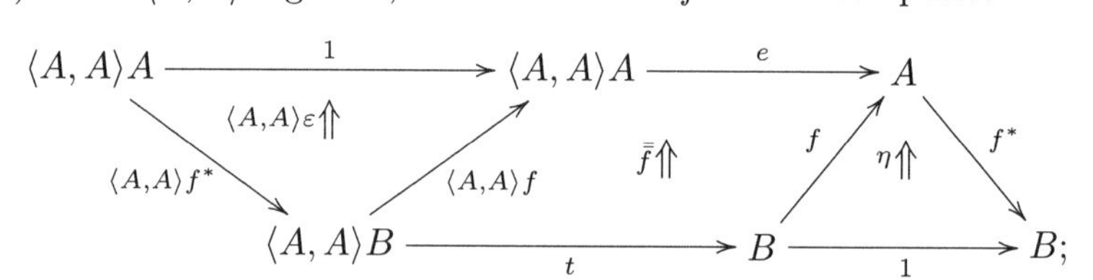

which, on substituting for $\bar{\bar{f}}$ its explicit value above and using one of the triangular equations, gives

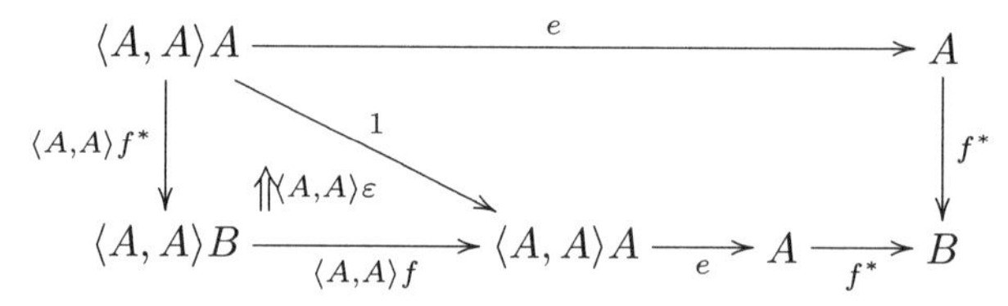

as the value of $\overline{f^*}$.

When $\bar{\bar{f}} = \varepsilon e.\langle A, A\rangle f$ is invertible — and so in particular when ε itself is invertible — we have a lax map $(f, \bar{f})$ of lax $\langle A, A\rangle$-algebras, where $\bar{f} = \bar{\bar{f}}^{-1}$, and by [8, Proposition 1.3] we have an adjunction $(f, \bar{f}) \dashv (f^*, \overline{f^*})$ in Lax-$\langle A, A\rangle$-Alg$_l$. We state this formally only in the most important case where ε is itself invertible; then both $\bar{f}$ and $\overline{f^*}$ are invertible so that $(f, \bar{f})$ and $(f^*, \overline{f^*})$ become pseudomorphisms:

Theorem 4.1 *Let the counit $\varepsilon : ff^* \to 1$ of the adjunction $\eta, \varepsilon : f \dashv f^* : A \to B$ be invertible. Then we have an adjunction $\eta, \varepsilon : (f, \bar{f}) \dashv (f^*, \overline{f^*}) : (A, e) \to (B, t, \hat{t}, \bar{t})$ in* Lax-$\langle A, A\rangle$-Alg, *where $\bar{f}$ is given by $\varepsilon^{-1}e.\langle A, A\rangle f$ and $\overline{f^*}$ is given by $f^*e.\langle A, A\rangle\varepsilon$. When η too is invertible, so that the original adjunction is an equivalence in $\mathcal{A}$, both $\hat{t}$ and $\bar{t}$ are invertible, so that the adjunction $(f, \bar{f}) \dashv (f^*, \overline{f^*})$ becomes an equivalence in* Ps-$\langle A, A\rangle$-Alg.

A somewhat different case that has been useful historically, as the motivation for introducing the concept of *flexibility* for 2-monads, is that where we suppose the invertibility only of $\overline{f^*} = f^*e.\langle A, A\rangle\varepsilon$, which gives us an adjunction

$$\eta, \varepsilon : (f, \overline{\overline{f}}) \dashv (f^*, (\overline{f^*})^{-1}) : (A, e) \to (B, t, \widehat{t}, \overline{t})$$

(with $\overline{t}$ too invertible) in the 2-category Lax-$\langle A, A\rangle$-Alg_c of lax $\langle A, A\rangle$-algebras and *colax* maps. The historical example supposed η too to be invertible — indeed, to be an identity — so that also $\widehat{t}$ was invertible, and we were dealing with pseudo $\langle A, A\rangle$-algebras. To regain lax maps instead of colax ones, we need only to pass to the dual case by supposing the original adjunction $\eta, \varepsilon : f \dashv f^* : A \to B$ to lie in $\mathcal{A}^{co}$ rather than $\mathcal{A}$. Leaving the reader to work through the simple dualizing process, we merely state the result (which essentially repeats [8, Theorem 3.2], itself a generalization of [5].)

Theorem 4.2 *With $\mathcal{V}$ and $\mathcal{A}$ as before, let $\eta, \varepsilon : f \dashv f^* : A \to B$ have ε invertible, and let $fe.\langle A, A\rangle\eta$ be invertible. Then we can enrich the adjunction $f \dashv f^*$ to an adjunction $\eta, \varepsilon : (f, \overline{f}) \dashv (f^*, \overline{f^*}) : (A, s, \widehat{s}, \overline{s}) \to (B, e)$ in* Ps-$\langle B, B\rangle$-Alg_l.

5 The monoid $\langle f, f^*\rangle : \langle A, A\rangle \to \langle B, B\rangle$ as an endo-object

Before turning to the applications of the above to transport of structure, we revisit our central results in Theorem 3.5 and in the prologue to Theorem 4.1, to cast a new light on them. The first of these asserts that, under the conditions of the theorem, the map $w = \langle f, f^*\rangle : \langle A, A\rangle \to \langle B, B\rangle$ underlies a lax monoid map $(w, w^\circ, \widetilde{w})$. By Section 2, such a lax monoid map is the same thing as a monoid in the monoidal 2-category Colax$[\mathbf{2}, \mathcal{V}]$. Now probably the most direct way of showing an object of a monoidal 2-category to admit a monoid structure is to exhibit it as an "object $\langle\langle A, A\rangle\rangle$ of endomorphisms" in the context of an action admitting the adjunction Φ of (1.1); in this way we saw $\langle A, A\rangle$ to be a monoid in Section 1, and $\{f, f\}$ in Section 2: the latter involving the action of $\mathcal{V}$ on Lax$[\mathbf{2}, \mathcal{V}]$ and the adjunction (2.1). Of course $\mathcal{V}$ also acts on Colax$[\mathbf{2}, \mathcal{A}]$ in the dual fashion, with a right adjoint say $\{f, g\}'$, so that a strict monoid map $T \to \{f, f\}'$ corresponds to a lax map $(f, \overline{f})$ of T-algebras. These three are all examples of monoids in $\mathcal{V}$; and the question suggests itself whether the monoid $(w, w^\circ, \widetilde{w})$ in Colax$[\mathbf{2}, \mathcal{V}]$, enriching $w = \langle f, f^*\rangle : \langle A, A\rangle \to \langle B, B\rangle$, is an object of endomorphisms for some suitable action.

To this end, we consider an action of Colax$[\mathbf{2}, \mathcal{V}]$ on Colax$[\mathbf{2}, \mathcal{A}]$ which extends the action above of $\mathcal{V}$ on Colax$[\mathbf{2}, \mathcal{A}]$. The 2-functor $* :$ Colax$[\mathbf{2}, \mathcal{V}] \times$ Colax$[\mathbf{2}, \mathcal{A}] \to$ Colax$[\mathbf{2}, \mathcal{A}]$ on objects sends $(\rho : T \to S, g : A \to B)$ to $\rho g : TA \to SB$; on morphisms it sends $((\alpha, \lambda, \beta), (a, \theta, b))$ to $(\alpha a, \lambda\theta, \beta b)$; and on 2-cells it sends $((\gamma, \delta), (\xi, \eta))$ to $(\gamma\xi, \delta\eta)$. That this is indeed an action is immediate. Consider now what it is to give a morphism $(a, \theta, b) : \rho g \to h$, as in

$$\begin{array}{ccc} TA & \xrightarrow{a} & C \\ {\scriptstyle \rho g}\downarrow & \overset{\theta}{\Rightarrow} & \downarrow{\scriptstyle h} \\ SB & \xrightarrow[b]{} & D. \end{array} \tag{5.1}$$

It comes to giving the images (α, φ, β) of (a, θ, b) under the isomorphism Φ, as in

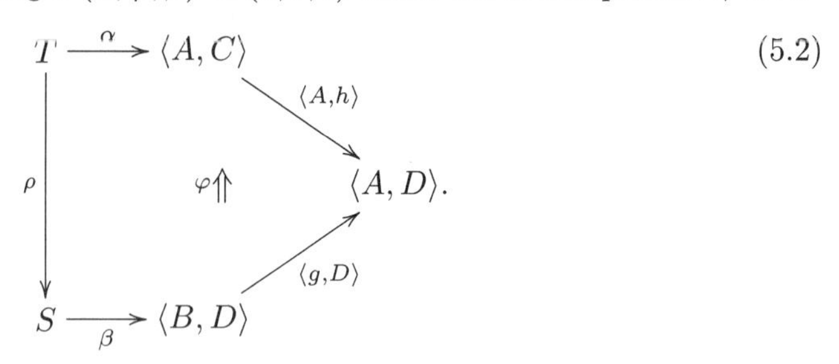

In general, this is not of the form $\rho \to \sigma$ for some object σ of Colax$[\mathbf{2}, \mathcal{V}]$: the present action does not admit a right adjoint like that in (1.1). Suppose however that the morphism g is a right adjoint — say $g = f^*$ where, as before, we have $\eta, \varepsilon : f \dashv f^* = g : A \to B$ in $\mathcal{A}$. These same data constitute, in $\mathcal{A}^{\mathrm{op}}$, an adjunction $\eta, \varepsilon : g \dashv f : A \to B$, which is sent by the 2-functor $\langle -, D \rangle : \mathcal{A}^{\mathrm{op}} \to \mathcal{V}$ to the adjunction

$$\langle \eta, D \rangle, \langle \varepsilon, D \rangle : \langle g, D \rangle \dashv \langle f, D \rangle : \langle A, D \rangle \to \langle B, D \rangle$$

in $\mathcal{V}$. Accordingly, to give the $\varphi : \langle g, D \rangle \beta \rho \to \langle A, h \rangle \alpha$ of (5.2) is equally (see [13]) to give a 2-cell $\psi : \beta\rho \to \langle f, D \rangle \langle A, h \rangle \alpha = \langle f, h \rangle \alpha$, where ψ is given in terms of φ as the pasting composite

$$\begin{array}{ccccc} T & \xrightarrow{\alpha} & \langle A, C \rangle & & \\ & & \searrow^{\langle A,h \rangle} & & \\ \rho \downarrow & \varphi \Uparrow & \langle A, D \rangle & & \\ & \nearrow^{\langle g,D \rangle} & \Uparrow \langle \eta, D \rangle & \searrow^{\langle f,D \rangle} & \\ S \xrightarrow{\beta} & \langle B, D \rangle & \xrightarrow{\quad 1 \quad} & & \langle B, D \rangle, \end{array} \tag{5.3}$$

with a similar formula giving φ in terms of ψ. The passage from φ to ψ is clearly 2-natural in the ρ and in the h of (5.1), so that we have a 2-natural isomorphism

$$\mathrm{Colax}[\mathbf{2}, \mathcal{A}](\rho g, h) \cong \mathrm{Colax}[\mathbf{2}, \mathcal{V}](\rho, \langle f, h \rangle).$$

Thus, although we have for a general g no adjunction of the form

$$\mathrm{Colax}[\mathbf{2}, \mathcal{A}](\rho g, h) \cong \mathrm{Colax}[\mathbf{2}, \mathcal{V}](\rho, [g, h]),$$

yet we do have such a $[g, h]$ when g is of the form f^*, it being given by $[g, h] = \langle f, h \rangle$; more succinctly, we have $[f^*, h] = \langle f, h \rangle$. In particular, $\langle f, f^* \rangle : \langle A, A \rangle \to \langle B, B \rangle$ is the value of $[f^*, f^*]$, which is a monoid in Colax$[\mathbf{2}, \mathcal{V}]$ because it has the form of an object of endomorphisms.

We now turn to our second main result, namely the observation in Section 4 that $f : B \to A$ underlies a colax map of lax $\langle A, A \rangle$-algebras, or equivalently that $f^* : A \to B$ underlies a lax map of such algebras. We can approach the latter, too, in terms of the present action of Colax$[\mathbf{2}, \mathcal{V}]$ on Colax$[\mathbf{2}, \mathcal{A}]$.

We may identify an object T of $\mathcal{V}$ with the object $1_T : T \to T$ of Colax$[\mathbf{2}, \mathcal{V}]$; and a monad structure on T gives rise to one on 1_T, with the same notation. To give a [lax] action of such a monad on an object $g : A \to B$ of Colax$[\mathbf{2}, \mathcal{A}]$ is clearly to give [lax] actions of T on A and on B, along with a lax map $(g, \overline{g}) : A \to B$ of such [lax] T-algebras; and the same is true when we omit each "[lax]". Accordingly to enrich $f^* : A \to B$ to a lax map of lax $\langle A, A\rangle$-algebras, we have only to provide in $\mathrm{Colax}[\mathbf{2}, \mathcal{A}]$ a lax action $\langle A, A\rangle f^* \to f^*$, or equivalently to provide in $\mathrm{Colax}[\mathbf{2}, \mathcal{V}]$ a lax monoid map $\langle A, A\rangle \to [f^*, f^*]$. Recall that $\langle A, A\rangle$ here stands for $1 : \langle A, A\rangle \to \langle A, A\rangle$, while $[f^*, f^*] = \langle f, f^*\rangle : \langle A, A\rangle \to \langle B, B\rangle$. We simplify now by writing C for $\langle A, A\rangle$ and D for $\langle B, B\rangle$, with $k : C \to D$ for $\langle f, f^*\rangle$; recall that $C = \langle A, A\rangle$ is a monoid (C, i, m) in $\mathcal{V}$, while $D = \langle B, B\rangle$ is a monoid (D, j, n), and $k : C \to D$ is a monoid in $\mathrm{Colax}[\mathbf{2}, \mathcal{V}]$, or equally a lax map $(k, k^\circ, \widetilde{k})$ of monoids in $\mathcal{V}$, where k° and $\widetilde{k}$ have the forms

$$\begin{array}{ccc} & \overset{i}{\nearrow} & C \\ 1 & \overset{k^\circ}{\Rightarrow} & \downarrow k \\ & \underset{j}{\searrow} & D, \end{array} \qquad \begin{array}{ccc} CC & \xrightarrow{m} & C \\ {\scriptstyle kk}\downarrow & \overset{\widetilde{k}}{\Rightarrow} & \downarrow{\scriptstyle k} \\ DD & \xrightarrow[n]{} & D. \end{array}$$

What we seek is a lax monoid map $(h, h^\circ, \widetilde{h}) : 1_C \to k$ in $\mathrm{Colax}[\mathbf{2}, \mathcal{V}]$. For h we take the map $(1, id, k)$ as in

$$\begin{array}{ccc} C & \xrightarrow{1} & C \\ {\scriptstyle 1}\downarrow & & \downarrow{\scriptstyle k} \\ C & \xrightarrow[k]{} & D. \end{array}$$

Next, h° has to be a 2-cell

$$\text{from}\quad \begin{array}{ccc} 1 & \xrightarrow{i} & C \\ {\scriptstyle 1}\downarrow & \overset{k^\circ}{\Rightarrow} & \downarrow{\scriptstyle k} \\ 1 & \xrightarrow[j]{} & D \end{array} \quad\text{to}\quad \begin{array}{ccccc} 1 & \xrightarrow{i} & C & \xrightarrow{1} & C \\ {\scriptstyle 1}\downarrow & & {\scriptstyle 1}\downarrow & & \downarrow{\scriptstyle k} \\ 1 & \xrightarrow[i]{} & C & \xrightarrow[k]{} & D, \end{array}$$

for which we take the pair (id, k°). Similarly the 2-cell $\widetilde{h}$

$$\text{from}\quad \begin{array}{ccccc} CC & \xrightarrow{11} & CC & \xrightarrow{m} & C \\ {\scriptstyle 11}\downarrow & & {\scriptstyle kk}\downarrow & \overset{\widetilde{k}}{\Rightarrow} & \downarrow{\scriptstyle k} \\ CC & \xrightarrow[kk]{} & DD & \xrightarrow[n]{} & D \end{array} \quad\text{to}\quad \begin{array}{ccccc} CC & \xrightarrow{m} & C & \xrightarrow{1} & C \\ {\scriptstyle 11}\downarrow & & {\scriptstyle 1}\downarrow & & \downarrow{\scriptstyle k} \\ CC & \xrightarrow[m]{} & C & \xrightarrow[k]{} & D \end{array}$$

is provided by the pair $(id, \widetilde{k})$. The easy verification that $(h, h^\circ, \widetilde{h})$ is indeed a lax map of monoids provides us with the desired enrichment $(f^*, \overline{f^*})$ of f^* to a lax map of lax $\langle A, A\rangle$-algebras.

We carry this analysis no further, since the calculations which give the explicit values of $\langle f, f^*\rangle : \langle A, A\rangle \to \langle B, B\rangle$ and of $(f^*, \overline{f^*}) : A \to B$ are no shorter if we begin with these present observations than were our calculations above based on the observations of Sections 3 and 4.

6 Transport of structure along an equivalence

We restrict ourselves here to the important case where the adjunction $\eta, \varepsilon : f \dashv f^* : A \to B$ is an *equivalence* in $\mathcal{A}$; the reader interested in the more general situations of the previous section will easily make the necessary extensions. We are used in universal algebra to the transport of structure along an isomorphism: if T is a monad on the mere category $\mathcal{A}$, if $a : TA \to A$ is an action of T on $A \in \mathcal{A}$, and if $f : B \to A$ is an isomorphism in $\mathcal{A}$ with inverse $f^* : A \to B$, there is a unique action $b : TB \to B$ of T on B for which f becomes an isomorphism of T-algebras — namely that given by $b = f^* a.Tf$. What replaces this result when $\mathcal{A}$ is a 2-category, the monad $T = (T, i, m)$ is a 2-monad, and the isomorphism $f : B \to A$ is replaced by the adjoint equivalence $f \dashv f^*$? We make use of the results above, taking for $\mathcal{V}$ the monoidal 2-category $[\mathcal{A}, \mathcal{A}]$, and supposing $\mathcal{A}$ complete and locally small, so that we have the adjunction $\Phi : \mathcal{A}(XA, B) \cong [\mathcal{A}, \mathcal{A}](X, \langle A, B\rangle)$. It is well known — see for instance [13] — that a strict map $\alpha : T \to S$ of monads on $\mathcal{A}$ (2-monads, of course, since $\mathcal{A}$ is a 2-category) induces a 2-functor $\alpha^* : S\text{-Alg} \to T\text{-Alg}$, commuting with the forgetful 2-functors to $\mathcal{A}$, and restricting to a 2-functor $S\text{-Alg}_s \to T\text{-Alg}_s$. We need a less strict analogue of this: we show that a pseudo map $\alpha = (\alpha, \alpha^\circ, \widetilde{\alpha}) : T \to S$ of monads induces a 2-functor $\alpha^* : \text{Ps-}S\text{-Alg} \to \text{Ps-}T\text{-Alg}$, again commuting with the forgetful 2-functors to $\mathcal{A}$, and again admitting a restriction $\alpha^*_s : \text{Ps-}S\text{-Alg}_s \to \text{Ps-}T\text{-Alg}_s$ to the *strict* maps of pseudo algebras. First, a pseudo action $s = (s, \widehat{s}, \overline{s})$ of S on B corresponds as in Section 2 to a pseudo map $\sigma = (\sigma, \sigma^\circ, \widetilde{\sigma}) : S \to \langle B, B\rangle$ of monads, which composes with $\alpha = (\alpha, \alpha^\circ, \widetilde{\alpha})$ to give a pseudo map $\rho = (\rho, \rho^\circ, \widetilde{\rho}) : T \to \langle B, B\rangle$, corresponding to a pseudo action $r = (r, \widehat{r}, \overline{r})$ of T on B; and $(B, r, \widehat{r}, \overline{r})$ is the image under α^* of $(B, s, \widehat{s}, \overline{s})$. Next, given a morphism $f : B \to B'$ where $(B, s, \widehat{s}, \overline{s})$ and $(B', s', \widehat{s}', \overline{s}')$ are pseudo S-algebras, to give f the structure of a morphism (that is, a pseudo morphism) of pseudo S-algebras is to give a pseudo monad map $\gamma : S \to \langle\langle f, f\rangle\rangle$ with $\chi\gamma = \sigma$ and $\chi'\gamma = \sigma'$, where

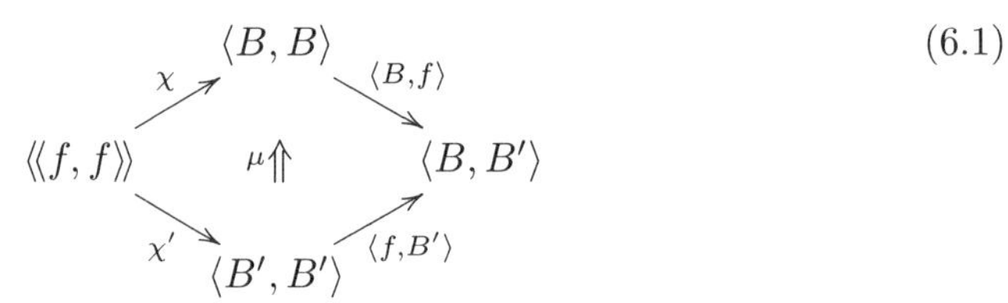

is the iso-comma object in $[\mathcal{A}, \mathcal{A}]$; and then the composite pseudo monad-map $\gamma\alpha : T \to \langle\langle f, f\rangle\rangle$ corresponds to a morphism $(f, \overline{\overline{f}}) : \alpha^*(B, s, \widehat{s}, \overline{s}) \to \alpha^*(B', s', \widehat{s}', \overline{s}')$ which is the desired $\alpha^*(f, \overline{f})$. The isomorphism $\overline{f}$ is an identity — that is, the morphism $(f, \overline{f})$ is strict — when $\gamma : S \to \langle\langle f, f\rangle\rangle$ factorizes through the canonical $\delta : (f, f) \to \langle\langle f, f\rangle\rangle$, where (f, f) is the pullback

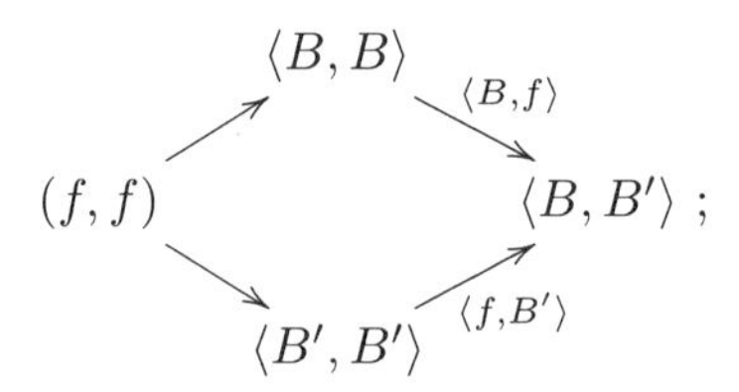

in which case $\gamma\alpha$ factorizes through δ. Thus α^* does indeed send strict morphisms to strict morphisms, and we have established the 2-functors $\alpha^* : \text{Ps-}S\text{-Alg} \to \text{Ps-}T\text{-Alg}$ and $\alpha^*_s : \text{Ps-}S\text{-Alg}_s \to \text{Ps-}T\text{-Alg}_s$.

With these tools at hand, we return to the question of transporting structure: let us have the adjoint equivalence $\eta, \varepsilon : f \dashv f^* : A \to B$ in $\mathcal{A}$, and a pseudo action $(a, \widehat{a}, \overline{a})$ of T on A. This corresponds to a pseudo monad-map $(\alpha, \alpha^\circ, \widetilde{\alpha}) : T \to \langle A, A\rangle$, where $\alpha : T \to \langle A, A\rangle$ is the image under Φ of $a : TA \to A$. This $\alpha = (\alpha, \alpha^\circ, \widetilde{\alpha})$ induces a 2-functor $\alpha^* : \text{Ps-}\langle A, A\rangle\text{-Alg} \to \text{Ps-}T\text{-Alg}$ which carries the adjoint equivalence $(f, \overline{f}) \dashv (f^*, \overline{f^*})$ in $\text{Ps-}\langle A, A\rangle\text{-Alg}$ of Theorem 4.1 to an adjoint equivalence

$$\eta, \varepsilon : \alpha^*(f, \overline{f}) \dashv \alpha^*(f^*, \overline{f^*}) : \alpha^*(A, e) \to \alpha^*(B, t, \widehat{t}, \overline{t})$$

in Ps-T-Alg. Since the strict action $e : \langle A, A\rangle A \to A$ corresponds to the identity morphism $\langle A, A\rangle \to \langle A, A\rangle$, the pseudo T-algebra $\alpha^*(A, e)$ is the $(A, a, \widehat{a}, \overline{a})$ we started with. Calculating $\alpha^*(B, t, \widehat{t}, \overline{t})$ is also straightforward, but we do it explicitly below only for the important case where A is a strict T-algebra, with $\widehat{a}$ and $\overline{a}$ identities. In fact there is a theoretical sense in which it suffices to study this case: it is shown in [3] that, under modest conditions on T, a pseudo T-algebra is just a T'-algebra for another monad T'; we shall return to this observation below, in connection with *flexible* monads.

We take (A, a), then, to be a strict T-algebra, observing that $\alpha : T \to \langle A, A\rangle$, as the image under Φ of $a : TA \to A$, satisfies $e.\alpha A = a$. (Note that we have earlier used i for the unit and m for the multiplication not only of $\langle A, A\rangle$ but also of T; but continuing to do so will lead to no confusion.) Let us write $(B, b, \widehat{b}, \overline{b})$ for the T-algebra $\alpha^*(B, t, \widehat{t}, \overline{t})$. Since $b : TB \to B$ is the composite $t.\alpha B$, while t is given by the composite

$$\langle A, A\rangle B \xrightarrow{\langle A,A\rangle f} \langle A, A\rangle A \xrightarrow{e} A \xrightarrow{f^*} B\ ,$$

the naturality of α along with the equation $e.\alpha A = a$ gives b as the composite

$$TB \xrightarrow{Tf} TA \xrightarrow{a} A \xrightarrow{f^*} B.$$

Similarly $\widehat{b}$ is simply $\eta : 1 \to f^*f = f^*a.Tf.iB$, while by the description (3.4) of $\overline{t}$ we see that $\overline{b}$ is given by

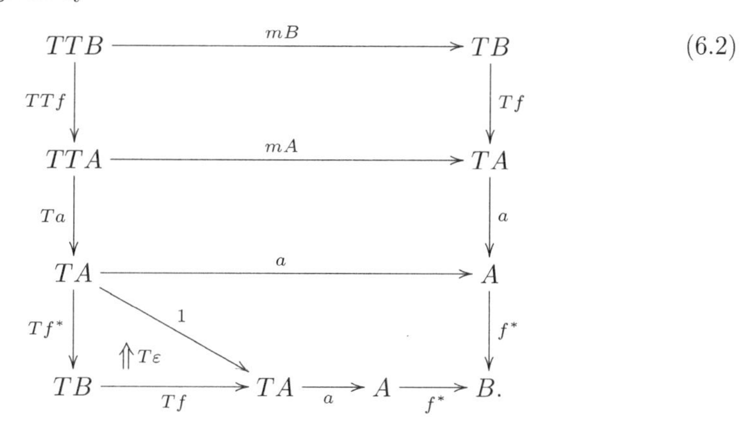

Similarly, $\alpha^*(f, \overline{f}) = (f, \overline{\overline{f}})$ and $\alpha^*(f^*, \overline{f^*}) = (f^*, \overline{\overline{f^*}})$, where $\overline{\overline{f}}$ is given by $\varepsilon^{-1}a.Tf : a.Tf \to ff^*a.Tf (= fb)$, and $\overline{\overline{f^*}}$ by $f^*a.T\varepsilon : (b.Tf^* =) f^*a.Tf.Tf^* \to f^*a$. Summing up, we have:

Theorem 6.1 *Given the equivalence $\eta, \varepsilon : f \dashv f^* : A \to B$ in the complete and locally small 2-category $\mathcal{A}$, and an algebra (A, a) for the monad $T = (T, i, m)$ on $\mathcal{A}$, the equivalence enriches to an equivalence*

$$\eta, \varepsilon : (f, \overline{\overline{f}}) \dashv (f^*, \overline{\overline{f^*}}) : (A, a) \to (B, b, \widehat{b}, \overline{b})$$

in Ps-T-Alg, *where $\widehat{b} = \eta$ and $\overline{b}$ is given by $f^*a.T\varepsilon.Ta.T^2f$ as in* (6.2), *and where $\overline{\overline{f}} = \varepsilon^{-1}a.Tf$ and $\overline{\overline{f^*}} = f^*a.T\varepsilon$.*

Consider the case where $\mathcal{A} = \mathbf{Cat}$ and $T = (T, i, m)$ is the 2-monad whose algebras are the strict monoidal categories. A consequence of the coherence theorem for monoidal categories is that for any monoidal category B there is a strict monoidal category A — that is, a strict T-algebra (A, a) — and an equivalence $\eta, \varepsilon : f \dashv f^* : A \to B$ with f and f^* strong monoidal functors. Now suppose that B is a skeleton of the category of countable sets, equipped with the cartesian monoidal structure; then the monoidal structure on A is again the cartesian one, for some choice of binary products and terminal object. If the equivalence $\eta, \varepsilon : f \dashv f^*$ underlay an equivalence in T-Alg, then the monoidal structure on B would be both cartesian and strict. By an argument due to Isbell [14, p.160], however, this is impossible.

In general, then, it is not possible to enrich an adjoint equivalence to one in T-Alg. However such an enrichment does exist when the monad T is *flexible* — a notion, originally introduced in [8], which we now recall. First note that, in the present case where $\mathcal{V} = [\mathcal{A}, \mathcal{A}]$, so that a monoid in $\mathcal{V}$ is a monad on $\mathcal{A}$, the 2-categories $\mathrm{Mon}_l\,\mathcal{V}$, $\mathrm{Mon}_p\,\mathcal{V}$, and $\mathrm{Mon}\,\mathcal{V}$ of Section 2 are conveniently renamed $\mathrm{Mnd}_l\,\mathcal{A}$, $\mathrm{Mnd}_p\,\mathcal{A}$, and $\mathrm{Mnd}\,\mathcal{A}$. In particular we have the inclusion 2-functor $J : \mathrm{Mnd}\,\mathcal{A} \to \mathrm{Mnd}_p\,\mathcal{A}$; and it was shown in Blackwell's thesis [3] that a partial left adjoint to J is defined at the monad T if $\mathcal{A}$ is cocomplete and T has some rank. (An endofunctor T of $\mathcal{A}$ is said to have rank κ, where κ is a regular cardinal, if T preserves κ-filtered colimits.) To say that the partial adjoint is defined at T means, of course, that there is a pseudo map $p : T \to T'$ of monads on $\mathcal{A}$ such that, for any monad S on $\mathcal{A}$, the 2-functor $\mathrm{Mnd}\,\mathcal{A}(T', S) \to \mathrm{Mnd}_p\,\mathcal{A}(T, S)$ given by composition with p is an isomorphism of 2-categories. In more elementary terms, every pseudo map $g : T \to S$ is of the form hp for a unique strict map $h : T' \to S$, and every monad 2-cell $\alpha : hp \to h'p$, where the monad maps h and h' are strict, is βp for a unique monad 2-cell $\beta : h \to h'$.

In particular, there is a unique strict monad map $q : T' \to T$ for which $qp = 1_T$. Even before Blackwell's result, Kelly had shown in [9] that, whenever the partial left adjoint is defined at T, there is an invertible 2-cell $\rho : 1_{T'} \cong pq$ with $\rho p = id$ and $q\rho = id$, so that we have in $\mathrm{Mnd}_p\,\mathcal{A}$ the equivalence

$$\rho, id : q \dashv p : T \to T'.$$

Taking $S = \langle B, B \rangle$ in the universal property of $p : T \to T'$ shows that to give a pseudo action of T on B is just to give a strict action of T' on B. And taking for S the $\langle\langle f, f \rangle\rangle$ of (6.1) shows that enriching $f : B \to B'$ to a morphism $(f, \overline{f})$ of pseudo T-algebras is the same as enriching it to a morphism of T'-algebras. Accordingly we have an isomorphism of 2-categories

$$\text{Ps-}T\text{-Alg} \cong T'\text{-Alg}$$

which commutes with the underlying 2-functors to $\mathcal{A}$, and which restricts to an isomorphism of 2-categories

$$\text{Ps-}T\text{-Alg}_s \cong T'\text{-Alg}_s;$$

moreover, by a similar argument, it extends to an isomorphism

$$\text{Ps-}T\text{-Alg}_l \cong T'\text{-Alg}_l.$$

This is the intent of our earlier remark that a pseudo T-algebra is just a T'-algebra for a certain monad T'. The strict monad map $q : T' \to T$ induces a 2-functor $q^* : T\text{-Alg} \to T'\text{-Alg}$, restricting to $T\text{-Alg}_s \to T'\text{-Alg}_s$ and extending to

$T\text{-Alg}_l \to T'\text{-Alg}_l$. If we identify T'-Alg with Ps-T-Alg via the isomorphism above, $q^* : T\text{-Alg} \to T'\text{-Alg}$ is of course nothing but the inclusion $T\text{-Alg} \to \text{Ps-}T\text{-Alg}$.

The notion of flexibility was introduced by Kelly in [8] as a property of 2-monads, which is the case of interest here; later it was generalized to be a property of algebras for a 2-monad, of which a 2-monad itself is a special case — see [4]; in another special case introduced there, flexibility is a property of a *weight* for **Cat**-enriched limits, the corresponding *flexible limits* being studied in [2].

Supposing the monad T on $\mathcal{A}$ to be such that $p : T \to T'$ exists as above, with $q : T' \to T$ the unique strict map for which $qp = 1$, we say that T is *flexible* if there is some *strict* map $r : T \to T'$ for which $qr = 1$. Since we have $\rho : 1 \cong pq$ as above, we have $\rho r : r \cong pqr = p$; so that besides the equivalence $\rho, id : q \dashv p : T \to T'$ in $\text{Mnd}_p\, \mathcal{A}$, we now have an equivalence

$$\sigma, id : q \dashv r : T \to T' \tag{6.3}$$

in $\text{Mnd}\, \mathcal{A}$ itself. One easily sees that (supposing the left adjoints to exist) the monad T' is always flexible, and in fact a monad S is flexible precisely when it is a retract in $\text{Mnd}\, \mathcal{A}$ of some T'; the details can be found in [4]. For a flexible T, the equivalence of 2-categories

$$q^* \dashv r^* : T'\text{-Alg} \cong \text{Ps-}T\text{-Alg} \to T\text{-Alg}$$

induced by the equivalence (6.3) restricts of course to an equivalence

$$q^* \dashv r^* : T'\text{-Alg}_s \cong \text{Ps-}T\text{-Alg}_s \to T\text{-Alg}_s$$

between the Eilenberg-Moore 2-categories for the monads.

We can now give our main result on flexible monads:

Theorem 6.2 *Let $\eta, \varepsilon : f \dashv f^* : A \to B$ be an equivalence in the complete, cocomplete, and locally-small 2-category $\mathcal{A}$, let $T = (T, i, m)$ be a flexible monad on $\mathcal{A}$ having some rank, let $a : TA \to A$ be an action (meaning a strict one) of T on A, and let $qr = 1_T$, where $q : T' \to T$ is as above and r is a strict monad map. Then the given equivalence has an enrichment to an equivalence*

$$\eta, \varepsilon : (f, \check{f}) \dashv (f^*, \check{f}^*) : (A, a) \to (B, \check{b})$$

in T-Alg.

Proof Identifying Ps-T-Alg with the isomorphic T'-Alg, we find the desired equivalence as the image under $r^* : \text{Ps-}T\text{-Alg} \to T\text{-Alg}$ of the equivalence of Theorem 6.1. Note here that the (A, a) in the equivalence of Theorem 6.1 really denotes $q^*(A, a)$ — the T-algebra (A, a) seen as a pseudo T-algebra — and that $r^* q^*(A, a)$ is (A, a) itself, since $qr = 1$. □

Remark 6.3 That the scope of the theorem is extremely broad will be clear from the forthcoming article [11], where it is shown that a monad T on **Cat** is flexible if the structure of a T-algebra can be presented by operations and equations, in the sense of [12], in such a way that there are no equations between objects, only between maps; with similar results for many other 2-categories in place of **Cat**.

References

[1] J. Bénabou, *Catégories avec multiplication*, C.R.Acad.Sci.Paris **256** (1963), 1887–1890.

[2] G.J. Bird, G.M. Kelly, A.J. Power, and R.H. Street, *Flexible limits for 2-categories*, J. Pure Appl. Alg. **61** (1989), 1–27.

[3] R. Blackwell, *Some existence theorems in the theory of doctrines*, Ph.D. Thesis, Univ. New South Wales, 1976.

[4] R. Blackwell, G.M. Kelly, and A.J. Power, *Two dimensional monad theory*, J. Pure Appl. Alg. **59** (1989), 1–41.

[5] B.J. Day, *A reflection theorem for closed categories*, J. Pure Appl. Alg. **2** (1972), 1–11.

[6] G. Janelidze and G.M. Kelly, *A note on actions of a monoidal category*, Theory Appl. Categ. **9** (2001), 61–91.

[7] André Joyal and Ross Street, *Braided tensor categories*, Adv. Math. **102** (1993), 20-78.

[8] G.M. Kelly, Doctrinal adjunction, in *Sydney Category Seminar*, Lecture Notes Math. 420, pp. 257–280, Springer, 1974.

[9] G.M. Kelly, *Coherence theorems for lax algebras and for distributive laws* in Sydney Category Seminar, Lecture Notes Math. vol. 420, Springer, Berlin-Heidelberg, 1974, pp. 281–375.

[10] G.M. Kelly and S. Lack, *On property-like structures* Theory Appl. Categ. **3** (1997), 213–250.

[11] G.M. Kelly, S. Lack, and A.J. Power, *A criterion for flexibility of a 2-monad*, in preparation.

[12] G.M. Kelly and A.J. Power, *Adjunctions whose counits are coequalizers and presentations of finitary enriched monads*, J. Pure Appl. Alg. **89** (1993), 163–179.

[13] G.M. Kelly and R. Street, *Review of the elements of 2-categories* in Sydney Category Seminar, Lecture Notes Math. vol. 420, Springer, Berlin-Heidelberg, 1974, pp. 75–103.

[14] S. Mac Lane, *Categories for the Working Mathematician*, Graduate Texts in Mathematics vol. 5, Springer, New York-Heidelberg-Berlin, 1971.

[15] S. Mac Lane and R. Paré, *Coherence for bicategories and indexed categories*, J. Pure Appl. Algebra **37** (1985), 59–80.

[16] R. Street, *Fibrations and Yoneda's lemma in a bicategory* in Sydney Category Seminar, Lecture Notes Math. vol. 420, Springer, Berlin-Heidelberg, 1974, pp. 104–133.

Received December 23, 2002; in revised form October 22, 2003

Fields Institute Communications
Volume **43**, 2004

On the Cyclic Homology of Hopf Crossed Products

M. Khalkhali
Department of Mathematics
University of Western Ontario
London, ON, Canada N6A 5B7
masoud@uwo.ca

B. Rangipour
Department of Mathematics
University of Western Ontario
London, ON, Canada N6A 5B7
brangipo@uwo.ca

Abstract. We consider Hopf crossed products of the the type $A\#_\sigma\mathcal{H}$, where $\mathcal{H}$ is a cocommutative Hopf algebra, A is an $\mathcal{H}$-module algebra and σ is a "numerical" convolution invertible 2-cocycle on $\mathcal{H}$. we give an spectral sequence that converges to the cyclic homology of $A\#_\sigma\mathcal{H}$ and identify the E^1 and E^2 terms of the spectral sequence.

1 Introduction

A celebrated problem in noncommutative geometry, more precisely in cyclic homology theory, is to compute the cyclic homology of a crossed product algebra. The interest in this problem stems from the fact that, according to a guiding principle in noncommutative geometry [3], crossed products play the role of "noncommutative quotients" in situations where the usual set theoretic quotients are ill behaved. For example, when a (locally compact) group G acts on a locally compact Hausdorff space X, the quotient space X/G may not be well behaved, e.g. may not be even a Hausdorff space. The crossed product (C^*-) algebra $C_0(X) \ltimes G$, however, is a good replacement for X/G [3]. In fact, if the action of G is free and proper, then by a theorem of Rieffel [16] the C^*-algebra of continuous functions vanishing at infinity on X/G, denoted by $C_0(X/G)$ is in a suitable C^*-algebraic sense, Morita equivalent to the crossed product algebra $C_0(X) \ltimes G$. Since K-Theory, Hochschild homology and cyclic homology are Morita invariant functors, replacing the commutative algebra $C_0(X/G)$ by the noncommutative algebra $C_0(X) \ltimes G$ results in no loss of information.

It is therefore natural and desirable to develop tools to compute the cyclic homology of crossed product algebras. Most of the results obtained so far are

2000 *Mathematics Subject Classification.* Primary 19D55 18G40; Secondary 16W30.

concerned with the action of groups on algebras [7, 8, 15]. For Hopf algebra crossed products, [9] gives a complete answer for Hochschild homology but it is not clear how to extend its method to cyclic homology. If one is only interested in smash products as opposed to crossed products, one can find a complete answer in [1] in terms of a spectral sequence converging to the cyclic homology of the smash product algebra $A\#\mathcal{H}$.

The goal of this article is to extend the results of [1] to Hopf algebra crossed products. Due to the fact that very complicated formulas appear in our constructions, however, we had to assume that the Hopf algebra is cocommutative and the 2-cocycle takes values in the ground field. Under these conditions, we give a spectral sequence for the cyclic homology of a crossed product algebra $A\underset{\sigma}{\#}\mathcal{H}$, when the cocycle σ is convolution invertible and takes values in the ground field k. The method of proof is similar to the one used in [1] which is based on the generalized cyclic Eilenberg-Zilber theorem of [8]. Though we think the same method should apply to arbitrary crossed products $A\underset{\sigma}{\#}\mathcal{H}$ (with convolution invertible cocycles), due to technical difficulties we are not able to verify this.

One of our main motivations to consider cocycle crossed products is to find simple methods to compute the cyclic cohomology of " noncommutative toroidal orbifolds" considered in [12, Sec. 9] and [6]. These examples are suggested by applications of noncommutative geometry to string theory and M(atrix) theory. In these examples one considers algebras of the type $B^d_{\theta,\sigma} = A^d_\theta \underset{\sigma}{\#}\mathbb{C}G$, where G is a finite group acting by automorphisms on the noncommutative d-dimensional torus A^d_θ and $\sigma \in H^2(G, U(1))$ is a group 2-cocycle on G. Corollary 3.8 shows that the cyclic cohomology of $B^d_{\theta,\sigma}$ can always be computed from a cyclic complex much simpler than the original cyclic complex of the algebra $B^d_{\theta,\sigma}$.

2 Preliminaries

In this paper we work over a fixed ground field k. All algebras are unital associative algebras over k and all modules are unitary. The unadorned tensor product $\otimes$ means tensor product over k. We denote the coproduct of a Hopf algebra by Δ, the counit by ϵ and the antipode by S. We use Sweedler's notation and write $\Delta(h) = h^{(1)} \otimes h^{(2)}$ to denote the coproduct, where summation is understood. Similarly, we write $\Delta^{(n)}(h) = h^{(1)} \otimes h^{(2)} \otimes \ldots \otimes h^{(n+1)}$ to denote the iterated coproducts defined by $\Delta^{(1)} = \Delta$ and $\Delta^{(n+1)} = (\Delta \otimes 1) \circ \Delta^{(n)}$.

We recall the concept of Hopf crossed product, introduced for the first time in [5], and independently [2]. A good reference for this notion is Chapter 7 of [14]. Let $\mathcal{H}$ be a Hopf algebra and A an algebra. Recall from [2] and [5] that a *weak action* of $\mathcal{H}$ on A is a linear map $\mathcal{H} \otimes A \longrightarrow A$, $h \otimes a \to h(a)$ such that, for all $h \in \mathcal{H}$, and $a, b \in A$

1) $h(ab) = h^{(1)}(a)h^{(2)}(b)$,
2) $h(1) = \epsilon(h)1$,
3) $1(a) = a$.

By an *action* of $\mathcal{H}$ on A, we mean a weak action such that A is an $\mathcal{H}$-module, i.e. for all $h, l \in \mathcal{H}$ and $a \in A$ we have $h(l(a)) = hl(a)$. In the latter case we say A is an $\mathcal{H}-$*module algebra.*

Let A be an $\mathcal{H}$-module algebra. The *smash product* $A\#\mathcal{H}$ of A and $\mathcal{H}$ is an associative algebra whose underlying vector space is $A \otimes \mathcal{H}$ and whose multiplication is defined by

$$(a \otimes h)(b \otimes l) = ah^{(1)}(b) \otimes h^{(2)}l.$$

If, on the other hand, we have only a weak action of $\mathcal{H}$ on A the above formula does not define an associative multiplication, and a modification is needed. Given a linear map $\sigma : \mathcal{H} \otimes \mathcal{H} \to A$ one defines a (not necessarily unital or associative) multiplication on $A \otimes \mathcal{H}$ by [2, 5]

$$(a \otimes h)(b \otimes l) = ah^{(1)}(b)\sigma(h^{(2)}, l^{(1)}) \otimes h^{(3)}l^{(2)}.$$

It can be shown that the above formula defines an associative product with $1 \otimes 1$ as its unit, if and only if σ and the weak action enjoy the following properties:

1) (Normality) For all $h \in \mathcal{H}$, $\sigma(h, 1) = \sigma(1, h) = \epsilon(h)1$.
2) (Cocycle property) For all $h, l, m \in \mathcal{H}$,
$\sum h^{(1)}(\sigma(l^{(1)}, m^{(1)}))\sigma(h^{(2)}, l^{(2)}m^{(2)}) = \sum \sigma(h^{(1)}, l^{(1)})\sigma(h^{(2)}l^{(2)}, m)$,
3) (Twisted module property) For all $h, l \in \mathcal{H}$ and $a \in A$,
$\sum h^{(1)}(l^{(1)}(a))\sigma(h^{(2)}, l^{(2)}) = \sum \sigma(h^{(1)}, l^{(1)})l^{(2)}h^{(2)}(a))$.

The cocycle σ is said to be *convolution invertible* if it is an invertible element of the convolution algebra $Hom_k(\mathcal{H} \otimes \mathcal{H}, A)$. Now assume the Hopf algebra $\mathcal{H}$ is cocommutative, $\sigma : \mathcal{H} \otimes \mathcal{H} \longrightarrow k1_A$ takes values in the ground field k, and σ is invertible. Then it follows that A is an $\mathcal{H}$-module algebra, i.e. the weak action in the above situation is in fact an action. To prove this let $a \in A$ be fixed. Define two functions in $Hom_k(\mathcal{H}\otimes\mathcal{H}, A)$ by $F(h, l) = \sum h^{(1)}(l^{(1)}(a))\sigma(h^{(2)}, l^{(2)})$ and $G(h, l) = \sum \sigma(h^{(1)}, l^{(1)})l^{(2)}h^{(2)}(a)$. Then $F = G$ by the twisted module property of σ, so $F * \sigma^{-1} = G * \sigma^{-1}$. In other word

$$h(l(a)) = F * \sigma^{-1}(h, l) = G * \sigma^{-1}(h, l) = hl(a).$$

One notes that the above proof remains valid when $\mathcal{H}$ is cocommutative and σ takes its values in the center of A, instead of k.

One of the main technical tools used in [1] to derive a spectral sequence for the cyclic homology of smash products is the generalized cyclic Eilenberg-Zilber theorem. This result was first stated in [8] but its first algebraic proof appeared in [11]. The idea of using an Eilenberg-Zilber type theorem to derive a spectral sequence for cyclic homology of smash products (for the action of groups) is due to Getzler and Jones [8]. We find it remarkable that the same idea works in the case of Hopf algebra crossed product (with convolution invertible cocycle). In the following we recall the definitions of (para)cyclic modules, cylindrical modules and state the Eilenberg-Zilber theorem for cylindrical modules.

Recall that a *paracyclic module* is a simplicial k-module $\{M_n\}_{n\geq 0}$, such that the following extra relations are satisfied [7, 8]:

$$\delta_i \tau = \tau \delta_{i-1}, \qquad \delta_0 \tau = \delta_n, \qquad 1 \leq i \leq n,$$
$$\sigma_i \tau = \tau \sigma_{i+1}, \qquad \sigma_0 \tau = \tau^2 \sigma_n, \qquad 1 \leq i \leq n,$$

where $\delta_i : M_n \to M_{n-1}$, $\sigma_i : M_n \to M_{n+1}$, $0 \leq i \leq n$, are faces and degeneracies of the simplicial module $\{M_n\}_{n\geq 0}$ and $\tau : M_n \to M_n$, $n \geq 0$ are k-linear maps. If furthermore we have $\tau^{n+1} = id_{M_n}$, for all $n \geq 0$, then we say that we have a *cyclic module*.

We denote the cyclic module of an associative unital k-algebra A, by $A^\natural$. It is defined by $A^\natural_n = A^{\otimes(n+1)}$, $n \geq 0$, and simplicial and cyclic operations defined by

$$\begin{aligned}
\delta_i(a_0 \otimes \cdots \otimes a_n) &= a_0 \otimes a_1 \otimes \cdots \otimes a_i a_{i+1} \cdots \otimes a_n, \quad 0 \leq i \leq n-1,\\
\delta_n(a_0 \otimes \cdots \otimes a_n) &= a_n a_0 \otimes a_1 \otimes \cdots \otimes a_{n-1},\\
\sigma_i(a_0 \otimes \cdots \otimes a_n) &= a_0 \otimes \cdots \otimes a_i \otimes 1 \cdots \otimes a_n, \quad 0 \leq i \leq n,\\
\tau_n(a_0 \otimes \cdots \otimes a_n) &= a_n \otimes a_0 \cdots \otimes a_{n-1}.
\end{aligned}$$

To any cyclic module one associates its cyclic homology groups [13, 4]. In particular, the cyclic homology groups of $A^\natural$ are denoted by $HC_n(A)$, $n \geq 0$, and are called cyclic homology of A.

By a *biparacyclic* module we mean a doubly graded sequence of k-modules $\{M_{p,q}\}_{p,q\geq 0}$ such that each row and each column is a paracyclic module and all vertical operators commute with all horizontal operators. In particular a *bicyclic* module is a biparacyclic module such that each row and each column is a cyclic module. We denote the horizontal and vertical operators of a biparacyclic module by $(\delta_i, \sigma_i, \tau)$ and (d_i, s_i, t) respectively. By a *cylindrical module* we mean a biparacyclic module such that for all $p, q \geq 0$,

$$\tau^{p+1} t^{q+1} = id_{M_{p,q}}. \tag{2.1}$$

Given a cylindrical module M, its *diagonal*, denoted by dM, is a cyclic module defined by $(dM)_n = M_{n,n}$ and with simplicial and cyclic operators given by $\delta_i d_i$, $\sigma_i s_i$, and τt. In view of (2.1), it is a cyclic module. The total complex of a cylindrical module, denoted by $Tot(M)$, is a mixed complex with operators given by $b + \bar{b}$ and $B + T\bar{B}$, where $T = 1 - (bB + Bb)$. Here b (resp. $\bar{b}$) and B (resp. $\bar{B}$) are the vertical (resp. horizontal) Hochschild and Connes boundary operators of cyclic modules. Note that it differs from the usual notion of total complex in that we use $B + T\bar{B}$ instead of $B + \bar{B}$. In fact the latter choice won't give us a mixed complex [8]. It can be checked that $Tot(M)$ is a mixed complex. Given a cylindrical or cyclic module M, we denote its normalization by $N(M)$.

The following theorem is the main technical result that enables us to derive spectral sequences for the cyclic homology of crossed product algebras.

Theorem 2.1 (Generalized cyclic Eilenberg-Zilber theorem ([11, 8])) *For any cylindrical module M there is a natural quasi-isomorphism of mixed complexes $f_0 + uf_1 : Tot(N(M)) \longrightarrow N(dM)$, where f_0 is the shuffle map.*

3 A Spectral sequence for Hopf crossed products

Let $\mathcal{H}$ be a cocommutative Hopf algebra, A a left $\mathcal{H}$- module algebra and $\sigma : \mathcal{H} \otimes \mathcal{H} \longrightarrow k1_A$ a two cocycle satisfying the cocycle conditions 1), 2), and 3) in Section 2. We further assume that σ is convolution invertible. We introduce a cylindrical module

$$A \natural_\sigma \mathcal{H} = \{\mathcal{H}^{\otimes(p+1)} \otimes A^{\otimes(q+1)}\}_{p,q\geq 0}$$

with vertical and horizontal simplicial and cyclic operators (δ,σ,τ) and (d,s,t), defined as follows

$$
\begin{aligned}
\tau(g_0,\ldots,g_p \mid a_0,\ldots,a_q) &= (g_0^{(2)},\ldots,g_p^{(2)} \mid \\
&\qquad S(g_0^{(1)}\ldots g_p^{(1)})(a_q),a_0,a_1,\ldots,a_{q-1})\\
\delta_i(g_0,\ldots,g_p \mid a_0,\ldots,a_q) &= (g_0,\ldots,g_p \mid a_0,\ldots,a_ia_{i+1},\ldots,a_q) \quad 0\le i<q\\
\delta_q^{p,q}(g_0,\ldots,g_p \mid a_0,\ldots,a_q) &= (g_0^{(2)},\ldots,g_p^{(2)} \mid \\
&\qquad S(g_0^{(1)}\ldots g_p^{(2)})(a_q)a_0,a_1\ldots,a_{q-1})\\
\sigma_i(g_0,\ldots,g_p \mid a_0,\ldots,a_q) &= (g_0,\ldots,g_p \mid a_0,\ldots,a_i,1,a_{i+1},\ldots,a_q) \quad 0\le i\le q\\
t(g_0,\ldots,g_p \mid a_0,\ldots,a_q) &= (g_p^{(q+2)},g_0,\ldots,g_{p-1} \mid g_p^{(1)}(a_0),\ldots,g_p^{(q+1)}(a_q))\\
d_i(g_0,\ldots,g_p \mid a_0,\ldots,a_q) &= (g_0,\ldots,g_i^{(1)}g_{i+1}^{(1)},\ldots,g_p \mid \\
&\qquad \sigma(g_i^{(2)},g_{i+1}^{(2)})a_0,\ldots,a_q) \quad 0\le i<p\\
d_p(g_0,\ldots,g_p \mid a_0,\ldots,a_q) &= (g_p^{(q+2)}g_0^{(1)},g_1,\ldots,g_{p-1} \mid \\
&\qquad \sigma(g_p^{(q+3)},g_0^{(2)})g_p^{(1)}(a_0),\ldots,g_p^{(q+1)}(a_q))\\
s_i(g_0,\ldots,g_p \mid a_0,\ldots,a_q) &= (g_0,\ldots,g_i,1,g_{i+1},\ldots,g_p \mid a_0,\ldots,a_q) \quad 0\le i\le p.
\end{aligned}
$$

Theorem 3.1 *Endowed with the above operators, $A\natural_\sigma\mathcal{H}$ is a cylindrical module.*

Proof We should check that every row and every column is a paracyclic module, and vertical operator commutes with each horizontal operator, and in addition the identity (2.1) holds. Since the weak action in our situation is actually an action and the vertical operators are the same as the vertical operators in ([1] Theorem 3.1), we refer the reader to [1] for the proof that the columns form paracyclic modules. To check that the rows are paracyclic modules we need to verify the following identities

$$
\begin{aligned}
d_id_j &= d_{j-1}d_i \qquad i<j\\
s_is_j &= s_{j+1}s_i \qquad i\le j\\
d_is_j &= \begin{cases} s_{j-1}d_i & i<j\\ \text{identity} & i=j \text{ or } i=j+1\\ s_jd_{i-1} & i>j+1.\end{cases}
\end{aligned}
$$

$$
\begin{aligned}
d_it_n = t_{n-1}d_{i-1} \quad 1\le i\le n\,, &\quad d_0t_n = d_n\\
s_it_n = t_{n+1}s_{i-1} \quad 1\le i\le n\,, &\quad s_0t_n = t_{n+1}^2s_n
\end{aligned}
$$

We just check $d_id_{i+1} = d_id_i$ and the cylindrical module condition (2.1). The rest can be proved by the same techniques. We have:

$$
\begin{aligned}
&d_id_{i+1}(g_0,\ldots,g_p \mid a_0,\ldots,a_q)\\
&= d_i(g_0,\ldots,g_{i+1}^{(1)}g_{i+2}^{(1)},\cdots g_p \mid \sigma(g_{i+1}^{(2)},g_{i+2}^{(2)})a_0,\ldots,a_q)\\
&= (g_0,\ldots,g_i^{(1)}g_{i+1}^{(1)}g_{i+2}^{(1)},\cdots g_p \mid \sigma(g_i^{(2)},g_{i+1}^{(2)}g_{i+2}^{(2)})\sigma(g_{i+1}^{(3)},g_{i+2}^{(3)})a_0\ldots,a_q),
\end{aligned}
$$

that by using the cocycle property 2) in Section 2 is equal to

$$
\begin{aligned}
&(g_0, \ldots, g_i^{(1)} g_{i+1}^{(1)} g_{i+2}^{(1)}, \ldots g_p \mid \sigma(g_i^{(2)}, g_{i+1}^{(2)}) \sigma(g_i^{(3)} g_{i+1}^{(3)}, g_{i+2}^{(3)}) a_0 \ldots, a_q) \\
&\qquad = d_i d_i (g_0, \ldots, g_p \mid a_0, \ldots, a_q).
\end{aligned}
$$

Next we check the cylindrical module condition (2.1). We have:

$$
\begin{aligned}
& t^{p+1} \tau^{q+1} (g_0, \ldots, g_p \mid a_0, \ldots, a_q) \\
= \;& t^{p+1} \tau^{q} (g_0^{(2)}, \ldots, g_p^{(2)} \mid S(g_0^{(1)} g_1^{(1)} \ldots g_p^{(1)}) \cdot a_q, a_0, \ldots, a_{q-1}) \\
= \;& t^{p+1} (g_0^{(q+1)}, \ldots, g_p^{(q+1)} \mid S(g_0^{(q)} \ldots g_p^{(q)}) \cdot a_0, S(g_0^{(q-1)} \ldots g_p^{(q-1)}) \cdot a_1, \\
& \qquad \ldots, S(g_0^{(0)} \ldots g_p^{(0)}) \cdot a_q) \\
= \;& t^{p} (g_p^{(2q+2)}, g_0^{(q+1)}, \ldots, g_{p-1}^{(q+1)} \mid (g_p^{(q+1)} S(g_0^{(q)} \ldots g_p^{(q)})) \cdot a_0, \\
& \qquad (g_p^{(q+2)} S(g_0^{(q-1)} \ldots g_p^{(q-1)}) \cdot a_1, \ldots, (g^{(2q+1)} S(g_0^{(0)} \ldots g_p^{(0)})) \cdot a_q) \\
= \;& t^{p} (g_p^{(2q)}, g_0^{(q+1)}, \ldots, g_{p-1}^{(q+1)} \mid (S(g_0^{(q)} \ldots g_{p-1}^{(q)})) \cdot a_0, \\
& \qquad (g_p^{(q)} S(g_0^{(q-1)} \ldots g_p^{(q-1)}) \cdot a_1, \ldots, (g_p^{(2q-1)} S(g_0^{(0)} \ldots g_p^{(0)})) \cdot a_q) \\
= \;& t^{p} (g_p, g_0^{(q+1)}, \ldots, g_{p-1}^{(q+1)} \mid (S(g_0^{(q)} \ldots g_{p-1}^{(q)})) \cdot a_0, \\
& \qquad (S(g_0^{(q-1)} \ldots g_{p-1}^{(q-1)}) \cdot a_1, \ldots, (S(g_0^{(0)} \ldots g_{p-1}^{(0)})) \cdot a_q) \\
= \;& (g_0, \ldots, g_p \mid a_0, \ldots, a_q).
\end{aligned}
$$

The theorem is proved. □

Next we show that the diagonal of the above cylindrical module, $d(A\natural_\sigma \mathcal{H})$, is isomorphic with the cyclic module $(A\#_\sigma \mathcal{H})^\natural$ associated with the crossed product algebra. To this end we define maps $\Phi : (A\#_\sigma \mathcal{H})^\natural \to d(A\natural_\sigma \mathcal{H})$ and $\Psi : d(A\natural_\sigma \mathcal{H}) \to (A\#_\sigma \mathcal{H})^\natural$ by the following formulas

$$
\begin{aligned}
& \Phi(a_0 \otimes g_0, \ldots, a_n \otimes g_n) \\
= \;& (g_0^{(2)}, g_1^{(3)}, \ldots, g_n^{(n+2)} \mid S(g_0^{(1)} g_1^{(2)} \ldots g_n^{(n+1)}) \cdot a_0, S(g_1^{(1)} g_2^{(2)} \ldots g_n^{(n)}) \cdot a_1, \ldots, \\
& \qquad S(g_{n-1}^{(1)} g_n^{(2)}) \cdot a_{n-1}, S(g_n^{(1)}) \cdot a_n),
\end{aligned}
$$

$$
\begin{aligned}
& \Psi(g_0, \ldots, g_n \mid a_0, \ldots, a_n) \\
= \;& ((g_0^{(1)} g_1^{(1)} \ldots g_n^{(1)}) \cdot a_0 \otimes g_0^{(2)}, (g_1^{(2)} \ldots g_n^{(2)}) \cdot a_1 \otimes g_1^{(3)}, \ldots, g_n^{(n+1)} \cdot a_n \otimes g_n^{(n+2)}).
\end{aligned}
$$

Theorem 3.2 *The above maps, Φ, Ψ, are morphisms of cyclic modules and are inverse to one another.*

Proof It is not hard to see that Φ and Ψ are inverse of each other. We just prove that Φ is a cyclic map. We first verify the commutativity of Φ and the cyclic operators, i.e., the relation $t\tau\Phi = \Phi\tau_{A\#_\sigma\mathcal{H}}$. We have:

$$
\begin{aligned}
&(t\tau)\Phi(a_0 \otimes g_0, \dots, a_n \otimes g_n) \\
&= t\tau(g_0^{(2)}, g_1^{(3)}, \dots, g_n^{(n+2)} \mid S(g_0^{(1)} g_1^{(2)} \dots g_n^{(n+1)}) \cdot a_0, S(g_1^{(1)} g_2^{(2)} \dots g_n^{(n)}) \cdot a_1, \\
&\qquad \dots, S(g_{n-1}^{(1)} g_n^{(2)}) \cdot a_{n-1}, S(g_n^{(1)}) \cdot a_n) \\
&= t(g_0^{(3)}, g_1^{(4)}, \dots, g_n^{(n+3)} \mid S(g_0^{(2)} g_1^{(3)} \dots g_n^{(n+2)}) S(g_n^{(1)}) \cdot a_n, \\
&\qquad S(g_0^{(1)} g_1^{(2)} \dots g_n^{(n+1)}) \cdot a_0 \dots, S(g_{n-1}^{(1)} g_n^{(2)}) \cdot a_{n-1}) \\
&= (g_n^{((2n+4)}, g_0^{(3)}, g_1^{(4)}, \dots, g_{n-1}^{(n+2)} \mid g_n^{(n+3)} S(g_0^{(2)} g_1^{(3)} \dots g_n^{(n+1)}) S(g_n^{(1)}) \cdot a_n, \\
&\qquad g_n^{(n+4)} S(g_0^{(1)} g_1^{(2)} \dots g_n^{(n+1)}) \cdot a_0 \dots, g_n^{(2n+3)} S(g_{n-1}^{(1)} g_n^{(2)}) \cdot a_{n-1}) \\
&= (g_n^{((2n+3)}, g_0^{(3)}, g_1^{(4)}, \dots, g_{n-1}^{(n+2)} \mid \epsilon(g_n^{(n+2)}) S(g_n^{(1)} g_0^{(2)} \dots g_{n-1}^{(n+1)}) \cdot a_n, \\
&\qquad g_n^{(n+3)} S(g_0^{(1)} g_1^{(2)} \dots g_n^{(n+1)}) \cdot a_0, \dots, g_n^{(2n+2)} S(g_{n-1}^{(1)} g_n^{(2)}) \cdot a_{n-1}) \\
&= (g_n^{((2n+2)}, g_0^{(3)}, g_1^{(4)}, \dots, g_{n-1}^{(n+2)} \mid S(g_n^{(1)} g_0^{(2)} \dots g_{n-1}^{(n+1)}) \cdot a_n, \\
&\qquad \epsilon(g_n^{(n+1)}) S(g_0^{(1)} g_1^{(2)} \dots g_{n-1}^{(n)}) \cdot a_0, \dots, g_n^{(2n)} S(g_{n-1}^{(1)} g_n^{(2)}) \cdot a_{n-1}) \\
&= (g_n^{(2)}, g_0^{(3)}, g_1^{(4)}, \dots, g_{n-1}^{(n+2)} \mid S(g_n^{(1)} g_0^{(2)} \dots g_{n-1}^{(n+1)}) \cdot a_n, \\
&\qquad S(g_0^{(1)} g_1^{(2)} \dots g_{n-1}^{(n)}) \cdot a_0, \dots, \dots, \epsilon(g_n^{(2)}) S(g_{n-1}^{(1)}) \cdot a_{n-1}) \\
&= (g_n^{(2)}, g_0^{(3)}, \dots, g_{n-1}^{(n+2)} \mid S(g_n^{(1)} g_0^{(2)} \dots g_{n-1}^{(n+1)}) \cdot a_n, \\
&\qquad S(g_0^{(1)} g_1^{(2)} \dots g_{n-1}^{(n)}) \cdot a_0, \dots, \dots, S(g_{n-1}^{(1)}) \cdot a_{n-1}) \\
&= \Phi(a_n \otimes g_n, a_0 \otimes g_0, \dots, a_{n-1} \otimes g_{n-1}) = \Phi(\tau_{A\#_\sigma\mathcal{H}}(a_0 \otimes g_0, \dots, a_n \otimes g_n)).
\end{aligned}
$$

Next we check the commutativity of Φ and the face operators, i.e. the relation $d_i\delta_i\Phi = \Phi d_i^{A\#_\sigma\mathcal{H}}$. For $0 \le i < n$, we have:

$$
\begin{aligned}
&d_i\delta_i\Phi(a_0 \otimes g_0, \dots, a_n \otimes g_n) \\
&= d_i\delta_i(g_0^{(2)}, g_1^{(3)}, \dots, g_n^{(n+2)} \mid S(g_0^{(1)} g_1^{(2)} \dots g_n^{(n+1)}) \cdot a_0, S(g_1^{(1)} g_2^{(2)} \dots g_n^{(n)}) \cdot a_1, \\
&\qquad \dots, S(g_{n-1}^{(1)} g_n^{(2)}) \cdot a_{n-1}, S(g_n^{(1)}) \cdot a_n) \\
&= d_i((g_0^{(2)}, g_1^{(3)}, \dots, g_n^{(n+2)} \mid S(g_0^{(1)} g_1^{(2)} \dots g_n^{(n+1)}) \cdot a_0, S(g_1^{(1)} g_2^{(2)} \dots g_n^{(n)}) \cdot a_1, \\
&\qquad \dots, S(g_{i+1}^{(1)}, \dots, g_n^{(n+1-i)}) \cdot (S(g_i^{(1)})(a_i)a_{i+1}), \\
&\qquad \dots, S(g_{n-1}^{(1)} g_n^{(2)}) \cdot a_{n-1}, S(g_n^{(1)}) \cdot a_n)) \\
&= ((g_0^{(2)}, g_1^{(3)}, \dots, g_i^{(i+2)} g_{i+1}^{(i+3)}, \dots g_n^{(n+2)} \mid \sigma(g_i^{(i+3)}, g_{i+1}^{(i+4)}) \\
&\quad S(g_0^{(1)} g_1^{(2)} \dots g_n^{(n+1)}) \cdot a_0, S(g_1^{(1)} g_2^{(2)} \dots g_n^{(n)}) \cdot a_1, \dots, S(g_{i+1}^{(1)} \dots g_n^{(n+1-i)}) \\
&\quad (S(g_i^{(1)})(a_i)a_{i+1}), \dots, S(g_{n-1}^{(1)} g_n^{(2)}) \cdot a_{n-1}, S(g_n^{(1)}) \cdot a_n)) \\
&= \Phi d_i^{A\#_\sigma\mathcal{H}}(a_0 \otimes g_0, \dots, a_n \otimes g_n).
\end{aligned}
$$

For $i=n$, we have:

$$
\begin{aligned}
& d_n\delta_n\Phi(a_0\otimes g_0,\ldots,a_n\otimes g_n)\\
&= d_n\delta_n(g_0^{(2)},g_1^{(3)},\ldots,g_n^{(n+2)}\mid S(g_0^{(1)}g_1^{(2)}\ldots g_n^{(n+1)})\cdot a_0,S(g_1^{(1)}g_2^{(2)}\ldots g_n^{(n)})\cdot a_1,\\
&\qquad \ldots,S(g_{n-1}^{(1)}g_n^{(2)})\cdot a_{n-1},S(g_n^{(1)})\cdot a_n)\\
&= d_n(g_0^{(3)},g_1^{(4)},\ldots,g_n^{(n+3)}\mid\\
&\qquad (S(g_0^{(2)}g_1^{(3)}\ldots g_n^{(n+2)})S(g_n^{(1)})\cdot a_n)(S(g_0^{(1)}g_1^{(2)}\ldots g_n^{(n+1)})\cdot a_0),\\
&\qquad S(g_1^{(1)}g_2^{(2)}\ldots g_n^{(n)})\cdot a_1,\ldots,S(g_{n-1}^{(1)}g_n^{(2)})\cdot a_{n-1})\\
&= (g_n^{(2n+4)}g_0^{(3)},g_1^{(4)},\ldots,g_{n-1}^{(n+2)}\mid\\
&\qquad \sigma(g_n^{(2n+5)},g_0^{(4)})g_n^{(n+3)}\cdot((S(g_0^{(2)}g_1^{(3)}\ldots g_n^{(n+2)})S(g_n^{(1)})\cdot a_n)\\
&\qquad (S(g_0^{(1)}g_1^{(2)}\ldots g_n^{(n+1)})\cdot a_0)),S(g_1^{(1)}g_2^{(2)}\ldots g_n^{(n)})\cdot a_1,\ldots,\\
&\qquad S(g_{n-1}^{(1)}g_n^{(2)})\cdot a_{n-1})\\
&= \Phi d_n^{A\#_\sigma\mathcal{H}}(a_0\otimes g_0,\ldots,a_n\otimes g_n).
\end{aligned}
$$

The commutativity of Φ and the degeneracies are easier to check and is left to the reader. The theorem is proved. □

Let $H_\sigma=k\underset{\sigma}{\#}\mathcal{H}$ denote the crossed product of $\mathcal{H}$ and k where $\mathcal{H}$ acts on k via the counit ϵ. One can check that the q-th row of the cylindrical module $A\underset{\sigma}{\natural}\mathcal{H}$ is the standard Hochschild complex of the algebra $\mathcal{H}_\sigma$ with coefficients in the bimodule $M_q=\mathcal{H}\otimes A^{\otimes(q+1)}$. Here H_σ acts on M_q on the left and right by

$$h\cdot(g\otimes a_0\otimes\ldots\otimes a_q)=\sigma(h^{(q+3)},g^{(2)})h^{(q+2)}g^{(1)}\otimes h^{(1)}a_0\otimes\ldots\otimes h^{(q+1)}a_q$$

$$(g\otimes a_0\otimes\ldots\otimes a_q)\cdot h=\sigma(g^{(2)},h^{(2)})g^{(1)}h^{(1)}\otimes a_0\otimes\ldots\otimes a_q.$$

For the proof of Theorem 3.4 we need an extension of Mac Lane's isomorphism, which relates group homology to Hochschild homology, to Hopf algebras.

We recall that the Hopf homology of a Hopf algebra $\mathcal{H}$ with coefficients in a left $\mathcal{H}$-module M is the homology of the following complex

$$M\overset{d_0}{\leftarrow}\mathcal{H}\otimes M\overset{d_1}{\leftarrow}\mathcal{H}\otimes\mathcal{H}\otimes M\overset{d_2}{\leftarrow}\ldots\mathcal{H}^{\otimes n}\otimes M\overset{d_n}{\leftarrow}\mathcal{H}^{\otimes(n+1)}\otimes M\leftarrow\ldots,$$

where the differential d_n is given by

$$
\begin{aligned}
& d_n(h_0\otimes h_1\otimes\ldots\otimes h_n\otimes m)=\epsilon(h_0)h_1\otimes\ldots\otimes h_n\otimes m+\\
& \sum_{1\le i\le n-1}(-1)^i h_0\otimes\ldots\otimes h_ih_{i+1}\otimes\ldots\otimes h_n\otimes m+(-1)^n h_0\otimes h_1\otimes\ldots\otimes h_{n-1}\otimes hm.
\end{aligned}
$$

We denote the nth Hopf homology group of $\mathcal{H}$ with coefficients in M by $H_n(\mathcal{H};M)$.

Let M be an $\mathcal{H}_\sigma$-bimodule. We can convert M to a new left $\mathcal{H}$-module, $\widetilde{M}=M$, where the action of $\mathcal{H}$ on $\widetilde{M}$ is defined by

$$h\blacktriangleright m=\sigma^{-1}(S(h^{(2)}),h^{(3)})\overline{h^{(4)}}m\overline{S(h^{(1)})},\tag{3.1}$$

where $\bar{h}$ denotes the image of h in $\mathcal{H}_\sigma$ under the map $h\to 1\#h$. Note that in the proof of the following lemma the cocommutativity of $\mathcal{H}$ is used.

Lemma 3.3 *Let M be an $\mathcal{H}_\sigma$-bimodule. Then by the above definition M is a left $\mathcal{H}$-module, i.e., $g\blacktriangleright(h\blacktriangleright m)=(gh)\blacktriangleright m$, for all $g,h\in\mathcal{H}$ and $m\in M$.*

Proof We have

$$
\begin{aligned}
& g \blacktriangleright (h \blacktriangleright m) \\
= \; & g \blacktriangleright (\sigma^{-1}(S(h^{(2)}), h^{(3)})\overline{h^{(4)}} m \overline{S(h^{(1)})}) \\
= \; & \sigma^{-1}(S(g^{(2)}), g^{(3)})\sigma^{-1}(S(h^{(2)}), h^{(3)})\overline{g^{(4)}}(\overline{h^{(4)}} m \overline{S(h^{(1)})})\overline{S(g^{(1)})} \\
= \; & \sigma^{-1}(S(g^{(3)}), g^{(4)})\sigma^{-1}(S(h^{(3)}), h^{(4)})\sigma(g^{(5)}, h^{(5)}) \\
& \sigma(S(h^{(2)}), S(g^{(2)}))\overline{g^{(5)}h^{(5)}} m \overline{S(g^{(1)}h^{(1)})} \\
= \; & \sigma^{-1}(S(g^{(3)}), g^{(4)})\sigma^{-1}(S(h^{(3)}), h^{(4)})\sigma(g^{(5)}, h^{(5)})\sigma(S(h^{(2)}), S(g^{(2)})) \\
& \sigma(S(h^{(9)})S(g^{(9)}), g^{(6)}h^{(6)})\sigma^{-1}(S(h^{(8)}S(g^{(8)}), g^{(7)}h^{(7)}) \\
& \overline{g^{(5)}h^{(5)}} m \overline{S(g^{(1)}h^{(1)})} \\
= \; & \sigma^{-1}(S(g^{(3)}), g^{(4)})\sigma^{-1}(S(h^{(3)}), h^{(4)})\sigma(g^{(5)}, h^{(5)})\sigma(S(h^{(2)}), S(g^{(2)})) \\
& \sigma(S(h^{(8)})S(g^{(8)}), g^{(6)}h^{(6)})\sigma^{-1}(S(h^{(9)}S(g^{(9)}), g^{(7)}h^{(7)}) \\
& \overline{g^{(5)}h^{(5)}} m \overline{S(g^{(1)}h^{(1)})} \\
= \; & \sigma(S(h^{(9)})S(g^{(9)}), g^{(5)})\sigma(S(h^{(8)}), h^{(5)})\sigma^{-1}(S(g^{(3)}), g^{(4)})\sigma^{-1}(S(h^{(3)}), h^{(4)}) \\
& \sigma(S(h^{(2)}), S(g^{(2)}))\sigma^{-1}(S(h^{(9)}S(g^{(9)}), g^{(7)}h^{(7)})\overline{g^{(5)}h^{(5)}} m \overline{S(g^{(1)}h^{(1)})} \\
= \; & \sigma(S(h^{(5)})S(g^{(5)}), g^{(8)})\sigma^{-1}(S(g^{(5)}), g^{(6)})\sigma(S(h^{(6)}), S(g^{(6)})) \\
& \sigma^{-1}(S(h^{(4)}S(g^{(4)}), g^{(3)}h^{(3)})\overline{g^{(2)}h^{(2)}} m \overline{S(g^{(1)}h^{(1)})} \\
= \; & \sigma^{-1}(S(g^{(8)}), g^{(7)})\sigma(S(g^{(6)}), g^{(5)})\sigma^{-1}(S(h^{(4)}S(g^{(4)}), g^{(3)}h^{(3)}) \\
& \overline{g^{(2)}h^{(2)}} m \overline{S(g^{(1)}h^{(1)})} \\
= \; & \sigma^{-1}(S(h^{(2)})S(g^{(2)}), g^{(3)}h^{(3)})\overline{g^{(4)}h^{(4)}} m \overline{S(g^{(1)}h^{(1)})} = (gh) \blacktriangleright m.
\end{aligned}
$$

□

The following result was first proved in [10] for σ a trivial cocycle.

Theorem 3.4 (Mac Lane Isomorphism for Hopf crossed products) *Let M be an $\mathcal{H}_\sigma$-bimodule and $\widetilde{M}$ be defined as above. Then the following map defines an isomorphism between Hochschild and Hopf homology complexes:*

$$
\begin{gathered}
\Theta : C_n(\mathcal{H}_\sigma, M) \longrightarrow C_n(\mathcal{H}; \widetilde{M}) \\
\Theta(\bar{h}_1 \otimes \ldots \otimes \bar{h}_n \otimes m) = h_1^{(2)} \otimes h_2^{(2)} \otimes \ldots \otimes h_n^{(2)} \otimes m\overline{h_1^{(1)}} \ldots \overline{h_n^{(1)}}.
\end{gathered}
$$

Proof We show more than what we need for the proof, namely we show that Θ is an isomorphisms of of simplicial modules. We have:

$$
\begin{aligned}
& \Theta\delta_0(\bar{h}_1 \otimes \ldots \otimes \bar{h}_n \otimes m) = \Theta(\bar{h}_2 \otimes \ldots \otimes \bar{h}_n \otimes m\bar{h}_1) \\
& \quad = h_2^{(2)} \otimes \ldots \otimes h_n^{(2)} \otimes m\overline{h_1}\overline{h_2^{(1)}} \ldots \overline{h_n^{(1)}} = \delta_0\Theta(\bar{h}_1 \otimes \ldots \otimes \bar{h}_n \otimes m).
\end{aligned}
$$

For $0 \leq i \leq n$, we have

$$
\begin{aligned}
& \Theta\delta_i(\bar{h}_1 \otimes \ldots \otimes \bar{h}_n \otimes m) \\
& \quad = \; \Theta(\bar{h}_1 \otimes \ldots \otimes (\overline{h_i})(\overline{h_{i+1}}) \otimes \ldots \otimes \bar{h}_n \otimes m) \\
& \quad = \; h_1^{(2)} \otimes h_2^{(2)} \otimes \ldots \otimes h_i^{(2)} h_{i+1}^{(2)} \otimes \ldots \otimes h_n^{(2)} \otimes m\overline{h_1^{(1)}} \ldots \overline{h_n^{(1)}} \\
& \quad = \; \delta_i\Theta(\bar{h}_1 \otimes \ldots \otimes \bar{h}_n \otimes m).
\end{aligned}
$$

We leave it to the reader to check the commutativity of Θ with the last face and the degeneracies.

To finish the proof one can check that the following map is the inverse of Θ

$$\mathfrak{T} : C_n(\mathcal{H}; \widetilde{M}) \longrightarrow C_n(\mathcal{H}_\sigma, M)$$

$$\begin{aligned}
&\mathfrak{T}(h_1 \otimes \ldots \otimes h_n \otimes m) \\
&= \sigma^{-1}(S(h_1^{(2)}), h_1^{(3)}) \ldots \sigma^{-1}(S(h_n^{(2)}), h_n^{(3)}) \overline{h_1^{(4)}} \otimes \ldots \otimes \overline{h_n^{(4)}} \\
&\quad \otimes m \overline{S(h_n^{(1)})} \ldots \overline{S(h_1^{(1)})}.
\end{aligned}$$

□

We apply the generalized cyclic Eilenberg-Zilber theorem (Theorem 2.1) to the cylindrical module $A\natural_\sigma\mathcal{H}$ to derive a spectral sequence for the cyclic homology of $A\#_\sigma\mathcal{H}$. We have

$$Tot(A\natural_\sigma\mathcal{H}) \to d(A\natural_\sigma\mathcal{H}) \cong (A\#_\sigma\mathcal{H})^\natural,$$

where the first map is a quasi-isomorphism of mixed complexes given in Theorem 2.1 and the second map is the isomorphism given in Theorem 3.2. We filter the mixed complex $Tot(A\natural_\sigma\mathcal{H})$ by sub mixed complexes

$$F^i(Tot(A\natural_\sigma\mathcal{H}))_n = \bigoplus_{\substack{p+q=n \\ q\leq i}} A^{\otimes(p+1)} \otimes \mathcal{H}^{\otimes(q+1)}.$$

This gives us an spectral sequence that converges to $HC_\bullet(A\#_\sigma\mathcal{H})$. We can then apply Theorem 3.3 to identify the E^1-term of this spectral sequence, i.e. the homology of rows, as Hopf homologies of $\mathcal{H}$, with coefficients in $M_q = \mathcal{H} \otimes A^{\otimes(q+1)}$, where $\mathcal{H}$ acts on M_q by

$$\begin{aligned}
&h \blacktriangleright (g \otimes a_0 \otimes a_1 \otimes \ldots \otimes a_q) = \sigma^{-1}(S(h^{(3)}), h^{(4)})\sigma(h^{(q+6)}, g^{(1)})\sigma(h^{(q+7)}g^{(2)}, S(h^2)) \\
&h^{(q+8)}g^{(3)}S(h^{(1)}) \otimes h^{(5)}(a_0) \otimes \ldots \otimes h^{(q+5)}(a_q). \qquad (3.2)
\end{aligned}$$

This proves the following theorem.

Theorem 3.5 *There is a spectral sequence that converges to $HC_{p+q}(A\#_\sigma\mathcal{H})$. The E^1-term of this spectral sequence is given by*

$$E^1_{p,q} = H_p(\mathcal{H}; M_q).$$

Given any cylindrical module $X = \{X_{p,q}\}_{p,q\geq 0}$, if we compute the Hochschild homologies of rows of X we obtain a new bigraded k-module $X' = \{X'_{p,q}\}_{p,q\geq 0}$. We claim that the columns of X', i.e. $\{X'_{p,q}\}_{q\geq 0}$ form a cyclic module for each $p \geq 0$. For some special cases one can find the proof in [8, 1]. The same proof, however, works in the general case. This observation proves the following proposition.

Proposition 3.6 *The pth column of E^1, i.e. $\{H_p(\mathcal{H}; M_q)\}_{q\geq 0}$ is a cyclic module for each $p \geq 0$.*

We denote the pth column of E^1 by N_p. One can observe that the induced differential d^1 on E^1 is simply the differential $b+B$ associated to the cyclic modules N_p. This finishes the proof of the following theorem.

Theorem 3.7 *The E^2 term of the spectral sequence in Theorem 3.5 is*

$$E^2_{p,q} = HC_q(N_p).$$

Recall that if $\mathcal{H}$ is a semisimple Hopf algebra, then for any $\mathcal{H}$-module M, $H_i(\mathcal{H}, M) = 0$ for $i \geq 1$. From this and Theorem 3.5, we obtain the following corollary.

Corollary 3.8 *Let $\mathcal{H}$ be semisimple. Then the above spectral sequence collapses and we obtain*

$$HC_q(A \underset{\sigma}{\#} \mathcal{H}) \cong HC_q(N_0).$$

Since $N_0 = H_0(\mathcal{H}, M_\bullet)$, we obtain

$$N_{0,q} = M_q^{\mathcal{H}} = (\mathcal{H} \otimes A^{\otimes(q+1)})^{\mathcal{H}},$$

where the action of $\mathcal{H}$ is defined by 3.2.

References

[1] Akbarpour, R. and Khalkhali, M. *Hopf Algebra Equivariant Cyclic Homology and Cyclic Homology of Crossed Product Algebras*, J. reine angew. Math. **559** (2003), 137-152.

[2] Blattner, R. J., Cohen, R. J. and Montgomery, S. *Crossed products and inner action of Hopf algebras*, Trans. Amer. Math. Soc. **298** (1986), 671-711.

[3] Connes, A. *Noncommutative geometry*, Academic Press, Inc., San Diego, CA, 1994.

[4] Connes, A. *Cohomologie cyclique et foncteurs* Ext^n , (French) (Cyclic cohomology and functors Ext^n) C. R. Acad. Sci. Paris Sr. I Math. **296** (1983), no. 23, 953–958.

[5] Doi, Y. and Takeuchi, M. *Cleft comodule algebras for a bialgebra*, Comm. in Alg. **14** (1986), 801-817.

[6] Douglas, M. *D-branes and discrete torsion I, II.* hep-th/9903031.

[7] Feĭgin, B. and Tsygan, B. L. *Additive K-theory,* K-theory, arithmetic and geometry (Moscow, 1984–1986), 67–209, Lecture Notes in Math., **1289**, Springer, Berlin, 1987.

[8] Getzler, E. and Jones, J. D. S. *The cyclic homology of crossed product algebras,* J. Reine Angew. Math. **445** (1993), 161–174.

[9] Guccione, J. A. and Guccione, J. J. *Hochschild (co)homology of Hopf crossed products,* math.KT/0104075.

[10] Khalkhali, M. and Rangipour, B. *A new cyclic module for Hopf algebras,* K-Theory **27**, 111-131, Nov. 2002.

[11] Khalkhali, M. and Rangipour, B. *On the generalized cyclic Eilenberg-Zilber theorem,* to appear in Canadian Math. Bulletin.

[12] Konechny, A. and Schwartz, A. *Introduction to M(atrix) theory and noncommutative geometry, part I,* hep-th/0107251.

[13] Loday, J. L. Cyclic Homology, Springer-Verlag, (1992).

[14] Montgomery, S. *Hopf algebras and their actions on rings, CBMS regional conference series in Mathematics,* **82**, AMS, 1993.

[15] Nistor, V. *Group cohomology and the cyclic cohomology of crossed products,* Invent. Math. **99**, (1989) 411-424.

[16] Rieffel, M. A. *Morita equivalence for C^*-algebras and W^*-algebras,* J. pure Apll. Algebras **5** (1974), 51-96.

[17] Sweedler, M. *Hopf algebras,* Mathematics Lecture Note Series W. A. Benjamin, Inc., New York 1969.

Received November 19, 2002; in revised form June 26, 2003

Fields Institute Communications
Volume **43**, 2004

On Sequentially h-complete Groups

Gábor Lukács
Department of Mathematics & Statistics
York University
4700 Keele Street
Toronto, ON, Canada M3J 1P3
Current address: FB 3 - Mathematik und Informatik
Universität Bremen
Bibliothekstrasse 1
28359 Bremen, Germany
lukacs@mathstat.yorku.ca

Abstract. A topological group G is *sequentially h-complete* if all the continuous homomorphic images of G are sequentially complete. In this paper we give necessary and sufficient conditions on a complete group for being compact, using the language of sequential h-completeness. In the process of obtaining such conditions, we establish a structure theorem for ω-precompact sequentially h-complete groups. As a consequence we obtain a reduction theorem for the problem of c-compactness.

All topological groups in this paper are assumed to be Hausdorff.

A topological group G is *sequentially h-complete* if all the continuous homomorphic images of G are sequentially complete (i.e., every Cauchy-sequence converges). G is called *precompact* if for any neighborhood U of the identity element there exists a finite subset F of G such that $G = UF$.

In [6, Theorem 3.6] Dikranjan and Tkačenko proved that nilpotent sequentially h-complete groups are precompact (also see [4]). Thus, if a group is nilpotent, sequentially h-complete and complete, then it is compact.

Inspired by this result, the aim of this paper is to give necessary and sufficient conditions on a complete group for being compact, using the language of sequential h-completeness. This aim is carried out in Theorem 6.

For an infinite cardinal τ, a topological group G is *τ-precompact* if for any neighborhood U of the identity element there exists $F \subset G$ such that $G = UF$ and $|F| \leq \tau$. In order to prove Theorem 6, we will first establish a strengthened version of the Guran's Embedding Theorem for ω-precompact sequentially h-completely groups (Theorem 5).

2000 *Mathematics Subject Classification.* Primary 22A05, 22C05; Secondary 54D30.
I gratefully acknowledge the financial support received from York University that enabled me to do this research.

A topological group G is *c-compact* if for any topological group H the projection $\pi_H : G \times H \to H$ maps closed subgroups of $G \times H$ onto closed subgroups of H (see [12], [5] and [2], as well as [3]). The problem of whether every c-compact topological group is compact has been an open question for more than ten years. As a consequence of Theorem 6, we obtain that the problem of c-compactness can be reduced to the second-countable case (Theorem 9).

The following Theorem is a slight generalization of Theorem 3.2 from [8]:

Theorem 1 *Let G be an ω-precompact sequentially h-complete topological group. Then every continuous homomorphism $f : G \to H$ onto a group H of countable pseudocharacter is open.*

In order to prove Theorem 1, we need the following three facts, two of which are due to Guran:

Fact A (Guran's Embedding Theorem) *A topological group is τ-precompact if and only if it is topologically isomorphic to a subgroup of a direct product of topological groups of weight $\leq \tau$.* (Theorem 4.1.3 in [14].)

Fact B (Banach's Open Map Theorem) *Any continuous homomorphism from a separable complete metrizable group onto a Baire group is open.* (Corollary V.4 in [10].)

Fact C *Let G be an ω-precompact topological group of countable pseudocharacter. Then G admits a coarser second countable group topology.* (Corollary 4 in [9].)

The proof below is just a slight modification of the proof of Theorem 3.2 from [8] mentioned above:

Proof First, suppose that H is metrizable. Let U be a neighborhood of e in G. Since, by Fact A, G embeds into a product of separable metrizable group, we may assume that $U = g^{-1}(V)$ for some continuous homomorphism $g : G \to M$ onto a separable metrizable group M and some neighborhood V of e in M. Let $h = (f, g) : G \to H \times M$, and put $L = h(G)$. Let $p : L \to H$ and $q : L \to M$ be the restrictions of the canonical projections $H \times M \to H$ and $H \times M \to M$. Clearly, one has $f = ph$ and $g = qh$. Since q is continuous, $W = q^{-1}(V)$ is open. We have $h(U) = h(g^{-1}(V)) = h(h^{-1}q^{-1}(V)) = W$ and $f(U) = ph(U) = p(W)$. Since G is sequentially h-complete, the groups L and H, being homomorphic images of G, are sequentially complete. Since H and L are metrizable, this means that they are simply complete. They are also separable, because they are metrizable and ω-precompact. Thus, by Fact B, $p : L \to H$ is open, and hence $f(U) = p(W)$ is open in H.

To show the general case, suppose that the topology $\mathcal{T}$ of the group H is of countable pseudocharacter; then H admits a coarser second countable topology $\mathcal{T}'$ (by Fact C); put $\iota : (H, \mathcal{T}) \to (H, \mathcal{T}')$ to be the identity map. Since $\mathcal{T}'$ is metrizable, the continuous homomorphism $\iota \circ \varphi : G \to (H, \mathcal{T}')$ is open by what we have already proved, and thus φ is also open. □

Corollary 2 *Let G be an ω-precompact sequentially h-complete group. The following statements are equivalent:*

(i) *G is second countable;*

(ii) *G contains a countable network;*

(iii) *G has a countable pseudocharacter;*
(iv) *G is metrizable.*

Proof The implications (i) $\Rightarrow$ (ii) $\Rightarrow$ (iii) are obvious, and so is the equivalence of (i) and (iv).

(iii) $\Rightarrow$ (iv): If $(G, \mathcal{T})$ is of countable pseudocharacter, it admits a coarser second countable topology $\mathcal{T}'$(by Fact C); put $\iota : G \to G_1$ to be the identity map. By Theorem 1, ι is open, and thus $\mathcal{T}$ is second countable, as desired. □

A topological group G is *minimal* if every continuous isomorphism $\varphi : G \to H$ is a homeomorphism, or equivalently, if the topology of G is a coarsest (Hausdorff) group topology on G. The group G is *totally minimal* if every continuous surjective homomorphism $f : G \to H$ is open; in other words, G is totally minimal if every quotient of G is minimal.

Corollary 3 *Every ω-precompact sequentially h-complete topological group of countable pseudocharacter is totally minimal and metrizable.*

Proof Let G be an ω-precompact sequentially h-complete group of countable pseudocharacter. By Corollary 2, G is metrizable and contains a countable network. Sequential h-completeness and the property of having a countable network are preserved under continuous homomorphic images, so for every continuous homomorphism $\varphi : G \to H$ onto a topological group H, the group H is sequentially h-complete and contains a countable network; in particular, H is ω-precompact. Thus, by Corollary 2, H is of countable pseudocharacter. Therefore, by Theorem 1, φ is open. □

Corollary 4 generalizes [8, 3.3] to the sequentially h-complete topological groups.

Corollary 4 *Every sequentially h-complete topological group with a countable network is totally minimal and metrizable.* □

A topological group is *h-complete* if all its continuous homomorphic images are complete (see [7]).

Using Theorem 1, we obtain the following strengthening of the Guran's Embedding Theorem for sequentially h-complete groups:

Theorem 5 *Every ω-precompact sequentially h-complete group G densely embeds into the (projective) limit of its metrizable quotients. In particular, if G is h-complete, then it is equal to the limit of its metrizable quotients.*

Proof By Fact A, since G is ω-precompact, it embeds into $\Sigma = \prod_{\alpha \in I} \Sigma_\alpha$, a product of second countable groups. Denote by $\pi_\alpha : G \to \Sigma_\alpha$ the restriction of the canonical projections to G; without loss of generality, we may assume that the π_α are onto. Since G is also sequentially h-complete, the π_α are open (by Theorem 1). Thus, one has $G/N_\alpha \cong \Sigma_\alpha$, where $N_\alpha = \ker \pi_\alpha$. We may also assume that *all* metrizable quotients of G appear in the product constituting Σ, because by adding factors we do not ruin the embedding.

Let ι be the embedding of G into the product of its metrizable quotients, and put $L = \varprojlim_{\alpha \in I} G/N_\alpha$. The image of ι is obviously contained in L. (We note that if G/N_α and G/N_β are metrizable quotients, then $G/N_\alpha N_\beta$ and $G/N_\alpha \cap N_\beta$ are also metrizable quotients; the latter one is metrizable, because the continuous homomorphism $G \to G/N_\alpha \times G/N_\beta$ is open onto its image, as the codomain is metrizable.) In order to show density, let $x = (x_\alpha N_\alpha)_{\alpha \in I} \in L$ and let

$$U = U_{\alpha_1} \times \cdots \times U_{\alpha_k} \times \prod_{\alpha \neq \alpha_i} G/N_\alpha$$

be a neighborhood of x. By the consideration above, the quotient $G/\bigcap_{i=1}^{k} N_{\alpha_i}$ is metrizable, so $\bigcap_{i=1}^{k} N_{\alpha_i} = N_\gamma$ for some $\gamma \in I$. Thus, $\pi_{\alpha_i}(x_\gamma) = x_{\alpha_i} N_{\alpha_i}$, therefore $\iota(G)$ intersects U, and hence $\iota(G)$ is dense in L. □

A topological group G is *maximally almost-periodic* (or briefly, *MAP*) if it admits a continuous monomorphism $m : G \to K$ into a compact group K, or equivalently, if the finite-dimensional unitary representations of G separate points.

The Theorem below is a far-reaching generalization of the main result of [11]:

Theorem 6 *Let G be a complete topological group. Then the following assertions are equivalent:*

(i) *the closed normal subgroups of closed separable subgroups of G are h-complete and MAP;*

(ii) *every closed separable subgroup H of G is sequentially h-complete, and its metrizable quotients H/N are MAP;*

(iii) *G is compact.*

The following easy consequence of a result by Dikranjan and Tkačenko plays a very important role in proving Theorem 6:

Fact D *G is precompact if and only if every closed separable subgroup of G is precompact.* (Theorem 3.5 in [6].)

Proof (i) ⇒ (ii): If H/N is a quotient described in (ii), then it is clearly h-complete, as the homomorphic image of H, which is assumed to be h-complete in (i). Let $m : H \to K$ be a continuous injective homomorphism into a compact group K. Since H is h-complete, $m(H)$ is closed in K, so we may assume that m is onto. The subgroup $m(N)$ is normal in K, because m is bijective, and it is closed because N is h-complete. Therefore, $\bar{m} : H/N \to K/m(N)$ is a continuous injective homomorphism, showing that H/N is MAP.

(ii) ⇒ (iii): Since G is complete, in order to show that G is compact, we show that it is also precompact. By Fact D, it suffices to show that every closed separable subgroup of G is precompact.

Let H be a closed separable subgroup of G. The group H is ω-precompact (because it is separable), and by (ii) H is sequentially h-complete. Applying Theorem 5, H densely embeds into the the product P of its metrizable quotient. In order to show that H is precompact, we show that each factor of the product P is precompact.

Let $Q = H/N$ be a metrizable quotient of H; since H is sequentially h-complete, so is Q. The group Q is second countable, because (being the continuous image of H) it is separable. Thus, by Corollary 4, Q is totally minimal, because it is sequentially h-complete. According to (ii), Q is MAP, and together with minimality this implies that Q is precompact. □

Remark The implication (i) $\Rightarrow$ (iii) can also be proved directly, without applying Theorem 5, by using Fact D and [8, 3.4].

Since subgroups and continuous homomorphic images of c-compact groups are c-compact again, and are in particular h-complete, we obtain:

Corollary 7 *If G is c-compact and MAP, then G is compact.*

Proof If G is MAP, then in particular all its subgroups are so, and all its closed subgroups are c-compact, thus h-complete. Hence Theorem 6 applies. □

A group is *minimally almost periodic* (or briefly, *m.a.p.*) if it has no non-trivial finite-dimensional unitary representations.

Corollary 8 *Every c-compact group G has a maximal compact quotient G/M, where M is a closed characteristic m.a.p. subgroup of G.*

Proof Let G be a c-compact group, and let $M = n(G)$, the von Neumann radical of G (the intersection of the kernels of all the finite-dimensional unitary representations of G). By its definition, G/M is the maximal MAP quotient of G, and according to Corollary 7 this is the same as the maximal compact quotient of G, because each quotient of G is c-compact.

The subgroup M is also c-compact, so by what we have proved so far, it has a maximal compact quotient M/L, where L is characteristic in M, thus L is normal in G. By the Third Isomorphism Theorem, $G/M \cong (G/L)/(M/L)$, and since both M/L and G/M are compact, by the three space property of compactness in topological groups, the quotient G/L is compact too. Thus, $M \subseteq L$, and therefore $M = L$. Hence, M is minimally almost periodic. □

We conclude with a reduction theorem. If G is c-compact, then so is every closed subgroup of G. Thus, each closed subgroup H and separable metrizable quotient H/N mentioned in (ii) of Theorem 6 is also c-compact. Therefore, the equivalence of (ii) and (iii) in Theorem 6 yields:

Theorem 9 *The following statements are equivalent:*

(i) *every c-compact group is compact;*
(ii) *every second countable c-compact group is MAP (and thus compact);*
(iii) *every second countable c-compact m.a.p. group is trivial.* □

It is not known whether (ii) is true, but by Corollary 4 every second-countable c-compact group is totally minimal, a fact that may help in proving or giving a counterexample to (ii).

Acknowledgements

I am deeply indebted to Prof. Walter Tholen, my PhD thesis supervisor, for giving me incredible academic and moral help.

I am thankful to Prof. Mikhail Tkačenko for sending me his papers [14] and [15] by e-mail.

I am grateful for the constructive comments of the anonymous referee that led to an improved presentation of the results in this paper.

Last but not least, special thanks to my students, without whom it would not be worth it.

References

[1] A. V. Arhangel'skiĭ. Cardinal invariants of topological groups. Embeddings and condensations. *Dokl. Akad. Nauk SSSR*, 247(4):779–782, 1979.

[2] M. M. Clementino, E. Giuli, and W. Tholen. Topology in a category: compactness. *Portugal. Math.*, 53(4):397–433, 1996.

[3] M. M. Clementino and W. Tholen. Tychonoff's theorem in a category. *Proc. Amer. Math. Soc.*, 124(11):3311–3314, 1996.

[4] D. N. Dikranjan. Countably compact groups satisfying the open mapping theorem. *Topology Appl.*, 98(1-3):81–129, 1999. II Iberoamerican Conference on Topology and its Applications (Morelia, 1997).

[5] D. N. Dikranjan and E. Giuli. Compactness, minimality and closedness with respect to a closure operator. In *Categorical topology and its relation to analysis, algebra and combinatorics (Prague, 1988)*, pages 284–296. World Sci. Publishing, Teaneck, NJ, 1989.

[6] D. N. Dikranjan and M. G. Tkačenko. Sequential completeness of quotient groups. *Bull. Austral. Math. Soc.*, 61(1):129–150, 2000.

[7] D. N. Dikranjan and A. Tonolo. On characterization of linear compactness. *Riv. Mat. Pura Appl.*, 17:95–106, 1995.

[8] D. N. Dikranjan and V. V. Uspenskij. Categorically compact topological groups. *J. Pure Appl. Algebra*, 126(1-3):149–168, 1998.

[9] I. Guran. On topological groups close to being lindelöf. *Soviet Math. Dokl.*, 23:173–175, 1981.

[10] T. Husain. *Introduction to topological groups.* W. B. Saunders Co., Philadelphia, Pa., 1966.

[11] G. Lukács. On locally compact *c*-compact groups. *Preprint*, 2003.

[12] E. G. Manes. Compact Hausdorff objects. *General Topology and Appl.*, 4:341–360, 1974.

[13] L. S. Pontryagin. *Selected works. Vol. 2.* Gordon & Breach Science Publishers, New York, third edition, 1986. Topological groups, Edited and with a preface by R. V. Gamkrelidze, Translated from the Russian and with a preface by Arlen Brown, With additional material translated by P. S. V. Naidu.

[14] M. G. Tkačenko. Topological groups for topologists. I. *Bol. Soc. Mat. Mexicana (3)*, 5(2):237–280, 1999.

[15] M. G. Tkačenko. Topological groups for topologists. II. *Bol. Soc. Mat. Mexicana (3)*, 6(1):1–41, 2000.

Received December 14, 2002; in revised form July 24, 2003

Fields Institute Communications
Volume **43**, 2004

Embeddings of Algebras

John L. MacDonald
Department of Mathematics
University of British Columbia
Vancouver, BC Canada V6T 1Z2
johnm@math.ubc.ca

This paper is dedicated to Saunders Mac Lane.

Abstract. Conditions are given for determining when the unit of an adjunction is monic when the domain of the left adjoint is a category **B** of algebras. Key necessary and sufficient conditions are given in terms of a graph assigned to object G of **B**. The general method is then applied to present a proof of a Clifford algebra embedding theorem which exactly parallels that of the Birkhoff-Witt theorem for Lie algebras. A contrasting application related to work of Schreier and Serre on amalgams of groups further illustrates the method.

1 Introduction

We consider here the general problem of determining when the unit morphisms $\eta_G : G \to UFG$ of an adjunction $(F, U, \eta, \varepsilon) : \mathbf{B} \rightharpoonup \mathbf{A}$ are monic when the domain **B** of F is a category of algebras.

When $\mathbf{B} = \mathbf{Sets}$ it is clear from the commutativity of the diagram

(1.1)

that η_G is always monic provided that there exists A in **A** such that UA has more than one element. There are, in fact, two monads on **Sets** which have only trivial algebras. On the empty set ϕ one of them takes value ϕ and the other 1. On $X \neq \phi$ both monads take value 1(cf. Manes [17]). Pareigis [18] makes similar remarks in his discussion of consistent algebraic theories.

2000 *Mathematics Subject Classification.* Primary 18B15, 18C15, 18C20; Secondary 11E88, 15A66, 17B35.

The author was supported in part by the Atlantis Consortium.

In the case when $\mathbf{B}$ is a more general algebraic category the problem becomes more intricate and has been a subject of research in such diverse areas as Lie algebras, Clifford algebras, amalgams of groups and categorical coherence (cf. [6], [8], [9], [12], [17] and [21]).

In earlier papers ([11], [12] and [13]) the author developed a general method for solving this problem and applied it to generalizations of the Birkhoff-Witt theorem, to coherence problems and to embeddings of partially defined algebras.

In this paper the general method has been reformulated in an isomorphism invariant way and a new application to the embedding of Clifford algebras has been made which demonstrates as well a strong parallel between the proof techniques used for Clifford and Lie algebra embeddings. It is shown explicitly here how embeddings involving group amalgams follow the same pattern as the Clifford algebra embedding. The Diamond, Embedding and Connectedness Principles used in support of the general method in earlier work have been reformulated in terms of costrict objects to ensure isomorphism invariance of results.

The method employed when looking at a unit morphism $\eta_G : G \to UFG$ is to associate to G a category $\mathbf{C}(S_V)$ called a *V picture of the adjoint F to U at G* and then to embed G as a costrict subcategory of $\mathbf{C}(S_V)$. The principles developed in Section 2 then will provide a set of necessary and sufficient test conditions to determine whether η_G is monic or not. $\mathbf{C}(S_V)$ is a certain quotient category of a free category generated by a graph S_V obtained by forgetting part of the structure of G using V and then taking a left adjoint to VU.

Section 2 considers various principles that ensure the embeddability of distinct isomorphism classes of objects of a subcategory $\mathbf{G}$ of $\mathbf{C}$ in the components of $\mathbf{C}$. For the purposes of applications in the later sections it is important to require that the objects of $\mathbf{G}$ are costrict in $\mathbf{C}$. Then diamond completeness for a pair of morphisms with common reducible domain is equivalent to the Diamond Principle, namely, that objects $X \to A$ of $X/\mathbf{C}$ with A in $\mathbf{G}$ are weakly terminal. There is another equivalent Embedding Principle and finally a Principle of Connectedness which is equivalent to the first two in the presence of a rank functor. Roughly speaking an Embedding Principle is one that ensures a monic η and the other equivalent principles are the ones tested on examples. If the objects of $\mathbf{G}$ satisfy the further property of being t-costrict (Section 2), then there is a corresponding set of equivalent principles.

Sections 3 and 4 translate these conditions into the context of adjoints with left adjoint domain a category of algebras. Instead of looking at the unit directly sufficiently many operations and equations are dropped until there is a monic unit. Then we take an adjoint and create a graph by reintroducing the operations and do a couple of other constructions to create an appropriate category for applying the results of Section 2. The starting object may then be viewed as a costrict subcategory of this newly created category and then the principles developed in Section 2 are applied.

The next two sections bring in the embedding of Clifford algebras as a new application of the results of the preceding sections. This is formulated in such a way that the close parallel with earlier development of a Lie algebra embedding result is apparent. This explicit comparison and introduction appears in Sections 5 and 6. Finally in Section 7 an application to group amalgams is given based on earlier work. This serves to give an application of the main theorem of a type contrasting with the Clifford algebra example. It should be noted that the method used here is based on structures with partially defined operations, thus providing a view of amalgamations of groups with common subgroup, distinct from the one appearing in Serre([21]).

2 Embedding principles for costrict subcategories

Let $\mathbf{G}$ be a subcategory of $\mathbf{C}$. In this section various equivalent principles are formulated which ensure that distinct isomorphism classes of objects of $\mathbf{G}$ appear in distinct components of $\mathbf{C}$. These details are developed for use in proving the embedding results of later sections.

Let $\mathbf{G}$ be a subcategory of a category $\mathbf{C}$ and let $\mathcal{P}(\mathbf{G})$ be the *power category* of $\mathbf{G}$. The objects of $\mathcal{P}(\mathbf{G})$ are the subclasses of objects of $\mathbf{G}$ and the morphisms are the inclusions.

The *reduction functor* $\mathcal{R}_{\mathbf{G}} : \mathbf{C}^{op} \to \mathcal{P}(\mathbf{G})$ is defined by

$$\mathcal{R}_{\mathbf{G}}X = \{A | A \text{ is an object of } \mathbf{G} \text{ and there exists a } \mathbf{C}\text{-morphism } X \to A\}$$

with the obvious definition on morphisms.

An object X of $\mathbf{C}$ is $\mathbf{G}$-*reducible* if $\mathcal{R}_{\mathbf{G}}X$ is nonempty.

An object A is *costrict* in $\mathbf{C}$ if each $\mathbf{C}$-morphism with domain A is an isomorphism. If the only such isomorphism is the identity, then we call A *t-costrict.*

We say that W is *weakly terminal* in a category $\mathbf{W}$ if for each object S of $\mathbf{W}$ there is a morphism $S \to W$ and any two such morphisms differ by an automorphism of the codomain.

Diamond Principle for G. Let $\mathbf{G}$ be a subcategory of $\mathbf{C}$, then the objects $\alpha : X \to A$ of $X/\mathbf{C}$ with A an object of $\mathbf{G}$ are weakly terminal in $X/\mathbf{C}$ for each object X of $\mathbf{C}$.

If we change the principle by requiring the objects α to be terminal and not just weakly terminal, then the result is called the Strong Diamond Principle.

Lemma 2.1 *Let* $\mathbf{G}$ *be a costrict subcategory of* $\mathbf{C}$. *Then the following statements are equivalent:*

(a) *The Diamond Principle for* $\mathbf{G}$.
(b) *Each pair* $Y \leftarrow X \to Z$ *of* $\mathbf{C}$*-morphisms with* X *a* $\mathbf{G}$*-reducible object can be completed to a commutative diamond in* $\mathbf{C}$.

Proof Suppose that (b) holds and let $\alpha : X \to A$ be an object of $X/\mathbf{C}$ with A in $\mathbf{G}$ and $\beta : X \to Y$ be any other object. Then, by hypothesis there exists a commutative diagram

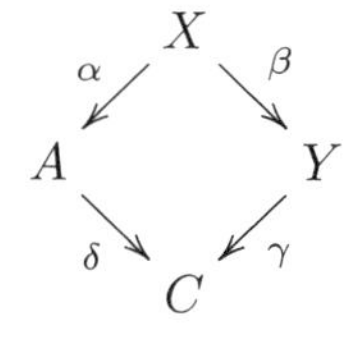

in $\mathbf{C}$. But A in $\mathbf{G}$ and $\mathbf{G}$ costrict implies that δ is an isomorphism. Thus $\delta^{-1}\gamma : \beta \to \alpha$. If $\gamma' : \beta \to \alpha$ then, by hypothesis, $A \overset{\delta^{-1}\gamma}{\longleftarrow} Y \overset{\gamma'}{\to} A$ can be completed to a commutative square since Y is G-reducible. Thus

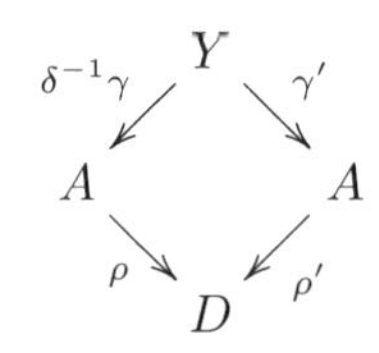

where ρ and ρ' are isomorphisms. Thus $\gamma' = ({\rho'}^{-1}\rho)(\delta^{-1}\gamma)$ and γ' differs from $\delta^{-1}\gamma$ by an automorphism. Thus α is weakly terminal in $X/\mathbf{C}$.

It is trivial to show that (a) implies (b). □

The following Lemma illustrates how the Strong Diamond Principle corresponds to t-costrict objects.

Lemma 2.2 *Let* $\mathbf{G}$ *be an t-costrict subcategory of* $\mathbf{C}$*, that is, the objects of* $\mathbf{G}$ *are t-costrict in* $\mathbf{C}$*. Then the following statements are equivalent:*

(a) *The Strong Diamond Principle for* $\mathbf{G}$*.*

(b) *Each pair* $Y \leftarrow X \rightarrow Z$ *of* $\mathbf{C}$*-morphisms with* X *a* $\mathbf{G}$*-reducible object can be completed to a commutative diamond in* $\mathbf{C}$*.*

The *component class* $[X]$ of an object X of a category $\mathbf{C}$ (or a graph $\mathbf{C}$) is the class of all objects Y which can be connected to X by a finite sequence of morphisms (e.g. $X \rightarrow X_1 \leftarrow X_2 \rightarrow Y$). We let $Comp\mathbf{C}$ denote the collection of component classes.

Embedding Principle for G. If $[X] = [Y]$ in $Comp\mathbf{C}$, then $\mathcal{R}_\mathbf{G}X = \mathcal{R}_\mathbf{G}Y$. Furthermore there is at most one morphism $X \rightarrow A$ up to an automorphism of A, for each pair (X, A) consisting of an object X of $\mathbf{C}$ and an object A of $\mathbf{G}$.

If we require there to be at most one morphism and not just at most one morphism up to automorphism, then we have the Strong Embedding Principle.

Theorem 2.3 *Let* $\mathbf{G}$ *be a costrict subcategory of* $\mathbf{C}$*. Then the following statements are equivalent:*

(a) *The Embedding Principle for* $\mathbf{G}$*.*

(b) *The Diamond Principle for* $\mathbf{G}$*.*

Proof Suppose (a) holds. Let $A \overset{\alpha}{\leftarrow} X \overset{\beta}{\rightarrow} Y$ be a diagram in $\mathbf{C}$ with A in $\mathbf{G}$. Thus $[A] = [Y] = [X]$ in $Comp\mathbf{C}$ and $\mathcal{R}_\mathbf{G}X = \mathcal{R}_\mathbf{G}Y$, by hypothesis. Hence A is in $\mathcal{R}_\mathbf{G}Y$ and there exists $\gamma : Y \rightarrow A$. But then $\gamma\beta$ and α are morphisms $X \rightarrow A$ and α and $\gamma\beta$ differ by an automorphism ρ of A, by hypothesis. Thus $\alpha = \rho\gamma\beta$ and $\rho\gamma : \beta \rightarrow \alpha$ in $X/\mathbf{C}$ and $\rho\gamma$ is unique up to automorphism of α since, by hypothesis, as a $\mathbf{C}$ morphism $Y \rightarrow A$ it is unique up to automorphism of A.

Conversely, suppose (b) holds and $[X] = [Y]$ in $Comp\mathbf{C}$. Then X and Y are connected by a finite sequence of morphisms. Thus to see that $\mathcal{R}_\mathbf{G}X = \mathcal{R}_\mathbf{G}Y$ it is sufficient to show that $\mathcal{R}_\mathbf{G}Z = \mathcal{R}_\mathbf{G}W$ for each morphism $\beta : Z \rightarrow W$. Clearly $\mathcal{R}_\mathbf{G}W \subseteq \mathcal{R}_\mathbf{G}Z$ since $\mathcal{R}_\mathbf{G}$ is a functor $\mathbf{C}^{op} \rightarrow \mathcal{P}(\mathbf{G})$. If $\mathcal{R}_\mathbf{G}Z$ is empty, then $\mathcal{R}_\mathbf{G}Z = \mathcal{R}_\mathbf{G}W$. Otherwise there is an object A of $\mathcal{R}_\mathbf{G}Z$ and a diagram $A \overset{\alpha}{\leftarrow} Z \overset{\beta}{\rightarrow} W$ for some $\alpha : Z \rightarrow A$. By (b) $\alpha : Z \rightarrow A$ is weakly terminal in $Z/\mathbf{C}$. Thus there exists a $\gamma : W \rightarrow A$ with $\alpha = \gamma\beta$. Thus A is in $\mathcal{R}_\mathbf{G}W$ and $\mathcal{R}_\mathbf{G}Z = \mathcal{R}_\mathbf{G}W$. Suppose $\alpha, \alpha' : X \rightarrow A$ with A in $\mathbf{G}$. We need to show that these are the same up to an automorphism of A. By (b) $\alpha : X \to A$ is weakly terminal in $X/\mathbf{C}$, thus there is a $\beta : A \to A$ with $\alpha = \beta\alpha'$. But A is costrict, thus β is an isomorphism, hence α and α' are the same up to isomorphism of A. □

We remark that for $\mathbf{G}$ a t-costrict subcategory of $\mathbf{C}$ the Strong Embedding Principle is equivalent to the Strong Diamond Principle.

Given an object X of $\mathbf{C}$ let $(X/\mathbf{C})_{\mathcal{P}}$ be the full subcategory of the slice category $X/\mathbf{C}$ obtained by omitting those objects which are isomorphisms in $\mathbf{C}$.

Principle of Connectedness for T. The categories $(X/\mathbf{C})_{\mathcal{P}}$ are connected, or empty, for each object X of $\mathbf{T}$, where $\mathbf{T}$ is a subcategory of $\mathbf{C}$.

Lemma 2.4 *If the Diamond Principle holds for an costrict subcategory* $\mathbf{G}$ *of* $\mathbf{C}$, *then the Principle of Connectedness holds for the full subcategory* $\mathbf{T}_{\mathbf{G}}$ *of* $\mathbf{C}$ *consisting of all* $\mathbf{G}$*-reducible objects of* $\mathbf{C}$.

Proof Let X be in $\mathbf{T}_{\mathbf{G}}$. Then there is a morphism $X \to A$ in $\mathbf{C}$ with A in $\mathbf{G}$. By the Diamond Principle for $\mathbf{G}$ the morphism $X \to A$ is weakly terminal in $X/\mathbf{C}$. Thus $(X/\mathbf{C})_{\mathcal{P}}$ is connected or empty. □

Let $\mathbb{P}$ be a preorder, considered as a category, and write $x \geq y$ when the hom set $\mathbb{P}(x, y)$ has exactly one member. We will use the following definition.

Definition 2.5 If $\mathbb{P}$ is a preorder, then a subset $\mathbb{S}$ is *inductive* iff for all $x \in \mathbb{P}$, if for all $y, (y < x) \Rightarrow y \in \mathbb{S}$, then $x \in \mathbb{S}$.
A preorder $\mathbb{P}$ is *inductive* iff for all $S \subset \mathbb{P}$, if S is inductive, then $S = \mathbb{P}$.

Definition 2.6 Let $\mathbf{C}$ be a category and $\mathbb{P}$ be an inductive preorder. A *rank functor* for $\mathbf{C}$ is a functor $R : \mathbf{C} \to \mathbb{P}$ with $R\alpha \neq 1$ whenever α is not an isomorphism, that is, R is strictly rank reducing on non isomorphisms.

Example 2.7 Let $\mathbb{N}$ be the preorder of nonnegative integers with $n \to m$ iff $n \geq m$. A functor $R : \mathbf{C} \to \mathbb{N}$ with $R\alpha \neq 1$ whenever α is not an isomorphism is then a rank functor.

The following theorem uses an inductive argument to show that (a) implies (b). The other implications follow from Theorem 2.3 and Lemma 2.4.

Theorem 2.8 *Let* $\mathbf{C}$ *be a category with a given rank functor and* $\mathbf{G}$ *a costrict subcategory. Then the following are equivalent.*

(a) *The Principle of Connectedness for the full subcategory* $\mathbf{T}_{\mathbf{G}}$ *of* $\mathbf{C}$ *of all* $\mathbf{G}$*-reducible objects of* $\mathbf{C}$.
(b) *The Diamond Principle for* $\mathbf{G}$.
(c) *The Embedding Principle for* $\mathbf{G}$.

Proof Suppose (a) holds. Let $\alpha : X \to A$ with $A \in \mathbf{G}$ and let $\beta : X \to Y$. We show, inductively that α is weakly terminal. If α is an isomorphism, then A costrict implies that $\beta\alpha^{-1}$ is an isomorphism. Let $\varphi : Y \to A$ be its inverse. Then $\varphi\beta = \alpha$. If $\varphi'\beta = \alpha$, then $\varphi'\beta\alpha^{-1} = 1$ Applying the inverse φ of $\beta\alpha^{-1}$ on the right we get $\varphi' = \varphi$. We now assume inductively that $Z \to A$ is weakly terminal in $Z/\mathbf{C}$ for all Z of lower rank than X. The case when β is an isomorphism is trivial since then $\gamma = \alpha\beta^{-1} : \beta \to \alpha$ and $\gamma'\beta = \gamma\beta$ implies that $\gamma' = \gamma$. It is sufficient to show that there is a morphism $\gamma : \beta \to \alpha$, unique up to automorphism of α whenever α and

β are not isomorphisms, that is, when they are objects of $(X/\mathbf{C})_{\mathcal{P}}$. By hypothesis the category $(X/\mathbf{C})_{\mathcal{P}}$ is connected, hence there is a diagram

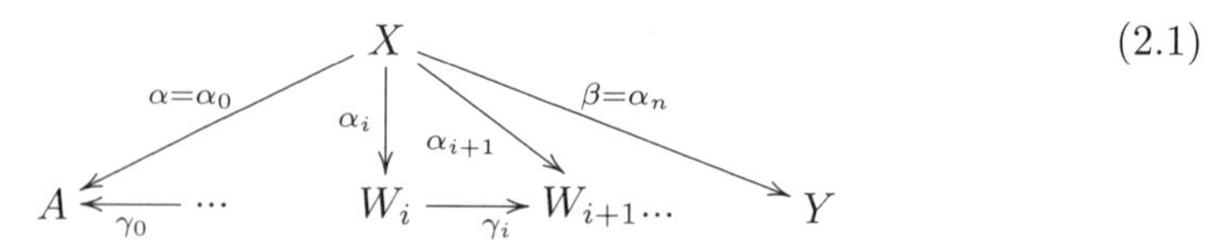

(2.1)

in $\mathbf{C}$ for which the α_is are not isomorphisms and $\gamma_i : W_i \to W_{i+1}$ for i odd and $\gamma_i : W_{i+1} \to W_i$ for i even. Note that γ_0 can always be regarded as having codomain A since the only morphisms with domain A are isomorphisms. Let $A = W_0$ and $Y = W_n$. If n = 1, then $\gamma_0 : \beta \to \alpha$. Assume inductively that $n > 1$ and that if the objects α and β are connected by a sequence of length $< n$, then there exists $\gamma : \beta \to \alpha$. Consider

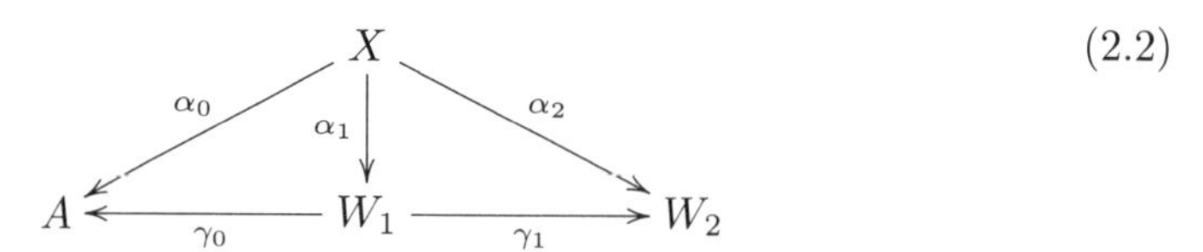

(2.2)

Rank W_1 is less than rank X since α_1 is not an isomorphism. Thus, by induction on rank, γ_0 is weakly terminal in $W_1/\mathbf{C}$. Thus there is a morphism $\delta : \gamma_1 \to \gamma_0$ unique up to automorphism of γ_0 and $\delta\gamma_1 = \gamma_0$ and $\delta\alpha_2 = \delta\gamma_1\alpha_1 = \gamma_0\alpha_1 = \alpha_0$. This connects α and β by a sequence of length $< n$ and thus there exists $\gamma : \beta \to \alpha$. Suppose γ, $\gamma' : \beta \to \alpha$. As $\mathbf{C}$ morphisms both are morphisms $Y \to A$. But β is not an isomorphism and thus rank of Y is less than the rank of X. Thus by the induction hypothesis $\gamma, \gamma' : Y \to A$ are weakly terminal. Thus there is a $\mathbf{C}$ morphism ξ such that $\gamma' = \xi\gamma$. Furthermore ξ is an isomorphism since A is costrict. Thus α is weakly terminal since any two morphisms $\gamma, \gamma' : \beta \to \alpha$ differ by ξ and ξ is an automorphism of α since $\xi\alpha = \xi\gamma\beta = \gamma'\beta = \alpha$.

□

3 Necessary and sufficient conditions

Let $\mathbf{Alg}(\Omega, E)$ denote the category of algebras defined by a set of operators Ω and identities E as described in [16]. In this section we describe a set of necessary and sufficient conditions for an adjunction of a certain type defined on a category $\mathbf{B}$ of algebras to have its unit a monomorphism. In particular, we see that this condition involves the Embedding Principle, which is actually the same principle used in case of classical coherence(cf. [11], [14], [15]).

Let

$$\mathbf{A} \xrightarrow{U} \mathbf{B} \xrightarrow{V} \mathbf{D} \tag{3.1}$$

be a diagram such that

(a) $\mathbf{A}$, $\mathbf{B}$ and $\mathbf{D}$ are the categories of (Ω, E), (Ω', E') and (Ω'', E'') algebras, respectively, with $\Omega'' \subseteq \Omega'$ and $E'' \subseteq E'$, and

(b) V is the forgetful functor on operators $\Omega' - \Omega''$ and identities $E' - E''$ and U is a functor commuting with the underlying set functors on $\mathbf{A}$ and $\mathbf{B}$.

Note that U is *not* necessarily a functor forgetting part of Ω and E.

We next describe a functor $C_V : \mathbf{B} \to \mathbf{Grph}$ associated to each pair consisting of a diagram (3.1) of algebras and an adjunction $(L, VU, \eta', \varepsilon') : \mathbf{D} \rightharpoonup \mathbf{A}$, where **Grph** is the category of directed graphs in the sense of [16].

Given an object G of $\mathbf{B}$ let the objects of the graph $C_V(G)$ be the elements of the underlying set $|LVG|$ of LVG.

Recursive definition of the arrows of $C_V(G)$:

$$\omega_{ULVG}(|\eta'_{VG}|x_1, \cdots, |\eta'_{VG}|x_n) \to |\eta'_{VG}|\omega_G(x_1, \cdots, x_n)$$

is an arrow if ω is in the set $\Omega' - \Omega''$ of operators forgotten by V and $(x_1, \cdots, x_n)$ is an n-tuple of elements of $|G|$ for which $\omega_G(x_1, \cdots, x_n)$ is defined.

If $d \to e$ is an arrow of $C_V(G)$, then so is

$$\rho_{LVG}(d_1, \cdots, d, \cdots, d_q) \to \rho_{LVG}(d_1, \cdots, e, \cdots, d_q)$$

for ρ an operator of arity q in Ω and $d_1, \cdots, d_{i-1}, d_{i+1}, \cdots, d_q$ arbitrary elements of $|LVG|$.

If $\beta : G \to G' \in \mathbf{B}$, then $C_V(\beta) : C_V(G) \to C_V(G')$ is the graph morphism which is just the function $|LV\beta| : |LVG| \to |LVG'|$ on objects and defined recursively on arrows in the obvious way.

In the following proposition note that if, in the diagram (3.1), $\mathbf{D}$ is the category of sets, then an adjunction $(L, VU, \eta', \varepsilon')$ is given by letting LX be the free algebra on the set X.

Proposition 3.1 *Suppose*

$$\mathbf{A} \underset{L}{\overset{U}{\rightleftarrows}} \mathbf{B} \xrightarrow{V} \mathbf{D}$$

is a diagram of algebra categories as in (3.1), with adjunction $(L, VU, \eta', \varepsilon') : \mathbf{D} \rightharpoonup \mathbf{A}$ *given. Then there is an adjunction* $(F, U, \eta, \varepsilon) : \mathbf{B} \rightharpoonup \mathbf{A}$ *with the following specific properties:*

(a) *The underlying set of* FG *is* $CompC_V(G)$.

(b) *If* ρ *is an operator of arity* n *in* Ω*, then* ρ_{FG} *is defined by*

$$\rho_{FG}([c_1], \cdots, [c_n]) = [\rho_{LVG}(c_1, \cdots, c_n)]$$

where $c_1, \cdots, c_n$ *are members of the set* $|LVG|$ *of objects of the graph* $C_V(G)$.

(c) *The unit morphism* $\eta_G : G \to UFG$ *of* $(F, U, \eta, \varepsilon)$ *has an underlying set map which is the composition* $[\,]\cdot \mid \eta'_{VG}|$*, where* $|\eta'_{VG}| : |G| \to |LVG| = ObjC_V(G)$ *is the set map underlying the unit* $\eta'_{VG} : VG \to VULVG$ *of the adjunction* $(L, VU, \eta', \varepsilon')$ *and* $[\ ] : ObjC_V(G) \to CompC_V(G)$ *is the function which sends each object to its component.*

Proof Suppose the hypotheses of the proposition hold. Let S_V be a subgraph of $C_V(G)$ having the same objects $|LVG|$ and the same components as $C_V(G)$. Then the proposition remains valid under substitution of S_V for $C_V(G)$ throughout. This allows us to "picture" the adjoint using a possibly smaller set of arrows than those present in $C_V(G)$. Accordingly, we define a *V picture of the adjoint F to U at $G \in |\mathbf{B}|$* to be any quotient category $\mathbf{C}(= \mathbf{C}(S_V))$ of the free category generated by such a subgraph S_V of $C_V(G)$. This proposition is then valid upon substitution of the underlying graph of a V picture $\mathbf{C}$ for $C_V(G)$ throughout. □

Theorem 3.2 *Let*

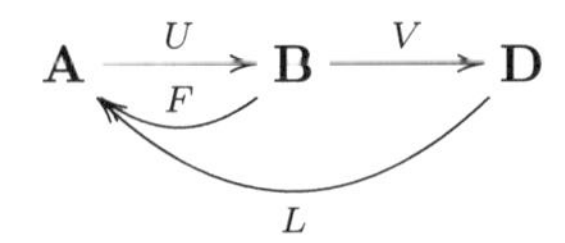

be given with adjunctions $(L, VU, \eta', \varepsilon') : \mathbf{D} \rightharpoonup \mathbf{A}$ *and* $(F, U, \eta, \varepsilon) : \mathbf{B} \rightharpoonup A$ *as described in Proposition* 3.1.

Given $G \in |\mathbf{B}|$ *let* $C(S_V)$ *be any* V *picture of the adjoint* F *to* U *at* G.

Then the unit morphism $\eta_G : G \to UFG$ *of the adjunction* $(F, U, \eta, \varepsilon)$ *is monic if and only if the following hold:*

(a) *The discrete subcategory* $\mathbf{G} = \eta'_{VG}(|G|)$ *is costrict in* $\mathbf{C}(S_V)$ *for* η'_{VG} *the unit of* $(L, VU, \eta', \varepsilon')$.
(b) *If* $[A] = [B]$ *in* $CompC(S_V)$, *then* $\mathcal{R}_{\mathbf{G}}A = \mathcal{R}_{\mathbf{G}}B$ *for all* $A, B \in |\mathbf{G}|$, *where* $\mathcal{R}_{\mathbf{G}} : C(S_V)^{op} \to \mathcal{P}(\mathbf{G})$ *is the reduction functor.*
(c) *The unit morphism* η'_{VG} *is monic.*

Strict Embedding Principle for G. If $[X] = [Y]$ in $Comp\mathbf{C}$, then $\mathcal{R}_{\mathbf{G}}X = \mathcal{R}_{\mathbf{G}}Y$.

Note that in Theorem 3.2 this principle is a restatement of condition (b).

4 Sufficient conditions

In the presence of a rank functor we have seen in Theorem 2.8 that the three principles of section 2 are equivalent. Applying this equivalence to Theorem 3.2 we obtain the following:

Theorem 4.1 *Let*

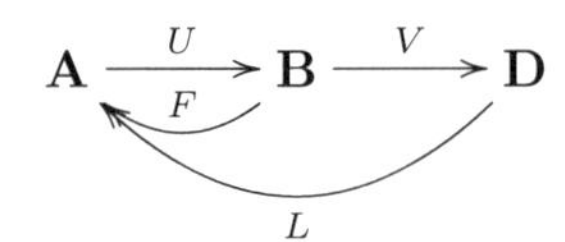

be given with hypotheses as in Theorem 3.2.

Let G *be an object of* $\mathbf{B}$. *Then the unit morphism* $\eta_G : G \to UFG$ *of* $(F, U, \eta, \varepsilon)$ *is monic provided there exists a* V *picture* $\mathbf{C}$ *of the adjoint* F *to* U *at* G *for which the following conditions hold:*

(a) $\mathbf{C}$ *has a rank functor.*
(b) *The discrete subcategory* $\mathbf{G} = \eta'_{VG}(|G|)$ *is costrict in* $\mathbf{C}$ *for* η'_{VG} *the unit of* $(L, VU, \eta', \varepsilon')$.
(c) *The categories* $(X/\mathbf{C})_{\mathcal{P}}$ *are connected, or empty, for each* $\mathbf{G}$*-reducible object* $X \in |\mathbf{C}|$.
(d) *The unit morphism* η'_{VG} *of* $(L, VU, \eta', \varepsilon')$ *is monic.*

Proof Condition (c) is the Principle of Connectedness which by Theorem 2.8 is equivalent to the Embedding Principle. This, in turn, implies the Strict Embedding Principle, which is condition (b) of Theorem 3.2. □

Remark 4.2 Assuming (a) in 4.1, then conditions (b), (c) and (d) are not in general equivalent to the monicity of η. See [12] for an example where η is monic and (a), (b) and (d) hold but not (c).

Remark 4.3 The example referred to in 4.2 shows that the Embedding Principle is stronger than the Strict Embedding Principle used in the necessary and sufficient conditions of Theorem 3.2 since, in the presence of a rank functor, condition (c) of 4.1 is equivalent to the Embedding Principle.

5 Associative embedding of Lie algebras

Given a commutative ring K, let $U : \mathbf{A} \to \mathbf{L}$ be the usual algebraic functor from associative algebras to Lie algebras which replaces the associative multiplication ab of A by the multiplication $[a, b] = ab - ba$ of UA.

The left adjoint $F : \mathbf{L} \to \mathbf{A}$ assigns to each Lie algebra L its universal enveloping algebra $\mathbf{F}L$ but the question as to whether the unit $\eta_L : L \to UFL$ is an embedding is a subtle one since not all Lie algebras can be so embedded(cf. Higgins [9]).

The Birkhoff-Witt theorem gives a positive answer to the embedding question for any K when L is projective as a K-module and for any L when K is a Dedekind domain as well as in various other special cases(cf. [6], [9], [20]). Generalizing from Lie algebras to colour Lie algebras leads to new positive answers in some cases as well as to "Non Birkhoff-Witt" results in other cases such as that of colour Lie algebras and their envelopes over the cyclic group with three elements. For a discussion of colour Lie algebras and special cases, including Lie superalgebras, see [2], [4], [5], [19] and [20].

In the following theorems 5.1 and 5.2 we apply general results of the preceding sections to the case of Lie algebras over a commutative ring K. The results are applied in the case when the underlying module is free.

We do not strive for maximum generality in the case of Lie algebras since our purpose here is rather to demonstrate the commonality of approach in different types of examples, which in this paper have been chosen to be group amalgams, Clifford algebras and Lie algebras.

We now consider specifics. Considering a Lie algebra as a K-module we have a diagram

$$\mathbf{A} \underset{L}{\overset{U}{\rightleftarrows}} \mathbf{L} \xrightarrow{V} K\text{-}\mathbf{Mod}$$

where the conditions (a) and (b) of (3.1) hold. The adjunction $(L, VU, \eta', \varepsilon')$: K-$\mathbf{Mod} \rightharpoonup \mathbf{A}$ can be described explicitly as follows.

Given $G \in |\mathbf{L}|$ it is known that LVG is the tensor algebra of VG. Thus

$$LVG = \oplus_{n \geq 0}(\otimes_{i=1}^{n}(VG)).$$

Furthermore $\eta'_{VG} : VG \to K \oplus VG \oplus (VG \otimes VG) \oplus \cdots$ is monic. In this section let $\mathbf{G}$ be the discrete subgraph $\eta'_{VG}(|G|)$ of $C_V(G)$. Thus $\mathbf{G}$ is a discrete subcategory of any V picture of F at G. Applying Theorem 3.2 the following embedding result holds.

Theorem 5.1 *Let G be a Lie-algebra and $\mathbf{C}$ any V picture of F at G. Then a Lie-algebra G can be embedded in its universal associative algebra FG if and only if the following hold:*

(a) $[A] = [B]$ *in CompC implies that $\mathcal{R}_{\mathbf{G}}A = \mathcal{R}_{\mathbf{G}}B$ for all $A, B \in |\mathbf{G}|$ where $\mathcal{R}_{\mathbf{G}} : \mathbf{C}^{op} \to \mathcal{P}(\mathbf{G})$ is the reduction functor.*

(b) $\mathbf{G}$ *is an costrict subcategory of* $\mathbf{C}$.

Similarly, by applying Theorem 4.1 and again noting that η'_{VG} is monic we have the following sufficient conditions:

Theorem 5.2 *A Lie-algebra G can be embedded in its universal associative algebra FG if there exists any V picture $\mathbf{C}$ of the adjoint F to U at G with the following properties.*

(a) *The categories $(X/\mathbf{C})_{\mathcal{P}}$ are connected for each $X \in |\mathbf{C}|$ which is $\mathbf{G}$-reducible.*
(b) *$\mathbf{C}$ has a rank functor.*
(c) *$\mathbf{G} = \eta'_{VG}(|G|)$ is a costrict subcategory of of $\mathbf{C}$.*

Theorem 5.3 (Birkhoff-Witt). *A Lie-algebra G whose underlying module VG is free can be embedded in its universal associative algebra FG.*

Proof The conditions of the previous theorem are to be verified for the following V picture of the adjoint F to U at G. Let $\mathbf{C}$ be the preorder which is a quotient of the free category on the following subgraph S_V of $C_V(G)$. The objects of S_V are the elements of the free K module LVG on all finite strings $x_{i_1}\cdots x_{i_n}$ of elements from a basis $(x_i)_{i\in I}$ of the free K module VG. Given a well ordering of I we let the arrows of S_V be those of the form

$$k_i x_{i_1}\cdots x_{i_n} + \alpha \to k_i x_{i_1}\cdots x_{i_{j+1}}x_{i_j}\cdots x_{i_n} + k_i x_{i_1}\cdots[x_{i_j}, x_{i_{j+1}}]\cdots x_{i_n} + \alpha$$

for $i_{j+1} < i_j$, $k_i \in K$, and α any element of LVG (not involving $x_{i_1}\cdots x_{i_n}$).

To show that the categories $(X/\mathbf{C})_{\mathcal{P}}$ are connected for each $\mathbf{G}$-reducible object in $\mathbf{C}$, it turns out that the key idea is to show that for $c < b < a$ in I the objects

$$\beta : x_a x_b x_c \to x_b x_a x_c + [x_a, x_b]x_c$$

and

$$\gamma : x_a x_b x_c \to x_a x_c x_b + x_a[x_b, x_c]$$

can be connected in $((x_a x_b x_c)/\mathbf{C})_{\mathcal{P}}$.

The next step in the process is to further reduce the ranges of β and γ.

From β we have

$$\begin{aligned} x_b x_a x_c + [x_a, x_b]x_c &\to x_b x_c x_a + x_b[x_a, x_c] + [x_a, x_b]x_c \\ &\to x_c x_b x_a + [x_b, x_c]x_a + x_b[x_a, x_c] + [x_a, x_b]x_c \end{aligned}$$

and from γ

$$\begin{aligned} x_a x_c x_b + x_a[x_b, x_c] &\to x_c x_a x_b + [x_a, x_c]x_b + x_a[x_b, x_c] \\ &\to x_c x_b x_a + x_c[x_a, x_b] + [x_a, x_c]x_b + x_a[x_b, x_c] \end{aligned}$$

Finally we use the Jacobi identity and the identity $[x, y] = -[y, x]$ to connect the arrows as required(cf. [12]).

The rank functor for $\mathbf{C}$ is given as follows. Given $X = kx_{a_1}\cdots x_{a_n}$ let $R(X) = (R_n(X))$ be a sequence of nonnegative integers defined by $R_n(X) = \sum_{i=1}^{n} p_{a_i}$ where p_{a_i} is the number of x_{a_j} to the right of x_{a_i} with $a_j < a_i$ and $R_s(X) = 0$ for $s \neq n$. We extend by linearity to all elements of $LVG = |\mathbf{C}|$. If $X \to Y$ is an arrow, then $R(Y) < R(X)$ where the latter inequality means that $R_n(Y) < R_n(X)$ for n the largest integer with $R_n(Y) \neq R_n(X)$. Thus R extends to a rank functor.

Finally, we verify condition (c) by observing that any element of $|G|$ may be expressed in the form $\sum_{i\in I} k_i x_i$ in terms of the basis $(x_i)_{i\in I}$ of G, which is regarded as a subset $\mathbf{G}$ of LVG via the embedding η'_{VG}. From the preceding description of arrows of S_V there is no arrow with domain an element of $\mathbf{G} = \eta'_{VG}(|G|)$. Thus $\mathbf{G}$ is costrict in $\mathbf{C}$. ☐

6 Embedding into Clifford algebras

In this section we show how the embedding of a vector space with symmetric bilinear form into its universal associative Clifford algebra follows the pattern of previous sections.

Clifford algebras are associative algebras generated using symmetric bilinear forms. They are important in mathematics because of their relationship to orthogonal Lie groups and in physics because they arise as the type of algebras generated by the Dirac matrices (cf. Hermann [8]).

Let X be a vector space over K with a symmetric bilinear form $\beta : X \times X \to K$. Then the Clifford algebra of the bilinear form β is the associative algebra $Cl(X, \beta)$ $= TX/I(\beta)$ where TX is the tensor algebra of X and $I(\beta)$ is the two-sided ideal generated by all elements

$$x_1x_2 + x_2x_1 - 2\beta(x_1, x_2)$$

for $x_1, x_2 \in X$ (cf. [8]).

The diagram used here is a generalization of the type used in section 3, namely,

$$\mathbf{A} \xleftarrow{\ I\ } \mathbf{Cl} \underset{F}{\overset{W}{\rightleftarrows}} \mathbf{B} \xrightarrow{\ R\ } \mathbf{D}, \qquad L : \mathbf{D} \to \mathbf{A}$$

where the functor $U : \mathbf{A} \to \mathbf{B}$ of section 3 is replaced by a relation $(I, W) : \mathbf{A} \leftarrow \mathbf{Cl} \to \mathbf{B}$ in **Cat** for I the inclusion.

For the rest of this section let **A** be the category of associative K-algebras, **Cl** the Clifford algebras $Cl(X, \beta)$, **B** the K-vector spaces X with bilinear form β and **D** the K-vector spaces. Furthermore I is the inclusion, W and R are forgetful, $F(X, \beta)$ $= Cl(X, \beta)$ and LX is the tensor algebra on X.

We show how results of earlier sections yield the following embedding.

Theorem 6.1 *A vector space X over K with symmetric bilinear form β : $X \times X \to K$ can be embedded in its universal associative Clifford algebra $Cl(X, \beta)$*

Proof We follow the pattern of Theorems 5.2 and 5.3. That is, we verify the same three conditions (a), (b) and (c) as in 5.2, only now for a V picture **C** of the adjoint F to W. In fact, for condition (c), we use the same tensor algebra monad $(L, VU, \eta', \varepsilon')$ and the same discrete subcategory **G**, only now arising from the vector space X $= R(X, \beta)$, instead of VG as in 5.2.

In this case the preorder **C** arises as a quotient of the free category on the following subgraph S_V of $C_V(G)$. The objects of S_V are exactly the same as in Theorem 5.3, namely, they are the elements of the vector space $LR(X, \beta)$ on all finite strings $x_{i_1} \cdots x_{i_n}$ of elements from a basis $(x_i)_{i \in I}$ of the space $R(X, \beta)$. The arrows of S_V are different from those of 5.3. Given a well ordering of I we let the arrows of S_V be those of the form

$$k_i x_{i_1} \cdots x_{i_n} + \alpha \to -k_i x_{i_1} \cdots x_{i_{j+1}} x_{i_j} \cdots x_{i_n} + k_i x_{i_1} \cdots 2\beta(x_{i_j}, x_{i_{j+1}}) \cdots x_{i_n} + \alpha$$

for $i_{j+1} < i_j$, $k_i \in K$, and α any element of $LR(X, \beta)$ (not involving $x_{i_1} \cdots x_{i_n}$).

To show that the categories $(X/\mathbf{C})_{\mathcal{P}}$ are connected for each **G**-reducible object in **C**, it turns out that the key idea is to show that for $c < b < a$ in I the objects

$$\delta : x_a x_b x_c \to - x_b x_a x_c + 2\beta(x_a, x_b) x_c$$

and

$$\gamma : x_a x_b x_c \to x_a x_c x_b + x_a 2\beta(x_b, x_c)$$

can be connected in $((x_a x_b x_c)/\mathbf{C})_{\mathcal{P}}$. This is done by further reduction of the ranges of δ and γ until they reach a common endpoint.

Following δ we have

$$-x_b x_a x_c + 2\beta(x_a, x_b)x_c \to x_b x_c x_a - x_b 2\beta(x_a, x_c) + 2\beta(x_a, x_b)x_c$$

$$\to - x_c x_b x_a + 2\beta(x_b, x_c)x_a - x_b 2\beta(x_a, x_c) + 2\beta(x_a, x_b)x_c$$

and following γ

$$-x_a x_c x_b + x_a 2\beta(x_b, x_c] \to x_c x_a x_b - 2\beta(x_a, x_c)x_b + x_a 2\beta(x_b, x_c)$$

$$\to - x_c x_b x_a + x_c 2\beta(x_a, x_b) - 2\beta(x_a, x_c)x_b + x_a 2\beta(x_b, x_c)$$

The category $\mathbf{G}$ is costrict for the same reasons as in 5.3 and rank on objects is defined the same way as well. Note the close parallel between the proof of this theorem and Theorem 5.2.

□

7 Amalgams of groups

We show how the following classical theorem follows from sections 3 and 4.

Theorem 7.1 (Schreier). *If S is a common subgroup of the groups X and Y and if*

$$\begin{array}{ccc} S & \xrightarrow{\subseteq} & X \\ {\scriptstyle\subseteq}\downarrow & & \downarrow{\scriptstyle\alpha} \\ Y & \xrightarrow[\beta]{} & P \end{array} \tag{7.1}$$

is the pushout in the category of groups, then α and β are monomorphisms. The group P is referred to as the free product of X and Y with amalgamated subgroup S.

Let $\mathbf{B}$ be the category of sets with a single partially defined binary operation. The diagram $X \longleftarrow S \longrightarrow Y$ of groups can be regarded as a diagram in $\mathbf{B}$ and can be completed to a diagram

$$\begin{array}{ccc} S & \xrightarrow{\subseteq} & X \\ {\scriptstyle\subseteq}\downarrow & & \downarrow{\scriptstyle\gamma} \\ Y & \xrightarrow[\delta]{} & Z \end{array} \tag{7.2}$$

commuting in $\mathbf{B}$ where Z is the disjoint union of X and Y with common subset S identified and ab is defined if both $a, b \in X$ or if both $a, b \in Y$, otherwise it is undefined. Clearly (7.2) is a pushout in **Set** and in $\mathbf{B}$. The morphisms γ and δ are the obvious monomorphisms. The next Lemma and Proposition describe how this approach yields the Schreier Theorem. (cf. Baer [1], Serre [21]).

Lemma 7.2 *The Schreier Theorem holds if in (7.2) the pushout codomain Z in $\mathbf{B}$ is embeddable in a group.*

Proof Let $\iota : Z \to P'$ be a monomorphism in $\mathbf{B}$ with P' a group. Then

$$\begin{array}{ccc} S & \xrightarrow{\subseteq} & X \\ {\scriptstyle\subseteq}\downarrow & & \downarrow{\scriptstyle\iota\gamma} \\ Y & \xrightarrow[\iota\delta]{} & P' \end{array}$$

commutes in groups. Thus for some group homomorphism ϕ we have $\iota\gamma = \phi\alpha$ and $\iota\delta = \phi\beta$ since (7.1) is a pushout in groups. Thus α, β are monic since ι, γ and δ are. $\square$

Proposition 7.3 *Let Z be as in (7.2). Then Z is embeddable in a group.*

Proof We embed Z in a particular semigroup which turns out to be a group. Begin with the diagram

$$\mathbf{A} \underset{L}{\overset{U}{\rightleftarrows}} \mathbf{B} \xrightarrow{V} \mathbf{Set}$$

where $\mathbf{A}$ is the category of semigroups (not necessarily with 1), U forgetful, V forgetful and $(L, VU, \eta', \varepsilon')$ an adjunction. We then have a preorder $\mathbf{C}_Z$ which is a quotient of the free category $\mathbf{F}_Z$ generated by $C_V(Z)$. Proposition 3.1 (which also holds for the category $\mathbf{B}$ of partial algebras, see [12]) shows that an adjunction $(F, U, \eta, \varepsilon) : \mathbf{B} \rightharpoonup \mathbf{A}$ exists and describes it. It is sufficient to show that the unit $\eta_Z : Z \to UFZ$ of the adjunction is a monomorphism. By Theorem 4.1 it is sufficient to verify conditions (a) through (d). These conditions are trivial except for (c), which requires that the categories $(X/\mathbf{C})_{\mathcal{P}}$ be connected for each $\mathbf{G}$-reducible object X of $\mathbf{C}$. The objects of $\mathbf{C}_Z$ are elements of the free semigroup LVZ on VZ. An object X may be written as a string $(a_1, \cdots, a_n)$ of length $n \geq 1$ where $a_i \in VZ$ for $i = 1, \cdots, n$. It is sufficient to show that C_V arrows

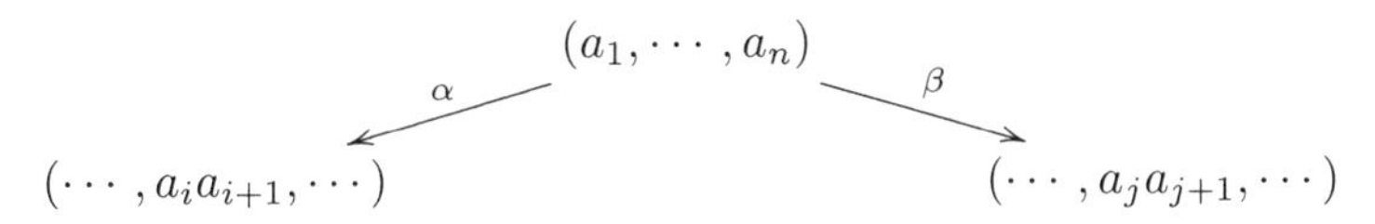

regarded as $(X/\mathbf{C}_Z)_{\mathcal{P}}$ objects can be connected by a finite sequence of morphisms in the same category. This requires a detailed argument when $i = j-1$ or $i = j+1$, otherwise it is trivial (cf. Baer [1], MacDonald [12]). $\square$

References

[1] R. Baer, *Free sums of groups and their generalizations*, Amer. J. Math. **71** (1949), 708-742.

[2] Y. Bahturin, et al., *Infinite-dimensional Lie superalgebras*, de Gruyter, 1992.

[3] M. Barr and C. Wells, *Toposes, Triples and Theories*, Springer-Verlag, 1985.

[4] E. J. Behr, *Enveloping algebras of Lie superalgebras*, Pacific J. Math. **130** (1987), 9-25.

[5] J. Bergen and D.S. Passman, *Delta methods in enveloping algebras of Lie superalgebras*, Trans. Amer. Math. Soc. **334** (1992) 259-280.

[6] G. Birkhoff, *Representability of Lie algebras and Lie groups by matrices*, Ann. of Math. II **38** (1937), 526-532.

[7] S. Eilenberg and J. C. Moore, *Adjoint functors and triples*, Illinois J. Math. **9** (1965), 231-244.

[8] R. Hermann, *Spinors, Clifford and Cayley Algebras*, Math. Sci. Press, 1974.

[9] P. Higgins, *Baer invariants and the Birkhoff-Witt theorem*, J. Algebra **11** (1969), 469-482.

[10] T. Lam, *A First Course in Noncommutative Rings*, Springer-Verlag, 1991.

[11] J. MacDonald, *Coherence of Adjoints, Associativities and Identities*, Arch. Math. **19**, (1968), 398-401.

[12] J. MacDonald, *Coherence and embedding of algebras*, Math. Zeitschrift **135** (1974), 185-220.
[13] J. MacDonald, *Conditions for a universal mapping of algebras to be a monomorphism*, Bull. A.M.S. **80**, (1974), 888-892.
[14] J. MacDonald and M. Sobral, *Aspects of Monads, Ch. V of Categorical Foundations. Special Topics in Order, Topology, Algebra and Sheaf Theory*, Cambridge University Press, 2003, to appear.
[15] S. Mac Lane, *Natural associativity and commutativity*, Rice Univ. Studies **49** (1963), 28-46.
[16] S. Mac Lane, *Categories for the Working Mathematician*, Springer-Verlag, 1971.
[17] E. Manes, *Algebraic Theories*, Springer-Verlag, 1976.
[18] B. Pareigis, *Categories and Functors*, Academic Press, 1970.
[19] M. Scheunert, *The theory of Lie superalgebras*, Springer Lecture Notes Math. **716**, 1979.
[20] J. Serre, *Lie algebras and Lie groups*, W.A. Benjamin, 1965.
[21] J. Serre, *Trees*, Springer-Verlag, 1980.
[22] M. C. Wilson, *Delta methods in enveloping algebras of Lie colour algebras*, J. Algebra, **175** (1995) 661-696.

Received December 14, 2002; November 1, 2003

Fields Institute Communications
Volume **43**, 2004

Universal Covers and Category Theory in Polynomial and Differential Galois Theory

Andy R. Magid
Department of Mathematics
University of Oklahoma
Norman, OK 73019 U.S.A.
amagid@ou.edu

Abstract. The category of finite dimensional modules for the proalgebraic differential Galois group of the differential Galois theoretic closure of a differential field F is equivalent to the category of finite dimensional F spaces with an endomorphism extending the derivation of F. This paper presents an expository proof of this fact modeled on a similar equivalence from polynomial Galois theory, whose proof is also presented as motivation.

1 Introduction

We begin by recalling some notation, definitions, and standard results:

k denotes a field.

$K \supset k$ is a *splitting field*, or *polynomial Galois* extension, for the degree n monic separable polynomial

$$p = X^n + a_{n-1}X^{n-1} + \cdots + a_1X + a_0, \quad a_i \in k$$

if:

1. K is a field extension of k generated over k by $W = \{y \in K \mid p(y) = 0\}$("generated by solutions"); and

2. p is a product of linear factors in $K[X]$ ("full set of solutions").

For polynomial Galois extensions, let $G(K/k) = \mathrm{Aut}_k(K)$; note that $G(K/k) \to S_n(W)$ is an injection.

Then we have the familiar Fundamental Theorem for polynomial Galois extensions:

2000 *Mathematics Subject Classification.* Primary 12H05; Secondary 12F10.

Research supported by the NSF under grant number DMS 0070748.

Theorem (***Fundamental Theorem for Polynomial Galois Extensions***) *Let $K \supset k$ be a polynomial Galois extension. Then $G = G(K/k)$ is a finite group and there is a one-one lattice inverting correspondence between subfields M, $K \supset M \supset k$, and subgroups H of G given by $M \mapsto G(K/M)$ and $H \mapsto K^H$. If M is itself a polynomial Galois extension, then the restriction map $G \to G(M/k)$ is a surjection with kernel $G(K/M)$. If H is normal in G, then K^H is a polynomial Galois extension.*

There is a completely analogous theory for differential fields:

F denotes a differential field of characteristic zero with derivation $D = D_F$ and algebraically closed field of constants C.

$E \supset F$ is a *Picard–Vessiot*, or *Differential Galois*, extension for an order n monic linear homogeneous differential operator

$$L = Y^{(n)} + a_{n-1}Y^{(n-1)} + \cdots + a_1 Y^{(1)} + a_0 Y, \quad a_i \in F$$

if:

1. E is a differential field extension of F generated over F by $V = \{y \in E \mid L(y) = 0\}$ ("generated by solutions").

2. The constants of E are those of F ("no new constants").

3. $\dim_C(V) = n$ ("full set of solutions").

For Picard–Vessiot extensions, let $G(E/F) = \mathrm{Aut}_F^{\mathrm{diff}}(E)$; then $G(E/F) \to GL(V)$ is an injection with Zariski closed image.

There is a "Fundamental Theorem" for differential Galois extensions:

Theorem (***Fundamental Theorem for Picard–Vessiot Extensions***) *Let $E \supset F$ be a Picard–Vessiot extension. Then $G = G(E/F)$ has a canonical structure of affine algebraic group and there is a one-one lattice inverting correspondence between differential subfields K, $E \supset K \supset F$, and Zariski closed subgroups H of G given by $K \mapsto G(E/K)$ and $H \mapsto E^H$. If K is itself a Picard–Vessiot extension, then the restriction map $G \to G(K/F)$ is a surjection with kernel $G(E/K)$. If H is normal in G, then E^H is a Picard–Vessiot extension.*

There are Fundamental Theorems for infinite extensions as well:

Theorem (***Fundamental Theorem for Infinite Polynomial Galois Extensions***) *Let k be a field and let $K \supseteq k$ be a directed union of polynomial Galois field extensions of k. Then the group of automorphisms $G = G(K/k)$ has a canonical structure of topological (in fact profinite) group and there is a bijection between the set of closed subgroups of G, and the set of subfields of K containing k, under which a subgroup H corresponds to the subfield K^H of K fixed element-wise by H and the subfield M corresponds to the subgroup $Aut_M(K)$ of G which fixes each element of M. If M is itself a union of polynomial Galois extensions, then the restriction map $G \to G(M/k)$ is a surjection with kernel $G(K/M)$. If H is (closed and) normal in G, then M^H is a union of polynomial Galois extensions.*

Theorem (***Fundamental Theorem for Infinite Picard–Vessiot Extensions***) *Let $E \supset F$ be a directed union of Picard–Vessiot extensions. Then the group of differential automorphisms $G = G(E/F)$ has a canonical structure of proaffine group and there is a one-one lattice inverting correspondence between differential subfields K, $E \supset K \supset F$, and Zariski closed subgroups H of G given by $K \mapsto G(E/K)$ and $H \mapsto E^H$. If K is itself an infinite Picard–Vessiot extension, then the restriction map $G \to G(K/F)$ is a surjection with kernel $G(E/K)$. If H is (Zariski closed and) normal in G, then K^H is an infinite Picard–Vessiot extension.* [5]

We shall call these theorems (and their finite dimensional versions stated previously) "Correspondence Theorem Galois Theory". These theorems are about the pair consisting of the base field and the extension . There is a another aspect of Galois theory, which we will call "Universal Cover Galois Theory", which focuses on the base field and hopes to understand all possible (polynomial or differential) Galois extensions of the base by constructing a closure (or universal cover) and looking at its group of automorphisms.

The field k has a *separable closure*, which can be defined as a union of polynomial Galois extensions of k such that every polynomial Galois extension of k has an isomorphic copy in it. (More generally, every algebraic separable extension of k embeds over k in a separable closure of k.)

For various reasons, the direct analogues of "algebraic closure" and its properties for differential Galois extensions do not hold. However, the following notion is of interest, and can be shown to exist [7]:

A *Picard–Vessiot closure* $E \supset F$ of F is a differential field extension which is a union of Picard–Vessiot extensions of F and such that every such Picard–Vessiot extension of F has an isomorphic copy in E.

For a differential field extension of F to embed in a Picard–Vessiot closure, it is necessary and sufficient that it have the same constants as F and be generated over F as a differential field by elements that satisfy monic linear homogeneous differential equations over F [8, Prop. 13].

As noted, the goal of what we are calling Universal Cover Galois Theory is produce the (profinite or proalgebraic) Galois group of the (polynomial or differential) universal cover of the base field. In the next section, we will see how this is done in the polynomial case, and then in the following how it is done in the differential case.

1.1 History and literature. This is an expository article. Section 2 is basically an account of the special (one point) case of A. Grothendieck's theory of Galois categories and the fundamental group. I learned this material from J. P. Murre's account of it [9] which is still an excellent exposition. Differential Galois theory is the work of Ellis Kolchin [4]. For a comprehensive modern introduction, see M. van der Put and M. Singer [10]. The survey article by Singer [11] is also a good introduction. Less advanced are the author's introductory expository lectures on the subject [6], which is a reference for much of the terminology used here. Section 3 is an account of a version of the Tannakian Categories methods in differential Galois Theory. This originated in work of P. Deligne [3]; there are explanations of this in both [11] and [10]. A compact explanation of the theory as well as how to do the Fundamental Theorem in this context is also found in D. Bertrand's article

[1]. Information about the Picard–Vessiot closure is found in [7] and [8]. And of course the standard reference for Galois theory is categories is F. Borceaux and G. Janelidze [2].

2 The Galois group of the separable closure

The Fundamental Theorems recalled above were called "Correspondence Theorem Galois Theory". Even in their infinite forms, they are just special cases of Categorical Galois Theory [2]. The point of view of Correspondence Theory is from the extension down to the base: somehow the extension has been constructed and the group of automorphisms obtained (if only in principle) and then the lattice of intermediate fields is equivalent to the lattice of subgroups.

Now the naive view of Galois theory, say the one often adopted by students, is often the opposite: the point of view is from the base up. Despite the fact that, as Aurelio Carboni for example has noted, Galois Theory is about *not* solving equations, not about solving equations, the base up point of view begins with the base field, the (polynomial or differential) equation and asks for the solutions. Even though this doesn't work, let us imagine how to conduct such a project. We will deal first with the polynomial situation.

Let $p = X^n + a_{n-1}X^{n-1} + \cdots + a_1X + a_0$ be a separable polynomial over k which we will assume has no repeated irreducible factors (and hence no multiple roots). The set W of roots of p in a separable closure K of k has, by itself, only the structure of a finite set. The elements of W, however, are not completely interchangeable (remember we are taking the point of view of the base k): the elements of W are grouped according to the irreducible factor of p of which they are a root. Within these groupings they are interchangeable, but there are still limitations, namely any multivariable polynomial relations (with k coefficients) should be preserved as well. Of course the "interchanging" we are talking about is the action of the group $\pi_1(k) = G(K/k)$ on W.

(The reader will note the obvious circularity here: we are trying to describe the set of roots of p from the point of view of k, and to do so we introduce the separable closure and its group of k automorphisms. But this means that be have in principle found the roots not only of p but of every (separable) polynomial over of k!).

It is easy to check that the action of $\pi_1(k)$ on W is topological (which means simply that the stabilizers of points are open). And some natural questions arise, for example, do all finite sets with continuous $\pi_1(k)$ action come from polynomials over k, and if so, how? To take up the first, one should consider all finite sets with $\pi_1(k)$ action, and therefore the category $\mathcal{M}(\pi_1(k))$ of all of them, morphisms in the category being $\pi_1(k)$ equivariant maps. (It is a theorem of Grothendieck [9] that $\pi_1(k)$ can be recovered from $\mathcal{M}(\pi_1(k))$, as we will recall later.)

Now let us ask about how finite sets with continuous $\pi_1(k)$ action might come from polynomials. For the set $W = p^{-1}(0) \subset K$ we considered above, we could find p as $\prod_{\alpha \in W}(X - \alpha)$. But suppose we start with an arbitrary finite set X with continuous $\pi_1(k)$ action. If we want to repeat the above construction, then the first thing that should be considered is how to embed X in K, $\pi_1(k)$ equivariantly of course . While we do not know if such a map even exists, it is clear that no one such should be privileged. Thus the natural thing is to consider the set of all $\pi_1(k)$

equivariant maps $X \to K$. This set is a ring, under pointwise operations on K, and is even a k algebra (the latter sitting in it as constant functions).

We use $C(X, K)$ to denote all the functions from X to K. Then the set $C(X, K)$ is a commutative k algebra under pointwise operations, and $\pi_1(k)$ acts on it via $\sigma \cdot \phi(x) = \sigma(\phi(\sigma^{-1}x))$. We consider the ring of invariants $C(X, K)^{\pi_1(k)}$, which is the ring of $\pi_1(k)$ equivariant functions from X to K. Suppose that ϕ is such a function and x is an element of X. Let $\{x = x_1, \ldots, x_n\}$ be the orbit of x and let H be the intersection of the stabilizers of the x_i. Note that H is closed and normal and of finite index in $\pi_1(k)$. Then all the $\phi(x_i)$ lie in $M = K^H$, which is a polynomial Galois extension of k. Since X is a finite union of orbits, it follows, by taking the compositum of such extensions for each orbit, that there is a finite, normal separable extension $N \supset k$ such that $C(X, K)^{\pi_1(k)} = C(X, N)^{\pi_1(k)}$. Now $C(X, N)$ is a finite product of finite, separable extensions of k, and it follows that its subalgebra $C(X, N)^{\pi_1(k)}$ is as well.

In other words, our search for a polynomial related to X led to a commutative k algebra which is a finite product of finite separable extensions of k. We consider the category of all such:

Let $\mathcal{A}(k)$ be the category whose objects are finite products of finite separable field extensions of k and whose morphisms are k algebra homomorphisms. From the discussion above, we have a contrafunctor

$$\mathcal{U} = C(\cdot, K)^{\pi_1(k)} : \mathcal{M}(\pi_1(k)) \to \mathcal{A}(k).$$

On the other hand, for any object $A = K_1 \times \cdots \times K_n$ in $\mathcal{A}(k)$, we can consider the set $\mathcal{V}(A) = \mathrm{Alg}_k(A, K)$. We have that $\mathcal{V}(A)$ is a finite set (its cardinality is the dimension of A as an k vector space) and there is a left $\pi_1(k)$ action on $\mathcal{F}(A)$, given by following an embeding by an automorphism of K. All the embedings of A into K lie in a fixed finite, separable, normal subextension $K_0 \supseteq k$ of K, and this implies that the action of $\pi_1(k)$ on $\mathcal{V}(A)$ is continuous. Thus we also have a contrafunctor

$$\mathcal{V} = \mathrm{Alg}_k(\cdot, K) : \mathcal{A}(k) \to \mathcal{M}(\pi_1(k))$$

to the category $\mathcal{M}(\pi_1(k))$ of finite sets with continuous $\pi_1(k)$ action.

We will see that $\mathcal{U}$ and $\mathcal{V}$ are equivalences of categories.

Here are some properties of the functor $\mathcal{V}$: if $K_0 \supseteq k$ is a finite separable extension, then $\mathcal{V}(K_0) = \mathrm{Alg}_k(K_0, K)$ has cardinality the dimension of K_0 over k. If $A = K_1 \times \cdots \times K_n$ is a finite product of finite separable extensions of k, then every homomorphism from A to a field must factor through a projection onto a K_i, so it follows that $\mathcal{V}(A)$ is the (disjoint) union $\mathcal{V}(K_1) \amalg \cdots \amalg \mathcal{V}(K_n)$ and hence has cardinality $\sum |\mathcal{V}(K_i)| = \sum \dim_k(K_i) = \dim_k(A)$. We also note that since K is the separable closure of k, $\mathcal{V}(A) = \mathrm{Alg}_k(A, K)$ is always non–empty.

And some properties of the functor $\mathcal{U}$: if X is a finite set with continuous $\pi_1(k)$ action and $X = X_1 \amalg X_2$ is a disjoint union of two proper $\pi_1(k)$ subsets, then the inclusions $X_i \to X$ induce maps $C(X, K) \to C(X_1, K)$ which in turn give an isomorphism $C(X, K) \to C(X_1, K) \times C(X_2, K)$. All these are $\pi_1(k)$ equivariant, and hence give an isomorphism $\mathcal{U}(X) \to \mathcal{U}(X_1) \times \mathcal{U}(X_2)$. Now suppose that X does not so decompose, which means that X is a single orbit, say of the element x with stabilizer H (which is, of course, closed and of finite index in $\pi_1(k)$). Then

$\pi_1(k)$ equivariant maps from X to K are determined by the image of x, which may be any element with stabilizer H. Thus $C(X,K)^{\pi_1(k)} \to K^H$ by $\phi \mapsto \phi(x)$ is a bijection. Combining this observation with the previous then yields the following formula for $\mathcal{U}$: If $X = X_1 \amalg \cdots \amalg X_n$ is a disjoint union of orbits with orbit representatives x_i with stabilizers H_i, then $\mathcal{U}(X) = \prod K^{H_i}$. The index of H_i in $\pi_1(k)$ is both the cardinality of X_i and the dimension of K^{H_i} over k, and it follows that $\dim_k(\mathcal{U}(X)) = |X|$. We also note that $\mathcal{U}(X) = \prod K^{H_i}$ is always non–zero.

Combining, we have cardinality/dimension equalities, $|\mathcal{U}(\mathcal{V}(X))| = |X|$ and $\dim_k(\mathcal{V}(\mathcal{U}(A)) = \dim_k(A)$

We also have "double dual" maps

$$A \to \mathcal{U}(\mathcal{V}(A)) = C(\mathrm{Alg}_k(A,K),K)^{\pi_1(k)} \text{ by } a \mapsto \hat{a}, \text{ where } \hat{a}(\tau) = \tau(a).$$

and

$$X \to \mathcal{V}(\mathcal{U}(X)) = \mathrm{Alg}_k(C(X,K)^{\pi_1(k)},K) \text{ by } x \mapsto \hat{x}, \text{ where } \hat{x}(\phi) = \phi(x).$$

It follows from the cardinality/dimension equalities that in the case that A is a field or X is a single orbit that the double dual maps are bijections, and then that they are bijections in general from the product formulae above.

The above remarks imply that the functors $\mathcal{U}$ and $\mathcal{V}$ give an (anti)equivalence of categories, a result which we now state as a theorem:

Theorem (*Categorical Classification Theorem*) *Let k be a field, let K be a separable closure of k and let $\pi_1(k) = Aut_k(K)$. Then $\pi_1(k)$ has a natural topological structure as a profinite group. Let $\mathcal{A}(k)$ denote the category of commutative k algebras which are finite products of finite separable field extensions of k and k algebra homomorphisms. Let $\mathcal{M}(\pi_1(k))$ denote the category of finite sets with continuous $\pi_1(k)$ action, and $\pi_1(k)$ equivariant functions. Consider the contravariant functors*

$$\mathcal{U} = C(\cdot,K)^{\pi_1(k)} : \mathcal{M}(\pi_1(k)) \to \mathcal{A}(k)$$

and

$$\mathcal{V} = Alg_k(\cdot,K) : \mathcal{A}(k) \to \mathcal{M}(\pi_1(k)).$$

Then both compositions $\mathcal{U}\circ\mathcal{V}$ and $\mathcal{V}\circ\mathcal{U}$ are naturally isomorphic to the identity using the double dual maps, and hence the categories $\mathcal{A}(k)$ and $\mathcal{M}(\pi_1(k))$ are equivalent.

The proofs of the assertions summarized as the Categorical Classification Theorem depended on the Fundamental Theorem of Galois Theory. Conversely, the Categorical Classification Theorem can be used to prove the Fundamental Theorem:

Suppose $K_0 \supset k$ is a finite, normal, separable extension, and that $\tau : K_0 \to K$ is an embeding over k. By normality, $\mathcal{V}(K_0) = \mathrm{Alg}_k(K_0,K)$ is a single orbit, and the stabilizer H of τ is a closed normal subgroup of $\pi_1(k)$ with $\pi_1(k)/H$ isomorphic to $G = \mathrm{Aut}_k(K_0)$. Also, $\mathcal{V}(k) = \{\mathrm{id}_k\}$ is a final object. The transitive $\pi_1(k)$ sets X which fit into a diagram $\mathcal{V}(K_0) \to X \to \mathcal{V}(k)$ are the $\pi_1(k)$ sets between $\pi_1(k)/H$ and $\pi_1(k)/\pi_1(k)$, namely those of the form $\pi_1(k)/K$ where K is a closed subgroup of $\pi_1(k)$ containing H, and thus correspond one–one to subgroups of $\pi_1(k)/H$. The quotients of $\mathcal{V}(K_0)$ correspond, by $\mathcal{U}$, to the subobjects of E. Thus once the Categorical Classification Theorem is available, the Fundamental Theorem of Galois Theory (for finite field extensions) translates into the simple correspondence between (homogeneous) quotients of a finite homogeneous space and the subgroups of the transformation group. We state this result, noting that it implies the Fundamental Theorem of Galois Theory:

Theorem (Fundamental Theorem for Faithful Transitive G Sets) *Let G be a finite group and regard G as a finite set on which G acts transitively and with trivial stabilizers, and let e be the identity of G. Let Z be a one point set and $p : X \to Z$ a map. Then transitive G sets Y and classes of G equivariant surjective maps $q : G \to Y$ which factor through p are in one–one correspondence with subgroups H of G as follows: to the subgroup H, make correspond the G set G/H and the map $G \to G/H$ by $g \mapsto gH$; to the surjective G map $q : G \to X$, make correspond the stabilizer H of $q(e)$.*

3 The Galois group of the Picard–Vessiot closure

In the preceding section, we saw how the category $\mathcal{M}(\pi_1(k))$ of finite sets on which the profinite Galois group of the separable closure of k acts continuously was (anti)equivalent to a category of k algebras. And we recalled that for any profinite group, the category of finite sets on which it acts continuously determines it. A similar statement is true about proalgebraic groups: such a group is determined by the category of vector spaces (or modules) on which it acts algebraically (this is the general Tannaka Duality Theorem [3]). The group of differential automorphisms $\Pi(F) = G(E/F)$ of the Picard–Vessiot closure E of the differential field F is a proalgebraic group, and it is therefore natural to consider the analogue of the functors of the preceding section in the differential case.

Thus we consider the category $\mathcal{M}(\Pi(F))$ of finite dimensional, algebraic, $\Pi(F)$ modules, and the functor $\mathrm{Hom}_{\Pi(F)}(\cdot, E)$ defined on it. The proalgebraic group $\Pi(F)$ is over the field C of constants of F, and the vector spaces in $\mathcal{M}(\Pi(F))$ are over C. The field E is not a $\Pi(F)$ module, although of course $\Pi(F)$ acts on it, since not every element in E has a $\Pi(F)$ orbit that spans a finite dimensional C vector space. Those elements that do form an F subalgebra S of E, which is additionally characterized by the fact that it consists of the elements of E which satisfy a linear homogeneous differential equation over F (see [6, Prop. 5.1, p.61], and [8]). Any $\Pi(F)$ equivariant homomorphism from an algebraic $\Pi(F)$ module to E must have image in S, so the functor to be considered is actually $\mathcal{V}(U) = \mathrm{Hom}_{\Pi(F)}(U, S)$.

It is clear that $\mathcal{V}(U)$, for U an object of $\mathcal{M}(\Pi(F))$, is an abelian group under pointwise addition of of functions. It is also true that $\mathcal{V}(U)$ is an F vector space via multiplication in the range of functions. We will see below that $\mathcal{V}(U)$ is finite dimensional over F. The derivation D of E preserves S, and this derivation of S defines an operator, which we also call D, on $\mathcal{V}(U)$ as follows: let $f \in \mathcal{V}(U)$ and let $u \in U$. Then $D(f)(u)$ is defined to be $D(f(u))$. It is easy to check that D on $\mathcal{V}(U)$ is additive and in fact is C linear. It is not F linear, but we do have the following formula: for $\alpha \in F$ and $f \in \mathcal{V}(U)$, $D(\alpha f) = D(\alpha)f + \alpha D(f)$.

This suggests we consider the category $\mathcal{M}(F \cdot D)$ of finite dimensional F vector spaces V equipped with C linear endomorphisms D_V (usually abbreviated D) such that for $\alpha \in F$ and $v \in V$, $D(\alpha v) = D(\alpha)v + \alpha D(v)$, morphisms being F linear maps which commute with D action. We call objects of $\mathcal{M}(F \cdot D)$ *$F \cdot D$ modules*, and morphisms of $\mathcal{M}(F \cdot D)$ *$F \cdot D$ morphisms.* (Sometimes $F \cdot D$ modules are known as connections [11, 2.4.1 p.536].) The contrafunctor $\mathcal{V}$ sends all objects and morphisms $\mathcal{M}(\Pi(F))$ to $\mathcal{M}(F \cdot D)$ (we still have to establish that $\mathcal{V}(U)$ is always finite dimensional over F). We note that, except for the finite dimensionality, S is like an object in $\mathcal{M}(F \cdot D)$ in that it has an operator D satisfying the appropriate relation, and for an object in $\mathcal{M}(F \cdot D)$ we will use $\mathrm{Hom}_{F \cdot D}(V, S)$ to denote the F linear D preserving homomorphisms from V to S.

It is clear that $\mathrm{Hom}_{F \cdot D}(V, S)$ is an abelian group under pointwise addition of functions, and a C vector space under the usual scalar multiplication operation.

The group $\Pi(F)$ acts on $\mathrm{Hom}_{F \cdot D}(V, S)$: for $\sigma \in \Pi(F), T \in \mathrm{Hom}_{F \cdot D}(V, S)$, and $v \in V$, define $\sigma(T)(v) = \sigma(T(v))$. We will see later that $\mathrm{Hom}_{F \cdot D}(V, S)$ is a finite dimensional C vector space, and that the action of $\Pi(F)$ on it is algebraic. Thus we will have a contrafunctor $\mathcal{U}(V) = \mathrm{Hom}_{F \cdot D}(V, S)$ from $\mathcal{M}(F \cdot D)$ to $\mathcal{M}(\Pi(F))$.

The pair of functors $\mathcal{U}$ and $\mathcal{V}$, therefore, are the analogues for the differential Galois case of the corresponding functors in the polynomial Galois case. And we will see that, as in the polynomial Galois case, both $\mathcal{V}(\cdot)$ and $\mathcal{U}(\cdot)$ are equivalences.

We begin by describing the $\Pi(F)$-module structure of S, and for this we now fix the following notation:

Notation 3.1 *Let Π denote $\Pi(F)$, the differential Galois group of the Picard–Vessiot closure E of F, and let Π^0 denote its identity component and $\overline{\Pi} = \Pi/\Pi^0$ the profinite quotient.*

We denote the algebraic closure of F by $\overline{F}$. We regard $\overline{F}$ as emdedded in S, where it is a Π submodule and, since also $\overline{F} = S^{\Pi^0}$, a $\overline{\Pi}$ module. Therefore, when we need to regard $\overline{F}$ as a trivial Π module we will denote it $\overline{F}_t$.

Proposition 3.2 *In Notation* (3.1),

1. $\overline{F}_t \otimes_F S \cong \overline{F}_t \otimes_C C[\Pi]$ *as $\overline{F}_t$ algebras and Π modules.*

2. $S \cong \overline{F} \otimes_C C[\Pi^0]$ *as $\overline{F}$ algebras and Π^0 modules.*

Proof Statement (1) is the infinite version of Kolchin's Theorem, [9, Thm. 5.12, p.67]. Since E is also a Picard–Vessiot closure of $\overline{F}$, whose corresponding ring is S as an $\overline{F}$ algebra, statement (2) is Kolchin's Theorem as well. □

We can analyze the functor $\mathcal{V} : \mathcal{M}(\Pi) \to \mathcal{M}(F \cdot D)$ using the structural description of the preceding theorem: since $\mathcal{V}(U) = \mathrm{Hom}_\Pi(U, S)$ we have

$$\begin{aligned} \mathcal{V}(U) = \mathrm{Hom}_\Pi(U, S) &= (\mathrm{Hom}_{\Pi^0}(U, S))^{\overline{\Pi}} \\ &= (\mathrm{Hom}_{\Pi^0}(U, \overline{F} \otimes_C C[\Pi^0]))^{\overline{\Pi}} \\ &= (\overline{F} \otimes_C \mathrm{Hom}_{\Pi^0}(U, C[\Pi^0]))^{\overline{\Pi}} \qquad (\mathcal{V}\text{ factor}) \end{aligned}$$

(For the third equality of ($\mathcal{V}$ factor) we used the isomorphism of Proposition (3.2)(2), and for the final equality of ($\mathcal{V}$ factor), we used the fact that U was finite dimensional.)

Using the decomposition ($\mathcal{V}$ factor), it is a simple matter to see that $\mathcal{V}$ is exact:

Proposition 3.3 *The contrafunctor $\mathcal{V} : \mathcal{M}(\Pi) \to \mathcal{M}(F \cdot D)$ is exact. Moreover, $\mathcal{V}(U)$ is F finite dimensional with $dim_F(\mathcal{V}(U)) = dim_C(U)$.*

Proof In ($\mathcal{V}$ factor), we have factored $\mathcal{V}$ as the composition of four functors: first the forgetful functor from Π modules to Π^0 modules, then $U \mapsto \mathrm{Hom}_{\Pi^0}(U, C[\Pi^0])$, $(\cdot) \mapsto \overline{F} \otimes_C (\cdot)$, and $(\cdot) \mapsto (\cdot)^{\overline{\Pi}}$. The first of these is obviously exact. For exactness of the second, we use that $C[\Pi^0]$ is an injective Π^0 module (in fact, for any finite dimensional algebraic Π^0 module W the map

$\mathrm{Hom}_{\Pi^0}(W, C[\Pi^0]) \to (W)^*$ by evaluation at the identity is a C isomorphism to the C linear dual of W). The third functor is also obviously exact. Since our modules are over a field of characteristic zero, taking invariants by a profinite group is also exact, and hence the final functor is exact as well.

To compute dimensions, we note that $\dim_F(\mathcal{V}(U)) = \dim_{\overline{F}}(\mathcal{V}(U) \otimes_F \overline{F})$, then that $\mathcal{V}(U) \otimes_F \overline{F} = \mathrm{Hom}_\Pi(U, S) \otimes_F \overline{F} = \mathrm{Hom}_\Pi(U, \overline{F}_t \otimes_F S)$, and by Proposition (3.2)(1), this latter is $\mathrm{Hom}_\Pi(U, \overline{F}_t \otimes_C C[\Pi]) = \mathrm{Hom}_\Pi(U, C[\Pi]) \otimes_C \overline{F} = U^* \otimes_C \overline{F}$, which has the same dimension over $\overline{F}$ as U does over C. □

Now we turn to the functor $\mathcal{U} = \mathrm{Hom}_{F \cdot D}(\cdot, S)$, and we will see that it also is exact and preserves dimensions. For both of these, we will need a few comments about cyclic $F \cdot D$ modules:

Remark 3.4 An $F \cdot D$ module W is *cyclic*, generated by x, if W is the smallest $F \cdot D$ submodule of W containing x. For any $F \cdot D$ module V and any element $x \in V$, the F span of its derivatives $\sum_{i \geq 0} FD^i(x)$ is a cyclic $F \cdot D$ module, generated by x. If $n = \dim_F(V)$, and $\{D^0x, D^1x, \ldots, D^{k-1}x\}$ is a maximal linearly independent set, then there are elements $\alpha_i \in F$ with $D^kx + \alpha_{k-1}D^{k-1}x + \cdots + \alpha_0 D^0 x = 0$. Note that $k \leq n$. We refer to the differential operator $L = Y^{(k)} + \alpha_{k-1}Y^{(k-1)} + \cdots + \alpha_0 Y$ as the *differential operator corresponding to* x in V.

Now we turn to exactness of $\mathcal{U}$

Proposition 3.5 *The contrafunctor* $\mathcal{U} : \mathcal{M}(F \cdot D) \to \mathcal{M}(\Pi)$ *is exact.*

Proof Since $\mathcal{U}$, being a "Hom into" functor, is right exact, what we need to show is that it carries monomorphisms $V_1 \to V_2$ into epimorphisms. We can assume that the monomorphism is an inclusion and that V_2 is generated over V_1 by a single element x (that is, that V_2 is the sum of V_1 and the cyclic submodule of V_2 generated by x.) We suppose given an $F \cdot D$ morphism $T_1 : V_1 \to S$. We consider the symmetric algebras over F on V_1 and V_2, which we denote $F[V_1]$ and $F[V_2]$. The D operators on the V_i extend to derivations of the $F[V_i]$, and T_1 extends to a differential homomorphism $h : F[V_1] \to S$. We have $F[V_1] \subset F[V_2]$ (this is split as a extension of polynomial algebras over F), and $F[V_2]$ is generated over $F[V_1]$ as a differential algebra by x, which is denoted $F[V_2] = F[V_1]\{x\}$. We will also use h for the extension of h to the quotient field E of S. Let P be the kernel of h, let $\overline{F[V_1]} = F[V_1]/P$ and let $\overline{F[V_2]} = F[V_2]/PF[V_2]$. (Since $PF[V_2]$ is a differential ideal, this latter is a differential algebra.) If $\overline{x}$ denotes the image of x in $\overline{F[V_2]}$, then $\overline{F[V_2]} = \overline{F[V_1]}\{\overline{x}\}$. Now we extend scalars to E:

$$R = E \otimes_{\overline{F[V_1]}} \overline{F[V_1]}\{\overline{x}\}.$$

Note that R is finitely generated as an algebra over E. This implies that if Q is any maximal differential ideal of R, then the quotient field K of R/Q is a differential field extension of E with the same constant field C [6, Cor. 1.18, p. 11]. By construction, K is generated over E as a differential field by the image y of $\overline{x}$. Now x, and hence $\overline{x}$ and y, is the zero of a linear differential operator L of order k, the operator corresponding to x defined above in Remark (3.4). On the other hand, E already contains a Picard–Vessiot extension of F for L, and hence a full set of zeros of L (that is, of dimension k over C). Since K has no new constants, the zero y of L must belong to this set and hence $y \in E$. But this then implies $K = E$. Thus we have a differential F algebra homomorphism $f : F[V_2] \to \overline{F[V_2]} \to R \to R/Q \to E$, and by construction f restricted to $F[V_1]$ is h. Moreover, the image y of x lies in S (since it satisfies a differential equation over F) and hence f has image in S. Finally, the restriction T_2 of f to V_2 is an $F \cdot D$ morphism from V_2 to S extending $T_1 : V_1 \to S$. It follows that $\mathcal{U}$ is left exact, as desired. □

Using exactness, we can also show how $\mathcal{U}$ preserves dimension:

Proposition 3.6 *$\mathcal{U}(V)$ is C finite dimensional with $dim_C(\mathcal{U}(V)) = dim_F(V)$.*

Proof By Proposition (3.5), $\mathcal{U}$ is exact, and since dimension is additive on exact sequences, we can reduce to the case that the $F \cdot D$ module V is cyclic with generator x. Then, by Remark (3.4), V has F basis $\{D^0x, D^1x, \ldots, D^{k-1}x\}$ and corresponding linear operator $L = Y^{(k)} + \alpha_{k-1}Y^{(k-1)} + \cdots + \alpha_0 Y$. Then an $F \cdot D$ morphism $V \to S$ is determined by the image of x, which is an element of S sent to zero by L, and every such element of S determines a morphism. Thus $\mathcal{U}(V)$ is the zeros of L in S, which is the same as the zeros of L in E. Since E contains a Picard–Vessiot extension of F for L, and hence a complete set of solutions, we have $\dim_C(\mathcal{U}(V)) = \dim_C(L^{-1}(0)) = k = \dim_F(V)$. □

Both $\mathcal{U}$ and $\mathcal{V}$ involve a "duality" into S, and hence a "double duality" which we now record, and use to prove that the functors are equivalences.

Theorem 3.7 1. *The function $V \to \mathcal{V}(\mathcal{U}(V)) = Hom_\Pi(Hom_{F \cdot D}(V, S), S)$ by $v \mapsto \hat{v}$, $\hat{v}(T) = T(v)$ is an $F \cdot D$ isomorphism natural in V.*

2. *The function $U \to \mathcal{U}(\mathcal{V}(U)) = Hom_{F \cdot D}(Hom_\Pi(U, S), S)$ by $u \mapsto \hat{u}$, $\hat{u}(\phi) = \phi(u)$ is a Π isomorphism natural in U.*

In particular, $\mathcal{U}$ and $\mathcal{V}$ are category equivalences between the categories $\mathcal{M}(F{\cdot}D)$ and $\mathcal{M}(\Pi(F))$.

Proof We leave to the reader to check that the maps $v \mapsto \hat{v}$ and $u \mapsto \hat{u}$ are well defined and natural in V and U. To see that they are isomorphisms, we use the fact that $\mathcal{V} \circ \mathcal{U}$ and $\mathcal{U} \circ \mathcal{V}$ are exact to reduce to the case of checking the isomorphism for non–zero simple modules, and then use the fact that $\mathcal{V} \circ \mathcal{U}$ and $\mathcal{U} \circ \mathcal{V}$ preserve dimension to reduce to showing that both double dual maps are non–zero. For (1), this means that there is a non–zero $T \in \mathrm{Hom}_{F \cdot D}(V, S) = \mathcal{U}(V)$. But since $\dim_C(\mathcal{U}(V)) = \dim_F(V) \neq 0$, this holds. For (2), this means that there is a non–zero $\phi \in \mathrm{Hom}_\Pi(U, S) = \mathcal{V}(U)$. Since $\dim_F(\mathcal{V}(U)) = \dim_C(U) \neq 0$, this holds as well. Thus the theorem is proved. □

Theorem (3.7) tells us that the category of $\Pi(F)$ modules is equivalent to the category of $F{\cdot}D$ modules. As we mentioned above, the proalgebraic group $\Pi(F)$ can be recovered from its category of modules $\mathcal{M}(\Pi(F))$ by the Tannaka Duality. We review this construction briefly: a *tensor automorphism* of $\mathcal{M}(\Pi(F))$ is a family of vector space automorphisms $\sigma_U, U \in |\mathcal{M}(\Pi(F))|$, one for each object in $\mathcal{M}(\Pi(F))$, such that

1. For any $\Pi(F)$ homomorphism $\phi : U \to U'$ we have $\sigma_{U'}\phi = \phi\sigma_U$, and

2. For any $\Pi(F)$ modules U and U', we have $\sigma_{U \otimes U'} = \sigma_U \otimes \sigma_{U'}$

An example of a tensor automorphism is $\mathrm{Id}_U, U \in |\mathcal{M}(\Pi(F))$.

The composition of tensor automorphisms are tensor automorphisms (composition of $\sigma_U, U \in |\mathcal{M}(\Pi(F))|$. and $\tau_U, U \in |\mathcal{M}(\Pi(F))|$ is $\sigma_U \tau_U, U \in |\mathcal{M}(\Pi(F))|$) and so are inverses, and the tensor automorphism $\mathrm{Id}_U, U \in |\mathcal{M}(\Pi(F))|$ is an identity for composition. Thus the tensor automorphisms form a group, denoted $\mathrm{Aut}_\otimes(\Pi(F))$. For notational convenience, we will denote the element of $\mathrm{Aut}_\otimes(\Pi(F))$ given by $\sigma_U, U \in |\mathcal{M}(\Pi(F))|$ simply as σ

If $U \in |\mathcal{M}(\Pi(F))|$, and $u \in U$ and $f \in U^*$, then we can define a function $m_{u,f}$ on $\mathrm{Aut}_\otimes(\Pi(F))$ by $m_{u,f}(\sigma) = f(\sigma_U(u))$. The C algebra of all such functions is denoted $C[\mathrm{Aut}_\otimes(\Pi(F))]$. One shows, as part of Tannaka Duality, that $C[\mathrm{Aut}_\otimes(\Pi(F))]$ is the coordinate ring of a proalgebraic group structure on $\mathrm{Aut}_\otimes(\Pi(F))$.

For any $g \in \Pi(F)$ and any $U \in |\mathcal{M}(\Pi(F))|$, let $L(g)_U$ denote the left action of g on U. Then $L(g)_U, U \in |\mathcal{M}(\Pi(F))|$ is a tensor automorphism, and $L : \Pi(F) \to \mathrm{Aut}_\otimes(\Pi(F))$ is a group homomorphism. Tannaka duality proves that L is actually a group isomorphism (of proalgebraic groups). Thus the proalgebraic group $\Pi(F)$ is recovered from the category $\mathcal{M}(\Pi(F))$ modules. (This procedure works for any proalgebraic group.)

Because of the importance of the tensor product in the Tannaka Duality, we record how the tensor products in $\mathcal{M}(F \cdot D)$ and $\mathcal{M}(\Pi)$ interact with the functors $\mathcal{U}$ and $\mathcal{V}$.

Proposition 3.8 *There are natural (and coherent) isomorphisms*

1. $\mathcal{V}(U_1) \otimes_F \mathcal{V}(U_2) \to \mathcal{V}(U_1 \otimes_C U_2)$, *and*

2. $\mathcal{U}(V_1) \otimes_C \mathcal{U}(V_2) \to \mathcal{U}(V_1 \otimes_F V_2)$.

Proof The map in (1) is defined as follows: if $\phi_i \in \mathrm{Hom}_\Pi(U_i, S)$, then $\phi_1 \otimes_F \phi_2$ is sent to the function in $\mathrm{Hom}_\Pi(U_1 \otimes_C U_2, S)$ given by $u_1 \otimes u_2 \mapsto \phi_1(u_1)\phi_2(u_2)$. Since $\mathcal{V}(U_1) \otimes_F \mathcal{V}(U_2)$ and $\mathcal{V}(U_1 \otimes_C U_2)$ have the same dimension over F (namely that of $U_1 \otimes_C U_2$ over C), to see that the map is an isomorphism it suffices to see that it is injective. To that end, we tensor it over F with $\overline{F}_t$. Then we consider successively:

$$(\mathrm{Hom}_\Pi(U_1, S) \otimes_F \mathrm{Hom}_\Pi(U_2, S)) \otimes_F \overline{F}_t \to \mathrm{Hom}_\Pi(U_1 \otimes_C U_2, S) \otimes_F \overline{F}_t$$

we distribute $\overline{F}_t$ over the tensors and inside the Hom's

$$(\mathrm{Hom}_\Pi(U_1, S) \otimes_F \overline{F}_t) \otimes_{\overline{F}_t} (\mathrm{Hom}_\Pi(U_2, S)) \otimes_F \overline{F}_t) \to \mathrm{Hom}_\Pi(U_1 \otimes_C U_2, S \otimes_F \overline{F}_t)$$

then we apply Proposition (3.2) (1)

$$\mathrm{Hom}_\Pi(U_1, \overline{F}_t \otimes_C C[\Pi]) \otimes_{\overline{F}_t} \mathrm{Hom}_\Pi(U_2, \overline{F}_t \otimes_C C[\Pi])) \to \mathrm{Hom}_\Pi(U_1 \otimes_C U_2, \overline{F}_t \otimes_C C[\Pi])$$

and finally use that spaces of Π maps into $C[\Pi]$ are duals

$$(\mathrm{Hom}_C(U_1, \overline{F}_t) \otimes_{\overline{F}_t} \mathrm{Hom}_C(U_2, \overline{F}_t) \to \mathrm{Hom}_C(U_1 \otimes_C U_2, \overline{F}_t).$$

And this final map is, of course, an isomorphism. This proves (1).

The map in (2) is defined similarly: if $T_i \in \mathrm{Hom}_{F \cdot D}(V_i, S)$ then $T_1 \otimes_F T_2$ is sent to the function in $\mathrm{Hom}_{F \cdot D}(V_1 \otimes V_2, S)$ given by $v_1 \otimes v_2 \mapsto T_1(v_1)T_2(v_2)$. To prove (2), we may assume that $V_i = \mathcal{V}(U_i)$, so we are trying to show that $\mathcal{U}(\mathcal{V}(U_1)) \otimes_C \mathcal{U}(\mathcal{V}(U_2)) \to \mathcal{U}(\mathcal{V}(U_1) \otimes_F \mathcal{V}(U_2))$ is an isomorphism. Note that the domain is, by Theorem (3.7) (2), $U_1 \otimes_C U_2$. On the other hand, if we apply $\mathcal{V}$ to

(1), then we have an isomorphism $\mathcal{U}(\mathcal{V}(U_1 \otimes_C U_2)) \to \mathcal{U}(\mathcal{V}(U_1) \otimes_F \mathcal{V}(U_2))$. Here again, by Theorem (3.7) (2) the domain is $U_1 \otimes_C U_2$. It is a simple matter to check that both maps are the same, and hence conclude (2). □

4 Conclusion

We try to set some of the above in perspective. We consider the problem of understanding the proalgebraic differential Galois group $\Pi(F)$ of a Picard–Vessiot closure of F. By Tannaka Duality, $\Pi(F)$ is determined by and recoverable from its category $\mathcal{M}(\Pi(F))$ of finite dimensional over C algebraic modules – the tensor product over C in $\mathcal{M}(\Pi(F))$ being an essential part of the structure. The (anti)equivalence $\mathcal{U}$ and its inverse $\mathcal{V}$ show that the category $\mathcal{M}(F \cdot D)$ of finite dimensional F spaces with an endomorphism compatible with the derivation on F is (anti)equivalent to the category $\mathcal{M}(\Pi(F))$. In other words, we might say that every finite dimensional $F \cdot D$ module has a "secret identity" as a $\Pi(F)$ module (more appropriately, perhaps, a "dual secret identity", since the equivalences are contravariant). And this identification includes converting tensors over F of $F \cdot D$ modules into tensors over C for $\Pi(F)$ modules. It follows, at least in principle, that the group $\Pi(F)$ is determined by, and determines, the category of $F \cdot D$ modules. So everything that could be learned about F from the group $\Pi(F)$ can be learned by studying $F \cdot D$ modules.

We would also like to tie this observation about differential Galois theory with our earlier discussion of polynomial Galois theory. In that case, we began by considering sets of solutions of polynomial equations in isolation, that is, simply as finite sets, and then found that the structure necessary to tell these "disembodied sets of solutions" from unstructured finite sets was an action of the Galois group of the separable closure. In the same way, modules for the differential Galois group are like "disembodied sets of solutions" for differential equations. But unlike the situation with the polynomial equations, where duality with respect to the closure leads from solution sets to extension fields (actually finite products of extension fields), in the differential case duality with respect to the closure lead from solution spaces to $F \cdot D$ modules, which are more like "disembodied differential equations" (see Remark (3.4)) than extensions. There is a way to pass from $F \cdot D$ modules to extensions (we saw some of this construction in the proof of Proposition (3.5)): for an $F \cdot D$ module V, we form the F symmetric algebra $F[V] = S_F[V]$. This F algebra has a derivation extending that of F, and if we mod out by a maximal differential ideal Q we obtain a differential F integral domain whose quotient field has the same constants as F. One can then show that this domain embeds in a Picard–Vessiot closure of F [8, Prop. 13], and in particular into a Picard–Vessiot extension of F. Different choices of Q are possible, of course. But each arises from a differential F algebra homomorphism from $F[V]$ to the Picard–Vessiot closure of F and hence from differential F algebra homomorphisms $F[V] \to S$. (Since these latter correspond to $F \cdot D$ module homomorphisms $V \to S$, we see our functor $\mathcal{U}$.) One should regard the whole collection of these homomorphisms, or at least all their images, as the corresponding object to the (finite product of) field extensions in the polynomial case.

References

[1] Bertrand, D. *Review of Lectures on Differential Galois Theory by A. Magid*, Bull. (New Series) Amer. Math. Soc. **33** (1966) 289–294.

[2] Borceau, F. and Janelidze, G. *Galois Theories*, Cambridge Studies in Advanced Mathematics **72**, Cambridge University Press, Cambridge, 2001.

[3] Deligne, P. *Catégories tannakiennes* in Carties P., et. al, eds, Grothendieck Festschrift, Vol. 2, Progress in Mathematics **87**, Birkhauser, Boston, 1990, 111–195.

[4] Kolchin, E. *Selected Works*, Amer. Math. Soc., Providence, 1999.

[5] Kovacic, J. *Pro-algebraic groups and the Galois theory of differential fields*, Amer. J. Math. **95** (1973), 507–536.

[6] Magid, A. *Lectures on Differential Galois Theory*, University Lecture Series **7**, American Mathematical Society, Providence RI, 1997 (second printing with corrections).

[7] Magid, A. *The Picard-Vessiot Antiderivative Closure*, J. of Algebra **244** (2001), 1–18.

[8] Magid, A. *The Picard–Vessiot closure in differential Galois theory* in Diferential Galois Theory, Banach Center Publications **58**, Polish Academy of Sciences, Warsaw, 2002, 157–164.

[9] Murre, J. P. *Lectures on an Introduction to Grothendieck's Theory of the Fundamental Group*, Tata Institute of Fundamental Research, Bombay, 1967.

[10] van der Put, M. and Singer, M. *Differential Galois Theory* Springer–Verlag, New York, 2003.

[11] Singer, M. *Direct and inverse problems in differential Galois theory* in H. Bass et al, eds, Selected Works of Ellis Kolchin with Commentary, Amer. Math. Soc., Providence, 1999, 527–554.

Received December 31, 2002; in revised form September 2, 2003

Fields Institute Communications
Volume **43**, 2004

Weak Categories in Additive 2-Categories with Kernels

N. Martins-Ferreira
Departamento de Matemática
ESTG de Leiria
Leiria, Portugal
nelsonmf@estg.ipleiria.pt
http://www.estg.ipleiria.pt

To my parents Ilídio and Bertina.

Abstract. We introduce a notion of weak category, define additive 2-categories and describe weak categories in them. We make this description more explicit in the case of the additive 2-category of morphisms of abelian groups. In particular we present internal bicategories in the category of abelian groups as presheaf categories.

1 Introduction

Consider the notion of monoidal category as an internal structure in (Cat, $\times$, 1). More generally consider it in an abstract 2-category, not necessarily Cat. For the notion emerging in this way we use the name *weak monoid.* Table 1 describes the notion of monoid and weak monoid (for the cases where it is applicable) in some concrete examples of ambient categories.

Table 1

Ambient Category	Monoids	Weak Monoids
Set	ordinary monoids	N/A
$\mathcal{O}$-graphs(Set)	ordinary categories (objects are the elements of $\mathcal{O}$)	N/A
Cat	strict monoidal categories	monoidal categories
$\mathcal{O}$-graphs(Cat)	double categories (the vertical structure is $\mathcal{O}$)	*Weak Categories*
$\mathcal{O}$-graphs(Cat) ($\mathcal{O}$ discrete)	2-categories (objects are the elements of $\mathcal{O}$)	bicategories (objects are the elements of $\mathcal{O}$)

The term *weak category* appears in this way as a generalization of double category and bicategory.

2000 *Mathematics Subject Classification.* Primary 18A05, 18D05; Secondary 18E05.
Key words and phrases. Weak category, additive 2-category, 2-V-category, internal bicategory, weak double category.
I wish to thank Professor G. Janelidze for the useful discussions about this work.
This paper is in final form and no version of it will be submitted for publication elsewhere.

In this work, after giving explicit description of weak monoids and weak categories, we analyze internal weak categories in additive 2-categories with kernels. We also introduce the notion of additive 2-category by defining 2-Ab-category and more generally 2-V-category where V is a monoidal category.

An example of additive 2-category with kernels is Mor(Ab) and we show that a weak category in Mor(Ab) is completely determined by four abelian groups A_1, A_0, B_1, B_0, together with four group homomorphisms $\partial, \partial', k_1, k_0$, such that the following square is commutative

$$\begin{array}{ccc} A_1 & \xrightarrow{\partial} & A_0 \\ {\scriptstyle k_1}\downarrow & & \downarrow{\scriptstyle k_0} \\ B_1 & \xrightarrow{\partial'} & B_0 \end{array}$$

and three more group homomorphisms

$$\lambda, \rho : A_0 \longrightarrow A_1 \ , \ \eta : B_0 \longrightarrow A_1,$$

such that

$$\begin{aligned} k_1\lambda &= 0 = k_1\rho, \\ k_1\eta &= 0. \end{aligned}$$

This result generalizes at the same time the description of internal double categories and of internal bicategories in Ab. Let the morphisms λ, ρ, η become zero morphisms; then we obtain the known description of internal double categories in Ab (for a similar description of (strict) n-categories see e.g. [5],[2],[6] and references there). Taking instead B_1 to be the trivial group, we obtain the description of internal bicategories in Ab [3].

2 Weak categories

This section begins with the formal definition of weak monoid and shows how the notion of weak category can be regarded as a weak monoid. Nevertheless, to consider the category of all weak categories (see [4]), an explicit definition of weak category is required. Last part of the section gives an explicit definition of weak category.

2.1 Weak monoids. An ordinary monoid in a monoidal category $(M, \Box, \mathbf{1})$ is a diagram

$$C\Box C \xrightarrow{m} C \xleftarrow{e} \mathbf{1}$$

(see [1]) in M such that the following diagrams are commutative

$$\begin{array}{ccc} C\Box C\Box C & \xrightarrow{1\Box m} & C\Box C \\ {\scriptstyle m\Box 1}\downarrow & & \downarrow{\scriptstyle m} \\ C\Box C & \xrightarrow{m} & C \end{array}, \tag{2.1}$$

$$
\begin{array}{ccccc}
C & \xrightarrow{e\Box 1} & C\Box C & \xleftarrow{1\Box e} & C \,. \\
 & \searrow & \Big\downarrow{\scriptstyle m} & \swarrow & \\
 & & C & &
\end{array}
\tag{2.2}
$$

In the monoidal category (Cat,$\times$, 1), the monoids are precisely the strict monoidal categories, whereas in the monoidal category ($\mathcal{O}$-Graphs, $\times_{\mathcal{O}}, \mathcal{O} = \mathcal{O}$) the monoids are all categories with the fixed set $\mathcal{O}$ of objects.

Consider a monoidal category $(M, \Box, \mathbf{1})$, in which M is a 2-category and $\Box$ is a 2-bifunctor. Replacing the commutativity of the diagrams (2.1) and (2.2) by the existence of suitable 2-cells satisfying the usual coherence conditions for monoidal categories (see [1]), we obtain the notion of *weak monoid.*

Definition 1 A weak monoid in a monoidal category $(M, \Box, \mathbf{1})$ (where M is a 2-category and $\Box$ is a 2-bifunctor) is a diagram of the form

$$C\Box C \xrightarrow{m} C \xleftarrow{e} \mathbf{1}$$

together with 2-cells

$$
\begin{aligned}
\alpha &: m\,(1\Box m) \longrightarrow m\,(m\Box 1)\,, \\
\lambda &: m\,(1\Box e) \longrightarrow 1, \\
\rho &: m\,(e\Box 1) \longrightarrow 1,
\end{aligned}
$$

that are isomorphisms satisfying the identity

$$\lambda \circ e = \rho \circ e,$$

and the commutativity of the following diagrams[1]

$$
\begin{array}{ccccc}
 & \cdot & \xrightarrow{m\circ(1\Box\alpha)} & \cdot & \\
{\scriptstyle \alpha\circ(1\Box 1\Box m)} \swarrow & & & & \searrow {\scriptstyle \alpha\circ(1\Box m\Box 1)} \\
\cdot & & & & \cdot \\
 & {\scriptstyle \alpha\circ(m\Box 1\Box 1)} \searrow & \cdot & \swarrow {\scriptstyle m\circ(\alpha\Box 1)} &
\end{array}
\;,
\tag{2.3}
$$

$$
\begin{array}{ccc}
\cdot & \xrightarrow{\alpha\circ(1\Box e\Box 1)} & \cdot \\
 & {\scriptstyle m\circ(1\Box\lambda)} \searrow & \Big\downarrow {\scriptstyle m\circ(\rho\Box 1)} \\
 & & \cdot
\end{array}
\;.
\tag{2.4}
$$

In the monoidal category (Cat,$\times$, 1) a weak monoid is precisely a monoidal category (not necessarily strict). A weak category (see Table 1) is obtained as a particular case of a weak monoid by considering a weak monoid in the monoidal category of ($\mathcal{O}$-Graphs(Cat), $\times_{\mathcal{O}}, \mathcal{O} = \mathcal{O}$) where $\mathcal{O}$-Graphs(Cat) is the category of

---

[1] The reader not familiar with internal constructions may think as if the object C had elements a, b, c, d, and write ab for $m\,(a, b)$ in order to obtain

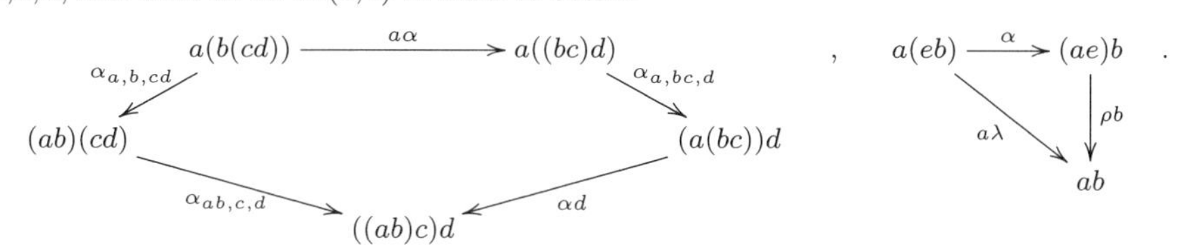

internal $\mathcal{O}$-graphs in Cat. In this case we have the notion of weak monoid written as

$$\begin{array}{ccccc} C\times_{\mathcal{O}}C & \xrightarrow{m} & C & \xleftarrow{e} & \mathcal{O} \\ {\scriptstyle c\pi_1}\downarrow\downarrow{\scriptstyle d\pi_2} & & {\scriptstyle c}\downarrow\downarrow{\scriptstyle d} & & \| \\ \mathcal{O} & = & \mathcal{O} & = & \mathcal{O} \end{array}$$

with

$$\begin{aligned} de &= 1_{\mathcal{O}} = ce, \\ dm &= d\pi_2 \ , \ cm = c\pi_1, \end{aligned}$$

and the commutativity of the diagrams for associativity and identities replaced by natural isomorphisms α, λ and ρ satisfying the usual coherence conditions. If $\mathbb{X}$ is a 2-category, then a weak monoid in $(\mathcal{O}\text{-Graphs}(\mathbb{X}), \times_{\mathcal{O}}, \mathcal{O} = \mathcal{O})$ is a weak category in $\mathbb{X}$.

2.2 Weak categories. For simplicity we introduce the notion of weak category in several steps. First we define the notion of *precategory,* which is just an internal reflexive graph with composition. Next we define precategory with associativity (up to isomorphism) and call it *associative precategory.* Afterwards we define *associative precategory with identity,* an associative precategory with (up to isomorphism) left and right identities.

With respect to coherence conditions we specify the usual *pentagon* and *triangle* (which generalize 2.3 and 2.4) but also consider an intermediate coherence condition (that we call *mixed* coherence condition). The mixed coherence condition is important since (in an additive 2-category with kernels) an associative precategory with identity, satisfying the triangle and the mixed coherence conditions, completely determines the structure of weak category.

Finally we will define the notion of weak category by saying that it is an associative precategory with identity satisfying the pentagon and the triangle coherence conditions.

Definition 2 An internal *precategory* in a category $\mathbb{C}$ is a diagram in $\mathbb{C}$ of the form

$$C_1\times_{C_0}C_1 \xrightarrow{m} C_1 \overset{\xrightarrow{d}}{\underset{\xrightarrow[c]{}}{\xleftarrow{e}}} C_0$$

with

$$\begin{aligned} de &= 1_{C_1} = ce, && (2.5)\\ dm &= d\pi_2 \ , \ cm = c\pi_1 && (2.6) \end{aligned}$$

and where $C_1\times_{C_0}C_1$ is defined via the pullback diagram

$$\begin{array}{ccc} C_1\times_{C_0}C_1 & \xrightarrow{\pi_2} & C_1 \\ {\scriptstyle \pi_1}\downarrow & & \downarrow{\scriptstyle c} \\ C_1 & \xrightarrow[d]{} & C_0 \end{array}.$$

Definition 3 An internal *associative precategory,* in a 2-category $\mathbb{C}$, is a system

$$(C_0, C_1, m, d, e, c, \alpha),$$

where (C_0, C_1, m, d, e, c) is a precategory, (internal to $\mathbb{C}$) and

$$\alpha : m\,(1 \times_{C_0} m) \longrightarrow m\,(m \times_{C_0} 1)$$

is an isomorphism with

$$d \circ \alpha = 1_{d\pi_3}, c \circ \alpha = 1_{c\pi_1}. \tag{2.7}$$

Definition 4 An internal *associative precategory with identity*, in a 2-category $\mathbb{C}$, is a system

$$(C_0, C_1, m, d, e, c, \alpha, \lambda, \rho)\,,$$

where $(C_0, C_1, m, d, e, c, \alpha)$ is an associative precategory, and

$$\lambda : m\,\langle ec, 1\rangle \longrightarrow 1_{C_1} \ , \ \rho : m\,\langle 1, ed\rangle \longrightarrow 1_{C_1}$$

are isomorphisms with

$$\begin{aligned} d \circ \lambda &= 1_d = d \circ \rho, \\ c \circ \lambda &= 1_c = c \circ \rho, \\ \lambda \circ e &= \rho \circ e. \end{aligned} \tag{2.8}$$

Definition 5 An internal *associative precategory with coherent identity*, in a 2-category $\mathbb{C}$, is a system

$$(C_0, C_1, m, d, e, c, \alpha, \lambda, \rho)\,,$$

forming an associative precategory with identity and satisfying the *triangle* and the *mixed* coherence conditions

$$(m \circ (\rho \times 1)) \cdot (\alpha \circ (1 \times \langle ec, 1\rangle)) = m \circ (1 \times \lambda)\,, \tag{2.9}$$

$$\rho \cdot (m \circ \langle \lambda, 1_{ed}\rangle) \cdot (\alpha \circ \langle ec, 1, ed\rangle) = \lambda \cdot (m \circ \langle 1_{ec}, \rho\rangle)\,. \tag{2.10}$$

Definition 6 An internal *weak category* in the 2-category $\mathbb{C}$ is a system

$$(C_0, C_1, m, d, e, c, \alpha, \lambda, \rho)\,,$$

forming an associative precategory with identity and satisfying the *triangle* and the *pentagon* coherence conditions

$$(m \circ (\rho \times 1)) \cdot (\alpha \circ (1 \times \langle ec, 1\rangle)) = m \circ (1 \times \lambda)\,,$$

$$\begin{aligned} &(\alpha \circ (m \times 1 \times 1)) \cdot (\alpha \circ (1 \times 1 \times m)) \\ &\quad = (m \circ (\alpha \times 1)) \cdot (\alpha \circ (1 \times m \times 1)) \cdot (m \circ (1 \times \alpha))\,. \end{aligned} \tag{2.11}$$

If the 2-cells α, λ, ρ were identities, then this would become nothing but the definition of internal category in $\mathbb{C}$. On the other hand, if we let the object C_0 be terminal, then the notion of internal monoidal category is obtained.

In the case where $\mathbb{C}$ is Cat, if the 2-cells are identities we get the definition of a double category; if the category C_0 is discrete (has only objects and the identity morphism for each object) then the definition of bicategory is obtained. More generally, if $\mathbb{C}$ is the category of internal categories in some category $\mathbb{X}$, i.e., $\mathbb{C} = \mathrm{Cat}(\mathbb{X})$ then we obtain the definition of double category in $\mathbb{X}$ on the one hand, and the definition of internal bicategory in $\mathbb{X}$ on the other hand.

In what follows, after defining additive 2-categories, we will give a complete description of the above structures inside (=internal to) them.

3 Additive 2-categories

In order to define additive 2-category we need the notion of 2-Ab-category. To do so, we give the general notion of a 2-V-category, a 2-category enriched in a monoidal category V.

3.1 2-V-categories. Let $\mathbb{V} = (\mathbb{V}, \Box, 1)$ be a monoidal category and O a fixed set of objects. A $\mathbb{V}$-category over the set of objects O is given by a system

$$(H, \mu, \varepsilon)$$

where H is a family of objects[2] of $\mathbb{V}$,

$$H = (H(A,B) \in \mathbb{V})_{A,B \in O},$$

μ is a family of morphisms of $\mathbb{V}$

$$\mu = (\mu_{A,B,C} : H(A,B) \Box H(B,C) \longrightarrow H(A,C))_{A,B,C \in O}$$

and ε is another family of morphisms of $\mathbb{V}$

$$\varepsilon = (\varepsilon_A : 1 \longrightarrow H(A,A))_{A \in O},$$

such that for every $A, B, C, D \in O$, the following diagrams commute

$$\begin{array}{ccc} H(A,B)\Box H(B,C)\Box H(C,D) & \xrightarrow{1_{H(A,B)}\Box \mu_{B,C,D}} & H(A,B)\Box H(B,D) \\ {\scriptstyle \mu_{A,B,C}\Box 1_{H(C,D)}} \downarrow & & \downarrow {\scriptstyle \mu_{A,B,D}} \\ H(A,C)\Box H(C,D) & \xrightarrow{\mu_{A,C,D}} & H(A,D) \end{array}, \qquad (3.1)$$

$$\begin{array}{ccc} H(A,B) & \xrightarrow{\epsilon_A \Box 1} & H(A,A)\Box H(A,B) \\ & \searrow & \downarrow {\scriptstyle \mu_{A,A,B}} \\ & & H(A,B) \end{array}, \quad \begin{array}{ccc} H(A,B)\Box H(B,B) & \xleftarrow{1\Box \epsilon_B} & H(A,B) \\ {\scriptstyle \mu_{A,B,B}} \downarrow & \swarrow & \\ H(A,B) & & \end{array}. \qquad (3.2)$$

A morphism φ between two $\mathbb{V}$-categories over the set of objects O

$$(H, \mu, \varepsilon) \xrightarrow{\varphi} (H', \mu', \varepsilon')$$

is a family of morphisms of $\mathbb{V}$

$$\varphi = (\varphi_{A,B} : H(A,B) \longrightarrow H'(A,B))_{A,B \in O}$$

such that for every $A, B, C \in O$ the following diagrams are commutative

$$\begin{array}{ccc} H(A,A) & \xrightarrow{\varphi_{A,A}} & H'(A,A) \\ {\scriptstyle \epsilon_A} \uparrow & \nearrow {\scriptstyle \epsilon'_A} & \\ 1 & & \end{array}, \qquad (3.3)$$

[2]The object $H(A,B) \in \mathbb{V}$ represents $\hom(A,B)$ of the $\mathbb{V}$-category that is being defined.

$$\begin{array}{ccc} H(A,B)\square H(B,C) & \xrightarrow{\varphi_{A,B}\square\varphi_{B,C}} & H'(A,B)\square H'(B,C) \\ {\scriptstyle \mu_{A,B,C}}\downarrow & & \downarrow{\scriptstyle \mu'_{A,B,C}} \\ H(A,C) & \xrightarrow{\varphi_{A,C}} & H'(A,C) \end{array} . \tag{3.4}$$

Defining composition in the usual way, the category of all $\mathbb{V}$-categories over the set of objects O, denoted by $(\mathbb{V}, O)$-Cat, can be formed.

Definition 7 A *2-$\mathbb{V}$-category* over the set of objects O is an internal category in the category $(\mathbb{V}, O)$-Cat.

A 2-*Ab*-category is obtained by considering the monoidal category $\mathbb{V}= (Ab, \otimes, Z)$.

3.2 2-Ab-categories. Following the previous definition, a 2-Ab-Category over the set of objects O, is an internal category in the category (Ab, O)-Cat, that is, a diagram of the form

$$C_1 \times_{C_0} C_1 \xrightarrow{m} C_1 \overset{\xrightarrow{d}}{\underset{\xrightarrow[c]{}}{\xleftarrow{e}}} C_0$$

satisfying the usual axioms for a category. In order to analyze the definition, it is convenient to think of the object C_0 as an ordinary Ab-category (not given in terms of hom objects) and to think of C_1 as given by a system $C_1 = (H, \mu, \varepsilon)$ (see previous section). Since m, d, e, c are morphisms between Ab-categories, for each two objects A, B of C_0 (note that the objects of C_0 are by definition the elements of O), we have the following diagram in the category of abelian groups

$$H(A,B) \times_{\hom(A,B)} H(A,B) \xrightarrow{m} H(A,B) \overset{\xrightarrow{d}}{\underset{\xrightarrow[c]{}}{\xleftarrow{e}}} \hom(A,B).$$

Using the well-known equivalence Cat(Ab)$\sim$Mor(Ab), the diagram can be presented as

$$\ker d_{A,B} \oplus \ker d_{A,B} \oplus \hom(A,B) \xrightarrow{m} \ker d_{A,B} \oplus \hom(A,B) \overset{\xrightarrow{(0\ 1)}}{\underset{\xrightarrow[(D\ 1)]{}}{\xleftarrow{\binom{0}{1}}}} \hom(A,B)$$

with $m = \begin{pmatrix} 1 & 1 & 0 \\ 0 & 0 & 1 \end{pmatrix}$.

The group homomorphism $D : \ker d_{A,B} \longrightarrow \hom(A,B)$ sends each 2-cell with zero domain to its codomain.

Applying the commutativity of (3.4) to the cases $\varphi = d, e, c$ we conclude that the horizontal composition is completely determined by the composition of the 2-cells in $\ker d$ and by the horizontal composition of each element in $\ker d$ with left and right identity 2-cells. In fact a 2-cell $\tau^* : f \longrightarrow g$ with $f, g : A \longrightarrow B$ may be decomposed into the sum

$$\tau^* = \tau + 1_f \ ,$$

where $\tau \in \ker d_{A,B}$ and $D(\tau) = g - f$ $(\tau : 0 \longrightarrow g - f)$. The horizontal composition (in C_1) of $\sigma^* : f' \longrightarrow g'$ $(: B \longrightarrow C)$ and $\tau^* : f \longrightarrow g$ $(: A \longrightarrow B)$ is given by

$$\mu(\tau^*, \sigma^*) = \sigma^* \circ \tau^* = (\sigma + 1_{f'}) \circ (\tau + 1_f)$$

and, since horizontal composition is bilinear, we obtain the following formula

$$\sigma^{*} \circ \tau^{*} = (\sigma \circ \tau + \sigma \circ 1_{f} + 1_{f'} \circ \tau) + 1_{f'f}.$$

Also, by condition (3.4) applied to $\varphi = c$, the homomorphism D must satisfy the following conditions

$$\begin{aligned} D(\tau \circ \sigma) &= D(\tau) D(\sigma), \\ D(\tau \circ 1_{f}) &= D(\tau) f, \\ D(1_{g} \circ \tau) &= gD(\tau). \end{aligned}$$

Moreover, requiring the commutativity of (3.4) for $\varphi = m$ is the same as to require the four middle interchange law. Consider a diagram of the form

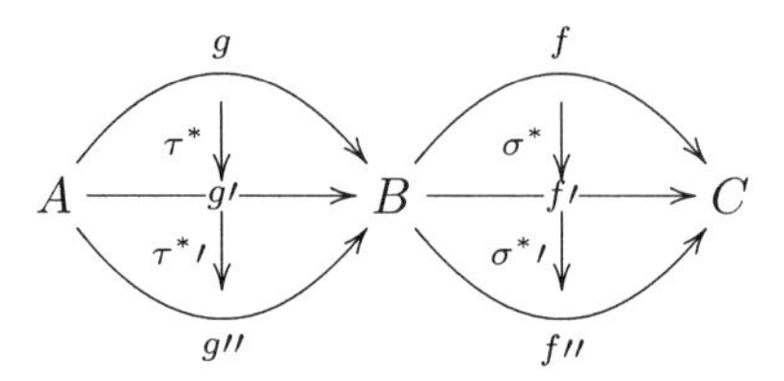

in C_1, the four middle interchange law states that

$$(\sigma^{*\prime} \cdot \sigma^{*}) \circ (\tau^{*\prime} \cdot \tau^{*}) = (\sigma^{*\prime} \circ \tau^{*\prime}) \cdot (\sigma^{*} \circ \tau^{*}).$$

Using the formulas obtained above, we have

$$\begin{aligned} ((\sigma' + \sigma + 1_{f}) \circ (\tau' + \tau + 1_{g})) &= ((\sigma' \circ \tau' + \sigma' \circ 1_{g'} + 1_{f'} \circ \tau') + 1_{f'g'}) \\ &\quad \cdot ((\sigma \circ \tau + \sigma \circ 1_{g} + 1_{f} \circ \tau) + 1_{fg}), \end{aligned}$$

which extends to

$$\begin{aligned} &\sigma' \circ \tau' + \sigma \circ \tau' + 1_{f} \circ \tau' + \sigma' \circ \tau + \sigma \circ \tau + 1_{f} \circ \tau + \sigma' \circ 1_{g} + \sigma \circ 1_{g} + 1_{fg} \\ &= ((\sigma' \circ \tau' + \sigma' \circ 1_{g'} + 1_{f'} \circ \tau' + \sigma \circ \tau + \sigma \circ 1_{g} + 1_{f} \circ \tau) + 1_{fg}), \end{aligned}$$

and then becomes

$$\sigma \circ \tau' + 1_{f} \circ \tau' + \sigma' \circ \tau + \sigma' \circ 1_{g} = \sigma' \circ 1_{g'} + 1_{f'} \circ \tau'.$$

By substituting

$$g' = D(\tau) + g \ , \ f' = D(\sigma) + f \ ,$$

in the formula above we obtain

$$\sigma \circ \tau' + \sigma' \circ \tau = \sigma' \circ 1_{D(\tau)} + 1_{D(\sigma)} \circ \tau'$$

which is equivalent to

$$\sigma \circ \tau = \sigma \circ 1_{D(\tau)} = 1_{D(\sigma)} \circ \tau.$$

Finally, by the commutativity of (3.2) we have

$$\begin{aligned} \tau \circ 1_{A} &= \tau \\ 1_{C} \circ \sigma &= \sigma. \end{aligned}$$

We may summarize the above calculations in the following proposition.

Proposition 1 *Giving a 2-Ab-category is the same as to give the following data:*

- *An Ab-category $\mathbb{A}$;*
- *An abelian group $K(A, B)$, for each pair of objects A, B of $\mathbb{A}$;*

- *A group homomorphism* $D_{A,B} : K(A,B) \longrightarrow \hom_{\mathbb{A}}(A,B)$, *for each pair of objects* A, B *of* $\mathbb{A}$;
- *Associative and bilinear laws of composition*

$$g\tau \ , \ \sigma\tau \ , \ \sigma f \in K(A,C)$$

for each $\tau \in K(A,B), \sigma \in K(B,C), f \in \hom_{\mathbb{A}}(A,B), g \in \hom_{\mathbb{A}}(B,C)$ *with* A, B, C *objects of* $\mathbb{A}$, *satisfying the following conditions*

$$\begin{aligned} \tau 1_A &= \tau, \\ 1_B \tau &= \tau, \end{aligned} \tag{3.5}$$

$$\begin{aligned} D(\sigma\tau) &= D(\sigma) D(\tau), \\ D(\sigma f) &= D(\sigma) f, \\ D(g\tau) &= g D(\tau), \end{aligned} \tag{3.6}$$

$$\sigma\tau = \sigma D(\tau) = D(\sigma)\tau. \tag{3.7}$$

The data given in the above proposition determines a 2-category structure in the Ab-category $\mathbb{A}$. Given two morphisms $f, g : A \longrightarrow B$ of $\mathbb{A}$, a 2-cell from f to g is a pair (τ, f) with τ in $K(A,B)$ and $D(\tau) = g - f$. Note that $K(A,B)$ plays the role of $\ker d_{A,B}$.

The vertical composition is given by the formula

$$(\sigma, g) \cdot (\tau, f) = (\sigma + \tau, f)$$

whereas the horizontal composition is given by

$$(\tau', f') \circ (\tau, f) = (\tau'\tau + \tau' f + f'\tau \ , \ f' f).$$

We always write the three different compositions $g\tau, \sigma\tau, \sigma f$ as justaposition, because it is clear from the context. We also use small letters like f, g, h, k to denote the morphisms of $\mathbb{A}$ and small greek letters, like $\alpha, \lambda, \rho, \eta$ to denote the elements of K. Sometimes the same greek letter is used to denote the element of K and the 2-cell itself, e.g. $\alpha = (\alpha, m(1 \times m))$.

Definition 8 An additive 2-category is a 2–Ab-category with:

- a zero object;
- all binary biproducts;

In the next section simple properties of additive 2-categories are presented.

3.3 Properties of additive 2-categories. Let $\mathbb{A}$ be an additive 2-category (with K and D as above). As is well known, in an additive category (see [1]), a morphism between iterated biproducts is described as a matrix of its components and composition is just the product of matrices. In an additive 2-category the same is true for the 2-cells since we are able to compose them with the projections and the injections of the biproducts. This means that if we have

$$\tau \in K(A_1 \oplus A_2, B_1 \oplus B_2),$$

then we can write

$$\tau = \begin{pmatrix} \tau_{11} & \tau_{12} \\ \tau_{21} & \tau_{22} \end{pmatrix}$$

with

$$\tau_{ij} \in K(A_j, B_i).$$

Let us recall:

Proposition 2 *A split epi $X \xrightarrow{u} Y$ (with splitting $Y \xrightarrow{v} X$) in an additive category with kernels is isomorphic to*

$$\ker u \oplus Y \underset{i}{\overset{p}{\rightleftarrows}} Y$$

where $p = (0\ ,\ 1)$ *and* $i = \begin{pmatrix} 0 \\ 1 \end{pmatrix}$.

Proposition 3 *Let $X \times_Y Z$ be the object of a pullback diagram in an additive category with kernels, where u is a split epi, with splitting v, as in the following diagram*

$$\begin{array}{ccc} X \times_Y Z & \xrightarrow{\pi_2} & Z \\ {\scriptstyle \pi_1}\downarrow & & \downarrow{\scriptstyle w} \\ X & \xrightarrow{u} & Y \end{array}.$$

Then $X \times_Y Z \cong \ker u \oplus Z$ and the pullback diagram becomes

$$\begin{array}{ccc} \ker u \oplus Z & \xrightarrow{(0\ 1)} & Z \\ {\scriptstyle \left(\begin{smallmatrix} 1 & 0 \\ 0 & w \end{smallmatrix}\right)}\downarrow & & \downarrow{\scriptstyle w} \\ \ker u \oplus Y & \xrightarrow{(0\ 1)} & Y \end{array}.$$

In the following section we will describe the notion of weak category in an additive 2-category with kernels.

4 Weak categories in additive 2-categories with kernels

Let $\mathbb{A}$ be an additive 2-category with kernels. We will identify $\mathbb{A}$ with the data $(\mathbb{A}, K, D)$ of Proposition 1. When it is clear from the context, we will refer to a 2-cell

$$(\tau, f) : f \longrightarrow f + D(\tau)$$

simply by τ.

4.1 Precategories. (See Definition 2).

Proposition 4 *An internal precategory in $\mathbb{A}$ is completely determined by four morphisms of $\mathbb{A}$,*

$$\begin{aligned} k &: A \longrightarrow B, \\ f, g &: A \longrightarrow A, \\ h &: B \longrightarrow A, \end{aligned}$$

with

$$kf = k = kg\ ,\ kh = 0, \tag{4.1}$$

and is given (up to an isomorphism) by

$$A\oplus A\oplus B \xrightarrow{m} A\oplus B \underset{(k\ 1)}{\overset{(0\ 1)}{\underset{\longrightarrow}{\overset{\longrightarrow}{\xleftarrow{\binom{0}{1}}}}}} B, \tag{4.2}$$

where $m = \begin{pmatrix} f & g & h \\ 0 & 0 & 1 \end{pmatrix}$.

Proof Since the morphism $d : C_1 \longrightarrow C_0$ in Definition 2 is a split epi, using Proposition 2 we conclude that the object C_1 is of the form $A \oplus B$ (considering A as the kernel of d and $C_0 = B$). This means that the underlying reflexive graph of our precategory is of the form

$$A\oplus B \underset{(k\ 1)}{\overset{(0\ 1)}{\underset{\longrightarrow}{\overset{\longrightarrow}{\xleftarrow{\binom{0}{1}}}}}} B.$$

The object $C_1 \times_{C_0} C_1$ is (by Proposition 3) isomorphic to

$$A \oplus A \oplus B$$

and the projections π_1 and π_2 are given by the diagram

$$A\oplus B \xleftarrow{\begin{pmatrix} 1 & 0 & 0 \\ 0 & k & 1 \end{pmatrix}} A\oplus A\oplus B \xrightarrow{\begin{pmatrix} 0 & 1 & 0 \\ 0 & 0 & 1 \end{pmatrix}} A\oplus B.$$

The composition $m : A \oplus A \oplus B \longrightarrow A \oplus B$ is a morphism satisfying $dm = d\pi_2$, which means that

$$m = \begin{pmatrix} f & g & h \\ 0 & 0 & 1 \end{pmatrix}$$

with $f, g : A \longrightarrow A$ and $h : B \longrightarrow A$ arbitrary morphisms of $\mathbb{A}$. Nevertheless, the condition $cm = c\pi_1$ yields

$$kf = k = kg\ ,\ kh = 0.$$

□

4.2 Associative precategories. In order to analyze the 2-cell

$$\alpha : m\,(1 \times_{C_0} m) \longrightarrow m\,(m \times_{C_0} 1)$$

(see Definition 3), the morphisms $m\,(1 \times_{C_0} m)$ and $m\,(m \times_{C_0} 1)$ have to be described. Having in mind (by Proposition 4) that $C_1 \times_{C_0} C_1$ is of the form $A \oplus A \oplus B$ and, using Proposition 3, we conclude that $C_1 \times_{C_0} C_1 \times_{C_0} C_1$ is of the form

$$A \oplus A \oplus A \oplus B.$$

The projections for $C_1 \times_{C_0} (C_1 \times_{C_0} C_1)$ are given as in the diagram

$$A\oplus B \xleftarrow{\begin{pmatrix} 1 & 0 & 0 & 0 \\ 0 & k & k & 1 \end{pmatrix}} A\oplus(A\oplus A\oplus B) \xrightarrow{(0\ 1)} (A\oplus A\oplus B)$$

and the projections for $(C_1 \times_{C_0} C_1) \times_{C_0} C_1$ are given by

$$A\oplus A\oplus B \xleftarrow{p} A\oplus A\oplus A\oplus B \xrightarrow{\begin{pmatrix} 0 & 0 & 1 & 0 \\ 0 & 0 & 0 & 1 \end{pmatrix}} A\oplus B$$

with

$$p = \begin{pmatrix} 1 & 0 & 0 & 0 \\ 0 & 1 & 0 & 0 \\ 0 & 0 & k & 1 \end{pmatrix}.$$

The reader may appreciate checking that

$$1 \times_{C_0} m = \begin{pmatrix} 1 & 0 & 0 & 0 \\ 0 & f & g & h \\ 0 & 0 & 0 & 1 \end{pmatrix}$$

and

$$m \times_{C_0} 1 = \begin{pmatrix} f & g & hk & h \\ 0 & 0 & 1 & 0 \\ 0 & 0 & 0 & 1 \end{pmatrix}.$$

Using matrix multiplication we have that

$$m\,(1 \times_{C_0} m) = \begin{pmatrix} f & gf & g^2 & gh+h \\ 0 & 0 & 0 & 1 \end{pmatrix}$$

and

$$m\,(m \times_{C_0} 1) = \begin{pmatrix} f^2 & fg & fhk+g & fh+h \\ 0 & 0 & 0 & 1 \end{pmatrix}.$$

The isomorphism

$$(\alpha, m\,(1 \times m)) : m\,(1 \times m) \longrightarrow m\,(m \times 1)$$

has α in $K\,(A \oplus A \oplus A \oplus B, A \oplus B)$ and

$$D\,(\alpha) = m\,(m \times 1) - m\,(1 \times m)\,. \tag{4.3}$$

Since α must satisfy

$$d \circ \alpha = 1_{d\pi_3},$$

which may be written as[3]

$$(d\alpha, d\pi_3) = (0, d\pi_3)\,,$$

we conclude that $d\alpha = 0$. Having in mind that $d = (0\ 1)$ and α is a 2×4 matrix, we have

$$\alpha = \begin{pmatrix} \alpha_1 & \alpha_2 & \alpha_3 & \alpha_0 \\ 0 & 0 & 0 & 0 \end{pmatrix}$$

with $\alpha_1, \alpha_2, \alpha_3 \in K\,(A, A)$ and $\alpha_0 \in K\,(B, A)$. Similarly, from $c \circ \alpha = 1_{c\pi_1}$, we conclude that $c\alpha = 0$. Furthermore, since $c = (k\ 1)$ we have

$$k\alpha_i = 0\ ,\ i = 0, 1, 2, 3.$$

In order to satisfy condition (4.3), we must also have

$$\begin{aligned} D\,(\alpha_1) &= f^2 - f, \\ D\,(\alpha_2) &= fg - gf, \\ D\,(\alpha_3) &= fhk + g - g^2, \\ D\,(\alpha_4) &= fh - gh. \end{aligned}$$

We are now ready to establish the following:

[3]Note that for any morphism $\varphi : A \longrightarrow A'$, the 2-cell 1_φ is of the form $(0, \varphi)$ and so, $1_{d\pi_3}$ is of the form $(0, d\pi_3)$. The composite $d \circ \alpha$ is of the form $(0, d) \circ (\alpha, m\,(1 \times m)) = (d\alpha, dm\,(1 \times m)) = (d\alpha, d\pi_3)\,.$

Proposition 5 *An internal associative precategory in $\mathbb{A}$ is completely determined by morphisms*

$$\begin{aligned} k &: A \longrightarrow B, \\ f, g &: A \longrightarrow A, \\ h &: B \longrightarrow A, \end{aligned}$$

with

$$kf = k = kg \ , \ kh = 0,$$

and objects $\alpha_1, \alpha_2, \alpha_3 \in K(A, A)$, $\alpha_0 \in K(B, A)$ with

$$k\alpha_i = 0 \ , \ i = 0, 1, 2, 3$$

$$\begin{aligned} D(\alpha_1) &= f^2 - f, \\ D(\alpha_2) &= fg - gf, \\ D(\alpha_3) &= fhk + g - g^2, \\ D(\alpha_4) &= fh - gh. \end{aligned}$$

4.3 Associative precategories with identity. In order to analyze the 2-cells for the left and right identities (see Definition 4), we have to describe the morphisms $m\langle ec, 1\rangle$ and $m\langle 1, ed\rangle$ from C_1 to C_1. Proposition 3 yields

$$\langle ec, 1\rangle = \begin{pmatrix} 0 & 0 \\ 1 & 0 \\ 0 & 1 \end{pmatrix} \ , \ \langle 1, ed\rangle = \begin{pmatrix} 1 & 0 \\ 0 & 0 \\ 0 & 1 \end{pmatrix}.$$

Hence,

$$m\langle ec, 1\rangle = \begin{pmatrix} g & h \\ 0 & 1 \end{pmatrix} \ , \ m\langle 1, ed\rangle = \begin{pmatrix} f & h \\ 0 & 1 \end{pmatrix}.$$

Since $(\lambda, m\langle ec, 1\rangle) : m\langle ec, 1\rangle \longrightarrow 1$ is a 2-cell from $A \oplus B$ to $A \oplus B$, we conclude that λ is in $K(A \oplus B, A \oplus B)$ and

$$D(\lambda) = \begin{pmatrix} 1 & 0 \\ 0 & 1 \end{pmatrix} - \begin{pmatrix} g & h \\ 0 & 1 \end{pmatrix}, \tag{4.4}$$

while ρ is in $K(A \oplus B, A \oplus B)$ and

$$D(\rho) = \begin{pmatrix} 1 & 0 \\ 0 & 1 \end{pmatrix} - \begin{pmatrix} f & h \\ 0 & 1 \end{pmatrix}. \tag{4.5}$$

In order to satisfy conditions (2.8) λ and ρ must be of the form

$$\begin{aligned} \lambda &= \begin{pmatrix} \lambda_1 & \lambda_0 \\ 0 & 0 \end{pmatrix}, \\ \rho &= \begin{pmatrix} \rho_1 & \rho_0 \\ 0 & 0 \end{pmatrix}, \end{aligned}$$

with $\lambda_1, \rho_1 \in K(A, A)$ and $\lambda_0, \rho_0 \in K(B, A)$ such that

$$\begin{aligned} k\lambda_1 &= 0 = k\rho_1, \\ k\lambda_0 &= 0 = k\rho_0. \end{aligned} \tag{4.6}$$

From the condition

$$\lambda \circ e = \rho \circ e,$$

we conclude that $\lambda_0 = \rho_0$. To simplify notation a new letter, η, is introduced to denote λ_0 and ρ_0. In this way, instead of having $\lambda_0, \rho_0 \in K(B, A)$ and one condition $\lambda_0 = \rho_0$ we simply have $\eta \in K(B, A)$. Since λ_0 and ρ_0 are not used anymore, we will write λ and ρ instead of λ_1 and ρ_1 respectively.

With this new notation, conditions (4.4) and (4.5) become

$$\begin{pmatrix} D(\lambda) & D(\eta) \\ 0 & 0 \end{pmatrix} = \begin{pmatrix} 1 & 0 \\ 0 & 1 \end{pmatrix} - \begin{pmatrix} g & h \\ 0 & 1 \end{pmatrix}$$

and

$$\begin{pmatrix} D(\rho) & D(\eta) \\ 0 & 0 \end{pmatrix} = \begin{pmatrix} 1 & 0 \\ 0 & 1 \end{pmatrix} - \begin{pmatrix} f & h \\ 0 & 1 \end{pmatrix}.$$

This means that λ and ρ completely determine the morphisms f, g, h and we have

$$\begin{aligned} g &= 1 - D(\lambda), \\ f &= 1 - D(\rho), \\ h &= -D(\eta). \end{aligned}$$

Next, we show that the conditions (4.1) are satisfied with f, g, h given as above. Since

$$kf = k - kD(\rho),$$

and $kD(\rho) = D(k\rho)$ (by condition (3.6)) and $k\rho = 0$, we have

$$kf = k.$$

The same argument shows that $k = kg$ and $kh = 0$, since $k\lambda = 0$ and $k\eta = 0$.

This suggests the following description of associative precategories with identity in an additive 2-category with kernels:

Proposition 6 *An associative precategory with identity in an additive 2-category with kernels is completely determined by a morphism*

$$A \xrightarrow{k} B$$

together with

$$\begin{aligned} \alpha_1, \alpha_2, \alpha_3, \lambda, \rho &\in K(A, A), \\ \alpha_0, \eta &\in K(B, A), \end{aligned}$$

subject to the following conditions

$$k\alpha_i = 0\ ,\ i = 0, 1, 2, 3$$

$$k\lambda = 0 = k\rho,\ k\eta = 0,$$

$$\begin{aligned} D(\alpha_1) &= f^2 - f, \\ D(\alpha_2) &= fg - gf, \\ D(\alpha_3) &= fhk + g - g^2, \\ D(\alpha_0) &= fh - gh, \end{aligned} \tag{4.7}$$

where f, g, h are defined as follows:

$$\begin{aligned} g &= 1 - D(\lambda), \\ f &= 1 - D(\rho), \\ h &= -D(\eta). \end{aligned} \tag{4.8}$$

4.4 Associative precategories with coherent identity. An associative precategory with coherent identity is an associative precategory with identity (see previous section) where the triangle and mixed coherent conditions are satisfied (see Definition 5). We proceed using the description of associative precategory with identity given as in the previous section to describe the triangle and the mixed coherence conditions in additive 2-categories with kernels.

4.4.1 *Triangle coherence condition.* In order to analyze the triangle coherence condition

$$(m \circ (\rho \times 1)) \cdot (\alpha \circ (1 \times \langle ec, 1\rangle)) = m \circ (1 \times \lambda)$$

the 2-cells $\rho \times_{C_0} 1$ and $1 \times_{C_0} \lambda$, have to be described. Since they are elements in $K(A \oplus A \oplus B, A \oplus A \oplus B)$, the reader is invited to show that

$$\rho \times_{C_0} 1 = \begin{pmatrix} \rho & \eta k & \eta \\ 0 & 0 & 0 \\ 0 & 0 & 0 \end{pmatrix}, 1 \times_{C_0} \lambda = \begin{pmatrix} 0 & 0 & 0 \\ 0 & \lambda & \eta \\ 0 & 0 & 0 \end{pmatrix}.$$

We have already seen that

$$\langle ec, 1\rangle = \begin{pmatrix} 0 & 0 \\ 1 & 0 \\ 0 & 1 \end{pmatrix}, \text{ so } (1 \times \langle ec, 1\rangle) = \begin{pmatrix} 1 & 0 & 0 \\ 0 & 0 & 0 \\ 0 & 1 & 0 \\ 0 & 0 & 1 \end{pmatrix}.$$

The definition of horizontal composition in an additive 2-category yields

$$\begin{aligned} m \circ (\rho \times 1) &= \begin{pmatrix} f\rho & f\eta k & f\eta \\ 0 & 0 & 0 \end{pmatrix}, \\ m \circ (1 \times \lambda) &= \begin{pmatrix} 0 & g\lambda & g\eta \\ 0 & 0 & 0 \end{pmatrix}, \\ \alpha \circ (1 \times \langle ec, 1\rangle) &= \begin{pmatrix} \alpha_1 & \alpha_3 & \alpha_0 \\ 0 & 0 & 0 \end{pmatrix}. \end{aligned}$$

Finally, the coherence condition may be written as

$$\begin{pmatrix} f\rho & f\eta k & f\eta \\ 0 & 0 & 0 \end{pmatrix} + \begin{pmatrix} \alpha_1 & \alpha_3 & \alpha_0 \\ 0 & 0 & 0 \end{pmatrix} = \begin{pmatrix} 0 & g\lambda & g\eta \\ 0 & 0 & 0 \end{pmatrix},$$

or, equivalently, as

$$\begin{aligned} \alpha_1 &= -f\rho, \\ \alpha_3 &= g\lambda - f\eta k, \\ \alpha_0 &= g\eta - f\eta. \end{aligned}$$

The components $\alpha_1, \alpha_3, \alpha_0$ are completely determined. In the next section we show that the component α_2 is also determined by the mixed coherence condition.

4.4.2 *Mixed coherence condition.* Consider the mixed coherence condition

$$\rho \cdot (m \circ \langle \lambda, 1_{ed}\rangle) \cdot (\alpha \circ \langle ec, 1, ed\rangle) = \lambda \cdot (m \circ \langle 1_{ec}, \rho\rangle).$$

We have already seen that

$$\langle 1, ed\rangle = \begin{pmatrix} 1 & 0 \\ 0 & 0 \\ 0 & 1 \end{pmatrix}. \text{ Thus } \langle ec, 1, ed\rangle = \begin{pmatrix} 0 & 0 \\ 1 & 0 \\ 0 & 0 \\ 0 & 1 \end{pmatrix}.$$

Furthermore, we have

$$\langle \lambda, 1_{ed} \rangle = \begin{pmatrix} \lambda & \eta \\ 0 & 0 \\ 0 & 0 \end{pmatrix}$$

and

$$\langle 1_{ec}, \rho \rangle = \begin{pmatrix} 0 & 0 \\ \rho & \eta \\ 0 & 0 \end{pmatrix}.$$

Using the definition of horizontal composition we obtain

$$\begin{aligned} m \circ \langle \lambda, 1_{ed} \rangle &= \begin{pmatrix} f\lambda & f\eta \\ 0 & 0 \end{pmatrix}, \\ \alpha \circ \langle ec, 1, ed \rangle &= \begin{pmatrix} \alpha_2 & \alpha_0 \\ 0 & 0 \end{pmatrix}, \\ m \circ \langle 1_{ec}, \rho \rangle &= \begin{pmatrix} g\rho & g\eta \\ 0 & 0 \end{pmatrix}, \end{aligned}$$

and the mixed coherence condition may be written as

$$\begin{pmatrix} \rho & \eta \\ 0 & 0 \end{pmatrix} + \begin{pmatrix} f\lambda & f\eta \\ 0 & 0 \end{pmatrix} + \begin{pmatrix} \alpha_2 & \alpha_0 \\ 0 & 0 \end{pmatrix} = \begin{pmatrix} \lambda & \eta \\ 0 & 0 \end{pmatrix} + \begin{pmatrix} g\rho & g\eta \\ 0 & 0 \end{pmatrix},$$

or, equivalently, as

$$\begin{aligned} \alpha_2 &= \lambda + g\rho - \rho - f\lambda, \\ \alpha_0 &= \eta + g\eta - \eta - f\eta. \end{aligned}$$

Therefor the 2-cell α is completely determined and it is a straightforward calculation checking that $k\alpha_i = 0,\ i = 0, 1, 2, 3$, and that conditions (4.7) are satisfied.

Hence, we have:

Proposition 7 *An associative precategory with coherent identity in an additive 2-category with kernels is completely determined by a morphism*

$$A \xrightarrow{k} B,$$

together with

$$\begin{aligned} \lambda, \rho &\in K(A, A), \\ \eta &\in K(B, A), \end{aligned}$$

subject to the conditions

$$\begin{aligned} k\lambda &= 0, \\ k\rho &= 0, \\ k\eta &= 0. \end{aligned}$$

It is given (up to an isomorphism) by

$$A \oplus A \oplus B \xrightarrow{m} A \oplus B \overset{(0\ 1)}{\underset{(k\ 1)}{\overset{\longrightarrow}{\underset{\longrightarrow}{\xleftarrow{\binom{0}{1}}}}}} B,$$

$$\alpha = \begin{pmatrix} \alpha_1 & \alpha_2 & \alpha_3 & \alpha_0 \\ 0 & 0 & 0 & 0 \end{pmatrix},$$

$$\lambda = \begin{pmatrix} \lambda & \eta \\ 0 & 0 \end{pmatrix}, \rho = \begin{pmatrix} \rho & \eta \\ 0 & 0 \end{pmatrix},$$

where

$$m = \begin{pmatrix} f & g & h \\ 0 & 0 & 1 \end{pmatrix},$$

$$\begin{aligned} g &= 1 - D(\lambda), \\ f &= 1 - D(\rho), \\ h &= -D(\eta), \end{aligned}$$

$$\begin{aligned} \alpha_1 &= \rho^2 - \rho, \\ \alpha_2 &= \rho\lambda - \lambda\rho, \\ \alpha_3 &= \lambda - \lambda^2 - f\eta k, \\ \alpha_0 &= \rho\eta - \lambda\eta. \end{aligned}$$

4.5 Weak categories. In this section we show that the pentagon coherence condition does not add new restrictions on the data involved in Proposition 7, i.e. the description of associative precategory with coherent identity is in fact the description of a weak category in an additive 2-category with kernels.

In order to analyze the pentagon coherence condition

$$\begin{aligned} &(\alpha \circ (m \times 1 \times 1)) \cdot (\alpha \circ (1 \times 1 \times m)) \\ &\quad = (m \circ (\alpha \times 1)) \cdot (\alpha \circ (1 \times m \times 1)) \cdot (m \circ (1 \times \alpha)), \end{aligned}$$

we need some preliminary calculations. Namely, all the arrows in the expression have to be described.

To describe the arrow $m \times_{C_0} 1 \times_{C_0} 1$, we have to analyze its domain

$$(C_1 \times_{C_0} C_1) \times_{C_0} C_1 \times_{C_0} C_1$$

and codomain

$$C_1 \times_{C_0} C_1 \times_{C_0} C_1.$$

Using the results obtained in the previous sections, we have that the domain, together with its three projections, is given by

$$\begin{array}{ccccc} A \oplus A \oplus B & \overset{\pi_1}{\longleftarrow} & A \oplus A \oplus A \oplus A \oplus B & \overset{\pi_3}{\longrightarrow} & A \oplus B \\ & & \downarrow{\scriptstyle \pi_2} & & \\ & & A \oplus B & & \end{array},$$

where

$$\begin{aligned} \pi_1 &= \begin{pmatrix} 1 & 0 & 0 & 0 & 0 \\ 0 & 1 & 0 & 0 & 0 \\ 0 & 0 & k & k & 1 \end{pmatrix}, \\ \pi_2 &= \begin{pmatrix} 0 & 0 & 1 & 0 & 0 \\ 0 & 0 & 0 & k & 1 \end{pmatrix}, \\ \pi_3 &= \begin{pmatrix} 0 & 0 & 0 & 1 & 0 \\ 0 & 0 & 0 & 0 & 1 \end{pmatrix}. \end{aligned}$$

The codomain, together with its projections, is given by

$$\begin{array}{ccccc} A \oplus B & \overset{\pi_1}{\longleftarrow} & A \oplus A \oplus A \oplus B & \overset{\pi_3}{\longrightarrow} & A \oplus B \\ & & \downarrow{\scriptstyle \pi_2} & & \\ & & A \oplus B & & \end{array},$$

where

$$\pi_1 = \begin{pmatrix} 1 & 0 & 0 & 0 \\ 0 & k & k & 1 \end{pmatrix},$$
$$\pi_2 = \begin{pmatrix} 0 & 1 & 0 & 0 \\ 0 & 0 & k & 1 \end{pmatrix},$$
$$\pi_3 = \begin{pmatrix} 0 & 0 & 1 & 0 \\ 0 & 0 & 0 & 1 \end{pmatrix}.$$

Having in mind that

$$m = \begin{pmatrix} f & g & h \\ 0 & 0 & 1 \end{pmatrix} : A \oplus A \oplus B \longrightarrow A \oplus B,$$
$$1 = \begin{pmatrix} 1 & 0 \\ 0 & 1 \end{pmatrix} : A \oplus B \longrightarrow A \oplus B,$$

we obtain

$$(m \times 1 \times 1) = \begin{pmatrix} f & g & hk & hk & h \\ 0 & 0 & 1 & 0 & 0 \\ 0 & 0 & 0 & 1 & 0 \\ 0 & 0 & 0 & 0 & 1 \end{pmatrix}.$$

By similar calculations we also have

$$(1 \times m \times 1) = \begin{pmatrix} 1 & 0 & 0 & 0 & 0 \\ 0 & f & g & hk & h \\ 0 & 0 & 0 & 1 & 0 \\ 0 & 0 & 0 & 0 & 1 \end{pmatrix}$$

and

$$(1 \times 1 \times m) = \begin{pmatrix} 1 & 0 & 0 & 0 & 0 \\ 0 & 1 & 0 & 0 & 0 \\ 0 & 0 & f & g & h \\ 0 & 0 & 0 & 0 & 1 \end{pmatrix}.$$

We remark that $\alpha \times 1$ is in fact an abbreviation of $\alpha \times_{C_0} 1_{1_{C_1}}$, where $1_{1_{C_1}}$ is the identity 2-cell of the arrow 1_{C_1}. So, it is the pair $(0, 1_{C_1})$ with 0 in $K(A \oplus B, A \oplus B)$.

The domain of $\alpha \times 1$ is given (together with its two projections) as in the diagram

$$A \oplus A \oplus A \oplus B \xleftarrow{\pi_1} A \oplus A \oplus A \oplus A \oplus B \xrightarrow{\pi_2} A \oplus B,$$

where

$$\pi_1 = \begin{pmatrix} 1 & 0 & 0 & 0 & 0 \\ 0 & 1 & 0 & 0 & 0 \\ 0 & 0 & 1 & 0 & 0 \\ 0 & 0 & 0 & k & 1 \end{pmatrix},$$
$$\pi_2 = \begin{pmatrix} 0 & 0 & 0 & 1 & 0 \\ 0 & 0 & 0 & 0 & 1 \end{pmatrix},$$

and the codomain is given as in

$$A \oplus B \xleftarrow{\pi_1} A \oplus A \oplus B \xrightarrow{\pi_2} A \oplus B,$$

where

$$\pi_1 = \begin{pmatrix} 1 & 0 & 0 \\ 0 & k & 1 \end{pmatrix},$$
$$\pi_2 = \begin{pmatrix} 0 & 1 & 0 \\ 0 & 0 & 1 \end{pmatrix}.$$

Hence, we obtain

$$\alpha \times 1 = \begin{pmatrix} \alpha_1 & \alpha_2 & \alpha_3 & \alpha_0 k & \alpha_0 \\ 0 & 0 & 0 & 0 & 0 \\ 0 & 0 & 0 & 0 & 0 \end{pmatrix}.$$

Similarly, we have get

$$1 \times \alpha = \begin{pmatrix} 0 & 0 & 0 & 0 & 0 \\ 0 & \alpha_1 & \alpha_2 & \alpha_3 & \alpha_0 \\ 0 & 0 & 0 & 0 & 0 \end{pmatrix}.$$

Now that we have all ingredients of our calculation, we begin the main part.

On the one hand, we have to describe

$$\alpha\,(m \times 1 \times 1) + \alpha\,(1 \times 1 \times m)$$

and the result is

$$\begin{pmatrix} \alpha_1 f + \alpha_1 & \alpha_1 g + \alpha_2 & \alpha_1 hk + \alpha_2 + \alpha_3 f & \alpha_1 hk + \alpha_3 + \alpha_3 g & \alpha_1 h + \alpha_0 + \alpha_3 h + \alpha_0 \\ 0 & 0 & 0 & 0 & 0 \end{pmatrix}. \tag{4.9}$$

On the other hand, we need

$$(m\,(\alpha \times 1)) + (\alpha\,(1 \times m \times 1)) + (m\,(1 \times \alpha))$$

and the result is

$$\begin{pmatrix} f\alpha_1 + \alpha_1 & f\alpha_2 + \alpha_2 f + g\alpha_1 & f\alpha_3 + \alpha_2 g + g\alpha_2 & * & ** \\ 0 & 0 & 0 & 0 & 0 \end{pmatrix}, \tag{4.10}$$

where $*$ is $f\alpha_0 k + \alpha_2 hk + \alpha_3 + g\alpha_3$ and $**$ is $f\alpha_0 + \alpha_2 h + \alpha_0 + g\alpha_0$.

To check whether (4.9) and (4.10) are equal is the same as checking whether the following identities hold

$$\alpha_1 f = f\alpha_1 \tag{4.11}$$
$$\alpha_1 g + \alpha_2 = f\alpha_2 + \alpha_2 f + g\alpha_1 \tag{4.12}$$
$$\alpha_1 hk + \alpha_2 + \alpha_3 f = f\alpha_3 + \alpha_2 g + g\alpha_2 \tag{4.13}$$
$$\alpha_1 hk + \alpha_3 g = f\alpha_0 k + \alpha_2 hk + g\alpha_3 \tag{4.14}$$
$$\alpha_1 h + \alpha_3 h + \alpha_0 = f\alpha_0 + \alpha_2 h + g\alpha_0. \tag{4.15}$$

We will use

$$\begin{aligned} f &= 1 - D(\rho)\ ,\ g = 1 - D(\lambda)\ ,\ h = -D(\eta)\,, \\ \alpha_1 &= \rho^2 - \rho\ ,\ \alpha_2 = \rho\lambda - \lambda\rho, \\ \alpha_3 &= \lambda - \lambda^2 - \eta k + \rho\eta k, \\ \alpha_0 &= \rho\eta - \lambda\eta. \end{aligned}$$

(see Proposition 7).

The condition (4.11) holds since we have $\rho f = f\rho$.

The condition (4.12) is equivalent to

$$\alpha_1 - \alpha_1\lambda + \alpha_2 = \alpha_2 - \rho\alpha_2 + \alpha_2 - \alpha_2\rho + \alpha_1 - \lambda\alpha_1,$$

which simplifies to

$$-\left(\rho^2-\rho\right)\lambda=-\rho\alpha_2+\alpha_2-\alpha_2\rho-\lambda\left(\rho^2-\rho\right)$$

and then becomes

$$-\rho^2\lambda+\rho\lambda=-\rho^2\lambda+\rho\lambda\rho+\rho\lambda-\lambda\rho-\rho\lambda\rho+\lambda\rho^2-\lambda\rho^2+\lambda\rho$$

which is a trivial condition.

Moreover, the condition (4.13) is equivalent to

$$\alpha_1hk+\alpha_2+\alpha_3-\alpha_3\rho=\alpha_3-\rho\alpha_3+\alpha_2-\alpha_2\lambda+\alpha_2-\lambda\alpha_2,$$

which extends to

$$\left(\rho^2-\rho\right)(-\eta)\,k-\alpha_3\rho=-\rho\alpha_3-(\rho\lambda-\lambda\rho)\,\lambda+\rho\lambda-\lambda\rho-\lambda\,(\rho\lambda-\lambda\rho)\,,$$

and also to

$$\begin{aligned}&\left(\rho^2-\rho\right)(-\eta)\,k-\left(\lambda-\lambda^2-\eta k+\rho\eta k\right)\rho\\&\qquad=-\rho\left(\lambda-\lambda^2-\eta k+\rho\eta k\right)-(\rho\lambda-\lambda\rho)\,\lambda+\rho\lambda-\lambda\rho-\lambda\,(\rho\lambda-\lambda\rho)\,.\end{aligned}$$

Since $k\rho=0$, this condition is also trivial.

The condition (4.14) is equivalent to

$$\alpha_1hk+\alpha_3-\alpha_3\lambda=f\alpha_0k+\alpha_2hk+\alpha_3-\lambda\alpha_3,$$

and then to

$$\begin{aligned}&\left(\rho^2-\rho\right)(-\eta)\,k-\left(\lambda-\lambda^2-\eta k+\rho\eta k\right)\lambda\\&\quad=(\rho\eta-\lambda\eta)\,k-\rho\,(\rho\eta-\lambda\eta)\,k+(\rho\lambda-\lambda\rho)\,(-\eta)\,k-\lambda\left(\lambda-\lambda^2-\eta k+\rho\eta k\right).\end{aligned}$$

Since $k\lambda=0$ it is trivial again.

The condition (4.15) is equivalent to

$$\alpha_1h+\alpha_3h=-\rho\alpha_0+\alpha_2h+\alpha_0-\lambda\alpha_0,$$

or

$$\begin{aligned}&-\rho^2\eta+\rho\eta-\lambda\eta+\lambda^2\eta+\eta k\eta-\rho\eta k\eta\\&\qquad\qquad=-\rho\,(\rho\eta-\lambda\eta)-\rho\lambda\eta+\lambda\rho\eta+\rho\eta-\lambda\eta-\lambda\,(\rho\eta-\lambda\eta)\,.\end{aligned}$$

Since $k\eta=0$, the condition is trivial.

Finally, we obtain:

Proposition 8 *An associative precategory with coherent identity in an additive 2-category with kernels is a weak category.*

5 Examples

In this section we consider internal weak categories in Ab and Mor(Ab) that are examples of additive 2-categories with kernels.

5.1 Abelian groups. According to Proposition 1, taking $\mathbb{A}$=Ab and $D = id : \hom_{\mathrm{Ab}}(A, B) \longrightarrow \hom_{\mathrm{Ab}}(A, B)$, the category Ab of abelian groups is an example of an additive 2-category.

The data describing a weak category in Ab consists of four morphisms of abelian groups

$$A \xrightarrow{k} B \ , \ \lambda, \rho : A \longrightarrow A \ , \ \eta : B \longrightarrow A$$

subject to the conditions

$$\begin{aligned} k\lambda &= 0 = k\rho, \\ k\eta &= 0. \end{aligned}$$

This information can be used to construct the corresponding weak category with the objects being the elements of B, the morphisms pairs $(a, b) \in A \oplus B$

$$b \xrightarrow{(a,b)} k(a) + b,$$

and the composition

$$b \xrightarrow{(a,b)} k(a) + b \xrightarrow{(a',k(a)+b)} k(a' + a) + b$$

given by

$$(a', k(a) + b)(a, b) = (a' - \rho(a') + a - \lambda(a) - \eta(b), b).$$

For every three composable morphisms

$$b \xrightarrow{(a,b)} k(a) + b = b' \xrightarrow{(a',b')} k(a') + b' = b'' \xrightarrow{(a'',b'')} k(a'') + b'',$$

the 2-cell

$$\alpha : (a'', b'')((a', b')(a, b)) \longrightarrow ((a'', b'')(a', b'))(a, b),$$

is given by

$$\begin{aligned} &\alpha(a'', a', a, b) \\ &= \left((\rho^2 - \rho)(a'') + (\rho\lambda - \lambda\rho)(a') + \left(\lambda - \lambda^2 - \eta k + \rho\eta k\right)(a) + (\rho\eta - \lambda\eta)(b)\right). \end{aligned}$$

The 2-cells λ and ρ for one morphism (a, b)

$$\begin{aligned} \lambda &: (0, b')(a, b) \longrightarrow (a, b) \\ \rho &: (a, b)(0, b) \longrightarrow (a, b) \end{aligned}$$

are given by

$$\begin{aligned} \lambda(a, b) &= (\lambda a + \eta b, 0), \\ \rho(a, b) &= (\rho a + \eta b, 0). \end{aligned}$$

Note that if $B = 0$, we obtain what we called a weak monoid. If the morphisms λ, ρ, η are the zero morphisms, then we obtain an internal category in Ab (which is well known to be just a group homomorphism).

5.2 Morphisms of Abelian groups. The category Mor(Ab) of morphisms of abelian groups is the category where the objects are morphisms of Ab, say

$$A = \left(A_1 \xrightarrow{\partial} A_0\right).$$

The arrows are pairs of morphisms of Ab $(f_1, f_0) : A \longrightarrow B$ such that the following square is commutative

$$\begin{array}{ccc} A_1 & \xrightarrow{\partial} & A_0 \\ {\scriptstyle f_1}\downarrow & & \downarrow{\scriptstyle f_0} \\ B_1 & \xrightarrow{\partial} & B_0 \end{array}.$$

For each two arrows $f = (f_1, f_0)$ and $g = (g_1, g_0)$ from A to B, a 2-cell from f to g is a pair $(\tau, f) : f \longrightarrow g$ where $\tau : A_0 \longrightarrow B_1$ is a homomorphism of abelian groups with

$$\begin{aligned} \tau\partial &= g_1 - f_1 \\ \partial\tau &= g_0 - f_0. \end{aligned}$$

In order to be able to see that this category is an example of an additive 2-category we note that, with respect to objects and arrows, it is in fact an additive category. Now, for each pair of objects, A and B, we define the abelian group $K(A, B)$ as

$$K(A, B) = \hom_{Ab}(A_0, B_1)$$

and the homomorphism D as

$$D(\tau) = (\tau\partial, \partial\tau).$$

It can be shown that D satisfies all the conditions in (3.6) and (3.7) if we define the following laws of composition

$$\begin{aligned} g\tau &= g_1\tau, \\ \sigma\tau &= \sigma\partial\tau, \\ \sigma f &= \sigma f_0, \end{aligned}$$

for every $\tau \in K(A, B), \sigma \in K(B, C), f \in \hom(A, B), g \in \hom(B, C)$.

A weak category C in the additive 2-category of Mor(Ab) is determined by a commutative square

$$\begin{array}{ccc} A_1 & \xrightarrow{\partial} & A_0 \\ {\scriptstyle k_1}\downarrow & & \downarrow{\scriptstyle k_0} \\ B_1 & \xrightarrow{\partial'} & B_0 \end{array}$$

together with three morphisms

$$\begin{aligned} \lambda, \rho &: A_0 \longrightarrow A_1, \\ \eta &: B_0 \longrightarrow A_1, \end{aligned}$$

satisfying the conditions

$$\begin{aligned} k_1\lambda &= 0 = k_1\rho, \\ k_1\eta &= 0. \end{aligned}$$

The objects of C are pairs (b,d) with $b \in B_0, d \in B_1$. The morphisms are of the form

$$\begin{pmatrix} b & x \\ d & y \end{pmatrix}$$

with $b \in B_0, d \in B_1, x \in A_0, y \in A_1$.

A weak category in Mor(Ab) may also be viewed as a structure with objects, vertical arrows, horizontal arrows and squares, in the following way

$$\begin{array}{ccc} b & \xrightarrow{(b,x)} & b + k_0(x) \\ \binom{b}{d}\downarrow & \begin{pmatrix} b & x \\ d & y \end{pmatrix} & \downarrow\binom{b+k_0(x)}{d+k_1(y)} \\ b + \partial'(d) & \xrightarrow[(b+\partial'(d),x+\partial(y))]{} & * \end{array},$$

where $*$ stands for $b + \partial'(d) + k_0(x + \partial(y)) = b + k_0(x) + \partial'(d + k_1(y))$.

The horizontal composition between squares is given by

$$\begin{pmatrix} b + k_0(x) & x' \\ d + k_1(y) & y' \end{pmatrix} \circ \begin{pmatrix} b & x \\ d & y \end{pmatrix} = \begin{pmatrix} b & f_0(x') + g_0(x) + h_0(b) \\ d & f_1(y') + g_1(y) + h_1(d) \end{pmatrix},$$

where

$$\begin{aligned} f_0 &= 1 - \partial\rho, \\ g_0 &= 1 - \partial\lambda, \\ h_0 &= -\partial\eta, \end{aligned}$$

$$\begin{aligned} f_1 &= 1 - \rho\partial, \\ g_1 &= 1 - \lambda\partial, \\ h_1 &= -\eta\partial. \end{aligned}$$

For each three horizontal arrows

$$b \xrightarrow{(b,x)} b' \xrightarrow{(b',x')} b'' \xrightarrow{(b'',x'')} b'' + k_0(x'')$$

with $b' = b + k_0(x)$ and $b'' = b' + k_0(x')$ the isomorphism for associativity is given by

$$\begin{array}{ccc} b & \xrightarrow{(b'',x'')\circ((b',x')\circ(b,x))} & b'' + k_0(x'') \\ \binom{b}{0}\downarrow & \begin{pmatrix} b & f_0(x'') + g_0(z) + h_0(b) \\ 0 & \alpha_1(x'') + \alpha_2(x') + \alpha_3(x) + \alpha_0(b) \end{pmatrix} & \downarrow\binom{b''+k_0(x'')}{0} \\ b & \xrightarrow[((b'',x'')\circ(b',x'))\circ(b,x)]{} & b'' + k_0(x'') \end{array}$$

where $z = f_0(x') + g_0(x) + h_0(b)$ and

$$\begin{aligned} \alpha_1 &= -f_1\rho, \\ \alpha_2 &= \lambda + g_1\rho - \rho - f_1\lambda, \\ \alpha_3 &= g_1\lambda - f_1\eta k_0, \\ \alpha_0 &= g_1\eta - f_1\eta. \end{aligned}$$

The left and right isomorphisms are given, respectively, by

$$\begin{array}{ccc} b & \xrightarrow{(b+k_0(x),0)\circ(b,x)} & b+k_0(x) \\ \binom{b}{0}\Big\downarrow & \begin{pmatrix} b & g_0(x)+h_0(b) \\ 0 & \lambda(x)+\eta(b) \end{pmatrix} & \Big\downarrow\binom{b+k_0(x)}{0} \\ b & \xrightarrow[(b,x)]{} & b+k_0(x) \end{array}$$

and

$$\begin{array}{ccc} b & \xrightarrow{(b,x)\circ(b,0)} & b+k_0(x) \\ \binom{b}{0}\Big\downarrow & \begin{pmatrix} b & f_0(x)+h_0(b) \\ 0 & \rho(x)+\eta(b) \end{pmatrix} & \Big\downarrow\binom{b+k_0(x)}{0} \\ b & \xrightarrow[(b,x)]{} & b+k_0(x) \end{array}.$$

References

[1] MacLane, Saunders, *Categories for the working Mathematician*, Springer-Verlag, 1998, 2nd edition.
[2] Crans, S.E, "Teisi in Ab", Homology, Homotopy and Applications 3 (2001) 87-100.
[3] Martins-Ferreira, N., "Internal Bicategories in Ab", Preprint CM03/I-24, Aveiro Universitiy, 2003
[4] Martins-Ferreira, N., "The Category of Weak Categories", in preparation.
[5] Bourn, D., "Another denormalization theorem for the abelian chain complexes", J. Pure and Applied Algebra 66, 1990, 229-249.
[6] Brown, R. and Higgins, Ph.J., "Cubical Abelian Groups with Connections are Equivalent to Chain Complexes", Homology, Homotopy and Applications vol. 5(1), 2003, 49-52.
[7] Paré, R. and Grandis, M. "Adjoints for Double Categories" Submitted to Cahiers Topologie Géom. Différentielle Catég.

Received December 20, 2002; in revised form July 21, 2003

Fields Institute Communications
Volume **43**, 2004

Dendrotopic Sets

Thorsten Palm
Department of Mathematics and Statistics
York University
Toronto, Ontario M3J 1P3, Canada
Current address: Institut für Theoretische Informatik
Technische Universität Braunschweig
38023 Braunschweig, Germany
palm@iti.cs.tu-bs.de

What is before the reader has arisen from research towards the thesis [7] on *weak higher-dimensional categories*. The thesis, to be more precise, will concern one of the many definitions of this notion that have been proposed so far, namely the one put forth by M. Makkai in his yet unpublished paper [5]. The present work concerns only the underlying geometric structures, which he calls *multitopic sets*.

The story of this definition begins with the work of J. Baez and J. Dolan, a good summary of which can be found in [1]. These authors suggest to define weak finite-dimensional categories as *opetopic sets* having universal fillers for certain configurations of cells. The striking feature of this definition is that the usually explicit algebraic operations — composition, associativity, coherence and so on — are implicit in the geometric structure. The aesthete, however, will find two minor flaws: the concept of universality cannot be generalized to the infinite-dimensional case, because it is defined by a reversed induction; and cells possess a non-geometric excess structure (an arbitrary total order on the source facets), which can be traced back to the definition of opetopic sets via *(sorted) operads* (whence the prefix 'ope-').

Both flaws have been eliminated by Makkai. As for the first, the object approach to universality is superseded by a represented-functor approach; the need for a notion of universality in the next dimension disappears. As for the second, symmetric operads are superseded by their non-symmetric counterparts, the so-called *multicategories* (whence the prefix 'multi-').

Multitopic sets were defined in the three-part paper [4]. The definition follows algebraic ideas: a multitopic set gives rise to a free multicategory in each dimension, which in turn allows for cells in the next dimension to be constructed in a globular fashion. In the more recent account [3], it is shown that an alternative definition is available: multitopic sets are precisely the *many-to-one computads*. Here 'computad' refers to a freely generating subset of a strict ∞-category, and 'many-to-one' indicates that this subset is closed under the target operations.

2000 *Mathematics Subject Classification.* Primary 18D05; Secondary 52B11.

The present account brings yet a third definition, under the more descriptive name '*dendrotopic set*'. It is briefer than the first and the most elementary of the three. It entirely follows geometric ideas.

Dendrotopic sets are built from *dendrotopes*, just as, say, globular sets are built from globes. Dendrotopes in turn are oriented polytopes satisfying certain simple conditions. Here a *polytope* is determined by the incidence structure of its faces. A 0-dendrotope is just a point,

$$\bullet.$$

A 1-dendrotope is just a directed line segment from a 0-dendrotope to a 0-dendrotope,

(So far the same can be said about globes.) A 2-dendrotope is a surface suspended from a path of 1-dendrotopes to a single 1-dendrotope; it is determined by the length of the path.

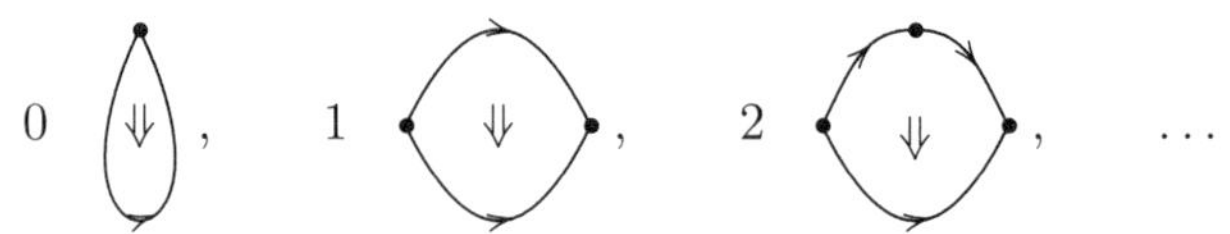

A 3-dendrotope is a solid filling the space from a tree of 2-dendrotopes to a single 2-dendrotope; it is determined by the bare planar tree. For example,

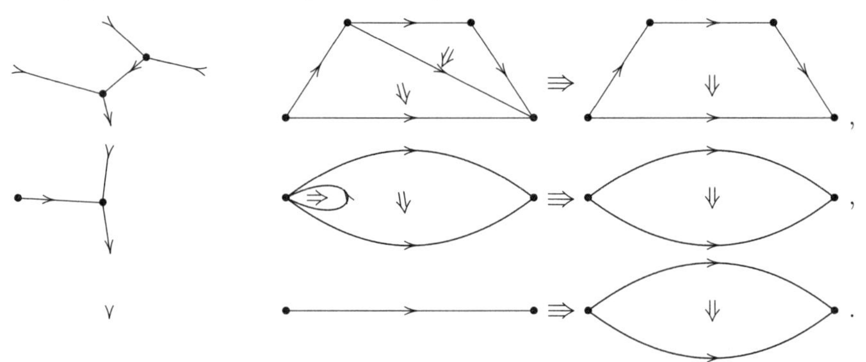

In general, a dendrotope is an oriented polytope whose facets are dendrotopes as well; furthermore

(Δ1) the target consists of a single facet,
(Δ2) the source inherits a tree structure (whence the prefix 'dendro-'),
(Δ3) each 3-codimensional face is a ridge of the target facet precisely once.

The paper is self-contained, to the point that it should be accessible to the general mathematical public. Only a few remarks are directed specifically at category theorists. Part I defines the notion of a dendrotopic set and related ones (with the latter accounting for much of its size). Part II shows that dendrotopic sets are precisely the many-to-one computads, whence they are precisely the multitopic sets as well.

Part I. The Definition

The definition of 'dendrotopic set' will be presented in three steps. The first two introduce notions which are interesting enough to deserve being named: *propolytopic sets* (Section 2) and *oriented propolytopic sets* (Section 4). The former can be specialized to obtain the notion of a *polytopic set*, that is, a combinatorial structure built from polytopes. This is done in Section 3, whose chief aim it is to compare the present concept of polytope with the standard one, as defined for example in the monograph [6]. Note that the concept considered standard here is *not* the one according to which a polytope is (the face lattice of) the convex hull of finitely many points in affine space.

In fact two definitions of 'dendrotopic set' will be presented, the first officially (Section 5), the second in the form of a theorem (Section 6). The first is the natural and practical one: the existence part of axiom ($\Delta 3$) is omitted. The second is the elegant one: the entire axiom ($\Delta 2$) is omitted. (In either case the omitted piece is a consequence of the remaining ones.) Part I is rounded off by the result that every finite arrangement of dendrotopes as a tree has the potential of being the source of another dendrotope (Section 7).

Section 1 supplies two auxiliary concepts.

Section 1. Higher-Dimensional Numbers and Hemigraphs

This preliminary section provides two new, yet basic notions for later reference. *Higher-dimensional numbers* will naturally enumerate the source facets of a dendrotope. The incidence structure of the overall source in codimensions 1 and 2 will be a particular kind of *hemigraph* (namely, a tree).

A *list of length* ℓ is simply an ℓ-tuple. We shall use angular brackets to denote lists. Thus, the list with entries $x_0, \ldots, x_{\ell-1}$ is denoted by $\langle x_0, \ldots, x_{\ell-1} \rangle$. Concatenation of lists is denoted by $+$; the empty list is denoted by 0. We shall differentiate between the element x and the list $\langle x \rangle$.

The lists with entries in a given set X form the monoid X^* freely generated by it. The product is $+$, and the neutral element is 0. Generators are inserted by means of the mapping $x \mapsto \langle x \rangle$. The smallest such monoid that is closed under the insertion of generators consists of the *higher-dimensional numbers*.

A more explicit description can be given as follows. For $n \geq 0$, we define an *n-dimensional number* inductively to be a list of $(n-1)$-dimensional numbers. The understanding is that -1-dimensional numbers do not exist. Thus, denoting the set of n-dimensional numbers by $\hat{\mathbf{N}}_n$, we have

$$\hat{\mathbf{N}}_{-1} = \varnothing \qquad \text{and} \qquad \hat{\mathbf{N}}_n = \hat{\mathbf{N}}^*_{n-1} \quad (n \geq 0).$$

The only 0-dimensional number is the empty list 0, and 1-dimensional numbers are lists $\langle 0, \ldots, 0 \rangle$. We can identify each of the latter with its length, so that $\hat{\mathbf{N}}_1$ becomes precisely the monoid $\mathbf{N}$ of natural numbers. A simple inductive argument shows that we have a chain of monoids $\hat{\mathbf{N}}_0 \subseteq \hat{\mathbf{N}}_1 \subseteq \cdots$. The union $\hat{\mathbf{N}}_\infty$ is the monoid of all higher-dimensional numbers.

The inclusions $\hat{\mathbf{N}}_{n-1} \subseteq \hat{\mathbf{N}}_n$ are convenient as of now, but it will sometimes be important exactly which dimension a number is considered in. We keep track of these dimensions by means of subscripts. We write $\langle \mathfrak{q}_0, \ldots, \mathfrak{q}_{\ell-1} \rangle_n$ for the specifically n-dimensional number whose entries are the specifically $(n-1)$-dimensional numbers $\mathfrak{q}_0, \ldots, \mathfrak{q}_{\ell-1}$. In the case $\ell = 0$ we write 0_n.

The concept of an n-dimensional number is equivalent to the concept of an *n-stage tree* considered by M. Batanin in [2]. In vivid terms, such a tree is an isomorphism class of finite planar rooted trees with all heights $\leq n$. (Here the term 'tree' is used in its standard meaning. A slightly refined notion of tree is introduced further below.) The proof of this equivalence is suggested by the diagrammatic equation

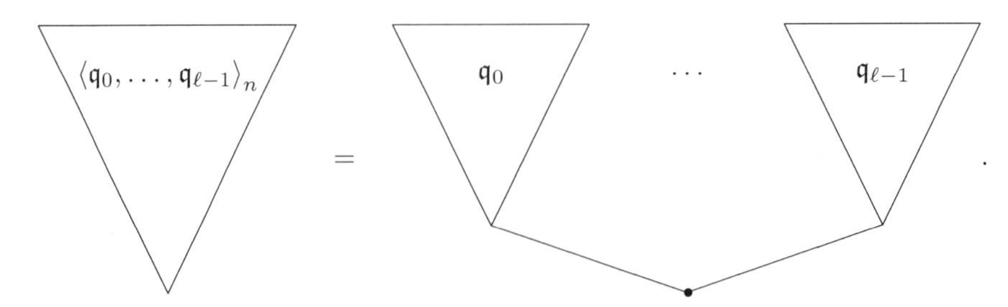

Despite the common goal of defining weak higher-dimensional categories, the present use of higher-dimensional numbers has no evident similarities to Batanin's use of trees.

Before we introduce hemigraphs, we recall some related notions.

A *(directed) graph* $\mathfrak{G}$ consists of two sets $\mathfrak{G}_0$ and $\mathfrak{G}_1$ and two mappings δ_-, $\delta_+ : \mathfrak{G}_1 \rightrightarrows \mathfrak{G}_0$. The set $\mathfrak{G}_0$ contains the *vertices* and the set $\mathfrak{G}_1$ contains the *arrows* of the graph. Given an arrow F, the vertices $F\delta_-$ and $F\delta_+$ are its *source* and its *target*. We may visualize F thus:

$$F\delta_- \bullet \xrightarrow{\quad F \quad} \bullet\, F\delta_+ .$$

A *path of length* ℓ from (= *starting* at = having *source*) a vertex A to (= *ending* at = having *target*) a vertex B consists of $\ell + 1$ vertices $A = A_0, A_1, \ldots, A_{\ell-1}, A_\ell = B$ and ℓ arrows $F_{1/2}, \ldots, F_{\ell-1/2}$ satisfying the *consecutiveness conditions* $F_i\delta_\eta = A_{i+\eta/2}$ ($i \in \{1/2, \ldots, \ell - 1/2\}$, $\eta \in \mathbf{Z}^\times$). (We take the signs $-$ and $+$ to be abbreviations for the numbers -1 and $+1$, which in turn form the group $\mathbf{Z}^\times$ of unit integers.) If $\ell > 0$ and $A = B$ we speak of a *cycle*. We slight the question of equality of cycles, which will not be of importance to us.

An *undirected graph* $\mathfrak{G}$ is a graph together with a direction-reversing involution fixing each vertex, but no arrow. That is, undirectedness is expressed by a mapping $\chi : \mathfrak{G}_1 \to \mathfrak{G}_1$ such that for each arrow F we have $F\chi \neq F$, $F\chi^2 = F$ and $F\chi\delta_\eta = F\delta_{-\eta}$. The orbits of χ are the *edges* of $\mathfrak{G}$. Given an edge $\{F, F\chi\}$, the vertices $F\delta_-$ and $F\delta_+$ are its *extremities*. We may visualize $\{F, F\chi\}$ thus:

$$F\delta_- \bullet \overset{\{F, F\chi\}}{\longrightarrow} \bullet\, F\delta_+ .$$

There is an evident way of specifying the structure of an undirected graph by listing vertices, edges, and the two extremities of each edge. (Specifying the extremities of all edges amounts to specifying a mapping of the set of edges into the set of unordered pairs of vertices.) We can in particular associate to a directed graph an underlying undirected graph, declaring each arrow to become an edge and its source and target to become its extremities. A path in this underlying structure is a *zigzag* in the original graph. Conversely, we can impose a direction on an undirected graph by choosing a representative arrow for each edge.

Hemigraphs are defined in the same way as graphs, the only difference being that the source and target mappings are allowed to be partial. Hence an arrow may lack a source or a target or both. These cases can be visualized thus:

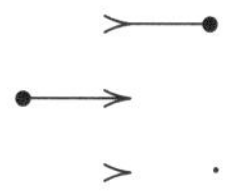

In this context it makes sense to allow paths to start or end at an arrow; lengths are measured naturally in half steps. For example, a path of length $3/2$ from a vertex A to an arrow G consists of two vertices A and B and two arrows F and G with $F\delta_-$ defined and equal to A and both $F\delta_+$ and $G\delta_-$ defined and equal to B.

A *tree* to us is a hemigraph with a distinguished arrow Ω, called *root*, such that from each vertex or arrow there is a unique path to Ω. The length of this path is the *height* of the vertex or arrow. (Hence the heights of arrows are natural numbers and the heights of vertices are natural numbers plus $1/2$.) Necessary conditions for a hemigraph to be a tree with root Ω are that each vertex has precisely one outgoing arrow and that Ω is the only arrow without a target. A hemigraph satisfying these conditions will be called *locally treelike* with *root* Ω. Conversely, in a locally treelike hemigraph the component containing the root is a tree. Each of the other components contains a *forward-infinite path*, that is: vertices A_0, A_1, ... and arrows $F_{1/2}$, $F_{3/2}$, ... satisfying the consecutiveness conditions. We may hence define trees equivalently by local treelikeness plus either connectedness or absence of forward-infinite paths. Now a forward-infinite path either contains a vertex twice, in which case the intermediate vertices and arrows form a cycle, or indicates that the hemigraph itself is infinite. Therefore a finite hemigraph is a tree if and only if it is locally treelike and cycle-free.

Use of the term 'tree' invokes a baggage of botanical vocabulary. For example, a *leaf* is an arrow without a source. This notion should be contrasted with that of a *stump*, which is a vertex without an incoming arrow. A tree is *empty* if it consists of the root only; that is: if the root is a leaf. In the other case the root has a source, called the *stem* of the tree. (Thus the stem is the only vertex with height $1/2$.)

Section 2. Propolytopic Sets

A polytope P can be constructed as follows. First take a finite family of polytopes of the previous dimension.

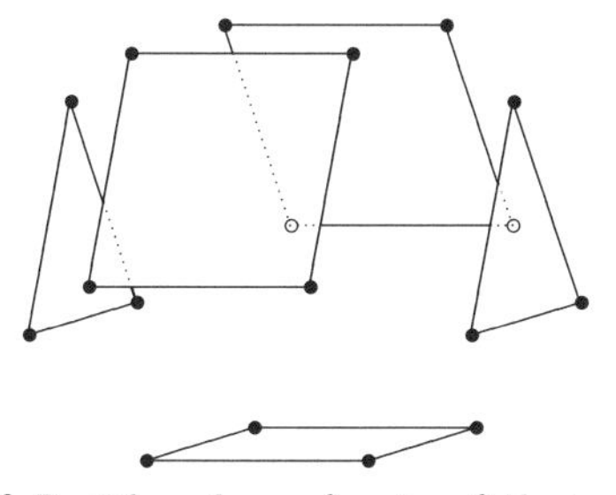

They become the *facets* of P. They have facets of their own, and thus give rise to a family of "facet facets". Now arrange the "facet facets" into pairs and glue them accordingly.

The joints become the *ridges* of P. Finally fill the interior of the (ideally) hollow figure that has arisen.

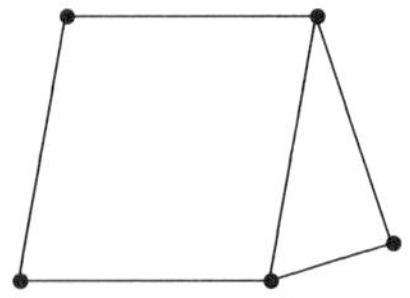

A formalization of the choices to be made in this construction leads to the definition of a *propolytopic set*.

Definition. A *propolytopic set* $\mathfrak{P}$ provides for each integer n the data (Π0), (Π1) and (Π2) below.

(Π0) a set $\mathfrak{P}_n$. We call the elements of $\mathfrak{P}_n$ the *cells of dimension* n of $\mathfrak{P}$, the n-*cells* for short.

(Π1) to each n-cell P, a finite family $\langle\, Pq \mid q \in P\mathrm{I}\,\rangle$ of $(n-1)$-cells. We call $q \in P\mathrm{I}$ a *facet location* of P and Pq its *resident*.

A *chamber of depth* d underneath an n-cell P, a d-*chamber* for short, is a d-tuple $\langle p^1, \ldots, p^d \rangle$ in which, for certain cells $P = P^0$, P^1, ... , P^d, the entry p^{i+1} is a facet location of P^i with resident P^{i+1}. The $(n-d)$-cell $P^d = Pp^1 \cdots p^d$ will be called the *floor* of the chamber. The set of all d-chambers underneath P will be denoted by $P\mathrm{I}^d$.

(Π2) to each n-cell P, a partition $P\diamondsuit$ of $P\mathrm{I}^2$ into 2-element subsets of 2-chambers with common floor. We call $\{\langle q_0, r_0\rangle, \langle q_1, r_1\rangle\} \in P\diamondsuit$ a *ridge location* of P and $Pq_0r_0 = Pq_1r_1$ its *resident*.

Instead of '$\{\langle q_0, r_0\rangle, \langle q_1, r_1\rangle\} \in P\diamondsuit$' we shall write more compactly '$P : \frac{q_1\ r_1}{q_0\ r_0}$'. Being given a partition into 2-element subsets is equivalent to being given an involution without fixed element; viewed as such, $P\diamondsuit$ will be called the *conjugation* underneath P. Consequently the relationship $\frac{q_1\ r_1}{q_0\ r_0}$ can be rendered by saying that $\langle q_0, r_0\rangle$ and $\langle q_1, r_1\rangle$ are *conjugates* of each other.

Cells of dimensions ≥ 0 will be called *proper*, cells of dimension < 0 will be called *improper*. A propolytopic set is *plain* if all of its cells are proper. In this case of course 0-cells have no facet locations and 1-cells have no ridge locations. Any propolytopic set has a *plain part* (that is: underlying plain propolytopic set), obtainable by removing improper cells and the locations where they reside. While our interest in propolytopic sets will be restricted to their plain parts, improper cells will serve a purpose in avoiding low-dimensional exceptions to general statements.

Examples. The plain propolytopic sets characterized below will be referred to as the *classical polytopic sets*.

- In a *globular set* each cell of dimension ≥ 1 has precisely 2 facet locations δ_- and δ_+, and each cell of dimension ≥ 2 has the 2 ridge locations

$$\frac{\delta_-\ \delta_\zeta}{\delta_+\ \delta_\zeta} \quad (\zeta \in \mathbf{Z}^\times).$$

- In a *simplicial set (without degeneracies!)* each cell of dimension $n \geq 1$ has precisely $n+1$ facet locations δ_0, ... , δ_n, and each cell of dimension $n \geq 2$ has the $(n+2)(n+1)/2$ ridge locations

$$\frac{\delta_j\ \delta_{i-1}}{\delta_i\ \delta_j} \quad (0 \leq j < i \leq n).$$

- In a *cubical set (without degeneracies!)* each cell of dimension $n \geq 1$ has precisely $2n$ facet locations $\delta_{0,-}, \ldots, \delta_{n-1,-}, \delta_{0,+}, \ldots, \delta_{n-1,+}$, and each cell of dimension $n \geq 2$ has the $2(n+1)n$ ridge locations

$$\frac{\delta_{j,\zeta}\ \delta_{i-1,\eta}}{\delta_{i,\eta}\ \ \delta_{j,\zeta}} \quad (0 \leq j < i < n;\ \eta, \zeta \in \mathbf{Z}^{\times}).$$

Facets and ridges are instances of *faces*, which we are now going to introduce in full generality.

A *panel of depth d* underneath an n-cell P, a *d-panel* for short, is a $(d-1)$-tuple

$$\langle p^1, \ldots, p^{\tau-1}, \frac{p_1^{\tau}\ p_1^{\tau+1}}{p_0^{\tau}\ p_0^{\tau+1}}, p^{\tau+2}, \ldots, p^d \rangle \tag{1}$$

in which, for certain cells $P = P^0, P^1, \ldots, P^{\tau-1}, P^{\tau+1}, \ldots, P^d$, the entry p^{i+1} is a facet location of P^i with resident P^{i+1} and the entry $\frac{p_1^{\tau}\ p_1^{\tau+1}}{p_0^{\tau}\ p_0^{\tau+1}}$ is a ridge location of $P^{\tau-1}$ with resident $P^{\tau+1}$. The number τ will be referred to as the *type* of the panel. We say that (1) *separates* the two chambers

$$\langle p^1, \ldots, p^{\tau-1}, p_j^{\tau}, p_j^{\tau+1}, p^{\tau+2}, \ldots, p^d \rangle \quad (j \in \{0, 1\}).$$

The $(n-d)$-cell P^d will be called the *floor* of the panel.

The *undirected depth-d chamber graph* $P\Gamma^d$ underneath P is defined as follows. Its vertices are the d-chambers and its edges are the d-panels underneath P. The extremities of a panel are the two chambers it separates. A path in this graph will also be called a *gallery of depth d* underneath P, a *d-gallery* for short. (The fancy terminology — 'chamber/panel/gallery' — goes back to Bourbaki and has been adapted for current needs.) Note that all the chambers and panels along a gallery have the same floor. This $(n-d)$-cell will be called the *floor* of the gallery.

We display a gallery of depth d and length ℓ as an (often somewhat repetitious) $(\ell+1)$-by-d matrix. Each row represents a chamber, and each two consecutive rows represent a panel, with the facet-location entries doubled and the ridge-location entry marked by a horizontal line as before. Thus the generic panel (1), viewed as a gallery of length 1, takes the following appearance:

$$\begin{array}{ccccccc} p^1 & \cdots & p^{\tau-1} & \underline{p_1^{\tau}\ p_1^{\tau+1}} & p^{\tau+2} & \cdots & p^d \\ p^1 & \cdots & p^{\tau-1} & p_0^{\tau}\ p_0^{\tau+1} & p^{\tau+2} & \cdots & p^d \end{array}.$$

Definition. The components of $P\Gamma^d$ are called the *d-face locations* of of P. Here 'd-face' is short for 'face of *codimension* d'. We write $P\Pi^d$ for the set of all d-face locations of P.

Thus each face location is represented by (= contains) a chamber, and two chambers represent the same face location if and only if they are connected by a gallery. It follows that all representatives of a face location f of P have the same floor. This cell is called the *resident* of f and denoted by Pf.

Examples. Let $0 < d \leq n$.

- Each n-cell in a globular set has precisely 2 d-face locations δ_{η}^d, where η ranges over the set of signs. Here each d-face location δ_{η}^d has 2^{d-1} representative chambers $\langle \delta_{(1)\theta}, \ldots, \delta_{(d-1)\theta}, \delta_{\eta} \rangle$, where θ ranges over the set of mappings $\{1, \ldots, d-1\} \twoheadrightarrow \mathbf{Z}^{\times}$.

- Each n-cell in a simplicial set has precisely $\binom{n+1}{d}$ d-face locations δ^d_Λ, where Λ ranges over the set of d-element subsets of $\{0,\dots,n\}$. Here each d-face location δ^d_Λ has $d!$ representative chambers $\langle \delta_{(1)\bar\lambda},\dots,\delta_{(d)\bar\lambda}\rangle$, where λ ranges over the set of bijective mappings $\{1,\dots,d\} \xrightarrow{\sim} \Lambda$ and

$$(i)\bar\lambda = (i)\lambda - \sum_{\substack{j<i \\ (j)\lambda<(i)\lambda}} 1.$$

- Each n-cell in a cubical set has precisely $2^d\binom{n}{d}$ d-face locations $\delta^d_{\Lambda,\theta}$, where Λ ranges over the set of d-element subsets of $\{0,\dots,n-1\}$ and θ ranges over the set of mappings $\Lambda \to \mathbf{Z}^\times$. Here each d-face location $\delta^d_{\Lambda,\theta}$ has $d!$ representative chambers $\langle \delta_{(1)\bar\lambda,(1)\lambda\theta},\dots,\delta_{(d)\bar\lambda,(d)\lambda\theta}\rangle$, where λ ranges over the set of bijective mappings $\{1,\dots,d\} \xrightarrow{\sim} \Lambda$ and $\bar\lambda$ is defined as before.

A *d-face* of an n-cell P is an $(n-d)$-cell P' together with a d-face location f such that $Pf = P'$. This almost empty definition is intended to solve the following linguistic puzzle: any one face is the resident of a face location (and hence a cell), two different faces are residents of two different face locations (but not necessarily different cells). Any term describing either a face location or the associated face will also be used to describe the other.

Let P' be an u-face of P with location f, and let v be a natural number. For every u-chamber $\langle p^1,\dots,p^u\rangle$ representing f, we have a canonical embedding

$$(\langle p^1,\dots,p^u\rangle + ?) : P'\Gamma^v \to P\Gamma^{u+v}$$

of graphs, which induces a mapping $P'\Pi^v \to P\Pi^{u+v}$ of the component sets. This mapping does not depend on the representative chamber, as the reader can easily verify. We denote the image of g under it by fg.

The depth-0 chamber graph underneath P contains but a single vertex, the empty list 0. The represented location is called *trivial* and denoted by 1; its resident is P itself. For $d > 0$, the depth-d chamber graph is, as a combinatorist would put it, $(d-1)$-*regular*, meaning that each vertex links precisely $d-1$ edges. In fact, each of the possible types $1, \dots, d-1$ occurs precisely once among the panels surrounding a chamber. We can readily classify the possible components for small depths.

1. Each component of $P\Gamma^1$ consists of one vertex and no edge. The three phrases 'q is a facet location of P with resident Q', '$\langle q\rangle$ is a 1-chamber underneath P with floor Q' and 'q is a 1-face location of P with resident Q' evidently mean the same. We identify $P\mathrm{I}$ and $P\Pi^1$ accordingly.
2. Each component of $P\Gamma^2$ consists of two vertices and one connecting edge. The three phrases '$\frac{q_1\ r_1}{q_0\ r_0}$ is a ridge location of P with resident R', '$\langle\frac{q_1\ r_1}{q_0\ r_0}\rangle$ is a 2-panel underneath P with floor R' and '$q_0r_0 = q_1r_1$ is a 2-face location of P with resident R (while $\langle q_0, r_0\rangle \neq \langle q_1, r_1\rangle$)' evidently mean the same. We identify $P\diamondsuit$ and $P\Pi^2$ accordingly.

3. Each component of $P\Gamma^3$ is a cycle. Its length is even, since panel types alternate between 1 and 2. Its general form is

$$\begin{array}{ccc} \cdots & \cdots & \cdots \\ q_0 & r_{2\ell-1} & s_{\ell-1} \\ \cline{1-2} q_{\ell-1} & r_{2\ell-2} & s_{\ell-1} \\ \cline{2-3} \vdots & \vdots & \vdots \\ \cline{2-3} q_1 & r_1 & s_0 \\ \cline{1-2} q_0 & r_0 & s_0 \\ & \cdots & \cdots \end{array} ,$$

where the two dotted lines together signify a single type-2 panel.

By now the category-theoretically educated reader will have noticed that a propolytopic set is nothing but a special kind of category presentation, introduced here with a weird terminology. Cells are the vertices/objects, facet locations are the arrows of the underlying graph (generators), ridge locations are the commutativity conditions (relators), and face locations in general are the morphisms of the presented category. Disagreement may arise about the direction of arrows/morphisms; the author decrees that they point upwards with respect to dimension. Thus a face location f of P is a morphism $f : Pf \twoheadrightarrow P$. (Traditionalists will be satisfied to see that the notation for composition, introduced in the previous paragraph but one, exhibits the confusing reversal of order that they have become so accustomed to. On a more serious note, the upward direction reflects the view that a face location is an inclusion of its resident by its owner. This view is made more precise in the following section, where *geometric realizations* are explained.) Dimensions amount to a *grading* of the underlying graph, by which we provisionally mean a homomorphism into the graph whose vertices as well as arrows are the integers, with the arrow n pointing from the vertex n to the vertex $n + 1$. All said, we can alternatively define a propolytopic set to be a category presentation with its underlying graph graded, satisfying the following two conditions.

- Each vertex has finitely many incoming arrows.
- The paths occurring in the commutativity conditions have length 2, and each length-2 path thus occurs precisely once. (A commutativity condition is taken to be an unordered pair of paths agreeing in source and target.)

A presentation with these properties can be recovered from the abstract category that it gives rise to. We can hence also define a propolytopic set as a special kind of category. Hints to this effect are given towards the end of the following section.

If the author has decided against the "categorical" approach outlined here, he did so for the following reasons. Firstly, he wanted to keep this work as elementary as possible. Secondly, making the graph structure explicit may have led to an overuse of graph terminology. And thirdly, hardly any feature of propolytopic sets discussed beyond Section 3 occurs in the context of category presentations at large. The reader may feel free to skip the parts that become trivial by translation into category-theoretic language.

Section 3. Polytopic Sets

Under a connectedness assumption, a propolytopic set may drop the prefix 'pro-', and its cells are deemed *polytopes*. In fact, the notion of 'polytope' thus defined is *almost* the standard (abstract) one. Arguably the former improves the latter, by allowing for multiple face incidences.

The section opens with three fundamental constructions on a propolytopic set $\mathfrak{P}$. Category theorists will recognize them as yielding slice categories, "coslice" categories and categories of factorizations.

Let A be an α-cell in $\mathfrak{P}$. The propolytopic set $A\mathfrak{P}$ is defined as follows. An n-cell of $A\mathfrak{P}$ consists of an n-cell P of $\mathfrak{P}$ and an $(\alpha - n)$-face location a of A with resident P. A face location of $(P; a)$ in $A\mathfrak{P}$ with resident $(P'; a')$ consists of a such face location f of P in $\mathfrak{P}$ with resident P' as satisfies $af = a'$. We sometimes identify $A\mathfrak{P}$ with A and therefore call the former a *cell* as well.

Let B be a β-cell in $\mathfrak{P}$. The propolytopic set $\mathfrak{P}B$ is defined as follows. An n-cell of $\mathfrak{P}B$ consists of a $(\beta+n+1)$-cell P of $\mathfrak{P}$ and an $(n+1)$-face location b of P with resident B. (The dimension shift is dictated by geometric considerations.) A face location of $(P; a)$ in $\mathfrak{P}B$ with resident $(P'; a')$ consists of a such face location f of P in $\mathfrak{P}$ with resident P' as satisfies $fb' = b$. We call $\mathfrak{P}B$ the *link* about B.

The third construction combines the other two. Let F be a face location of an α-cell A of $\mathfrak{P}$ with resident a β-cell B. The link about $(B; F)$ of the cell A is isomorphic to the cell $(A; F)$ of the link about B. Let us denote either of the two by $A\mathfrak{P}_F B$. The propolytopic set $A\mathfrak{P}_F B$ can also be defined directly as follows. An n-cell of $A\mathfrak{P}_F B$ consists of a $(\beta+n+1)$-cell P of $\mathfrak{P}$ and two face locations, a of A with resident P (and codimension $\alpha - \beta - n - 1$) and b of P with resident B (and codimension $n+1$), such that $ab = F$. A face location of $(P; a, b)$ in $A\mathfrak{P}_F B$ with resident $(P'; a', b')$ consists of a such face location f of P in $\mathfrak{P}$ with resident P' as satisfies $af = a'$ and $fb' = b$. We call $A\mathfrak{P}_F B$ the *link* about F.

By a *co-d-face* of an n-cell P we mean the link about an $(n-d+1)$-face location of P. A *cofacet* is a co-1-face, and a *coridge* is a co-2-face.

The three constructions yield the principal examples for the three notions to be introduced next: a cell is a *propolytope*, the link about a cell is a *polytopic set*, and the link about a face location is a *polytope*.

A *propolytope of dimension n*, an *n-propolytope* for short, is a propolytopic set with an n-cell $\top$ that has each cell as a face precisely once. The same definition with 'propolytope/propolytopic' replaced by any similar noun/adjective pair is implicit whenever the adjective as an attribute to 'set' refers to a variant of a propolytopic set. Thus it is clear what is meant by the terms '*globe*', '*simplex*', '*cube*', and it will be clear what is meant by the term '*polytope*' as soon as the following statement has been issued.

Definition. A *polytopic set* is a propolytopic set with a -1-cell $\perp$ that is a face of each cell precisely once.

It follows that each 0-cell has precisely one facet location, which we shall denote by ε, and that each 1-cell has precisely one ridge location. The first observation implies that the facet locations t of a 1-cell S are in one-to-one correspondence with the 2-chambers $\langle t, \varepsilon \rangle$ underneath S, whence the second observation implies that S has precisely two facet locations.

A propolytopic set $\mathfrak{P}$ each of whose 1-cells has precisely two facets will be called *augmentable*. In this situation the 0-cells and 1-cells of $\mathfrak{P}$ are in an obvious manner the vertices and edges of an undirected graph. We can modify any such $\mathfrak{P}$ (the case of interest being the one where $\mathfrak{P}$ is plain) by affixing a -1-cell $\perp$, for each 0-cell a facet location ε with resident $\perp$ and for each 1-cell with facet locations t_0 and t_1 a ridge location $\frac{t_1\ \varepsilon}{t_0\ \varepsilon}$. This process is known as *trivial augmentation*. Thus, any polytopic set can be obtained from a certain augmentable plain propolytopic set (namely, its plain part) by trivial augmentation.

Examples. Classical polytopic sets as defined in the previous section are not polytopic sets as defined here, but they become polytopic sets by trivial augmentation.

Why trivially augmented classical polytopic sets are indeed polytopic sets can be seen most conveniently by taking the following alternative approach to the new concept, which also has the advantage of being more intuitive.

We now clarify the geometric content of the notions discussed in Section 2. Let $\mathfrak{P}$ be a propolytopic set. We assign to each cell P a topological space $P\Gamma$, to be called its *(geometric) realization*, and to each of its facet locations q a continuous mapping $q\Gamma : Pq\Gamma \twoheadrightarrow P\Gamma$. If P is improper, its realization is the empty space. The case of a proper cell is handled by induction on its dimension. The induction step was illustrated at the beginning of Section 2. It is divided into three smaller steps. In the first, we form the topological sum $P\Gamma^{\bullet\bullet}$ of the realizations $Pq\Gamma$ of all facets. In the second, we form the quotient space $P\Gamma^{\bullet}$ of $P\Gamma^{\bullet\bullet}$ that arises from identifying corresponding image points under the two mappings $r_j\Gamma : P\frac{q_1\ r_1}{q_0\ r_0}\Gamma \twoheadrightarrow Pq_j\Gamma$ ($j \in \{0, 1\}$), for all ridges. We call $P\Gamma^{\bullet}$ the *boundary* of the realization of P. In the third step, we form the (unreduced) cone $P\Gamma$ over $P\Gamma^{\bullet}$. (The cone is obtained from the cylinder by collapsing one of the two frontiers to a point; it contains at least this point.) The mapping $q\Gamma$ is the composite

$$Pq\Gamma \twoheadrightarrow P\Gamma^{\bullet\bullet} \twoheadrightarrow P\Gamma^{\bullet} \twoheadrightarrow P\Gamma$$

of summand inclusion, projection, base inclusion.

If P has dimension n and $P\Gamma^{\bullet}$ is an $(n-1)$-sphere, we call P *spherical*. In this case $P\Gamma$ itself is an n-ball, while otherwise it is not even an n-manifold with the indicated boundary. Note that for $n > 0$, the space $P\Gamma^{\bullet}$ is a (closed) $(n-1)$-manifold if and only if all facets and cofacets of P are spherical.

Let us call the propolytopic set $\mathfrak{P}$ *spherical* if all of its proper cells are. Ideally sphericality is implied by additional axioms (and in this sense the present use of the term 'cell' is justified). Making it a separate axiom would be useless, since it is too hard to verify. All classical polytopic sets are well known to be spherical.

In low dimensions, the following picture emerges.

0. Let T be a 0-cell. Then $T\Gamma^{\bullet}$ is empty, whence T is automatically spherical.
1. Let S be a 1-cell. Then $S\Gamma^{\bullet}$ is a discrete space, with each point corresponding to a facet of S. Hence S is spherical if and only if it has precisely two facets.
2. Assume all 1-cells of $\mathfrak{P}$ spherical (that is, assume $\mathfrak{P}$ augmentable), and let R be a 2-cell. Then $R\Gamma^{\bullet}$ is a disjoint union of circles. Hence R is spherical if and only if $R\Gamma^{\bullet}$ is connected.

More generally, for every cell of dimension ≥ 2 sphericality implies connectedness of the boundary of the geometric realization. This property in turn is equivalent to connectedness of the combinatorial structures considered next.

Let P be a cell of $\mathfrak{P}$. The *undirected dual graph* $P\mathrm{D}$ of P is defined as follows. Vertices are the facet locations of P; edges are the ridge locations of P. The extremities of a ridge location $\frac{q_1\ r_1}{q_0\ r_0}$ are q_0 and q_1.

The condition we have alluded to appears in the following result.

Proposition 1. *Let $\mathfrak{P}$ be a propolytopic set with precisely one improper cell, in dimension -1. Then $\mathfrak{P}$ is a polytopic set if and only if the dual graphs of all proper cells of $\mathfrak{P}$ are connected.*

Proof. We only show sufficiency. The necessity proof reaches its climax with the observation that the first entries of the type-1 panels of a gallery are the edges of a path in the dual graph.

Suppose the dual graphs of all proper cells of $\mathfrak{P}$ are connected, and let P be an n-cell of $\mathfrak{P}$. We show by induction on n that P has precisely one $(n+1)$-face location. The result is obvious for $n = -1$, so let us suppose that $n \geq 0$.

Since PD is connected, it contains a vertex q. By the existence part of the induction hypothesis, Pq has an n-face location f. Now qf is an $(n+1)$-face location of P, and this settles the existence part of the claim.

Now consider an arbitrary $(n+1)$-face location of P. Being represented by a chamber, it can be written $q'f'$, where q' is a facet location of P and f' is an n-face location of Pq'. Since the graph PD is connected, it contains a path

$$q = q_0 \underset{\frac{q_1\ r'_0}{q_0\ r_0}}{\bullet\!\!-\!\!-\!\!-\!\!-\!\!\bullet} q_1 \quad \cdots \quad q_{\ell-1} \underset{\frac{q_\ell\ r'_{\ell-1}}{q_{\ell-1}\ r_{\ell-1}}}{\bullet\!\!-\!\!-\!\!-\!\!-\!\!\bullet} q_\ell = q'.$$

(If $n = 0$ then $\ell = 0$.) By the existence part of the induction hypothesis, the resident of $\frac{q_{i+1}\ r'_i}{q_i\ \ r_i}$ has an $(n-1)$-face location g_i. By the uniqueness part of the induction hypothesis, applied to Pq_i, we have $r'_{i-1}g_{i-1} = r_i g_i$, where the reader substitutes 'f' or 'f'' for an undefined expression '$r'_{-1}g_{-1}$' or '$r_\ell g_\ell$'. The situation is summarized in the commutative diagram (containing some abbreviations with obvious meaning)

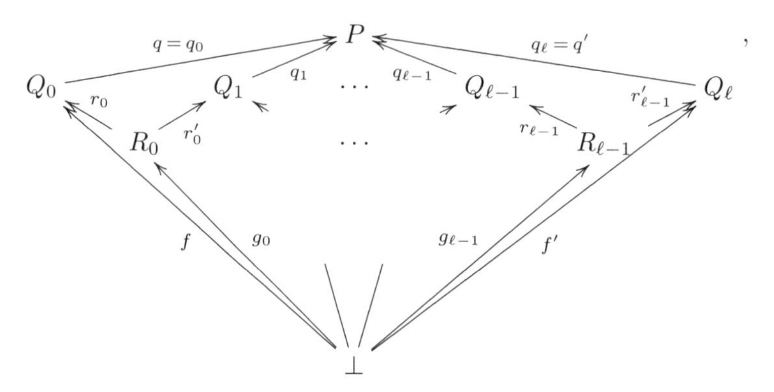

which can be read

$$qf = q_0 r_0 g_0 = q_1 r'_0 g_0 = \cdots = q_{\ell-1} r_{\ell-1} g_{\ell-1} = q_\ell r'_{\ell-1} g_{\ell-1} = q'f'.$$

This settles the uniqueness part of the claim. □

In an augmentable plain propolytopic set, the dual graph of a 0-cell is empty, and the dual graph of a 1-cell consists of two vertices. Trivial augmentation adds a vertex to the former and an edge to the latter, making them connected; the dual graphs of all other cells remain unchanged. Consequently trivial augmentation makes a spherical plain propolytopic set — the classical polytopic sets being examples — into a (spherical) polytopic set.

The section closes with a (very condensed) comparison of "our" polytopes with the standard ones. The attributes 'present-type' and 'standard' will make the distinction clear; the bare term 'polytope' is reserved for a notion unifying the other two. Namely, *we pretend for the remainder of this section that the word 'finite' has been omitted from item* (Π1) *and that the subsequent text has been changed accordingly*. These changes do not concern the main ideas and results; the most significant one accounts for the fact that a 3-face location (as a graph) can now be a path infinite in both directions (rather than a cycle of even length).

It may be interesting to note that the new polytopes are still modest in size: a simple inductive argument using the connectedness result from Proposition 1 shows that they are countable, that is, the cells of each of them together have countably many facet locations.

The result of our comparison will be as follows: *present-type polytopes are precisely the finite polytopes; standard polytopes are precisely the monic polytopes.* Here a polytope is called *monic* if for any two of its cells one is a face of the other at most once. While the first statement is obviously valid, the second needs an explanation. We start with a definition of standard polytopes, as given with more explicitness in [6].

A *standard polytope* $\mathscr{P}$ *of dimension* n is a (partially) ordered set satisfying conditions (i)–(iv) below.

(i) $\mathscr{P}$ has a smallest element $\perp$ and a largest element $\top$.

(ii) All maximal chains in $\mathscr{P}$ have length $n+1$.

Here the *length* of a chain is the number of its elements minus one. For $B < A$ in $\mathscr{P}$, we consider the segments

$$[B, A] = \{ P \mid B \leq P \leq A \} \qquad \text{and} \qquad]B, A[= \{ P \mid B < P < A \}.$$

as ordered subsets of $\mathscr{P}$. Condition (ii) implies:

(ii)* All maximal chains in $[B, A]$ have the same, finite length.

We denote this length by $d_{B,A}$.

(iii) If $d_{B,A} > 2$, then $]B, A[$ is connected.

(iv) If $d_{B,A} = 2$, then $]B, A[$ contains precisely two elements.

It turns out that polytopes in general can be given the same definition, with only a few words and symbols changed. To achieve this end, we define a *multiordered set* to be a category in which all isomorphisms and all endomorphisms are identities, that is, a category generated by a cycle-free graph. The terms 'element', 'smallest', 'largest' will stand for 'object', 'initial', 'terminal'. A *chain* is a subcategory which is a totally ordered set. For a non-identity morphism $F : B \longrightarrow A$, we denote by $[F]$ the category of factorizations and by $]F[$ the category of proper factorizations ('proper' excluding the two identities as factors) of F.

Now a polytope is (after interpretation as a category and up to isomorphism) exactly a multiordered set $\mathscr{P}$ satisfying conditions (i)–(iv) above, with the symbol string 'B, A' replaced by 'F'. Let us see why this is true, without going through the details of an actual proof. First, let us call the number d_F the *codimension* of F. The two parts of condition (i) evidently correspond, respectively, to the definitions of polytopic sets and propolytopes as particular propolytopic sets. The remaining conditions, with (ii)* replacing (ii) to account for the potential absence of $\perp$ or $\top$, characterize propolytopic sets. More precisely, they imply that $\mathscr{P}$ has a presentation in which the generators are the 1-codimensional morphisms (condition (ii)), the relators are pairs of length-2 paths of generators (length ≥ 2: condition (ii); length ≤ 2: condition (iii)), and each length-2 path of generators occurs in precisely one relator (condition (iv)).

We have considered three different notions of 'polytope'. The differences can be well illustrated in dimension 2. Here a polytope is a *polygon*, more precisely the *ℓ-gon*, where ℓ is the number of its facets. The possible values for ℓ are the integers ≥ 1 and ∞. The ∞-gon (= *apeirogon*) is not a present-type polytope, for it is not finite. The 1-gon (= *monogon*) is not a standard polytope, for it is (ironically) not monic: its only vertex is twice a facet of its only edge.

Section 4. Orientation

An important feature of ∞-categorical diagrams has not yet been accounted for: the arrows. An arrow distinguishes the boundary of a cell into a source and a target. For the present — and most other — purposes it is enough to know how the facets are distributed over the two parts. This information is in each case encoded by a sign: '$-$' for 'source' and '$+$' for 'target'. Now for any given ridge, only half of the possible combinations of signs about it can actually occur. Either the ridge is properly contained in either the source or the target; then the two facets have opposite (matching) directions relative to it.

Or the ridge is part of the common boundary of the overall source and the overall target; then the two facets have the same direction relative to it.

There finally is a third, degenerate case, which agrees in signs with the first. Thus, of the four signs involved, an odd number is $-$ and an odd number is $+$.

Definition. An *orientation* on a propolytopic set provides the data (O1) below and satisfies condition (O2) below.

(O1) for each facet location q, a sign $\{\!\!\{q\}\!\!\}$. The set of facet locations of sign η of a cell P is denoted by $P\mathrm{I}_\eta$.

To a chamber $\langle p^1, \ldots, p^d \rangle$, we associate the sign

$$\{\!\!\{p^1, \ldots, p^d\}\!\!\} = \{\!\!\{p^1\}\!\!\} \cdots \{\!\!\{p^d\}\!\!\}.$$

Put differently: a chamber is positive or negative according to whether an even or an odd number of its entries is negative. (The adjectives 'positive' and 'negative' apply in the obvious way.) The set of all d-chambers of sign η underneath a cell P is denoted by $P\mathrm{I}^d_\eta$.

(O2) *Sign-flip axiom.* For each ridge location $\{\langle q_0, r_0\rangle, \langle q_1, r_1\rangle\}$, we have

$$\{\!\!\{q_0, r_0\}\!\!\} \neq \{\!\!\{q_1, r_1\}\!\!\}.$$

Hence we may view $P\Diamond$ as a one-to-one correspondence between $P\mathrm{I}^2_+$ and $P\mathrm{I}^2_-$.

If an orientation is specified, the line notation for ridge locations serves the additional purpose of keeping track of signs. To this end we decree: the negative chamber always goes above the line; the positive chamber always goes below the line. Thus if we write $\frac{q_1\ r_1}{q_0\ r_0}$, we imply that $\{\!|q_1, r_1|\!\} = -$ and $\{\!|q_0, r_0|\!\} = +$. For the sake of flexibility, the two chambers may be permuted, as long as the sign of the permutation is indicated. Thus if we write $\eta\frac{q_1\ r_1}{q_0\ r_0}$, we imply that $\{\!|q_1, r_1|\!\} = -\eta$ and $\{\!|q_0, r_0|\!\} = +\eta$.

Examples. The classical polytopic sets can be oriented. In fact, each of them carries a particular orientation suggested by the indices denoting facet locations:

- in a globular set, $\{\!|\delta_\eta|\!\} = \eta$.
- in a simplicial set, $\{\!|\delta_i|\!\} = -^i$.
- in a cubical set, $\{\!|\delta_{i,\eta}|\!\} = -^i\eta$.

The displays of Section 2 showing the corresponding ridge locations now have to be accompanied by a sign each, so as to read $\varsigma\frac{\delta_-\ \delta_\varsigma}{\delta_+\ \delta_\varsigma}$ in the globular case, $-^{i+j}\frac{\delta_j\ \delta_{i-1}}{\delta_i\ \delta_j}$ in the simplicial case and $-^{i+j}\eta\varsigma\frac{\delta_{j,\varsigma}\ \delta_{i-1,\eta}}{\delta_{i,\eta}\ \delta_{j,\varsigma}}$ in the cubical case.

Let $\mathfrak{P}$ be an oriented propolytopic set (that is, a propolytopic set together with an orientation). We take the chamber graphs $X\Gamma^d$ of $\mathfrak{P}$ to be *directed* as follows. Note that the two chambers separated by a panel carry opposite signs. (Hence chamber graphs are *bipartite*, as a combinatorist would put it.) A type-τ panel

$$\langle p^1, \ldots, p^{\tau-1}, \frac{p^\tau_-\ p^{\tau+1}_-}{p^\tau_+\ p^{\tau+1}_+}, p^{\tau+2}, \ldots, p^d\rangle \tag{2}$$

points towards that one of its adjacent chambers that agrees in sign with its initial $(\tau-1)$-subchamber $\langle p^1, \ldots, p^{\tau-1}\rangle$. We then use the term 'gallery' only to refer to directed paths (that is, paths in the directed chamber graphs). Thus we may no more say that each two chambers representing the same face location are connected by a gallery; but of course they remain connected by a zigzag in the chamber graph.

Let us take a closer look at the case $d = 3$. The components of 3-chamber graphs can no longer be described as cycles; rather, they are closed zigzags. The directions along these zigzags, however, are not distributed entirely arbitrarily. Type-1 panels point from negative to positive chambers (this is true for all depths) and so determine a preferred direction along the zigzag. A type-2 panel points in the preferred direction if and only if its first entry is negative. Thus, a 3-face location whose chambers have only negative first entries remains a cycle; a 3-face location whose chambers have only positive first entries becomes a zigzag with alternating directions.

The rules spelled out above for the line notation of ridge incidences extend accordingly to the matrix notation of galleries. Thus, galleries usually run downwards; a change of direction must be indicated by its sign. The generic panel (2), viewed as a gallery of length 1, can be written

$$\begin{matrix} p^1 & \cdots & p^{\tau-1} & p^\tau_{-\eta} & p^{\tau+1}_{-\eta} & p^{\tau+2} & \cdots & p^d \\ \hline p^1 & \cdots & p^{\tau-1} & p^\tau_{+\eta} & p^{\tau+1}_{+\eta} & p^{\tau+2} & \cdots & p^d \end{matrix}$$

or

$$\eta\begin{cases} p^1 & \cdots & p^{\tau-1} & p^\tau_- & p^{\tau+1}_- & p^{\tau+2} & \cdots & p^d \\ p^1 & \cdots & p^{\tau-1} & p^\tau_+ & p^{\tau+1}_+ & p^{\tau+2} & \cdots & p^d \end{cases},$$

where η is the sign of the final $(d-\tau-1)$-subchamber $\langle p^{\tau+2}, \ldots, p^d\rangle$.

The choice of direction may be a bit surprising. An explanation is given in [7], where the present author shows that in the dendrotopic case, the chosen directions induce a total order on the chambers representing a common face location. (Thus two such chambers are still connected by a path — rather than by a mere zigzag —, but only in one of the two possible ways.) A favourable argument is the fact that the canonical embedding $(\langle p^1, \ldots, p^u \rangle + \,?\,)$ remains a homomorphism of graphs (that is, preserves the chosen directions), independent of the sign of $\langle p^1, \ldots, p^u \rangle$. (Of course this argument may be countered by the observation that the dual statement is wrong: if we consider graphs of chambers and panels with common floors, then the analogous embedding $(\,?\, + \langle p'^1, \ldots, p'^v \rangle)$ preserves or reverses direction according to whether $\langle p'^1, \ldots, p'^v \rangle$ is positive or negative.)

Use of the term 'orientation' seems sufficiently justified by the following fact. For a spherical polytopic set $\mathfrak{P}$, providing an orientation on $\mathfrak{P}$ is equivalent to providing an orientation on the geometric realization of each proper cell of $\mathfrak{P}$. The sphericality assumption is actually redundant, if one agrees that for the purpose of orienting an arbitrary space, points where the manifold axiom of local euclidianness is violated may be cut off.

The above statement refers to a specific one-to-one correspondence. The forward mapping can be described inductively as follows. Suppose $\mathfrak{P}$ is oriented. For a cell P of dimension 0, the orientation on the point $P\Gamma$ is the sign of the only facet location of P. For a cell P of dimension > 0, the orientation on the manifold $P\Gamma$ is determined by the orientation induced on its boundary $P\Gamma^\bullet$, which in turn is determined by demanding that for each facet location q of P, the canonical mapping $Pq\Gamma \twoheadrightarrow P\Gamma^\bullet$ preserve or reverse orientation according to whether q is positive or negative. The sign-flip axiom ensures that the orientations thus described on the facets match up on the ridges.

The idea that an individual cell carries one of two possible orientations leads to a construction that applies to oriented propolytopic sets in general. Let Δ be a bipartition of the set of all cells in an oriented propolytopic set $\mathfrak{P}$. We can modify $\mathfrak{P}$ to obtain an oriented propolytopic set $\mathfrak{P}_\Delta$ by reversing the signs of those facet locations $q \in P\mathrm{I}$ for which P and Pq belong to different parts of Δ. For instance, if one part of Δ comprises all cells of dimension $\geq n$, then the sign reversal occurs precisely for the facet locations of n-cells (duality!). When applying this construction to a polytopic set, the part of Δ containing $\perp$ is the one where cell orientations remain unchanged.

It may counter intuition, or be unwanted for concrete reasons, that the points of an (although) oriented polytopic set carry orientations. In this case we may demand axiomatically that $\perp$ not occur as a negative facet. Put affirmatively, this means that the only facet location ε of each 0-cell is positive. The sign-flip axiom then yields that the two facet locations of each 1-cell have opposite signs. Any oriented polytopic set $\mathfrak{P}$ has a variant with "disoriented" points, namely $\mathfrak{P}_\Delta$, where one part of Δ comprises the 0-cells for which ε is negative.

An *oriented* propolytopic set $\mathfrak{P}$ will be called *augmentable* if each of its 1-cells has precisely one facet of each sign. In this situation the 0-cells and 1-cells of $\mathfrak{P}$ are in an obvious manner the vertices and arrows of a directed graph. For any such $\mathfrak{P}$, the process of trivial augmentation works as it did in the absence of orientations; the affixed facet locations ε are taken to be positive.

A polytope P can be represented by drawing the realizations of all its facets Pf as they appear under the mappings $f\Gamma$. An orientation can — in the cases of interest for us — be represented by adding arrows, one for each $>$ 0-dimensional face, pointing from the totality of negative facets to the totality of positive facets and having as many "tails" as the dimension indicates. Here are a picture of a 2-globe and a picture of a 2-cube.

If the dimension of the polytope exceeds the dimension of the paper, source and target of its arrow may be drawn separately in a "double Schlegel diagram". Here is a picture of a 3-simplex.

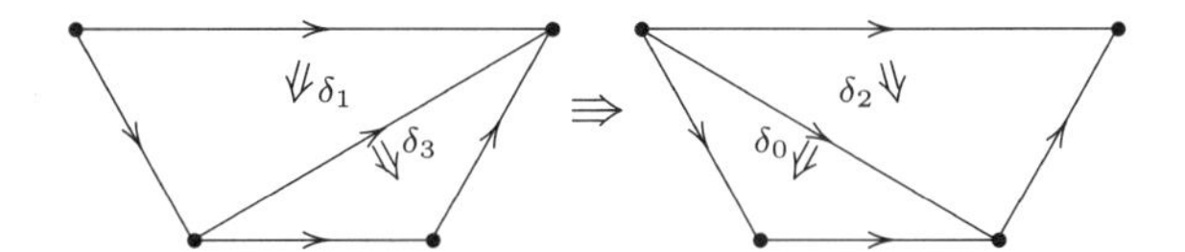

Similar pictures occur in the theory of higher-dimensional categories, where they may be viewed as freely generating subsets of strict ∞-categories (see Sections 1 and 3 in Part II). Let us for this and the following two paragraphs call such a gadget an *(∞-categorical) diagram scheme.* (The artificial term 'computad' is the standard.) The following two questions naturally arise.

(i) Which oriented polytopic sets can be viewed as ∞-categorical diagram schemes?

(ii) Which ∞-categorical diagram schemes can be viewed as oriented polytopic sets?

A partial answer to both these questions is given by Theorem 10 of Part II. This work is far from pursuing general answers, but a few remarks may be of interest to the reader.

As for (i), it is not difficult to come up with certain geometric conditions satisfied by the classical polytopic sets and apparently sufficient in order for a polytopic set to be a diagram scheme. The envisioned result has already been obtained in two instances. In [2], Batanin explicitly constructs the cells of the free ∞-category associated to a globular set as trees of the kind mentioned in Section 1, labelled in a certain way by the generators. In [8], R. Street gives an implicit description of the free ∞-category associated to a simplicial set, and more generally of the ∞-category associated to what may be described as a presentation by simplices. The question has been fully answered under assumptions of global cycle freeness, reducing the ∞-categorical operations to plain set-theoretic unions. The best-known work of this kind is Street's [9] on his *parity complexes*.

As for (ii), two shortcomings of the notion of a polytopic set ought to be stressed. First, the "gluing" step in the construction of a polytope is restricted to codimension 2. Thus, for example, in an attempt to interpret the ∞-categorical diagram

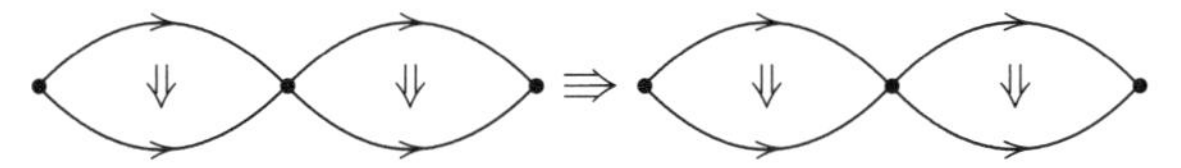

as a 3-(pro-)polytope, the middle one of the three 0-dimensional faces would break into two. (Parity complexes, by contrast, are subject to no such restriction.) Second, the "face pick" step in the construction of a polytope is restricted to codimension 1. Thus, for example, the ∞-categorical diagram

showing a 2-cell looping about a degenerate 1-cell, cannot be interpreted as a 2-(pro-)polytope at all. Apart from that, an oriented polytope lacking facets of one of the two signs may be ∞-categorically underdetermined: compare the two diagrams

This last problem will not arise for dendrotopic sets: here a $>$ 0-dimensional cell lacking negative facets will have precisely one ridge.

As have been the chamber graphs, so will be the dual graphs of a propolytopic set in presence of an orientation: *directed*. We let a ridge location $\frac{q_- \; r_-}{q_+ \; r_+}$ point from the facet location q_- to the facet location q_+. The following diagram shows the directed dual graph of a 3-simplex.

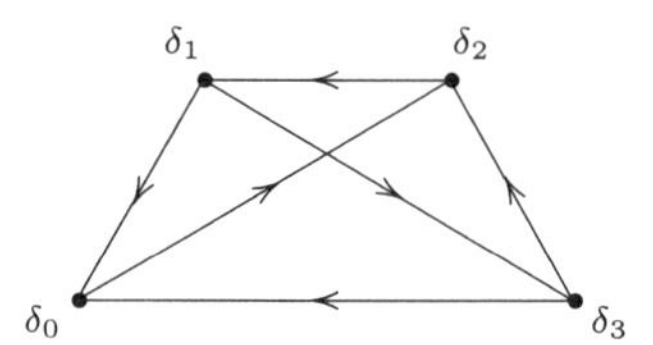

We shall be not so much interested in the dual graph itself than in certain two substructures (in fact, only in one of the two). We define the *sign-η dual hemigraph* $P\mathrm{D}_\eta$ of P as follows. Vertices are the facet locations of P of sign η; arrows are the ridge locations of P. A source or a target of an arrow is its source or its target in the dual graph if possible, and otherwise does not exist. In brief, $P\mathrm{D}_\eta$ is obtained from $P\mathrm{D}$ by removing the vertices of sign $-\eta$. Note that in general this operation leaves isolated arrows behind. (It does so for a purpose.) The following two diagrams show the negative and the positive dual hemigraphs of a 3-simplex.

Section 5. Dendrotopic Sets

The time has come to state the main definition of this work.

Definition. Let $\mathfrak{P}$ be an oriented propolytopic set with precisely one improper cell, $\bot$. Let $\bot$ have dimension -1 and not occur as a negative facet. Then $\mathfrak{P}$ is a *dendrotopic set* provided conditions $(\Delta 1)$, $(\Delta 2)$ and $(\Delta 3)_!$ below are satisfied.

(Δ1) Each cell of dimension ≥ 0 has precisely one positive facet. The associated location will be denoted by ω.

It follows that in the negative dual graph of a cell P of dimension ≥ 1, each vertex q has precisely one outgoing arrow, namely $q\omega$, and there is precisely one arrow without target, namely $\omega\omega$. Thus PD_- is locally treelike with root $\omega\omega$.

(Δ2) *Tree axiom.* The negative dual hemigraph of each cell of dimension ≥ 1 is a tree with root $\omega\omega$.

By the preceding remark, this condition can be expressed equivalently (assuming (Δ1)) as either connectedness or cycle freeness of the (finite) hemigraphs in question.

$(\Delta 3)_!$ *Normalization, uniqueness part.* Each 3-face location is represented by at most one panel of the form $\langle \omega, ** \rangle$.

We call *normal* such a panel, as well as the two chambers it separates. (Hence the name 'normalization' for the axiom.) The latter can be referred to individually as the positive normal chamber and the negative normal chamber.

The qualifier 'uniqueness part' suggests that we shall consider an existence part as well. In fact it occurs in the following proposition, which will be proven at the end of this section.

Proposition 2. *Any dendrotopic set satisfies condition* $(\Delta 3)_{\mathsf{i}}$ *below.*

$(\Delta 3)_{\mathsf{i}}$ *Normalization, existence part.* Each 3-face location is represented by at least one panel of the form $\langle \omega, ** \rangle$.

We may more generally define a *normal panel* to be one of the form

$$\langle \omega, \dots, \omega, ** \rangle.$$

This is done in [7], where the corresponding existence and uniqueness statements are proven for all depths/codimensions ≥ 2.

We know that a 3-face location (as a graph) is a closed zigzag with a preferred direction, given by its type-1 representatives. The normal panel, say $\langle \omega, \frac{r_- \; s_-}{r_+ \; s_+} \rangle$, is the only one that opposes the preferred direction. If we remove it, we are left with a gallery from $\langle \omega, r_-, s_- \rangle$ to $\langle \omega, r_+, s_+ \rangle$, which we call the *Hamilton gallery* of the location. We can display the entire location as its Hamilton gallery, with the normal panel indicated by two dotted lines:

$$\begin{array}{ccc} \omega & r_- & s_- \\ \hline * & * & * \\ \vdots & \vdots & \vdots \\ * & * & * \\ \hline \omega & r_+ & s_+ \end{array} .$$

Let us take a closer look at low dimensions.

0. Let T be a 0-cell. Since $\perp$ does not occur as a negative facet, ω is the only facet location of T. We denote it also by ε. Here is a picture of T.

1. Let S be a 1-cell. All 2-chambers underneath S are of the form $\langle *, \varepsilon \rangle$. Hence the negative facet locations of S are precisely the solutions of $S : \frac{* \; \varepsilon}{\omega \; \varepsilon}$, of which there is of course precisely one. We denote the sign-η facet location of

S by δ_η (so that $\delta_+ = \omega$). The negative dual hemigraph of S automatically takes the form

$$\bullet\,\delta_- \qquad \frac{\delta_-\ \varepsilon}{\delta_+\ \varepsilon}\downarrow \quad .$$

Thus the tree axiom is redundant for dimension 1. Here is a picture of S.

$$\delta_-\ \bullet\!\longrightarrow\!\bullet\ \delta_+$$

2. Let R be a 2-cell. Each vertex s of its negative dual hemigraph has precisely one incoming arrow, namely $s\delta_-$. Therefore $R\mathrm{D}_-$, being a tree, consists of a single branch

$$\xrightarrow[\frac{\alpha_{\ell-1}\ \delta_-}{}]{\omega\ \ \delta_-}\overset{\alpha_{\ell-1}}{\bullet}\xrightarrow[\frac{\alpha_{\ell-2}\ \delta_-}{}]{\alpha_{\ell-1}\ \delta_+}\cdots\xrightarrow[\frac{\alpha_0\ \delta_-}{}]{\alpha_1\ \delta_+}\overset{\alpha_0}{\bullet}\xrightarrow[\frac{\omega\ \delta_+}{}]{\alpha_0\ \delta_+}\ .$$

(The naming of the vertices is explained further below in this section.) The 3-chamber graph underneath R automatically takes the form

$$\begin{array}{ccc}
\omega & \delta_- & \varepsilon \\ \hline
\alpha_{\ell-1} & \delta_- & \varepsilon \\
\alpha_{\ell-1} & \delta_+ & \varepsilon \\ \hline
\vdots & \vdots & \vdots \\ \hline
\alpha_0 & \delta_- & \varepsilon \\
\alpha_0 & \delta_+ & \varepsilon \\ \hline
\omega & \delta_+ & \varepsilon
\end{array}\ .$$

Thus the normalization axiom is redundant for dimension 2. Here is a picture of R.

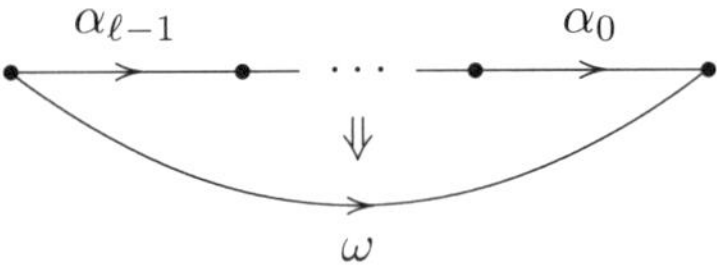

3. Let Q be a 3-cell. The incoming arrows of each vertex r of $Q\mathrm{D}_-$ can be listed $r\alpha_0, \ldots, r\alpha_{\ell_r-1}$; they thus carry a total order. These orders make $Q\mathrm{D}_-$ into a *planar tree* and thus induce a total order on the $\ell_\omega = 1+\sum_r(\ell_r - 1)$ leafs. The leafs in turn can be listed $\omega\alpha_0, \ldots, \omega\alpha_{\ell_\omega-1}$; they thus again carry a total order. The two orders agree, as one can show by using the normalization property.

Examples. Every globular set becomes a dendrotopic set by trivial augmentation.

Let us relate the above conditions to some of the notions discussed previously. In the following statements we assume that $\mathfrak{P}$ satisfies the hypothesis of the Definition along with condition $(\Delta 1)$.

- If $\mathfrak{P}$ has property $(\Delta 2)$, then $\mathfrak{P}$ is a polytopic set. This can easily be checked in view of Proposition 1.
- If $\mathfrak{P}$ has properties $(\Delta 2)$ and $(\Delta 3)_!$, then $\mathfrak{P}$ is spherical. This should intuitively be clear; the major ingredients of the proof can be found in that of Theorem 3. The theorem itself will tell us that under assumption of $(\Delta 3)_!$, conditions $(\Delta 2)$ and $(\Delta 3)_¡$ are equivalent.

The converses of these two statements are false. In fact, there are oriented spherical 3-polytopes satisfying $(\Delta 1)$, but none of the other conditions. For instance, consider the oriented 3-propolytopes with one facet location q_η of each sign η both of whose residents are dendrotopes with two negative facets. There are (up to isomorphism) six of them, all of which are polytopes. The two with a ridge location $\frac{q_- \ \omega}{q_+ \ \omega}$ are not of interest here. (One is a dendrotope, the other satisfies $(\Delta 2)$ and $(\Delta 3)_{\mathrm{i}}$ and has a realization whose boundary is a torus.) The remaining four meet the description.

Let X be an $(n+1)$-cell, and let p' be a negative facet location of X. We are going to assign to p' an n-dimensional number $|p'|$, which we shall call its *index*. The definition is by a double induction, first on the dimension of X, then on the height of p' in the tree $X\mathrm{D}_-$. If $\frac{p' \ \omega}{\omega \ \omega}$, we put $|p'| = 0$. If $\frac{p' \ \omega}{p \ q}$ with $\langle p, q\rangle \neq \langle \omega, \omega\rangle$, then p is below p' in $X\mathrm{D}_-$ and we put $|p'| = |p| + \langle |q| \rangle$. We may think of $|p'|$ as encoding the route along which to climb the tree $X\mathrm{D}_-$ in order to reach the vertex p'.

Before we proceed, we introduce an order $\leq^1$ on the set of higher-dimensional numbers. We decree that $\mathfrak{p}_0 \leq^1 \mathfrak{p}_1$ if and only if $\mathfrak{p}_0$ is an initial segment of $\mathfrak{p}_1$, that is, $\mathfrak{p}_1 = \mathfrak{p}_0 + \mathfrak{p}'$. Note that for two facet locations p_0 and p_1 of X, we have $|p_0| \leq^1 |p_1|$ if and only if there is a path from p_1 to p_0 in the negative dual hemigraph of X. We have thus represented the order induced by the tree $X\mathrm{D}_-$ on its vertex set $X\mathrm{I}_-$. We can conclude in particular that different negative facet locations of X have different indices. We may hence write, for example, $p = \alpha_{|p|}$. In this way, the negative facet locations at a given cell bear *a posteori* names $\alpha_{\mathfrak{p}}$, with the indices $\mathfrak{p}$ forming a downward-closed set of higher-dimensional numbers. In this regard dendrotopic sets become roughly similar to simplicial sets, for which the facet locations of a given cell bear *a priori* names δ_i, with the indices i forming a downward-closed set of natural numbers.

For an arbitrary positive 2-chamber $\langle p, q\rangle$ underneath the $(n+1)$-cell X, let us put

$$|p, q| = \begin{cases} 0 & \text{if } \langle p, q\rangle = \langle \omega, \omega\rangle, \\ |p| + \langle |q| \rangle & \text{otherwise.} \end{cases}$$

We call this n-dimensional number the *index* of $\langle p, q\rangle$. Expressed with the new notation, the definition of the index of $p' \in X\mathrm{I}_-$ becomes

$$|p'| = |p, q| \quad \text{if} \quad \frac{p' \ \omega}{p \ q}.$$

Thus, each vertex of $X\mathrm{D}_-$ carries the same index as the positive representative of its outgoing arrow. By contrast, the positive representatives of the leafs have indices that do not occur elsewhere in $X\mathrm{D}_-$.

We are almost set to prove Proposition 2. To do so, we need another order, denoted by $\leq^2$, of higher-dimensional numbers, defined to be lexicographical with respect to $\leq^1$. More explicitly, we decree that $\mathfrak{p}_0 \leq^2 \mathfrak{p}_1$ if and only if either $\mathfrak{p}_0$ is an initial segment of $\mathfrak{p}_1$ or $\mathfrak{p}_i = \mathfrak{p} + \langle \mathfrak{q}_i \rangle + \mathfrak{p}'_i$ with $\mathfrak{q}_0 <^1 \mathfrak{q}_1$. (Usually lexicographical orders are constructed with respect to total orders, in which case they are total themselves. The reader may want to convince himself of the redundancy of the totality assumption.)

Proof of Proposition 2. Let us define the *index* of a non-normal negative 3-chamber $\langle p, q, r\rangle$ to be

$$|p, q, r| = |p| + \langle |q, r|\rangle.$$

(Note that since p is negative, $\langle q, r\rangle$ is positive.) We show that for two such chambers successive in $X\Gamma^3$ (that is, with a length-2 gallery from the first to the second), the index of the first is strictly bigger than the index of the second with respect to $\leq^2$. Put more formulaicly, the assertion states that if

$$\begin{array}{ccc} \underline{p'} & \underline{q''} & r' \\ p & \underline{q'} & \underline{r'} \\ p & q & r \end{array} \qquad \text{with } p,\, p' \neq \omega,$$

then $|p', q'', r'| >^2 |p, q, r|$. Consider the sign of r', which determines the signs of q' and q''. If $r' \neq \omega$, then $q' = \omega$ and $q'' \neq \omega$, and so

$$|p', q'', r'| = |p'| + \langle |q''| + \langle |r'|\rangle\rangle >^2 |p'| + \langle |q''|, |q, r|\rangle = |p| + \langle |q, r|\rangle = |p, q, r|.$$

If $r' = \omega$, then $q' \neq \omega$ and $q'' = \omega$, and so

$$|p', q'', r'| = |p'| + \langle 0\rangle >^2 |p'| = |p| + \langle |q'|\rangle = |p| + \langle |q, r|\rangle = |p, q, r|.$$

If there were a 3-face location without a normal chamber, it would be a (directed) cycle. But along this cycle, the indices of negative vertices would strictly decrease, and this is impossible. □

Section 6. Normalization

The existence part of the normalization condition, proven to be satisfied in all dendrotopic sets, can in fact replace the tree condition as an axiom. The present section is devoted to proving this fact.

Theorem 3. *Let $\mathfrak{P}$ be an oriented propolytopic set with precisely one improper cell, $\perp$. Let $\perp$ have dimension -1 and not occur as a negative facet. Then $\mathfrak{P}$ is a dendrotopic set if and only if it satisfies conditions* ($\Delta 1$) *above and* ($\Delta 3$) *below.*

($\Delta 3$) *Normalization.* Each 3-face location is represented by exactly one normal panel.

Proof. Let us temporarily call a propolytopic set satisfying the conditions *normalizing*. Thus we have to show that in a normalizing propolytopic set $\mathfrak{N}$, the tree condition ($\Delta 2$) is satisfied. To this end, it suffices to show that the negative dual hemigraph of each cell A of dimension ≥ 1 is cycle-free. We do so by induction on the dimension n of A.

By a previous observation, the statement is automatically true for $n = 1$. So let us consider the case $n \geq 2$. We proceed by another induction, this time on the number ℓ of negative facets of A.

If $\ell = 0$, then AD_- consists of the arrow $\omega\omega$ only and is hence clearly cycle-free. In order to handle the general case, we reduce the problem by means of the following two constructions. (The diagrams will illustrate an example with $n = 3$.)

Claim A. *Let B_0 and B_1 be two $(n-1)$-cells of $\mathfrak{N}$, and let c be a negative facet location of B_0 with resident C equal to the positive facet of B_1.*

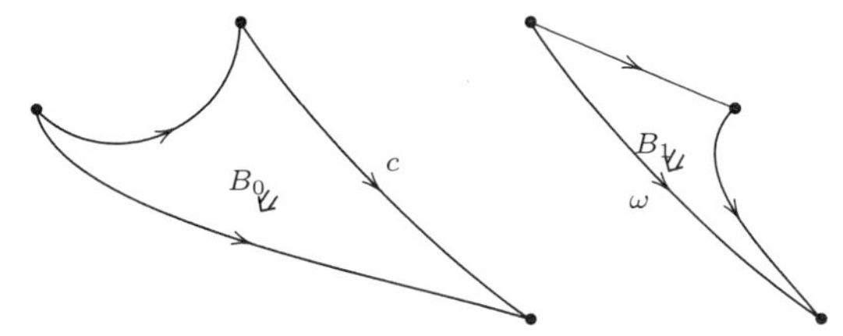

We obtain a new normalizing propolytopic set $\mathfrak{N}'$ by adding to $\mathfrak{N}$ an $(n-1)$-cell B' with incidence structure prescribed as follows. We let the facets of B' be those of B_0 except the one with location c and those of B_1 except the one with location ω. (Without loss of generality disjointness of $B_0\mathrm{I}$ and $B_1\mathrm{I}$ is assumed. The signs are kept, so that the ω of B' is the ω of B_0.) The ridge locations of B' are defined so that (with the evident subscript convention)

$$B' : \frac{r_0^- \; s_0^-}{r_0^+ \; s_0^+} \quad \textit{if and only if}$$

$$\textit{either} \quad B_0 : \frac{r_0^- \; s_0^-}{r_0^+ \; s_0^+} \quad \textit{or} \quad \left\{ \begin{array}{l} B_0 : \dfrac{r_0^- \; s_0^-}{c \quad s} \\[2ex] B_1 : \dfrac{\omega \quad s}{\omega \quad \omega} \\[2ex] B_0 : \dfrac{c \quad \omega}{r_0^+ \; s_0^+} \end{array} \right\} \quad \textit{for some } s \in C\mathrm{I}_-;$$

$$B' : \{\!\!\{ s_1 \}\!\!\} \frac{r_1 \; s_1}{r_0 \; s_0} \quad \textit{if and only if} \quad \left\{ \begin{array}{l} B_1 : \{\!\!\{ s_1 \}\!\!\} \dfrac{r_1 \; s_1}{\omega \quad s} \\[2ex] B_0 : \{\!\!\{ s_1 \}\!\!\} \dfrac{c \quad s}{r_0 \; s_0} \end{array} \right\} \quad \textit{for some } s \in C\mathrm{I}_{\{\!\!\{ s_1 \}\!\!\}};$$

$$B' : \frac{r_1^- \quad \omega}{r_1^+ \; s_1^+} \quad \textit{if and only if} \quad B_1 : \frac{r_1^- \quad \omega}{r_1^+ \; s_1^+}.$$

Proof. We can describe $B'\diamondsuit$ more vividly as follows. Think of a ridge location of B_0 or B_1 as a domino, with its two representative 2-chambers being the faces. The face $\langle c, s\rangle$ of a B_0-domino is taken to match the face $\langle \omega, s\rangle$ of a B_1-domino. Now form the longest possible rows of dominoes. We can think of each row as a ridge location of B', with its two unmatched faces being the representative 2-chambers. Indeed all possible rows are mentioned in the statement of the claim, as a consequence of the outer induction hypothesis: a ridge location $B_0 : \frac{c\;\omega}{c\;s}$ would be a loop in $B_0\mathrm{D}_-$. This alternative description conveniently shows that we have indeed a one-to-one correspondence between $B'\mathrm{I}^2_-$ and $B'\mathrm{I}^2_+$.

The only part of the claim that remains difficult to check is that the normalization condition is satisfied underneath B'. We do so by noting that $B'\Gamma^3$ can be obtained from $B_0\Gamma^3$ by replacing every occurrence of a type-2 panel with first entry c, the separated chambers included,

$$\left.\begin{array}{ccc} * & * & * \\ \hline c & s_- & t_- \\ \cline{2-3} c & s_+ & t_+ \\ \hline * & * & * \end{array}\right\} \text{(this part)}$$

by the non-normal segment of the corresponding 3-face location of B_1.

$$\left.\begin{array}{ccc} \omega & s_- & t_- \\ \hline * & * & * \\ \vdots & \vdots & \vdots \\ * & * & * \\ \hline \omega & s_+ & t_+ \end{array}\right\} \text{(this part)}$$

□

Claim B. *Let b_0 and b_1 be two different negative facet locations of A, and let c be a negative facet location of Ab_0 with*

$$A : \frac{b_1 \ \omega}{b_0 \ c}. \tag{3}$$

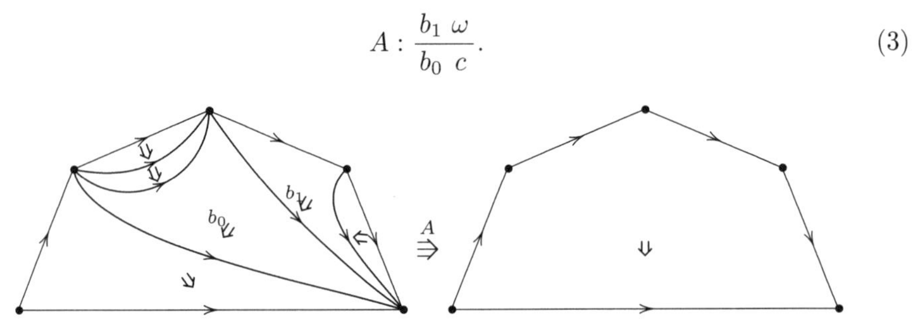

We obtain a new normalizing propolytopic set $\mathfrak{N}''$ by adding two cells to $\mathfrak{N}$: first B' of dimension $n-1$ as in Claim A for $B_i = Ab_i$; then A' of dimension n with incidence structure prescribed as follows. We let the facets of A' be those of A, minus the two with locations b_0 and b_1, plus B' with negative location b'. The ridge locations of A' are defined to be the same as those of A, with the understanding that b' replaces both b_0 and b_1 and that (3) is absent.

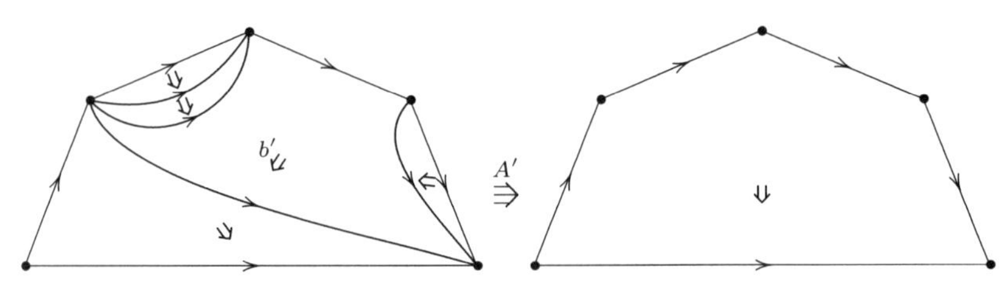

Proof. Again the only difficult part is to check the normalization condition. We do so by noting that $A'\Gamma^3$ can be obtained from $A\Gamma^3$ by replacing segments

$$\begin{array}{ccc} \hline b_0 & r_0^- & s_0^- \\ \cline{2-3} b_0 & r_0^+ & s_0^+ \\ \hline \end{array} \quad \text{or} \quad \begin{array}{ccc} \hline b_0 & r_0^- & s_0^- \\ \cline{2-3} b_0 & c & s \\ \hline b_1 & \omega & s \\ \cline{2-3} b_1 & \omega & \omega \\ \hline b_0 & c & \omega \\ \cline{2-3} b_0 & r_0^+ & s_0^+ \\ \hline \end{array} \quad \text{by} \quad \begin{array}{ccc} \hline b' & r_0^- & s_0^- \\ \cline{2-3} b' & r_0^+ & s_0^+ \\ \hline \end{array},$$

$$\{\!\!|s|\!\!\}\left\{\begin{array}{ccc} b_1 & r_1 & s_1 \\ b_1 & \omega & s \\ b_0 & c & s \\ b_0 & r_0 & s_0 \end{array}\right. \quad \text{by} \quad \{\!\!|s|\!\!\}\left\{\begin{array}{ccc} b' & r_1 & s_1 \\ b' & r_0 & s_0 \end{array}\right. ,$$

$$\begin{array}{ccc} b_1 & r_1^- & s_1^- \\ b_1 & r_1^+ & s_1^+ \end{array} \quad \text{by} \quad \begin{array}{ccc} b' & r_1^- & s_1^- \\ b' & r_1^+ & s_1^+ \end{array} . \qquad \square$$

Back to the theorem. In the inner induction step, we analyse the situation of Claim B in a way similar to its proof, but with regard to negative dual hemigraphs rather than depth-3 chamber graphs, and in the opposite direction. Our findings will be the following. We can obtain $A\mathrm{D}_-$ from $A'\mathrm{D}_-$ by "stretching out" the vertex b' to become the arrow (3) with extremities b_0 and b_1. The outgoing arrow of b' becomes the outgoing arrow of b_0, while an incoming arrow of b' becomes an incoming arrow of either b_0 or b_1. It is clear that no essentially new cycle is created by this modification.

The crucial part of the proof, however, is to make sure that Claim B can actually be applied when needed. This is where the following result comes in. In order to state it conveniently, let us agree to write

$$A : \frac{q \ \ \omega}{\beta_q \ \gamma_q} \qquad (q \in A\mathrm{I}_-).$$

Claim C. *If there is $b \in A\mathrm{I}_-$ such that $\langle \beta_b, \gamma_b \rangle \neq \langle \omega, \omega \rangle$, then there is $b_1 \in A\mathrm{I}_-$ such that $\langle \beta_{b_1}, \gamma_{b_1} \rangle \neq \langle \omega, \omega \rangle$ and $\beta_{b_1} \neq b_1$.*

Proof. Let us assume the contrary. Pick b as indicated, whence $\beta_b = b$. By the outer induction hypothesis, $Ab\mathrm{D}_-$ is a tree with root $\omega\omega$, whence there is a path

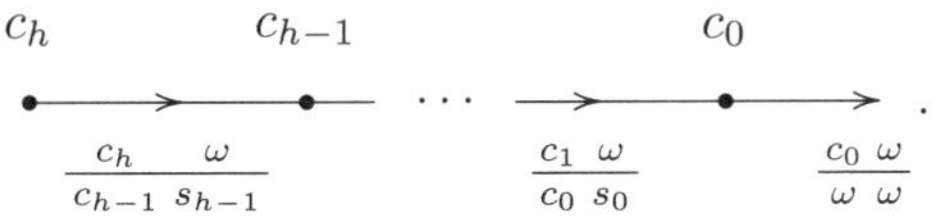

$$\begin{array}{ccccccc} \gamma_b = c_h & & c_{h-1} & \cdots & c_0 & & \\ \bullet & \xrightarrow{\ \frac{c_h \ \ \omega}{c_{h-1} \ s_{h-1}}\ } & \bullet & \cdots \xrightarrow{\ \frac{c_1 \ \omega}{c_0 \ s_0}\ } & \bullet & \xrightarrow{\ \frac{c_0 \ \omega}{\omega \ \omega}\ } & . \end{array}$$

We have $h > 0$, for otherwise there would be a cycle

$$\begin{array}{ccc} b & \omega & \omega \\ b & c_0 & \omega \end{array}$$

underneath A, contradicting the existence part of the normalization property. For each $i < h$, the positive 2-chamber $\langle b, c_i \rangle$ is not the conjugate of any $\langle q, \omega \rangle$ with $\{\!\!|q|\!\!\} = -$, since otherwise by assumption $q = \beta_q = b$, but $\gamma_q = c_i \neq c_h = \gamma_b$. We can hence select $r_i \in A\omega\mathrm{I}_-$ by demanding that $A : \frac{\omega \ r_i}{b \ c_i}$.

If $i < h - 1$, there is a gallery

$$\begin{array}{ccc} \omega & r_{i+1} & \omega \\ b & c_{i+1} & \omega \\ b & c_i & s_i \\ \omega & r_i & s_i \end{array}$$

underneath A. By the uniqueness part of the normalization property, $A\omega : \frac{r_{i+1}\ \omega}{r_i\ \ s_i}$. There also is a gallery

$$\begin{array}{ccc} \omega & r_0 & \omega \\ \hline b & c_0 & \omega \\ \hline b & \omega & \omega \\ \hline b & c_h & \omega \\ \hline b & c_{h-1} & s_{h-1} \\ \hline \omega & r_{h-1} & s_{h-1} \end{array}$$

underneath A. For the same reason as before, $A\omega : \frac{r_0 \quad \omega}{r_{h-1}\ s_{h-1}}$. We have thus constructed a cycle in $A\omega\mathrm{D}_-$,

$$\cdots \xrightarrow[\frac{r_1\ \omega}{r_0\ s_0}]{} \underset{r_0}{\bullet} \xrightarrow[\frac{r_0\quad \omega}{r_{h-1}\ s_{h-1}}]{} \underset{r_{h-1}}{\bullet} \cdots,$$

contradicting the outer induction hypothesis. □

Now consider the case $\ell = 1$. Let us denote the only vertex of the negative dual hemigraph of A by b. The outgoing arrow of b must be $\frac{b\ \omega}{\omega\ \omega}$, since by Claim C, $\beta_b = b$ is impossible. But this arrow has no target in $A\mathrm{D}_-$, whence there is no cycle.

Finally consider the general case $\ell > 1$. As there is at most one $b \in A\mathrm{I}_-$ with $A : \frac{b\ \omega}{\omega\ \omega}$, the hypothesis of Claim C is satisfied. We can hence pick $b_1 \in A\mathrm{I}_-$ as in the conclusion. We then apply Claim B with $\langle b_0, c\rangle = \langle \beta_{b_1}, \gamma_{b_1}\rangle$. The resulting n-cell A' of the extension $\mathfrak{N}''$ has $\ell - 1$ negative facets, whence by inner induction hypothesis, $A'\mathrm{D}_-$ is cycle-free. It follows that $A\mathrm{D}_-$ is cycle-free as well, as we have seen above. □

Section 7. Cell Trees

An $(n+1)$-cell in a dendrotopic set has a positive half boundary, which is an n-cell, and a negative half boundary, which is an *n-cell tree*. An n-cell tree can be viewed as a partitioned cell of sorts and has a boundary of its own. An n-cell and an n-cell tree whose two boundaries agree together are potentially the positive and the negative halves of the boundary of an $(n+1)$-cell. Somewhat hidden in this statement are the globular laws.

Let $\mathfrak{D}$ be a dendrotopic set, and let P be a cell of $\mathfrak{D}$. The *boundary* $P^\bullet$ of P consists of the set $P^\bullet\mathrm{I} = P\mathrm{I}$ of its facet locations, the mappings associating a sign $\{\!|q|\!\}$ and a resident $P^\bullet q = Pq$ to each $q \in P\mathrm{I}$, and the conjugation $P^\bullet\Diamond = P\Diamond$. In brief, the boundary of P is the collection of the atomic data that depend directly on P, except for the element P itself.

A *frame* is a collection of data that has the potential of being the boundary of a cell. To be precise, without reiterating the entire definition of 'dendrotopic set', we define an n-frame of $\mathfrak{D}$ to be the boundary of an n-cell of a dendrotopic set obtained by extending $\mathfrak{D}$ by just this cell. We usually denote a frame as we would denote such a cell. The boundary of an (actual) n-cell of $\mathfrak{D}$ is an example of an n-frame of $\mathfrak{D}$.

We identify two frames if they can be made equal by renaming their facet locations. (That is, we shall be dealing with frames-of-old and frames-of-new, the latter being isomorphism classes of the former.) Explicitly, given two n-frames P_0

and P_1, the equation $P_0 = P_1$ will mean that there is a one-to-one correspondence $P_0\mathrm{I} \rightleftarrows P_1\mathrm{I}$ with the following properties: whenever $q_0 \mapsto\!\!\!\!\leftarrow q_1$, we have $P_0 q_0 = P_1 q_1$ and $\{\!|q_0|\!\} = \{\!|q_1|\!\}$; whenever $q_0^\eta \mapsto\!\!\!\!\leftarrow q_1^\eta$ ($\eta \in \mathbf{Z}^\times$), we have $P_0 : \frac{q_0^- \ r^-}{q_0^+ \ r^+}$ if and only if $P_1 : \frac{q_1^- \ r^-}{q_1^+ \ r^+}$. From these properties it follows that whenever $q_0 \mapsto\!\!\!\!\leftarrow q_1$ and $\{\!|q_0|\!\}$, $\{\!|q_1|\!\} = -$, we have also $|q_0| = |q_1|$. We conclude that there exists at most one correspondence as stated. (That is, there is at most one isomorphism between two frames-of-old. In particular, all automorphisms are trivial.) We conclude further that every frame will be identified with one for which $q = |q|$ for all negative facet locations. (Readers who doubt the soundness of our identification can replace this paragraph by an additional axiom, demanding that each negative facet location agree with its index and that each positive facet location agree with some element they have chosen once and for all. They then have to revise all the constructions of frames yet to come.) Thus the frames of $\mathfrak{D}$ will form a set within the universe of discourse.

We may ascribe a geometric realization $P\Gamma$ to an n-frame P. Namely, we may take $P\Gamma$ to be the boundary of the geometric realization of P as a cell in a defining extension of $\mathfrak{D}$. This space is an $(n-1)$-sphere. Thus the prefix 'n-' does not denote geometric dimension. The reason is that we always think of a frame as a cell-to-be. The most general notion of a cell-to-be is to be introduced next.

We inductively define a dendrotopic set $\mathfrak{D}^\dagger$ extending $\mathfrak{D}$ by demanding that those n-cells of $\mathfrak{D}^\dagger$ that are not already in $\mathfrak{D}$ are precisely the n-frames of $\mathfrak{D}^\dagger$, each being its own boundary. (Note that the n-frames of a dendrotopic set, here $\mathfrak{D}^\dagger$, are precisely the n-frames of the dendrotopic subset consisting of cells of dimensions up to $n-1$.) We call the cells of $\mathfrak{D}^\dagger$ the *frameworks* of $\mathfrak{D}$. The terminology for cells will be applied for frameworks accordingly.

A face of a framework may be either a cell or not; we call it (and hence the associated location as well) *complete* or *incomplete* accordingly. Of course all faces of a cell are cells and hence complete. It follows that for a framework P and locations f of P and g of Pf, completeness of f implies completeness of fg, whence (contraposition) incompleteness of fg implies incompleteness of f.

A framework with no incomplete face location is a cell. A framework whose only incomplete face location is the trivial one, 1, is a frame. A framework whose only incomplete locations are 1 and the positive facet location, ω, will be called an *outniche*. (The term 'frame' is adopted from Baez and Dolan's pioneering work. What we call an outniche, these authors would call a punctured frame; what they would call a niche, we should call an *inniche* given an opportunity.)

Definition. Let $n \geq 0$. An *n-cell tree* P conveys the following data:

- an $(n-1)$-cell $P{\uparrow}\omega\omega$, called the *root ridge* of P;
- a finite family $\langle\, P{\uparrow}p \mid p \in P{\uparrow}\mathrm{I}_- \,\rangle$ of n-cells, called the *facets* of P;
- a tree $P{\uparrow}\mathrm{D}_-$ with vertex set $P{\uparrow}\mathrm{I}_-$ and arrow set

$$P{\uparrow}\mathrm{I}_+^2 = \{\langle\omega,\omega\rangle\} \cup \{\, \langle p,q\rangle \mid p \in P{\uparrow}\mathrm{I}_-;\ q \in P{\uparrow}p\mathrm{I}_- \,\}$$

such that

— the arrow $\langle\omega,\omega\rangle$ is the root, and in the case $n = 0$ not a leaf;

— the target of an arrow $\langle p,q\rangle \neq \langle\omega,\omega\rangle$ is p;

— if p' is the source of an arrow $\langle p,q\rangle$ — a situation we render by $P{\uparrow} : \frac{p' \ \omega}{p \ \ q}$ — then $P{\uparrow}p'\omega = P{\uparrow}pq$.

A *facet location* of an n-cell tree P is an element of $P{\uparrow}\mathrm{I}_-$. We identify two n-cell trees if they can be made equal by renaming their facet locations.

For $n = 0$, the tree $P{\uparrow}\mathrm{D}_-$ has a single vertex δ_-. Thus, a 0-cell tree P is given by a 0-cell $P{\uparrow}\delta_-$. For $n > 0$, the tree $P{\uparrow}\mathrm{D}_-$ either is empty or has a stem α_0. In the former case we call P itself *empty*. In the latter case the path from a vertex $p \neq \alpha_0$ to the root reaches the stem via an arrow $\langle \alpha_0, \beta_p \rangle$, for a certain $\beta_p \in P{\uparrow}\alpha_0\mathrm{I}_-$. For a given negative facet location q_0 of $P{\uparrow}\alpha_0$, we can form a new n-cell tree $P_{(q_0)}$ by putting

$$P_{(q_0)}{\uparrow}\omega\omega = P{\uparrow}\alpha_0 q_0, \qquad P_{(q_0)}{\uparrow}\mathrm{I}_- = \{\, p \in P{\uparrow}\mathrm{I}_- \mid \beta_p = q_0 \,\}, \qquad P_{(q_0)}{\uparrow}p = P{\uparrow}p$$

and $P_{(q_0)}{\uparrow} : \frac{p' \;\; \omega}{p \;\; q}$ if and only if either $\langle p, q \rangle \neq \langle \omega, \omega \rangle$ and $P{\uparrow} : \frac{p' \;\; \omega}{p \;\; q}$ or $\langle p, q \rangle = \langle \omega, \omega \rangle$ and $P{\uparrow} : \frac{p' \;\; \omega}{\alpha_0 \;\; q_0}$. A converse construction is also available and leads to the following result. For $n > 0$, an n-cell tree P with root ridge Q is given either by the information that $P{\uparrow}\mathrm{I}_- = \varnothing$ or by an n-cell $P{\uparrow}\alpha_0$ with $P{\uparrow}\alpha_0\omega = Q$ and an n-cell tree $P_{(q)}$ with root ridge $P{\uparrow}\alpha_0 q$ for each $q \in P{\uparrow}\alpha_0\mathrm{I}_-$. This statement is a recursive definition of n-cell trees.

The notation makes clear that every $(n + 1)$-cell X has an underlying n-cell tree, which we call the *negative half boundary* of X and denote by $X\alpha$. (There is only a minor subtlety: an arrow $\frac{p_- \;\; q_-}{p_+ \;\; q_+}$ of $X\mathrm{D}_-$ occurs in the guise of its positive representative $\langle p_+, q_+ \rangle$ in $X\alpha{\uparrow}\mathrm{D}_-$.)

What has been said for $(n+1)$-cells is equally true for $(n+1)$-frames and $(n+1)$-outniches: they have negative half boundaries which are n-cell trees. An arbitrary $(n+1)$-framework has a negative half boundary which is an n-framework tree (that is, an n-cell tree in the dendrotopic set $\mathfrak{D}^\dagger$ of frameworks of $\mathfrak{D}$). Roughly speaking, what distinguishes $(n + 1)$-outniches from n-cell trees is the presence of positive facets (which are n-frames). But, as the following theorem tells us, a positive facet is implicit in the structure of a cell tree.

Theorem 4. *Let $n \geq 0$. Each n-cell tree is the negative half boundary of a unique $(n + 1)$-outniche.*

From here we can immediately derive the more general result that every n-framework tree is the negative half boundary of a unique $(n + 1)$-framework with incomplete face locations 1 and ω.

Proof. Let P be an n-cell tree. We are going to construct an $(n + 1)$-outniche X with negative half boundary P. The uniqueness part of the claim can eventually be verified by inspection: the outniche X has to be constructed in the way stated.

We let the negative facet locations of $X\omega$ be the leafs of $P{\uparrow}\mathrm{D}_-$, each written ${}^{p}q$ rather than $\langle p, q \rangle$. In the case $n = 0$ the root $\langle \omega, \omega \rangle$ is the only arrow of $P{\uparrow}\mathrm{D}_-$, but by definition not a leaf, whence $X\omega$ has no negative facet location, as required. We complete the definition of conjugation underneath X by putting $\frac{\omega \;\; {}^{p}q}{p \;\; q}$, whence we have to let $X\omega\,{}^{p}q = Xpq$. We have thus constructed all of the chamber graph $X\Gamma^3$, except for its normal panels. They are determined by the normalization condition on X. Thus conjugation underneath $X\omega$ is also defined. We now have to examine whether $X\omega$ meets the dendrotopic axioms.

First we verify the tree property. For the cell tree P, we define indices $|p|$ ($p \in P{\uparrow}\mathrm{I}_-$) and $|p, q|$ ($\langle p, q \rangle \in P{\uparrow}\mathrm{I}_+^2$) and even $|p, q, r|$ ($p \in P{\uparrow}\mathrm{I}_-$; $\langle q, r \rangle \in P{\uparrow}p\mathrm{I}_+^2$)

just as we have done for cells. We now show that whenever we have an arrow in $X\omega\mathrm{D}_-$ possessing source and target,

$$X\omega : \frac{{}^{p'}q' \;\; \omega}{{}^{p}q \;\; r}, \tag{4}$$

indices decrease from source to target, $|p',q'| >^2 |p,q|$; from here it follows that the negative dual hemigraph of $X\omega$ has no cycles. Consider the Hamilton gallery underneath X that has led to the definition of (4),

$$\begin{array}{ccc} \omega & {}^{p'}q' & \omega \\ \hline p' & q' & \omega \\ \hline * & * & * \\ \vdots & \vdots & \vdots \\ * & * & * \\ \hline p & q & r \\ \hline \omega & {}^{p}q & r \end{array}.$$

By the argument employed in the proof of Proposition 2, the indices of non-normal negative chambers decrease along this gallery. In particular, a comparison of the third chamber from above with the second chamber from below yields $|p',q'| \geq^2 |p,q,r|$. Since $|p,q,r| >^2 |p,q|$, the result follows.

Now we verify the uniqueness part of the normalization property. Let $\langle r^0_\eta, s^0_\eta\rangle$ ($\eta \in \mathbf{Z}^\times$) be two different 2-chambers underneath $X\omega\omega$. Suppose that there is a zigzag from $\langle \omega, r^0_-, s^0_-\rangle$ to $\langle \omega, r^0_+, s^0_+\rangle$ in $X\omega\Gamma^3$; we want to show that $\langle r^0_\eta, s^0_\eta\rangle$ are conjugates of each other. If we disallow backtracking, the only vertices along the zigzag where the direction changes are normal chambers. Hence we may as well assume that the zigzag is a path, that is, a gallery underneath $X\omega$. Via the canonical embedding $(\langle\omega\rangle + ?) : X\omega\Gamma^3 \rightarrow X\Gamma^4$ we obtain a gallery from $\langle \omega, \omega, r^0_-, s^0_-\rangle$ to $\langle \omega, \omega, r^0_+, s^0_+\rangle$ underneath X. Unravelling the definition of $X\omega\Diamond$, we can replace each arrow of the form

$$\text{normal panel} + \langle s\rangle$$

by a path of the form

$$\text{Hamilton gallery} + \langle s\rangle.$$

There is hence a gallery from $\langle \omega, \omega, r^0_-, s^0_-\rangle$ to $\langle \omega, \omega, r^0_+, s^0_+\rangle$ containing no type-2 panel with first entry ω. Of all those galleries we pick a shortest one, say $\mathscr{G}$.

The gallery $\mathscr{G}$ contains no chamber of the form $\langle \omega, {}^{p}q, *, *\rangle$: if there were one, $\mathscr{G}$ would have a subgallery

$$\begin{array}{cccc} p & q & r_- & s_- \\ \hline \omega & {}^{p}q & r_- & s_- \\ \hline \omega & {}^{p}q & r_+ & s_+ \\ \hline p & q & r_+ & s_+ \end{array},$$

which could be replaced by the shorter

$$\begin{array}{cccc} p & q & r_- & s_- \\ \hline p & q & r_+ & s_+ \end{array}.$$

Now pick a negative facet location p' which has $\leq^1$-maximal index among those occurring as first entries of chambers of $\mathscr{G}$. There is a subgallery $\langle p'\rangle + \mathscr{G}'$ of $\mathscr{G}$ bordered by type-1 panels. Since the previous and the next chamber can be of neither of the forms $\langle p'', \omega, *, *\rangle$ with p'' negative (maximality of p') or $\langle \omega, {}^{p'}q', *, *\rangle$

(just excluded), it is of the form $\langle p, q, *, *\rangle$, where $P{\uparrow} : \frac{p' \;\omega}{p \;\; q}$. It follows that source and target of $\mathscr{G}'$ are normal chambers $\langle \omega, r_\eta, s_\eta \rangle$:

$$\left.\begin{array}{cccc}
\underline{p} & \underline{q} & r_- & s_- \\
p' & \omega & \underline{r_-} & \underline{s_-} \\
p' & * & * & * \\
\vdots & \vdots & \vdots & \vdots \\
p' & \underline{*} & \underline{*} & \underline{*} \\
\underline{p'} & \underline{\omega} & r_+ & s_+ \\
p & q & r_+ & s_+
\end{array}\right\} = \langle p' \rangle + \mathscr{G}' \,. \tag{5}$$

(The long lines indicate uncertainty as to the types of the panels they symbolize.) Since the normalization condition is satisfied underneath Xp', the 2-chambers $\langle r_\eta, s_\eta \rangle$ underneath $Xp'\omega = Xpq$ are conjugate. We can hence replace the subgallery (5) of $\mathscr{G}$ by the shorter

$$\begin{array}{cccc}
p & q & \underline{r_-} & \underline{s_-} \\
p & q & r_+ & s_+
\end{array} \;;$$

a contradiction. Hence the implicit assumption that a negative first entry occurs at all among the chambers of $\mathscr{G}$ was wrong. We have thus ruled out all occurrences of chambers with either of the first two entries negative: all chambers of $\mathscr{G}$ are of the form $\langle \omega, \omega, *, * \rangle$. Since type-2 panels with first entry ω are also ruled out, $\mathscr{G}$ consists of nothing but a type-3 panel:

$$\begin{array}{cccc}
\omega & \omega & \underline{r^0_-} & \underline{s^0_-} \\
\omega & \omega & r^0_+ & s^0_+
\end{array} \;.$$

Thus indeed $\langle r^0_\eta, s^0_\eta \rangle$ are conjugates of each other. □

The frame $X\omega$ constructed in this proof will be called the *boundary* of P and denoted by $P^\bullet$. Note that for an $(n+1)$-cell X, the uniqueness part of the theorem yields

$$X\alpha^\bullet = X\omega^\bullet . \tag{6}$$

Let us denote by $\mathfrak{D}^{1*}_n$ the set of n-cell trees of $\mathfrak{D}$. A single n-cell P can be viewed as an n-cell tree with a single facet location δ_-: put $P{\uparrow}\delta_- = P$. Thus we obtain an inclusion $\mathfrak{D}_n \subseteq \mathfrak{D}^{1*}_n$. The boundary of the cell P is independent of the way it is viewed: there is an $(n+1)$-frame X with two facets $X\delta_\eta = P$ $(\eta \in \mathbf{Z}^\times)$ and ridge locations $\{\!\!\{q\}\!\!\} \frac{\delta_- \; q}{\delta_+ \; q}$ $(q \in P\mathrm{I})$.

Now let P be an n-cell tree. If $n \geq 1$, we put $P\omega = P^\bullet\omega$ $(= P{\uparrow}\omega\omega)$ and $P\alpha = P^\bullet\alpha$. If $n \geq 2$, we can conclude from (6) that $P\alpha\omega = P\omega\omega$ and $P\alpha\alpha = P\omega\alpha$. Thus, we have constructed a globular set $\mathfrak{D}^{1*}$ with positive facet locations ω and negative facet locations α. The situation may be summarized by the partially commutative diagram

$$\begin{array}{ccccccccc}
\cdots & \to \mathfrak{D}^{1*}_{n+1} & \xrightarrow{\alpha} & \mathfrak{D}^{1*}_n & - & \cdots & \to \mathfrak{D}^{1*}_1 & \xrightarrow{\alpha} & \mathfrak{D}^{1*}_0 \,. \\
 & \cup\!\uparrow & \searrow^{\omega}\!\!\nearrow_{\alpha} & \cup\!\uparrow & & & \cup\!\uparrow & \searrow^{\omega}\!\!\nearrow_{\alpha} & \| \\
\cdots & \to \mathfrak{D}_{n+1} & \xrightarrow[\omega]{} & \mathfrak{D}_n & - & \cdots & \to \mathfrak{D}_1 & \xrightarrow[\omega]{} & \mathfrak{D}_0
\end{array}$$

Part II. Correctness

The task ahead is to show that that the definition of dendrotopic sets is correct, in the sense that it is equivalent to the Hermida–Makkai–Power definition of multitopic sets. There is bound to be a direct proof of this fact; the proof presented here involves a detour via ∞-categories. More precisely, it is shown (in Section 5) that the Harnik–Makkai–Zawadowski characterization is valid here too: dendrotopic sets are precisely the freely generating subsets, closed under the target operation, of strict ∞-categories. To this end, a lemma is proven (in Section 4) which seems to be of interest in its own right. To keep the exposition self-contained, definitions of strict ∞-categories (Section 1) and their freedom (Section 3) are supplied. The elegant way to represent the n-cells of the ∞-category to be generated by a dendrotopic set is as certain $(n+1)$-frameworks of the latter. These frameworks are called *roofs*; they are introduced in Section 2.

Section 1. Strict ∞-Categories

We discuss various ways to define the term 'strict ∞-category' (about which there can be no dispute, contrary to the 'weak' situation).

Put into more familiar terms, a globular set $\mathfrak{G}$ is given by the following items:

- for each $n \geq 0$, the set $\mathfrak{G}_n$ of n-cells;
- for each $n \geq 1$ and $\eta \in \mathbf{Z}^\times$, the *facet operator* ${}_n\delta_\eta : \mathfrak{G}_n \twoheadrightarrow \mathfrak{G}_{n-1}$, $A \mapsto A\delta_\eta$;
- (*globular law*) for each $n \geq 2$, $\zeta \in \mathbf{Z}^\times$ and $A \in \mathfrak{G}_n$, satisfaction of the equation $A\delta_+\delta_\zeta = A\delta_-\delta_\zeta$.

It follows that the expression '$A\delta_{\theta^{(1)}} \cdots \delta_{\theta^{(u)}}$', whenever defined, takes a value independent of $\theta^{(1)}, \ldots, \theta^{(u-1)}$. This value will be denoted by $A\delta^u_{\theta^{(u)}}$. We are thus led to yet another definition. A globular set is given by the following items: for each $n \geq 0$,

- the set $\mathfrak{G}_n$ of n-cells,
- the *face operators* ${}_n\delta^u_\eta : \mathfrak{G}_n \twoheadrightarrow \mathfrak{G}_{n-u}$, $A \mapsto A\delta^u_\eta$ $(0 < u \leq n;\ \eta \in \mathbf{Z}^\times)$,
- (*generalized globular law*) satisfaction of the equations

$$A\delta^u_\eta\delta^v_\zeta = A\delta^{u+v}_\zeta \qquad (0 < u, v;\ u + v \leq n;\ \eta, \zeta \in \mathbf{Z}^\times;\ A \in \mathfrak{G}_n).$$

In view of the "enumeration" below, the generalized globular law will also be referred to by the symbol $(\delta\delta2)^*$.

A *subset* $\mathfrak{g}$ of a globular set $\mathfrak{G}$ consists of a subset $\mathfrak{g}_n$ of each $\mathfrak{G}_n$. A subset $\mathfrak{g}$ of $\mathfrak{G}$ is called *sign-η semiglobular* if $n > 0$ and $A \in \mathfrak{g}_n$ imply $A\delta_\eta \in \mathfrak{g}_{n-1}$. In absence of further qualification the sign is understood to be $+$. For instance, at the end of Part I we have seen how a dendrotopic set $\mathfrak{D}$ is viewed as a semiglobular subset of the globular set $\mathfrak{D}^{1*}$ of its cell trees. A *globular* subset of $\mathfrak{G}$ is one that is both positively and negatively semiglobular.

A *system of constants* for $\mathfrak{G}$ consists of maps $\sigma^u_n : \mathfrak{G}_{n-u} \twoheadrightarrow \mathfrak{G}_n$ $(n \geq u > 0)$, written $A' \mapsto A'\sigma^u$, or just $A' \mapsto A'\sigma$ in the case $u = 1$, satisfying the equations

$$(\sigma\delta1) \qquad A'\sigma\delta_\eta = A',$$

$$(\sigma\delta2) \qquad A''\sigma^{v+1}\delta_\eta = A''\sigma^v.$$

We can immediately infer validity of the more general equations

$$(\sigma\delta 2)^* \qquad A''\sigma^{u+v}\delta^u_\eta = A''\sigma^v,$$
$$(\sigma\delta 1)^* \qquad A'\sigma^u\delta^u_\eta = A',$$
$$(\delta\sigma 2)^* \qquad A'\sigma^u\delta^{u+v}_\zeta = A'\delta^v_\zeta.$$

In view of imminent developments, we call $A'\sigma^u$ the *u-identity* of A'. A cell can be the u-identity of at most one other cell. In fact, if $A = A'\sigma^u$, then by $(\sigma\delta 1)^*$ we have $A' = A\delta^u_\eta$, independent of $\eta \in \mathbf{Z}^\times$. In this case we denote the cell A' suggestively by $A\delta^u$.

We say that two n-cells A and B of a globular set $\mathfrak{G}$ are *u-consecutive*, in symbols $A \circ^u B$, if $A\delta^u_+ = B\delta^u_-$. We allow $\circ$ as a shorthand notation for $\circ^1$. The set of all pairs $\langle A, B\rangle$ of u-consecutive n-cells of $\mathfrak{G}$ will be denoted by $\mathfrak{G}_n \times^u \mathfrak{G}_n$.

A *system of binary operations* for $\mathfrak{G}$ consists of maps $\mu^u_n : \mathfrak{G}_n \times^u \mathfrak{G}_n \to \mathfrak{G}_n$ $(n \geq u > 0)$, written $\langle A, B\rangle \mapsto A \cdot^u B$, or just $\langle A, B\rangle \mapsto A \cdot B$ in the case $u = 1$, satisfying the equations

$$(\mu\delta 1) \qquad (A\cdot B)\delta_- = A\delta_- \quad\text{and}\quad (A\cdot B)\delta_+ = B\delta_+ \quad (A \circ B),$$
$$(\mu\delta 2) \qquad (A \cdot^{v+1} B)\delta_\eta = A\delta_\eta \cdot^v B\delta_\eta \qquad \big(A \circ^{v+1} B\big).$$

(Note that $A \circ^{v+1} B$ implies $A\delta_\eta \circ^v B\delta_\eta$.) We can immediately infer validity of the more general equations

$$(\mu\delta 2)^* \qquad (A \cdot^{u+v} B)\delta^u_\eta = A\delta^u_\eta \cdot^v B\delta^u_\eta \qquad (A \circ^{u+v} B),$$
$$(\mu\delta 1)^* \qquad (A \cdot^u B)\delta^u_- = A\delta^u_- \quad\text{and}\quad (A\cdot^u B)\delta^u_+ = B\delta^u_+ \quad (A \circ^u B),$$
$$(\delta\mu 2)^* \qquad (A \cdot^u B)\delta^{u+v}_\zeta = A\delta^{u+v}_\zeta = B\delta^{u+v}_\zeta \qquad (A \circ^u B).$$

In view of imminent developments, we call $A \cdot^u B$ the *u-composite* of A and B.

Definition. A *(strict) ∞-category* $\mathfrak{K}$ is a globular set with a system of constants and **a** system of binary operations, satisfying the following equations.

$$(\sigma\mu 1)^* \qquad A\delta^u_-\sigma^u \cdot^u A = A = A \cdot^u A\delta^u_+\sigma^u,$$
$$(\mu\mu 1)^* \qquad (A\cdot^u B)\cdot^u C = A \cdot^u (B\cdot^u C) \qquad (A \circ^u B \circ^u C),$$
$$(\mu\sigma 2)^* \qquad (A' \cdot^v B')\sigma^u = A'\sigma^u \cdot^{u+v} B'\sigma^u \qquad (A' \circ^v B'),$$
$$(\mu\mu 2)^* \; (A \cdot^{u+v} C)\cdot^u (B\cdot^{u+v} D) = (A\cdot^u B)\cdot^{u+v}(C\cdot^u D) \qquad \begin{pmatrix} A & \circ^u & B \\ \circ^{u+v} & & \circ^{u+v} \\ C & \circ^u & D\end{pmatrix}.$$

(In $(\sigma\mu 1)^*$, the composites exist because of $(\sigma\delta 1)^*$. In $(\mu\mu 1)^*$, the outer composites exist because of $(\mu\delta 1)^*$. In $(\mu\sigma 2)^*$, the right-hand composite exists because of $(\delta\sigma 2)^*$. In $(\mu\mu 2)^*$, the outer composites exist because of $(\mu\delta 2)^*$ and $(\delta\mu 2)^*$.)

The equations $(**1)^*$ state that we have a category $\mathfrak{K}^u_n$ with object set $\mathfrak{K}_{n-u}$, morphism set $\mathfrak{K}_n$ and structure maps ${}_n\delta^u_-$, ${}_n\delta^u_+$, σ^u_n, μ^u_n. The equations $(\delta*2)^*$ state that for each $\eta \in \mathbf{Z}^\times$ we have a functor of $\mathfrak{K}^u_n$ into the discrete category $\mathfrak{K}_{n-u-v}$ with object map ${}_{n-u}\delta^v_\eta$ and morphism map ${}_n\delta^{u+v}_\eta$. It follows that we have a category $\mathfrak{K}^u_n \times^{u+v} \mathfrak{K}^u_n$ with object set $\mathfrak{K}_{n-u} \times^v \mathfrak{K}_{n-u}$, morphism set $\mathfrak{K}_n \times^{u+v} \mathfrak{K}_n$ and structure maps ${}_n\delta^u_- \times^{u+v} {}_n\delta^u_-$, ${}_n\delta^u_+ \times^{u+v} {}_n\delta^u_+$, $\sigma^u_n \times^{u+v} \sigma^u_n$, $\mu^u_n \times^{u+v} \mu^u_n$. The equations $(\mu*2)^*$ state that we have a functor of $\mathfrak{K}^u_n \times^{u+v} \mathfrak{K}^u_n$ into $\mathfrak{K}^u_n$ with object map μ^v_{n-u} and morphism map μ^{u+v}_n.

By a standard argument (Eckmann–Hilton), the n-cells A of $\mathfrak{K}$ with $A\delta_-^u = A' = A\delta_+^u$ for some fixed $A' = A''\sigma^v$ form a commutative monoid. The operations are equally those of $\mathfrak{K}_n^u$ and those of $\mathfrak{K}_n^{u+v}$. For the neutral element in particular we find that

$$(\sigma\sigma 2)^* \qquad A''\sigma^v\sigma^u = A''\sigma^{u+v}.$$

Let still $\mathfrak{G}$ be a globular set. A system of constants for $\mathfrak{G}$ satisfying $(\sigma\sigma 2)^*$ is given by maps $\sigma_n : \mathfrak{G}_{n-1} \twoheadrightarrow \mathfrak{G}_n$ $(n > 0)$ satisfying $(\sigma\delta 1)$. Such a system makes $\mathfrak{G}$ into a *globular set with degeneracies*. (The phrase 'with degeneracies' stresses the analogy with the other classical polytopic sets. Usually the term 'reflexive' is employed.) We are thus led to a slightly slimmer version of the Definition: an ∞-category $\mathfrak{K}$ is precisely a globular set with degeneracies that bears a system of binary operations, satisfying $(\sigma\mu 1)^*$, $(\mu\mu 1)^*$, $(\mu\mu 2)^*$ and

$$(\mu\sigma 2) \qquad (A' \cdot^v B')\sigma = A'\sigma \cdot^{v+1} B'\sigma \quad (A' \circ^v B').$$

Example. It has been mentioned that higher-dimensional numbers occur in the work of Batanin in the guise of trees. They in fact form an ∞-category $\hat{\mathbf{N}}$, which we are now going to describe. The n-cells of $\hat{\mathbf{N}}$ are the n-dimensional numbers. We define the facet operators by putting

$$\langle \mathfrak{q}_0, \ldots, \mathfrak{q}_{\ell-1}\rangle_n \delta_\eta = \begin{cases} 0_0 & \text{if } n = 1, \\ \langle \mathfrak{q}_0\delta_\eta, \ldots, \mathfrak{q}_{\ell-1}\delta_\eta\rangle_{n-1} & \text{if } n > 1. \end{cases}$$

Note that ${}_n\delta_+$ and ${}_n\delta_-$ have exactly the same effect. We define the degeneracies by putting

$$\langle \mathfrak{r}_0, \ldots, \mathfrak{r}_{\ell-1}\rangle_{n-1}\sigma = \langle \mathfrak{r}_0\sigma, \ldots, \mathfrak{r}_{\ell-1}\sigma\rangle_n.$$

The consecutive pairs can be characterized as follows: we always have

$$\langle \mathfrak{q}_0, \ldots, \mathfrak{q}_{\ell-1}\rangle_n \circ^n \langle \mathfrak{q}'_0, \ldots, \mathfrak{q}'_{\ell'-1}\rangle_n,$$

and for $d < n$ we have

$$\langle \mathfrak{q}_0, \ldots, \mathfrak{q}_{\ell-1}\rangle_n \circ^d \langle \mathfrak{q}'_0, \ldots, \mathfrak{q}'_{\ell'-1}\rangle_n \text{ if and only if } \ell = \ell' \text{ and } \mathfrak{q}_i \circ^d \mathfrak{q}'_i \ (0 \leq i < \ell).$$

We define the composition operators by putting

$$\langle \mathfrak{q}_0, \ldots, \mathfrak{q}_{\ell-1}\rangle_n \cdot^d \langle \mathfrak{q}'_0, \ldots, \mathfrak{q}'_{\ell'-1}\rangle_n = \begin{cases} \langle \mathfrak{q}_0, \ldots, \mathfrak{q}_{\ell-1}, \mathfrak{q}'_0, \ldots, \mathfrak{q}'_{\ell'-1}\rangle_n & \text{if } d = n, \\ \langle \mathfrak{q}_0 \cdot^d \mathfrak{q}'_0, \ldots, \mathfrak{q}_{\ell-1} \cdot^d \mathfrak{q}'_{\ell-1}\rangle_n & \text{if } d < n. \end{cases}$$

Often the one-set approach to defining strict ∞-categories (and more generally strict ω-categories; see [8]) is taken. Here each cell of $\mathfrak{K}$ is identified with all its identities. To avoid confusion, the notation for the category operations ${}_n\delta_\eta^u$ and μ_n^u provides the object dimension $n - u$, while the object codimension u can be omitted. Thus either side of $(\delta\sigma 2)^*$ is written, say, $A'\delta_{n'',\eta}$, and either side of $(\mu\sigma 2)^*$ is written, say, $A' \cdot_{n''} B'$. Here $n'' = n' - v$, where n' is the dimension of the "primed" cells. The one-set approach is impractical for our purposes, but it may increase the reader's appreciation of the Example.

The axioms of a system of constants or binary operations prescribe precisely the boundaries of identities or composites. In the axioms of an ∞-category, the boundaries of the equated cells agree automatically. These two statements have to be taken with care: they refer to an inductive construction of an ∞-category. To clarify them, we have to go a bit farther afield.

Let n be an integer. Given any concept $\mathscr{C}$ derived in this paper from that of a propolytopic set, the attribute '*n-dimensional*' will be used to name the corresponding concept $\mathscr{C}_{\leq n}$ that can be defined in place of $\mathscr{C}$ by omitting all references to data in dimensions $> n$. Any model $\mathfrak{X}$ of the concept $\mathscr{C}$ has an underlying model $\mathfrak{X}_{\leq n}$ of the concept $\mathscr{C}_{\leq n}$. We call $\mathfrak{X}_{\leq n}$ the *n-skeleton* of $\mathfrak{X}$. Given a model $\mathfrak{O}$ of $\mathscr{C}_{\leq n}$, we say that $\mathfrak{X}$ is a model *over* $\mathfrak{O}$ if $\mathfrak{X}_{\leq n} = \mathfrak{O}$. Analogous statements apply to the passage from $\mathscr{C}_{\leq N}$ to $\mathscr{C}_{\leq n}$ for $n \leq N < \infty$. We write $\mathfrak{X}_{<N}$ instead of $\mathfrak{X}_{\leq N-1}$.

A model of $\mathscr{C}$ can be constructed inductively by constructing its skeleta. More precisely, a model $\mathfrak{X}$ of $\mathscr{C}$ is given by a sequence consisting of one model $\mathfrak{X}_{\leq N}$ of each $\mathscr{C}_{\leq N}$ and satisfying the condition that $\mathfrak{X}_{\leq N}$ is over $\mathfrak{X}_{<N}$. Thus, in order to understand models of $\mathscr{C}$, it suffices to understand models of some $\mathscr{C}_{<N_0}$ as well as models of $\mathscr{C}_{\leq N}$ over fixed models of $\mathscr{C}_{<N}$ for all $N \geq N_0$.

It is high time to be more concrete. Let $N \geq -1$. An N-dimensional ∞-category is simply called an *N-category*. To be sure, here is a somewhat more explicit definition. First, an N-dimensional globular set $\mathfrak{X}$ consists of sets $\mathfrak{X}_0, \ldots,$ $\mathfrak{X}_N$ and mappings ${}_1\delta_+, \ldots, {}_n\delta_+, {}_1\delta_-, \ldots, {}_n\delta_-$ satisfying the globular law. An N-category $\mathfrak{K}$ is an N-dimensional globular set together with mappings σ_n^u and μ_n^u $(0 < u \leq n \leq N)$ satisfying the laws of an ∞-category. We furthermore define a *pre-N-category* $\mathfrak{k}$ to be an N-dimensional globular set together with mappings σ_n^u and μ_n^u $(0 < u \leq n < N)$ satisfying the laws of an ∞-category. Every N-category has an underlying pre-N-category (drop the mappings σ_N^u and μ_N^u), and every pre-N-category has an underlying $(N-1)$-category, its $(N-1)$-skeleton.

Let $N \geq 0$ and let $\mathfrak{O}$ be an $(N-1)$-dimensional globular set. A *(globular) N-frame* a of $\mathfrak{O}$ provides for each integer u with $0 < u \leq N$ and each sign η an $(N-u)$-cell $a\delta_\eta^u$, where

$$a\delta_\eta^u \delta_\zeta = a\delta_\zeta^{u+1} \quad (0 < u < N;\ \eta, \zeta \in \mathbf{Z}^\times).$$

If $N = 0$, there is precisely one N-frame (which provides no information). If $N = 1$, an N-frame a is just a pair of 0-cells $a\delta_+$ and $a\delta_-$. If $N \geq 2$, an N-frame a is given by two $(N-1)$-cells $a\delta_+$ and $a\delta_-$ with $a\delta_+\delta_\zeta = a\delta_-\delta_\zeta$.

An N-dimensional globular set $\mathfrak{G}$ over $\mathfrak{O}$ is given by an N-cell set $\mathfrak{G}_N$ and face operators ${}_N\delta_\eta^u : \mathfrak{G}_N \longrightarrow \mathfrak{O}_{N-u}$ satisfying the generalized globular laws. A particular N-dimensional globular set $\mathfrak{O}[\mathbf{1}]$ over $\mathfrak{O}$ can be obtained by taking $\mathfrak{O}[\mathbf{1}]_N$ to be the set of N-frames of $\mathfrak{O}$ and ${}_N\delta_\eta^u$ as suggested by the notation for frames. Back to the general case. Having a family of maps ${}_N\delta_\eta^u : \mathfrak{G}_N \longrightarrow \mathfrak{O}_{N-u}$ satisfying the generalized globular laws is equivalent to having a single map $\mathfrak{G}_N \longrightarrow \mathfrak{O}[\mathbf{1}]_N$. Indeed, given the former, we can assign to each N-cell A of $\mathfrak{G}$ its *boundary* $A^\bullet$, defined by $A^\bullet\delta_\eta^u = A\delta_\eta^u$. (We have constructed $\mathfrak{O}[\mathbf{1}]_N$ as a certain projective limit in the category of sets.) But any map can be replaced by the family of its fibres. An N-cell A of $\mathfrak{G}$ with boundary a will also be called a *filler* of a. The set of all fillers of a will be denoted by $\mathfrak{G}_a$. Thus, an N-dimensional globular set $\mathfrak{G}$ over $\mathfrak{O}$ is given by a family of disjoint sets $\mathfrak{G}_a$ $(a \in \mathfrak{O}[\mathbf{1}]_N)$. Taking this last statement as a definition, we can drop the disjointness condition without gain of generality.

Now suppose that $\mathfrak{O}$ is an $(N-1)$-category. The previous paragraph can be repeated with the 'N-dimensional globular set $\mathfrak{G}$' replaced by a 'pre-N-category $\mathfrak{k}$'. Now an N-category $\mathfrak{K}$ over $\mathfrak{O}$ is given by pre-N-category over $\mathfrak{O}$ together with identity operators σ^u_N and composition operators μ^u_N satisfying the relevant laws. In the case of $\mathfrak{O}[\mathbf{1}]$ there is a unique way to define such operators $\sigma^u_N = \sigma^{u}_N{}^{\bullet}$ and $\mu^u_N = \mu^{u}_N{}^{\bullet}$: the u-identity of an $(N-u)$-cell is determined by the laws of a system of constants; the u-composite of two u-consecutive N-frames is determined by the laws for a system of binary operations. One can verify that all the laws of an ∞-category are thus satisfied. Back to the general case. Having a map $\sigma^u_N : \mathfrak{O}_{N-u} \longrightarrow \mathfrak{K}_N$ satisfying the relevant laws is equivalent to having maps (constants) $\sigma^u_{A'} : 1 \longrightarrow \mathfrak{K}_{A'\sigma^{u\bullet}}$, one for each $(N-u)$-cell A' of $\mathfrak{O}$. Having a map $\mu^u_N : \mathfrak{K}_N \cdot^u \mathfrak{K}_N \longrightarrow \mathfrak{K}_N$ satisfying the relevant laws is equivalent to having maps (binary operations) $\mu^u_{a,b} : \mathfrak{K}_a \times \mathfrak{K}_b \longrightarrow \mathfrak{K}_{a\cdot^u b}$, one for each pair of u-consecutive N-frames a and b of $\mathfrak{O}$. The remaining laws can similarly be broken down for frames. They assert commutativity of the following diagrams (with the obvious conditions on the frames).

$(\sigma\mu 1)^*$

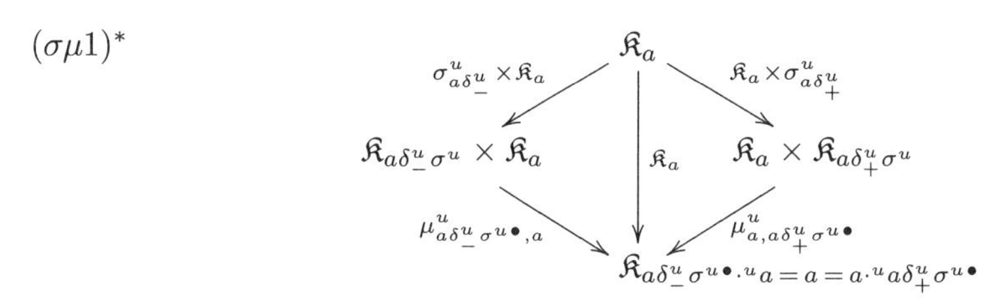

$(\mu\mu 1)^*$

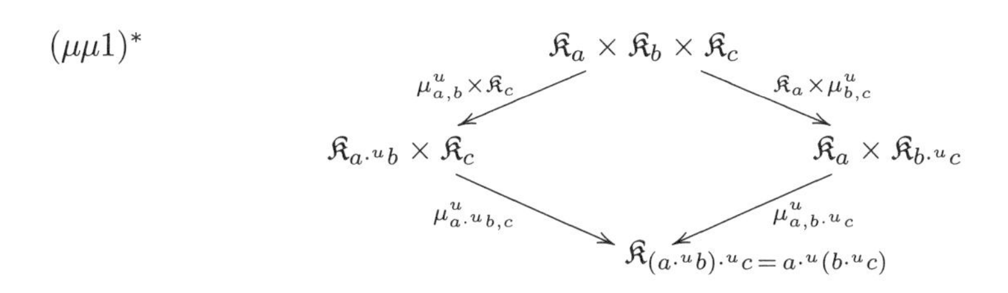

$(\mu\sigma 2)^*$

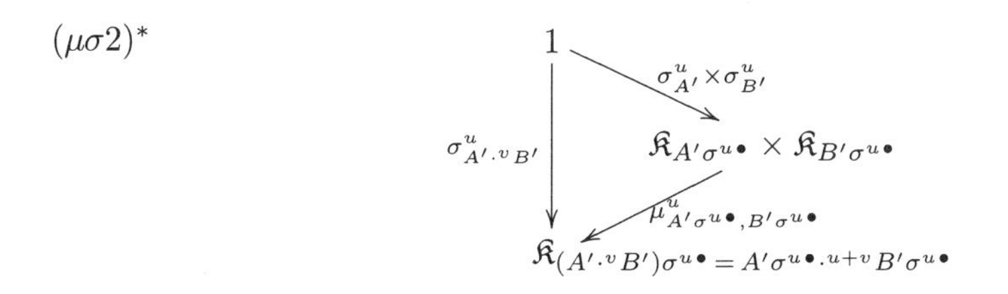

$(\mu\mu 2)^*$

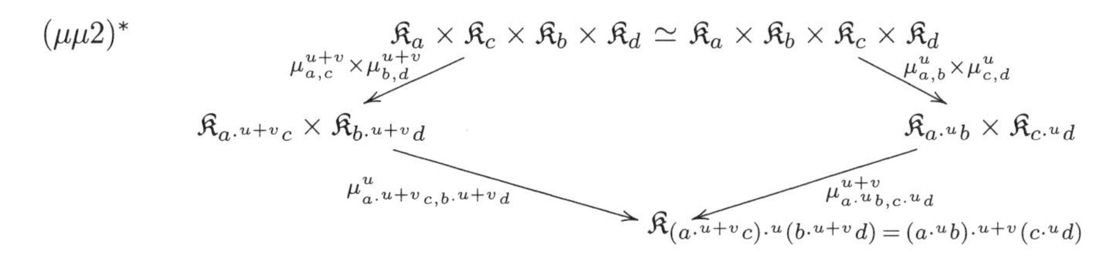

(Thus the N-categories over a fixed $(N-1)$-category $\mathfrak{O}$ form an $\mathfrak{O}[\mathbf{1}]_N$-sorted variety.)

Section 2. Roofs

We construct the ∞-category generated by a dendrotopic set. The cells of the former will be certain frameworks of the latter, called *roofs*.

Let $\mathfrak{D}$ be a dendrotopic set. We want to define those finite configurations of cells in $\mathfrak{D}$ which have an overall direction in every dimension. By their appearance we may call them *cell forests*. Each n-cell forest is equipped with a finite family of n-cells, its facets. A 0-cell forests consists of just a single facet. For $n > 0$, an n-cell forest has an underlying $(n-1)$-cell forest, to each of whose facets it associates an n-cell tree rooted there. Its facets are the facets of these trees.

This description, while being very natural, turns out to be cumbersome. An important piece of data is shielded from direct access: the leafs of the n-cell trees that constitute an n-cell forest form an $(n-1)$-cell forest of their own. To have all the relevant structure at hand, we collect the data of an n-cell forest into an $(n+1)$-framework of a particular kind.

Definition. Let $n \geq 0$, and let X be an $(n+1)$-framework of a dendrotopic set $\mathfrak{D}$. We call X an *$(n+1)$-roof* under the circumstances described inductively as follows. If $n = 0$, the negative facet is a cell, the positive facet is not. If $n \geq 1$, there is precisely one negative facet that is not a cell, and this facet is an n-roof.

Let us temporarily denote the incomplete negative facet location of a roof by τ. In order fully to appreciate the definition, the reader may need the following supplementary information.

Proposition 1. *Let X be an $(n+1)$-roof. If $n \geq 1$, then $X\omega$ is an n-roof (too). If $n \geq 2$, then*

$$X : \frac{\tau\ \omega}{\omega\ \omega} \qquad \text{and} \qquad X : \frac{\omega\ \tau}{\tau\ \tau}. \tag{1}$$

Proof. We use induction on n.

First consider the case $n = 1$. The floor of $\langle p, \delta_\zeta \rangle \in X\mathrm{I}^2$ is a cell if either $p \notin \{\omega, \tau\}$ or $p = \tau$ and $\zeta = -$. The floor of $\langle \tau, \delta_+ \rangle$ is not a cell. Hence the conjugate of this negative 2-chamber is the only remaining positive 2-chamber, namely $\langle \omega, \delta_+ \rangle$, and so $X\omega\delta_+ = X\tau\delta_+$ is not a cell. Now the conjugate of the other remaining 2-chamber $\langle \omega, \delta_- \rangle$ must be among those 2-chambers whose floors are cells, and so $X\omega\delta_-$ is a cell. Thus $X\omega$ is a roof.

Now consider the case $n \geq 2$. The floor of $\langle p, q \rangle \in X\mathrm{I}^2$ is a cell if either $p \notin \{\omega, \tau\}$ or $p = \tau$ and $q \notin \{\omega, \tau\}$. The floors of $\langle \tau, \omega \rangle$ and $\langle \tau, \tau \rangle$ are roofs, but these two 2-chambers cannot be conjugates of each other because of their common negative first entry. Hence the conjugate of the former, which is negative, is the only remaining positive 2-chamber, namely $\langle \omega, \omega \rangle$ — this proves the left part of (1) —, and the conjugate of the latter, which is positive, is among the remaining negative 2-chambers, namely $\langle \omega, q \rangle$ $(q \in X\omega\mathrm{I}_-)$ — this means $X : \frac{\omega\ q_0}{\tau\ \tau}$ for some negative facet location q_0 of $X\omega$. In particular, $X\omega q_0 = X\tau\tau$ is a roof. Now let q be any other negative facet location of $X\omega$. The conjugate of $\langle \omega, q \rangle$ must be among those 2-chambers whose floors are cells, and so $X\omega q$ is a cell. Thus $X\omega$ is a roof with $\tau = q_0$. This last equation settles the right part of (1). □

Let us come back to the introductory remarks about cell forests. The negative half boundary of an $(n+1)$-roof is an n-framework tree with the following property. If $n = 0$, its only facet location is complete; if $n \geq 1$, it has precisely one incomplete

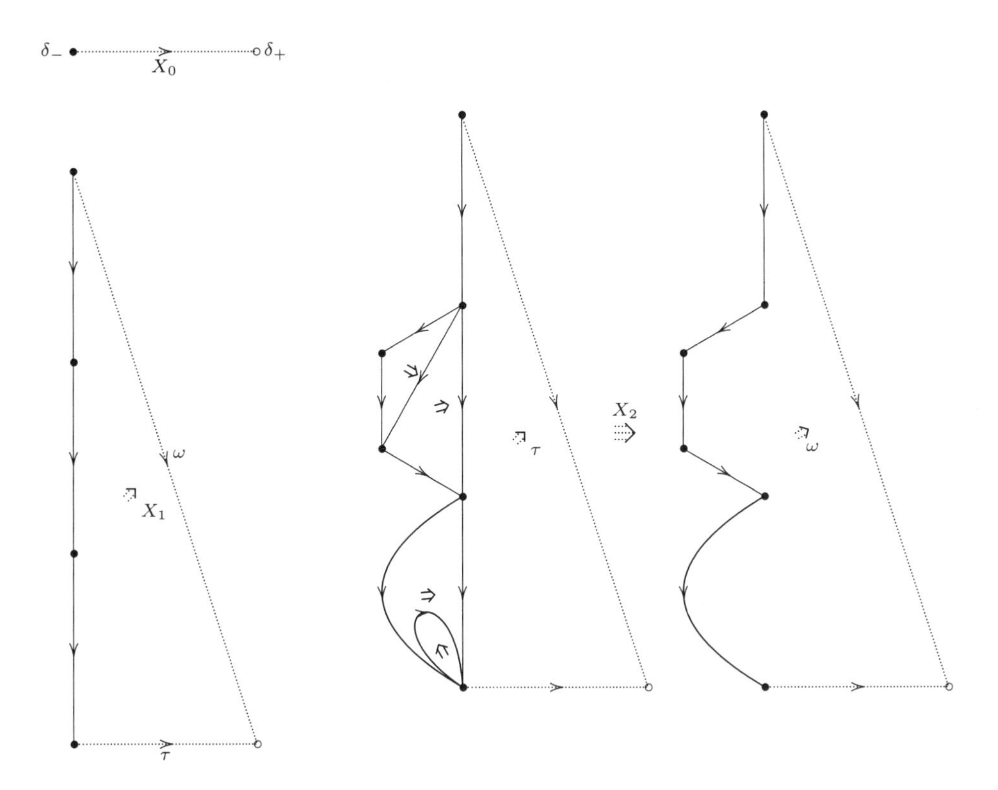

Figure. A 1-roof X_0, a 2-roof X_1 with $X_1\tau = X_0$, a 3-roof X_2 with $X_2\tau = X_1$. (The unfilled circle represents the 0-frame; dotted lines mark incomplete $>$ 0-frameworks.)

facet location, whose resident is a roof. Let us call such a framework tree *roofed*, for lack of a better word. By an adapted version of Theorem 4 of Part I, every roofed n-framework tree is the negative half boundary of exactly one $(n+1)$-roof.

Examining the proof of Proposition 1 again, we find that it in fact shows $X : \frac{\tau\ \omega}{\omega\ \omega}$ for all $n \geq 1$. Thus the incomplete negative facet location of any roof has index 0. We conclude that for roofed framework trees, the roof is the stem. Hence, according to the recursive definition of cell trees, a roofed n-framework tree P consists of an n-roof $P{\uparrow}\tau$ and, for each complete facet location q of $P{\uparrow}\tau$, an n-cell tree $P_{(q)}$ with root $P{\uparrow}\tau q$. (The n-cell tree $P_{(\tau)}$, for $n \geq 2$, is automatically empty, as shown by the left part of (1).) This observation leads to an inductive definition of roofs directly via cell trees, namely the one given above, with the term 'n-cell forest' replacing '$(n+1)$-roof'.

The incomplete facet locations of roofs of dimensions ≥ 1 exhibit a certain symmetry. It will be convenient to express this symmetry in the notation: let us write $\tau = \partial_+$ and $\omega = \partial_-$. Note that ∂_η has the sign $-\eta$, opposite to what the index indicates. The reason becomes clear from the forest point of view: the resident of τ stands for the target of a forest (consisting of its roots), and the resident of ω stands for the source of a forest (consisting of its leafs). Now (1) can be instantly recognized as the globular relationship

$$X : \eta \frac{\partial_- \; \partial_\eta}{\partial_+ \; \partial_\eta}.$$

This tells us that we can define a globular set $\mathfrak{D}^*$ as follows: an n-cell A of $\mathfrak{D}^*$ is an $(n+1)$-roof $A^\natural$ of $\mathfrak{D}$, and the sign-η facet of A, for $n \geq 1$, is specified by

$$A\delta_\eta{}^\natural = A^\natural \partial_\eta.$$

We now want to add degeneracies and a system of binary operations to $\mathfrak{D}^*$. As was remarked before, the boundary of an identity $A'\sigma$ or a composite $A \cdot^u B$ is determined by the operands via the axioms. That is, if we write $X = A'\sigma^\natural$ or $X = (A \cdot^u B)^\natural$, we do not need to specify the facets $X\partial_\eta$. (We may mention them nevertheless, for the convenience of the reader.) The incidence structure of incomplete faces (in particular the incomplete 2-face and 3-face locations) are determined by X being a roof. Thus, we can restrict our attention to the complete part of the boundary. To this end, the subscript '$_{\mathfrak{D}}$' will denote the restriction of any set of face locations to the complete ones.

Let $n \geq 1$, and let A' be an $(n-1)$-cell of $\mathfrak{D}^*$. We want to construct the 1-identity $A'\sigma$ of A'. That is, we want to construct the $(n+1)$-roof $A'\sigma^\natural = X$. We let X have no complete facet. The incomplete ones are $X\partial_\eta = A'^\natural$. We let the complete ridge locations of X be

$$\frac{\partial_- \; q}{\partial_+ \; q} \qquad \left(q \in A'^\natural \mathrm{I}_{\mathfrak{D}} \right).$$

Then the complete 3-face locations of X are

$$\begin{array}{l} \partial_- \; q_- \; r_- \\ \hline \partial_+ \; q_- \; r_- \\ \hline \partial_+ \; q_+ \; r_+ \\ \hline \partial_- \; q_+ \; r_+ \end{array} \qquad \left(\tfrac{q_- \; r_-}{q_+ \; r_+} \in A'^\natural \Diamond_{\mathfrak{D}} \right).$$

For $u > 1$, the facets of the u-identity $A'\sigma^u$ of A' are identities themselves. In particular, $A'\sigma^{u\,\natural}\partial_+ = A'\sigma^{u-1\,\natural}$ has no complete facets. From the cell-tree approach it follows that given an n-roof Y without complete facets, there is only one $(n+1)$-roof X with $X\partial_+ = Y$. Hence whenever $A\delta_+ = A'\sigma^{u-1}$, we can conclude that $A = A'\sigma^u$.

Let $u \geq 1$, and let A and B be two u-consecutive n-cells. Let us denote by A' the intermediate $(n-u)$-cell $A\delta_+^u = B\delta_-^u$. We want to construct the u-composite $A \cdot^u B$ of A and B. That is, we want to construct the $(n+1)$-roof $(A \cdot^u B)^\natural = X$. This construction has to be carried out by induction on u. In all cases, we let the complete facets of X be those of $A^\natural$ and those of $B^\natural$.

In the case $u = 1$, the incomplete facets of X are $X\partial_- = A^\natural \partial_-$ and $X\partial_+ = B^\natural \partial_+$. We let the complete ridge locations of X be those of $A^\natural$ and those of $B^\natural$, except that

$$A^\natural : \frac{p_- \; q_-}{\partial_+ \; q} \quad \text{and} \quad B^\natural : \frac{\partial_- \; q}{p_+ \; q_+}$$

are replaced by

$$X : \frac{p_- \; q_-}{p_+ \; q_+}$$

for each $q \in A'^{\natural}\,\mathrm{I}_{\mathfrak{D}}$. Let us say that the chambers $\langle \partial_+, q\rangle$ of $A^{\natural}$ and $\langle \partial_-, q\rangle$ of $B^{\natural}$ *cancel* each other. The construction is an instance of the one discussed in Claim A in the proof of Theorem 3 of Part I (with $A^{\natural} : \partial_+$ and $B^{\natural} : \partial_-$ here playing the roles of $B_0 : c$ and $B_1 : \omega$ there). Note that the ridge location $\eta\frac{\partial_-\ \partial_\eta}{\partial_+\ \partial_\eta}$ of X replaces the corresponding ones of $A^{\natural}$ and $B^{\natural}$.

In the case $u > 1$, the incomplete facets of X are $X\partial_\eta = (A\delta_\eta \cdot^{u-1} B\delta_\eta)^{\natural}$. We let the complete ridge locations of X be those of $A^{\natural}$ and those of $B^{\natural}$. Then the complete 3-face locations of X are also those of $A^{\natural}$ and those of $B^{\natural}$, with the following exceptions in the case $u = 2$: here for each $r \in A'^{\natural}\,\mathrm{I}_{\mathfrak{D}}$, the two locations

$$A^{\natural} : \{\!\!\{ r \}\!\!\} \left\{ \begin{array}{ccc} \partial_- & \partial_+ & r \\ \partial_+ & \partial_+ & r \\ \left.\begin{array}{ccc} \partial_+ & q_{+,+} & r_{+,+} \\ * & * & * \\ \vdots & \vdots & \vdots \\ * & * & * \\ \partial_- & q_{-,+} & r_{-,+} \end{array}\right\} = \mathscr{G}_+ \end{array}\right. \qquad \text{and} \qquad B^{\natural} : \{\!\!\{ r \}\!\!\} \left\{ \begin{array}{ccc} \left.\begin{array}{ccc} \partial_- & q_{-,-} & r_{-,-} \\ * & * & * \\ \vdots & \vdots & \vdots \\ * & * & * \\ \partial_+ & q_{+,-} & r_{+,-} \end{array}\right\} = \mathscr{G}_- \\ \partial_+ & \partial_- & r \\ \partial_- & \partial_- & r \end{array}\right.$$

are relinked and thus replaced by

$$X : \{\!\!\{ r \}\!\!\} \left\{ \begin{array}{l} \left.\begin{array}{ccc} \partial_- & q_{-,-} & r_{-,-} \\ * & * & * \\ \vdots & \vdots & \vdots \\ * & * & * \\ \partial_+ & q_{+,-} & r_{+,-} \end{array}\right\} = \mathscr{G}_- \\ \left.\begin{array}{ccc} \partial_+ & q_{+,+} & r_{+,+} \\ * & * & * \\ \vdots & \vdots & \vdots \\ * & * & * \\ \partial_- & q_{-,+} & r_{-,+} \end{array}\right\} = \mathscr{G}_+ \end{array}\right. .$$

For $u > 1$, the cell-tree approach yields a different way to describe $(A \cdot^u B)^{\natural} = X$. By induction hypothesis, we know that $X\partial_+\mathrm{I}_{\mathfrak{D}}$ is the disjoint union of $A^{\natural}\,\mathrm{I}_{\mathfrak{D}}$ and $B^{\natural}\,\mathrm{I}_{\mathfrak{D}}$. For each $q \in A^{\natural}\partial_+\mathrm{I}_{\mathfrak{D}}$ we have put $X\alpha_{(q)} = A^{\natural}\alpha_{(q)}$, and for each $q \in B^{\natural}\partial_+\mathrm{I}_{\mathfrak{D}}$ we have put $X\alpha_{(q)} = B^{\natural}\alpha_{(q)}$. This description does not, however, directly reveal the fact that $X\partial_- = (A\delta_- \cdot^{u-1} B\delta_-)^{\natural}$.

We now want to verify that $\mathfrak{D}^*$, including the structure just introduced, is indeed an ∞-category. We have to consider four equations of cells in $\mathfrak{D}^*$, each of which can be equivalently expressed as an equation of frameworks in $\mathfrak{D}$. Following the construction steps above, we easily obtain a common descriptions of both — or in case of $(\sigma\mu 1)^*$ all three — sides of each of the latter equations. We only give this description here. According to an earlier remark, the globular boundaries in the original equations agree as a consequence of the axioms on identities and composites. Hence we can again restrict our attention to the complete parts of the dendrotopic boundaries.

$(\sigma\mu 1)^*$ $(A\delta_-^u\sigma^u \cdot^u A)^{\natural} = A^{\natural} = (A \cdot^u A\delta_+^u\sigma^u)^{\natural}$. The complete facets as well as the complete ridge locations are precisely those of $A^{\natural}$. To evaluate the left-hand side ($\eta = -$) and the right-hand side ($\eta = +$) in the case $u = 1$, note that $A\delta_\eta\sigma^{\natural} : \frac{\partial_-\ q}{\partial_+\ q}$ ($q \in A\delta_\eta{}^{\natural}\,\mathrm{I}_{\mathfrak{D}}$) act as neutral elements for cancellation.

$(\mu\mu 1)^*$ $((A \cdot^u B) \cdot^u C)^\natural = (A \cdot^u (B \cdot^u C))^\natural$. The complete facets are precisely those of $A^\natural$, $B^\natural$, $C^\natural$. The same goes for the complete ridge locations, with the following exceptions in the case $u = 1$. Let us denote by A' the cell $A\delta_+ = B\delta_-$ and by B' the cell $B\delta_+ = C\delta_-$. For each $q \in A'^\natural \mathrm{I}_{\mathfrak{D}}$ the chambers $\langle \partial_+, q \rangle$ underneath $A^\natural$ and $\langle \partial_-, q \rangle$ underneath $B^\natural$ cancel each other, and so do for each $q \in B'^\natural \mathrm{I}_{\mathfrak{D}}$ the chambers $\langle \partial_+, q \rangle$ underneath $B^\natural$ and $\langle \partial_-, q \rangle$ underneath $C^\natural$.

$(\mu\sigma 2)$ $(A \cdot^u B)\sigma^\natural = (A\sigma \cdot^{u+1} B\sigma)^\natural$. There are no complete facets. The complete ridge locations are precisely $\frac{\partial_- \, q}{\partial_+ \, q}$ for all complete facet locations q of $A^\natural$ and of $B^\natural$.

$(\mu\mu 2)^*$ $((A \cdot^u B) \cdot^{u+v} (C \cdot^u D))^\natural = ((A \cdot^{u+v} C) \cdot^u (B \cdot^{u+v} D))^\natural$. The complete facets are precisely those of $A^\natural$, $B^\natural$, $C^\natural$, $D^\natural$. The same goes for the complete ridge locations, with the following exceptions in the case $u = 1$. Let us denote by A' the cell $A\delta_+ = B\delta_-$ and by C' the cell $C\delta_+ = D\delta_-$. For each $q \in A'^\natural \mathrm{I}_{\mathfrak{D}}$ the chambers $\langle \partial_+, q \rangle$ underneath $A^\natural$ and $\langle \partial_-, q \rangle$ underneath $B^\natural$ cancel each other, and so do for each $q \in C'^\natural \mathrm{I}_{\mathfrak{D}}$ the chambers $\langle \partial_+, q \rangle$ underneath $C^\natural$ and $\langle \partial_-, q \rangle$ underneath $D^\natural$. (Note once again that $(A \cdot^{v+1} C)\delta_+ = A' \cdot^v C' = (B \cdot^{v+1} D)\delta_-$.)

The achievement of this section ought to be stated for the record.

Proposition 2. *Let $\mathfrak{D}$ be a dendrotopic set. The roofs of $\mathfrak{D}$ form an ∞-category $\mathfrak{D}^*$ as defined above.*

Not surprisingly, a quick comparison of the ∞-category construction of the Hermida–Makkai–Zawadowski paper [3] with the present one uncovers many parallels, hidden not least by different terminologies. The translations

pasting diagram = cell tree,

indeterminate = incomplete framework,

good pasting diagram (in the superstructure with indeterminates) = roofed framework tree

suggest themselves. Of course these "equations" gain formal meaning only after the concepts of 'multitopic set' and of 'dendrotopic set' have been suitably related. Then the following assessments, delivered here without proofs, can be made.

The first equation clearly holds. The second equation holds too, but to evaluate the left-hand side one has to go beyond the cited paper, which (in its present form) is silent on how many 0-dimensional indeterminates there are. In avoiding such objects, the authors unnecessarily give special treatment to the case of dimension 1. Here they take a "good pasting diagram" to be a pasting diagram in the original structure, while a "roofed framework tree" is a cell tree extended by incomplete root and stem. Thus the third equation fails, even though the steps in the inductive definitions of the two sides amount to the same.

Section 3. Freedom

We say what it means for a strict ∞-category to be *free*. Two concepts of an n-category being freely generated by a pre-n-category naturally arise, and we show that they are equivalent.

Oddly enough, we have so far been discussing only objects, not their accompanying *morphisms*. With the introduction of the latter, some essentially category-theoretical arguments will enter the scene.

The objects defined in Section 1 have obvious (homo-)morphisms, sending cells to cells of the same dimension and preserving the operators. Here the term 'preservation' takes its usual sense (meaning "on the nose"). The morphisms of globular sets are also called *globular maps*, the morphisms of ∞-categories are also called ∞*-functors*, and so on. Terminology and notation for morphisms follows that for objects as far as possible.

To be explicit in at least one case, here is what we mean by a globular map $\Phi : \mathfrak{G} \to \mathfrak{H}$: it consists of maps $\Phi_n : \mathfrak{G}_n \to \mathfrak{H}_n$, written $A \mapsto A\Phi$, that satisfy the equations $A\Phi\delta_\eta = A\delta_\eta\Phi$. Just as we can construct a globular set $\mathfrak{G}$ inductively by specifying its filler sets $\mathfrak{G}_a$, we can construct a globular map $\Phi : \mathfrak{G} \to \mathfrak{H}$ inductively by specifying its filler maps $\Phi_a : \mathfrak{G}_a \to \mathfrak{H}_{a\Phi}$. (Note that an $(N-1)$-dimensional globular map sends N-frames to N-frames.)

There is an even more obvious way to obtain identity morphisms and composite morphisms. As a result, we have a notion of *isomorphism*. The identity for an object is denoted in the same way as the object itself. The composite of two morphisms is denoted by writing them next to each other, first the one to be carried out first, second the one to be carried out second.

There are two sensible concepts of an N-category being freely generated by a pre-N-category $\mathfrak{k}$. We call an N-category $\mathfrak{K}$ and a pre-N-functor $e : \mathfrak{k} \to \mathfrak{K}$ *universal* if for every N-category $\mathfrak{L}$ and every pre-N-functor $f : \mathfrak{k} \to \mathfrak{L}$, there is a unique N-functor $F : \mathfrak{K} \to \mathfrak{L}$ such that

$$f = eF. \tag{2}$$

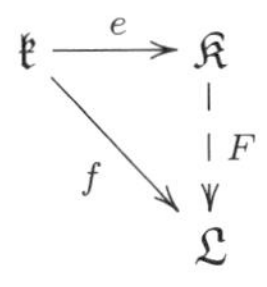

Now write $\mathfrak{D}$ for the $(N-1)$-skeleton of $\mathfrak{k}$. In order to avoid overuse of the word 'over', let us agree that the prefix 'N-' and the qualifier 'over $\mathfrak{D}$' together can be replaced by the prefix '$\mathfrak{D}$-'. Thus, $\mathfrak{k}$ is a pre-$\mathfrak{D}$-category. We call an $\mathfrak{D}$-category $\mathfrak{K}$ and a pre-$\mathfrak{D}$-functor $e : \mathfrak{k} \to \mathfrak{K}$ *universal over* $\mathfrak{D}$ if for every $\mathfrak{D}$-category $\mathfrak{L}$ and every pre-$\mathfrak{D}$-functor $f : \mathfrak{k} \to \mathfrak{L}$, there is a unique $\mathfrak{D}$-functor $F : \mathfrak{K} \to \mathfrak{L}$ such that (2). Of course we could have omitted the explicit requirement that F be over $\mathfrak{D}$, since it is implicit in (2).

The two concepts are equivalent in the following sense.

Proposition 3. *If* $(\mathfrak{K}; e)$ *is universal over* $\mathfrak{D}$*, then* $(\mathfrak{K}; e)$ *is universal.*

Proposition 4. *If* $(\mathfrak{K}; e)$ *is universal, then* $e_{<N} : \mathfrak{k}_{<N} \to \mathfrak{K}_{<N}$ *is an isomorphism of* $(N-1)$*-categories. (Thus, when we identify* $\mathfrak{K}_{<N}$ *with* $\mathfrak{D} = \mathfrak{k}_{<N}$ *accordingly,* $(\mathfrak{K}; e)$ *becomes universal over* $\mathfrak{D}$*.)*

Readers familiar with category theory know that these statements are consequences of the fact that in the commutative triangle of categories and functors

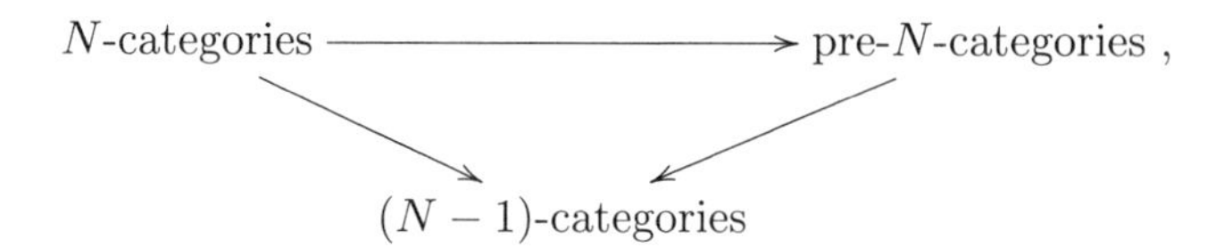

the two lateral ones are fibrations, and the top one preserves cartesian liftings. Of course this fact itself would still deserve a proof. We use a weaker fact instead, which suffices to obtain the intended conclusions in a categorical manner.

Lemma 5. *Let $g : \mathfrak{k} \to \mathfrak{M}$ be a pre-N-functor into an N-category $\mathfrak{M}$. There is a decomposition of g into a pre-$\mathfrak{O}$-functor $f : \mathfrak{k} \to \mathfrak{L}$ and an N-functor $H : \mathfrak{L} \to \mathfrak{M}$ with the following universal property. For any decomposition of g into a pre-$\mathfrak{O}$-functor $f' : \mathfrak{k} \to \mathfrak{L}'$ and an N-functor $H' : \mathfrak{L}' \to \mathfrak{M}$ there is a unique $\mathfrak{O}$-functor $D : \mathfrak{L}' \to \mathfrak{L}$ such that $f'D = f$ and $DH = H'$.*

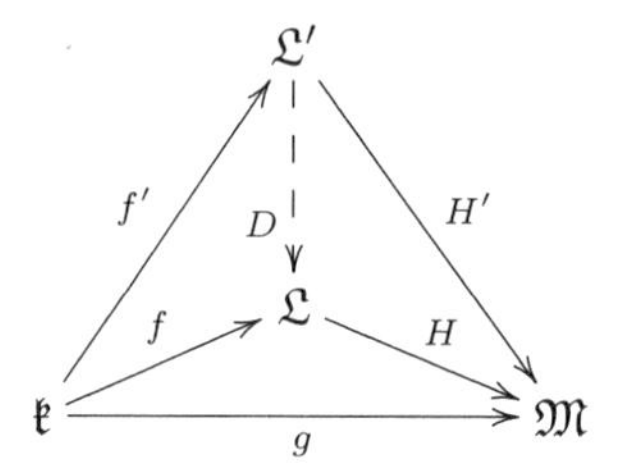

Again we could have omitted the explicit requirement that the fill-in N-functor (in this case D) be over $\mathfrak{O}$.

Proof. The straight-forward translation of the conditions into constructions is left to the reader. The only slightly difficult step is the specification of the set $\mathfrak{L}_N$ and the map $H_N : \mathfrak{L}_N \to \mathfrak{M}_N$. We demand (and have to demand) that for each N-frame a of $\mathfrak{L}_{<N} = \mathfrak{k}_{<N}$, the filler map $H_a : \mathfrak{L}_a \to \mathfrak{M}_{aH=ag}$ be bijective. □

Proof of Proposition 3. Suppose $(\mathfrak{K}; e)$ is universal over $\mathfrak{O}$. Let $\mathfrak{M}$ be an arbitrary N-category, and let $g : \mathfrak{k} \to \mathfrak{L}$ be a pre-N-functor. We pick a decomposition $g = fH$ as in the lemma. By universality of $(\mathfrak{K}; e)$, there is a unique $\mathfrak{O}$-functor $F : \mathfrak{K} \to \mathfrak{L}$ such that $eF = f$. Now $G = FH$ is an N-functor $\mathfrak{K} \to \mathfrak{M}$ such that $eG = g$.

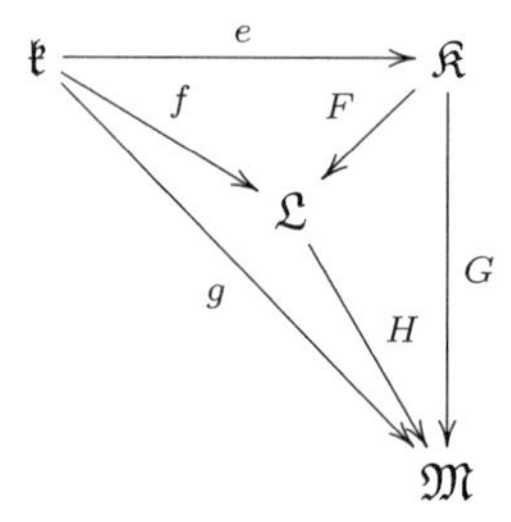

Conversely, suppose that $G' : \mathfrak{K} \to \mathfrak{M}$ is an N-functor such that $eG' = g$. By universality of the decomposition, there is an $\mathfrak{O}$-functor $F' : \mathfrak{K} \to \mathfrak{L}$ such that $eF' = f$ and $F'H = G$. Of these two equations, the former yields $F' = F$ by uniqueness of F, whence the latter yields $G' = G$. □

Proof of Proposition 4. Suppose $(\mathfrak{K}; e)$ is universal. We pick a decomposition of e into $e^0 : \mathfrak{k} \to \mathfrak{K}^0$ and $E : \mathfrak{K}^0 \to \mathfrak{K}$ as in the lemma. By universality of $\mathfrak{K}$, there is an N-functor $E^0 : \mathfrak{K} \to \mathfrak{K}^0$ such that $eE^0 = e^0$.

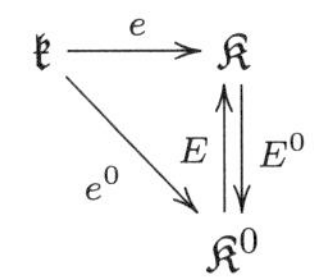

Now $eE^0E = e^0E = e$, whence by universality of e the endomorphism E^0E of $\mathfrak{K}$ is the identity. It follows that $E^0EE^0 = E^0$. Also $e^0EE^0 = eE^0 = e^0$,

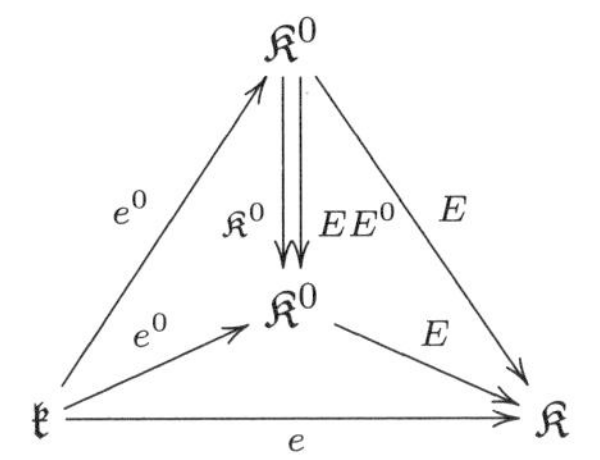

whence by universality of the decomposition the endomorphism EE^0 of $\mathfrak{K}^0$ is the identity as well. Thus E and E^0 are isomorphisms. Applying this result to the $(N-1)$-skeleton of the situation, we find that $e_{<N} = E_{<N}$ is also an isomorphism. (It is furthermore clear that $(\mathfrak{K}^0; e^0)$ is universal over $\mathfrak{D}$.) □

Now let us view things from the perspective of a given $\mathfrak{D}$-category $\mathfrak{K}$. Providing a pre-$\mathfrak{D}$-functor $e : \mathfrak{k} \to \mathfrak{K}$ reduces to providing a map $e_N : \mathfrak{k}_N \to \mathfrak{K}_N$. Indeed, the boundaries for the N-cells of $\mathfrak{k}$ are determined by the homomorphy requirement. If e_N is moreover injective, we can identify $\mathfrak{k}_N$ with a subset of $\mathfrak{K}_N$. The following result shows that the assumption is satisfied if $(\mathfrak{K}; e)$ is universal (generally or over $\mathfrak{D}$).

Proposition 6. *Let $(\mathfrak{K}; e)$ be universal over $\mathfrak{D}$. Then $e_N : \mathfrak{k}_N \to \mathfrak{K}_N$ is injective.*

Proof. We construct an $\mathfrak{D}$-category $\mathfrak{D}[\mathbf{N}]$ as follows. We let the filler set of each N-frame of $\mathfrak{D}$ be $\mathbf{N}$, the set of natural numbers. In dimension N, we let every identity be zero and every composite be the sum of the operands. The equations of an ∞-category clearly hold. (An analogous construction can be carried out for any commutative monoid in place of $\mathbf{N}$.)

Now let a be an N-frame in $\mathfrak{D}$, and let $A_0, A_1 \in \mathfrak{k}_a$ such that $A_0e = A_1e$. We define a pre-$\mathfrak{D}$-functor $\chi^{(A_1)} : \mathfrak{k} \to \mathfrak{D}[\mathbf{N}]$ by sending A_1 to $1 \in \mathfrak{D}[\mathbf{N}]_a$ and all other elements $A \in \mathfrak{k}_N$ to 0. By universality of e, there is an $\mathfrak{D}$-functor $X^{(A_1)} : \mathfrak{K} \to \mathfrak{D}[\mathbf{N}]$ such that $eX^{(A_1)} = \chi^{(A_1)}$. Now $A_0\chi^{(A_1)} = A_0eX^{(A_1)} = A_1eX^{(A_1)} = A_1\chi^{(A_1)} = 1$, whence $A_0 = A_1$. □

We say that a subset of $\mathfrak{K}_N$ *freely generates* $\mathfrak{K}$ *over* $\mathfrak{D}$ if $\mathfrak{K}$ together with the pre-$\mathfrak{D}$-functor arising from the inclusion is universal over $\mathfrak{D}$. We can now submit this section to its main purpose.

Definition. Let $\mathfrak{K}$ be an ∞-category, and let $\mathfrak{k}$ be a subset of $\mathfrak{K}$. We say that $\mathfrak{k}$ *freely generates* $\mathfrak{K}$ if for each $n \geq 0$ the subset $\mathfrak{k}_n$ of $\mathfrak{K}_n$ freely generates $\mathfrak{K}_{\leq n}$ over $\mathfrak{K}_{<n}$.

Section 4. Factoriality and Exact Size Functions

We prove an important lemma, called Proposition 8, with the help of which we can recognize certain ∞-categories as being freely generated by a semiglobular subset.

Throughout this section, we fix an ∞-category $\mathfrak{K}$. For convenience we put $A\delta_\eta^0 = A$. (Note, however, that the generalized globular law $A\delta_\eta^u\delta_\zeta^v = A\delta_\zeta^{u+v}$ cannot be extended to the case $v = 0$.)

We call $\mathfrak{K}$ *(positively) factorial* if for each of its cells C the following two conditions are satisfied. Let us write $C\delta_+ = C'$.

(i) If $C' = C'\delta\sigma$, then $C = C\delta^2\sigma^2$.

(ii) Whenever $C' = A' \cdot^v B'$, there are unique cells A and B such that $A\delta_+ = A'$, $B\delta_+ = B'$ and $A \cdot^{v+1} B = C$.

By induction on u we can infer, respectively, the following more general conditions. Here we write $C\delta_+^u = C'$.

(i)* If $C' = C'\delta\sigma$, then $C = C\delta^{u+1}\sigma^{u+1}$.

(ii)* Whenever $C' = A' \cdot^v B'$, there are unique cells A and B such that $A\delta_+^u = A'$, $B\delta_+^u = B'$ and $A \cdot^{u+v} B = C$.

A *size function* on $\mathfrak{K}$ assigns to each cell A a natural number $|A|$, called the *size* of A, such that the following two conditions are satisfied.

(iii) For each cell A', we have $|A'\sigma| = 0$.

(iv) Whenever $A \circ^u B$, we have $|A \cdot^u B| = |A| + |B|$.

Thus a size function provides for each natural number n an n-functor $\mathfrak{K}_{\leq n} \to \mathfrak{K}_{<n}[\mathbf{N}]$ (see the proof of Proposition 6) over $\mathfrak{K}_{<n}$.

In presence of a size function, we may proof facts on n-cells A by $(n+1)$-fold induction: first on $|A|$, then on $|A\delta_+|$, and so on, and finally on $|A\delta_+^n|$. We refer to this proof method as *δ_+-induction on A*. To be more systematic, we may express this method as induction on the set $\mathfrak{K}_n$ with respect to the (strict) well-founding order $\prec$ according to which

$$A \prec B \quad \text{if and only if} \quad \langle |A|, |A\delta_+|, \ldots, |A\delta_+^n| \rangle <_{\text{lex}} \langle |B|, |B\delta_+|, \ldots, |B\delta_+^n| \rangle,$$

where $<_{\text{lex}}$ stands for (strict) lexicographical order. Note that if $C = A \cdot^u B$, then existence of a natural number $i < u$ with $|A\delta_+^i| < |C\delta_+^i|$ (or, equivalently, $|B\delta_+^i| > 0$) implies $A \prec C$, and similarly existence of a natural number $j < u$ with $|B\delta_+^j| < |C\delta_+^j|$ (or, equivalently, $|A\delta_+^j| > 0$) implies $B \prec C$.

We keep considering a size function on $\mathfrak{K}$. We call an n-cell A of $\mathfrak{K}$ *(positively) simple* if $|A|$, $|A\delta_+|$, ... , $|A\delta_+^n| = 1$. Note that the simple cells form a semiglobular subset of $\mathfrak{K}$. The size function will be called *(positively) exact* if for every cell A such that $A\delta_+$ is simple, the following two conditions are satisfied.

(v) If $|A| = 0$, then $A = A\delta\sigma$.

(vi) If $|A| > 1$, then there are unique cells A_+ and A_0 such that A_0 is simple and $A = A_+ \cdot A_0$.

(These statements imply that every 0-cell is simple.) If $\mathfrak{K}$ is also factorial we can infer, respectively, that for every $u > 0$ and for every cell A such that $A\delta_+^u$ is simple, the following more general conditions are satisfied.

(v)* If $|A\delta_+^{u-1}| = 0$, then $A = A\delta^u\sigma^u$.

(vi)* If $|A\delta_+^{u-1}| > 1$, then there are unique cells A_+ and A_0 such that $A_0\delta_+^{u-1}$ is simple and $A = A_+ \cdot^u A_0$.

We refer to the cells A_+ and A_0 of condition (vi)* as forming the *principal decomposition* of A. Note that A_+, $A_0 \prec A$. Note also that the codimension u that gives rise to the principal decomposition does indeed depend only on A. In general, the minimal number u for which $A\delta_+^u$ is simple will be called the *complexity* of A. Thus a cell is simple if and only if its complexity is 0.

In order to deal conveniently with degeneracies in the main lemma, we prove the following preliminary lemma. In the application we have in mind, its conclusion is evidently satisfied.

Lemma 7. *Let $\mathfrak{K}$ be a factorial strict ∞-category with an exact size function, and let A be a cell in $\mathfrak{K}$. If $|A| = 0$, then $A = A\delta\sigma$.*

As a consequence, we have $A = A\delta^u\sigma^u$ if and only if $|A\delta_+^{u-1}| = 0$. No simplicity assumption is needed.

Proof. Suppose $|A| = 0$; we want to show that $A = A'\sigma$ for some cell A'. We do so by δ_+-induction on A. The complexity k of A is clearly > 0. If $|A\delta_+^{k-1}| = 0$, then $A = A\delta^k\sigma^k$, in particular $A = A\delta\sigma$. So let us suppose that $|A\delta_+^{k-1}| > 1$, whence $k > 1$.

Let $A = A_+ \cdot^k A_0$ be the principal decomposition. Then $|A_+| + |A_0| = |A| = 0$ and hence $|A_+|$, $|A_0| = 0$. By induction hypothesis, we have $A_+ = A'_+\sigma$ and $A_0 = A'_0\sigma$, say. From $A_+ \circ^k A_0$ we infer $A'_+ \circ^{k-1} A'_0$ and then put $A' = A'_+ \cdot^{k-1} A'_0$. Now $A = A'_+\sigma \cdot^k A'_0\sigma = A'\sigma$. □

If A is simple, we can put $A_+ = A\delta_-\sigma$ and $A_0 = A$, thus obtaining a decomposition $A_+ \cdot A_0$ of A that looks very much like the one in condition (vi). In fact, — supposing exactness and factoriality — Lemma 7 implies that it is the only decomposition $A_+ \cdot A_0$ of A with A_0 being simple. This observation yields a slightly generalized version of property (vi)*: whenever $A\delta_+^u$ is simple and $|A\delta_+^{u-1}| \geq 1$, there are unique cells A_+ and A_0 such that $A_0\delta_+^{u-1}$ is simple and $A = A_+ \cdot^u A_0$.

We have arrived at the main lemma itself.

Proposition 8. *A factorial strict ∞-category with an exact size function is freely generated by the simple cells.*

Proof. Fix $n \geq 0$. We write $\mathfrak{k}_n$ for the set of simple n-cells of $\mathfrak{K}$, which gives rise to a pre-n-category $\mathfrak{k}$ over $\mathfrak{K}_{<n}$. Let φ be a pre-n-functor of $\mathfrak{k}$ into an n-category $\mathfrak{L}$. We show that there is a unique n-functor Φ extending φ to all of $\mathfrak{K}_{\leq n}$. (We directly show the stronger of the two universal properties because it makes the algebraic manipulations clearer — though slightly longer — without causing actual extra work.)

First we put $\Phi_{<n} = \varphi_{<n}$. Then we consider an n-cell A of $\mathfrak{K}$. By δ_+-induction on A we define an n-cell $A\Phi$ of $\mathfrak{L}$ with $A\Phi\delta_\eta = A\delta_\eta\Phi$ as follows. If A is simple, we put $A\Phi = A\varphi$. Otherwise A has complexity $u > 0$. If $A = A\delta^u\sigma^u = A\delta\sigma$, we put $A\Phi = A\delta^u\Phi\sigma^u = A\delta\Phi\sigma$. If A has a principal decomposition $A_+ \cdot^u A_0$, we use the induction hypothesis in order to put $A\Phi = A_+\Phi \cdot^u A_0\Phi$. The reader will find no difficulty in verifying the condition on the boundary of $A\Phi$.

We have thus defined a pre-n-functor $\Phi : \mathfrak{K} \twoheadrightarrow \mathfrak{L}$. Moreover every n-functor $\mathfrak{K} \twoheadrightarrow \mathfrak{L}$ extending $\varphi : \mathfrak{k} \twoheadrightarrow \mathfrak{L}$ clearly has to agree with Φ, showing that the uniqueness part of the universal property is satisfied. All that is left to be verified is that Φ is indeed an n-functor.

Let $C = C'\sigma \in \mathfrak{K}_n$; we want to show that $C\Phi = C'\Phi\sigma$. We do so by δ_+-induction on C. The complexity k of C is clearly > 0. If $C = C\delta^k\sigma^k$, then we obtain directly $C\Phi = C\delta\Phi\sigma = C'\sigma\delta\Phi\sigma = C'\Phi\sigma$. Otherwise $k > 1$, and C has a principal decomposition $C_+ \cdot^k C_0$. Let us put $C'_i = C_i\delta_+$ $(i \in \{0, +\})$. Then $C' = C\delta_+ = C'_+ \cdot^{k-1} C'_0$ and hence $C = C'_+\sigma \cdot^k C'_0\sigma$, where $C'_0\sigma\delta_+^{k-1} = C_0\delta_+^{k-1}$ is simple. Since the principal decomposition is unique, we can conclude that $C_i = C'_i\sigma$ $(i \in \{0, +\})$. (This argument can be simplified slightly by using Lemma 7.) Now we obtain

$$\begin{aligned} C\Phi &= C_+\Phi \cdot^k C_0\Phi && \text{(definition of } C\Phi) \\ &= C'_+\sigma\Phi \cdot^k C'_0\sigma\Phi \\ &= C'_+\Phi\sigma \cdot^k C'_0\Phi\sigma && \text{(induction hypothesis)} \\ &= (C'_+\Phi \cdot^{k-1} C'_0\Phi)\sigma \\ &= (C'_+ \cdot^{k-1} C'_0)\Phi\sigma \\ &= C'\Phi\sigma. \end{aligned}$$

Let $C = A \cdot^u B \in \mathfrak{K}_n$; we want to show that $C\Phi = A\Phi \cdot^u B\Phi$. We do so by δ_+-induction on C. The case that either $A = A\delta^u\sigma^u = B\delta_-^u\sigma^u$ or $B = B\delta^u\sigma^u = A\delta_+^u\sigma^u$ is easy and left to the reader. We may henceforth suppose that neither A nor B is a u-identity. Lemma 7 allows us to conclude that $|A\delta^{u-1}|, |B\delta^{u-1}| \geq 1$, whence we further have $|C\delta^{u-1}| \geq 2$. The complexity k of C is consequently $\geq u$. If we had $C = C\delta^k\sigma^k$, then $C\delta_+^{u-1} = C\delta^u\sigma$, and hence $|C\delta^{u-1}| = 0$, contradicting a previous statement. Thus C has a principal decomposition $C_+ \cdot^k C_0$.

Note that $B\delta_+^k = C\delta_+^k$ is simple. If we had $|B\delta_+^{k-1}| = 0$, then $B = B\delta^k\sigma^k$ by Lemma 7, and in particular $B = B\delta^u\sigma^u$, contradicting our assumption. We therefore have $|B\delta_+^{k-1}| \geq 1$, whence there is some decomposition $B_+ \cdot^k B_0$ of B with $B_0\delta^{k-1}$ being simple.

First consider the case $k = u$. Here $C = A \cdot^u B_+ \cdot^u B_0$, and since $B_0\delta^{u-1}$ is simple, uniqueness of the principal decomposition yields $C_0 = B_0$ and $C_+ = A \cdot^u B_+$. We conclude that

$$\begin{aligned} C\Phi &= C_+\Phi \cdot^u C_0\Phi && \text{(definition of } C\Phi) \\ &= (A\Phi \cdot^u B_+\Phi) \cdot^u C_0\Phi && \text{(induction hypothesis for } C_+) \\ &= A\Phi \cdot^u (B_+\Phi \cdot^u B_0\Phi) \\ &= A\Phi \cdot^u B\Phi && \text{(induction hypothesis for } B). \end{aligned}$$

Now consider the case $k > u$. Note that $A\delta_+^u = B\delta_-^u = B_+\delta_-^u \cdot^{k-u} B_0\delta_-^u$, whence by factoriality we have $A = A_+ \cdot^k A_0$ with $A_i\delta_+^u = B_i\delta_-^u$ $(i \in \{0, +\})$. It follows that $C = (A_+ \cdot^k A_0) \cdot^u (B_+ \cdot^k B_0) = (A_+ \cdot^u B_+) \cdot^k (A_0 \cdot^u B_0)$. Since $(A_0 \cdot^u B_0)\delta_+^{k-1} = B_0\delta_+^{k-1}$ is simple, uniqueness of the principal decomposition yields $C_0 = A_0 \cdot^u B_0$ and $C_+ = A_+ \cdot^u B_+$. We conclude that

$$\begin{aligned} C\Phi &= C_+\Phi \cdot^k C_0\Phi && \text{(definition of } C\Phi) \\ &= (A_+\Phi \cdot^u B_+\Phi) \cdot^k (A_0\Phi \cdot^u B_0\Phi) && \text{(induction hypothesis for } C_+ \text{ and } C_0) \\ &= (A_+\Phi \cdot^k A_0\Phi) \cdot^u (B_+\Phi \cdot^k B_0\Phi) \\ &= A\Phi \cdot^u B\Phi && \text{(induction hypothesis for } A \text{ and } B). \quad \square \end{aligned}$$

Example. The ∞-category $\hat{\mathbf{N}}$ of higher-dimensional numbers is factorial. It can be equipped with an exact size function, namely by putting

$$|0_0| = 1; \qquad |\langle \mathfrak{q}_0, \ldots, \mathfrak{q}_{\ell-1} \rangle_n| = |\mathfrak{q}_0| + \cdots + |\mathfrak{q}_{\ell-1}| \quad (n > 0).$$

There is precisely one simple cell $\mathfrak{e}_n$ in each dimension n. We can define $\mathfrak{e}_n$ inductively by putting

$$\mathfrak{e}_0 = 0_0; \qquad \mathfrak{e}_n = \langle \mathfrak{e}_{n-1} \rangle_n \quad (n > 0).$$

Since δ_+ and δ_- act in the same way, factoriality and exactness hold not only positively, but also negatively (in the obvious sense), and the simple cells are the same in both instances.

Section 5. The Makkai Equivalence

We show that dendrotopic sets are (essentially) precisely the freely generating semiglobular subsets of strict ∞-categories. Thus, according to [3], the concept of a dendrotopic set introduced here is equivalent to the Makkai concept of a multitopic set.

We are going to show that the ∞-category $\mathfrak{D}^*$ generated by a dendrotopic set $\mathfrak{D}$ meets the assumptions of Proposition 8, with the cells of $\mathfrak{D}$ playing the role of the simple cells.

We first show that $\mathfrak{D}^*$ is factorial. To this end, let C be a cell of $\mathfrak{D}^*$, and let $C\delta_+ = C'$. In terms of frameworks of $\mathfrak{D}$, this means that $C^\natural \partial_+ = C'^\natural$.

(i) Suppose that $C' = C'\delta\sigma$; we want to show that $C = C'\sigma$. Since $C'^\natural = C'\delta\sigma^\natural$ has no complete facet, there is precisely one special framework with ∂_+-facet $C'^\natural$. Since $C^\natural \partial_+ = C'^\natural$ as well as $C'\sigma^\natural \partial_+ = C'^\natural$, we can conclude that $C^\natural = C'\sigma^\natural$.

(ii) Let $C' = A' \cdot^v B'$; we want to show that there uniquely exist two cells A and B such that $A\delta_+ = A'$, $B\delta_+ = B'$ and $C = A \cdot^{v+1} B$. Since $C'^\natural = (A' \cdot^v B')^\natural$, the complete facets of $C'^\natural$ are the complete facets of $A'^\natural$ and $B'^\natural$. Assuming $A^\natural \partial_+ = A'^\natural$ and $B^\natural \partial_+ = B'^\natural$, the frameworks $A^\natural$ and $B^\natural$ are determined by the cell trees $A^\natural \alpha_{(q)}$ $(q \in A'^\natural \mathrm{I}_{\mathfrak{D}})$ and $B^\natural \alpha_{(q)}$ $(q \in B'^\natural \mathrm{I}_{\mathfrak{D}})$, and assuming further $C^\natural = (A \cdot^{v+1} B)^\natural$, these cell trees are given by $A^\natural \alpha_{(q)} = C^\natural \alpha_{(q)}$ and $B^\natural \alpha_{(q)} = C^\natural \alpha_{(q)}$, respectively. Along the same lines we find a construction of $A^\natural$ and $B^\natural$.

Now we want to put an exact size function on $\mathfrak{D}^*$. To this end, we let $|A|$ be the number of complete facets of $A^\natural$. The homomorphy conditions (iii) and (iv) are clearly satisfied. Before we proceed to show exactness, we want to understand the structure of the critical cells.

Let A be a simple n-cell of $\mathfrak{D}^*$. The corresponding roof $A^\natural$ has precisely one complete facet, whose location we shall denote by ι. If $n = 0$, then $A^\natural : \iota = \delta_-$. Suppose now that $n > 0$. Then $A^\natural : \frac{\iota \ \omega}{\partial_+ \ *}$ is complete, and since $A\delta_+$ itself is simple, the yet unknown complete facet location is the respective ι:

$$A^\natural : \frac{\iota \quad \omega}{\partial_+ \quad \iota}. \tag{3}$$

(Hence, the complete facet location of the roof corresponding to a simple n-cell is $\mathfrak{e}_n$; see the Example in the previous section.) As a consequence we have the identity

$$A\delta_+{}^\natural \iota = A^\natural \iota\omega, \tag{4}$$

which states that the assignment $A \mapsto A^\natural \iota$ preserves positive facets. The complete ridge locations of $A^\natural$ apart from (3) have to be of the form $\frac{\partial_- \;*}{\iota \;\; *}$. Thus $A^\natural \Diamond$ induces a one-to-one correspondence between the complete facet locations of $A^\natural \partial_- = A\delta_-{}^\natural$ on the one hand and the negative facet locations of $A^\natural \iota$ on the other hand. We express this correspondence by an identification according to the rule

$$A^\natural : \frac{\partial_- \; q}{\iota \;\; q}.$$

If the dimension of A is 1, we thus identify the two facet locations already both known as δ_-. Suppose now that the dimension of A is > 1. By removing the type-2 panels with first entry ∂_- or ι, each complete 3-face location of $A^\natural$ can be decomposed into an "upper" gallery

$$\begin{array}{ccc} \partial_- & \partial_- & r \\ \hline \partial_+ & \partial_- & r \\ \hline \partial_+ & \iota & r \\ \hline \iota & \omega & r \end{array} \quad \bigl(r \in A\delta_+{}^\natural \partial_- \mathrm{I}_{\mathfrak{D}} = A\delta_+{}^\natural \iota \mathrm{I}_-\bigr) \qquad \text{or}$$

$$\frac{\partial_- \; q \; \omega}{\iota \;\; q \; \omega} \qquad \bigl(q \in A^\natural \partial_- \mathrm{I}_{\mathfrak{D}} = A^\natural \iota \mathrm{I}_-\bigr)$$

and a "lower" gallery

$$\begin{array}{ccc} \iota & \omega & \omega \\ \hline \partial_+ & \iota & \omega \\ \hline \partial_+ & \partial_+ & \iota \\ \hline \partial_- & \partial_+ & \iota \end{array} \qquad \text{or} \qquad \frac{\iota \;\; q \; r}{\partial_- \; q \; r} \qquad \bigl(q \in A^\natural \partial_- \mathrm{I}_{\mathfrak{D}} = A^\natural \iota \mathrm{I}_-;\; r \neq \omega\bigr).$$

Thus $A^\natural \Gamma^3$ induces a one-to-one correspondence between the complete ridge locations of $A^\natural \partial_- = A\delta_-{}^\natural$ on the one hand and the ridge locations of $A^\natural \iota$ on the other hand. The best way to put this result is the following. First note that since A is simple, so is $A\delta_+^2 = A\delta_-\delta_+$. Therefore the complete part of the boundary of $A\delta_-{}^\natural$ makes up precisely the $(n-1)$-cell tree $A\delta_-{}^\natural \alpha_{(\iota)}$. What we have shown here is that this cell tree is precisely the negative half boundary of $A^\natural \iota$:

$$A\delta_-{}^\natural \alpha_{(\iota)} = A^\natural \iota \alpha. \tag{5}$$

The assignment $A \mapsto A^\natural \iota$ is a bijection of the set of simple n-cells of $\mathfrak{D}^*$ onto the set of n-cells of $\mathfrak{D}$. We prove this statement by induction on n. Let P be an n-cell in $\mathfrak{D}$; we want to show that there is a unique simple n-cell A in $\mathfrak{D}^*$ such that $A^\natural \iota = P$. Our reasoning for uniqueness will also produce a construction. If $n = 0$, then the cell A is determined by $A^\natural \iota = A^\natural \delta_-$, which is required to be P. Now suppose that $n > 0$. Then the cell A is determined by $A^\natural \iota$ and $A\delta_+$. The former is required to be P; the latter has to be a simple $(n-1)$-cell A' such that $A'^\natural \iota = A^\natural \partial_+ \iota = P\omega$. By induction hypothesis, there is only one such cell A'.

We now want to understand the structure of those cells of $\mathfrak{D}^*$ whose positive facets are simple. Let A be such a cell. Then $A\delta_+{}^\natural$ has precisely one complete facet location ι. Hence A determines and is determined by the cell tree $A^\natural \alpha_{(\iota)}$, which we shall abbreviate by $A^\natural \iota$ as well. If A itself is simple, then the cell $A^\natural \iota$ defined earlier can be viewed as a one-facet cell tree, which is precisely the cell tree $A^\natural \iota$ defined here. For all $n \geq 0$, the assignment $A \mapsto A^\natural \iota$ is evidently a bijection of the set of n-cells of $\mathfrak{D}^*$ with complexity ≤ 1 onto the set of n-cells trees of $\mathfrak{D}$.

Using the new notation, we may rewrite the identity (5) as follows:

$$A\delta_-{}^{\natural}\,\iota = A^{\natural}\,\iota\alpha. \tag{6}$$

This result also holds if the dimension of A is 1, since then $A^{\natural} : \frac{\partial_-\ \ \delta_-}{\iota\ \ \delta_-}$. We now want to generalize the identities (4) and (6) by proving them for all cells A of dimension ≥ 1 and complexity ≤ 1 (not just the simple ones). The anticipated result can also be phrased as follows.

Proposition 9. *The assignment $A \longmapsto A^{\natural}\,\iota$ is an isomorphism of the globular subset of $\mathfrak{D}^*$ constituted by cells of complexity ≤ 1 onto the globular set $\mathfrak{D}^{1*}$ of cell trees of $\mathfrak{D}$.*

Proof. We have to show that in the case in question, the two equations (4) and (6) are satisfied. As for the first, we find on both sides the root of the cell tree $A^{\natural}\,\alpha_{(\iota)}$. As for the second, we take recourse to a trick, to avoid the tiresome fiddling with 3-face locations.

The problem remains unchanged if new cells are added to $\mathfrak{D}$. We may hence work with the $(n+1)$-outniche X corresponding to the n-cell tree $A^{\natural}\,\iota$ as if it were an actual cell of $\mathfrak{D}$. Thus, without loss of generality we pick an $(n+1)$-cell X with $X\alpha = A^{\natural}\iota$. There is a corresponding simple cell F in $\mathfrak{D}^*$. The following calculations will make use of the results (4) and (6) for the simple cells F and $F\delta_+$. First note that since

$$F\delta_-{}^{\natural}\,\iota = F^{\natural}\,\iota\alpha = X\alpha = A^{\natural}\,\iota,$$

we have $F\delta_- = A$. We conclude that

$$A\delta_-{}^{\natural}\,\iota = F\delta_-\delta_-{}^{\natural}\,\iota = F\delta_+\delta_-{}^{\natural}\,\iota = F\delta_+{}^{\natural}\,\iota\alpha = F^{\natural}\,\iota\omega\alpha = F^{\natural}\,\iota\alpha\alpha = F\delta_-{}^{\natural}\,\iota\alpha = A^{\natural}\,\iota\alpha.$$

□

We henceforth view $\mathfrak{D}^{1*}$ as a globular subset of $\mathfrak{D}^*$ as suggested by the proposition. Given the earlier identification of dendrotopic cells with one-facet cell trees, we view in particular $\mathfrak{D}$ as a semiglobular subset of $\mathfrak{D}^*$.

We now show that the given size function is exact. To this end, let A be a cell of $\mathfrak{D}^*$ for which $A\delta_+$ is simple.

(v) Suppose that $|A| = 0$; we want to show that $A = A\delta\sigma$. The result follows from the fact that both $A^{\natural}$ and $A\delta_+\sigma^{\natural}$ have the ∂_+-facet $A\delta_+{}^{\natural}$ and no complete facet.

(vi) Suppose that $|A| > 1$; we want to show that there uniquely exist two cells A_+ and A_0 such that A_0 is simple and $A = A_+ \cdot A_0$. We show uniqueness by assuming the conditions and then expressing A_0 and A_+ in terms of A. The complete facet location ι of $A_0{}^{\natural}$ corresponds to a facet location ι' of $A^{\natural}$ determined by the requirement that

$$A^{\natural} : \frac{\iota'\ \ \ \omega}{\partial_+\ \ \iota}. \tag{7}$$

In turn A_0 is determined by $A_0{}^{\natural}\,\iota = A^{\natural}\,\iota'$. Now $A_+{}^{\natural}\,\partial_+ = A_0{}^{\natural}\,\partial_-$, and the complete facets of $A_+{}^{\natural}$ are those of $A^{\natural}$ except the one with location ι'. The complete ridge locations of $A_+{}^{\natural}$ are the ones of $A^{\natural}$ except (7) and except for the replacement of each $\frac{p_-\ \ q_-}{\iota'\ \ q}$ with $\frac{p_-\ \ q_-}{\partial_+\ \ q}$ (corresponding to a ridge location $\frac{\partial_-\ \ q}{\iota\ \ q}$ of $A_0{}^{\natural}$). Thus A_+ is also determined. To show existence, we first note that since $|A| > 1$, there is in particular some complete facet location ι' of $A^\natural$ such that (7). Now we construct A_0 and A_+ along the lines of the uniqueness proof and convince ourselves that $A_+{}^\natural$ has the tree property.

We have arrived at the main theorem of this work.

Theorem 10. *Let $\mathfrak{K}$ be a strict ∞-category, and let $\mathfrak{k}$ be a subset of $\mathfrak{K}$. The following three conditions are equivalent.*

(i) *$\mathfrak{k}$ is semiglobular and freely generates $\mathfrak{K}$.*

(ii) *$\mathfrak{K}$ is factorial and carries an exact size function with respect to which $\mathfrak{k}$ is the set of simple cells.*

(iii) *$(\mathfrak{K};\mathfrak{k})$ is isomorphic to $(\mathfrak{D}^*;\mathfrak{D})$ for a dendrotopic set $\mathfrak{D}$.*

As for condition (iii), the notion of an isomorphism is the obvious one: an invertible ∞-functor for the ∞-categories inducing a bijective mapping for the distinguished subsets.

Proof. The implication of (i) by (ii) is Proposition 8. The implication of (ii) by (iii) has just been demonstrated. We hence know that every dendrotopic set generates a free ∞-category. It remains to show the implication of (iii) by (i).

So suppose that $\mathfrak{k}$ is semiglobular and freely generates $\mathfrak{K}$. We want to construct a dendrotopic set $\mathfrak{D}$ and an isomorphism $\Phi : (\mathfrak{K};\mathfrak{k}) \twoheadrightarrow (\mathfrak{D}^*;\mathfrak{D})$. We do so by inductively constructing the skeleta $\mathfrak{D}_{\leq n}$ and $\Phi_{\leq n}$ of $\mathfrak{D}$ and Φ. So let us suppose that we have an isomorphism $\Phi_{<n} : (\mathfrak{K}_{<n};\mathfrak{k}_{<n}) \twoheadrightarrow (\mathfrak{D}^*{}_{<n};\mathfrak{D}_{<n})$ of $(n-1)$-categories with distinguished subsets. We let $\mathfrak{D}_n$ contain one copy $A\varphi$ of each $A \in \mathfrak{k}_n$. We now have to specify a boundary for each $A\varphi$ so that the assignment $A \mapsto A\varphi$ specifies a pre-n-functor φ over $\Phi_{<n}$. Then the desired isomorphism $\Phi_{\leq n}$ of N-categories arises due to the universal properties of the respective inclusions $\mathfrak{k}_n \subseteq \mathfrak{K}_n$ and $\mathfrak{D}_n \subseteq \mathfrak{D}^*{}_n$.

We explicitly treat the case $n \geq 2$; the constructions are even simpler in the cases $n = 1$ and $n = 0$. Since $\mathfrak{k}$ is semiglobular, $A\delta_+$ is in $\mathfrak{k}_{n-1}$, and is hence sent to an $(n-1)$-cell Q_+ in $\mathfrak{D}_{<n}$ by $\Phi_{<n}$. Similarly, $A\delta_-\delta_+ = A\delta_+\delta_+ \in \mathfrak{k}_{n-2}$, hence $A\delta_-\delta_+\Phi \in \mathfrak{D}_{n-2}$, and we conclude that $A\delta_-$ is sent to an $(n-1)$-cell tree Q_- in $\mathfrak{D}_{<n}$. Preservation of the globular operations in dimension $n-1$ ensures that $Q_+^\bullet = Q_-^\bullet$, so that we can form an n-frame P of $\mathfrak{D}_{<n}$ by putting $P\omega = Q_+$ and $P\alpha = Q_-$. We take this frame to be the boundary of $A\varphi$. Then φ clearly preserves the globular operations in dimension n. $\square$

The theorem can be extended to assert the equivalence of three (ordinary) categories:

(i) the category of ∞-categories with freely generating semiglobular subsets,

(ii) the category of factorial ∞-categories with exact size functions,

(iii) the category of dendrotopic sets.

References

[1] J. Baez: "An introduction to n-categories", *7th Conference on Category Theory and Computer Science* (Lecture Notes in Computer Science **1290**), Springer (1997)

[2] M. Batanin: "Monoidal Globular Categories as a Natural Environment for the Theory of Weak n-Categories", *Advances in Mathematics* **136**, 39–103 (1997)

[3] V. Harnik, M. Makkai, M. Zawadowski: "Multitopic sets are the same as many-to-one computads", `www.math.mcgill.ca/makkai` (2002)

[4] C. Hermida, M. Makkai, A. Power: "On weak higher dimensional categories", *Journal of Pure and Applied Algebra.* Part 1: **153**, 221–246 (2000); Part 2: **157**, 247–277 (2001); Part 3: **166**, 83–104 (2002)

[5] M. Makkai: "The multitopic ω-category of all multitopic ω-categories", www.math.mcgill.ca/makkai (1999)
[6] P. McMullen, E. Schulte: *Abstract Regular Polytopes*, Cambridge University Press (2003)
[7] T. Palm: *Dendrotopic Sets for Weak ∞-Categories*, Ph. D. Thesis, York University (2003)
[8] R. Street: "The algebra of oriented simplexes", *Journal of Pure and Applied Algebra* **49**, 283–335 (1987)
[9] R. Street: "Parity complexes", *Cahiers de topologie et géométrie différentielle catégoriques* **32**, 315–343 (1991). Corrigenda: **35**, 359–361 (1994)

Received December 16, 2002; in revised form May 20, 2003

Fields Institute Communications
Volume **43**, 2004

On Factorization Systems and Admissible Galois Structures

Ana Helena Roque
Departamento de Matemática
Universidade de Aveiro
3810-193 Aveiro, Portugal
a.h.roque@mat.ua.pt

Abstract. Full reflective subcategories of congruence-modular regular categories which are closed under subobjects and quotients are admissible in the sense of G. Janelidze's categorical Galois theory, with respect to regular epimorphisms.

1 Introduction

As defined in [3], a *Galois structure* Γ in a category $\mathbb{C}$ with finite limits, is an adjunction $(I, H, \eta, \varepsilon) : \mathbb{C} \to \mathbb{X}$ together with two classes of morphisms (called fibrations) F and Φ in $\mathbb{C}$ and $\mathbb{X}$, respectively, such that

1. F and Φ are pullback stable and pullbacks along morphisms of F or Φ exist,
2. F and Φ are closed under composition and contain all isomorphisms,
3. $I(F) \subseteq \Phi$ and $H(\Phi) \subseteq F$.

Γ is *admissible* if for every object C in $\mathbb{C}$ and every fibration $\phi : X \to I(C)$ in $\mathbb{X}$, the composite of canonical morphisms $I(C \times_{HI(C)} H(X)) \to IH(X) \to X$ is an isomorphism.

When the adjunction is a full reflection, admissibility is equivalent to preservation by I of pullbacks of the form

$$\begin{array}{ccc} C & \xrightarrow{t} & X \\ {\scriptstyle s}\downarrow & & \downarrow{\scriptstyle \varphi \in \Phi} \\ B & \xrightarrow[\eta_B]{} & I(B) \end{array} \tag{1.1}$$

with X in $\mathbb{X}$, which coincides with semi-left exactness with respect to the class Φ.

In [4] it is shown that full reflective subcategories of congruence-modular exact categories closed under subobjects and quotients, are admissible with respect to regular epimorphisms. Here, we show that the same result holds when the

2000 *Mathematics Subject Classification.* Primary 18A32, 18A40; Secondary 08C99, 03C99, 12F10.

This work is a part of author's Ph.D. thesis under supervision of Prof. G. Janelidze.

exactness assumption is relaxed to regularity. In fact, we show that modularity of $\mathcal{E}-Quot(A)$ for each A in $\mathbb{C}$, implies admissibility of the Galois structure $(I, H, \eta, \varepsilon) : \mathbb{C} \to \mathbb{X}$, $(\mathcal{E}, \mathcal{E} \cap mor\mathbb{X})$ when $\mathbb{X}$ is a full reflective subcategory of $\mathbb{C}$, closed under $\mathcal{M}$-subobjects and $\mathcal{E}$-quotients, $(\mathcal{E}, \mathcal{M})$ is a stable factorization system and $\mathcal{E}$ is contained in the class of epimorphisms.

We will use the following terminology and notation, for any factorization system $(\mathcal{E}, \mathcal{M})$ in a category $\mathbb{C}$ with finite limits:

- an $\mathcal{M}$-subobject of an object A in $\mathbb{C}$ is an isomorphism class of morphisms in $\mathcal{M}$ with codomain A;
- an $\mathcal{E}$-quotient of an object A in $\mathbb{C}$ is an isomorphism class of morphisms in $\mathcal{E}$ with domain A, the set of which we denote by $\mathcal{E}-Quot(A)$.

$\mathcal{E} - Quot(A)$ is preordered by $[s] \le [r]$ if $s = t \circ r$ for some morphism t, with greatest element 1_A; suprema of pairs exist and the supremum of $\{[p], [q]\}$ (denoted $[p] \vee [q]$) is the $\mathcal{E}$ factor of the $(\mathcal{E}, \mathcal{M})$ factorization of (p, q); the infimum of $\{[p], [q]\}$ (denoted $[p] \wedge [q]$) exists whenever a pushout of p and q exists and these conditions are equivalent if $\mathcal{E}$ is contained in the class of epimorphisms.

The properties of factorization systems that we use can be found in [2].

2 Full reflective subcategories closed under $\mathcal{M}$-subobjects and $\mathcal{E}$-quotients, for a factorization system $(\mathcal{E}, \mathcal{M})$

Let $\mathbb{C}$ be a category with finite limits, and $(\mathcal{E},\mathcal{M})$ any factorization system in $\mathbb{C}$.

Let $\mathbb{X}$ be a full replete reflective subcategory of $\mathbb{C}$, and let H denote the inclusion functor and I the reflector. Since $\mathbb{X}$ is full and $\mathbb{C}$ has finite limits, $\mathbb{X}$ has finite limits, and these are exactly the same as in $\mathbb{C}$.

It is well known (see e.g. [1], Theorem 16.8) that

Proposition 2.1 *$\mathbb{X}$ is closed under $\mathcal{M}$-subobjects iff for each C in $\mathbb{C}$, $\eta_C \in \mathcal{E}$.*

Corollary 2.2 *If $\mathbb{X}$ is closed under $\mathcal{M}$-subobjects, for any f in $\mathcal{E}$, $If \in \mathcal{E}$.*

Proof Let $f : A \to B$ be in $\mathcal{E}$. From Proposition 2.1 and, as $\mathcal{E}$ is closed under composition, $\eta_A \in \mathcal{E}$ and $If \circ \eta_A = \eta_B \circ f \in \mathcal{E}$. Then, If is in $\mathcal{E}$, since $(\mathcal{E}, \mathcal{M})$ is a factorization system. □

Lemma 2.3 *If $\mathbb{X}$ is closed under both $\mathcal{M}$-subobjects and $\mathcal{E}$-quotients, for each C in $\mathbb{C}$, η_C is an epimorphism.*

Proof : Let $f, g : I(C) \to B$ be such that $f \circ \eta_C = g \circ \eta_C$ and assume these factorize as $m \circ e$; let f and g factorize as $f = m_1 \circ e_1$ and $g = m_2 \circ e_2$ as in the commutative diagram:

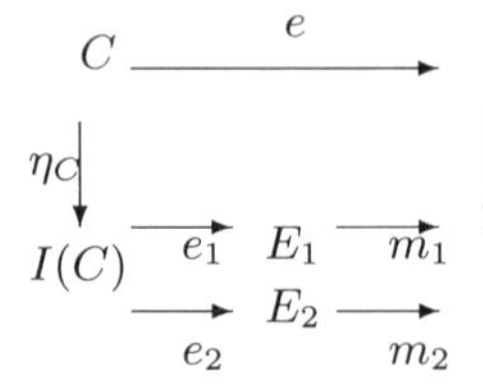

$$\begin{array}{ccccc} C & & \xrightarrow{\quad e \quad} & & A \\ \downarrow{\scriptstyle \eta_C} & & & & \downarrow{\scriptstyle m} \\ I(C) & \xrightarrow{e_1} & E_1 & \xrightarrow{m_1} & B \\ & \xrightarrow[e_2]{} & E_2 & \xrightarrow[m_2]{} & \end{array}$$

Since $\mathbb{X}$ is closed under $\mathcal{M}$-subobjects, we know that η_C is in $\mathcal{E}$. Then there are isomorphisms $i : E_1 \to A$ and $j : E_2 \to A$ such that

$$i \circ e_1 \circ \eta_C = e \quad \text{and} \quad m \circ i = m_1, \qquad j \circ e_2 \circ \eta_C = e \quad \text{and} \quad m \circ j = m_2.$$

As $\mathbb{X}$ is closed under $\mathcal{E}$-quotients, A is in $\mathbb{X}$. Therefore, by the universal property of the adjunction, the unique morphism w satisfying $Hw \circ \eta_C = e$, must be $i \circ e_1 = j \circ e_2$. But then, $f = m_1 \circ e_1 = m \circ i \circ e_1 = m \circ j \circ e_2 = m_2 \circ e_2 = g$, and so, η_C is an epimorphism. $\square$

Proposition 2.4 *If $\mathbb{X}$ is closed under $\mathcal{M}$-subobjects then, $\mathbb{X}$ is closed under $\mathcal{E}$-quotients if and only if for each $f : A \to B$, $f \in \mathcal{E}$, the diagram*

$$\begin{array}{ccc} A & \xrightarrow{\eta_A} & I(A) \\ {\scriptstyle f}\downarrow & & \downarrow{\scriptstyle If} \\ B & \xrightarrow[\eta_B]{} & I(B) \end{array}$$

is a pushout.

Proof If diagrams as in the statement are pushouts, in particular they are so for A in $\mathbb{X}$ and $f \in \mathcal{E}$. Then η_B is an isomorphism, since η_A is, and therefore B is in $\mathbb{X}$.

Conversely, assume that $\mathbb{X}$ is closed under $\mathcal{E}$-quotients. Let $f : A \to B$ be in $\mathcal{E}$, and suppose that $u : I(A) \to D$ and $v : B \to D$ are morphisms such that $u \circ \eta_A = v \circ f$. Consider the factorizations $u = m \circ q$, $v = m' \circ q'$ with $m, m' \in \mathcal{M}$ and $q, q' \in \mathcal{E}$. We have then the commutative diagram:

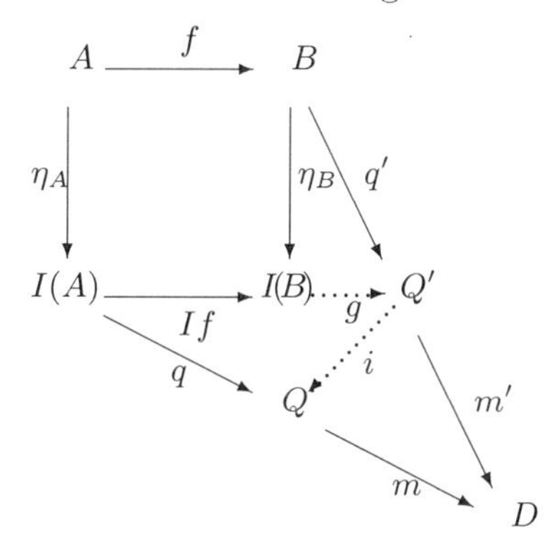

As $\eta_A, \eta_B \in \mathcal{E}$ and $\mathcal{E}$ is closed under composition, $q \circ \eta_A$, $\eta_B \circ f \in \mathcal{E}$. By the uniqueness of the factorization there exists an isomorphism $i : Q' \to Q$ such that $i \circ q' \circ f = q \circ \eta_A$ and $m \circ i = m'$. Then, as $\mathbb{X}$ is closed under $\mathcal{E}$-quotients, Q is in $\mathbb{X}$ and therefore Q' is in $\mathbb{X}$. By the universal property, there is then a morphism $g : I(B) \to Q'$ satisfying $q' = g \circ \eta_B$. Hence, $m' \circ g \circ \eta_B = m' \circ q' = v$, and

$$u \circ \eta_A = v \circ f = m' \circ q' \circ f = m' \circ g \circ \eta_B \circ f = m' \circ g \circ If \circ \eta_A.$$

By the Lemma, as η_A is an epimorphism, $m' \circ g \circ If = u$. To show uniqueness, suppose that $w \circ \eta_B = v$; then $w \circ \eta_B = m' \circ g \circ \eta_B$, and as η_B is an epimorphism, $w = m' \circ g$. $\square$

Corollary 2.5 *Let $\mathbb{C}$ be a category with finite limits, and $\mathbb{X}$ a full, replete and reflective subcategory of $\mathbb{C}$ with the reflection given by $(I, H, \eta, \varepsilon) : \mathbb{C} \to \mathbb{X}$. Let $(\mathcal{E}, \mathcal{M})$ be a stable factorization system in $\mathbb{C}$. Then, $(I, H, \eta, \varepsilon) : \mathbb{C} \to \mathbb{X}$ together with the pair of classes of morphisms $(\mathcal{E}, \mathcal{E} \cap mor\mathbb{X})$ is a Galois structure.*

Theorem 2.6 *Let $(\mathcal{E},\mathcal{M})$ be a stable factorization system in a category $\mathbb{C}$ with finite limits, with $\mathcal{E}$ contained in the class of epimorphisms. Let $\mathbb{X}$ be a full subcategory of $\mathbb{C}$, reflective through $(I,H,\eta,\varepsilon):\mathbb{C}\to\mathbb{X}$, and closed under $\mathcal{M}$-subobjects and $\mathcal{E}$-quotients.*

Then, if for each $C\in\mathbb{C}$ and each $s,t\in\mathcal{E}$, with $t\leq\eta_C$

$$\eta_C\wedge(s\vee t)=(\eta_C\wedge s)\vee t,$$

the Galois structure $(I,H,\eta,\varepsilon):\mathbb{C}\to\mathbb{X}$, $(\mathcal{E},\mathcal{E}\cap mor\mathbb{X})$ is admissible.

Proof Since ε_X is an isomorphism for each X in $\mathbb{X}$, we only need to show that under the stated conditions, for a pullback as in (1.1) It is an isomorphism. Note that as $\mathcal{E}$ is pullback stable and $\eta_B\in\mathcal{E}$, both s and t are in $\mathcal{E}$. As X is in $\mathbb{X}$, by the universal property, there exists $\beta: I(C)\to X$ such that $t=\beta\circ\eta_C$. Then, $\varepsilon_X\circ It=\varepsilon_X\circ IH\beta\circ I\eta_C=\beta\circ\varepsilon_{I(C)}\circ I\eta_C=\beta$ and therefore, It is invertible if and only if so is β; since $\eta_C\in\mathcal{E}$ is an epimorphism, this happens precisely when in $\mathcal{E}-Quot(C)$, the classes $[\eta_C]$ and $[t]$, coincide. In the diagram

$$\begin{array}{ccccccc}
C & \xrightarrow{\eta_C} & I(C) & \xrightarrow{It} & I(X) & \xrightarrow{\varepsilon_X} & X \\
{\scriptstyle s}\downarrow & & \downarrow{\scriptstyle Is} & & \downarrow{\scriptstyle I\varphi} & & \downarrow{\scriptstyle\varphi} \\
B & \xrightarrow[\eta_B]{} & I(B) & \xrightarrow[I\eta_B]{} & I(B) & \xrightarrow[\varepsilon_{IB}]{} & I(B)
\end{array}$$

the square on the left is a pushout by assumption, so that $\eta_C\wedge s$ exists. Since the exterior diagram is a pullback, the morphism $(s,t): C\to B\times_{I(B)}X$, induced by s and t, is in $\mathcal{M}$, and therefore $s\vee t=1_C$. Also,

$$s\wedge\eta_C=Is\circ\eta_C=\varphi\circ\varepsilon_X\circ It\circ\eta_C=\varphi\circ t,$$

and thus $\eta_C\wedge s\leq t$. Hence, $t=(\eta_C\wedge s)\vee t\leq\eta_C\wedge(s\vee t)=\eta_C\wedge 1_C=\eta_C$.

Therefore, β is invertible if and only if $(\eta_C\wedge s)\vee t=\eta_C\wedge(s\vee t)$. □

Corollary 2.7 *If $\mathbb{C}$ is regular and for each object C in $\mathbb{C}$, $Cong(C)$ is a modular lattice, any full and reflective subcategory of $\mathbb{C}$ closed under mono-subobjects and regular epi-quotients is admissible with respect to regular epimorphisms.*

3 Examples

In the category of structures for a first order language regular epis are the strong surjective homomorphisms that is, surjective homomorphisms $h:A\longrightarrow B$, such that, for each natural n, and each n-ary predicate symbol P, $h^n(P^A)\subseteq P^B$; varieties are regular categories (not necessarily exact).

1. A variety of Ω-groups with relations is congruence modular. Any of its full and reflective subcategories, closed under subobjects and strong homomorphic images is, by Corollary 2.7, admissible with respect to the class of strong surjective homomorphisms. Examples of such are axiomatizable subclasses of varieties of Ω-groups with relations, whose axioms, besides those that characterize the variety, are equivalent to either universal atomic sentences or strict universal Horn sentences $\forall x_1...\forall x_n(\theta_1\wedge...\wedge\theta_k\to\delta)$ which satisfy:

i) for each $1\leq i\leq k$, θ_i is of the form $Px_1^i...x_{l_i}^i$ where P is a predicate symbol and the x_j^i's are distinct variables

ii) for each $1\leq i,j\leq k$, $i\neq j$, θ_i and θ_j have disjoint sets of variables,

iii) δ is of the form $t_1\approx t_2$.

2. Let $\mathbb{X}$ be a subvariety of a variety $\mathbb{V}$, in a *relational* language. The pair $(\mathcal{E}, \mathcal{M})$ where $\mathcal{E}$ is the class of surjective homomorphisms and $\mathcal{M}$ the class of substructure monomorphisms, is a factorization system in $\mathbb{V}$ ([5]). $(\mathcal{E}, \mathcal{M})$ together with the reflection $(I, H, \eta, \varepsilon) : \mathbb{V} \to \mathbb{X}$, $\eta_C = id_C$, $\varepsilon_X = 1_X$, is a Galois structure. Assume that $\mathbb{X}$ is axiomatizable by (universal atomic) $T' = T \cup \Sigma$ where T is an axiomatization of $\mathbb{V}$ and Σ consists of sentences with atomic part $Px_1 ... x_n$ where $x_i \neq x_j$ for $i \neq j$, that is to say, the interpretation in a structure A of each predicate symbol P in Σ is A^k where k is the arity of P (*). Then, for $s, t \in \mathcal{E}$, $t : C \to B$ with $t \leq \eta_C$, B is in $\mathbb{X}$ and it follows that $\eta_C \wedge (s \vee t) = (\eta_C \wedge s) \vee t$, so that, by Theorem 2.6, the above Galois structure is admissible. In fact, condition (*) is necessary ([5]).

3. Consider the adjunction $\mathbb{C}at \rightleftarrows Preord$ formed by the inclusion of the category $Preord$ of preorders into the category $\mathbb{C}at$ of small categories, and its left adjoint. We define the fibrations in $\mathbb{C}at$ as all full functors bijective on objects, and accordingly the fibrations in $Preord$ are just the isomorphisms. The admissibility of this Galois structure follows from Theorem 2.6; however, since the isomorphisms are involved, this case is in fact trivial. The same is true if we replace $Preord$ with the category of (partially) ordered sets. Note that those admissibilities also hold for the *(All functors, Isomorphisms)* factorization system, as shown in [6].

References

[1] J. Adámek, H: Herrlich, G. Strecker *Abstract and Concrete Categories*; John Wiley & Sons, Inc. 1990.

[2] A. Carboni, G. Janelidze, G. M. Kelly, R. Paré, *On Localization and Stabilization for Factorization Systems*; Appl. Cat. Struct. **5** (1997), 1-58.

[3] G. Janelidze, *Categorical Galois Theory: revision and some recent developments*; Proc. Int. Conf. on Galois Connections, Potsdam 2001.

[4] G. Janelidze, G. M. Kelly, *Galois theory and a general notion of central extension*; J. Pure and Appl. Algebra **97** (1994), 135-161.

[5] A. H. Roque, *PhD Thesis*, in preparation.

[6] J. Xarez, *The monotone-light factorization for categories via preorders* submitted to the Proceedings of the "Workshop on Categorical Structures for Descent and Galois Theory, Hopf Algebras and Semiabelian Categories", September 2002.

Received December 20, 2002; in revised form June 14, 2003

Fields Institute Communications
Volume **43**, 2004

Hopf-Galois and Bi-Galois Extensions

Peter Schauenburg
Mathematisches Institut der Universität München
Theresienstr. 39
80333 München, Germany
schauen@mathematik.uni-muenchen.de

Contents

1 Introduction

Hopf-Galois extensions were introduced by Chase and Sweedler [8] (in the commutative case) and Kreimer and Takeuchi [25] (in the case of finite dimensional Hopf algebras) by axioms directly generalizing those of a Galois extension of rings, replacing the action of a group on the algebra by the coaction of a Hopf algebra H; the special case of an ordinary Galois extension is recovered by specializing H to be the dual of a group algebra. Hopf-Galois extensions also generalize strongly graded algebras (here H is a group algebra) and certain inseparable field extensions (here the Hopf algebra is the restricted envelope of a restricted Lie algebra, or, in more general cases, generated by higher derivations). They comprise twisted group rings $R * G$ of a group G acting on a ring R (possibly also twisted by a cocycle), and similar constructions for actions of Lie algebras. If the Hopf algebra involved is the coordinate ring of an affine group scheme, faithfully flat Hopf-Galois extensions are precisely the coordinate rings of affine torsors or principal homogeneous spaces. By analogy, Hopf-Galois extensions with Hopf algebra H the coordinate ring of a quantum group can be considered as the noncommutative analog of a principal homogeneous space, with a quantum group as its structure group. Apart from this noncommutative-geometric interpretation, and apart from their role as a unifying

2000 *Mathematics Subject Classification.* Primary 16W30; Secondary 18D10, 14L30.
Key words and phrases. Hopf algebra, Hopf-Galois extension, bi-Galois extension.

language for many examples of good actions of things on rings, Hopf-Galois extensions are frequently used as a tool in the investigation of the structure of Hopf algebras themselves.

In this paper we try to collect some of the basic facts of the theory of Hopf-Galois extensions and (see below) bi-Galois extensions, offering alternative proofs in some instances, and proving new facts in very few instances.

In the first part we treat Hopf-Galois extensions and discuss various properties by which they can, to some extent, be characterized. After providing the necessary definitions, we first treat the special case of cleft extensions, repeating (with some more details) a rather short proof from [38] of their characterization, due to Blattner, Cohen, Doi, Montgomery, and Takeuchi [15, 5, 6]. Cleft extensions are the same as crossed products, which means that they have a combinatorial description that specializes in the case of cocommutative Hopf algebras to a cohomological description in terms of Sweedler cohomology [46].

In Section 2.3 we prove Schneider's structure theorem for Hopf modules, which characterizes faithfully flat Hopf-Galois extensions as those comodule algebras A that give rise to an equivalence of the category of Hopf modules $\mathcal{M}_A^H$ with the category of modules of the ring of coinvariants under the coaction of H. The structure theorem is one of the most ubiquitous applications of Hopf-Galois theory in the theory of Hopf algebras. We emphasize the role of faithfully flat descent in its proof.

A more difficult characterization of faithfully flat Hopf-Galois extensions, also due to Schneider, is treated in Section 2.4. While the definition of an H-Galois extension A of B asks for a certain canonical map $\beta\colon A \otimes_B A \to A \otimes H$ to be bijective, it is sufficient to require it to be surjective, provided we work over a field and A is an injective H-comodule. When we think of Hopf-Galois extensions as principal homogeneous spaces with structure quantum group, this criterion has a geometric meaning. We will give a new proof for it, which is more direct than that in [44]. The new proof has two nice side-effects: First, it is more parallel to the proof that surjectivity of the canonical map is sufficient for finite-dimensional Hopf algebras (in fact so parallel that we prove the latter fact along with Schneider's result). Secondly, it yields without further work the fact that an H-Galois extension A/B that is faithfully flat as a B-module is always projective as a B-module [1].

Section 2.5 treats (a generalized version of) a characterization of Hopf-Galois extensions due to Ulbrich: An H-Galois extension of B is (up to certain additional conditions) the same thing as a monoidal functor ${}^H\mathcal{M} \to {}_B\mathcal{M}_B$ from the monoidal category of H-comodules to the category of B-bimodules.

In Section 2.6 we deal with another characterization of Hopf-Galois extensions by monoidal functors: Given any H-comodule algebra A with coinvariants B, we can define a monoidal category ${}_A\mathcal{M}_A^H$ of Hopf bimodules (monoidal with the tensor product over A), and a *weak* monoidal functor from this to the category of B-bimodules. Again up to some technical conditions, the functor is monoidal if and only if A is an H-Galois extension of B.

In Section 2.8 we show how to characterize Hopf-Galois extensions without ever mentioning a Hopf algebra. The axioms of a torsor we give here are a simplified

[1]Since the present paper was submitted, the new proof has been developed further in joint work with H.-J. Schneider, in particular to also prove some results on Q-Galois extensions for a quotient coalgebra and one sided module of H; this type of extensions will not be considered in the present paper.

variant of axioms recently introduced by Grunspan. A crucial ingredient in the characterization is again the theory of faithfully flat descent.

The second part of the paper deals with bi-Galois objects. This means, first of all, that we restrict our attention to Galois extensions of the base ring k rather than of an arbitrary coinvariant subring. Contrary, as it were, to the theory of torsors that can do without any Hopf algebras, the theory of bi-Galois extensions exploits the fact that any Hopf-Galois object has *two* rather than only one Hopf algebra in it. More precisely, for every H-Galois extension A of k there is a uniquely determined second Hopf algebra L such that A is a left L-Galois extension of A and an L-H-bicomodule. We will give an account of the theory and several ways in which the new Hopf algebra L can be applied. Roughly speaking, this may happen whenever there is a fact or a construction that depends on the condition that the Hopf algebra H be cocommutative (which, in terms of bi-Galois theory, means $L \cong H$). If this part of the cocommutative theory does not survive if H fails to be cocommutative, then maybe L can be used to replace H. Our approach will stress a very general universal property of the Hopf algebra L in an L-H-Galois extension. Several versions of this were already used in previous papers, but the general version we present here appears to be new. The construction of L was invented in the commutative case by Van Oystaeyen and Zhang to repair the failing of the fundamental theorem of Galois theory for Hopf-Galois extensions. We will discuss an application to the computation of Galois objects over tensor products, and to the problem of reducing the Hopf algebra in a Hopf-Galois object to a quotient Hopf algebra (here, however, L arises because of a lack of commutativity rather than cocommutativity). Perhaps the most important application is that bi-Galois extensions classify monoidal category equivalences between categories of comodules over Hopf algebras.

Some conventions and background facts can be found in an appendix. Before starting, however, let us point out a general notational oddity: Whenever we refer to an element $\xi \in V \otimes W$ of the tensor product of two modules, we will take the liberty to "formally" write $\xi = v \otimes w$, even if we know that the element in question is not a simple tensor, or, worse, has to be chosen from a specific submodule that is not even generated by simple tensors. Such formal notations are of course widely accepted under the name Sweedler notation for the comultiplication $\Delta(c) = c_{(1)} \otimes c_{(2)} \in C \otimes C$ in a coalgebra C, or $\delta(v) = v_{(0)} \otimes v_{(1)}$ for a right comodule, or $\delta(v) = v_{(-1)} \otimes v_{(0)}$ for a left comodule.

For a coalgebra C and a subspace $V \subset C$ we will write $V^+ = V \cap \operatorname{Ker}(\varepsilon)$. C^{cop} denotes the coalgebra C with coopposite comultiplication, A^{op} the algebra A with opposite multiplication. Multiplication in an algebra A will be denoted by $\nabla \colon A \otimes A \to A$.

2 Hopf-Galois theory

2.1 Definitions. Throughout this section, H is a k-bialgebra, flat over k. A (right) H-comodule algebra A is by definition an algebra in the monoidal category of right H-comodules, that is, a right H-comodule via $\delta \colon A \ni a \mapsto a_{(0)} \otimes a_{(1)}$ and an algebra, whose multiplication $\nabla \colon A \otimes A \to A$ is a colinear map, as well as the unit $\eta \colon k \to A$. These conditions mean that the unit $1_A \in A$ is a coinvariant element, $1_{(0)} \otimes 1_{(1)} = 1 \otimes 1$, and that $\delta(xy) = x_{(0)} y_{(0)} \otimes x_{(1)} y_{(1)}$ holds for all $x, y \in A$. Equivalently, A is an algebra and an H-comodule in such a way that the comodule structure is an algebra homomorphism $\delta\colon A \to A \otimes H$. For any H-comodule M we let $M^{\operatorname{co} H} := \{m \in M | \delta(m) = m \otimes 1\}$ denote the subset of H-coinvariants. It is straightforward to check that $A^{\operatorname{co} H}$ is a subalgebra of A.

Definition 2.1.1 The right H-comodule algebra A is said to be an H-Galois extension of $B := A^{\operatorname{co} H}$, if the Galois map

$$\beta\colon A \underset{B}{\otimes} A \ni x \otimes y \mapsto xy_{(0)} \otimes y_{(1)} \in A \otimes H$$

is a bijection. More precisely we should speak of a right H-Galois extension; it is clear how a left H-Galois extension should be defined.

We will use the term "(right) Galois object" as shorthand for a right H-Galois extension A of k which is a faithfully flat k-module.

The first example that comes to mind is the H-comodule algebra H itself:

Example 2.1.2 Let H be a bialgebra. Then H is an H-comodule algebra, with $H^{\operatorname{co} H} = k$. The Galois map $\beta\colon H \otimes H \to H \otimes H$ is the map $T(id)$, where

$$T\colon \operatorname{Hom}(H, H) \to \operatorname{End}_{H-}^{-H}(H \otimes H)$$

is the anti-isomorphism from Lemma 4.4.1. Thus, H is a Hopf algebra if and only if the identity on H is convolution invertible if and only if the Galois map is bijective if and only if H is an H-Galois extension of k.

The notion of a Hopf-Galois extension serves to unify various types of extensions. These are recovered as we specialize the Hopf algebra H to one of a number of special types:

Example 2.1.3 Let A/k be a Galois field extension, with (finite) Galois group G. Put $H = k^G$, the dual of the group algebra. Then A is an H-Galois extension of k. Bijectivity of the Galois map $A \otimes A \to A \otimes H$ is a consequence of the independence of characters.

The definition of a Galois extension A/k of commutative rings in [9] requires (in one of its many equivalent formulations) precisely the bijectivity of the Galois map $A \otimes A \to A \otimes k^G$, beyond of course the more obvious condition that k be the invariant subring of A under the action of a finite subgroup G of the automorphism group of A. Thus Hopf-Galois extensions of commutative rings are direct generalizations of Galois extensions of commutative rings.

Example 2.1.4 Let $A = \bigoplus_{g \in G} A_g$ be a k-algebra graded by a group G. Then A is naturally an H-comodule algebra for the group algebra kG, whose coinvariant subring is $B = A_e$, the homogeneous component whose degree is the neutral element. The Galois map $A \otimes_B A \to A \otimes H$ is surjective if and only if $A_g A_h = A_{gh}$ for all $g, h \in G$, that is, A is strongly graded [10, 52]. As we shall see in Corollary 2.4.9, this condition implies that A is an H-Galois extension of B if k is a field.

We have seen already that a bialgebra H is an H-Galois extension of k if and only if it is a Hopf algebra. The following more general observation is the main result of [34]; we give a much shorter proof that is due to Takeuchi [51].

Lemma 2.1.5 *Let H be a k-flat bialgebra, and A a right H-Galois extension of $B := A^{\operatorname{co} H}$, which is faithfully flat as k-module. Then H is a Hopf algebra.*

Proof H is a Hopf algebra if and only if the map $\beta_H \colon H \otimes H \ni g \otimes h \mapsto gh_{(1)} \otimes h_{(2)}$ is a bijection. By assumption the map $\beta_A \colon A \otimes_B A \ni x \otimes y \mapsto xy_{(0)} \otimes y_{(1)} \in A \otimes H$ is a bijection. Now the diagram

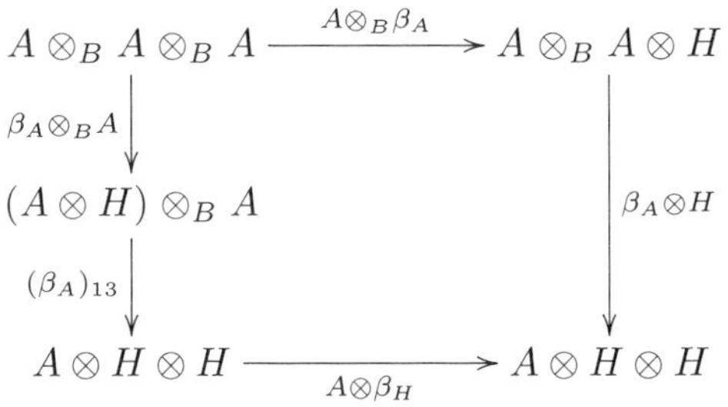

$$\begin{array}{ccc}
A \otimes_B A \otimes_B A & \xrightarrow{A \otimes_B \beta_A} & A \otimes_B A \otimes H \\
\downarrow{\scriptstyle \beta_A \otimes_B A} & & \\
(A \otimes H) \otimes_B A & & \downarrow{\scriptstyle \beta_A \otimes H} \\
\downarrow{\scriptstyle (\beta_A)_{13}} & & \\
A \otimes H \otimes H & \xrightarrow[A \otimes \beta_H]{} & A \otimes H \otimes H
\end{array}$$

commutes, where $(\beta_A)_{13}$ denotes the map that applies β_A to the first and third tensor factor, and leaves the middle factor untouched. Thus $A \otimes \beta_H$, and by faithful flatness of A also β_H, is a bijection. □

The Lemma also shows that if A is an H-Galois extension and a flat k-module, then A^{op} is never an H^{op}-Galois extension, unless the antipode of H is bijective. On the other hand (see [44]):

Lemma 2.1.6 *If the Hopf algebra H has bijective antipode and A is an H-comodule algebra, then A is an H-Galois extension if and only if A^{op} is an H^{op}-Galois extension.*

Proof The canonical map $A^{\mathrm{op}} \otimes_{B^{\mathrm{op}}} A^{\mathrm{op}} \to A^{\mathrm{op}} \otimes H^{\mathrm{op}}$ identifies with the map $\beta' \colon A \otimes_B A \to A \otimes H$ given by $\beta'(x \otimes y) = x_{(0)} y \otimes x_{(1)}$. One checks that the diagram

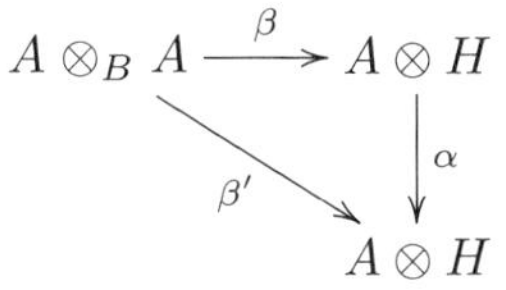

$$\begin{array}{ccc}
A \otimes_B A & \xrightarrow{\beta} & A \otimes H \\
& \searrow^{\beta'} & \downarrow{\scriptstyle \alpha} \\
& & A \otimes H
\end{array}$$

commutes, where $\alpha \colon A \otimes H \ni a \otimes h \mapsto a_{(0)} \otimes a_{(1)} S(h)$ is bijective with $\alpha^{-1}(a \otimes h) = a_{(0)} \otimes S^{-1}(h) a_{(1)}$. □

Lemma 2.1.7 *Let A be an H-Galois extension of B. For $h \in H$ we write $\beta^{-1}(1 \otimes h) =: h^{[1]} \otimes h^{[2]}$. For $g, h \in H$, $b \in B$ and $a \in A$ we have*

$$h^{[1]} h^{[2]}{}_{(0)} \otimes h^{[2]}{}_{(1)} = 1 \otimes h \tag{2.1.1}$$

$$h^{[1]} \otimes h^{[2]}{}_{(0)} \otimes h^{[2]}{}_{(1)} = h_{(1)}{}^{[1]} \otimes h_{(1)}{}^{[2]} \otimes h_{(2)} \tag{2.1.2}$$

$$h^{[1]}{}_{(0)} \otimes h^{[2]} \otimes h^{[1]}{}_{(1)} = h_{(2)}{}^{[1]} \otimes h_{(2)}{}^{[2]} \otimes S(h_{(1)}) \tag{2.1.3}$$

$$h^{[1]} h^{[2]} = \varepsilon(h) 1_A \tag{2.1.4}$$

$$(gh)^{[1]} \otimes (gh)^{[2]} = h^{[1]} g^{[1]} \otimes g^{[2]} h^{[2]} \tag{2.1.5}$$

$$b h^{[1]} \otimes h^{[2]} = h^{[1]} \otimes h^{[2]} b \tag{2.1.6}$$

$$a_{(0)} a_{(1)}{}^{[1]} \otimes a_{(1)}{}^{[2]} = 1 \otimes a \tag{2.1.7}$$

We will omit the proof, which can be found in [45, (3.4)].

Definition 2.1.8 Let H be a Hopf algebra, and A an H-Galois extension of B. The Miyashita-Ulbrich action of H on the centralizer A^B of B in A is given by $x \leftharpoonup h = h^{[1]}xh^{[2]}$ for $x \in A^B$ and $h \in H$.

The expression $h^{[1]}xh^{[2]}$ is well-defined because $x \in A^B$, and it is in A^B again because $h^{[1]} \otimes h^{[2]} \in (A \otimes_B A)^B$. The following properties of the Miyashita-Ulbrich action can be found in [52, 16] in different language.

Lemma 2.1.9 *The Miyashita-Ulbrich action makes A^B an object of $\mathcal{YD}^H_H$, and thus the weak center of the monoidal category $\mathcal{M}^H$ of right H-comodules. A^B is the center of A in the sense of Definition 4.2.1.*

Proof It is trivial to check that A^B is an subcomodule of A. It is a Yetter-Drinfeld module by (2.1.3) and (2.1.2). Now the inclusion $A^B \hookrightarrow A$ is central in the sense of Definition 4.2.1, since $a_{(0)}(x \leftharpoonup a_{(1)}) = a_{(0)}a_{(1)}{}^{[1]}xa_{(1)}{}^{[2]} = xa$ for all $a \in A$ and $x \in A^B$ by (2.1.7). Finally let us check the universal property in Definition 4.2.1: Let V be a Yetter-Drinfeld module, and $f\colon V \to A$ an H-colinear map with $a_{(0)}f(v \leftharpoonup a_{(1)}) = f(v)a$ for all $v \in V$ and $a \in A$. Then we see immediately that f takes values in A^B. Moreover, we have $f(v) \leftharpoonup h = h^{[1]}f(v)h^{[2]} = h^{[1]}h^{[2]}{}_{(0)}f(v \leftharpoonup h^{[2]}{}_{(1)}) = f(v \leftharpoonup h)$ for all $h \in H$ by (2.1.1). □

Much of the "meaning" of the Miyashita-Ulbrich action can be guessed from the simplest example $A = H$. Here we have $h^{[1]} \otimes h^{[2]} = S(h_{(1)}) \otimes h_{(2)} \in H \otimes H$, and thus the Miyashita-Ulbrich action is simply the adjoint action of H on itself.

2.2 Cleft extensions and crossed products. Throughout the section, H is a k-bialgebra.

Definition 2.2.1 Let B be a k-algebra. A map $\rightharpoonup\colon H \otimes B \to B$ is a measuring if $h \rightharpoonup (bc) = (h_{(1)} \rightharpoonup b)(h_{(2)} \rightharpoonup c)$ and $h \rightharpoonup 1 = 1$ hold for all $h \in H$ and $b, c \in B$.

Let H be a bialgebra, and B an algebra. A crossed product $B\#_\sigma H$ is the structure of an associative algebra with unit $1\#1$ on the k-module $B\#_\sigma H := B \otimes H$, in which multiplication has the form

$$(b\#g)(c\#h) = b(g_{(1)} \rightharpoonup c)\sigma(g_{(2)} \otimes h_{(1)})\#g_{(3)}h_{(2)}$$

for some measuring $\rightharpoonup\colon H \otimes B \to B$ and some linear map $\sigma\colon H \otimes H \to B$.

We have quite deliberately stated the definition without imposing any explicit conditions on σ. Such conditions are implicit, however, in the requirement that multiplication be associative and have the obvious unit. We have chosen the definition above to emphasize that the explicit conditions on σ are never used in our approach to the theory of crossed products. They are, however, known and not particularly hard to derive:

Proposition 2.2.2 *Let H be a bialgebra, $\rightharpoonup\colon H \otimes B \to B$ a measuring, and $\sigma\colon H \otimes H \to B$ a k-linear map. The following are equivalent:*

1. *$A = B\#H := B \otimes H$ is an associative algebra with unit $1\#1$ and multiplication*

$$(b\#g)(c\#h) = b(g_{(1)} \rightharpoonup c)\sigma(g_{(2)} \otimes h_{(1)})\#g_{(3)}h_{(2)}.$$

2. (a) *$\rightharpoonup$ is a twisted action, that is $(g_{(1)} \rightharpoonup (h_{(1)} \rightharpoonup b))\sigma(g_{(2)} \otimes h_{(2)}) = \sigma(g_{(1)} \otimes h_{(1)})(g_{(2)}h_{(2)} \rightharpoonup b)$ and $1 \rightharpoonup b = b$ hold for all $g, h \in H$ and $b \in B$.*

(b) *σ is a two-cocycle, that is* $(f_{(1)} \rightharpoonup \sigma(g_{(1)} \otimes h_{(1)}))\sigma(f_{(2)} \otimes g_{(2)}h_{(2)}) = \sigma(f_{(1)} \otimes g_{(1)})\sigma(f_{(2)} \otimes g_{(2)} \otimes h)$ *and* $\sigma(h \otimes 1) = \sigma(1 \otimes h) = 1$ *hold for all* $f, g, h \in H$.

Not only are the conditions on σ known, but, more importantly, they have a cohomological interpretation in the case where H is cocommutative and B is commutative. In this case a twisted action is clearly simply a module algebra structure. Sweedler [46] has defined cohomology groups $H^\bullet(H, B)$ for a cocommutative bialgebra H and commutative H-module algebra B, and it turns out that a convolution invertible map σ as above is precisely a two-cocycle in this cohomology. Sweedler's paper also contains the construction of a crossed product from a two-cocycle, and the fact that his second cohomology group classifies cleft extensions (which we shall define below) by assigning the crossed product to a cocycle. Group cohomology with coefficients in the unit group of B as well as (under some additional conditions) Lie algebra cohomology with coefficients in the additive group of B are examples of Sweedler cohomology, and the cross product costruction also has precursors for groups (twisted group rings with cocycles, which feature in the construction of elements of the Brauer group from group cocycles) and Lie algebras. Thus, the crossed product construction from cocycles can be viewed as a nice machinery producing (as we shall see shortly) Hopf-Galois extensions *in the case of cocommutative Hopf algebras and commutative coinvariant subrings.* In the general case, the equations do not seem to have any reasonable cohomological interpretation, so while cleft extensions remain an important special class of Hopf-Galois extensions, it is rarely possible to construct them by finding cocycles in some conceptually pleasing way.

We now proceed to prove the characterization of crossed products as special types of comodule algebras, which is due to Blattner, Cohen, Doi, Montgomery, and Takeuchi:

Definition 2.2.3 Let A be a right H-comodule algebra, and $B := A^{\operatorname{co} H}$.

1. A is cleft if there exists a convolution invertible H-colinear map $j\colon H \to A$ (also called a cleaving).
2. A normal basis for A is an H-colinear and B-linear isomorphism $\psi\colon B \otimes H \to A$.

If $\tilde{j}$ is a cleaving, then $\tilde{j}(1)$ is a unit in B, and thus $j(h) = \tilde{j}(1)^{-1}\tilde{j}(h)$ defines another cleaving, which, moreover, satisfies $j(1) = 1$.

It was proved by Doi and Takeuchi [15] that A is H-Galois with a normal basis if and only if it is cleft, and in this case A is a crossed product $A \cong B\#_\sigma H$ with an invertible cocycle $\sigma\colon H \otimes H \to B$. Blattner and Montgomery [6] have shown that crossed products with an invertible cocycle are cleft.

Clearly a crossed product is always an H-comodule algebra with an obvious normal basis.

Lemma 2.2.4 *Assume that the H-comodule algebra A has a normal basis $\psi\colon B \otimes H \to A$ satisfying $\psi(1 \otimes 1) = 1$. Then A is isomorphic (via ψ) to a crossed product.*

Proof In fact we may as well assume $B \otimes H = A$ as B-modules and H-comodules. Define $h \rightharpoonup b = (B \otimes \varepsilon)((1 \otimes h)(b \otimes 1))$ and $\sigma(g \otimes h) = (B \otimes \varepsilon)((1 \otimes$

$g)(1 \otimes h))$. Since multiplication is H-colinear, we find

$$(1 \otimes g)(c \otimes 1) = (B \otimes \varepsilon \otimes H)(B \otimes \Delta)((1 \otimes g)(c \otimes 1))$$
$$= (B \otimes \varepsilon \otimes H)((1 \otimes g_{(1)})(b \otimes 1) \otimes g_{(2)}) = g_{(1)} \rightharpoonup b \otimes g_{(2)},$$

$$(1 \otimes g)(1 \otimes h) = (B \otimes \varepsilon \otimes H)(B \otimes \Delta)((1 \otimes g)(1 \otimes h))$$
$$= (B \otimes \varepsilon \otimes H)((1 \otimes g_{(1)})(1 \otimes h_{(1)}) \otimes g_{(2)}h_{(2)}) = \sigma(g_{(1)} \otimes h_{(1)}) \otimes g_{(2)}h_{(2)},$$

and finally

$$(b \otimes g)(c \otimes h) = (b \otimes 1)(1 \otimes g)(c \otimes 1)(1 \otimes h) = (b \otimes 1)(g_{(1)} \rightharpoonup c \otimes g_{(2)})(1 \otimes h)$$
$$= b(g_{(1)} \rightharpoonup c)\sigma(g_{(2)} \otimes h_{(1)}) \otimes g_{(3)}h_{(2)}.$$

□

To prove the remaining parts of the characterization, we will make heavy use of the isomorphisms T_A^C from Lemma 4.4.1, for various choices of algebras A and coalgebras C.

Lemma 2.2.5 *Let $j\colon H \to A$ be a cleaving. Then there is a normal basis $\psi\colon B \otimes H \to A$ with $j = \psi(\eta_B \otimes H)$. If $j(1) = 1$, then $\psi(1 \otimes 1) = 1$.*

Proof We claim that $\psi\colon B \otimes H \ni b \otimes h \mapsto bj(h) \in A$ is a normal basis.

Since the comodule structure $\delta\colon A \to A \otimes H$ is an algebra map, δj is convolution invertible. Moreover $\delta j = (j \otimes H)\Delta$ by assumption. For $a \in A$, we have

$$T^H_{A\otimes H}(\delta j)(\delta(a_{(0)}j^{-1}(a_{(1)})) \otimes a_{(2)}) = T^H_{A\otimes H}(\delta j)T^H_{A\otimes H}(\delta j^{-1})(\delta(a_{(0)}) \otimes a_{(1)})$$
$$= a_{(0)}j^{-1}(a_{(1)})j(a_{(2)}) \otimes a_{(3)} \otimes a_{(4)} = T^H_{A\otimes H}((j \otimes H)\Delta)(a_{(0)}j^{-1}(a_{(1)}) \otimes 1 \otimes a_{(2)}),$$

hence $\delta(a_{(0)}j^{-1}(a_{(1)})) \otimes a_{(2)} = a_{(0)}j^{-1}(a_{(1)}) \otimes 1 \otimes a_{(2)}$, and further $\delta(a_{(0)}j^{-1}(a_{(1)})) = a_{(0)}j^{-1}(a_{(1)}) \otimes 1$. Thus $A \ni a \mapsto a_{(0)}j^{-1}(a_{(1)}) \otimes a_{(2)} \in B \otimes H$ is well defined and easily checked to be an inverse for ψ. □

Lemma 2.2.6 *Let A be an H-comodule algebra with a normal basis. Put $B := A^{\mathrm{co}\,H}$. The following are equivalent:*

1. *A is an H-Galois extension of B.*
2. *A is cleft.*

Proof We can assume that $A = B\#_\sigma H$ is a crossed product, and that $j(h) = 1 \otimes h$.

The map $\alpha\colon B \otimes H \otimes H \to A \otimes_B A$ with $\alpha(b \otimes g \otimes h) = b \otimes g \otimes 1 \otimes h$ is an isomorphism. For $b \in B$ and $g, h \in H$ we have

$$\beta_A\alpha(b \otimes g \otimes h) = \beta_A(b \otimes g \otimes j(h)) = (b \otimes g)j(h_{(1)}) \otimes h_{(2)} = T_A^H(j)(b \otimes g \otimes h),$$

that is $\beta_A\alpha = T_A^H(j)$. In particular, β_A is an isomorphism if and only if $T_A^H(j)$ is, if and only if j is convolution invertible. □

In particular, if B is faithfully flat over k, then cleft extensions can only occur if H is a Hopf algebra. If this is the case, we find:

Theorem 2.2.7 *Let H be a Hopf algebra and A a right H-comodule algebra with $B := A^{\mathrm{co}\,H}$. The following are equivalent:*

1. *A is H-cleft.*
2. *A is H-Galois with a normal basis.*

3. *A is isomorphic to a crossed product* $B\#_\sigma H$ *such that the cocycle* $\sigma\colon H\otimes H\to B$ *is convolution invertible.*

Proof We have already shown that under any of the three hypotheses we can assume that $A\cong B\#_\sigma H=B\otimes H$ is a crossed product, with $j(h)=1\otimes h$, and we have seen that (1) is equivalent to (2), even if H does not have an antipode.

Now, for $b\in B$, $g,h\in H$ we calculate

$$\begin{aligned}T_A^H(j)(b\otimes g\otimes h)&=bj(g)j(h_{(1)})\otimes h_{(2)}=b\sigma(g_{(1)}\otimes h_{(1)})\otimes g_{(2)}h_{(2)}\otimes h_{(3)}\\&=(B\otimes\beta_H)(b\sigma(g_{(1)}\otimes h_{(1)})\otimes g_{(2)}\otimes h_{(2)})=(B\otimes\beta_H)T_B^{H\otimes H}(\sigma)(b\otimes g\otimes h)\end{aligned}$$

that is, $T_A^H(j)=(B\otimes\beta_H)T_B^{H\otimes H}(\sigma)$. Since we assume that β_H is a bijection, we see that j is convolution invertible if and only if $T_A^H(j)$ is bijective, if and only if $T_B^{H\otimes H}(\sigma)$ is bijective, if and only if σ is convolution invertible. □

The reader that has seen the proof of (3)⇒(1) in [6] may be worried that we have lost some information: In [6] the convolution inverse of j is given explicitly, while we only seem to have a rather roundabout existence proof. However, we see from our arguments above that

$$j^{-1}=(T_A^H)^{-1}\left(T_B^{H\otimes H}(\sigma^{-1})(B\otimes\beta_H^{-1})\right),$$

that is,

$$\begin{aligned}j^{-1}(h)&=(A\otimes\varepsilon)T_B^{H\otimes H}(\sigma^{-1})(B\otimes\beta_H^{-1})(1\otimes 1\otimes h)\\&=(A\otimes\varepsilon)T_B^{H\otimes H}(\sigma^{-1})(1\otimes S(h_{(1)})\otimes h_{(2)})\\&=(A\otimes\varepsilon)(\sigma^{-1}(S(h_{(2)})\otimes h_{(3)})\otimes S(h_{(1)})\otimes h_{(4)})\\&=\sigma^{-1}(S(h_{(2)})\otimes h_{(3)})\#S(h_{(1)}).\end{aligned}$$

2.3 Descent and the structure of Hopf modules.

Definition 2.3.1 Let A be a right H-comodule algebra. A Hopf module $M\in\mathcal{M}_A^H$ is a right A-module in the monoidal category of H-comodules. That is, M is a right H-comodule and a right A-module such that the module structure is an H-colinear map $M\otimes A\to M$. This in turn means that $\delta(ma)=m_{(0)}a_{(0)}\otimes m_{(1)}a_{(1)}$ holds for all $m\in M$ and $a\in A$.

For any comodule algebra, one obtains a pair of adjoint functors between the category of Hopf modules and the category of modules over the coinvariant subalgebra.

Lemma 2.3.2 *Let H be a k-flat Hopf algebra, A a right H-comodule algebra, and $B=A^{\mathrm{co}\,H}$. Then the functor*

$$\mathcal{M}_A^H\ni M\mapsto M^{\mathrm{co}\,H}\in\mathcal{M}_B$$

is right adjoint to

$$\mathcal{M}_B\ni N\mapsto N\underset{B}{\otimes}A\in\mathcal{M}_A^H$$

Here, both the A-module and H-comodule structures of $N\otimes_B A$ are induced by those of A. The unit and counit of the adjunction are

$$N\ni n\mapsto n\otimes 1\in(N\underset{B}{\otimes}A)^{\mathrm{co}\,H}$$

$$M^{\mathrm{co}\,H}\underset{B}{\otimes}A\ni m\otimes a\mapsto ma\in M$$

If the adjunction in the Lemma is an equivalence, then we shall sometimes say that the structure theorem for Hopf modules holds for the extension. A theorem of Schneider [44] characterizes faithfully flat Hopf-Galois extensions as those comodule algebras for which the adjunction above is an equivalence. The proof in [44] uses faithfully flat descent; we rewrite it to make direct use of the formalism of faithfully flat descent of modules that we recall in Section 4.5. This approach was perhaps first noted in my thesis [32], though it is certainly no surprise; in fact, one of the more prominent special cases of the structure theorem for Hopf modules over Hopf-Galois extensions that is one direction of the characterization goes under the name of Galois descent.

Example 2.3.3 Let A/k be a Galois field extension with Galois group G. A comodule structure making an A-vector space into a Hopf module $M \in \mathcal{M}_A^{kG}$ is the same as an action of the Galois group G on M by semilinear automorphisms, i.e. in such a way that $\sigma \cdot (am) = \sigma(a)(\sigma \cdot m)$ holds for all $m \in M$, $a \in A$ and $\sigma \in G$. Galois descent (see for example [23]) says, most of all, that such an action on M forces M to be obtained from a k-vector space by extending scalars. This is (part of) the content of the structure theorem for Hopf modules.

Remark 2.3.4 Let A be an H-comodule algebra; put $B := A^{\operatorname{co} H}$. As a direct generalization of the Galois map $\beta\colon A \otimes_B A \to A \otimes H$, we have a right A-module map

$$\beta_M\colon M \underset{B}{\otimes} A \ni m \otimes a \mapsto ma_{(0)} \otimes a_{(1)} \in M. \otimes H.$$

which is natural in $M \in \mathcal{M}_A$. Of course, the Galois map is recovered as $\beta = \beta_A$. Note that β_M can be identified with $M \otimes_A \beta_A$, so that all β_M are bijective once β_A is bijective.

Lemma 2.3.5 *Let A be a right H-comodule algebra, and $B := A^{\operatorname{co} H}$. For each descent data $(M, \theta) \in \mathcal{D}(A \downarrow B)$, the map*

$$\delta := \left(M \xrightarrow{\theta} M \underset{B}{\otimes} A \xrightarrow{\beta_M} M \otimes H\right)$$

is a right H-comodule structure on M making $M \in \mathcal{M}_A^H$.

Thus, we have defined a functor $\mathcal{D}(A \downarrow B) \to \mathcal{M}_A^H$.

If A is an H-Galois extension of B, then the functor is an equivalence.

Proof Let $\theta\colon M \to M \otimes_B A$ be a right A-module map, and $\delta := \beta_M \theta$. Of course δ is a right A-module map, so that M is a Hopf module if and only if it is a comodule.

Now we have the commutative diagrams

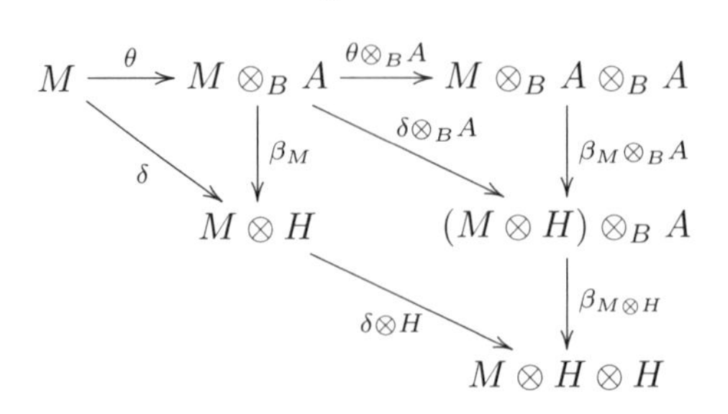

using naturality of β with respect to the right A-module map δ, and

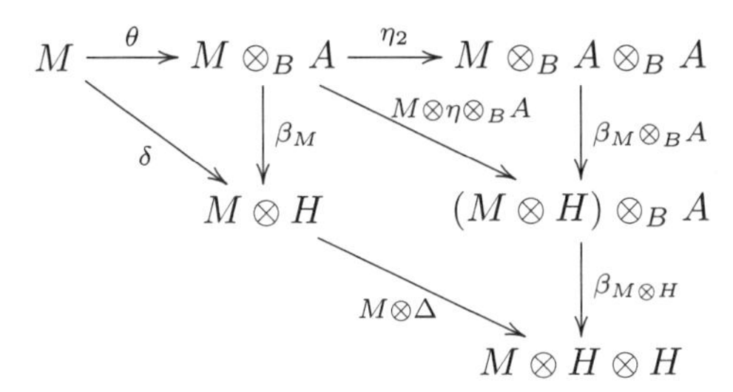

using $\beta_{M\otimes H}(m\otimes 1\otimes a) = (m\otimes 1)a_{(0)}\otimes a_{(1)} = ma_{(0)}\otimes a_{(1)}\otimes a_{(2)} = (M\otimes\Delta)\beta_M(m\otimes a)$. Moreover

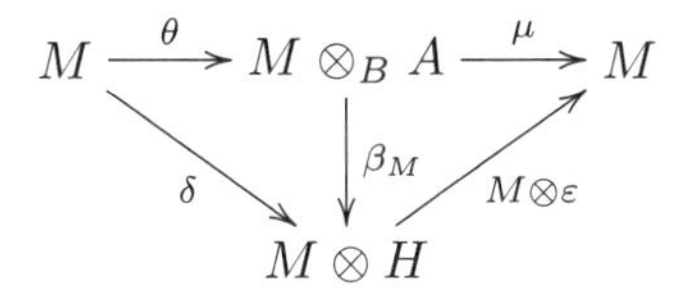

also commutes. Thus, if θ is a descent data, then δ is a comodule.

Conversely, if β is bijective, then the natural transformation β_M is an isomorphism. In particular the formula $\delta = \beta_M\theta$ defines a bijective correspondence between A-module maps $\theta\colon M \to M \otimes_B A$ and $\delta\colon M \to M_{\bullet} \otimes H_{\bullet}$. The same diagrams as above show that δ is a comodule structure if and only if θ is a descent data. □

Schneider's structure theorem for Hopf modules is now an immediate consequence of faithfully flat descent:

Corollary 2.3.6 *The following are equivalent for an H-comodule algebra A:*

1. *A is an H-Galois extension of $B := A^{\mathrm{co}\,H}$, and faithfully flat as left B-module.*
2. *The functor $\mathcal{M}_B \ni N \mapsto N \otimes_B A \in \mathcal{M}_A^H$ is an equivalence.*

Proof (1)⇒(2): We have established an equivalence $\mathcal{D}(A \downarrow B) \to \mathcal{M}_A^H$, and it is easy to check that the diagram

$$\begin{array}{ccc} \mathcal{D}(A\downarrow B) & \xrightarrow{\quad\sim\quad} & \mathcal{M}_A^H \\ & {\scriptstyle (-)^{\theta}}\searrow \quad \swarrow{\scriptstyle (-)^{\mathrm{co}\,H}} & \\ & \mathcal{M}_B & \end{array} \tag{2.3.1}$$

commutes. Thus the coinvariants functor is an equivalence by faithfully flat descent.

(2)⇒(1): Since $(-)^{\mathrm{co}\,H}\colon \mathcal{M}_A^H \to \mathcal{M}_B$ is an equivalence, and $\beta\colon A \otimes_B A^{\bullet}_{\bullet} \to A_{\bullet} \otimes H^{\bullet}_{\bullet}$ is a Hopf module homomorphism, β is an isomorphism if and only if $\beta^{\mathrm{co}\,H}$ is. But

$$A \cong A \underset{B}{\otimes} A^{\mathrm{co}\,H} \xrightarrow{\beta^{\mathrm{co}\,H}} A \otimes H^{\mathrm{co}\,H} \cong A$$

is easily checked to be the identity. Thus, A is an H-Galois extension of B. It is faithfully flat since $(-) \otimes_B A\colon \mathcal{M}_B \to \mathcal{M}_A^H$ is an equivalence. □

To shed some further light on the connection between descent data and the Galois map, it may be interesting to prove a partial converse to Lemma 2.3.5:

Proposition 2.3.7 *Let H be a bialgebra, A a right H-comodule algebra, and $B = A^{\mathrm{co}\, H}$.*

If the natural functor $\mathcal{D}(A \downarrow B) \to \mathcal{M}_A^H$ is an equivalence, then the Galois map $\beta\colon A \otimes_B A \to A \otimes H$ is surjective.

If, moreover, A is flat as left B-module, then A is an H-Galois extension of B.

Proof By assumption there is an A-module map $\theta = \theta_M\colon M \to M \otimes_B A$, natural in $M \in \mathcal{M}_A^H$, such that the H-comodule structure of M is given by $\delta_M = \beta_M \theta_M$.

Specializing $M = V^{\bullet} \otimes A^{\bullet}_{\bullet}$ for $V \in \mathcal{M}^H$, we obtain a natural A-module map $\tilde{\theta}_V\colon V \otimes A \to V \otimes A \otimes_B A$, which, being an A-module map, is determined by $\phi_V\colon V \to V \otimes A \otimes_B A$. Finally, since ϕ_V is natural, it has the form

$$\phi_V(v) = v_{(0)} \otimes {v_{(1)}}^{[1]} \otimes {v_{(1)}}^{[2]}$$

for the map $\gamma\colon H \ni h \mapsto h^{[1]} \otimes h^{[2]} \in A \otimes_B A$ defined by $\gamma = (\varepsilon \otimes A \otimes_B A)\phi_H$. In particular we have $\tilde{\theta}_V(v \otimes a) = v_{(0)} \otimes {v_{(1)}}^{[1]} \otimes {v_{(1)}}^{[2]}a$, and hence, specializing $V = H$ and $a = 1$:

$$\begin{aligned} h_{(1)} \otimes 1 \otimes h_{(2)} &= \delta_{H\otimes A}(h \otimes 1) \\ &= \beta_{H\otimes A}\theta_{H\otimes A}(h \otimes 1) \\ &= \beta_{H\otimes A}(h_{(1)} \otimes {h_{(2)}}^{[1]} \otimes {h_{(2)}}^{[2]}) \\ &= h_{(1)} \otimes {h_{(2)}}^{[1]}{{h_{(2)}}^{[2]}}_{(0)} \otimes {{h_{(2)}}^{[2]}}_{(1)} \\ &= h_{(1)} \otimes \beta_A(h^{[1]} \otimes h^{[2]}) \end{aligned}$$

for all $h \in H$, and thus $\beta(ah^{[1]} \otimes h^{[2]}) = a \otimes h$ for all $a \in A$.

If A is left B-flat, then $\theta_M(m) \in M^{\theta} \otimes_B A \subset M^{\mathrm{co}\, H} \otimes_B A$ implies, in particular, that $a_{(0)} \otimes {a_{(1)}}^{[1]} \otimes {a_{(1)}}^{[2]} \in (A \otimes A)^{\mathrm{co}\, H} \otimes_B A$, and thus $a_{(0)}{a_{(1)}}^{[1]} \otimes {a_{(1)}}^{[2]} \in B \otimes A$. Hence $\beta^{-1}(a \otimes h) = ah^{[1]} \otimes h^{[2]}$ is actually (not only right) inverse to β by the calculation $\beta^{-1}\beta(x \otimes y) = \beta^{-1}(xy_{(0)} \otimes y_{(1)}) = xy_{(0)}{y_{(1)}}^{[1]} \otimes {y_{(1)}}^{[2]} = x \otimes y_{(0)}{y_{(1)}}^{[1]}{y_{(1)}}^{[2]} = x \otimes y$. □

2.4 Coflat Galois extensions. A faithfully flat H-Galois extension is easily seen to be a faithfully coflat H-comodule:

Lemma 2.4.1 *Let H be a k-flat Hopf algebra, and A an H-Galois extension of B. If A_B is faithfully flat and A is a faithfully flat k-module, then A is a faithfully coflat H-comodule.*

Proof If A_B is flat, then we have an isomorphism, natural in $V \in {}^H\mathcal{M}$:

$$A \underset{B}{\otimes} (A \underset{H}{\square} V) \cong (A \underset{B}{\otimes} A) \underset{H}{\square} V \cong (A \otimes H) \underset{H}{\square} V \cong A \otimes V.$$

If A_B is faithfully flat and A is faithfully flat over k, then it follows that the functor $A \square_H -$ is exact and reflects exact sequences. □

The converse is trivial if $B = k$, for then any (faithfully) coflat comodule is a (faithfully) flat k-module by the definition we chose for coflatness. This is not at all clear if B is arbitrary. However, it is true if k is a field. In this case much more can be said. Schneider [44] has proved that a coflat H-comodule algebra A is already a faithfully flat (on either side) Hopf-Galois extension if we only assume that the Galois map is surjective, and the antipode of H is bijective. We will give a different proof of this characterization of faithfully flat Hopf-Galois extensions. Like the original, it is based on Takeuchi's result that coflatness and injectivity coincide for comodules if k is a field, and on a result of Doi on injective comodule algebras (for which, again, we will give a slightly different proof). Our proof of Schneider's criterion will have a nice byproduct: In the case that k is a field and the Hopf algebra H has bijective antipode, every faithfully flat H-Galois extension is a projective module (on either side) over its coinvariants.

Before going into any details, let us comment very briefly on the algebro-geometric meaning of Hopf-Galois extensions and the criterion. If H is the (commutative) Hopf algebra representing an affine group scheme G, A the algebra of an affine scheme X on which H acts, and Y the affine scheme represented by $A^{\mathrm{co}\,H}$, then A is a faithfully flat H-Galois extension of B if and only if the morphism $X \to Y$ is faithfully flat, and the map $X \times G \to X \times_Y X$ given on elements by $(x, g) \mapsto (x, xg)$ is an isomorphism of affine schemes. This means that X is an affine scheme with an action of G and a projection to the invariant quotient Y which is locally trivial in the faithfully flat topology (becomes trivial after a faithfully flat extension of the base Y). This is the algebro-geometric version of a principal fiber bundle with structure group G, or a G-torsor [11]. If we merely require the canonical map $A \otimes A \to A \otimes H$ to be surjective, this means that we require the map $X \times G \to X \times X$ given by $(x, g) \mapsto (x, xg)$ to be a closed embedding, or that we require the action of G on X to be free. Thus, the criterion we are dealing with in this section says that under the coflatness condition on the comodule structure freeness of the action is sufficient to have a principal fiber bundle. Note in particular that surjectivity of the canonical map is trivial in the case where H is a quotient Hopf algebra of a Hopf algebra A (or G is a closed subgroup scheme of an affine group scheme X), while coflatness in this case is a representation theoretic condition (the induction functor is exact). See [44] for further literature.

For the rest of this section, we assume that k is a field.

We start by an easy and well-known observation regarding projectivity of modules over a Hopf algebra.

Lemma 2.4.2 *Let H be a Hopf algebra and $M, P \in {}_H\mathcal{M}$ with P projective. Then $\bullet P \otimes \bullet M \in {}_H\mathcal{M}$ is projective. In particular, H is semisimple if and only if the trivial H-module is projective.*

Proof The second statement follows from the first, since every module is its own tensor product with the trivial module. The diagonal module $\bullet H \otimes \bullet M$ is free by the structure theorem for Hopf modules, or since

$$\bullet H \otimes M \ni h \otimes m \mapsto h_{(1)} \otimes h_{(2)} m \in \bullet H \otimes \bullet M$$

is an isomorphism. Since any projective P is a direct summand of a direct sum of copies of H, the general statement follows. □

If H has bijective antipode, then in the situation of the Lemma also $M \otimes P$ is projective.

For our proof, we will need the dual variant. To prepare, we observe:

Lemma 2.4.3 *Let C be a coalgebra and $M \in \mathcal{M}^C$. Then M is injective if and only if it is a direct summand of $V \otimes C^{\bullet} \in \mathcal{M}^C$ for some $V \in \mathcal{M}_k$. In particular, if M is injective, then so is every $V \otimes M^{\bullet} \in \mathcal{M}^C$ for $V \in \mathcal{M}_k$.*

Lemma 2.4.4 *Let H be a Hopf algebra and $M, I \in \mathcal{M}^H$ with I injective. Then $M^\bullet \otimes I^\bullet \in \mathcal{M}^H$ is injective.*

Proof Since I is a direct summand of some $V \otimes H^\bullet$, it is enough to treat the case $I = H$. But then

$$M^\bullet \otimes H^\bullet \ni m \otimes h \mapsto m_{(0)} \otimes m_{(1)}h \in M \otimes H^\bullet$$

is a colinear bijection, and $M \otimes H^\bullet$ is injective. □

We come to a key property of comodule algebras that are injective comodules, which is due to Doi [13]:

Proposition 2.4.5 *Let H be a Hopf algebra and A an H-comodule algebra that is an injective H-comodule. Then every Hopf module in $\mathcal{M}_A^H$ is an injective H-comodule. If H has bijective antipode, then also every Hopf module in ${}_A\mathcal{M}^H$ is an injective H-module.*

Proof Let $M \in \mathcal{M}_A^H$. Since the module structure $\mu\colon M^\bullet \otimes A^\bullet \to M$ is H-colinear, and splits as a colinear map via $M \ni m \mapsto m \otimes 1 \in M \otimes A$, the comodule M is a direct summand of the diagonal comodule $M \otimes A$. The latter is injective, since A is. The statement on Hopf modules in ${}_A\mathcal{M}^H$ follows since H^{op} is a Hopf algebra and we can identify ${}_A\mathcal{M}^H$ with $\mathcal{M}_{A^{\mathrm{op}}}^{H^{\mathrm{op}}}$. □

Lemma 2.4.6 *The canonical map $\beta_0\colon A \otimes A \to A \otimes H$ is a morphism of Hopf modules in ${}_A\mathcal{M}^H$ if we equip its source and target with the obvious left A-module structures, the source with the comodule structure coming from the left tensor factor, and its target with the comodule structure given by $(a \otimes h)_{(0)} \otimes (a \otimes h)_{(1)} = a_{(0)} \otimes h_{(2)} \otimes a_{(1)}S(h_{(1)})$. The latter can be viewed as a codiagonal comodule structure, if we first endow H with the comodule structure restricted along the antipode. Thus we may write briefly that*

$$\beta_0\colon .A^\bullet \otimes A \to .A^\bullet \otimes H^S$$

is a morphism in ${}_A\mathcal{M}^H$.

Proposition 2.4.7 *Let H be a Hopf algebra, and A a right H-comodule algebra; put $B := A^{\mathrm{co}\,H}$. Assume there is an H-comodule map $\gamma\colon H^S \to A^\bullet \otimes A$ such that $\beta(\gamma(h)) = 1 \otimes h$ for all $h \in H$ (where we abuse notations and also consider $\gamma(h) \in A \otimes_B A$).*

Then the counit $M^{\mathrm{co}\,H} \otimes_B A \to M$ of the adjunction in Lemma 2.3.2 is an isomorphism for every $M \in \mathcal{M}_A^H$. Its inverse lifts to a natural transformation $M \to M^{\mathrm{co}\,H} \otimes A$ (with the tensor product over k).

In particular A is an H-Galois extension of B, and a projective left B-module.

Proof We shall write $\gamma(h) =: h^{[1]} \otimes h^{[2]}$. This is to some extent an abuse of notations, since the same symbol was used for the map $H \to A \otimes_B A$ induced by the inverse of the canonical map in a Hopf-Galois extension. However, the abuse is not so bad, because in fact the map we use in the present proof will turn out to induce that inverse. Our assumptions on γ read ${h_{(2)}}^{[1]} \otimes {h_{(2)}}^{[2]} \otimes S(h_{(1)}) = {h^{[1]}}_{(0)} \otimes h^{[2]} \otimes {h^{[1]}}_{(1)}$ and $h^{[1]}{h^{[2]}}_{(0)} \otimes {h^{[2]}}_{(1)} = 1 \otimes h \in A \otimes H$ for all $h \in H$. The latter implies in particular that $h^{[1]}h^{[2]} = \varepsilon(h)1_A$.

It follows for all $m \in M \in \mathcal{M}_A^H$ that $m_{(0)}m_{(1)}{}^{[1]} \otimes m_{(1)}{}^{[2]} \in M^{\operatorname{co} H} \otimes A$; indeed $\rho(m_{(0)}m_{(1)}{}^{[1]}) \otimes m_{(1)}{}^{[2]} = m_{(0)}m_{(3)}{}^{[1]} \otimes m_{(1)}S(m_{(2)}) \otimes m_{(3)}{}^{[2]} = m_{(0)}m_{(1)}{}^{[1]} \otimes 1 \otimes m_{(1)}{}^{[2]}$.

Now we can write down the natural transformation $\psi\colon M \ni m \mapsto m_{(0)}m_{(1)}{}^{[1]} \otimes m_{(1)}{}^{[2]} \in M^{\operatorname{co} H} \otimes A$, and define $\vartheta\colon M \to M^{\operatorname{co} H} \otimes_B A$ as the composition of ψ with the canonical surjection.

We claim that ϑ is inverse to the adjunction map $\phi\colon M^{\operatorname{co} H} \otimes_B A \to M$.

Indeed

$$\phi\vartheta(m) = \phi(m_{(0)}m_{(1)}{}^{[1]} \otimes m_{(1)}{}^{[2]}) = m_{(0)}m_{(1)}{}^{[1]}m_{(1)}{}^{[2]} = m$$

and

$$\vartheta\phi(n \otimes a) = \phi^{-1}(na) = na_{(0)}a_{(1)}{}^{[1]} \otimes a_{(1)}{}^{[2]} = n \otimes a_{(0)}a_{(1)}{}^{[1]}a_{(1)}{}^{[2]} = n \otimes a,$$

using that $a_{(0)}a_{(1)}{}^{[1]} \otimes a_{(1)}{}^{[2]} \in B \otimes A$.

Since the adjunction map is an isomorphism, A is an H-Galois extension of B.

The instance $\psi_A\colon A \ni a \mapsto a_{(0)}a_{(1)}{}^{[1]} \otimes a_{(2)}{}^{[2]} \in B \otimes A$ of ψ splits the multiplication map $B \otimes A \to A$, so that A is a direct summand of $B \otimes A$ as left B-module, and hence a projective B-module. □

Corollary 2.4.8 *Let H be a Hopf algebra and A a right H-comodule algebra such that that the canonical map $\beta_0\colon A \otimes A \to A \otimes H$ is a surjection.*

Assume in addition that $\beta_0\colon A^\bullet \otimes A \to A^\bullet \otimes H^S$ splits as a comodule map for the indicated H-comodule structures. Then A is a right H-Galois extension of B and a projective left B-module.

In particular, the assumption can be verified in the following cases:

1. *H is finite dimensional.*
2. *A is injective as H-comodule, and H has bijective antipode.*

Proof First, if β_0 splits as indicated via a map $\alpha\colon A^\bullet \otimes H^S \to A^\bullet \otimes A$ with $\beta_0\alpha = id$, then the composition

$$\gamma = \left(H \xrightarrow{\eta \otimes H} A \otimes H \xrightarrow{\alpha} A \otimes A\right)$$

satisfies the assumptions of Proposition 2.4.7.

If A is an injective comodule, and H has bijective antipode, then every Hopf module in ${}_A\mathcal{M}^H$ is an injective comodule by Proposition 2.4.5. Thus the (kernel of the) Hopf module morphism β_0 splits as a comodule map. Finally, if H is finite dimensional, then we take the view that β_0 should split as a surjective H^*-module map. But H^S is projective as H^*-module, and hence $A \otimes H^S$ is projective as well, and thus the map splits. □

As a corollary, we obtain Schneider's characterization of faithfully flat Hopf-Galois extensions from [44] (and in addition projectivity of such extensions).

Corollary 2.4.9 *Let H be a Hopf algebra with bijective antipode over a base field k, A a right H-comodule algebra, and $B := A^{\operatorname{co} H}$. The following are equivalent:*

1. *The Galois map $A \otimes A \to A \otimes H$ is onto, and A is injective as H-comodule.*
2. *A is an H-Galois extension of B, and right faithfully flat as B-module.*
3. *A is an H-Galois extension of B, and left faithfully flat as B-module.*

In this case, A is a projective left and right B-module.

Proof We already know from the beginning of this section that 2⇒1. Assume 1. Then Corollary 2.4.8 implies that A is Galois and a projective left B-module, and that the counit of the adjunction in Lemma 2.3.2 is an isomorphism. By Corollary 2.3.6 it remains to prove that the unit $N \to (N \otimes_B A)^{\operatorname{co} H}$ is also a bijection for all $N \in \mathcal{M}_B$. But $N \otimes_B A$ is defined by a coequalizer

$$N \otimes B \otimes A \rightrightarrows N \otimes A \to N \underset{B}{\otimes} A \to 0,$$

which is a coequalizer in the category $\mathcal{M}_A^H$. Since every Hopf module is an injective comodule, every short exact sequence in $\mathcal{M}_A^H$ splits colinearly, so the coinvariants functor $\mathcal{M}_A^H \to \mathcal{M}_B$ is exact, and applying it to the coequalizer above we obtain a coequalizer

$$N \otimes B \otimes A \rightrightarrows N \otimes B \to (N \underset{B}{\otimes} A)^{\operatorname{co} H}$$

which says that $(N \otimes_B A)^{\operatorname{co} H} \cong N \otimes_B B \cong N$.

The equivalence of 1 and 3 is proved by applying that of 1 and 2 to the H^{op} comodule algebra A^{op}. □

2.5 Galois extensions as monoidal functors. In this section we prove the characterization of Hopf-Galois extensions as monoidal functors from the category of comodules due to Ulbrich [53, 54]. We are somewhat more general in allowing the invariant subring to be different from the base ring. In this general setting, we have proved one direction of the characterization in [35], but the proof is really no different from Ulbrich's. Some details of the reverse direction (from functors to extensions) are perhaps new. It will turn out that in fact suitably exact *weak* monoidal functors on the category of comodules are the same as comodule algebras, while being monoidal rather than only weak monoidal is related to the Galois condition.

Proposition 2.5.1 *Let H be a bialgebra, and $A \in \mathcal{M}^H$ coflat.*
If A is an H-comodule algebra, then

$$\xi\colon (A \underset{H}{\Box} V) \otimes (A \underset{H}{\Box} W) \ni (x \otimes v) \otimes (y \otimes w) \mapsto xy \otimes v \otimes w \in A \underset{H}{\Box} (V \otimes W)$$

and $\xi_0\colon k \ni \alpha \mapsto 1 \otimes \alpha \in A \Box_H k$ define the structure of a weak monoidal functor on $A \Box_H -\colon {}^H\mathcal{M} \to \mathcal{M}_k$.

Conversely, every weak monoidal functor structure on $A \Box_H -$ has the above form for a unique H-comodule algebra structure on A.

Proof The first claim is easy to check. For the second, given a monoidal functor structure ξ, define multiplication on A as the composition

$$A \otimes A \cong (A \underset{H}{\Box} H) \otimes (A \underset{H}{\Box} H) \xrightarrow{\xi} A \underset{H}{\Box} (H \otimes H) \xrightarrow{A \Box \nabla} A \underset{H}{\Box} H \cong A.$$

By naturality of ξ in its right argument, applied to $\Delta\colon H \to {}^{\bullet}H \otimes H$, we have a commutative diagram

$$\begin{array}{ccc}
(A \square_H H) \otimes (A \square_H H) & \xrightarrow{\xi} & A \square_H (H \otimes H) \\
\downarrow{\scriptstyle A\square_H H \otimes A \square_H \Delta} & & \downarrow{\scriptstyle A\square_H (H\otimes\Delta)} \\
(A \square_H H) \otimes (A \square_H (H \otimes H)) & & A \square_H (H \otimes H \otimes H) \\
\downarrow & & \downarrow \\
(A \square_H H) \otimes (A \square_H H) \otimes H & \xrightarrow{\xi \otimes H} & (A \square_H (H \otimes H)) \otimes H
\end{array}$$

In other words, $\xi\colon (A \square_H H) \otimes (A \square_H H^{\bullet}) \to A \square_H (H \otimes H^{\bullet})$ is an H-comodule map with respect to the indicated structures. Similarly (though a little more complicated to write), $\xi\colon (A \square_H H^{\bullet}) \otimes (A \square_H H) \to A \square_H (H^{\bullet} \otimes H)$ is also colinear, and from both we deduce that $\xi\colon (A \square_H H^{\bullet}) \otimes (A \square_H H^{\bullet}) \to A \square_H (H^{\bullet} \otimes H^{\bullet})$ is colinear. Hence the muliplication on A is colinear. Associativity of multiplication follows from coherence of ξ, so that A is a comodule algebra. □

Corollary 2.5.2 *Let H be a bialgebra, A a right H-comodule algebra, and $\iota\colon B \to A^{\mathrm{co}\,H}$ a subalgebra.*

Then for each $V \in {}^H\mathcal{M}$ we have $A \square_H V \in {}_B\mathcal{M}_B$ with bimodule structure induced by that of A (induced in turn by ι). The weak monoidal functor structure in Proposition 2.5.1 induces a weak monoidal functor structure on $A \square_H (-)\colon {}^H\mathcal{M} \to {}_B\mathcal{M}_B$, which we denote again by

$$\xi\colon (A \underset{H}{\square} V) \underset{B}{\otimes} (A \underset{H}{\square} W) \ni x \otimes v \otimes y \otimes w \mapsto xy \otimes v \otimes w \in A \underset{H}{\square} (V \otimes W)$$

and $\xi_0\colon B \ni b \mapsto b \otimes 1 \in A \square_H k$. If A is a (faithfully) coflat H-comodule, the functor is (faithfully) exact

Every exact weak monoidal functor ${}^H\mathcal{M} \to {}_B\mathcal{M}_B$ commuting with arbitrary direct sums, for a k-algebra B, has this form.

Proof Again, it is not hard to verify that every comodule algebra A and homomorphism ι gives rise to a weak monoidal functor as stated. For the converse, note that a weak monoidal functor ${}^H\mathcal{M} \to {}_B\mathcal{M}_B$ can be composed with the weak monoidal underlying functor ${}_B\mathcal{M}_B \to \mathcal{M}_k$ to yield a weak monoidal functor ${}^H\mathcal{M} \to \mathcal{M}_k$. The latter is exact by assumption, so has the form $V \mapsto A \square_H V$ for some coflat H-comodule A by Lemma 4.3.3, and A is an H-comodule algebra by Proposition 2.5.1. One ingredient of the weak monoidal functor structure that we assume to exist is a B-B-bimodule map $\xi_0\colon B \to A \square_H k$ with $\xi_0(1) = 1$, which has the form $\xi_0(b) = \iota(b) \otimes 1$ for some map $\iota\colon B \to A^{\mathrm{co}\,H}$ that also satisfies $\iota(b) = 1$. By coherence of the weak monoidal functor, the left B-module structure of $A \square_H V$, which is also one of the coherence isomorphisms of the monoidal category of B-B-bimodules, is given by

$$B \underset{B}{\otimes} (A \underset{H}{\square} V) \xrightarrow{\xi_0 \otimes id} (A \underset{H}{\square} k) \underset{B}{\otimes} (A \underset{H}{\square} V) \xrightarrow{\xi} A \underset{H}{\square} V.$$

Thus $b \cdot (x \otimes v) = \iota(b)x \otimes v$ holds for all $b \in B$ and $x \otimes v \in A \square_H V$. If we specialize $V = H$ and use the isomorphism $A \square_H H$, we see that ι is an algebra homomorphism, and for general V we see that $A \square_H V$ has the claimed B-B-bimodule structure. □

Theorem 2.5.3 *Let H be a k-flat Hopf algebra, and B a k-algebra.*

1. *Every exact monoidal functor $\mathcal{F}\colon {}^H\mathcal{M} \to {}_B\mathcal{M}_B$ that commutes with arbitrary colimits has the form $\mathcal{F}(V) = A \Box_H V$ for some right coflat H-Galois extension A of B, with monoidal functor structure given as in Corollary 2.5.2.*
2. *Assume that A is a right faithfully flat H-Galois extension of B. Then the weak monoidal functor $A \Box_H -$ as in Corollary 2.5.2 is monoidal.*

If we assume that k is a field, and H has bijective antipode, then a Hopf-Galois extension is coflat as H-comodule if and only if it is faithfully flat as right (or left) B-module. Also, if k is arbitrary, then a Hopf-Galois extension of k is faithfully coflat as H-comodule if and only if it is faithfully flat as k-module. Thus we have:

Corollary 2.5.4 *Let H be a Hopf algebra and B a k-algebra. Assume either of the following conditions:*

1. *k is a field and the antipode of H is bijective.*
2. *$B = k$.*

Then Corollary 2.5.2 establishes a bijective correspondence between exact monoidal functors ${}^H\mathcal{M} \to {}_B\mathcal{M}_B$ and faithfully flat H-Galois extensions of B.

Closing the section, let us give two curious application of the monoidal functor associated to a Galois object.

If H is a Hopf algebra, then any $V \in {}^H\mathcal{M}$ that is a finitely generated projective k-module has a right dual object in the monoidal category ${}^H\mathcal{M}$. Monoidal functors preserve duals. Thus, whenever A is a right faithfully flat H-Galois extension of B, the B-bimodule $A \Box_H V$ will have a right dual in the monoidal category of B-bimodules. This in turn means that $A \Box_H V$ is finitely generated projective as a left B-module. We have proved:

Corollary 2.5.5 *Let A be a right H-Galois extension of B and a right faithfully flat B-module. Then for every $V \in {}^H\mathcal{M}$ which is a finitely generated projective k-module, the left B-module $A \Box_H V$ is finitely generated projective. If H has bijective antipode, the right B-module $A \Box_H V$ is also finitely generated projective.*

The corollary (which has other proofs as well) has a conceptual meaning when we think of A as a principal fiber bundle with structure quantum group H. Then $A \Box_H V$ is analogous to the module of sections in an associated vector bundle with fiber V, and it is of course good to know that such a module of sections is projective, in keeping with the classical Serre-Swan theorem.

Definition 2.5.6 Let H be a k-flat Hopf algebra, and B a k-algebra. We define $\mathrm{Gal}_B(H)$ to be the set of all isomorphism classes of H-Galois extensions of B that are faithfully flat as right B-modules and (faithfully) flat as k-modules. We write $\mathrm{Gal}(H) = \mathrm{Gal}_B(H)$.

Proposition 2.5.7 *$\mathrm{Gal}_B(-)$ is a contravariant functor. For a Hopf algebra map $f\colon F \to H$ between k-flat Hopf algebras, the map $\mathrm{Gal}_B(f)\colon \mathrm{Gal}_B(H) \to \mathrm{Gal}_B(F)$ maps the isomorphism class of A to that of $A \Box_H F$.*

Proof In fact, f defines an exact monoidal functor ${}^F\mathcal{M} \to {}^H\mathcal{M}$, which composes with the monoidal functor $A \Box_H (-)\colon {}^H\mathcal{M} \to {}_B\mathcal{M}_B$ defined by A to give the functor $(A \Box_H F) \Box_F -$, since $A \Box_H V = A \Box_H (F \Box_F V) \cong (A \Box_H F) \Box_F V$ by k-flatness of A. This implies that $A \Box_H F$ is a right F-Galois extension of B.

It is faithfully flat on the right since A is, and for any left B-module M we have $A \otimes_B (A \Box_H F) \otimes_B N \cong ((A \otimes_B A) \Box_H F) \otimes_B N \cong (A \otimes H \Box_H F) \otimes_B N \cong A \otimes_B N \otimes H$. $\square$

2.6 Hopf bimodules. Let A be an H-comodule algebra. Since A is an algebra in the monoidal category of H-comodules, we can consider the category of bimodules over A in the monoidal category $\mathcal{M}^H$. Such a bimodule $M \in {}_A\mathcal{M}^H_A$ is an A-bimodule fulfilling both Hopf module conditions for a Hopf module in $\mathcal{M}^H_A$ and ${}_A\mathcal{M}^H$. By the general theory of modules over algebras in monoidal categories, the category ${}_A\mathcal{M}^H_A$ is a monoidal category with respect to the tensor product over A. Now without further conditions, taking coinvariants gives a *weak* monoidal functor:

Lemma 2.6.1 *Let A be an H-comodule algebra, and let $B \subset A^{\operatorname{co} H}$ be a subalgebra. Then*

$$ {}_A\mathcal{M}^H_A \ni M \mapsto M^{\operatorname{co} H} \in {}_B\mathcal{M}_B $$

is a weak monoidal functor with structure maps

$$\xi_0 \colon M^{\operatorname{co} H} \underset{B}{\otimes} N^{\operatorname{co} H} \ni m \otimes n \mapsto m \otimes n \in (M \underset{A}{\otimes} N)^{\operatorname{co} H}$$

and $\xi_0 \colon B \to A^{\operatorname{co} H}$ the inclusion.

The proof is straightforward. The main result of this section is that the functor from the Lemma is monoidal rather than only weak monoidal if and only if A is an H-Galois extension. The precise statement is slightly weaker:

Proposition 2.6.2 *Let H be a Hopf algebra, A a right H-comodule algebra, and $B := A^{\operatorname{co} H}$.*

If A is a left faithfully flat H-Galois extension of B, then the weak monoidal functor from Lemma 2.6.1 is monoidal.

Conversely, if the weak monoidal functor from Lemma 2.6.1 is monoidal, then the counit of the adjunction 2.3.2 is an isomorphism, and in particular, A is an H-Galois extension of B.

Proof If A is a left faithfully flat H-Galois extension of B, then ξ is an isomorphism if and only if $\xi \otimes_B A$ is. But via the isomorphisms

$$M \underset{A}{\otimes} N \cong M^{\operatorname{co} H} \underset{B}{\otimes} A \underset{A}{\otimes} N \cong M^{\operatorname{co} H} \underset{B}{\otimes} N \cong M^{\operatorname{co} H} \underset{B}{\otimes} N^{\operatorname{co} H} \underset{B}{\otimes} A$$

and $(M \otimes_A N)^{\operatorname{co} H} \otimes_B A \cong M \otimes_A N$, the map $\xi \otimes_B A \colon M^{\operatorname{co} H} \otimes_B N^{\operatorname{co} H} \otimes_B A \to (M \otimes_A N)^{\operatorname{co} H} \otimes_B A$ identifies with the identity on $M \otimes_A N$.

Conversely, if ξ is an isomorphism, we can specialize $N := {}_{\bullet}A_{\bullet} \otimes {}_{\bullet}H^{\bullet}_{\bullet} \in {}_A\mathcal{M}^H_A$. We have $N^{\operatorname{co} H} \cong A$. In ${}_A\mathcal{M}^H$ we have an isomorphism

$${}_{\bullet}A^{\bullet} \otimes H^{\bullet} \cong {}_{\bullet}A \otimes {}_{\bullet}H^{\bullet}; a \otimes h \mapsto a_{(0)} \otimes a_{(1)}h.$$

Thus we have an isomorphism

$$M \underset{A}{\otimes} N \cong M \underset{A}{\otimes} ({}_{\bullet}A^{\bullet} \otimes H^{\bullet}) \cong M^{\bullet} \otimes H^{\bullet}; m \otimes a \otimes h \mapsto ma_{(0)} \otimes S(a_{(1)})h.$$

composing with $M^{\bullet} \otimes H^{\bullet} \ni m \otimes h \mapsto m_{(0)} \otimes m_{(1)}h \in M \otimes H^{\bullet}$ yields the isomorphism

$$M \underset{A}{\otimes} N \cong M \otimes H^{\bullet}; m \otimes a \otimes h \mapsto m_{(0)}a \otimes m_{(1)}h.$$

Thus we find that

$$M^{\operatorname{co} H} \underset{B}{\otimes} A \cong M^{\operatorname{co} H} \underset{B}{\otimes} N^{\operatorname{co} H} \xrightarrow{\xi} (M \underset{A}{\otimes} N)^{\operatorname{co} H} \cong (M \otimes H^{\bullet})^{\operatorname{co} H} \cong M$$

maps $m \otimes a$ to ma, hence is the adjunction counit in question. □

2.7 Reduction. We have already seen that $\mathrm{Gal}_B(—)$ is a functor. In particular, we have a map $\mathrm{Gal}_B(Q) \to \mathrm{Gal}_B(H)$ for any (suitable) quotient Hopf algebra Q of H. In this section we will be concerned with the image and fibers of this map. The question has a geometric interpretation when we think of Galois extensions as principal fiber bundles: It is then the question under what circumstances a principal bundle with structure group G can be reduced to a principal bundle whose structure group is a prescribed subgroup of G.

The results in this section were proved first in [37] for the case of conormal quotients Q (i.e. normal subgroups, when we think of principal homogeneous spaces). The general case was obtained in [20, 21]. The proof we give here was essentially given in [43]; we rewrite it here with (yet) more emphasis on its background in the theory of algebras in monoidal categories. We begin with a Theorem of Takeuchi [49] on Hopf modules for a quotient of a Hopf algebra. We prove a special case in a new way here, which we do not claim to be particularly natural, but which only uses category equivalences that we have already proved above.

Theorem 2.7.1 *Let H be a k-flat Hopf algebra, and $H \to Q$ a quotient Hopf algebra of H which is also k-flat and has bijective antipode. Assume that H is a left Q-Galois extension of $K := {}^{\mathrm{co}\,Q}H$, and faithfully flat as left as well as right K-module.*

Then $\mathcal{M}^H_K \ni M \mapsto M/MK^+ \in \mathcal{M}^Q$ is a category equivalence. The inverse equivalence maps $N \in \mathcal{M}^Q$ to $N \,\Box_Q\, H$ with the K-module and H-comodule structures induced by those of H.

Remark 2.7.2 As we learned in Corollary 2.4.9, our list of requirements on the quotient $H \to Q$ is fulfilled if k is a field, Q has bijective antipode, and H is a coflat left Q-comodule.

Proof By the structure theorem for Hopf modules over a Hopf-Galois extension, we have an equivalence $\mathcal{F}\colon \mathcal{M}_K \to {}^Q\mathcal{M}_H$ given by $\mathcal{F}(N) = N \otimes_K H$, with quasi-inverse $\mathcal{F}(M) = {}^{\mathrm{co}\,Q}M$. We claim that $\mathcal{F}$ induces an equivalence $\hat{\mathcal{F}}\colon \mathcal{M}^H_K \to {}^Q\mathcal{M}^H_H$. Indeed, if $N \in \mathcal{M}^H_K$, then $N \otimes_K H$ is an object of ${}^Q\mathcal{M}^H_H$ when endowed with the diagonal right H-module structure (which is well-defined since K is an H-comodule subalgebra of H). Conversely, if $M \in {}^Q\mathcal{M}^H_H$, then ${}^{\mathrm{co}\,Q}M$ is a right H-subcomodule of M and in this way a Hopf module in $\mathcal{M}^H_K$. It is straightforward to check that the adjunction morphisms for $\mathcal{F}$ and $\mathcal{F}^{-1}$ are compatible with these additional structures.

Next, we have the category equivalence $\mathcal{M}^H_H \cong \mathcal{M}_k$, which induces an equivalence $\mathcal{G}\colon {}^Q\mathcal{M} \to {}^Q\mathcal{M}^H_H$. Indeed, if $V \in {}^Q\mathcal{M}$, then $V \otimes H \in {}^Q\mathcal{M}^H_H$ with the codiagonal left Q-comodule structure, and conversely, if $M \in {}^Q\mathcal{M}^H_H$, then $M^{\mathrm{co}\,H}$ is a left Q-subcomodule of M. Now consider the composition

$$\mathcal{T} := \left(\mathcal{M}^Q \to {}^Q\mathcal{M} \xrightarrow{\mathcal{G}} {}^Q\mathcal{M}^H_H \xrightarrow{\hat{\mathcal{F}}^{-1}} \mathcal{M}^H_K\right)$$

where the first functor is induced by the inverse of the antipode. We have $\mathcal{T}(V) = {}^{\mathrm{co}\,Q}(V^{S^{-1}} \otimes H) = V \,\Box_Q\, H$. We leave it to the reader to check that the module and comodule structure of $\mathcal{T}(V)$ are indeed those induced by H.

Since it is, finally, easy to check that the $\mathcal{M}^H_K \ni M \to M/MK^+ \in \mathcal{M}^Q$ is left adjoint to $\mathcal{T}$, it is also its quasi-inverse. □

Let H and Q be k-flat Hopf algebras, and $\nu\colon H \to Q$ a Hopf algebra map. Then $K := {}^{\mathrm{co}\,Q}H$ is stable under the right adjoint action of H on itself defined by $x \leftharpoonup h = S(h_{(1)})xh_{(2)}$, since for $x \in K$ we have $\nu((x \leftharpoonup h)_{(1)}) \otimes (x \leftharpoonup h)_{(2)} = \nu(S(h_{(2)})x_{(1)}h_{(3)}) \otimes S(h_{(1)})x_{(2)}h_{(4)} = \nu(S(h_{(2)})h_{(3)}) \otimes S(h_{(1)})xh_{(4)} = 1 \otimes S(h_{(1)})xh_{(2)}$ for all $h \in H$. Thus K is a subalgebra of the (commutative) algebra H in the category $\mathcal{YD}_H^H$ of right-right H-Yetter-Drinfeld modules, which in turn is the center of the monoidal category $\mathcal{M}^H$ of right H-comodules.

As a corollary, the category $\mathcal{M}_K^H$ is equivalent to the monoidal subcategory $\mathcal{S} \subset {}_K\mathcal{M}_K^H$ of symmetric bimodules in $\mathcal{M}^H$, that is, the category of those $M \in {}_K\mathcal{M}_K^H$ for which $xm = m_{(0)}(x \leftharpoonup m_{(1)})$ holds for all $m \in M$ and $x \in K$. The equivalence is induced by the underlying functor ${}_K\mathcal{M}_K^H \to \mathcal{M}_K^H$. Since now the source and target of the equivalence in Theorem 2.7.1 are monoidal functors, the following Theorem answers an obvious question:

Theorem 2.7.3 *The category equivalence from Theorem 2.7.1 is a monoidal category equivalence with respect to the isomorphisms*

$$\xi\colon (V \underset{Q}{\Box} H) \underset{K}{\otimes} (W \underset{Q}{\Box} H) \ni v \otimes g \otimes w \otimes h \mapsto v \otimes w \otimes gh \in (V \otimes W) \underset{Q}{\Box} H$$

Proof We have already seen (with switched sides) in Section 2.5 that ξ makes $(\text{—}) \Box_Q H\colon \mathcal{M}^Q \to {}_K\mathcal{M}_K$ a monoidal functor. Quite obviously, $V \Box_Q H$ has the structure of a right H-comodule in such a way that $V \Box_Q H \in {}_K\mathcal{M}_K^H$, and ξ is an H-comodule map. Thus, we have a monoidal functor $(\text{—}) \Box_Q H\colon \mathcal{M}^Q \to {}_K\mathcal{M}_K^H$. For $t = v \otimes h \in V \Box_Q H$ we have $xt = v \otimes xh = v \otimes h_{(1)}(x \leftharpoonup h_{(2)}) = t_{(0)}(x \leftharpoonup t_{(1)})$, so that the monoidal functor $(\text{—}) \Box_Q H$ takes values in the subcategory $\mathcal{S}$. Observe, finally, that it composes with the underlying functor to $\mathcal{M}_K^H$ to give the equivalence of categories from Theorem 2.7.1. From the commutative triangle

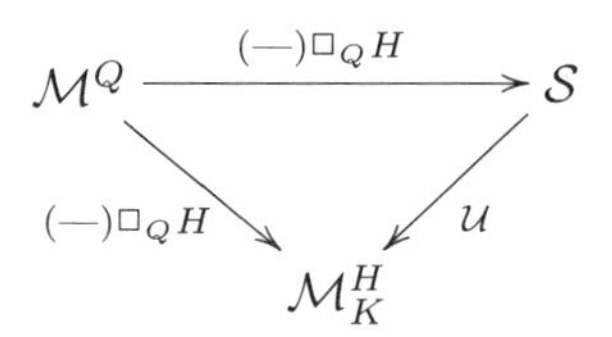

of functors, in which we already know the slanted arrows to be equivalences, we deduce that the top arrow is an equivalence. $\Box$

Corollary 2.7.4 *Assume the hypotheses of Theorem 2.7.1.*

The categories of right Q-comodule algebras, and of algebras in the category $\mathcal{M}_K^H$ are equivalent. The latter consists of pairs (A, f) in which A is a right H-comodule algebra, and $f\colon K \to A$ is a right H-comodule algebra map satisfying $f(x)a = a_{(0)}f(x \leftharpoonup a_{(1)})$ for all $x \in K$ and $a \in A$.

Suppose given an algebra $A \in \mathcal{M}_K^H$, corresponding to an algebra $\overline{A} \in \mathcal{M}^Q$. Then the categories of right A-modules in $\mathcal{M}_K^H$ and of right $\overline{A}$-modules in $\mathcal{M}^Q$ are equivalent. The former is the category of right A-modules in $\mathcal{M}_K^H$, so we have a

commutative diagram of functors

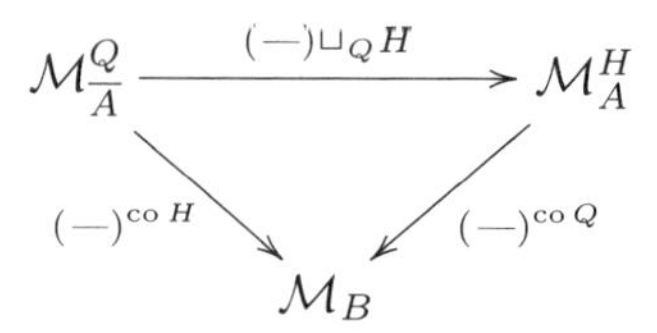

where $B = A^{\mathrm{co}\,H} \cong \overline{A}^{\mathrm{co}\,Q}$.

Corollary 2.7.5 *Assume the hypotheses of Theorem 2.7.1. Then we have a bijection between*

1. *isomorphism classes of left faithfully flat* Q*-Galois extensions of* B*, and*
2. *equivalences of pairs* (A, f) *in which* A *is a left faithfully flat* H*-Galois extension of* B*, and* $f\colon K \to A^B$ *is a homomorphism of algebras in* $\mathcal{YD}^H_H$*. Here, two pairs* (A, f) *and* (A', f') *are equivalent if there is a* B*-linear* H*-comodule algebra map* $t\colon A \to A'$ *such that* $tf = f'$.

Proof We know that Q-comodule algebras $\overline{A}$ correspond to isomorphism classes of pairs (A, f) in which A is an H-comodule algebra and $f\colon K \to A$ an H-comodule algebra map which is central in the sense of Definition 4.2.1. Since faithfully flat Galois extensions are characterized by the structure theorem for Hopf modules, see Corollary 2.3.6, the diagram in Corollary 2.7.4 shows that $\overline{A}$ is faithfully flat Q-Galois if and only if A is faithfully flat H-Galois. But if A is faithfully flat H-Galois, then every central H-comodule algebra map factors through a Yetter-Drinfeld module algebra map to A^B by Lemma 2.1.9. □

The preceding corollary can be restated as follows:

Corollary 2.7.6 *Assume the hypotheses of Theorem 2.7.1. Consider the map* $\pi\colon \mathrm{Gal}_B(Q) \to \mathrm{Gal}_B(H)$ *given by* $\pi(\overline{A}) = A \,\square_Q\, H$*, and let* $A \in \mathrm{Gal}_B(H)$*. Then*

$$\pi^{-1}(A) \cong \mathrm{Alg}^{-H}_{-H}(K, A^B) / \mathrm{Aut}^H_B(A),$$

where $\mathrm{Aut}^H_B(A)$ *acts on* $\mathrm{Alg}^{-H}_{-H}(K, A^B)$ *by composition.*

2.8 Hopf Galois extensions without Hopf algebras. Cyril Grunspan [19] has revived an idea that appears to have been known in the case of commutative Hopf-Galois extensions (or torsors) for a long time, going back to a paper of Reinhold Baer [1]: It is possible to write down axioms characterizing a Hopf-Galois extension without mentioning a Hopf algebra.

This approach to (noncommutative) Hopf-Galois extensions begins in [19] with the definition of a quantum torsor (an algebra with certain additional structures) and the proof that every quantum torsor gives rise to two Hopf algebras over which it is a bi-Galois extension of the base field. The converse was proved in [41]: Every Hopf-Galois extension of the base field is a quantum torsor in the sense of Grunspan. Then the axioms of a quantum torsor were simplified in [42] by showing that a key ingredient of Grunspan's definition (a certain endomorphism of the torsor) is actually not needed to show that a torsor is a Galois object. The simplified version of the torsor axioms admits a generalization to general Galois extensions (not only of the base ring or field).

Definition 2.8.1 Let B be a k-algebra, and $B \subset T$ an algebra extension, with T a faithfully flat k-module. The centralizer $(T \otimes_B T)^B$ of B in the (obvious) B-B-bimodule $T \otimes_B T$ is an algebra by $(x \otimes y)(a \otimes b) = ax \otimes yb$ for $x \otimes y, a \otimes b \in (T \otimes_B T)^B$.

A B-torsor structure on T is an algebra map $\mu\colon T \to T \otimes (T \otimes_B T)^B$; we denote by $\mu_0\colon T \to T \otimes T \otimes_B T$ the induced map, and write $\mu_0(x) = x^{(1)} \otimes x^{(2)} \otimes x^{(3)}$.

The torsor structure is required to fulfill the following axioms:

$$x^{(1)}x^{(2)} \otimes x^{(3)} = 1 \otimes x \in T \underset{B}{\otimes} T \tag{2.8.1}$$

$$x^{(1)} \otimes x^{(2)}x^{(3)} = x \otimes 1 \in T \otimes T \tag{2.8.2}$$

$$\mu(b) = b \otimes 1 \otimes 1 \quad \forall b \in B \tag{2.8.3}$$

$$\mu(x^{(1)}) \otimes x^{(2)} \otimes x^{(3)} = x^{(1)} \otimes x^{(2)} \otimes \mu(x^{(3)}) \in T \otimes T \underset{B}{\otimes} T \otimes T \underset{B}{\otimes} T \tag{2.8.4}$$

Note that (2.8.4) makes sense since μ is a left B-module map by (2.8.3).

Remark 2.8.2 If $B = k$, then the torsor axioms simplify as follows: They now assume the existence of an algebra map $\mu\colon T \to T \otimes T^{\mathrm{op}} \otimes T$ such that the diagrams

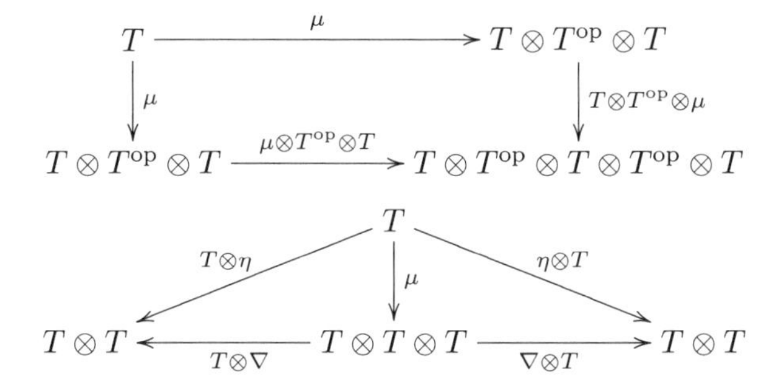

commute.

The key observation is now that a torsor provides a descent data. Here we use left descent data, i.e. certain S-linear maps $\theta\colon M \to S \otimes_R M$ for a ring extension $R \subset S$ and a left S-module M, as opposed to the right descent data in Section 4.5. For a left descent data $\theta\colon M \to S \otimes_R M$ from S to R on a left S-module M we will write ${}^{\theta}M := \{m \in M | \theta(m) = 1 \otimes m\}$.

Lemma 2.8.3 *Let T be a B-torsor. Then a descent data D from T to k on $T \otimes_B T$ is given by $D(x \otimes y) = xy^{(1)} \otimes y^{(2)} \otimes y^{(3)}$. It satisfies $(T \otimes D)\mu(x) = x^{(1)} \otimes 1 \otimes x^{(2)} \otimes x^{(3)}$.*

Proof Left T-linearity of D is obvious. We have

$$\begin{aligned}
(T \otimes D)\mu(x) &= x^{(1)} \otimes D(x^{(2)} \otimes x^{(3)}) \\
&= x^{(1)} \otimes (\nabla \otimes T \otimes T)(x^{(2)} \otimes \mu(x^{(3)})) \\
&= (T \otimes \nabla \otimes T \otimes T)(\mu(x^{(1)}) \otimes x^{(2)} \otimes x^{(3)}) \\
&= x^{(1)} \otimes 1 \otimes x^{(2)} \otimes x^{(3)}
\end{aligned}$$

and thus

$$\begin{aligned}(T\otimes D)D(x\otimes y) &= xy^{(1)}\otimes D(y^{(2)}\otimes y^{(3)})\\ &= xy^{(1)}\otimes 1\otimes y^{(2)}\otimes y^{(3)}\\ &= (T\otimes\eta\otimes T\otimes T)D(x\otimes y).\end{aligned}$$

Finally $(\nabla\otimes T\otimes T)D(x\otimes y) = xy^{(1)}y^{(2)}\otimes y^{(3)} = x\otimes y$. □

Note that $D(T\otimes_B T)\subset T\otimes(T\otimes_B T)^B$. Since T is faithfully flat over k, then faithfully flat descent implies that ${}^D(T\otimes_B T)\subset(T\otimes_B T)^B$.

Theorem 2.8.4 *Let T be a B-torsor, and assume that T is a faithfully flat right B-module.*

Then $H := {}^D(T\otimes_B T)$ is a k-flat Hopf algebra. The algebra structure is that of a subalgebra of $(T\otimes_B T)^B$, the comultiplication and counit are given by

$$\Delta(x\otimes y) = x\otimes y^{(1)}\otimes y^{(2)}\otimes y^{(3)},$$
$$\varepsilon(x\otimes y) = xy$$

for $x\otimes y\in H$. The algebra T is an H-Galois extension of B under the coaction $\delta\colon T\to T\otimes H$ given by $\delta(x)=\mu(x)$.

Proof H is a subalgebra of $(T\otimes_B T)^B$ since for $x\otimes y, a\otimes b\in H$ we have

$$\begin{aligned}D((x\otimes y)(a\otimes b)) &= D(ax\otimes yb)\\ &= ax(yb)^{(1)}\otimes(yb)^{(2)}\otimes(yb)^{(3)}\\ &= axy^{(1)}b^{(1)}\otimes b^{(2)}y^{(2)}\otimes y^{(3)}b^{(3)}\\ &= ab^{(1)}\otimes b^{(2)}x\otimes yb^{(3)}\\ &= 1\otimes ax\otimes yb\\ &= 1\otimes(x\otimes y)(a\otimes b).\end{aligned}$$

To see that the coaction δ is well-defined, we have to check that the image of μ is contained in $T\otimes H$, which is, by faithful flatness of T, the equalizer of

$$T\otimes T\otimes_B T \underset{T\otimes\eta\otimes T\otimes_B T}{\overset{T\otimes D}{\rightrightarrows}} T\otimes T\otimes T\otimes_B T\,.$$

But $(T\otimes D)\mu(x) = (T\otimes\eta\otimes T\otimes_B T)\mu(x)$ was shown in Lemma 2.8.3. Since μ is an algebra map, so is the coaction δ, for which we employ the usual Sweedler notation $\delta(x)=x_{(0)}\otimes x_{(1)}$. Note that (2.8.3) implies that $\delta(b)=b\otimes 1$ for all $b\in B$; in other words, δ is left B-linear.

The Galois map $\beta\colon T\otimes_B T\to T\otimes H$ for the coaction δ is given by $\beta(x\otimes y) = xy_{(0)}\otimes y_{(1)} = xy^{(1)}\otimes y^{(2)}\otimes y^{(3)} = D(x\otimes y)$. Thus it is an isomorphism by faithfully flat descent, Theorem 4.5.2. It follows that H is faithfully flat over k.

Since δ is left B-linear,

$$\Delta_0\colon T\underset{B}{\otimes}T\ni x\otimes y\mapsto x\otimes y^{(1)}\otimes y^{(2)}\otimes y^{(3)}\in T\underset{B}{\otimes}T\otimes H$$

is well-defined. To prove that Δ is well-defined, we need to check that $\Delta_0(H)$ is contained in $H\otimes H$, which, by faithful flatness of H, is the equalizer of

$$T\otimes_B T\otimes H \underset{\eta\otimes T\otimes T\otimes H}{\overset{D\otimes H}{\rightrightarrows}} T\otimes T\otimes_B T\otimes H\,.$$

Now for $x \otimes y \in H$ we have

$$\begin{aligned}(D \otimes H)\Delta_0(x \otimes y) &= (D \otimes H)(x \otimes y^{(1)} \otimes y^{(2)} \otimes y^{(3)}) \\ &= xy^{(1)(1)} \otimes y^{(1)(2)} \otimes y^{(1)(3)} \otimes y^{(2)} \otimes y^{(3)} \\ &= xy^{(1)} \otimes y^{(2)} \otimes \mu(y^{(3)}) \\ &= (T \otimes T \otimes \mu)D(x \otimes y) \\ &= (T \otimes T \otimes \mu)(1 \otimes x \otimes y) \\ &= 1 \otimes \Delta_0(x \otimes y)\end{aligned}$$

Δ is an algebra map since μ is, and coassociativity follows from the coassociativity axiom of the torsor T.

For $x \otimes y \in H$ we have $xy \otimes 1 = xy^{(1)} \otimes y^{(2)}y^{(3)} = 1 \otimes xy$, whence $xy \in k$ by faithful flatness of T. Thus, ε is well-defined. It is straightforward to check that ε is an algebra map, that it is a counit for Δ, and that the coaction δ is counital. Thus, H is a bialgebra.

We may now write the condition $\delta(b) = b \otimes 1$ for $b \in B$ simply as $B \subset T^{\operatorname{co} H}$. Conversely, $x \in T^{\operatorname{co} H}$ implies $x \otimes 1 = x^{(1)}x^{(2)} \otimes x^{(3)} = 1 \otimes x \in T \otimes_B T$, and thus $x \in B$ by faithful flatness of T as a B-module. Since we have already seen that the Galois map for the H-extension $B \subset T$ is bijective, T is an H-Galois extension of B, and from Lemma 2.1.5 we deduce that H is a Hopf algebra. □

Lemma 2.8.5 *Let H be a k-faithfully flat Hopf algebra, and let T be a right faithfully flat H-Galois extension of $B \subset T$. Then T is a B-torsor with torsor structure*

$$\mu \colon T \ni x \mapsto x_{(0)} \otimes x_{(1)}{}^{[1]} \otimes x_{(1)}{}^{[2]} \in T \otimes (T \underset{B}{\otimes} T)^B,$$

where $h^{[1]} \otimes h^{[2]} = \beta^{-1}(1 \otimes h) \in T \otimes_B T$, with $\beta \colon T \otimes_B T \to T \otimes H$ the Galois map.

3 Hopf-bi-Galois theory

3.1 The left Hopf algebra. Let A be a faithfully flat H-Galois object. Then A is a torsor by Lemma 2.8.5 By the left-right switched version of Theorem 2.8.4, there exists a Hopf algebra $L := L(A, H)$ such that A is a left L-Galois extension of k. Moreover, since the torsor structure $\mu \colon A \to A \otimes A^{\mathrm{op}} \otimes A$ is right H-colinear, we see that A is an L-H-bicomodule.

Definition 3.1.1 An L-H-bi-Galois object is a k-faithfully flat L-H-bicomodule algebra A which is simultaneously a left L-Galois object and a right H-Galois object.

We have seen that every right H-Galois object can be endowed with a left L-comodule algebra structure making it an L-H-bi-Galois object. We shall prove uniqueness by providing a universal property shared by every L that makes a given H-Galois object into an L-H-bi-Galois object.

Proposition 3.1.2 *Let H and L be k-flat Hopf algebras, and A an L-H-bi-Galois object.*

Then for all $n \in \mathbb{N}$ and k-modules V, W we have a bijection

$$\Phi := \Phi_{V,W,n} \colon \operatorname{Hom}(V \otimes L^{\otimes n}, W) \cong \operatorname{Hom}^{-H}(V \otimes A^{\otimes n}, W \otimes A)$$

(where $A^{\otimes n}$ carries the codiagonal comodule structure), given by $\Phi(f)(v\otimes x_1\otimes\ldots\otimes x_n) = f(v\otimes x_{1(-1)}\otimes\ldots\otimes x_{n(-1)})\otimes x_{1(0)}\cdot\dots\cdot x_{n(0)}$.

In particular, for every k-module we have the universal property that every right H-colinear map $\phi\colon A\to W\otimes A$ factors uniquely in the form $\phi = (f\otimes A)\delta_\ell$ as in the diagram

$$\begin{array}{ccccc} A & \xrightarrow{\delta_\ell} & L\otimes A & \qquad & L \\ & \searrow^{\phi} & \downarrow{\scriptstyle f\otimes A} & & \downarrow{\scriptstyle f} \\ & & W\otimes A & & W \end{array}$$

Proof Note first that the left Galois map $\beta_\ell\colon A^\bullet\otimes A^\bullet_\bullet\to L\otimes A^\bullet_\bullet$ is evidently a map of Hopf modules in $\mathcal{M}^H_A$ with the indicated structures. We deduce that for any $M\in\mathcal{M}^H$ we have

$$M^\bullet\otimes L^{\otimes m}\otimes A^\bullet_\bullet\cong M^\bullet\otimes L^{\otimes(m-1)}\otimes A^\bullet\otimes A^\bullet_\bullet\cong M^\bullet\otimes A^\bullet\otimes L^{\otimes(m-1)}\otimes A^\bullet_\bullet$$

in $\mathcal{M}^H_A$, and hence by induction

$$V\otimes L^{\otimes n}\otimes A^\bullet_\bullet\cong V\otimes (A^\bullet)^{\otimes n}\otimes A^\bullet_\bullet\in\mathcal{M}^H_A$$

We can now use the structure theorem for Hopf modules, Corollary 2.3.6, to compute

$$\begin{aligned}\operatorname{Hom}(V\otimes L^{\otimes n},W) &\cong \operatorname{Hom}^{-H}_{-A}(V\otimes L^{\otimes n}\otimes A^\bullet_\bullet, W\otimes A^\bullet_\bullet)\\ &\cong \operatorname{Hom}^{-H}_{-A}(V\otimes (A^\bullet)^{\otimes n}\otimes A^\bullet_\bullet, W\otimes A^\bullet_\bullet)\cong \operatorname{Hom}^{-H}(V\otimes A^{\otimes n}, W\otimes A).\end{aligned}$$

We leave it to the reader to verify that the bijection has the claimed form. □

Corollary 3.1.3 *Let A be an L-H-bi-Galois object, B a k-module, $f\colon L\to B$, and $\lambda = \Phi(f)\colon A\to B\otimes A$.*

1. *Assume B is a coalgebra. Then f is a coalgebra map if and only if λ is a comodule structure.*
2. *Assume B is an algebra. Then f is an algebra map if and only if λ is.*
3. *In particular, assume B is a bialgebra. Then f is a bialgebra map if and only if λ is a comodule algebra structure. In particular, the bialgebra L in an L-H-bi-Galois object is uniquely determined by the H-Galois object A.*

Proof We have $\Delta f = (f\otimes f)\Delta\colon L\to B\otimes B$ if and only if $\Phi(\Delta f) = \Phi((f\otimes f)\Delta)\colon A\to B\otimes B\otimes A$. But $\Phi(\Delta f)(a) = (\Delta\otimes A)\lambda$, and $\Phi((f\otimes f)\Delta)(a) = (f\otimes f)\Delta(a_{(-1)})\otimes a_{(0)} = f(a_{(-2)})\otimes f(a_{(-1)})\otimes a_{(0)} = f(a_{(-1)})\otimes\lambda(a_{(0)}) = (B\otimes\lambda)\lambda(a)$, proving (1).

We have $\nabla(f\otimes f) = f\nabla\colon L\otimes L\to B$ if and only if $\Phi(\nabla(f\otimes f)) = \Phi(f\nabla)\colon A\otimes A\to B\otimes A$. But $\Phi(\nabla(f\otimes f))(x\otimes y) = f(x_{(-1)})f(y_{(-1)})\otimes x_{(0)}y_{(0)} = \lambda(x)\lambda(y)$ and $\Phi(f\nabla)(x\otimes y) = f(x_{(-1)}y_{(-1)})\otimes x_{(0)}y_{(0)} = \lambda(xy)$, proving (2).

(3) is simply a combination of (1) and (2), since L as a bialgebra is uniquely determined once it fulfills a universal property for bialgebra maps. □

Corollary 3.1.4 *Let A be an L-H-bi-Galois object. Then*

$$\operatorname{Alg}(L,k)\ni\varphi\mapsto(a\mapsto\varphi(a_{(-1)})a_{(0)})\in\operatorname{Aut}^{-H}(A)$$

is an isomorphism from the group of algebra maps from L to k (i.e. the group of grouplikes of L^* if L is finitely generated projective) to the group of H-colinear algebra automorphisms of A.

If H is cocommutative, then every H-Galois object is trivially an H-H-bi-Galois object, so:

Corollary 3.1.5 *If H is cocommutative and A is an L-H-bi-Galois object, then $L \cong H$.*

It is also obvious that $L(H, H) = H$. There is a more general important case in which $L(A, H)$ can be computed in some sense (see below, though), namely that of cleft extensions:

Proposition 3.1.6 *Let $A = k\#_\sigma H$ be a crossed product with invertible cocycle σ. Then $L(A, H) = H$ as coalgebras, while multiplication in $L(A, H)$ is given by*

$$g \cdot h = \sigma(g_{(1)} \otimes h_{(1)}) g_{(2)} h_{(2)} \sigma^{-1}(g_{(3)} \otimes h_{(3)}).$$

We will say that $L(A, H) := H^\sigma$ is a cocycle double twist of H. The construction of a cocycle double twist is dual to the construction of a Drinfeld twist [17], and was considered by Doi [14]. We have said already that the isomorphism $L(k\#_\sigma H, H) = H^\sigma$ computes the left Hopf algebra in case of cleft extensions in some sense. In applications, this may rather be read backwards: Cocycles in the non-cocommutative case are not easy to compute for lack of a cohomological interpretation, while it may be easier to guess a left Hopf algebra from generators and relations of A. In this sense the isomorphism may be used to compute the Hopf algebra H^σ helped by the left Hopf algebra construction. This is quite important in the applications we will cite in Section 3.2.

We will give a different proof from that in [33] of Proposition 3.1.6. It has the advantage not to use the fact that H^σ is a Hopf algebra — checking the existence of an antipode is in fact one of the more unpleasant parts of the construction.

Proof of Proposition 3.1.6 We will not check here that H^σ is a bialgebra. Identify $A = k\#_\sigma H = H$, with multiplication $g \circ h = \sigma(g_{(1)} \otimes h_{(1)}) g_{(2)} h_{(2)}$. Then it is straightforward to verify that comultiplication in H is an H^σ-comodule algebra structrure $A \to H^\sigma \otimes A$ which, of course, makes A an H^σ-H-bicomodule algebra. One may now finish the proof by appealing to Lemma 3.2.5 below, but we will stay more elementary. We shall verify that H^σ fulfills the universal property of $L(A, H)$. Of course it does so as a coalgebra, since the left coaction is just the comultiplication of H. Thus a B-H-bicomodule algebra structure $\lambda\colon A \to B \otimes A$ gives rise to a unique coalgebra map $f\colon H^\sigma \to B$ by $f(h) = (B \otimes \varepsilon)\lambda(h) = h_{(-1)}\varepsilon(h_{(0)})$. We have to check that f is an algebra map:

$$\begin{aligned} f(g \cdot h) &= \sigma(g_{(1)} \otimes h_{(1)}) f(g_{(2)} h_{(2)}) \sigma^{-1}(g_{(3)} \otimes h_{(3)}) = f(g_{(1)} \circ h_{(1)}) \sigma^{-1}(g_{(2)} \otimes h_{(2)}) \\ &= (B \otimes \varepsilon)(\lambda(g_{(1)})\lambda(h_{(1)})) \sigma^{-1}(g_{(2)} \otimes h_{(2)}) \\ &= g_{(1)(-1)} h_{(1)(-1)} \varepsilon(g_{(1)(0)} \circ h_{(1)(0)}) \sigma^{-1}(g_{(2)} \otimes h_{(2)}) \\ &= g_{(-1)} h_{(-1)} \varepsilon(g_{(0)(1)} \circ g_{(0)(1)}) \sigma^{-1}(g_{(0)(2)} \otimes h_{(0)(2)}) = g_{(-1)} h_{(-1)} \varepsilon(g_{(0)} \cdot h_{(0)}) \\ &= g_{(-1)} \varepsilon(g_{(0)}) h_{(-1)} \varepsilon(h_{(0)}) = f(g) f(h) \qquad \square \end{aligned}$$

Remark 3.1.7 Let A be an L-H-bi-Galois object. The left Galois map $\beta_\ell\colon A^\bullet \otimes A^\bullet \to L \otimes A^\bullet$ is right H-colinear as indicated, and thus induces an isomorphism $(A \otimes A)^{\operatorname{co} H} \cong (L \otimes A)^{\operatorname{co} H} \cong L$, where the coinvariants of $A \otimes A$ are taken with respect to the codiagonal comodule structure. Let us check that the isomorphism is an algebra map to a subalgebra of $A \otimes A^{\mathrm{op}}$: If $x \otimes y, x' \otimes y' \in A \otimes A$

are such that $x_{(-1)} \otimes x_{(0)}y = \ell \otimes 1$ and $x'_{(-1)} \otimes x'_{(0)}y = \ell' \otimes 1$ for $\ell, \ell' \in L$, then $\beta_{\ell'}(xx' \otimes y'y) = x_{(-1)}x'_{(-1)} \otimes x_{(0)}x'_{(0)}y'y = x_{(-1)}\ell' \otimes x_{(0)}y = \ell\ell' \otimes 1$.

3.2 Monoidal equivalences and the groupoid of bi-Galois objects. Let H be a Hopf algebra, and A an L-H-bi-Galois object. Then the monoidal functor $(A \,\square_H\, —, \xi)$ considered in Section 2.5 also defines a monoidal functor $A \,\square_H\, —: {}^H\mathcal{M} \to {}^L\mathcal{M}$. If B is an H-R-bi-Galois object, then $A \,\square_H\, B$ is an L-H-bicomodule algebra, and since the functor ${}^R\mathcal{M} \ni V \mapsto (A \,\square_H\, B) \,\square_R\, V \in {}^L\mathcal{M}$ is the composition of the two monoidal functors $(B \,\square_R\, —)$ and $A \,\square_H\, —$, it is itself monoidal, so that $A \,\square_H\, B$ is an R-Galois object by Corollary 2.5.4. By symmetric arguments, $A \,\square_H\, B$ is also a left L-Galois and hence an L-R-bi-Galois object. Thus, without further work, we obtain:

Corollary 3.2.1 *k-flat Hopf algebras form a category $\underline{BiGal}$ when we define a morphism from a Hopf algebra H to a Hopf algebra L to be an isomorphism class of L-H-bi-Galois objects, and if we define the composition of bi-Galois objects as their cotensor product.*

On the other hand we can define a category whose objects are Hopf algebras, and in which a morphism from H to L is an isomorphism class of monoidal functors ${}^H\mathcal{M} \to {}^L\mathcal{M}$.

A functor from the former category to the latter is described by assigning to an L-H-bi-Galois object A the functor $A \,\square_H\, —: {}^H\mathcal{M} \to {}^L\mathcal{M}$.

The purpose of the Corollary was to collect what we can deduce without further effort from our preceding results. The following Theorem gives the full information:

Theorem 3.2.2
1. *The category $\underline{BiGal}$ is a groupoid; that is, for every L-H-bi-Galois object A there is an H-L-bi-Galois object A^{-1} such that $A \,\square_H\, A^{-1} \cong L$ as L-L-bicomodule algebras and $A^{-1} \,\square_L\, A \cong H$ as H-H-bicomodule algebras.*
2. *The category $\underline{BiGal}$ is equivalent to the category whose objects are all k-flat Hopf algebras, and in which a morphism from H to L is an isomorphism class of monoidal category equivalences ${}^H\mathcal{M} \to {}^L\mathcal{M}$.*

If k is a field, there is a short conceptual proof for the Theorem, in which the second claim is proved first, and the first is an obvious consequence. If k is arbitrary, there does not seem to be a way around proving the first claim first. This turns out to be much easier if we assume all antipodes to be bijective. We will sketch all approaches below, but we shall comment first on the main application of the result.

Definition 3.2.3 Let H, L be two k-flat Hopf algebras. We call H and L monoidally Morita-Takeuchi equivalent if there is a k-linear monoidal equivalence ${}^H\mathcal{M} \to {}^L\mathcal{M}$.

Since the monoidal category structure of the comodule category of a Hopf algebra is one of its main features, it should be clear that monoidal Morita-Takeuchi equivalence is an interesting notion of equivalence between two Hopf algebras, weaker than isomorphy. Theorem 3.2.2 immediately implies:

Corollary 3.2.4 *For two k-flat Hopf algebras H and L, the following are equivalent:*

1. *H and L are monoidally Morita-Takeuchi equivalent.*
2. *There exists an L-H-bi-Galois object.*

3. *There is a k-linear monoidal category equivalence $\mathcal{M}^H \to \mathcal{M}^L$.*

As a consequence of Corollary 3.2.4 and Proposition 3.1.6, Hopf algebras are monoidally Morita-Takeuchi equivalent if they are cocycle double twists of each other (one should note, though, that it is quite easy to give a direct proof of this fact). Conversely, if H is a finite Hopf algebra over a field k, then every H-Galois object is cleft. Thus every Hopf algebra L which is monoidally Morita-Takeuchi equivalent to H is a cocycle double twist of H.

In many examples constructing bi-Galois objects has proved to be a very practicable way of constructing monoidal equivalences between comodule categories. This is true also in the finite dimensional case over a field. The reason seems to be that it is much easier to construct an associative algebra with nice properties, than to construct a Hopf cocycle (or, worse perhaps, a monoidal category equivalence). I will only very briefly give references for such applications: Nice examples involving the representation categories of finite groups were computed by Masuoka [27]. In [28] Masuoka proves that certain infinite families of non-isomorphic pointed Hopf algebras collapse under monoidal Morita-Takeuchi equivalence. That paper also contains a beautiful general mechanism for constructing Hopf bi-Galois objects for quotient Hopf algebras of a certain type. This was applied further, and more examples of families collapsing under monoidal Morita-Takeuchi equivalence were given, in Daniel Didt's thesis [12]. Bichon [4] gives a class of infinite-dimensional examples that also involve non-cleft extensions.

Now we return to the proof of Theorem 3.2.2. First we state and prove (at least sketchily) the part that is independent of k and any assumptions on the antipode.

Lemma 3.2.5 *Let L and H be k-flat Hopf algebras. Then every k-linear equivalence $\mathcal{F}\colon {}^H\mathcal{M} \to {}^L\mathcal{M}$ has the form $\mathcal{F}(V) = A \,\square_H\, V$ for some L-H-bi-Galois object A.*

More precisely, every exact k-linear functor $\mathcal{F}\colon {}^H\mathcal{M} \to {}^L\mathcal{M}$ commuting with arbitrary colimits has the form $\mathcal{F}(V) = A \,\square_H\, V$ for an L-H-bicomodule algebra that is an H-Galois object, and if $\mathcal{F}$ is an equivalence, then A is an L-Galois object.

Proof Let B be a k-flat bialgebra, and $\mathcal{F}\colon {}^H\mathcal{M} \to {}^L\mathcal{M}$ an exact functor commuting with colimits. We already know that the composition $\mathcal{F}_0\colon {}^H\mathcal{M} \to \mathcal{M}_k$ of $\mathcal{F}$ with the underlying functor has the form $\mathcal{F}_0(V) = A \,\square_H\, V$ for an H-Galois object A. It is straightforward to check that $\mathcal{F}$ has the form $\mathcal{F}(V) = A \,\square_H\, V$ for a suitable L-comodule algebra structure on A making it an L-H-bicomodule algebra (just take the left L-comodule structure of $A = A \,\square_H\, H = \mathcal{F}_0(A)$, and do a few easy calculations). Conversely, every B-H-bicomodule algebra structure on A for some flat bialgebra B lifts $\mathcal{F}_0$ to a monoidal functor $\mathcal{G}\colon {}^H\mathcal{M} \to {}^B\mathcal{M}$. If $\mathcal{F}$ is an equivalence, we can fill in the dashed arrow in the diagram

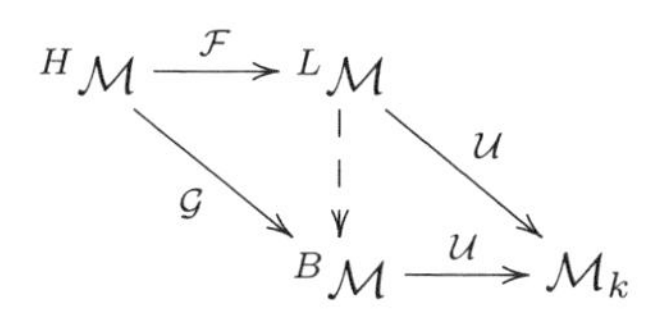

by a monoidal functor. To see this, simply note that every L-module is by assumption naturally isomorphic to one of the form $A \,\square_H\, V$ with $V \in {}^H\mathcal{M}$, and thus it is also a B-module. Now a monoidal functor ${}^L\mathcal{M} \to {}^B\mathcal{M}$ that commutes with the underlying functors has the form ${}^f\mathcal{M}$ for a unique bialgebra map $f\colon L \to B$. We have shown that L has the universal property characterizing the left Hopf algebra $L(A, H)$. $\square$

Now what is left of the proof of Theorem 3.2.2 is to provide a converse to Lemma 3.2.5.

In the case that k is a field, we can argue by the general principles of reconstruction theory for quantum groups, which also go back to work of Ulbrich [55]; see e.g. [31]. Assume given an H-Galois object A. The restriction $A \,\square_H\, —\colon {}^H\mathcal{M}_{\mathrm{f}} \to \mathcal{M}_k$ of the functor $A \,\square_H\, —$ to the category of finite-dimensional H-comodules takes values in finite dimensional vector spaces (see Corollary 2.5.5). Thus there exists a Hopf algebra L such that the functor factors over an equivalence ${}^H\mathcal{M}_{\mathrm{f}} \to {}^L\mathcal{M}_{\mathrm{f}}$; by the finiteness theorem for comodules this also yields an equivalence ${}^H\mathcal{M} \to {}^L\mathcal{M}$. By Lemma 3.2.5 we see that this equivalence comes from an L-H-bi-Galois structure on A, and in particular that cotensoring with a bi-Galois extension A is an equivalence ${}^H\mathcal{M} \to {}^{L(A,H)}\mathcal{M}$.

The general technique of reconstruction behind this proof is to find a Hopf algebra from a monoidal functor $\omega\colon \mathcal{C} \to \mathcal{M}_k$ by means of a coendomorphism coalgebra construction. More generally, one can construct a cohomomorphism object $\mathrm{cohom}(\omega, \nu)$ for every pair of functors $\omega, \nu\colon \mathcal{C} \to \mathcal{M}_k$ taking values in finite dimensional vector spaces. Ulbrich in fact reconstructs a Hopf-Galois object from a monoidal functor ${}^H\mathcal{M}_{\mathrm{f}} \to \mathcal{M}_k$ by applying this construction to the monoidal functor in question on one hand, and the underlying functor on the other hand. It is clear that the left Hopf algebra of a Hopf-Galois object A can be characterized as the universal Hopf algebra reconstructed as a coendomorphism object from the functor $A \,\square_H\, —$. Bichon [3] has taken this further by reconstructing a bi-Galois object, complete with both its Hopf algebras, from a pair of monoidal functors $\omega, \nu\colon \mathcal{C} \to \mathcal{M}_k$ taking values in finite dimensional vector spaces. He also gives an axiom system (called a Hopf-Galois system, and extended slightly to be symmetric by Grunspan [19]) characterizing the complete set of data arising in a bi-Galois situation: An algebra coacted upon by two bialgebras, and in addition another bicomodule algebra playing the role of the inverse bi-Galois extension.

In the case where k is not a field, reconstruction techniques as the ones used above are simply not available, and we have to take a somewhat different approach. If we can show that $\underline{BiGal}$ is a groupoid, then the rest of Theorem 3.2.2 follows: The inverse of the functor $A \,\square_H\, —{}^H\mathcal{M} \to {}^L\mathcal{M}$ can be constructed as $A^{-1} \,\square_L\colon {}^L\mathcal{M} \to {}^H\mathcal{M}$ when A^{-1} is the inverse of A in the groupoid $\underline{BiGal}$.

Now let A be an L-H-bi-Galois object. By symmetry it is enough to find a right inverse for A. For this in turn it is enough to find some left H-Galois object B such that $A \,\square_H\, B \cong L$ as left L-comodule algebras. For B is an H-R-bi-Galois object for some Hopf algebra R, and $A \,\square_H\, B$ is then an L-R-bi-Galois object. But if $A \,\square_H\, B \cong L$ as left L-comodule algebra, then $R \cong L$ by the uniqueness of the right Hopf algebra in the bi-Galois extension L. More precisely, there is an automorphism of the Hopf algebra L such that $A \,\square_H\, B \cong L^f$, where L^f has the right L-comodule algebra structure induced along f. But then $A \,\square_H\, (B^{f^{-1}}) \cong L$ as L-bicomodule algebras.

We already know that $L \cong (A \otimes A)^{\mathrm{co}\,H}$, a subalgebra of $A \otimes A^{\mathrm{op}}$. From the way the isomorphism was obtained in Remark 3.1.7, it is obviously left L-colinear, with the left L-comodule structure on $(A \otimes A)^{\mathrm{co}\,H}$ induced by that of the left tensor factor A. Thus it finally remains to find some left H-Galois object B such that

$A \,\square_H\, B \cong (A \otimes A)^{\mathrm{co}\,H}$. If the antipode of H is bijective, we may simply take $B := A^{\mathrm{op}}$, with the left comodule structure $A \ni a \mapsto S^{-1}(a_{(1)}) \otimes a_{(0)}$. If the antipode of H is not bijective, we can take $B := (H \otimes A)^{\mathrm{co}\,H}$, where the coinvariants are taken with respect to the diagonal comodule structure, the algebra structure is that of a subalgebra of $H \otimes A^{\mathrm{op}}$, and the left H-comodule structure is induced by that of H. By contrast to the case where the antipode is bijective, it is not entirely trivial to verify that B is indeed a left H-Galois object. We refer to [33] for details at this point. However, it is easy to see that the obvious isomorphism $A \,\square_H\, B = A \,\square_H\, (H \otimes A)^{\mathrm{co}\,H} \cong (A \,\square_H\, H \otimes A)^{\mathrm{co}\,H} \cong (A \otimes A)^{\mathrm{co}\,H}$ is a left L-comodule algebra map.

3.3 The structure of Hopf bimodules. Let A be an L-H-bi-Galois object. We have studied already in Section 2.6 the monoidal category ${}_A\mathcal{M}_A^H$ of Hopf bimodules, which allows an underlying functor to the category $\mathcal{M}_k$ which is monoidal. The result of this section is another characterization of the left Hopf algebra L: It is precisely that Hopf algebra for which we obtain a commutative diagram of monoidal functors

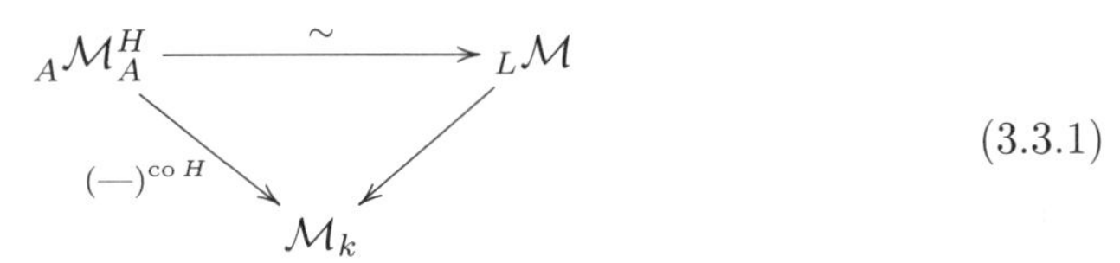

(3.3.1)

in which the top arrow is an equivalence, and the unmarked arrow is the underlying functor.

Theorem 3.3.1 *Let A be an L-H-bi-Galois object. Then a monoidal category equivalence ${}_L\mathcal{M} \to {}_A\mathcal{M}_A^H$ is defined by sending $V \in {}_L\mathcal{M}$ to $V \otimes A$ with the obvious structure of a Hopf module in $\mathcal{M}_A^H$, and the additional left A-module structure $x(v \otimes y) = x_{(-1)} \cdot v \otimes x_{(0)} y$. The monoidal functor structure is given by the canonical isomorphism $(V \otimes A) \otimes_A (V \otimes A) \cong V \otimes W \otimes A$.*

Proof We know that every Hopf module in $\mathcal{M}_A^H$ has the form $V \otimes A$ for some k-module V. It remains to verify that left A-module structures on $V \otimes A$ making it a Hopf module in ${}_A\mathcal{M}_A^H$ are classified by left L-module structures on V. A suitable left A-module structure is a colinear right A-module map $\mu\colon A \otimes V \otimes A \to V \otimes A$, and such maps are in turn in bijection with colinear maps $\sigma\colon A \otimes V \to V \otimes A$. By the general universal property of L, such maps σ are in turn classified by maps $\mu_0\colon L \otimes V \to V$ through the formula $\sigma(a \otimes v) = a_{(-1)} \cdot v \otimes a_{(0)}$, with $\mu_0(\ell \otimes v) =: \ell \cdot v$. Now it only remains to verify that μ_0 is an L-module structure if and only if μ, which is now given by $\mu(x \otimes v \otimes y) = x_{(-1)} \cdot v \otimes x_{(0)} y$, is an A-module structure. We compute

$$x(y(v \otimes z)) = x(y_{(-1)} \cdot v \otimes y_{(0)} z) = x_{(-1)} \cdot (y_{(-1)} \cdot v) \otimes x_{(0)} y_{(0)} z$$
$$(xy)(v \otimes z) = (xy)_{(-1)} \cdot v \otimes (xy)_{(0)} z = (x_{(-1)} y_{(-1)}) \cdot v \otimes x_{(0)} y_{(0)} z$$

so that the associativity of μ and μ_0 is equivalent by another application of the universal property of L. We skip unitality.

We have seen that the functor in consideration is well-defined and an equivalence. To check that it is monoidal, we should verify that the canonical isomorphism

$f\colon V\otimes W\otimes A\to (V\otimes A)\otimes_A (W\otimes A)$ is a left A-module map for $V,W\in L$. Indeed

$$\begin{aligned}f(x(v\otimes w\otimes y)) &= f(x_{(-1)}\cdot(v\otimes w)\otimes x_{(0)}y) = f(x_{(-2)}\cdot v\otimes x_{(-1)}\cdot w\otimes x_{(0)}y)\\ &= x_{(-2)}\cdot v\otimes 1\otimes x_{(-1)}\cdot w\otimes x_{(0)}y = x_{(-1)}\cdot v\otimes 1\otimes x_{(0)}(w\otimes y)\\ &= x_{(-1)}\cdot v\otimes x_{(0)}\otimes (w\otimes y) = x(v\otimes 1)\otimes 1\otimes y = xf(v\otimes w\otimes y)\end{aligned}$$

for all $x,y\in A$, $v\in V$ and $w\in W$. □

Corollary 3.3.2 *Let A be an L-H-bi-Galois object. Then there is a bijection between isomorphism classes of*

1. *Pairs (T,f), where T is an H-comodule algebra, and $f\colon A\to T$ is an H-comodule algebra map, and*
2. *L-module algebras R*

It is given by $R := T^{\operatorname{co} H}$, and $T := R\#A := R\otimes A$ with multiplication given by $(r\#x)(s\#y) = r(x_{(-1)}\cdot s)\#x_{(0)}y$.

Note in particular that every T as in (1) is a left faithfully flat H-Galois extension of its coinvariants.

Remark 3.3.3 If A is a faithfully flat H-Galois extension of B, then ${}_A\mathcal{M}_A^H$ is still a monoidal category, and the coinvariants functor is still a monoidal functor to ${}_B\mathcal{M}_B$. It is a natural question whether there is still some L whose modules classify Hopf modules in the same way as we have shown in this section for the case $B=k$, and whether L is still a Hopf algebra in any sense. This was answered in [35] by showing that $L=(A\otimes A)^{\operatorname{co} H}$ still yields a commutative diagram (3.3.1), and that L now has the structure of a $\times_B$-bialgebra in the sense of Takeuchi [48]. These structures have been studied more recently under the name of quantum groupoids or Hopf algebroids. They have the characteristic property that modules over a $\times_B$-bialgebra still form a monoidal category, so that it makes sense to say that (3.3.1) will be a commutative diagram of monoidal functors. The $\times_B$-bialgebra L can step in in some cases where the left Hopf algebra L is useful, but $B\neq k$. Since the axiomatics of $\times_B$-bialgebras are quite complicated, we will not pursue this matter here.

3.4 Galois correspondence. The origin of bi-Galois theory is the construction in [18] of certain separable extensions of fields that are Hopf-Galois with more than one possibility for the Hopf algebra. The paper [18] also contains information about what may become of the classical Galois correspondence between subfields and subgroups in this case. In particular, there are examples of classically Galois field extensions that are also H-Galois in such a way that the quotient Hopf algebras of H correspond one-to-one to the *normal* intermediate fields, that is, to the intermediate fields that are stable under the coaction of the dual Hopf algebra k^G of the group algebra of the Galois group. Van Oystaeyen and Zhang [56] then constructed what we called $L(A,H)$ above for the case of commutative A (and H), and proved a correspondence between quotients of $L(A,H)$ and H-costable intermediate fields in case A is a field. The general picture was developed in [33, 36]. We will not comment on the proof here, but simply state the results.

Theorem 3.4.1 *Let A be an L-H-bi-Galois object for k-flat Hopf algebras L,H with bijective antipodes.*

A bijection between

- *coideal left ideals $I \subset L$ such that L/I is k-flat and L is a faithfully coflat left (resp. right) L/I-comodule, and*
- *H-subcomodule algebras $B \subset A$ such that B is k-flat and A is a faithfully flat left (resp. right) B-module*

is given as follows: To a coideal left ideal $I \subset L$ we assign the subalgebra $B := {}^{\mathrm{co}\,A}L/I$. To an H-subcomodule algebra $B \subset A$ we assign the coideal left ideal $I \subset L \cong (A \otimes A)^{\mathrm{co}\,H}$ such that $L/I \cong (A \otimes_B A)^{\mathrm{co}\,H}$.

Let $I \subset L$ and $B \subset A$ correspond to each other as above. Then

1. *I is a Hopf ideal if and only if B is stable under the Miyashita-Ulbrich action of H on A.*
2. *I is stable under the left coadjoint coaction of L on itself if and only if B is stable under the coaction of L on A.*
3. *I is a conormal Hopf ideal if and only if B is stable both under the coaction of L and the Miyashita-Ulbrich action of H on A.*

As the special case $A = H$, the result contains the quotient theory of Hopf algebras, that is, the various proper Hopf algebra analogs of the correspondence between normal subgroups and quotient groups of a group. See [50, 26].

3.5 Galois objects over tensor products. Let H_1, H_2 be two Hopf algebras. If both H_i are cocommutative, then $\mathrm{Gal}(H_i)$ are groups under cotensor product, as well as $\mathrm{Gal}(H_1 \otimes H_2)$. If both H_i are also commutative, then we have the subgroups of these three groups consisting of all commutative Galois extensions. If, in particular, we take both H_i to be the duals of group algebras of abelian groups, then $H_1 \otimes H_2$ is the group algebra of the direct sum of those two groups, and the groups of commutative Galois objects are the Harrison groups. It is an old result that the functor "Harrison group" is additive. This means that $\mathrm{Har}(H_1 \otimes H_2) \cong \mathrm{Har}(H_1) \oplus \mathrm{Har}(H_2)$ as (abelian) groups. The same result holds true unchanged if we consider general commutative and cocommutative Hopf algebras. However, the same is not true for the complete $\mathrm{Gal}(—)$ groups. A result of Kreimer [24] states very precisely what is true instead: For two commutative cocommutative finitely generated projective Hopf algebras, we have an isomorphism of abelian groups

$$\mathrm{Gal}(H_1) \oplus \mathrm{Gal}(H_2) \oplus \mathrm{Hopf}(H_2, H_1^*) \to \mathrm{Gal}(H_1 \otimes H_2),$$

where $\mathrm{Hopf}(H_2, H_1^*)$ denotes the set of all Hopf algebra maps from H_2 to H_1^*, which is a group under convolution because H_2 is cocommutative and H_1^* is commutative. The assumption that both Hopf algebras H_i are commutative is actually not necessary. One can also drop the assumption that they be finitely generated projective, if one replaces the summand $\mathrm{Hopf}(H_2, H_1^*)$ by the group (under convolution) $\mathrm{Pair}(H_2, H_1)$ of all Hopf algebra pairings between H_2 and H_1; this does not change anything if H_1 happens to be finitely generated projective.

One cannot, however, get away without the assumption of cocommutativity: First of all, of course, we do not have any groups in the case of general H_i. Secondly, some of the information in the above sequence does survive on the level of pointed sets, but not enough to amount to a complete description of $\mathrm{Gal}(H_1 \otimes H_2)$.

As we will show in this section (based on [39]), bi-Galois theory can come to the rescue to recover such a complete description. Instead of pairings between the Hopf algebras H_i, one has to take into account pairings between the left Hopf algebras L_i in certain H_i-Galois objects.

Lemma 3.5.1 *Let H_1, H_2 be two k-flat Hopf algebras, and A a right H-comodule algebra for $H = H_1 \otimes H_2$.*

We have $A_1 := A^{\operatorname{co} H_2} \cong A \,\square_H\, H_1$ and $A_2 := A^{\operatorname{co} H_1} \cong A \,\square_H\, H_2$.

A is an H-Galois object if and only if A_i is an H_i-Galois object for $i = 1, 2$.

If this is the case, then multiplication in A induces an isomorphism $A_1 \# A_2 \to A$, where the algebra structure of $A_1 \# A_2$ is a smash product as in Corollary 3.3.2 for some L_2-module structure on A_1, where $L_2 := L(A_2, H_2)$; the H-comodule structure is the obvious one.

Proof It is straightforward to check that $A^{\operatorname{co} H_i} \cong A \,\square_H\, H_j$ for $i \neq j$. We know from Proposition 2.5.7 that A_i are Hopf-Galois objects if A is one.

Now assume that A_i is a faithfully flat H_i-Galois extension of k for $i = 1, 2$. By Corollary 3.3.2 we know that multiplication in A induces an isomorphism $A_1 \# A_2 \to A$ for a suitable L_2-module algebra structure on A_1. We view the Galois map $A \otimes A \to A \otimes H$ as a map of Hopf modules in $\mathcal{M}^{H_2}_{A_2}$. Its H_2-coinvariant part is the map $A \otimes A_1 \to A \otimes H_1$ given by $x \otimes y \mapsto xy_{(0)} \otimes y_{(1)}$, which we know to be a bijection. Thus the canonical map for A is a bijection, and A is faithfully flat since it is the tensor product of A_1 and A_2. □

To finish our complete description of $H_1 \otimes H_2$-Galois objects, we need two more consequences from the universal property of the left Hopf algebra:

Lemma 3.5.2 *Let A be an L-H-bi-Galois object, and let R be an L-module algebra and F-comodule algebra for some k-flat bialgebra F. Then $R \# A$ as in Corollary 3.3.2 is an $F \otimes H$-comodule algebra if and only if it is an F-comodule algebra, if and only if R is an L-F-dimodule in the sense that $(\ell \cdot r)_{(0)} \otimes (\ell \cdot r)_{(1)} = \ell \cdot r_{(0)} \otimes r_{(1)}$ holds for all $r \in R$ and $\ell \in L$.*

Proof Clearly $R \# A$ is an $F \otimes H$-comodule algebra if and only if it is an F-comodule algebra, since we already know it to be an H-comodule algebra.

Now (ignoring the unit conditions) $R \# A$ is an F-comodule algebra if and only if

$$r_{(0)}(x_{(-1)} \cdot s)_{(0)} \# x_{(0)} y \otimes r_{(1)}(x_{(-1)} \cdot s)_{(1)}$$
$$r_{(0)}(x_{(-1)} \cdot s_{(0)}) \# x_{(0)} y \otimes r_{(1)} s_{(1)}$$

agree for all $r, s \in R$ and $x, y \in A$. By the universal property of L, this is the same as requiring

$$r_{(0)}(\ell \cdot s)_{(0)} \otimes r_{(1)}(\ell \cdot s)_{(1)} = r_{(0)}(\ell \cdot s_{(0)}) \otimes r_{(1)} s_{(1)}$$

for all $r, s \in R$ and $\ell \in A$, which in turn is the same as requiring the dimodule condition for R. □

Lemma 3.5.3 *Let A be an L-H-bi-Galois object, B a k-module, and $\mu\colon B \otimes A \to A$ an H-colinear map. Then $\mu = \Phi(\tau)$, that is, $\mu(b \otimes a) = \tau(b \otimes a_{(-1)}) a_{(0)}$ for a unique $\tau\colon B \otimes L \to A$. Moreover,*

1. *Assume that B is a coalgebra. Then μ is a measuring if and only if $\tau(b \otimes \ell m) = \tau(b_{(1)} \otimes \ell)\tau(b_{(2)} \otimes m)$ and $\tau(b \otimes 1) = \varepsilon(b)$ hold for all $b \in C$ and $\ell, m \in L$.*
2. *Assume that B is an algebra. Then μ is a module structure if and only if $\tau(bc \otimes \ell) = \tau(b \otimes \ell_{(2)})\tau(c \otimes \ell_{(1)})$ and $\tau(1 \otimes \ell) = \varepsilon(\ell)$ hold for all $b, c \in B$ and $\ell \in L$.*

3. *Assume that B is a bialgebra. Then μ makes A a B-module algebra if and only if τ is a skew pairing between B and L, in the sense of the following definition:*

Definition 3.5.4 Let B and L be two bialgebras. A map $\tau\colon B\otimes L\to k$ is called a skew pairing if

$$\begin{aligned} \tau(b\otimes \ell m) &= \tau(b_{(1)}\otimes \ell)\tau(b_{(2)}\otimes m), & \tau(b\otimes 1) &= \varepsilon(b)\\ \tau(bc\otimes \ell) &= \tau(b\otimes \ell_{(2)})\tau(c\otimes \ell_{(1)}), & \tau(1\otimes \ell) &= \varepsilon(\ell) \end{aligned}$$

hold for all $b,c\in B$ and $\ell,m\in L$. Note that if B is finitely generated projective, then a skew pairing is the same as a bialgebra morphism $L^{\mathrm{cop}}\to B^*$

Proof We write $\mu(b\otimes a)=b\cdot a$. We have $\mu=\Phi(\tau)$ for $\tau\colon B\otimes L\to k$ as a special case of the universal properties of L.

If B is a coalgebra, then

$$\begin{aligned} b\cdot(xy) &= \tau(b\otimes x_{(-1)}y_{(-1)})x_{(0)}y_{(0)}\\ (b_{(1)}\cdot x)(b_{(2)}\cdot y) &= \tau(b_{(1)}\otimes x_{(-1)})\tau(b_{(2)}\otimes y_{(-1)})x_{(0)}y_{(0)} \end{aligned}$$

are the same for all $b\in B$, $x,y\in A$ if and only if $\tau(b\otimes \ell m)=\tau(b_{(1)}\otimes \ell)\tau(b_{(2)}\otimes m)$ for all $b\in B$, $\ell,m\in L$, by the universal property again. We omit treating the unit condition for a measuring

If B is an algebra then

$$\begin{aligned} (bc)\cdot x &= \tau(bc\otimes x_{(-1)})x_{(0)}\\ b\cdot c\cdot x = \tau(c\otimes x_{(-1)})b\cdot x_{(0)} &= \tau(c\otimes x_{(-2)})\tau(b\otimes x_{(-1)})x_{(0)} \end{aligned}$$

agree for all $b,c\in B$, $x\in A$ if and only if $\tau(bc\otimes \ell)=\tau(c\otimes \ell_{(1)})\tau(b\otimes \ell_{(2)})$ holds for all $b,c\in B$ and $\ell\in L$. Again, we omit treating the unit condition for a module structure.

Since a module algebra structure is the same as a measuring that is a module structure, we are done. □

Now we merely need to put together all the information obtained so far to get the following theorem.

Theorem 3.5.5 *Let H_1,H_2 be two k-flat Hopf algebras, and put $H=H_1\otimes H_2$. The map*

$$\pi\colon \operatorname{Gal}(H_1\otimes H_2)\to \operatorname{Gal}(H_1)\times \operatorname{Gal}(H_2);\, A\mapsto (A\mathbin{\underset{H}{\square}} H_1, A\mathbin{\underset{H}{\square}} H_2)$$

is surjective. For $A_i\in \operatorname{Gal}(H_i)$ let $L_i:=L(A_i,H_i)$. The Hopf algebra automorphism groups of L_i act on the right on the set of all skew pairings between L_1 and L_2. We have a bijection

$$\operatorname{SPair}(L_1,L_2)/\operatorname{CoInn}(L_1)\times \operatorname{CoInn}(L_2)\to \pi^{-1}(A_1,A_2),$$

given by assigning to the class of a skew pairing τ the algebra $A_1\#_\tau A_2:=A_1\otimes A_2$ with multiplication $(r\#x)(s\#y)=r\tau(s_{(-1)}\otimes x_{(-1)})s_{(0)}\#x_{(0)}y$.

In particular, we have an exact sequence

$$\operatorname{CoInn}(H_1)\times \operatorname{CoInn}(H_2)\to \operatorname{SPair}(H_2,H_1)\to \operatorname{Gal}(H_1\otimes H_2)\to \operatorname{Gal}(H_1)\times \operatorname{Gal}(H_2)$$

Proof Since $\pi(A_1 \otimes A_2) = (A_1, A_2)$, the map π is onto. Fix $A_i \in \mathrm{Gal}(H_i)$. Then the inverse image of $A_1 \otimes A_2$ under π consists of all those H-Galois objects A for which $A \,\square_H\, H_i \cong A_i$. By the discussion preceding the theorem, every such A has the form $A = A_1 \# A_2$, with multiplication given by an L_2-module algebra structure on A_1, which makes A_1 an L_2-H_1-dimodule, and is thus given by a skew pairing between L_1 and L_2.

Assume that for two skew pairings τ, σ we have an isomorphism $f\colon A_1 \#_\tau A_2 \to A_1 \#_\sigma A_2$. Then f has the form $f = f_1 \otimes f_2$ for automorphisms f_i of the H_i-comodule algebra A_i, which are given by $f_i(x) = u_i(x_{(-1)})x_{(0)}$ for algebra maps $u_i\colon L_i \to k$. The map f is an isomorphism of algebras if and only if $f((1\#x)(r\#1)) = f(1\#x)f(r\#1)$ for all $r \in A_1$ and $x \in A_2$. Now

$$
\begin{aligned}
f((1\#x)(r\#1)) &= \tau(r_{(-1)} \otimes x_{(-1)}) f(r_{(0)} \# x_{(0)}) \\
&= \tau(r_{(-2)} u_1(r_{(-1)}) \otimes x_{(-2)} u_2(x_{(-1)})) r_{(0)} \# x_{(0)}
\end{aligned}
$$

and on the other hand

$$
f(1\#x)f(r\#1) = u_1(r_{(-2)}) u_2(x_{(-2)}) \tau(r_{(-1)} \otimes x_{(-1)}) r_{(0)} \# x_{(0)}
$$

These two expressions are the same for all r, x if and only if τ and σ agree up to composition with $\mathrm{coinn}(u_1) \otimes \mathrm{coinn}(u_2)$, by yet another application of the universal properties of L_1 and L_2. □

3.6 Reduction. We take up once again the topic of reduction of the structure group, or the question of when an H-Galois extension reduces to a Q-Galois extension for a quotient Hopf algebra Q of H. We treated the case of a general base B of the extension in Section 2.7. Here, we treat some aspects that are more or less special to the case of a trivial coinvariant subring k, and involve the left Hopf algebra L.

We start by a simple reformulation of the previous results, using Corollary 3.1.4:

Corollary 3.6.1 *Assume the hypotheses of Theorem 2.7.1. Consider the map* $\pi\colon \mathrm{Gal}(Q) \to \mathrm{Gal}(H)$ *given by* $\pi(\overline{A}) = A \,\square_Q\, H$*, and let* $A \in \mathrm{Gal}(H)$*. Then* $\pi^{-1}(A) \cong \mathrm{Hom}_{-H}^{-H}(K, A)/\mathrm{Alg}(L, k)$*, where* $L = L(A, H)$.

The criterion we have given above for reducibility of the structure quantum group (i.e. the question when an H-Galois extension comes from a Q-Galois extension) is "classical" in the sense that analogous results are known for principal fiber bundles: If we take away the Miyashita-Ulbrich action on A^B, which is a purely noncommutative feature, we have to find a colinear algebra map $K \to A$, which is to say an equivariant map from the principal bundle (the spectrum of A) to the coset space of the structure group under the subgroup we are interested in. Another criterion looks even simpler in the commutative case: According to [11, III §4, 4.6], a principal fiber bundle, described by a Hopf-Galois extension A, can be reduced if and only if the associated bundle with fiber the coset space of the subgroup in the structure group admits a section. In our terminology, this means that there is an algebra map $(A \otimes K)^{\mathrm{co}\,H} \to B$ of the obvious map $B \to A$; alternatively, one may identify the associated bundle with $A^{\mathrm{co}\,Q} \cong (A \otimes K)^{\mathrm{co}\,H}$, see below. As it turns out, this criterion can be adapted to the situation of general Hopf-Galois extensions as well. In the noncommutative case, there are, again, extra requirements on the map $A^{\mathrm{co}\,Q} \to B$. In fact suitable such conditions were spelled out in [7, Sec.2.5], although the formulas there seem to defy a conceptual interpretation. As it turns out, the extra conditions can be cast in a very simple form using the left Hopf algebra L: The relevant map $A^{\mathrm{co}\,Q} \to B$ should simply be L-linear with respect to the Miyashita-Ulbrich action of L. Since the result now involves the left Hopf algebra, it can only be formulated like that in the case $B = k$; we note, however, that the result, as well as its proof, is still valid for the general case — one only has to take the $\times_B$-bialgebra L, see Remark 3.3.3, in place of the ordinary bialgebra L.

Theorem 3.6.2 *Assume the hypotheses of Theorem 2.7.1, and let A be an L-H-bi-Galois object. $A^{\mathrm{co}\,Q} \subset A$ is a submodule with respect to the Miyashita-Ulbrich action of L on A.*

The following are equivalent:

1. *$A \in \mathrm{Gal}(H)$ is in the image of $\mathrm{Gal}(Q) \ni \overline{A} \mapsto \overline{A} \,\square_Q\, H \in \mathrm{Gal}(H)$.*
2. *There is an L-module algebra map $A^{\mathrm{co}\,Q} \to k$.*

Proof The inverse of the Galois map of the left L-Galois extension A maps L to $(A \otimes A)^{\mathrm{co}\,H}$, so it is straightforward to check that $A^{\mathrm{co}\,Q}$ is invariant under the Miyashita-Ulbrich action of L.

We have an isomorphism $\theta_0\colon A \to (A \otimes H)^{\mathrm{co}\,H}$ with $\theta(a) = a_{(0)} \otimes S(a_{(1)})$ and $\theta_0^{-1}(a \otimes h) = a\varepsilon(h)$. One checks that $\theta_0(a) \in (A \otimes K)^{\mathrm{co}\,H}$ if and only if $a \in A^{\mathrm{co}\,Q}$, so that we have an isomorphism $\theta\colon A^{\mathrm{co}\,Q} \to (A \otimes K)^{\mathrm{co}\,H}$ given by $\theta(a) = a_{(0)} \otimes S(a_{(1)})$. It is obvious that θ is an isomorphism of algebras, if we regard $(A \otimes K)^{\mathrm{co}\,H}$ as a subalgebra of $A \otimes K^{\mathrm{op}}$, but we would like to view this in a more complicated way: Since K is an algebra in the center of $\mathcal{M}^H$, we can endow $A \otimes K$ with the structure of an algebra in $\mathcal{M}^H$ by setting $(a \otimes x)(b \otimes y) = ab_{(0)} \otimes (x \leftharpoonup b_{(1)})y$. If $x \otimes a \in (A \otimes K)^{\mathrm{co}\,H}$, then $(a \otimes x)(b \otimes y) = ab_{(0)} \otimes (x \leftharpoonup b_{(1)})y = ab_{(0)} \otimes y_{(0)}(x \leftharpoonup b_{(1)}y_{(1)}) = ab \otimes yx$, so $(A \otimes K)^{\mathrm{co}\,H}$ is a subalgebra of $A \otimes K^{\mathrm{op}}$.

Now the obvious map $A \to A \otimes K$ is an H-colinear algebra map, so $A \otimes K$ is an algebra in the monoidal category ${}_A\mathcal{M}^H_A$ and hence $(A \otimes K)^{\mathrm{co}\,H}$ is an L-module algebra by Corollary 3.3.2. Writing $\ell^{(1)} \otimes \ell^{(2)}$ for the preimage of $\ell \otimes 1$ under the Galois map for the left L-Galois extension A, we can compute the relevant L-module structure as $\ell \triangleright (a \otimes x) = \ell^{(1)}(a \otimes x)\ell^{(2)} = \ell^{(1)}a\ell^{(2)}{}_{(0)} \otimes x \leftharpoonup \ell^{(2)}{}_{(1)}$. It is immediate that θ^{-1} is L-linear.

Now we have a bijection between H-colinear maps $f\colon K \to A$ and H-colinear and left A-linear maps $\hat{f}\colon A \otimes K \to A$ given by $f(x) = \hat{f}(1 \otimes x)$ and $\hat{f}(a \otimes x) = af(x)$. Let us check that f is a right H-module algebra map if and only if $\hat{f}$ is an A-ring map in $\mathcal{M}^H$. First, assume that f is an H-module algebra map. Then

$$\begin{aligned}
\hat{f}((a \otimes x)(b \otimes y)) &= \hat{f}(ab_{(0)} \otimes (x \leftharpoonup b_{(1)})y) \\
&= ab_{(0)} f(x \leftharpoonup b_{(1)}) f(y) \\
&= ab_{(0)} (f(x) \leftharpoonup b_{(1)}) f(y) \\
&= af(x)bf(y) = \hat{f}(a \otimes x)\hat{f}(b \otimes y),
\end{aligned}$$

and $\hat{f}(1 \otimes a) = a$. Conversely, assume that $\hat{f}$ is an A-ring morphism in $\mathcal{M}^H$. Then f is trivially an algebra map, and

$$f(x)a = \hat{f}(1 \otimes x)\hat{f}(a \otimes 1) = \hat{f}((1 \otimes x)(a \otimes 1)) = \hat{f}(a_{(0)} \otimes x \leftharpoonup a_{(1)}) = a_{(0)} f(x \leftharpoonup a_{(1)})$$

for all $a \in A$ and $x \in K$ implies that f is H-linear.

Finally, we already know that A-ring morphisms $\hat{f}\colon A \otimes K \to A$ in $\mathcal{M}^H$, that is, algebra maps in ${}_A\mathcal{M}^H_A$, are in bijection with L-module algbra maps $g\colon (A \otimes K)^{\operatorname{co} H} \to k$. □

If we want to reduce the right Hopf algebra in an L-H-bi-Galois extension, it is of course also a natural question what happens to the left Hopf algebra in the process:

Lemma 3.6.3 *Assume the situation of Theorem 2.7.1. Let $\overline{A}$ be a Q-Galois object, and A the corresponding H-Galois object. Then for any $V \in \mathcal{M}^H$ the map $(V \otimes A)^{\operatorname{co} H} \to (V \otimes \overline{A})^{\operatorname{co} Q}$ induced by the surjection $A \to \overline{A}$ is an isomorphism.*

Proof It is enough to check that $\alpha\colon (V \otimes A)^{\operatorname{co} H} \to (V \otimes \overline{A})^{\operatorname{co} Q}$ is bijective after tensoring with $\overline{A}$. We compose $\alpha \otimes \overline{A}$ with the isomorphism $(V \otimes \overline{A})^{\operatorname{co} Q} \otimes \overline{A} \to V \otimes \overline{A}$ from the structure theorem for Hopf modules in $\mathcal{M}^Q_{\overline{A}}$, and have to check that

$$(V \otimes A)^{\operatorname{co} H} \otimes \overline{A} \to (V \otimes \overline{A})^{\operatorname{co} Q} \otimes \overline{A} \to V \otimes \overline{A}; \quad v \otimes a \otimes \overline{b} \mapsto v \otimes \overline{ab}$$

is bijective. But this is the image under the equivalence $\mathcal{M}^H_K \to \mathcal{M}^Q$ of the isomorphism

$$(V \otimes A)^{\operatorname{co} H} \otimes A \to V \otimes A; \quad v \otimes a \otimes b \to v \otimes ab$$

from the structure theorem for Hopf modules in $\mathcal{M}^H_A$. □

Theorem 3.6.4 *Assume the situation of Theorem 2.7.1, let A be an H-Galois object, $f\colon K \to A$ a Yetter-Drinfeld algebra map, and $\overline{A} = A/Af(K^+)$ the corresponding Q-Galois object.*

Using the identification $L := L(A, H) = (A \otimes A)^{\operatorname{co} H}$, the left Hopf algebra of $\overline{A}$ is given by

$$L(\overline{A}, Q) = (A \underset{K}{\otimes} A)^{\operatorname{co} H},$$

where the K-module structure of A is induced via f.

Proof We have to verify that the surjection $A \to \overline{A}$ induces an isomorphism $(A \otimes_K A)^{\operatorname{co} H} \to (\overline{A} \otimes \overline{A})^{\operatorname{co} Q}$. Since $(—)^{\operatorname{co} H}\colon \mathcal{M}^H_A \to \mathcal{M}_k$ is an equivalence, this amounts to showing that we have a coequalizer

$$(A \otimes K \otimes A)^{\operatorname{co} H} \rightrightarrows (A \otimes A)^{\operatorname{co} H} \to (\overline{A} \otimes \overline{A})^{\operatorname{co} Q} \to 0.$$

Using Lemma 3.6.3 this means a coequalizer

$$(A \otimes K \otimes \overline{A})^{\operatorname{co} Q} \rightrightarrows (A \otimes \overline{A})^{\operatorname{co} Q} \to (\overline{A} \otimes \overline{A})^{\operatorname{co} Q} \to 0.$$

Since $(—)^{\operatorname{co} Q}\colon \mathcal{M}^Q_{\overline{A}} \to \mathcal{M}_k$ is an equivalence, we may consider this before taking the Q-coinvariants, when it is just the definition of $\overline{A} = A/Af(K^+)$ tensored with $\overline{A}$. □

4 Appendix: Some tools

4.1 Monoidal category theory. A monoidal category $\mathcal{C} = (\mathcal{C}, \otimes, \Phi, I, \lambda, \rho)$ consists of a category $\mathcal{C}$, a bifunctor $\otimes\colon \mathcal{C} \times \mathcal{C} \to \mathcal{C}$, a natural isomorphism $\Phi\colon (X \otimes Y) \otimes Z \to X \otimes (Y \otimes Z)$, an object I, and natural isomorphisms $\lambda\colon I \otimes X \to X$ and $\rho\colon X \otimes I \to X$, all of which are coherent. This means that all diagrams that one can compose from Φ (which rearranges brackets), λ, ρ (which cancel instances of the unit object I) and their inverses commute. By Mac Lane's coherence theorem, it is actually enough to ask for one pentagon of Φ's, and one triangle with λ, ρ, and Φ, to commute in order that all diagrams commute. A monoidal category is called strict if Φ, λ, and ρ are identities.

The easiest example of a monoidal category is the category $\mathcal{M}_k$ of modules over a commutative ring, with the tensor product over k and the canonical isomorphisms expressing associativity of tensor products. Similarly, the category ${}_R\mathcal{M}_R$ of bimodules over an arbitrary ring R is monoidal with respect to the tensor product over R. We are interested in monoidal category theory because of its very close connections with Hopf algebra theory. If H is a bialgebra, then both the category of, say, left H-modules, and the category of, say, right H-comodules have natural monoidal category structures. Here, the tensor product of $V, W \in {}_H\mathcal{M}$ (resp. $V, W \in \mathcal{M}^H$) is $V \otimes W$, the tensor product over k, equipped with the diagonal module structure $h(v \otimes w) = h_{(1)}v \otimes h_{(2)}w$ (resp. the codiagonal comodule structure $\delta(v \otimes w) = v_{(0)} \otimes w_{(0)} \otimes v_{(1)}w_{(1)}$. The unit object is the base ring k with the trivial module (resp. comodule) structure induced by the counit ε (resp. the unit element of H). Since the associativity and unit isomorphisms in all of these examples are "trivial", it is tempting never to mention them at all, practically treating all our examples as if they were strict monoidal categories; we will do this in all of the present paper. In fact, this sloppiness is almost justified by the fact that every monoidal category is monoidally equivalent (see below) to a strict one. For the examples in this paper, which are categories whose objects are sets with some algebraic structure, the sloppiness is even more justified [40].

A weak monoidal functor $\mathcal{F} = (\mathcal{F}, \xi, \xi_0)\colon \mathcal{C} \to \mathcal{D}$ consists of a functor $\mathcal{F}\colon \mathcal{C} \to \mathcal{D}$, a natural transformation $\xi\colon \mathcal{F}(X) \otimes \mathcal{F}(Y) \to \mathcal{F}(X \otimes Y)$ and a morphism $\xi_0\colon \mathcal{F}(I) \to I$ making the diagrams

$$\begin{array}{ccccc}
(\mathcal{F}(X) \otimes \mathcal{F}(Y)) \otimes \mathcal{F}(Z) & \xrightarrow{\xi\otimes 1} & \mathcal{F}(X \otimes Y) \otimes \mathcal{F}(Z) & \xrightarrow{\xi} & \mathcal{F}((X \otimes Y) \otimes Z) \\
\downarrow{\scriptstyle\Phi} & & & & \downarrow{\scriptstyle\mathcal{F}(\Phi)} \\
\mathcal{F}(X) \otimes (\mathcal{F}(Y) \otimes \mathcal{F}(Z)) & \xrightarrow[1\otimes\xi]{} & \mathcal{F}(X) \otimes \mathcal{F}(Y \otimes Z) & \xrightarrow[\xi]{} & \mathcal{F}(X \otimes (Y \otimes Z))
\end{array}$$

commute and satisfying

$$\mathcal{F}(\lambda)\xi(\xi_0 \otimes id) = \lambda\colon I \otimes \mathcal{F}(X) \to \mathcal{F}(X)$$
$$\mathcal{F}(\rho)\xi(id \otimes \xi_0) = \rho\colon \mathcal{F}(X) \otimes I \to \mathcal{F}(X).$$

A standard example arises from a ring homomorphism $R \to S$. The restriction functor ${}_S\mathcal{M}_S \to {}_R\mathcal{M}_R$ is a weak monoidal functor, with $\xi\colon M \otimes_R N \to M \otimes_S N$ for $M, N \in {}_S\mathcal{M}_S$ the canonical surjection.

A monoidal functor is a weak monoidal functor in which ξ and ξ_0 are isomorphisms. Typical examples are the underlying functors ${}_H\mathcal{M} \to \mathcal{M}_k$ and $\mathcal{M}^H \to \mathcal{M}_k$

for a bialgebra H. In this case, the morphisms ξ, ξ_0 are even identities; we shall say that we have a strict monoidal functor.

A prebraiding for a monoidal category $\mathcal{C}$ is a natural transformation $\sigma_{XY}: X \otimes Y \to Y \otimes X$ satisfying

$$\sigma_{X,Y\otimes Z} = (Y \otimes \sigma_{XZ})(\sigma_{XY} \otimes Z): X \otimes Y \otimes Z \to Y \otimes Z \otimes X$$
$$\sigma_{X\otimes Y,Z} = (\sigma_{XZ} \otimes Y)(X \otimes \sigma_{YZ}): X \otimes Y \otimes Z \to Z \otimes X \otimes Y$$
$$\sigma_{XI} = \sigma_{IX} = id_X$$

A braiding is a prebraiding that is an isomorphism. A symmetry is a braiding with $\sigma_{XY} = \sigma_{YX}^{-1}$. The notion of a symmetry captures the properties of the monoidal category of modules over a commutative ring. For the topological flavor of the notion of braiding, we refer to Kassel's book [22]. We call a (pre)braided category a category with a (pre)braiding.

The (weak) center construction produces a (pre)braided monoidal category from any monoidal category: Objects of the weak center $\mathcal{Z}_0(\mathcal{C})$ are pairs $(X, \sigma_{X,-})$ in which $X \in \mathcal{C}$, and $\sigma_{XY}: X \otimes Y \to Y \otimes X$ is a natural transformation satisfying

$$\sigma_{X,Y\otimes Z} = (Y \otimes \sigma_{XZ})(\sigma_{XY} \otimes Z): X \otimes Y \otimes Z \to Y \otimes Z \otimes X$$

for all $Y, Z \in \mathcal{C}$, and $\sigma_{XI} = id_X$. The weak center is monoidal with tensor product $(X, \sigma_{X,-}) \otimes (Y, \sigma_{Y,-}) = (X \otimes Y, \sigma_{X\otimes Y,-})$, where

$$\sigma_{X\otimes Y,Z} = (\sigma_{XZ} \otimes Y)(X \otimes \sigma_{YZ}): X \otimes Y \otimes Z \to Z \otimes X \otimes Y$$

for all $Z \in \mathcal{C}$, and with neutral element $(I, \sigma_{I,-})$, where $\sigma_{IZ} = id_Z$. The weak center is prebraided with the morphism σ_{XY} as the prebraiding of X and Y. The center $\mathcal{Z}(\mathcal{C})$ consists of those objects $(X, \sigma_{X-}) \in \mathcal{Z}_0(\mathcal{C})$ in which all σ_{XY} are isomorphisms.

The main example of a (pre)braided monoidal category which we use in this paper is actually a center. Let H be a Hopf algebra. The category $\mathcal{Z}_0(\mathcal{M}^H)$ is equivalent to the category $\mathcal{YD}_H^H$ of right-right Yetter-Drinfeld modules, whose objects are right H-comodules and right H-modules V satisfying the condition

$$(v \leftharpoonup h)_{(0)} \otimes (v \leftharpoonup h)_{(1)} = v_{(0)} \leftharpoonup h_{(2)} \otimes S(h_{(1)})v_{(1)}h_{(3)}$$

for all $v \in V$ and $h \in H$. A Yetter-Drinfeld module V becomes an object in the weak center by

$$\sigma_{VW}(v \otimes w) = w_{(0)} \otimes v \leftharpoonup w_{(1)}$$

for all $v \in V$ and $w \in W \in \mathcal{M}^H$. It is an object in the center if and only if H has bijective antipode, in which case $\sigma_{VW}^{-1}(w \otimes v) = v \leftharpoonup S^{-1}(w_{(1)}) \otimes w_{(0)}$.

4.2 Algebras in monoidal categories. At some points in this paper we have made free use of the notion of an algebra within a monoidal category, modules over it, and similar notions. In this section we will spell out (without the easy proofs) some of the basic facts. It is possible that the notion of center that we define below is new.

Let $\mathcal{C}$ be a monoidal category, which we assume to be strict for simplicity. An algebra in $\mathcal{C}$ is an object A with a multiplication $\nabla: A \otimes A \to A$ and a unit $\eta: I \to A$ satisfying associativity and the unit condition that $A \cong A \otimes I \xrightarrow{A\otimes\eta} A \otimes A \xrightarrow{\nabla} A$ (and a symmetric construction) should be the identity. It should be clear what a morphism of algebras is. A left A-module in $\mathcal{C}$ is an object M together with a module structure $\mu: A \otimes M \to M$ which is associative and fulfills an obvious unit condition. It is clear how to define right modules and bimodules in a monoidal category.

An algebra in the (pre)braided monoidal category $\mathcal{C}$ is said to be commutative if $\nabla_A = \nabla_A\sigma\colon A \otimes A \to A$. Obviously we can say that a subalgebra B in A (or an algebra morphism $\iota\colon B \to A$) is central in A if

$$\nabla_A(\iota \otimes A) = \nabla_A\sigma_{AA}(\iota \otimes A) = \nabla_A(A \otimes \iota)\sigma_{BA}\colon B \otimes A \to A.$$

Since the last notion needs only the braiding between B and A to be written down, it can be generalized as follows:

Definition 4.2.1 Let A be an algebra in $\mathcal{C}$. A morphism $f\colon V \to A$ in $\mathcal{C}$ from an object $V \in \mathcal{Z}_0(\mathcal{C})$ is called central, if $\nabla_A(f \otimes A) = \nabla_A(A \otimes f)\sigma_{VA}\colon V \otimes A \to A$.

A center of A is a couniversal central morphism $c\colon C \to A$, that is, an object $C \in \mathcal{Z}_0(\mathcal{C})$ with a morphism $c\colon C \to A$ in $\mathcal{C}$ such that every central morphism $f\colon V \to A$ factors through a morphism $g\colon V \to C$ in $\mathcal{Z}_0(\mathcal{C})$.

It is not clear whether a center of an algebra A exists, or if it does, if it is a subobject in $\mathcal{C}$, though this is true in our main application Lemma 2.1.9. However, the following assertions are not hard to verify:

Remark 4.2.2 Let A be an algebra in $\mathcal{C}$, and assume that A has a center (C, c).

1. Any center of A is isomorphic to C.
2. C is a commutative algebra in $\mathcal{Z}_0(\mathcal{C})$.
3. If R is an algebra in $\mathcal{Z}_0(\mathcal{C})$ and $f\colon R \to A$ is central and an algebra morphism in $\mathcal{C}$, then its factorization $g\colon R \to C$ is an algebra morphism.

Let A and B be algebras in the prebraided monoidal category $\mathcal{C}$. Then $A \otimes B$ is an algebra with multiplication

$$A \otimes B \otimes A \otimes B \xrightarrow{A \otimes \sigma_{BA} \otimes B} A \otimes A \otimes B \otimes B \xrightarrow{\nabla_A \otimes \nabla_B} A \otimes B.$$

Again, this is also true if we merely assume B to be an algebra in the weak center of $\mathcal{C}$.

If B is a commutative algebra in the weak center of $\mathcal{C}$, then every right B-module M has a natural left B-module structure

$$B \otimes M \xrightarrow{\sigma_{BM}} M \otimes B \xrightarrow{\mu} M$$

which makes it a B-B-bimodule.

Provided that the category $\mathcal{C}$ has coequalizers, one can define the tensor product of a right A-module M and a left A-module N by a coequalizer

$$M \otimes A \otimes N \rightrightarrows M \otimes N \to M \underset{A}{\otimes} N.$$

If M is an L-A-bimodule, and N is an A-R-bimodule, then $M \otimes_A N$ is an L-R-bimodule provided that tensoring on the left with L and tensoring on the right with R preserves coequalizers. The extra condition is needed to show, for example, that $L \otimes (M \otimes_A N) \to M \otimes_A N$ is well-defined, using that $L \otimes (M \otimes_A N) \cong (L \otimes M) \otimes_A N$, which relies on $L \otimes -$ preserving the relevant coequalizer.

Some more technicalities are necessary to assure that the tensor product of three bimodules is associative. Assume given in addition an S-R-bimodule T such that $T \otimes -$ and $S \otimes -$ preserve coequalizers. Since colimits commute with colimits, $T \otimes_S -$ also preserves coequalizers, and we have in particular a coequalizer

$$T \underset{S}{\otimes} (M \otimes A \otimes N) \rightrightarrows T \underset{S}{\otimes} (M \otimes N) \to T \underset{S}{\otimes} (M \underset{A}{\otimes} N).$$

To get the desired isomorphism

$$T \underset{L}{\otimes} (M \underset{A}{\otimes} N) \cong (T \underset{L}{\otimes} M) \underset{A}{\otimes} N,$$

we need to compare this to the coequalizer

$$(T \underset{S}{\otimes} M) \otimes A \otimes N \rightrightarrows (T \underset{S}{\otimes} M) \otimes N \to (T \underset{S}{\otimes} M) \underset{A}{\otimes} N,$$

which can be done if we throw in the extra condition that the natural morphism $(T \otimes_S M) \otimes X \to T \otimes_S (M \otimes X)$ is an isomorphism for all $X \in \mathcal{C}$.

4.3 Cotensor product. To begin with, the cotensor product of comodules is nothing but a special case of the tensor product of modules in monoidal categories: A k-coalgebra C is an algebra in the opposite of the category of k-modules, so the cotensor product of a right C-comodule M and a left C-comodule N (two modules in the opposite category) is defined by an equalizer

$$0 \to M \underset{C}{\Box} N \to M \otimes N \rightrightarrows M \otimes C \otimes N.$$

We see that the cotensor product of a B-C-bicomodule M and a C-D-bicomodule N is a B-D-bicomodule provided that B and C are flat k-modules. Since flatness of C is even needed to make sense of equalizers within the category of C-comodules, it is assumed throughout this paper that all coalgebras are flat over k.

A right C-comodule V is called C-coflat if the cotensor product functor $V \Box_C -: {}^C\mathcal{M} \to \mathcal{M}_k$ is exact. Since $V \Box_C (C \otimes W) = V \otimes W$ for any k-module W, this implies that V is k-flat. If V is k-flat, it is automatic that $V \Box_C -$ is left exact. Also, $V \Box_C -$ commutes with (infinite) direct sums. From this we can deduce

Lemma 4.3.1 *If V is a coflat right C-comodule, then for any k-module X and any left C-comodule W the canonical map $(V \Box_C W) \otimes X \to V \Box_C (W \otimes X)$ is a bijection.*

In particular, if D is another k-flat coalgebra, W is a C-D-bicomodule, and U is a left D-comodule, then cotensor product is associative:

$$(V \underset{C}{\Box} W) \underset{D}{\Box} U \cong V \underset{C}{\Box} (W \underset{D}{\Box} U).$$

Proof The second claim follows from the first and the discussion at the end of the preceding section. For the first, observe first that cotensor product commutes with direct sums, so that the canonical map is bijective with a free module $k^{(I)}$ in place of X. Now we choose a presentation $k^{(I)} \to k^{(J)} \to X \to 0$ of X. Since $V \Box_C -$ commutes with this coequalizer, we see that the canonical map for X is also bijective. □

It is a well-known theorem of Lazard that a module is flat if and only if it is a direct limit of finitely generated projective modules. It is well-known, moreover, that a finitely presented module is flat if and only if it is projective. If k is a field, and C a k-coalgebra, every C-comodule is the direct limit of its finite dimensional subcomodules. Thus the following remarkable characterization of Takeuchi [47, A.2.1] may seem plausible (though of course far from obvious):

Theorem 4.3.2 *Let k be a field, C a k-coalgebra, and V a C-comodule. Then V is coflat if and only if C is injective (that is, an injective object in the category of comodules).*

We refer to [47] for the proof.

In Section 2.5 we have made use of a comodule version of Watts' theorem (which, in the original, states that every right exact functor between module categories is tensor product by a bimodule). For the sake of completeness, we prove the comodule version here:

Lemma 4.3.3 *Let C be a k-flat coalgebra, and $\mathcal{F}\colon {}^C\mathcal{M} \to \mathcal{M}_k$ an exact additive functor that commutes with arbitrary direct sums.*

Then there is an isomorphism $\mathcal{F}(M) \cong A \,\Box_C\, M$, natural in $M \in {}^C\mathcal{M}$, for some comodule $A \in \mathcal{M}^C$ which is k-flat and C-coflat.

Proof We first observe that $\mathcal{F}$ is an $\mathcal{M}_k$-functor. That is to say, there is an isomorphism $\mathcal{F}(M\otimes V) \cong \mathcal{F}(M)\otimes V$, natural in $V \in \mathcal{M}_k$, which is coherent (which is to say, the two obvious composite isomorphisms $\mathcal{F}(M\otimes V\otimes W) \cong \mathcal{F}(M)\otimes V\otimes W$ coincide, and $\mathcal{F}(M\otimes k) \cong \mathcal{F}(M)\otimes k$ is trivial). We only sketch the argument: To construct $\zeta\colon \mathcal{F}(M)\otimes V \to \mathcal{F}(M\otimes V)$, choose a presentation $k^{(I)} \xrightarrow{p} k^{(J)} \to V$. The map p can be described by a column-finite matrix, which can also be used to define a morphism $\hat{p}\colon \mathcal{F}(M)^{(I)} \to \mathcal{F}(M)^{(J)}$, which has both $\mathcal{F}(M)\otimes V$ and (since $\mathcal{F}$ commutes with cokernels) $\mathcal{F}(M\otimes V)$ as its cokernel, whence we get an isomorphism ζ between them. Clearly ζ is natural in M. Naturality in V is proved along with independence of the presentation: Let $k^{(K)} \to k^{(L)} \to W$ be a presentation of another k-module W, and $f\colon V \to W$. By the Comparison Theorem for projective resolutions, f can be lifted to a pair of maps $f_1\colon k^{(J)} \to k^{(L)}$ and $f_2\colon k^{(I)} \to k^{(K)}$. Since the maps of free k-modules can be described by matrices, they give rise to a diagram

$$\begin{array}{ccc} \mathcal{F}(M)^{(I)} & \longrightarrow & \mathcal{F}(M)^{(J)} \\ \downarrow & & \downarrow \\ \mathcal{F}(M)^{(K)} & \longrightarrow & \mathcal{F}(M)^{(L)} \end{array}$$

which can be filled to the right both by $\mathcal{F}(M)\otimes f\colon \mathcal{F}(M)\otimes V \to \mathcal{F}(M)\otimes W$ and by $\mathcal{F}(M\otimes f)\colon \mathcal{F}(M\otimes V) \to \mathcal{F}(M\otimes W)$. If $W = V$ and $f = id$, this proves independence of ζ of the resolution, and for a general choice of W and f it proves naturality of ζ. Coherence is now easy to check.

The rest of our claim is now Pareigis' version [30, Thm. 4.2] of Watts' theorem [57]. For completeness, we sketch the proof: Put $A := \mathcal{F}(C)$. Then A is a C-comodule via

$$A = \mathcal{F}(C) \xrightarrow{\mathcal{F}(\Delta)} \mathcal{F}({}^{\bullet}C\otimes C) \cong \mathcal{F}(C)\otimes C = A\otimes C.$$

The functors $\mathcal{F}$ and $A \,\Box_C\, -$ are isomorphic, since for $M \in {}^C\mathcal{M}$ we have $M \cong C \,\Box_C\, M$, that is, we have an equalizer

$$M \to {}^{\bullet}C\otimes M \rightrightarrows {}^{\bullet}C\otimes C\otimes M,$$

which is preserved by $\mathcal{F}$, and hence yields an equalizer

$$\mathcal{F}(M) \to A\otimes M \rightrightarrows A\otimes C\otimes M.$$

□

Finally, let us note the following associativity between tensor and cotensor product:

Lemma 4.3.4 *Let C be a coalgebra, A an algebra, M a right A-module, N a left A-module and right C-comodule satisfying the dimodule condition $(am)_{(0)} \otimes (am)_{(1)} = am_{(0)} \otimes m_{(1)}$ for all $a \in M$ and $c \in C$; finally let W be a left C-comodule. There is a canonical map*

$M \otimes_A (N \,\square_C\, V) \to (M \otimes_A N) \,\square_C\, V$, given by $m \otimes (n \otimes v) \mapsto (m \otimes n) \otimes v$.

If M is flat as A-module, or V is coflat as left C-comodule, then the canonical map is a bijection.

In fact, if M is flat, then $M \otimes_A -$ preserves the equalizer defining the cotensor product. If V is coflat, we may argue similarly using Lemma 4.3.1.

4.4 Convolution and composition. Let C be a k-coalgebra and A a k-algebra. The convolution product

$$f * g = \nabla_A(f \otimes g)\Delta_C : C \to a$$

defined for any two k-linear maps $f, g: C \to A$ is ubiquitous in the theory of coalgebras and bialgebras. It makes $\mathrm{Hom}(C, A)$ into an algebra, with the k-dual C^* as a special case. A lemma due to Koppinen (see [29, p.91] establishes a correspondence of convolution with composition (which, of course, is an even more ubiquitous operation throughout all of mathematics):

Lemma 4.4.1 *Let C be a k-coalgebra, and A a k-algebra. Then*

$$T = T_A^C : \mathrm{Hom}(C, A) \ni \varphi \mapsto (a \otimes c \mapsto a\varphi(c_{(1)}) \otimes c_{(2)}) \in \mathrm{End}_{A-}^{-C}(A \otimes C)$$

is an anti-isomorphism of k-algebras, with inverse given by $T^{-1}(f) = (A \otimes \varepsilon_C) f(\eta_A \otimes C)$.

In particular, $\varphi: C \to A$ is invertible with respect to convolution if and only if $T(\varphi)$ is bijective.

The assertions are straightforward to check. Let us point out that bijectivity of T is a special case of the following Lemma, which contains the facts that $A \otimes V$ is the free A-module over the k-module V, and $W \otimes C$ is the cofree C-comodule over the k-module W:

Lemma 4.4.2 *Let A be a k-algebra, C a k-coalgebra, and V a right C-comodule, and W a left A-module. Then we have an isomorphism*

$$\tilde{T}: \mathrm{Hom}(V, W) \ni \varphi \mapsto (a \otimes v \mapsto a\varphi(v_{(0)}) \otimes v_{(1)}) \in \mathrm{Hom}_{A-}^{-C}(A \otimes V, W \otimes C)$$

4.5 Descent. In this section we very briefly recall the mechanism of faithfully flat descent for extensions of noncommutative rings. This is a very special case of Beck's theorem; a reference is [2].

Definition 4.5.1 Let $\eta: R \subset S$ be a ring extension. A (right) descent data from S to R is a right S-module M together with an S-module homomorphism $\theta: M \to M \otimes_R S$ (also called a descent data on the module M) making the diagrams

$$\begin{array}{ccc} M & \xrightarrow{\theta} & M \otimes_R S \\ \downarrow{\scriptstyle\theta} & & \downarrow{\scriptstyle\theta \otimes_R S} \\ M \otimes_R S & \xrightarrow{M \otimes_R \eta \otimes_R S} & M \otimes_R S \otimes_R S \end{array} \qquad \begin{array}{ccc} M & \xrightarrow{\theta} & M \otimes_R S \\ & \searrow\!\!\searrow & \downarrow{\scriptstyle m} \\ & & M \end{array}$$

commute (where m is induced by the S-module structure of M). Descent data (M, θ) from S to R form a category $\mathcal{D}(S \downarrow R)$ in an obvious way.

If N is a right R-module, then the induced S-module $N \otimes_R S$ carries a natural descent data, namely the map $\theta\colon N \otimes_R S \ni n \otimes s \mapsto n \otimes 1 \otimes s \in N \otimes_R S \otimes_R S$. This defines a functor from $\mathcal{M}_R$ to the category of descent data from S to R.

Theorem 4.5.2 (Faithfully flat descent) *Let $\eta\colon R \subset S$ be an inclusion of rings, such that S is faithfully flat as a left R-module.*

Then the canonical functor from $\mathcal{M}_R$ to the category of descent data from S to R is an equivalence of categories. The inverse equivalence maps a descent data (M, θ) to

$$M^\theta := \{m \in M | \theta(m) = m \otimes 1\}.$$

In particular, for every descent data (M, θ), the map $f\colon (M^\theta) \otimes_R S \ni m \otimes s \mapsto ms \in M$ is an isomorphism with inverse induced by θ, i.e. $f^{-1}(m) = \theta(m) \in M^\theta \otimes_R S \subset M \otimes_R S$.

References

[1] Baer, R. Zur Einführung des Scharbegriffs. *J. Reine Angew. Math. 160* (1929), 199–207.

[2] Benabou, J., and Roubaud, J. Monades et descente. *C. R. Acad. Sci. Paris, Série A 270* (1970), 96–98.

[3] Bichon, J. Hopf-Galois systems. *J. Algebra 264* (2003), 565–581.

[4] Bichon, J. The representation category of the quantum group of a non-degenerate bilinear form. *Comm. Algebra 31* (2003), 4831–4851.

[5] Blattner, R. J., Cohen, M., and Montgomery, S. Crossed products and inner actions of Hopf algebras. *Trans. AMS 298* (1986), 671–711.

[6] Blattner, R. J., and Montgomery, S. Crossed products and Galois extensions of Hopf algebras. *Pacific J. of Math. 137* (1989), 37–54.

[7] Brzeziński, T., and Hajac, P. Galois type extensions and noncommutative geometry. *preprint* (2003).

[8] Chase, S., and Sweedler, M. *Hopf algebras and Galois theory*, vol. 97 of *Lecture Notes in Math.* Springer, Berlin, Heidelberg, New York, 1969.

[9] Chase, S. U., Harrison, D. K., and Rosenberg, A. Galois theory and cohomology of commutative rings. *Mem. AMS 52* (1965).

[10] Dade, E. C. Group graded rings and modules. *Math. Zeitschrift 174* (1980), 241–262.

[11] Demazure, M., and Gabriel, P. *Groupes Algébriques I.* North Holland, Amsterdam, 1970.

[12] Didt, D. *Linkable Dynkin diagrams and quasi-isomorphisms for finite-dimensional pointed Hopf algebras.* PhD thesis, Universität München, 2002.

[13] Doi, Y. Algebras with total integrals. *Comm. in Alg. 13* (1985), 2137–2159.

[14] Doi, Y. Braided bialgebras and quadratic bialgebras. *Comm. in Alg. 21* (1993), 1731–1749.

[15] Doi, Y., and Takeuchi, M. Cleft comodule algebras for a bialgebra. *Comm. in Alg. 14* (1986), 801–817.

[16] Doi, Y., and Takeuchi, M. Hopf-Galois extensions of algebras, the Miyashita-Ulbrich action, and Azumaya algebras. *J. Algebra 121* (1989), 488–516.

[17] Drinfel'd, V. G. Quasi-Hopf algebras. *Leningrad Math. J. 1* (1990), 1419–1457.

[18] Greither, C., and Pareigis, B. Hopf Galois theory for separable field extensions. *J. Algebra 106* (1987), 239–258.

[19] Grunspan, C. Quantum torsors. *J. Pure Appl. Algebra 184* (2003), 229–255.

[20] Günther, R. *Verschränkte Produkte für punktierte Hopfalgebren.* PhD thesis, Universität München, 1998.

[21] Günther, R. Crossed products for pointed Hopf algebras. *Comm. Algebra 27*, 9 (1999), 4389–4410.

[22] Kassel, C. *Quantum Groups*, vol. 155 of *GTM.* Springer, 1995.

[23] Knus, M.-A., and Ojanguren, M. *Théorie de la descente et algèbres d'Azumaya.* Springer-Verlag, Berlin, 1974. Lecture Notes in Mathematics, Vol. 389.

[24] Kreimer, H. F. Hopf-Galois theory and tensor products of Hopf algebras. *Comm. in Alg. 23* (1995), 4009–4030.

[25] Kreimer, H. F., and Takeuchi, M. Hopf algebras and Galois extensions of an algebra. *Indiana Univ. Math. J. 30* (1981), 675–692.

[26] Masuoka, A. Quotient theory of Hopf algebras. In *Advances in Hopf Algebras*, J. Bergen and S. Montgomery, Eds. Marcel Dekker Inc., 1994, pp. 107–133.

[27] Masuoka, A. Cocycle deformations and Galois objects for some cosemisimple Hopf algebras of finite dimension. In *New trends in Hopf algebra theory (La Falda, 1999)*. Amer. Math. Soc., Providence, RI, 2000, pp. 195–214.

[28] Masuoka, A. Defending the negated Kaplansky conjecture. *Proc. Amer. Math. Soc. 129*, 11 (2001), 3185–3192.

[29] Montgomery, S. *Hopf algebras and their actions on rings*, vol. 82 of *CBMS Regional Conference Series in Mathematics*. AMS, Providence, Rhode Island, 1993.

[30] Pareigis, B. Non-additive ring and module theory II. $\mathcal{C}$-categories, $\mathcal{C}$-functors and $\mathcal{C}$-morphisms. *Publ. Math. Debrecen 24* (1977), 351–361.

[31] Schauenburg, P. Tannaka duality for arbitrary Hopf algebras. *Algebra Berichte 66* (1992).

[32] Schauenburg, P. *Zur nichtkommutativen Differentialgeometrie von Hauptfaserbündeln—Hopf-Galois-Erweiterungen von de Rham-Komplexen*, vol. 71 of *Algebra Berichte [Algebra Reports]*. Verlag Reinhard Fischer, Munich, 1993.

[33] Schauenburg, P. Hopf Bigalois extensions. *Comm. in Alg 24* (1996), 3797–3825.

[34] Schauenburg, P. A bialgebra that admits a Hopf-Galois extension is a Hopf algebra. *Proc. AMS 125* (1997), 83–85.

[35] Schauenburg, P. Bialgebras over noncommutative rings and a structure theorem for Hopf bimodules. *Appl. Categorical Structures 6* (1998), 193–222.

[36] Schauenburg, P. Galois correspondences for Hopf bigalois extensions. *J. Algebra 201* (1998), 53–70.

[37] Schauenburg, P. Galois objects over generalized Drinfeld doubles, with an application to $u_q(\mathfrak{sl}_2)$. *J. Algebra 217*, 2 (1999), 584–598.

[38] Schauenburg, P. A generalization of Hopf crossed products. *Comm. Algebra 27* (1999), 4779–4801.

[39] Schauenburg, P. Bi-Galois objects over the Taft algebras. *Israel J. Math. 115* (2000), 101–123.

[40] Schauenburg, P. Turning monoidal categories into strict ones. *New York J. Math. 7* (2001), 257–265 (electronic).

[41] Schauenburg, P. Quantum torsors and Hopf-Galois objects. *preprint* (math.QA/0208047).

[42] Schauenburg, P. Quantum torsors with fewer axioms. *preprint* (math.QA/0302003).

[43] Schauenburg, P., and Schneider, H.-J. Galois type extensions and Hopf algebras. *preprint* (2003).

[44] Schneider, H.-J. Principal homogeneous spaces for arbitrary Hopf algebras. *Israel J. of Math. 72* (1990), 167–195.

[45] Schneider, H.-J. Representation theory of Hopf-Galois extensions. *Israel J. of Math. 72* (1990), 196–231.

[46] Sweedler, M. E. Cohomology of algebras over Hopf algebras. *Trans. AMS 133* (1968), 205–239.

[47] Takeuchi, M. Formal schemes over fields. *Comm. Algebra 5*, 14 (1977), 1483–1528.

[48] Takeuchi, M. Groups of algebras over $A \otimes \overline{A}$. *J. Math. Soc. Japan 29* (1977), 459–492.

[49] Takeuchi, M. Relative Hopf modules — equivalence and freeness criteria. *J. Algebra 60* (1979), 452–471.

[50] Takeuchi, M. Quotient spaces for Hopf algebras. *Comm. Algebra 22*, 7 (1994), 2503–2523.

[51] Takeuchi, M. Comments on Schauenburg's construction $L(A, H)$. unpublished manuscript, 1995.

[52] Ulbrich, K.-H. Galoiserweiterungen von nicht-kommutativen Ringen. *Comm. in Alg. 10* (1982), 655–672.

[53] Ulbrich, K.-H. Galois extensions as functors of comodules. *manuscripta math. 59* (1987), 391–397.

[54] Ulbrich, K.-H. Fiber functors of finite dimensional comodules. *manuscripta math. 65* (1989), 39–46.

[55] ULBRICH, K.-H. On Hopf algebras and rigid monoidal categories. *Israel J. Math. 72* (1990), 252–256.
[56] VAN OYSTAEYEN, F., AND ZHANG, Y. Galois-type correspondences for Hopf Galois extensions. *K-Theory 8* (1994), 257–269.
[57] WATTS, C. E. Intrinsic characterizations of some additive functors. *Proc. AMS 11* (1960), 5–8.

Received February 7, 2003; in revised form June 10, 2003

Fields Institute Communications
Volume **43**, 2004

Extension Theory in Mal'tsev Varieties

Jonathan D.H. Smith
Department of Mathematics
Iowa State University
Ames, Iowa 50011 U.S.A.
jdhsmith@math.iastate.edu

Abstract. The paper provides a brief survey of extension theory for Mal'tsev varieties based on centrality and monadic cohomology. Extension data are encoded in the form of a *seeded simplicial map*. Such a map yields an extension if and only if it is unobstructed. Second cohomology groups classify extensions, and third cohomology groups classify obstructions.

1 Introduction

Extension theory for Mal'tsev varieties was developed in [9, Chapter 6], generalising earlier treatment of special cases such as groups (cf. [4], [7]), commutative algebras (cf. [1]), loops [6], and other "categories of interest" [8]. Because of renewed attention being paid to the topic, a brief survey of the theory appears timely. With the exception of parts of Section 2, the context throughout the paper is that of a Mal'tsev variety $\mathfrak{V}$. An *extension* of a $\mathfrak{V}$-algebra R is considered as a $\mathfrak{V}$-algebra T equipped with a congruence α such that R is isomorphic to the quotient T^α of T by the congruence α.

The essential properties of centrality in Mal'tsev varieties are recalled in Section 2. Section 3 describes the "seeded simplicial maps" which provide a concise encoding of the raw material required for constructing an extension (analogous to the "abstract kernels" of [7]). Section 4 gives a brief, algebraic description of the rudiments of monadic cohomology, culminating in the Definition 4.2 of the obstruction of a seeded simplicial map as a cohomology class. Theorem 5.2 then shows that a seeded simplicial map yields an extension if and only if it is unobstructed. The final section discusses the classification of extensions by second cohomology groups, and of obstructions by third cohomology groups. Against this background, the pessimism expressed in [5] ("To classify the extensions ... is too big a project to admit of a reasonable answer") appears unwarranted.

For concepts and conventions not otherwise explained in the paper, readers are referred to [10]. In particular, note the general use of postfix notation, so that

2000 *Mathematics Subject Classification.* Primary 18G50; Secondary 08B10.

composites are read in natural order from left to right. For a binary relation ρ on a set X, and an element x of X, write $x^\rho = \{y \mid (x,y) \in \rho\}$.

2 Centrality in Mal'tsev varieties

Recall that a variety $\mathfrak{V}$ of universal algebras is a *Mal'tsev variety* if there is a derived ternary *parallelogram* operation P such that the identities

$$(x,x,y)P = y = (y,x,x)P$$

are satisfied. Equivalently, the relation product of two congruences is commutative, and thus agrees with their join. Moreover, reflexive subalgebras of direct squares are congruences [9, Proposition 143].

Consider two congruences γ and β on a general algebra, not necessarily in a Mal'tsev variety. Then γ is said *to centralise* β if there is a congruence $(\gamma|\beta)$ on β, called a *centreing congruence*, such that the following conditions are satisfied [9, Definition 211]:

(C0): $(x,y)\ (\gamma|\beta)\ (x',y') \Rightarrow x\ \gamma\ x'$;
(C1): $\forall (x,y) \in \beta,\ \pi^0 : (x,y)^{(\gamma|\beta)} \to x^\gamma; (x',y') \mapsto x'$ bijects;
(C2): **(RR):** $\forall (x,y) \in \gamma,\ (x,x)\ (\gamma|\beta)\ (y,y)$;
(RS): $(x,y)\ (\gamma|\beta)\ (x',y') \Rightarrow (y,x)\ (\gamma|\beta)\ (y',x')$;
(RT): $(x,y)\ (\gamma|\beta)\ (x',y')$ and $(y,z)\ (\gamma|\beta)\ (y',z') \Rightarrow (x,z)\ (\gamma|\beta)\ (x',z')$.

Example 2.1 Suppose that A is a (not necessarily associative) ring. For congruences β and γ on A, consider the ideals $B = 0^\beta$ and $C = 0^\gamma$. Then γ centralises β iff $BC + CB = \{0\}$ [9, pp.27-8].

In a Mal'tsev variety $\mathfrak{V}$, centreing congruences are unique [9, Proposition 221]. Moreover, for each congruence α on an algebra A in $\mathfrak{V}$, there is a unique largest congruence $\eta(\alpha)$, called the *centraliser* of α, which centralises α [9, 228]. Note that $\alpha \circ \eta(\alpha)$ centralises $\alpha \cap \eta(\alpha)$ [9, Corollary 227].

If R is a member of a variety $\mathfrak{V}$ of universal algebras, then following Beck [2] the *category of R-modules* is the category of abelian groups in the slice category $\mathfrak{V}/R$. For example, if A is an algebra in a Mal'tsev variety $\mathfrak{V}$ having nested congruences $\beta \leq \gamma$ such that γ centralises β, then $\beta^{(\gamma|\beta)} \to A^\gamma$ is an A^γ-module. Indeed, given $a_1\ \beta\ a_0\ \gamma\ b_0\ \beta\ b_2$ in A, one has

$$(a_0,a_1)^{(\gamma|\beta)} + (b_0,b_1)^{(\gamma|\beta)} = (a_0,a_3)^{(\gamma|\beta)}$$

for a_2 given by $(a_0,a_2)(\gamma|\beta)(b_0,b_2)$ using (C1) and then for a_3 given similarly by $(a_0,a_1)(\gamma|\beta)(a_2,a_3)$.

3 Seeded simplicial maps

The data used for the construction of extensions are most succinctly expressed in terms of simplicial maps. These are described using the direct algebraic approach of [9], to which the reader is referred for fuller detail. Compare also [3].

Let ε_n^i be the operation which deletes the $(i+1)$-th letter from a non-empty word of length n. Let δ_n^i be the operation which repeats the $(i+1)$-th letter in a non-empty word of length n. These operations, for all positive integers n and natural numbers $i < n$, generate (the morphisms of) a category Δ called the *simplicial category*. A *simplicial object* B^* in $\mathfrak{V}$ is (the image of) a functor from Δ to $\mathfrak{V}$. A *simplicial map* is (the set of components of) a natural transformation between such functors. Generically, the morphisms of a simplicial object B^* are denoted by their preimages in Δ, namely as $\varepsilon_n^i : B^n \to B^{n-1}$ and $\delta_n^i : B^n \to B^{n+1}$. (In other words, one treats simplicial objects as heterogeneous algebras in $\mathfrak{V}$.)

Given $(\theta^0, \ldots, \theta^{n-1}) \in \mathfrak{V}(X,Y)^n$, the *simplicial kernel* $\ker(\theta^0, \ldots, \theta^{n-1})$ is the largest subalgebra K of the power X^{n+1} for which the θ^i and the restrictions of the projections from the power respectively model the identities satisfied by the simplicial ε_n^i and ε_{n+1}^i. For example, the simplicial kernel of a single $\mathfrak{V}$-morphism $\theta^0 : X \to Y$ is $K = \{(x_0, x_1) \in X^2 \mid x_0\theta^0 = x_1\theta^0\}$, namely the usual kernel of θ^0, which models the single simplicial identity $\varepsilon_2^0\varepsilon_1^0 = \varepsilon_2^1\varepsilon_1^0$ by $\pi^0\theta^0 = \pi^1\theta^0$ for $\pi^i : K \to X; (x_0, x_1) \mapsto x_i$. Similarly, one has

$$\ker(\theta^0, \theta^1) = \{(x_0, x_1, x_2) \in X^3 \mid x_0\theta^0 = x_1\theta^0, x_1\theta^1 = x_2\theta^1, x_2\theta^0 = x_0\theta^1\},$$

on which for instance $\pi^2\theta^0 = \pi^0\theta^1$ models $\varepsilon_3^2\varepsilon_2^0 = \varepsilon_3^0\varepsilon_2^1$ by virtue of the condition $x_2\theta^0 = x_0\theta^1$.

For each positive integer n, removing all operations from Δ that involve words of length greater than n leaves the *simplicial category* Δ_n *truncated at* n. Functors from Δ_n are called *simplicial objects truncated at dimension* n. Truncated simplicial objects may be extended to full simplicial objects by successively tacking on simplicial kernels. In such cases one may omit the epithet "truncated," speaking merely of simplicial objects, even when one has only specified the lower-dimensional part.

Definition 3.1 A simplicial object B^* is said to be *seeded* if:

1. It is truncated at dimension 2;
2. $(\varepsilon_2^0, \varepsilon_2^1) : B^2 \to \ker(\varepsilon_1^0)$ surjects;
3. $\varepsilon_1^0 : B^1 \to B^0$ surjects;
4. $\ker(\varepsilon_2^0 : B^2 \to B^1) = \eta(\ker(\varepsilon_2^1 : B^2 \to B^1))$.

Lemma 3.2 *In a seeded simplicial object B, let C be the equalizer of the pair $(\varepsilon_2^0, \varepsilon_2^1)$. Define V on C by*

$$c \ V \ c' \Leftrightarrow ((c\varepsilon_2^0\delta_1^0, c), (c'\varepsilon_2^0\delta_1^0, c')) \in (\ker\varepsilon_2^0 \circ \ker\varepsilon_2^1 | \ker\varepsilon_2^0 \cap \ker\varepsilon_2^1).$$

Then

$$C^V \to B^0; c^V \mapsto c\varepsilon_2^0\varepsilon_1^0 \tag{3.1}$$

is a module over B^0, isomorphic to $(\ker\varepsilon_2^0 \cap \ker\varepsilon_2^1)^{(\ker\varepsilon_2^0 \circ \ker\varepsilon_2^1 | \ker\varepsilon_2^0 \cap \ker\varepsilon_2^1)}$.

The module (3.1) of Lemma 3.2 is called the module *grown* by the seeded simplicial object B^*. If α is a congruence on a $\mathfrak{V}$-algebra T, then

$$\alpha^{(\eta(\alpha)|\alpha)} \rightrightarrows T^{\eta(\alpha)} \to T^{\alpha\circ\eta(\alpha)} \tag{3.2}$$

is a seeded simplicial object with $\varepsilon_2^i : (t_0, t_1)^{(\eta(\alpha)|\alpha)} \mapsto t_i^{\eta(\alpha)}$, growing the module

$$(\alpha \cap \eta(\alpha))^{(\alpha\circ\eta(\alpha)|\alpha\cap\eta(\alpha))} \to T^{\alpha\circ\eta(\alpha)}; (t_0, t_1)^{(\alpha\circ\eta(\alpha)|\alpha\cap\eta(\alpha))} \mapsto t_0^{\alpha\circ\eta(\alpha)}.$$

The seeded simplicial object (3.2) is said to be *planted* by the congruence α on the algebra T.

Definition 3.3 A simplicial map $p^* : A^* \to B^*$ is said to be *seeded* if the codomain object B^* is seeded in the sense of Definition 3.1, and if $p^0 : A^0 \to B^0$ surjects.

4 Obstructions

Along with the simplicial theory outlined in Section 3, the second tool used for studying extensions of Mal'tsev algebras is monadic cohomology. Once again, full details may be found in [3] and [9]; the summary given here follows the direct approach of the latter reference.

For each $\mathfrak{V}$-algebra A, let AG denote the free $\mathfrak{V}$-algebra over the generating set $\{\{a\} \mid a \in A\}$. Given a $\mathfrak{V}$-algebra R, let $\varepsilon_n^j : RG^n \to RG^{n-1}$ denote the uniquely defined $\mathfrak{V}$-morphism deleting the j-th layer of braces, where $j = 0$ corresponds to the inside layer and $j = n-1$ to the outside. Let $\delta_n^j : RG^n \to RG^{n+1}$ insert the j-th layer of braces. One obtains a simplicial object RG^*, known as the *free resolution* of A. Each RG^n projects to R by a composition

$$\varepsilon_n^0 \dots \varepsilon_1^0 : RG^n \to R. \tag{4.1}$$

An R-module $E \to R$ becomes an RG^n-module by pullback along (4.1). Write $\mathrm{Der}(RG^n, E)$ for the abelian group $\mathfrak{V}/R(RG^n \to R, E \to R)$ of *derivations.* Define *coboundary homomorphisms*

$$d_n : \mathrm{Der}(RG^n, E) \to \mathrm{Der}(RG^{n+1}, E); f \mapsto \sum_{i=0}^{n} (-)^i \varepsilon_{n+1}^i f$$

for each natural number n. For each positive integer n, define

$$\mathrm{H}^n(R, E) = \mathrm{Ker}(d_n)/\mathrm{Im}(d_{n-1}), \tag{4.2}$$

the so-called n-th *monadic cohomology group of R with coefficients in E.* [Note that [3] and [9] use $\mathrm{H}^{n-1}(R, E)$ for (4.2).] The cosets forming (4.2) are known as *cohomology classes.* Elements of $\mathrm{Ker}(d_n)$ are known as *cocycles*, and elements of $\mathrm{Im}(d_{n-1})$ are *coboundaries.*

Lemma 4.1 *Let $p^* : RG^* \to B^*$ be a seeded simplicial map whose codomain grows module M. Pull M from B^0 back to R along p^0. Then*

$$p^3(\varepsilon_3^0, \varepsilon_3^1, \varepsilon_3^2)P^V : RG^3 \to M \tag{4.3}$$

is a cocycle in $\mathrm{Der}(B^3, M)$.

Definition 4.2 The cohomology class of (4.3) is called the *obstruction* of the seeded simplicial map p^*. The simplicial map is said to be *unobstructed* if this class is zero.

Lemma 4.3 *The obstruction of a seeded simplicial map $p^* : RG^* \to B^*$ is uniquely determined by its bottom component $p^0 : R \to B^0$.*

The diagram-chasing proofs of Lemmas 4.1 and 4.3 are given in [9, pp.124–7].

5 Constructing extensions

Definition 5.1 A seeded simplicial map $p^* : RG^* \to B^*$ is said to be *realised* by an algebra T if there is a congruence α on T planting B^* such that p^0 is the natural projection $T^\alpha \to T^{\alpha \circ \eta(\alpha)}$.

Theorem 5.2 *A seeded simplicial map $p^* : RG^* \to B^*$ is unobstructed iff it is realised by an algebra T.*

Proof (Sketch.) "**If:**" Consider the diagram

$$\begin{array}{ccccccc}
\Rrightarrow & RG^2 & \Rrightarrow & RG & \rightarrow & R & \\
& \downarrow \sigma^2 & & \downarrow \sigma^1 & & \downarrow \sigma^0 & \\
\Rrightarrow & \alpha & \Rrightarrow & T & \rightarrow & R & \\
& \downarrow & & \downarrow & & \downarrow p^0 & \\
\Rrightarrow & \alpha^{(\eta(\alpha)|\alpha)} & \Rrightarrow & T^{\eta(\alpha)} & \rightarrow & T^{\alpha\circ\eta(\alpha)} &
\end{array} \tag{5.1}$$

in which σ^0 is the identity on $R = T^\alpha$, σ^1 is given by the freeness of RG, and σ^2 exists since $\alpha = \ker(T \to R)$. Take p^2, p^1, p^0 to be the composites down the respective columns of (5.1), the second factors of these composites all being natural projections. Writing $\pi^i : \alpha \to T; (t_0, t_1) \mapsto t_i$, one has

$$(\varepsilon_3^0\sigma^2, \varepsilon_3^1\sigma^2, \varepsilon_3^2\sigma^2)P\pi^0 = (\varepsilon_3^0\varepsilon_2^0, \varepsilon_3^1\varepsilon_2^0, \varepsilon_3^2\varepsilon_2^0)P\sigma^1 = \varepsilon_3^2\varepsilon_2^0\sigma^1$$
$$= \varepsilon_3^0\varepsilon_2^1\sigma^1 = (\varepsilon_3^0\varepsilon_2^1, \varepsilon_3^1\varepsilon_2^1, \varepsilon_3^2\varepsilon_2^1)P\sigma^1 = (\varepsilon_2^0\sigma^2, \varepsilon_2^1\sigma^2, \varepsilon_2^2\sigma^2)P\pi^1,$$

so the obstruction of p^* is the zero element $(\varepsilon_3^0 p^2, \varepsilon_3^1 p^2, \varepsilon_3^2 p^2)P^V$ of the group $\mathrm{Der}(RG^3, (\alpha \cap \eta(\alpha))^{(\alpha\circ\eta(\alpha)|\alpha\cap\eta(\alpha))})$, as required.

"**Only if:**" If p^* is unobstructed, then as shown in [9, p.129], one may assume without loss of generality that (4.3) itself is zero, and not just in the zero cohomology class. Let Q be a pullback in

$$\begin{array}{ccc}
Q & \rightarrow & RG \\
\downarrow & & \downarrow p_1 \\
B^2 & \underset{\varepsilon_2^1}{\rightarrow} & B^1
\end{array}$$

realised, say, by $Q = \{(b, w) \in B^2 \times RG \mid wp^1 = b\varepsilon_2^1\}$. Define a congruence W on Q by $(b, w)\ W\ (b', w')$ iff $w\varepsilon_1^0 = w'\varepsilon_1^0$ and

$$(b, b')\ (\ker \varepsilon_2^1 | \ker \varepsilon_2^0)\ (\{w\}p^2, \{w'\}p^2).$$

Set $T = Q^W$, and take α on T to be the kernel of $T \to R; (b, w)^W \mapsto w\varepsilon_1^0$. For the details of the verification that T realises p^*, with α planting B^*, see [9, pp.129–132]. In particular, note that $\eta(\alpha)$ is the kernel of $T \to B^1; (b, w)^W \mapsto b\varepsilon_2^0$.

□

6 Classifying extensions and obstructions

Let $p^* : RG^* \to B^*$ be a seeded simplicial map whose codomain grows a module $M \to B^0$. Pull M back along $p^0 : R \to B^0$ to an R-module. An extension $\alpha \rightrightarrows T \to R$ is said to be *singular for* p^* if its kernel α is self-centralising, with an R-module isomorphism $\alpha^{(\alpha|\alpha)} \to M$. (Note that the central extensions of [5] form a special case of the singular extensions, in which α centralises all of $T \times T$.) Let p^*S be the set of $\mathfrak{V}/R$-isomorphism classes of extensions that are singular for p^*. This set becomes an abelian group, with the class of the split extension $M \to R$ as zero. The addition operation on p^*S is known as the *Baer sum*. To obtain a representative of the Baer sum of the isomorphism classes of two extensions

$\alpha_i \rightrightarrows T_i \to R$, with module isomorphism $\theta : \alpha_1^{(\alpha_1|\alpha_1)} \to \alpha_2^{(\alpha_2|\alpha_2)}$, take the quotient of the pullback $T_1 \times_R T_2$ by the congruence

$$\{((t_1, t_2), (t_1', t_2')) \mid (t_i, t_i') \in \alpha_i,\ (t_1, t_1')^{(\alpha_1|\alpha_1)}\theta = (t_2, t_2')^{(\alpha_2|\alpha_2)}\}.$$

Singular extensions are then classified as follows [9, Theorem 632], cf. [2], [3].

Theorem 6.1 *The groups p^*S and $\mathrm{H}^2(R, M)$ are isomorphic.*

Now assume additionally that the seeded simplicial map $p^* : RG^* \to B^*$ is unobstructed. An extension $\alpha \rightrightarrows T \to R$ is said to be *non-singular for p^** if T realises p^*. Let p^*N denote the set of $\mathfrak{V}/R$-isomorphism classes of extensions that are non-singular for p^*. By Theorem 5.2, p^*N is non-empty. Non-singular extensions are then classified as follows [9, Theorem 634].

Theorem 6.2 *The abelian group p^*S acts regularly on p^*N, so the sets p^*N and $\mathrm{H}^2(R, M)$ are isomorphic.*

Let $\beta \rightrightarrows S \to R$ be singular for p^*, and let $\alpha \rightrightarrows T \to R$ be non-singular for p^*. To obtain a representative for the image of the class of α under the action of the class of β, assuming an R-module isomorphism $\theta : (\alpha \cap \eta(\alpha))^{(\alpha \circ \eta(\alpha)|\alpha \cap \eta(\alpha))} \to \beta^{(\beta|\beta)}$, take the quotient of the pullback $T \times_R S$ by the congruence

$$\{((t, s), (t', s')) \mid (t, t') \in \alpha \cap \eta(\alpha),\ (s', s) \in \beta,\ (t, t')^{(\alpha|\alpha \cap \eta(\alpha))}\theta = (s', s)^{(\beta|\beta)}\}.$$

The final result [9, Theorem 641] shows how obstructions may be classified by elements of the third monadic cohomology groups. Note that for non-trivial R, the hypothesis on R is always satisfied in varieties $\mathfrak{V}$, such as the variety of all groups, where free algebras have little centrality. On the other hand, it is not satisfied, for example, by the three-element group in the variety of commutative Moufang loops.

Theorem 6.3 *Let R be a $\mathfrak{V}$-algebra for which $\eta(\ker(\varepsilon_1^0 : RG \to R)) = \widehat{RG}$. Let $M \to R$ be an R-module, and let $\xi \in \mathrm{H}^3(R, M)$. Then ξ is the obstruction to a seeded simplicial map $p^* : RG^* \to B^*$ whose codomain grows a module that pulls back to $M \to R$ along p^0.*

References

[1] Barr, M. *Cohomology and obstructions: commutative algebras*, Springer Lecture Notes in Mathematics No. 80, (ed. B. Eckmann), Springer-Verlag, Berlin, 1969, pp. 357–375.

[2] Beck, J. *Triples, Algebras, and Cohomology*, Ph.D. thesis, Columbia University, New York, NY, 1967, Theory and Applications of Categories Reprint No. 2.

[3] Duskin, J. *Simplicial Methods and the Interpretation of "Triple" Cohomology*, Memoirs of the American Mathematical Society No. 163, American Mathematical Society, Providence, RI, 1975.

[4] Gruenberg, K. *Cohomological Topics in Group Theory*, Springer Lecture Notes in Mathematics No. 143, Springer-Verlag, Berlin, 1970.

[5] Janelidze, G. and Kelly, G.M. *Central extensions in universal algebra: a unification of three notions*, Alg. Univ. **44** (2000), 123–128.

[6] Johnson, K.W. and Leedham-Green, C.R. *Loop cohomology*, Czech. Math. J. **40** (**115**) (1990), 182–195.

[7] Mac Lane, S. *Homology*, Springer-Verlag, Berlin, 1963.

[8] Orzech, G.G. *Obstruction Theory in Algebraic Categories*, Ph.D. thesis, University of Illinois, Urbana, IL, 1970.

[9] Smith, J.D.H. *Mal'cev Varieties*, Springer Lecture Notes in Mathematics No. 554, Springer-Verlag, Berlin, 1976.

[10] Smith, J.D.H. and Romanowska, A.B. *Post-Modern Algebra*, Wiley, New York, NY, 1999.

Received October 28, 2002; in revised form June 7, 2003

Fields Institute Communications
Volume **43**, 2004

On Projective Generators Relative to Coreflective Classes

Lurdes Sousa
Department of Mathematics
School of Technology
Polytechnic Institute of Viseu
3504-510 Viseu, Portugal
sousa@mat.estv.ipv.pt

Abstract. Projective $\mathcal{E}$-generators, for $\mathcal{E}$ a coreflective class of morphisms, are studied. Under mild conditions, it is shown that, for cocomplete categories $\mathbf{A}$ with a projective $\mathcal{E}$-generator P, the colimit-closure of P is the smallest $\mathcal{E}$-coreflective subcategory of $\mathbf{A}$, and, furthermore, it is premonadic over **Set** via the functor $\hom(P,-)$. A variety of examples is given.

Introduction

Several generalizations of the concept of factorization system for morphisms have appeared in the literature; here we work with one of them, the notion of coreflective class (see, e.g., [7], [12], [17], [18]). Pushout-stable coreflective classes of morphisms (in a category $\mathbf{A}$ with pushouts) are just those which, regarded as subcategories[1] of $\mathrm{Mor}(\mathbf{A})$, are coreflective. A significant role is played by the stabilization of a coreflective class $\mathcal{E}$, denoted by $\mathrm{St}(\mathcal{E})$, and defined as being the subclass of $\mathcal{E}$ which consists of all morphisms whose pullbacks along any morphism belong to $\mathcal{E}$ (see [7]). By a projective $\mathcal{E}$-generator it is meant an $\mathcal{E}$-generator which is projective with respect to $\mathrm{St}(\mathcal{E})$. Many examples of categories with a projective $\mathcal{E}$-generator P are given. In all of them projectivity does not hold with respect to the whole $\mathcal{E}$ unless $\mathcal{E}$ coincides with its stabilization. In fact, $\mathrm{St}(\mathcal{E})$ is shown to be precisely the class of all those morphisms to which P is projective.

Subcategories which are "colimit-generated" by a projective $\mathcal{E}$-generator of their cocomplete supercategories are shown to have special properties. Namely, let $\mathbf{A}$ be an $\mathcal{E}$-cocomplete category with pullbacks and $\mathcal{E}$ closed under composition with split epimorphisms. We show that if P is a projective $\mathcal{E}$-generator of $\mathbf{A}$, the colimit closure of P, denoted by $\mathbb{C}(P)$, is the smallest $\mathcal{E}$-coreflective subcategory of $\mathbf{A}$; furthermore, if the stabilization of $\mathcal{E}$ is closed under coequalizers of kernel pairs,

---

2000 *Mathematics Subject Classification.* Primary 18A20, 18A22; Secondary 18C15, 18G05.
The author was supported in part by Center of Mathematics of Coimbra University.
[1]Throughout, by subcategory we mean a full and isomorphism-closed subcategory.

then P is a regular generator of $\mathbb{C}(P)$, and thus $\mathbb{C}(P)$ is equivalent to a reflective subcategory of the category of Eilenberg-Moore algebras of the monad induced by the functor $\hom(P,-)$, that is, $\mathbb{C}(P)$ is premonadic over **Set**. Analogous results are obtained when (under the cocompleteness of $\mathbf{A}$) the coreflectiveness of $\mathbb{C}(P)$ replaces the $\mathcal{E}$-cocompleteness of $\mathbf{A}$. These properties are illustrated with many examples.

1 Projective $\mathcal{E}$-generators

Definition 1.1 A class $\mathcal{E}$ of epimorphisms of a category $\mathbf{A}$ (closed under composition with isomorphisms) is said to be a *coreflective class* whenever, for each $B \in \mathbf{A}$, the embedding $\mathcal{E}(B) \to B \downarrow \mathbf{A}$, where $\mathcal{E}(B)$ denotes de subcategory of $B \downarrow \mathbf{A}$ whose objects are $\mathcal{E}$-morphisms, is a left adjoint; that is, each $\mathbf{A}$-morphism f has a factorization $m \cdot e$ with $e \in \mathcal{E}$, and such that if $m' \cdot e'$ is another such a factorization of f then there is a (unique) morphism t with $t \cdot e' = e$ and $m \cdot t = m'$. We say that $m \cdot e$ is the *$\mathcal{E}$-factorization of f*. (cf. [12], [18] and [7].)

Remark 1.2 If $\mathbf{A}$ has pushouts and $\mathcal{E}$ is a coreflective class of $\mathcal{A}$, the following facts are easy consequences of the above definition:

1. $\mathcal{E}$ is pushout-stable if and only if the "local coreflections" from $B \downarrow \mathbf{A}$ to $\mathcal{E}(B)$ determine a "global coreflection" from Mor($\mathbf{A}$) to $\mathcal{E}$, where $\mathcal{E}$ is regarded as a subcategory of Mor($\mathbf{A}$). (The fact that $\mathbf{A}$ has a pushout-stable coreflective class $\mathcal{E}$ means in the terminology of [12] that $\mathbf{A}$ has a locally orthogonal $\mathcal{E}$-factorization.)
2. $\mathcal{E}$ determines a factorization system for morphisms if and only if it is pushout-stable and closed under composition.

We add another property which will play a role throughout:

Lemma 1.3 *If $\mathcal{E}$ is a pushout-stable coreflective class in a category with pushouts, then the following conditions are equivalent:*

(i) Any split epimorphism m which is part of an $\mathcal{E}$-factorization $m \cdot e$ is an isomorphism.

(ii) $\mathcal{E}$ is closed under composition with split epimorphisms from the left.

(iii) $\mathcal{E}$ is closed under composition with split epimorphisms (from the left and from the right).

Proof (i) $\Rightarrow$ (iii): Let $r \cdot s$ be defined with $s \in \mathcal{E}$ and r a split epi, and let $m \cdot e$ be the $\mathcal{E}$-factorization of $r \cdot s$. Then there is t such that $m \cdot t = r$ and, since r is a split epi, so is m, thus m is an isomorphism and $r \cdot s$ belongs to $\mathcal{E}$.

Take $r \cdot s$ with $r \in \mathcal{E}$ and s a split epi, let u be such that $s \cdot u = 1$, and let $m \cdot e$ be the $\mathcal{E}$-factorization of $r \cdot s$. From 1.2.1 and the equality $1 \cdot r = m \cdot e \cdot u$, we get a morphism t such that $mt = 1$ and $tr = eu$. Condition (i) ensures that m is an iso, and $r \cdot s \in \mathcal{E}$.

(ii) $\Rightarrow$ (i): Let $m \cdot e$ be the $\mathcal{E}$-factorization of some f with m a split epi. Then $f \in \mathcal{E}$ and so f has an $\mathcal{E}$-factorization of the form $1 \cdot f$. Therefore, m is iso. $\square$

Remark 1.4 The property of "$\mathcal{E}$ being closed under composition with split epimorphisms" will be used several times along the paper. If $\mathcal{E}$ is part of a proper factorization system, then the property holds. But the converse is not true as it is shown by the following example (G. Janelidze, private communication): Let $\mathbf{A}$ be the ordered set $0 \to 1 \to 2$ regarded as a category. Let $\mathcal{E}$ be the set of all maps in $\mathbf{A}$ except $0 \to 2$. Then $\mathcal{E}$ is a pushout-stable coreflective class closed under composition with split epimorphisms but not closed under composition. Another example of the fact that the closedness of a pushout-stable coreflective class $\mathcal{E}$ under composition with split epimorphisms does not imply that $\mathcal{E}$ is part of a factorization system for morphisms is given in [5], for $\mathcal{E} = \{\text{regular epimorphisms}\}$.

Definition 1.5 (cf. [6]) An object P is an *$\mathcal{E}$-generator* of the category $\mathbf{A}$ with copowers of P if, for each $A \in \mathbf{A}$, the canonical morphism ε_A from the coproduct $\coprod_{\hom(P,A)} P$ to A belongs to $\mathcal{E}$.

Assumption 1.6 From now on we assume that the category $\mathbf{A}$ has pullbacks and pushouts, and $\mathcal{E}$ is a pushout-stable coreflective class contained in $\mathrm{Epi}(\mathbf{A})$ which is closed under composition with split epimorphisms.

A morphism is said to be *stably in $\mathcal{E}$* if its pullback along any morphism belongs to $\mathcal{E}$. The *stabilization of $\mathcal{E}$* is the class of all morphisms that are stably in $\mathcal{E}$; it is denoted by $\mathrm{St}(\mathcal{E})$ and it is clearly contained in $\mathcal{E}$ (see [7]).

Lemma 1.7 (see [5]) *$\mathcal{E}$ and St($\mathcal{E}$) are strongly right-cancellable.*

Proof Given $r \cdot s \in \mathcal{E}$, we want to show that $r \in \mathcal{E}$. Let $m \cdot e$ be the $\mathcal{E}$-factorization of r. Then the equality $1 \cdot r \cdot s = m \cdot e \cdot s$ determines, by 1.2.1, the existence of a morphism t such that $mt = 1$ and $t \cdot (r \cdot s) = e \cdot s$. Since m is a split epimorphism and $m \cdot e$ is an $\mathcal{E}$-factorization, then, by 1.3, m is an isomorphism; so $r \in \mathcal{E}$. The strong right-cancellability of $\mathrm{St}(\mathcal{E})$ follows easily from the strong right-cancellability of $\mathcal{E}$. □

Definition 1.8 An object P is said to be a *projective $\mathcal{E}$-generator* if it is an $\mathcal{E}$-generator which is $\mathrm{St}(\mathcal{E})$-projective, that is, for each $f \in \mathrm{St}(\mathcal{E})$, the function $\hom(P, f)$ is surjective.

Notation 1.9 For A an $\mathbf{A}$-object, $\mathrm{Proj}(A)$ denotes the class of all $\mathbf{A}$-morphisms f such that A is f-projective. It is easily seen that $\mathrm{Proj}(A)$ is pullback-stable.

Assumption 1.10 In the following, besides the assumptions stated in 1.6, we also assume that $\mathbf{A}$ is cocomplete.

Proposition 1.11 *If P is a projective $\mathcal{E}$-generator, then St($\mathcal{E}$) = Proj(P).*

Proof By the assumption on P, the inclusion $\mathrm{St}(\mathcal{E}) \subseteq \mathrm{Proj}(P)$ is trivial. It remains to show that if P is $(f : X \to Y)$-projective then $f \in \mathrm{St}(\mathcal{E})$, i.e., any pullback of f along any morphism belongs to $\mathcal{E}$. Let $(\bar{f} : W \to Z, \bar{g} : W \to X)$ be the pullback of (f, g). Since P is f-projective, any coproduct of P is also f-projective; thus, from the pullback-stability of any class $\mathrm{Proj}(A)$, $\bar{f} \in \mathrm{Proj}(\coprod_{\hom(P,Z)} P)$. Let s be a morphism fulfilling $\bar{f} \cdot s = \varepsilon_Z$ and let $m \cdot e$ be the $\mathcal{E}$-factorization of $\bar{f}$. We get the equality $1_Z \cdot \varepsilon_Z = m \cdot e \cdot s$, which, since $\varepsilon_Z \in \mathcal{E}$, implies the existence of some t such that $t \cdot \varepsilon_Z = e \cdot s$ and $m \cdot t = 1_Z$. Then, from 1.3, and in view of 1.6, m is an iso and $\bar{f} \in \mathcal{E}$. □

Corollary 1.12 *If P is a projective $\mathcal{E}$-generator, then it is a projective St($\mathcal{E}$)-generator.*

Several examples of categories with a projective $\mathcal{E}$-generator are described in 2.5 below.

One question arises: When is the stabilization of a corefletive class $\mathcal{E}$ of the form Proj(P), or, at least, when is it of the form Proj$(\mathbf{B})$ for some subcategory $\mathbf{B}$ of $\mathbf{A}$? The next proposition gives a partial answer.

We are going to make use of the following lemma.

Lemma 1.13 *If* $\mathbf{B}$ *is an* $\mathcal{E}$*-coreflective subcategory of* $\mathbf{A}$*, then it is* $(St(\mathcal{E}) \cap Mono(\mathbf{A}))$*-coreflective.*

Proof Let $s_X : S(X) \to X$ be a coreflection of X in $\mathbf{B}$ (with S the coreflector functor). By hypothesis, s_X lies in $\mathcal{E}$; in order to show that it belongs to St$(\mathcal{E})$, let $(\bar{s} : W \to Y, \bar{g} : W \to S(X))$ be the pullback of (s_X, g), for some morphism g. To conclude that $\bar{s} \in \mathcal{E}$, let $m \cdot e$ be the $\mathcal{E}$-factorization of $\bar{s}$. Since $s_X \cdot S(g) = g \cdot s_Y$, there is a unique morphism v such that $\bar{s} \cdot v = s_Y$ and $\bar{g} \cdot v = S(g)$. Then we have $1_Y \cdot s_Y = m \cdot e \cdot v$; since $m \cdot e$ is an $\mathcal{E}$-factorization and $s_Y \in \mathcal{E}$, there is a morphism t such $m \cdot t = 1_Y$, and, taking into account 1.3 and 1.6, we get that $\bar{s} \in \mathcal{E}$. It remains to show that s_X is a monomorphism: Let $a, b : Y \to S(X)$ be such that $s_X \cdot a = s_X \cdot b$; then the equality $s_X \cdot a \cdot s_Y = s_X \cdot b \cdot s_Y$ implies that $a \cdot s_Y = b \cdot s_Y$, and thus, since $\mathcal{E} \subseteq$ Epi$(\mathbf{A})$, $a = b$. □

Let us recall that, if $\mathcal{F}$ is a class of morphisms of a category $\mathbf{A}$ containing all isomorphisms and closed under composition with isomorphisms, an $\mathcal{F}$-morphism $f : B \to A$ is said to be $\mathcal{F}$*-coessential* whenever any composition $f \cdot g$ belongs to $\mathcal{F}$ only if $g \in \mathcal{F}$. We say that the category $\mathbf{A}$ has *enough* $\mathcal{F}$*-projectives* if, for each $A \in \mathbf{A}$, there is some $\mathcal{F}$-morphism $f : B \to A$ with B $\mathcal{F}$-projective; if, in addition, f can be chosen to be $\mathcal{F}$-coessential, we say that $\mathbf{A}$ has $\mathcal{F}$*-projective hulls.*

Proposition 1.14 *1. If* $\mathbf{A}$ *has enough* $St(\mathcal{E})$*-projectives, then* $St(\mathcal{E}) = Proj(\mathbf{B})$ *for some subcategory* $\mathbf{B}$ *of* $\mathbf{A}$*.*

2. If $St(\mathcal{E}) = Proj(\mathbf{B})$ *for some* $\mathcal{E}$*-coreflective subcategory* $\mathbf{B}$ *of* $\mathbf{A}$*, then* $\mathbf{B}$ *has* $St(\mathcal{E})$*-projective hulls.*

Proof 1. Let $\mathbf{B}$ consist of all objects of $\mathbf{A}$ which are St$(\mathcal{E})$-projective; clearly St$(\mathcal{E}) \subseteq$ Proj$(\mathbf{B})$. In order to show the converse inclusion, let $f : A \to B$ belong to Proj$(\mathbf{B})$, and let $(\bar{f} : D \to C, \bar{g} : D \to A)$ be the pullback of f and g for some $g : C \to B$. Since $\mathbf{A}$ has enough St$(\mathcal{E})$-projectives, there is some St$(\mathcal{E})$-morphism $q : E \to C$ with $E \in \mathbf{B}$. Now, using with the morphism q the same technique used in the proof of 1.11 with the morphism ε_Z, we get that $\bar{f}$ belongs to $\mathcal{E}$.

2. By Lemma 1.13, each coreflection $s_A : S(A) \to A$ into $\mathbf{B}$ belongs to St$(\mathcal{E})$. It remains to show that s_A is a St$(\mathcal{E})$-coessential morphism. Let g be such that $s_A \cdot g$ belongs to St$(\mathcal{E})$, and let $\bar{g}$ be the pullback of g along any h. By 1.13, s_A is a monomorphism, then $\bar{g}$ is also the pullback of $s_A \cdot g$ along $s_A \cdot h$. Thus $\bar{g} \in \mathcal{E}$, and so g belongs to St$(\mathcal{E})$. □

2 The colimit-closure of a projective $\mathcal{E}$-generator

In this section we study properties of the colimit-closure of a projective $\mathcal{E}$-generator.

Remark 2.1 Let us recall that, for $\mathcal{E}$ a class containing all isomorphisms and closed under composition with isomorphisms, $\mathbf{A}$ is said to be $\mathcal{E}$-cocomplete if every pushout of any $\mathcal{E}$-morphism exists and belongs to $\mathcal{E}$ and any family of morphisms of $\mathcal{E}$ has a cointersection belonging to $\mathcal{E}$. The $\mathcal{E}$-cocompletness of $\mathbf{A}$ implies that $\mathcal{E} \subseteq$ Epi$(\mathbf{A})$ and that $\mathcal{E}$ is a coreflective class (see [17]). On the other hand, it is easily seen that if $\mathbf{A}$ is a cocomplete and $\mathcal{E}$-cowellpowered category with $\mathcal{E}$ a pushout-stable coreflective class then $\mathbf{A}$ is $\mathcal{E}$-cocomplete.

Notation 2.2 $\mathbb{C}(P)$ denotes the colimit-closure of P in $\mathbf{A}$, that is, the smallest subcategory of $\mathbf{A}$ containing P and closed under all colimits in $\mathbf{A}$.

Theorem 2.3 *Let* $\mathbf{A}$ *be* $\mathcal{E}$*-cocomplete and let* P *be a projective* $\mathcal{E}$*-generator of* $\mathbf{A}$*. Then* $\mathbb{C}(P)$ *consists of all* $\mathbf{A}$*-objects* A *such that the function* $hom(f, A)$ *is bijective for each* $f \in St(\mathcal{E}) \cap Mono(\mathbf{A})$*. Furthermore,* $\mathbb{C}(P)$ *is the smallest* $\mathcal{E}$*-coreflective subcategory of* $\mathbf{A}$*.*

Proof In order to prove that $\mathbb{C}(P)$ is coreflective in $\mathbf{A}$, it suffices to show the solution set condition for every $A \in \mathbf{A}$. Let $A \in \mathbf{A}$ and consider the following family, indexed by I:

$$\mathcal{S}_A = \{e_i : \coprod_{\hom(P,A)} P \to S_i \,|\, e_i \in \mathcal{E},\ S_i \in \mathbb{C}(P) \text{ and } t_i \cdot e_i = \varepsilon_A \text{ for some } t_i\}.$$

Let $(e : \coprod_{\hom(P,A)} P \to S;\ d_i : S_i \to S)$ be the cointersection of the family $\mathcal{S}_A$, and let $t : S \to A$ be the unique morphism such that $t \cdot e = \varepsilon_A$ and $t \cdot d_i = t_i$. We show that any morphism $f : B \to A$ with B in $\mathbb{C}(P)$ factorizes through S. Let $f^\star : \coprod_{\hom(P,B)} P \to \coprod_{\hom(P,A)} P$ be the morphism determined by $f : B \to A$, and let $(\bar{e} : \coprod_{\hom(P,A)} P \to \bar{S};\ \bar{f} : B \to \bar{S})$ be the pushout of $(\varepsilon_B, f^\star)$. Then $\bar{e}$ belongs to $\mathcal{S}_A$: The equality $f \cdot \varepsilon_B = \varepsilon_A \cdot f^\star$ gives rise to the existence of a unique morphism $w : \bar{S} \to A$ such that $w \cdot \bar{e} = \varepsilon_A$ and $w \cdot \bar{f} = f$. The morphism $\bar{e}$ is equal to e_i for some $i \in I$. Then $w \cdot e_i = \varepsilon_A = t \cdot e = t \cdot d_i \cdot e_i$, and since e_i is an epimorphism, $w = t \cdot d_i$. Therefore $f = t \cdot (d_i \cdot \bar{f})$.

It is well known that, being coreflective, $\mathbb{C}(P)$ coincides with the co-orthogonal closure of P in $\mathbf{A}$, that is, $\mathbb{C}(P)$ consists of all those $\mathbf{A}$-objects A such that for any morphism f, $\hom(A, f)$ is a bijection whenever $\hom(P, f)$ is so. Consequently, in order to conclude that $\mathbb{C}(P)$ is the subcategory of $\mathbf{A}$ of those objects A such that the function $\hom(f, A)$ is bijective for each $f \in \mathrm{St}(\mathcal{E}) \cap \mathrm{Mono}(\mathbf{A})$, it suffices to show that

$$\mathrm{St}(\mathcal{E}) \cap \mathrm{Mono}(\mathbf{A}) = \{f \in \mathrm{Mor}(\mathbf{A}) \,|\, \hom(P, f) \text{ is an iso}\}. \tag{2.1}$$

From 1.11, the inclusion "$\subseteq$" is trivial. Let $\hom(P, f)$ be an iso. Then, by 1.11, f belongs to $\mathrm{St}(\mathcal{E})$. In order to conclude that f is a mono, let $a, b : S \to X$ be morphisms such that $fa = fb$. Then for any $t : P \to S$, we have $fat = fbt$, which implies, since $\hom(P, f)$ is an iso, that $at = bt$. Thus $a = b$, because P is a generator.

By 1.11, each coreflection $r_A : RA \to A$ into $\mathbb{C}(P)$ belongs to $\mathrm{St}(\mathcal{E}) \subseteq \mathcal{E}$, because P is r_A-projective. In order to show that $\mathbb{C}(P)$ is the smallest $\mathcal{E}$-coreflective subcategory of $\mathbf{A}$, let $\mathbf{B}$ be another $\mathcal{E}$-coreflective subcategory of $\mathbf{A}$. Let $C \in \mathbb{C}(P)$ and let $s_C : S(C) \to C$ be the coreflection of C in $\mathbf{B}$. By Lemma 1.13, s_C belongs to $\mathrm{St}(\mathcal{E}) \cap \mathrm{Mono}(\mathbf{A})$. Therefore, by the equality (2.1), we get a morphism $t : C \to S(C)$ such that $s_C \cdot t = 1_C$, and thus s_C is an isomorphism. □

Remark 2.4 In the above proof, the only role of the $\mathcal{E}$-cocompleteness of $\mathbf{A}$ is to assure that $\mathbb{C}(P)$ is corefletive in $\mathbf{A}$. So, in Theorem 2.3 (in the presence of the conditions of Assumption 1.10), we can replace "$\mathbf{A}$ is $\mathcal{E}$-cocomplete" by "$\mathbb{C}(P)$ is coreflective".

We have seen that the existence of a projective $\mathcal{E}$-generator P gives a characterization of the stabilization of $\mathcal{E}$, $\mathrm{St}(\mathcal{E}) = \mathrm{Proj}(P)$, and, in case $\mathbf{A}$ is $\mathcal{E}$-cocomplete, it guarantees that P "generates" the smallest $\mathcal{E}$-coreflective subcategory of $\mathbf{A}$. In the following we give several examples of this situation.

Examples 2.5 1. For any monadic category $\mathbf{A}$ over $\mathbf{Set}$ and $\mathcal{E} = \mathrm{RegEpi}(\mathbf{A})$, let $P = FS$ for some $S \neq \emptyset$, where F is the corresponding left adjoint. Then P is an $\mathcal{E}$-generator such that $\mathrm{St}(\mathcal{E}) = \mathcal{E} = \mathrm{Proj}(P)$, and $\mathbf{A} = \mathbb{C}(P)$.

We point out that, under the conditions of Theorem 2.3, $\mathbf{A}$ and $\mathbb{C}(P)$ are identical whenever $\mathrm{St}(\mathcal{E}) \cap \mathrm{Mono}(\mathbf{A}) = \mathrm{Iso}(\mathbf{A})$, since $\{$ coreflections of $\mathbf{A}$ in $\mathbb{C}(P) \} \subseteq \mathrm{St}(\mathcal{E}) \cap \mathrm{Mono}(\mathbf{A})$.

2. For $\mathbf{A} = \mathbf{Cat}$, $\mathcal{E}$ the class of extremal epimorphisms of $\mathbf{A}$, and $\mathbf{2} = \{0 \to 1\}$ the category given by the ordered set 2, we have that $\mathbf{2}$ is an $\mathcal{E}$-generator, $\mathrm{St}(\mathcal{E}) = \mathrm{Proj}(\mathbf{2})$, and $\mathcal{E}$ does not coincides with its stabilization (see [8], [10]). The equality $\mathbf{A} = \mathbb{C}(\mathbf{2})$ also holds.
3. Let **PreOrd** be the category whose objects are preordered sets (i.e., sets with a reflexive and transitive binary relation), and whose morphisms are preorder-preserving maps. For $\mathcal{E} = \{$ regular epimorphisms $\} = \{$ extremal epimorphisms $\}$, the object $\mathbf{2}$ as in the above example is a projective $\mathcal{E}$-generator, in particular $\mathrm{St}(\mathcal{E}) = \mathrm{Proj}(\mathbf{2})$, although $\mathrm{St}(\mathcal{E}) \neq \mathcal{E}$. The colimit closure of $\mathbf{2}$ coincides with **PreOrd**.

In the following examples, it is more convenient to consider the dual situation. That is, now P is an injective $\mathcal{M}$-cogenerator of $\mathbf{A}$, with $\mathbf{A}$ and $\mathcal{M}$ fulfilling the dual conditions of 1.10. A morphism belongs to $\mathrm{St}(\mathcal{M})$ whenever its pushout along any morphism lies in $\mathcal{M}$. It holds the equality $\mathrm{St}(\mathcal{M})=\mathrm{Inj}(P)$, and the limit closure of P in $\mathbf{A}$, $\mathbb{L}(P)$, is the smallest $\mathcal{M}$-reflective subcategory of $\mathbf{A}$.

4. In the category $\mathbf{Set}$, the pushout-stable class $\mathcal{M}$ of monomorphisms coincides with $\mathrm{Inj}(P)$ for P the $\mathcal{M}$-cogenerator set $\{0,1\}$, and $\mathbb{L}(P) = \mathbf{Set}$.
5. For $\mathbf{A}$ the category **Top** of topological spaces and continuous maps, let $\mathcal{M} = \{\text{embeddings}\}$, and let P be the topological space $\{0,1,2\}$ whose only non trivial open is $\{0\}$. Then P is an $\mathcal{M}$-cogenerator and $\mathrm{Inj}(P)=\mathrm{St}(\mathcal{M})=\mathcal{M}$. The subcategory $\mathbb{L}(P)$ is the whole category **Top**.
6. For the category $\mathbf{Top_0}$ of T_0-topological spaces, $\mathcal{M} = \{\text{embeddings}\}$, the Sierpiński space S is an $\mathcal{M}$-cogenerator which fulfils $\mathrm{Inj}(S)=\mathrm{St}(\mathcal{M})=\mathcal{M}$. Here $\mathbb{L}(S)$ is the subcategory of sober spaces.
7. If $\mathbf{A}$ is the subcategory of **Top** of all 0-dimensional spaces, and $\mathcal{M}$ consists of all embeddings, then $\mathcal{M} \neq \mathrm{St}(\mathcal{M})$, but again $\mathrm{St}(\mathcal{M}) = \mathrm{Inj}(P)$, where P is the space $\{0,1,2\}$ whose only non trivial opens are $\{0\}$ and $\{1,2\}$. (The morphisms of $\mathrm{St}(\mathcal{M})$ are just those embeddings $m : X \to Y$ such that for each clopen set G of X there is some clopen H in Y such that $G = m^{-1}(H)$ (see [15]).) We have that $\mathbb{L}(P) = \mathbf{A}$ (by the dual of 2.3). A similar situation occurs for the category of 0-dimensional Hausdorff spaces and $\mathcal{M}$ the class of embeddings, which again is not pushout-stable, if we choose P as being the space $\{0,1\}$ with the discrete topology (see [15]).
8. For **Tych** the category of Tychonoff spaces, the class $\mathcal{M}$ of embeddings is not stable under pushouts, and $\mathrm{St}(\mathcal{M}) = \mathrm{Inj}(I)$, where I is the unit interval,

with the usual topology. The $\mathrm{St}(\mathcal{M})$-morphisms are just the C^*-embeddings (see [15]) and $\mathbb{L}(I)$ is the subcategory of compact Hausdorff spaces.

9. For $\mathbf{Vect}_{\mathbb{K}}$ the category of vector spaces and linear maps over the field $\mathbb{K}$, the class $\mathcal{M}$ of all monomorphisms is stable under pushouts and $\mathbb{K}$ is an injective $\mathcal{M}$-cogenerator.
10. In the category $\mathbf{Ab}$ of abelian groups and homomorphisms of group, let $\mathcal{M}$ be the class of all monomorphisms. It is well known that $\mathbb{Q}/\mathbb{Z}$ (where $\mathbb{Q}$ and $\mathbb{Z}$ are the groups of rational numbers and of integer numbers, respectively) is an $\mathcal{M}$-cogenerator, and it holds that $\mathrm{St}(\mathcal{M}) = \mathcal{M} = \mathrm{Inj}(\mathbb{Q}/\mathbb{Z})$. The limit closure of $\mathbb{Q}/\mathbb{Z}$ is the whole category $\mathbf{Ab}$, taking into account the dual of 2.3 and that in $\mathbf{Ab}$ any monomorphism is regular.
11. In the category of torsion-free abelian groups and homomorphisms of group, for $\mathcal{M}$ the class of all monomorphisms, the group of rational numbers $\mathbb{Q}$ is an $\mathcal{M}$-cogenerator and $\mathrm{St}(\mathcal{M}) = \mathcal{M} = \mathrm{Inj}(\mathbb{Q})$. $\mathbb{L}(\mathbb{Q})$ is the subcategory of torsion-free divisible abelian groups.

Next we are going to study the premonadicity of the colimit-closure of a projective $\mathcal{E}$-generator.

Definition 2.6 A functor $U : \mathbf{A} \to \mathbf{Set}$ is said to be *premonadic* if it is a right-adjoint whose comparison functor of the induced monad is full and faithful.

Remark 2.7 In [13] it is proven that for a faithful right adjoint $U : \mathbf{A} \to \mathbf{X}$ the following are equivalent:

(i) U is premonadic;

(ii) U reflects split coequalizers;

(iii) for morphisms $f : C \to A$ and $g : C \to B$ in $\mathbf{A}$ and $h : UA \to UB$ in $\mathbf{X}$, if $Ug = h \cdot Uf$ and Uf is a split epimorphism, then there is some $h' : A \to B$ in $\mathbf{A}$ such that $Uh' = h$.

The equivalence (i) $\Leftrightarrow$ (iii) was obtained in [2] for $\mathbf{X} = \mathbf{Set}$.

It is clear that each monadic functor is premonadic. It is known that there exists a monadic functor $U : \mathbf{A} \to \mathbf{Set}$ (see e.g. [4] or [11]) if and only if (i) $\mathbf{A}$ has finite limits, (ii) $\mathbf{A}$ is exact, (iii) $\mathbf{A}$ has copowers of P, for P such that (iv) P is a regular generator and (v) P is projective. In this classical case, the assumption (ii) ensures that $\mathrm{St}(\mathcal{E}) = \mathcal{E}$ for $\mathcal{E}$ the class of regular epimorphisms; thus (iv) and (v) mean that P is a projective $\mathcal{E}$-generator in the sense of Definition 1.8.

For any cocomplete category $\mathbf{A}$ and a generator P of $\mathbf{A}$, the functor $U = \hom(P, -)$ is a right adjoint with the counit consisting of all canonical morphisms of the form $\varepsilon_A : \coprod_{\hom(P,A)} P \to A$. It is well known that the comparison functor of the monad induced by U is full and faithful if and only if the morphisms ε_A are regular epimorphisms, that is, P is a regular generator of $\mathbf{A}$. Moreover, it is proven in [1] that full reflective subcategories of monadic categories over $\mathbf{Set}^S$ (for some set S) are just the cocomplete categories with a regular generator. (Here by a regular generator of the category $\mathbf{A}$ is meant a set $\{G_s,\ s \in S\}$ of $\mathbf{A}$-objects such that, for each object A of $\mathbf{A}$, the canonical morphism $\coprod_S \left(\coprod_{\hom(G_s,A)} G_s\right) \to A$ is a regular epimorphism.) As a consequence, if $U = \hom(P, -)$ is premonadic then $\mathbf{A}$ is just $\mathbb{C}(P)$. Indeed, given $A \in \mathbf{A}$, let $\varepsilon_A = \mathrm{coeq}(f, g)$, with $f,\ g : B \to \coprod_{\hom(P,A)} P$; it is trivially seen that, since ε_B is an epimorphism, ε_A is also the coequalizer of $f \cdot \varepsilon_B$ and $g \cdot \varepsilon_B$. Thus $\mathbb{C}(P)$ is the only colimit-closed subcategory of $\mathbf{A}$ containing

P which is candidate to be equivalent to a reflective subcategory of the category of algebras of the monad induced by U via the corresponding comparison functor. The next theorem (2.11), where projectivity has a relevant role, gives conditions under which such a candidate satisfies that property. The required conditions have a certain parallel with (i)–(v) above for monadic functors.

A significant role is played by the following notion:

Definition 2.8 We say that a class $\mathcal{F}$ of epimorphisms is *saturated* if, for each $f \in \mathcal{F}$, the coequalizer of the kernel pair of f belongs to $\mathcal{F}$.

Lemma 2.9 *If the class $\mathcal{E}$ admits a projective $\mathcal{E}$-generator, then the following two assertions are equivalent:*

1. *$St(\mathcal{E})$ is saturated.*
2. *For each $e \in St(\mathcal{E})$, the unique morphism d such that $d \cdot c = e$, for c the coequalizer of the kernel pair of e, is a monomorphism.*

Proof By Proposition 1.11, $\mathrm{St}(\mathcal{E}) = \mathrm{Proj}(P)$, where P is a projective $\mathcal{E}$-generator. Let $\mathrm{St}(\mathcal{E})$ be saturated, let $e \in \mathrm{St}(\mathcal{E})$, let c be the coequalizer of the kernel pair (u, v) of e, and let d be the unique morphism d such that $d \cdot c = e$. Let a and b be morphisms such that $d \cdot a = d \cdot b$; in order to show that $a = b$, since P is a generator, we may assume without loss of generality that P is the domain of a and b. Consequently, since $c \in \mathrm{St}(\mathcal{E}) = \mathrm{Proj}(P)$, there are $\bar{a}$ and $\bar{b}$ such that $c \cdot \bar{a} = a$ and $c \cdot \bar{b} = b$; then $e \cdot \bar{a} = d \cdot c \cdot \bar{a} = d \cdot a = d \cdot b = d \cdot c \cdot \bar{b} = e \cdot \bar{b}$. Since (u, v) is the kernel pair of e, this implies the existence of a unique morphism t such that $u \cdot t = \bar{a}$ and $v \cdot t = \bar{b}$. Therefore, $a = c \cdot \bar{a} = c \cdot u \cdot t = c \cdot v \cdot t = c \cdot \bar{b} = b$.

Conversely, if d is a monomorphism, for each morphism f with domain P and codomain in the codomain of c, we have some morphism f' such that $e \cdot f' = d \cdot f$, because $e \in \mathrm{St}(\mathcal{E}) = \mathrm{Proj}(P)$. Hence, $d \cdot c \cdot f' = d \cdot f$, and $c \cdot f' = f$, proving that $c \in \mathrm{Proj}(P) = \mathrm{St}(\mathcal{E})$. □

Remark 2.10 If $\mathcal{E} \subseteq \{\text{regular epimorphisms}\}$, $\mathrm{St}(\mathcal{E})$ is trivially saturated, since a regular epimorphism is the coequalizer of its kernel pair. The saturation of $\mathrm{St}(\mathcal{E})$ is also clear when $\mathcal{E}$ is pullback stable and $\mathcal{E}$ contains all regular epimorphisms.

In fact, non-trivial classes $\mathcal{E}$ with saturated $\mathrm{St}(\mathcal{E})$ do exist in everyday categories. Indeed, in all examples of 2.5, for the considered class $\mathcal{E}$ or $\mathcal{M}$, the corresponding stabilization is saturated, except in the second example. In this last case, where $\mathcal{E}$ is the class of extremal epimorphisms in the category **Cat**, $\mathrm{St}(\mathcal{E})$ is not saturated. To see that, consider the functor $F : A \to B$ where $A = \mathbf{2}$ and B is the category with a unique object and with an only non-identity morphism f such that $f \cdot f = f$. Then the coequalizer of the kernel pair of F is $G : A \to C$ where C is the monoid of natural numbers regarded as one-object category, and G sends the non-identity morphism of $\mathbf{2}$ to the generator of this monoid. Clearly the functor G does not belong to $\mathrm{St}(\mathcal{E})$ (see [10]).

Theorem 2.11 *Let P be a projective $\mathcal{E}$-generator of $\mathbf{A}$ and let $St(\mathcal{E})$ be saturated. Then, assuming that $\mathbb{C}(P)$ is coreflective in $\mathbf{A}$, the functor $hom(P, -) : \mathbb{C}(P) \to \mathbf{Set}$ is premonadic, and $\mathbb{C}(P)$ is equivalent to a reflective subcategory of the corresponding category of Eilenberg-Moore algebras.*

Proof Since $\mathrm{hom}(P, -)$ is a right adjoint and $\mathbb{C}(P)$ has coequalizers, we know that the comparison functor is a right adjoint. In order to show that it is full and faithful, it suffices to prove that the counits of the right adjoint $\mathrm{hom}(P, -)$:

$\mathbb{C}(P) \to \mathbf{Set}$ are regular epimorphisms. For each $B \in \mathbb{C}(P)$, let us consider the corresponding co-unit

$$\varepsilon_B : \coprod_{\hom(P,B)} P \longrightarrow B .$$

Since P is ε_B-projective, ε_B belongs to $\mathrm{St}(\mathcal{E})$, by 1.11. Let (u, v) be the kernel pair of ε_B and let c be the coequalizer of u and v. Since $\mathrm{St}(\mathcal{E})$ is saturated, the morphism d such that $d \cdot c = \varepsilon_B$ is a monomorphism. But $\mathrm{St}(\mathcal{E})$ is strongly right-cancellable, by 1.7, so $d \in \mathrm{St}(\mathcal{E})$. Then $d \in \mathrm{St}(\mathcal{E}) \cap \mathrm{Mono}(\mathbf{A})$ and, thus, by Theorem 2.3 (see also Remark 2.4), there is some morphism t such that $d \cdot t = 1_B$. Therefore d is an isomorphism and ε_B is a regular epimorphism. □

Remark 2.12 In the proof of Theorem 2.11 the saturation of $\mathrm{St}(\mathcal{E})$, that is, the fact that $\mathrm{St}(\mathcal{E})$ is closed under coequalizers of kernel pairs, is crucial for concluding that P is a regular generator of $\mathbb{C}(P)$. All examples of 2.5 fulfil the conditions of the above theorem, except the second one. As seen in 2.10, in this case, the saturation of $\mathrm{St}(\mathcal{E})$ fails. And, curiously, the corresponding functor into **Set** is not premonadic: It is easily seen that the corresponding counits are not necessarily regular epimorphisms (cf. [10]).

Acknowledgements I would like to thank George Janelidze for very valuable discussions on the subject of this paper. I also thank Walter Tholen and the referee for drawing my attention to the papers [5] and [13] respectively.

References

[1] Adámek, J., Sousa, L. *On reflective subcategories of varieties*, Jounal of Algebra (to appear).

[2] Barankovic, T. M. *O kategorijach strukturno ekvivalentnych nekotorym kategorijam algebr*, Mat. Sbornik 83 (125), 1(9)(1970), 3-14.

[3] Bednarczyk, M.A., Borzyszkowski, A., Pawlowski, W. *Generalized congruences – epimorphisms in* **Cat**, Theory Appl. Categories 5 (1999) 266–280.

[4] Borceux, F. *Handbook of categorical Algebra*, Vol. 1 and 2, Cambridge University Press 1994.

[5] Börger, R. *Making factorizations compositive*, Comment. Math. Univ. Carolin. 32 (1991), no. 4, 749–759.

[6] Börger, R., Tholen, W. *Strong, regular and dense generators*, Cahiers Topologie Géom. Différentielle Catég. 32 (1991), no. 3, 257–276.

[7] Carboni, A., Janelidze, G., Kelly, G. M., and Paré, R. *On localization and stabilization for factorization systems*, Appl. Categorical Structures **5** (1997), 1–58.

[8] Giraud, J. *Méthode de la descente*, Bull. Soc. Math. France Mém. 2 (1964).

[9] Janelidze, G. and Sobral, M. *Finite preorders and topological descent*, J. Pure Appl. Algebra (to appear)

[10] Janelidze, G., Sobral, M., and Tholen, W. *Beyond Barr Exactness: Effective Descent Morphisms*, preprint.

[11] MacDonald, J. and Sobral, M. *Aspects of Monads*, preprint.

[12] MacDonald, J. and Tholen, W. *Decomposition of morphisms into infinitely many factors*, Category Theory Proceedings Gummersbch 1981, Lecture Notes in Mathematics 962, Springer, Berlin, 1982, 175–189.

[13] Rosický, J. *Strong embeddings into categories of algebras over a monad II*, Comment. Math. Univ. Carolinae 15 (1974), 131-147.

[14] Sobral, M. *Absolutely closed spaces and categories of algebras*, Portugaliae Math. 47 (1990), 341–351.

[15] Sousa, L. *Orthogonal and Reflective Hulls*, PhD. thesis, University of Coimbra, 1997.

[16] Sousa, L. *Pushout stability of embeddings, injectivity and categories of algebras*, Proceedings of the Ninth Prague Topological Symposium (2001), Topol. Atlas, North Bay, ON, 2002, 295–308.

[17] Tholen, W. *Semi-topological functors I*, J. Pure App. Algebra 15 (1979), 53–73.

[18] Tholen, W. *Factorizations, localizations, and the orthogonal subcategory problem*, Math. Nachr. **114** (1983), 63–85.

Received December 16, 2002; in revised form August 12, 2003

Fields Institute Communications
Volume **43**, 2004

The Monotone-light Factorization for Categories Via Preorders

João J. Xarez
Departamento de Matemática
Universidade de Aveiro.
Campus Universitário de Santiago
3810-193 Aveiro, Portugal
jxarez@mat.ua.pt

Abstract. It is shown that the reflection **Cat** $\rightarrow$ **Preord** of the category of all categories into the category of preorders determines a monotone-light factorization system on **Cat** and that the light morphisms are precisely the faithful functors.

1 Introduction

1.1 Every map $\alpha : A \rightarrow B$ of compact Hausdorff spaces has a factorization $\alpha = me$ such that $m : C \rightarrow B$ has totally disconnected fibres and $e : A \rightarrow C$ has only connected ones. This is known as the classical monotone-light factorization of S. Eilenberg [3] and G. T. Whyburn [7].

Consider now, for an arbitrary functor $\alpha : A \rightarrow B$, the factorization $\alpha = me$ such that m is a faithful functor and e is a full functor bijective on objects. We shall show that this familiar factorization for categories is as well monotone-light, meaning that both factorizations are special and very similar cases of the categorical monotone-light factorization in an abstract category $\mathbb{C}$, with respect to a full reflective subcategory $\mathbb{X}$, as was studied in [1].

It is well known that any full reflective subcategory $\mathbb{X}$ of a category $\mathbb{C}$ gives rise, under mild conditions, to a factorization system $(\mathcal{E}, \mathcal{M})$. Hence, each of the two reflections **CompHaus** $\rightarrow$ **Stone**, of compact Hausdorff spaces into Stone spaces, and **Cat** $\rightarrow$ **Preord**, of categories into preorders, yields its own reflective factorization system.

Moreover, the process of simultaneously stabilizing $\mathcal{E}$ and localizing $\mathcal{M}$, in the sense of [1], was already known to produce a new non-reflective and stable factorization system $(\mathcal{E}', \mathcal{M}^*)$ for the adjunction **CompHaus** $\rightarrow$ **Stone**, which is just the

2000 *Mathematics Subject Classification.* Primary 18A32, 12F10; Secondary 54C10, 18A40, 18A20.

The author is a PhD student at the University of Aveiro, under scientific supervision of Prof. G. Janelidze. This research was supported by the program 4/5.3/PRODEP/2000.

(**Monotone, Light**)-factorization mentioned above. But this process does not work in general, the monotone-light factorization for the reflection **CompHaus** $\to$ **Stone** being just one of a few known examples where it does. Nevertheless, we shall prove that the (**Full and Bijective on Objects**, **Faithful**)-factorization for categories is another instance of a successful simultaneous stabilization and localization.

What guarantees the success is the following pair of conditions, which hold in both cases:

1. the reflection $I : \mathbb{C} \to \mathbb{X}$ has stable units (in the sense of [2]);
2. for each object B in $\mathbb{C}$, there is a monadic extension[1] (E, p) of B such that E is in the full subcategory $\mathbb{X}$.

Indeed, the two conditions 1 and 2 trivially imply that the $(\mathcal{E}, \mathcal{M})$-factorization is locally stable, which is a necessary and sufficient condition for $(\mathcal{E}', \mathcal{M}^*)$ to be a factorization system (cf. the central result of [1]).

Actually, we shall prove that the reflection **Cat** $\to$ **Preord** also has stable units, as the reflection **CompHaus** $\to$ **Stone** was known to have. And, for the reflection **Cat** $\to$ **Preord**, the monadic extension (E, p) of B may be chosen to be the obvious projection from the coproduct $E = \mathbf{Cat}(\mathbf{4}, B) \cdot \mathbf{4}$ of sufficiently many copies of the ordinal number **4**, one copy for each triple of composable morphisms in B. As for **CompHaus** $\to$ **Stone**, it was chosen to be the canonical surjection from the Stone-Čech compactification $E = \beta|B|$ of the underlying set of B.

In both cases these monadic extensions are precisely the counit morphisms of the following adjunctions from **Set**: the unique (up to an isomorphism) adjunction $\mathbf{Cat}(\mathbf{4}, -) \dashv (-) \cdot \mathbf{4} : \mathbf{Set} \to \mathbf{Cat}$ which takes the terminal object 1 to the ordinal number **4**, and the adjunction $|\cdot| \dashv \beta : \mathbf{Set} \to \mathbf{CompHaus}$, where the standard forgetful functor $|\cdot|$ is monadic, respectively.

Notice that this perfect matching exists in spite of the fact that **CompHaus** is an exact category and **Cat** is not even regular[2] (the reader may even extend the analogy, to the characterizations of the classes in the factorization systems involved, by simply making the following naive correspondence between some concepts of spaces and categories: "point"/"arrow"; "connected component"/"hom-set"; "fibre"/"inverse image of an arrow"; "connected"/"in the same hom-set"; "totally disconnected"/"every two arrows are in distinct hom-sets").

1.2 The two reflections may be considered as admissible Galois structures, in the sense of categorical Galois theory, since having stable units implies admissibility.

Therefore, in both cases, for every object B in $\mathbb{C}$, one knows that the full subcategory $TrivCov(B)$ of $\mathbb{C}/B$, determined by the trivial coverings of B (i.e., the morphisms over B in $\mathcal{M}$), is equivalent to $\mathbb{X}/I(B)$.

Moreover, the categorical form of the fundamental theorem of Galois theory gives us even more information on each $\mathbb{C}/B$ using the subcategory $\mathbb{X}$. It states that the full subcategory $Spl(E, p)$ of $\mathbb{C}/B$, determined by the morphisms split by the monadic extension (E, p) of B, is equivalent to the category $\mathbb{X}^{Gal(E,p)}$ of internal actions of the Galois pregroupoid of (E, p).

---

[1] It is said that (E, p) is a monadic extension of B, or that p is an effective descent morphism, if the pullback functor $p^* : \mathbb{C}/B \to \mathbb{C}/E$ is monadic.

[2] A monadic extension in **CompHaus** is just an epimorphism, i.e., a surjective mapping, while on **Cat** epimorphisms, regular epimorphisms and monadic extensions are distinct classes.

In fact, conditions 1 and 2 above imply that $Gal(E,p)$ is really an internal groupoid in $\mathbb{X}$ (cf. section 5.3 of [1]).

And, as all the monadic extensions (E,p) of B described above are projective[3], one has in both cases that $Spl(E,p) = Cov(B)$, the full subcategory of $\mathbb{C}/B$ determined by the coverings of B (i.e., the morphisms over B in $\mathcal{M}^*$).

Condition 1 implies as well that any covering over an object which belongs to the subcategory is just a trivial covering.

An easy consequence of this last statement, condition 2, and of the fact that coverings are pullback stable, is that a morphism $\alpha : A \to B$ is a covering over B if and only if, for every morphism $\phi : X \to B$ with X in the subcategory $\mathbb{X}$, the pullback $X \times_B A$ of α along ϕ is also in $\mathbb{X}$.

In particular, when the reflection has stable units, a monadic extension (E,p) as in condition 2 is a covering if and only if the kernel pair of p is in the full subcategory $\mathbb{X}$ of $\mathbb{C}$.

Thus, since the monadic extensions considered for the two cases are in fact coverings, one concludes that $Gal(\mathbf{Cat}(\mathbf{4},B)\cdot\mathbf{4},p)$ and $Gal(\beta|B|,p)$ are not just internal groupoids, but internal equivalence relations in **Preord** and **Stone**, respectively.

In symbols, specifically for the reflection **Cat** $\to$ **Preord**:

- $Faithful(B) \simeq \mathbf{Preord}^{Gal(\mathbf{Cat}(\mathbf{4},B)\cdot\mathbf{4},p)}$, for a general category B, and
- $Faithful(X) \simeq \mathbf{Preord}/X$, when X is a preorder.

As for **CompHaus** $\to$ **Stone**:

- $Light(B) \simeq \mathbf{Stone}^{Gal(\beta|B|,p)}$, for a general compact Hausdorff space B, and
- $Light(X) \simeq \mathbf{Stone}/X$, when X is a Stone space.

1.3 The fact that $Gal(\beta|B|,p)$ is an internal equivalence relation in **Stone** was already stated in [5].

Actually, the Stone spaces constitute what was defined there to be a generalized semisimple class of objects in **CompHaus**, and such that every compact Hausdorff space is a quotient of a Stone space.

In this way, the equivalence $Light(B) \simeq \mathbf{Stone}^{Gal(\beta|B|,p)}$ is just a special case of its main Theorem 3.1. Which can be easily extended to non-exact categories, by using monadic extensions instead of regular epis, and in such a manner that the equivalence $Faithful(B) \simeq \mathbf{Preord}^{Gal(\mathbf{Cat}(\mathbf{4},B)\cdot\mathbf{4},p)}$ is also a special case of it.

Hence, the faithful functors coincide with the locally semisimple coverings[4], $Gal(\mathbf{Cat}(\mathbf{4},B)\cdot\mathbf{4},p)$ is an internal equivalence relation in **Preord**, and the reflection **Cat** $\to$ **Preord** stands now as an interesting non-exact example of the case studied in [5].

2 The reflection of Cat into Preord has stable units

Consider the adjunction

$$(I,H,\eta,\epsilon) : \mathbf{Cat} \to \mathbf{Preord}, \tag{2.1}$$

where:

---

[3] I.e., for each monadic extension (A,f) of B there exists a morphism $g : E \to A$ with $fg = p$.

[4] On condition that one replaces regular epis by monadic extensions in the Definition 2.1 of [5].

- $H(X)$ is the preordered set X regarded as a category;
- $I(A) = A_0$ is the preordered set of objects a in A, in which $a \leq a'$ if and only if there exists a morphism from a to a';
- $\eta_A : A \to HI(A)$ is the unique functor with $\eta_A(a) = a$ for each object a in A;
- $\epsilon : IH \to 1$ is the identity natural transformation.

The following obvious lemma will be used many times below:

Lemma 2.1 *A commutative diagram*

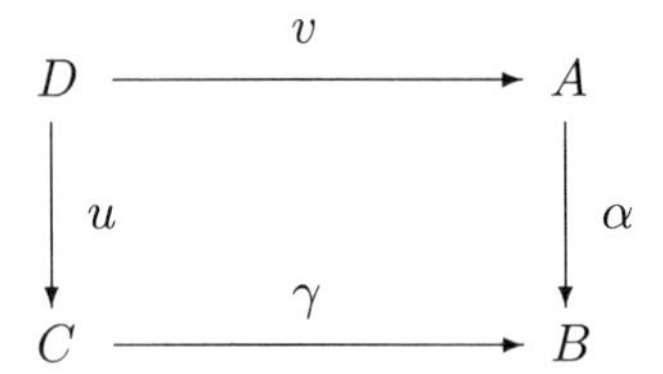

in **Cat** *is a pullback square if and only if its object version*

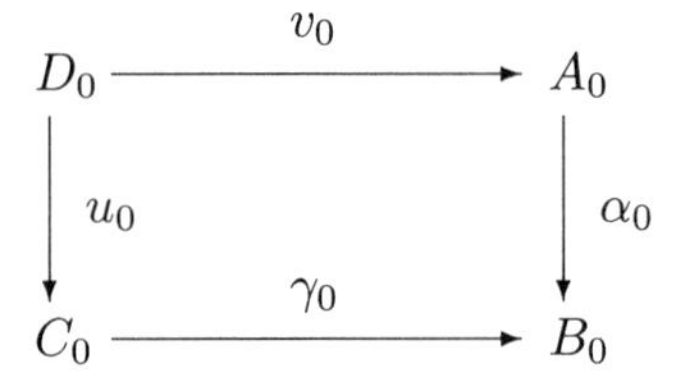

is a pullback square in **Set**, *and also its hom-set version*

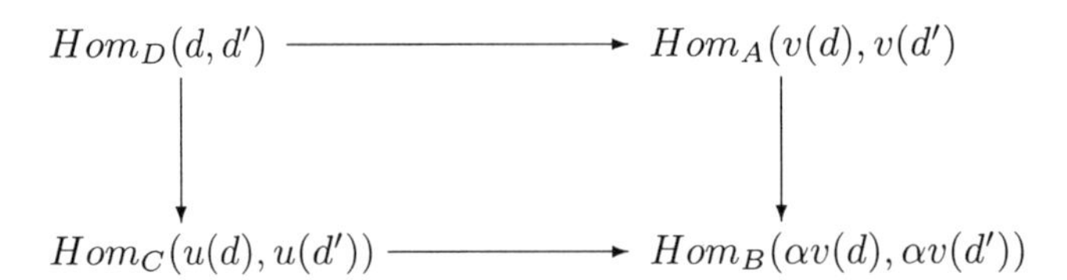

for arbitrary objects d and d' in D, where the maps are induced by the arrow functions of the functors above.

Proposition 2.2 *The adjunction 2.1 has stable units in the sense of* [2] *and* [1]; *that is, the functor I :* **Cat** $\to$ **Preord** *preserves every pullback square of the form*

$$\begin{array}{ccc} A \times_{H(X)} B & \xrightarrow{\pi_2} & B \\ \downarrow{\scriptstyle \pi_1} & & \downarrow{\scriptstyle \beta} \\ A & \xrightarrow{\alpha} & H(X) \end{array} .$$

Proof Since the reflector I does not change the sets of objects (i.e., the underlying set of $I(A)$ is the same as the set of objects in A), the underlying sets of the two preorders $I(A \times_{H(X)} B)$ and $I(A) \times_{IH(X)} I(B)$ are both equal to $A_0 \times_{H(X)_0} B_0$.

Moreover, for any pair of objects (a, b) and (a', b') in $A_0 \times_{H(X)_0} B_0$, we observe that:

$(a, b) \leq (a', b')$ in $I(A \times_{H(X)} B) \Leftrightarrow$
there exist two morphisms $f : a \to a'$ in A and $g : b \to b'$ in B
such that $\alpha(f) = \beta(g) \Leftrightarrow$
(since $H(X)$ has no parallel arrows!)
there exist two morphisms $f : a \to a'$ in A and $g : b \to b'$ in $B \Leftrightarrow$
$a \leq a'$ in $I(A)$ and $b \leq b'$ in $I(B) \Leftrightarrow$
$(a, b) \leq (a', b')$ in $I(A) \times_{IH(X)} I(B)$.

□

3 Trivial coverings

Consider the two classes of functors $\mathcal{E}$ and $\mathcal{M}$:

- $\mathcal{E}$ is the class of all functors inverted by I : **Cat** $\to$ **Preord**, i.e., of all morphisms $e : A \to C$ in **Cat** such that $I(e) : I(A) \to I(C)$ is a preorder isomorphism;
- $\mathcal{M}$ is the class of all trivial coverings with respect to the adjunction 2.1, i.e., of all morphisms $m : C \to B$ in **Cat** such that the following diagram

$$\begin{array}{ccc} C & \xrightarrow{\eta_C} & HI(C) \\ {\scriptstyle m}\downarrow & & \downarrow{\scriptstyle HI(m)} \\ B & \xrightarrow{\eta_B} & HI(B) \end{array} \qquad (3.1)$$

  is a pullback square.

The fact that the reflection of **Cat** into **Preord** has stable units is known to imply that

- $(\mathcal{E}, \mathcal{M})$ is a factorization system on **Cat**, and that
- the $(\mathcal{E}, \mathcal{M})$-factorization of an arbitrary functor $\alpha : A \to B$ is given by $\alpha = u \circ \langle \alpha, \eta_A \rangle$ in the commutative diagram of Figure 1, whose square part is a pullback.

Proposition 3.1 *A functor $m : C \to B$ belongs to $\mathcal{M}$ if and only if for every two objects c and c' in C with $Hom_C(c, c')$ nonempty, the map $Hom_C(c, c') \to Hom_B(m(c), m(c'))$ induced by m is a bijection.*

We will also express this by saying that m is a trivial covering with respect to the adjunction 2.1 if and only if m is a faithful and "almost full" functor.

Proof According to Lemma 2.1, the diagram 3.1 is a pullback square in **Cat** if and only if the diagram of Figure 2 and the diagrams of Figure 3, for arbitrary

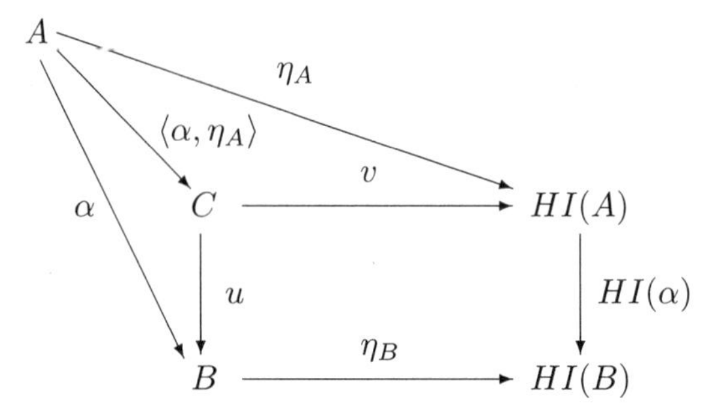

Figure 1 The $(\mathcal{E}, \mathcal{M})$-factorization of $\alpha : A \to B$

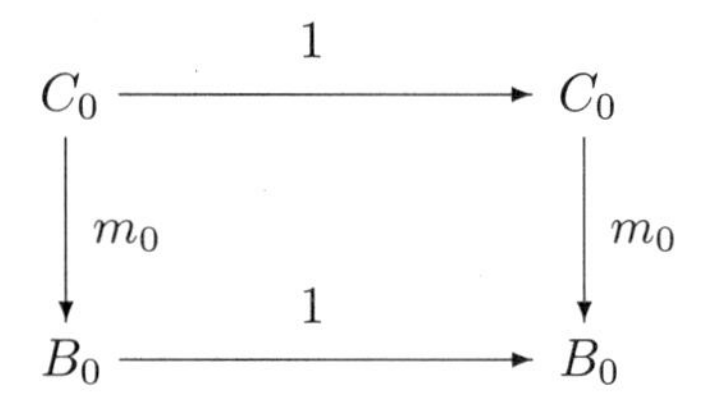

Figure 2 The object version diagram of the adjunction unit $\eta : 1 \to HI$

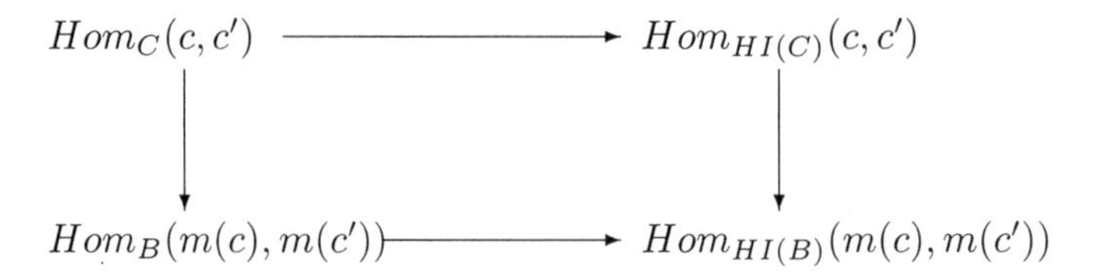

Figure 3 The arrow function diagrams of the adjunction unit $\eta : 1 \to HI$

objects c and c' in C, whose maps are induced by the arrow functions of the functors in diagram 3.1, are all pullback squares in **Set**.

We then observe:

the functor $m : C \to B$ belongs to $\mathcal{M} \Leftrightarrow$
(since the diagram of Figure 2 is obviously a pullback square)
for every two objects c and c' in C,
the diagram of Figure 3 is a pullback square $\Leftrightarrow$
(if $Hom_C(c, c')$ is empty then $Hom_{HI(C)}(c, c')$ is also empty!)
for every two objects c and c' in C, provided $Hom_C(c, c')$ is nonempty,
the diagram of Figure 3 is a pullback square $\Leftrightarrow$
(if $Hom_C(c, c')$ is nonempty then $Hom_{HI(C)}(c, c') \cong 1 \cong Hom_{HI(B)}(m(c), m(c')))$
for every two objects c and c' in C, provided $Hom_C(c, c')$ is nonempty,
the induced map $Hom_C(c, c') \to Hom_B(m(c), m(c'))$ is a bijection.

$\square$

Proposition 3.2 *A functor $\alpha : A \to B$ belongs to $\mathcal{E}$ if and only if the following two conditions hold:*

1. *the functor α is bijective on objects;*
2. *for every two objects a and a' in A, if $Hom_B(\alpha(a), \alpha(a'))$ is nonempty then so is $Hom_A(a, a')$.*

Proof The condition 2 reformulated in terms of the functor I becomes:

- *for every two objects a and a' in $I(A)$, $a \leq a'$ in $I(A)$ if and only if $\alpha(a) \leq \alpha(a')$ in $I(B)$.*

Therefore, the two conditions together are satisfied if and only if $I(\alpha)$ is an isomorphism, i.e., α is in $\mathcal{E}$.

□

4 Coverings

Starting from the given factorization system $(\mathcal{E}, \mathcal{M})$, we define in the following manner two new classes $\mathcal{E}'$ and $\mathcal{M}^*$ of functors:

- $\mathcal{E}'$ is the class of all functors $e' : A \to C$ in **Cat** such that every pullback of e' is in $\mathcal{E}$, i.e., $\mathcal{E}'$ is the largest pullback-stable class contained in $\mathcal{E}$;
- $\mathcal{M}^*$ is the class of all coverings with respect to the adjunction 2.1, i.e., of all functors $m^* : C \to B$ in **Cat** such that some pullback of m^* along a monadic extension (E, p) of B is in $\mathcal{M}$.

The next Lemma 4.1 is needed to prove the following Lemma 4.2, from which the characterization of coverings in Proposition 4.3 becomes an easy task.

Lemma 4.1 *Consider a Galois structure $\mathbf{\Gamma} = ((I, H, \eta, \varepsilon), \mathbf{F}, \mathbf{\Phi})$*[5] *on a category $\mathbb{C}$ with finite limits, such that its trivial coverings are pullback stable and the left adjoint $I : \mathbb{C} \to \mathbb{X}$ preserves every pullback square like the one in Proposition 2.2, provided its right-hand edge is a fibration.*

If a fibration (A, α) over B is a covering and $\varphi : H(X) \to B$ is any morphism in $\mathbb{C}$ with X an object in $\mathbb{X}$, then the pullback $(H(X) \times_B A, \varphi^*(\alpha))$ of (A, α) along φ is a trivial covering of $H(X)$.

Proof The lemma follows immediately from the next two facts:

- coverings are pullback stable, whenever trivial coverings are also so;
- any covering over an object of the form $H(X)$ is a trivial covering, whenever the left adjoint $I : \mathbb{C} \to \mathbb{X}$ preserves the pullback squares as above.

The proofs of which are given in §6.1 and §5.4 of [1], respectively.

□

Lemma 4.2 *A functor $\alpha : A \to B$ in* **Cat** *is a covering with respect to the adjunction 2.1 if and only if, for every functor $\varphi : X \to B$ over B from any preorder X, the pullback $X \times_B A$ of α along φ is also a preorder.*

Proof Consider the adjunction 2.1 as a Galois structure in which all morphisms are fibrations. One just has to show that

- the adjunction 2.1 satisfies the preceding lemma, and that

---

[5] In the sense of categorical Galois theory as presented in [4].

- for every category B in **Cat**, there is a monadic extension (X, p) of B with X a preorder.

We already know that the reflection 2.1 has stable units, by Proposition 2.2. Therefore, our reflection is an *admissible* Galois structure[6], in which the trivial coverings are known to be pullback stable.

Thus, we complete the proof by presenting, for each category B in **Cat**, a monadic extension (X, p) of B with X a preorder:[7]

make X the coproduct of all composable triples in B,
and then let p be the obvious projection of X into B.

□

Proposition 4.3 *A functor $\alpha : A \to B$ in* **Cat** *is a covering with respect to the adjunction 2.1 if and only if it is faithful.*

Proof We have:

the functor $\alpha : A \to B$ in **Cat** is a covering $\Leftrightarrow$ (by Lemma 4.2)

for every functor $\varphi : X \to B$ from a preorder X, the pullback $X \times_B A$ is a preorder $\Leftrightarrow$

for every functor $\varphi : X \to B$ from a preorder X, for any (x, a) and (x', a') in $X \times_B A$, $Hom_{X \times_B A}((x, a), (x', a'))$ has at most one element $\Leftrightarrow$

for every functor $\varphi : X \to B$ from a preorder X, if f is the unique morphism from x to x' in X, and if any two morphisms $g : a \to a'$ and $h : a \to a'$ in A are such that $\alpha(g) = \varphi(f) = \alpha(h)$, then $g = h \Leftrightarrow$

the functor $\alpha : A \to B$ is faithful.

□

Proposition 4.4 *A functor $\alpha : A \to B$ belongs to $\mathcal{E}'$ if and only if it is a full functor bijective on objects.*

Proof We have:

a functor $\alpha : A \to B$ belongs to $\mathcal{E}' \Leftrightarrow$

(according to the above definitions of $\mathcal{E}'$ and $\mathcal{E}$)
for every pullback u of α, $I(u)$ is an isomorphism $\Leftrightarrow$

$I(\alpha)$ is an isomorphism and I preserves every pullback of $\alpha \Leftrightarrow$

(according to Proposition 3.2)
α is bijective on objects
and
$Hom_A(a, a')$ is empty if and only if $Hom_B(\alpha(a), \alpha(a'))$ is so,
for arbitrary a and a' in A
and
(by Lemma 2.1, and since the reflector I does not change the sets of objects)
for every pullback

---

[6]A Galois structure, like the one in Lemma 4.1, is said to be admissible if for every object C in $\mathbb{C}$ and every fibration $\varphi : X \to I(C)$ in $\mathbb{X}$, the composite of canonical morphisms $I(C \times_{HI(C)} H(X)) \to IH(X) \to X$ is an isomorphism.

[7]A monadic extension in **Cat** is just a functor surjective on composable triples.

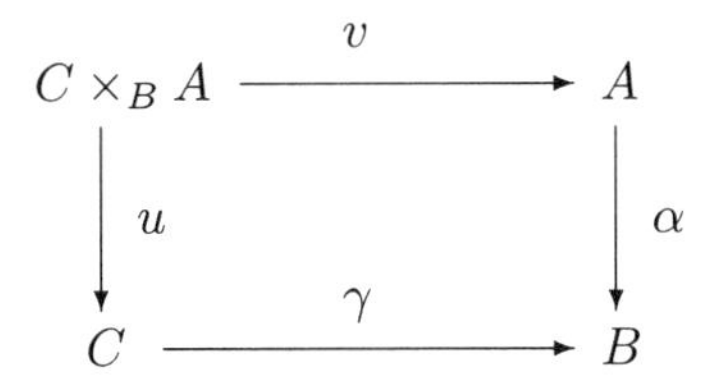

of α, the hom-set version

$$\begin{array}{ccc} Hom_{I(C\times_B A)}((c,a),(c',a')) & \longrightarrow & Hom_{I(A)}(a,a') \\ \downarrow & & \downarrow \\ Hom_{I(C)}(c,c') & \longrightarrow & Hom_{I(B)}(\alpha(a),\alpha(a')) \end{array}$$

of its image by I is also a pullback square in **Set**,
for arbitrary objects (c,a) and (c',a') in $C \times_B A \Leftrightarrow$

α is bijective on objects
and
$Hom_{I(A)}(a,a') \cong Hom_{I(B)}(\alpha(a),\alpha(a'))$, for arbitrary a and a' in A
and
for every pullback $C \times_B A$ of α, $Hom_{I(C\times_B A)}((c,a),(c',a')) \cong Hom_{I(C)}(c,c')$
for arbitrary objects (c,a) and (c',a') in $C \times_B A \Leftrightarrow$

α is bijective on objects
and
α is full.

□

Conclusion 4.5 *As follows from the previous results (and the results of* [1]*), $(\mathcal{E}', \mathcal{M}^*)$ is a factorization system. Moreover, Propositions 4.3 and 4.4 also tell us that it is a well-known one.*

References

[1] Carboni, A., Janelidze, G., Kelly, G. M., Paré, R. *On localization and stabilization for factorization systems.* App. Cat. Struct. **5**, (1997) 1–58.

[2] Cassidy, C., Hébert, M., Kelly, G. M. *Reflective subcategories, localizations and factorization systems.* J. Austral. Math. Soc. **38A** (1985) 287–329.

[3] Eilenberg, S. *Sur les transformations continues d'espaces métriques compacts.* Fundam. Math. **22** (1934) 292–296.

[4] Janelidze, G. *Categorical Galois theory: revision and some recent developments*, Manuscript (2001).

[5] Janelidze, G., Márki, L., Tholen, W. *Locally semisimple coverings.* J. Pure Appl. Algebra **128** (1998) 281–289.

[6] Mac Lane, S. *Categories for the Working Mathematician*, 2nd ed., Springer, 1998.

[7] Whyburn, G. T. *Non-alternating transformations.* Amer. J. Math. **56** (1934) 294–302.

Received December 9, 2002; in revised form June 16, 2003

Fields Institute Communications
Volume **43**, 2004

Separable Morphisms of Categories Via Preordered Sets

João J. Xarez
Departamento de Matemática
Universidade de Aveiro.
Campus Universitário de Santiago
3810-193 Aveiro, Portugal
jxarez@mat.ua.pt

Abstract. We give explicit descriptions of separable and purely inseparable morphisms with respect to the reflection **Cat** $\to$ **Preord** of the category of all categories into the category of preorders. It follows that the monotone-light, concordant-dissonant and inseparable-separable factorizations on **Cat** do coincide in this case.

1 Introduction

1.1 The monotone-light factorization. The classical monotone-light factorization of S. Eilenberg [3] and G. T. Whyburn [8], for maps of compact Hausdorff spaces, is a special case of the abstract categorical process studied by A. Carboni, G. Janelidze, G. M. Kelly, and R. Paré [1].

The reflection **CompHaus** $\to$ **Stone**, of compact Hausdorff spaces into Stone spaces, induces a reflective factorization system $(\mathcal{E}, \mathcal{M})$ on **CompHaus**. The $(\mathcal{E}, \mathcal{M})$-factorization of an arbitrary map $\alpha : A \to B$ is given by $\alpha = u \circ \langle \alpha, \eta_A \rangle$ in the commutative diagram of Figure 1, whose square part is a pullback, I is the reflector, H the inclusion functor and η the unit of the adjunction.

Then, by stabilizing $\mathcal{E}$ and localizing $\mathcal{M}$, a new monotone-light factorization system $(\mathcal{E}', \mathcal{M}^*)$ is obtained.

Indeed, the reflection **CompHaus** $\to$ **Stone** has stable units (in the sense of [2]), and for each object B in **CompHaus**, there is an effective descent morphism $p : E \to B$ such that E is in the full subcategory **Stone**: the canonical surjection from the Stone-Čech compactification $E = \beta|B|$ of the underlying set of B. These two conditions trivially imply that the $(\mathcal{E}, \mathcal{M})$-factorization is locally stable, which is a necessary and sufficient condition for $(\mathcal{E}', \mathcal{M}^*)$ to be a factorization system by the central result of [1].

We showed in [9] that the process outlined above for compact Hausdorff spaces is also successful when applied to the reflection **Cat** $\to$ **Preord** of categories into

2000 *Mathematics Subject Classification.* Primary 18A32, 12F10; Secondary 54C10, 18A40, 18A20.

The author is a PhD student at the University of Aveiro, under scientific supervision of Prof. G. Janelidze. This research was supported by the program 4/5.3/PRODEP/2000.

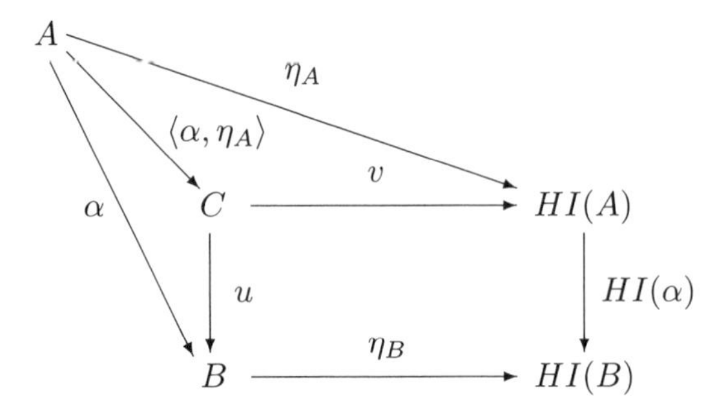

Figure 1 The $(\mathcal{E}, \mathcal{M})$-factorization of $\alpha : A \to B$

preorders, which takes a category to the preorder obtained by identifying all morphisms in the same hom-set.

Hence, the $(Full\ and\ Bijective\ on\ Objects, Faithful)$-factorization of functors can be viewed as another non-trivial example of the monotone-light factorization: **Cat** $\to$ **Preord** has also stable units and for each category B there is an effective descent morphism $p : E \to B$ with E a preorder. Notice that there are not many other examples where this process produces a non-reflective factorization system.

Having stable units is a condition stronger than admissibility (also called semi-left-exactness), and so the reflection **Cat** $\to$ **Preord**, as before **CompHaus** $\to$ **Stone**, provides a new application for categorical Galois theory. In this context the light morphisms are called the coverings, which are classified for each object B with the category of internal actions of the precategory $Gal(E,p)$ (reflection of the equivalence relation associated to the effective descent morphism $p : E \to B$ mentioned above) in the full subcategory:

$$Faithful(B) \simeq \mathbf{Preord}^{Gal(E,p)},$$

for a general category B.

In fact, the reflection **Cat** $\to$ **Preord** is an interesting non-exact example of the case studied in [5], and therefore $Gal(E,p)$ is really an equivalence relation for both adjunctions that we are comparing.

One sees that there is a strong analogy between **CompHaus** $\to$ **Stone** and **Cat** $\to$ **Preord** in what concerns categorical Galois theory and monotone-light factorization.

1.2 The concordant-dissonant factorization. Moreover, both reflections have concordant-dissonant $(Conc, Diss)$ factorization systems, in the sense of [6, §2.11], where $Conc$ is the class $\mathcal{E} \cap RegEpi$ of regular epimorphisms in the left-hand side $\mathcal{E}$ of the reflective factorization system $(\mathcal{E}, \mathcal{M})$.

One concludes from Corollary 2.11 in [6] that $(Conc, Diss)$ is a factorization system on **CompHaus**, since **CompHaus** is an exact category and so it has a regular epi-mono factorization system $(RegEpi, Mono)$.

On the other hand, the existence of an extremal epi-mono factorization system $(ExtEpi, Mono)$ on **Cat** implies that $(\mathcal{E} \cap ExtEpi, Diss)$ is also a factorization system on **Cat** (cf. [1, §3.9] which generalizes Corollary 2.11 in [6]), where $\mathcal{E} \cap ExtEpi$

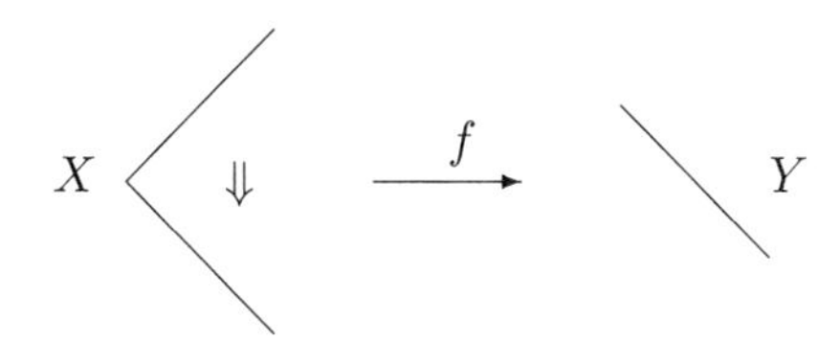

Figure 2 The concordant map $f : X \to Y$

is the class of extremal epis in $\mathcal{E}$.

Finally, remark that the two classes $\mathcal{E} \cap ExtEpi$ and $Conc = \mathcal{E} \cap RegEpi$ coincide on **Cat**.

This follows easily from the known characterizations of extremal epis and regular epis on **Cat** and from Lemma 2.3 below, since in this case the functors in $\mathcal{E}$ are known to be exactly the vertical ones in the sense of Definition 2.1 below (cf. [1]).

So far, the analogy between the two reflections continues.

1.3 The analogy ends. Now notice that for **Cat** $\to$ **Preord** the concordant morphisms are exactly the monotone morphisms, i.e., the full functors bijective on objects (cf. [9]), but for **CompHaus** $\to$ **Stone** it is not so.

Indeed, consider the map in Figure 2 which bends a closed segment in the Euclidean plane through its middle point, identifying in this way its two halves.

It is a concordant map, i.e., a surjection whose fibres are contained in connected components[1], since X has only one component. But it is not monotone, i.e., a map whose fibres are all connected: every point of Y, excepted one of the vertices, has disconnected two-point fibres.

Hence, we have:

- $(\mathcal{E}', \mathcal{M}^*) = (Conc, Diss)$, for **Cat** $\to$ **Preord**;
- $\mathcal{M}^*$ contains strictly the maps in $Diss$, i.e., the maps whose fibres meet the connected components in at most one point, for **CompHaus** $\to$ **Stone**.

1.4 The inseparable-separable factorization of functors. One also knows from [6, §4.1] that

$$Pin \cap RegEpi = Pin^* \subseteq Ins \subseteq Conc \ ,$$

where Ins and Pin are respectively the classes of inseparable and purely inseparable morphisms on **Cat**.

By Proposition 2.5 a functor is in Pin if and only if it is injective on objects. So, one easily concludes that

$$Pin^* = Ins = Conc = \mathcal{E}' \ .$$

---

[1]Every description, given at this Introduction, of a class of maps of compact Hausdorff spaces, was either taken from [1, §7] or obtained from the Example 5.1 in [6]. At the latter, those descriptions were stated for the reflection of topological spaces into hereditarily disconnected ones, which extends **CompHaus** $\to$ **Stone**.

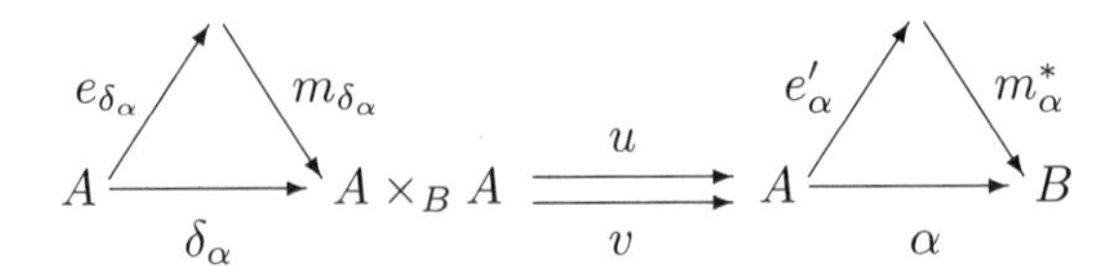

Figure 3 The inseparable-separable factorization

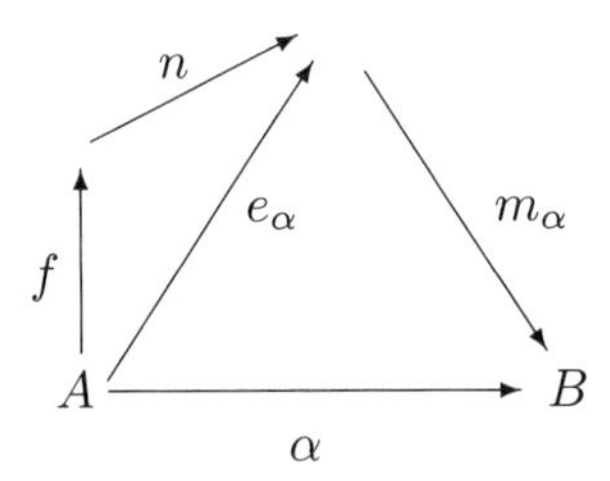

Figure 4 The concordant-dissonant factorization

And, by Theorem 4.4 in [6], the monotone-light factorization on **Cat**, besides being also a concordant-dissonant factorization, is in addition an inseparable-separable factorization:

$$(\mathcal{E}', \mathcal{M}^*) = (Conc, Diss) = (Pin^*, Sep) = (Ins, Sep) .$$

Proposition 2.4 gives a direct proof of this fact by stating that the separable morphisms on **Cat** are just the faithful functors, i.e., the light morphisms on **Cat** (cf. [9]).

We see that the reflection of categories into preorders is very well-behaved, in the sense that it equalizes three factorization systems.

1.5 Two procedures for computing the monotone-light factorization of a functor. For an inseparable-separable factorization $m^*_\alpha \cdot e'_\alpha$ of any functor $\alpha : A \to B$ (see Figure 3), e'_α is just the coequalizer of $(u \cdot m_{\delta_\alpha}, v \cdot m_{\delta_\alpha})$, where (u, v) is the kernel pair of α and $m_{\delta_\alpha} \cdot e_{\delta_\alpha}$ is the reflective $(\mathcal{E}, \mathcal{M})$-factorization of the fibred product $\delta_\alpha : A \to A \times_B A$ (see [6, §3.2]).

Hence, one has two procedures for obtaining the monotone-light factorization of a functor via preorders:

- the one just given in Figure 3, corresponding to the inseparable-separable factorization;
- another one in Figure 4, associated with the concordant-dissonant factorization considered above: $\alpha = (m_\alpha \cdot n) \cdot f$, such that $e_\alpha = n \cdot f$ is the extremal epi-mono factorization of e_α, and $\alpha = m_\alpha \cdot e_\alpha$ is the reflective $(\mathcal{E}, \mathcal{M})$-factorization of α.

1.6 The not so good behaviour of monotone-light factorization of maps of compact Hausdorff spaces. As for the reflection **CompHaus** $\to$ **Stone**, the classes Pin and Pin^* are not closed under composition, and so they cannot be part of a factorization system. This was shown in [6, §5.1] with a counterexample. It also follows from that same counterexample that $Ins \neq Pin^*$.[2]

Remark that, as far as separable and dissonant morphisms are concerned, and unlike the reflection **CompHaus** → **Stone**, the reflection **Top** → $\mathbf{T_0}$ of topological spaces into T_0 – spaces is analogous to the reflection **Cat** → **Preord**. Indeed, one has for the reflection **Top** → $\mathbf{T_0}$ that (see [6, §5.4]):

$$Conc = Ins = Pin^* \quad \text{and} \quad Diss = Sep\ .$$

2 Separable and purely inseparable morphisms

Definition 2.1 Consider the adjunction $(I, H, \eta, \epsilon) : \mathbb{C} \to \mathbb{X}$ and the morphism $\alpha : A \to B$ in $\mathbb{C}$ with kernel pair $(u, v) : A \times_B A \to A$.

We call the morphism α separable with respect to the given adjunction if the diagram

$$\begin{array}{ccc} A & \xrightarrow{\eta_A} & HI(A) \\ \downarrow{\scriptstyle \delta_\alpha} & & \downarrow{\scriptstyle HI(\delta_\alpha)} \\ A \times_B A & \xrightarrow{\eta_{A \times_B A}} & HI(A \times_B A) \end{array}$$

is a pullback square, where δ_α is the fibred product $\langle 1_A, 1_A \rangle$ of the identity morphism 1_A by itself.

That is, α is a separable morphism if δ_α is a trivial covering[3], with respect to the same adjunction.

Similarly, α is called purely inseparable with respect to the given adjunction if there exists a morphism d making the diagram

$$\begin{array}{ccc} A \times_B A & \xrightarrow{d} & HI(A) \\ \downarrow{\scriptstyle 1_{A \times_B A}} & & \downarrow{\scriptstyle HI(\delta_\alpha)} \\ A \times_B A & \xrightarrow{\eta_{A \times_B A}} & HI(A \times_B A) \end{array}$$

a pullback square.

That is, α is a purely inseparable morphism if δ_α is vertical[4], with respect to the same adjunction.

Consider the adjunction

$$(I, H, \eta, \epsilon) : \mathbf{Cat} \to \mathbf{Preord}, \tag{2.1}$$

where:

[2]See the last sentence in the previous footnote.
[3]In the sense of categorical Galois theory as presented in [4].
[4]In the sense of [6, §2.4].

- $H(X)$ is the preordered set X regarded as a category;
- $I(A) = A_0$ is the preordered set of objects a in A,
  in which $a \leq a'$ if and only if there exists a morphism from a to a';
- $\eta_A : A \to HI(A)$ is the unique functor with $\eta_A(a) = a$
  for each object a in A;
- $\epsilon : IH \to 1$ is the identity natural transformation.

With respect to the adjunction 2.1, which is certainly a reflection, it is known that a functor α is a trivial covering or vertical exactly when it belongs respectively to the right-hand or left-hand side of the associated reflective factorization system $(\mathcal{E}, \mathcal{M})$ (see [9]).

The next two lemmas were proved in [9].

Lemma 2.2 *A functor $\alpha : A \to B$ is a trivial covering with respect to the adjunction 2.1 if and only if, for every two objects a and a' in A with $Hom_A(a, a')$ nonempty, the map $Hom_A(a, a') \to Hom_B(\alpha(a), \alpha(a'))$ induced by α is a bijection.*

We will also express this by saying that α is a trivial covering with respect to the adjunction 2.1 if and only if α is a faithful and "almost full" functor.

Lemma 2.3 *A functor $\alpha : A \to B$ is vertical with respect to the adjunction 2.1 if and only if the following two conditions hold:*

1. *the functor α is bijective on objects;*
2. *for every two objects a and a' in A, if $Hom_B(\alpha(a), \alpha(a'))$ is nonempty then so is $Hom_A(a, a')$.*

Proposition 2.4 *A functor $\alpha : A \to B$ in* **Cat** *is a separable morphism with respect to the adjunction 2.1 if and only if it is faithful.*

Proof According to Definition 2.1 and Lemma 2.2 one has to show that the functor $\delta_\alpha : A \to A \times_B A$ is faithful and "almost full" if and only if α is faithful.

We observe that for all objects a and a' in A, the map

$$Hom_A(a, a') \to Hom_{A \times_B A}((a, a), (a', a'))$$

induced by δ_α is injective. Hence, this map is bijective if and only if it is surjective.

Since the domain of the map is empty if and only if the codomain is empty, the surjectivity condition for all of these maps amounts to asking that any two morphisms f and g in A with the same domain and codomain and with $\alpha(f) = \alpha(g)$ must coincide, which is to say that α must be a faithful functor.

□

Proposition 2.5 *A functor $\alpha : A \to B$ in* **Cat** *is a purely inseparable morphism with respect to the adjunction 2.1 if and only if its object function is injective.*

Proof According to Definition 2.1 and Lemma 2.3, $\alpha : A \to B$ is purely inseparable if and only if the following two conditions hold:

1. the functor $\delta_\alpha : A \to A \times_B A$ is bijective on objects;
2. for every two objects a and a' in A, if $Hom_{A \times_B A}(\delta_\alpha(a), \delta_\alpha(a'))$ is nonempty then so is $Hom_A(a, a')$.

Of these condition 2 holds trivially, since $\delta_\alpha(a) = (a, a)$ for every object a in A, and the morphisms in $A \times_B A$ are just the ordered pairs (f, g) of morphisms in

A such that $\alpha(f) = \alpha(g)$.

Furthermore, since the object function of δ_α is always injective on objects, it is bijective on objects if and only if it is surjective on objects, which is to say that the functor α must be injective on objects.

$\square$

References

[1] Carboni, A., Janelidze, G., Kelly, G. M., Paré, R. *On localization and stabilization for factorization systems.* App. Cat. Struct. **5**, (1997) 1–58.

[2] Cassidy, C., Hébert, M., Kelly, G. M. *Reflective subcategories, localizations and factorization systems.* J. Austral. Math. Soc. **38A** (1985) 287–329.

[3] Eilenberg, S. *Sur les transformations continues d'espaces métriques compacts.* Fundam. Math. **22** (1934) 292–296.

[4] Janelidze, G. *Categorical Galois theory: revision and some recent developments*, Manuscript (2001).

[5] Janelidze, G., Márki, L., Tholen, W. *Locally semisimple coverings.* J. Pure Appl. Algebra **128** (1998) 281–289.

[6] Janelidze, G., Tholen, W. *Functorial factorization, well-pointedness and separability* J. Pure Appl. Algebra **142** (1999) 99–130.

[7] Mac Lane, S. *Categories for the Working Mathematician*, 2nd ed., Springer, 1998.

[8] Whyburn, G. T. *Non-alternating transformations.* Amer. J. Math. **56** (1934) 294–302.

[9] Xarez, J. J. *The monotone-light factorization for categories via preorders*, submitted (2002).

Received December 30, 2002; in revised form October 25, 2003

Fields Institute Communications
Volume **43**, 2004

Frobenius Algebras in Tensor Categories and Bimodule Extensions

Shigeru Yamagami
Department of Mathematical Sciences
Ibaraki University
Mito, 310-8512, Japan
yamagami@mx.ibaraki.ac.jp

Introduction

By recent research developments, the notion of tensor category has been recognized as a fundamental language in describing quantum symmetry, which can replace the traditional method of groups for investigating symmetry.

The terminology of tensor category is used here as a synonym of linear monoidal category and hence it has a good affinity with semigroup. One way to incorporate the invertibility axiom of groups is to impose rigidity (or duality) on tensor categories, which will be our main standpoint in what follows.

When a tensor category bears a finite group symmetry inside, it is an interesting problem to produce a new tensor category by taking quotients with respect to this inner symmetry. For quantum symmetries of rational conformal field theory, this kind of constructions are worked out in a direct and individual way with respect to finite cyclic groups.

In our previous works, these specific constructions are organized by interpreting them as bimodule tensor categories for the symmetry of finite groups with a satisfactory duality on bimodule extensions [12]. The construction is afterward generalized to the symmetry of tensor categories governed by finite-dimensional Hopf algebras [13].

We shall present in this paper a further generalization to symmetries described by categorical Frobenius algebras, which are formulated and utilized by J. Fuchs and C. Schweigert for a mathematical description of boundary conditions in conformal field theory [3] (see [5] for earlier studies on categorical Frobenius structures). A similar notion has been introduced under the name of Q-systems by R. Longo in connection with subfactory theory ([6], cf. also [9]). More precisely, a Q-system, if it is algebraically formulated, is equivalent to giving a Frobenius algebra satisfying a certain splitting condition, which is referred to as a special Frobenius algebra according to the terminology in [3].

2000 *Mathematics Subject Classification.* Primary 18D10; Secondary 46L37.
Key words and phrases. Frobenius algebra, tensor category, bimodule.

Since our viewpoint here is that Q-systems (or special Frobenius algebras) should play the role of group algebras in classical symmetries, we first give an autonomic status to categorical Frobenius algebras as algebraic systems, which enables us to introduce the dual Frobenius algebras without assuming background tensor categories, together with a satisfactory duality on Frobenius algebras.

On the other hand, if Frobenius algebras are realized inside a tensor category $\mathcal{T}$, it is fundamental to consider bimodule extensions of $\mathcal{T}$ and we shall generalize the duality result on bimodule extensions to symmetries specified by categorical Frobenius algebras.

More precisely, given a special Frobenius algebra A realized inside a tensor category $\mathcal{T}$, we show the existence of a natural imbedding of the dual Frobenius algebra B of A into the tensor category ${}_A\mathcal{T}_A$ of A-A bimodules in $\mathcal{T}$. The duality for bimodule extensions is then formulated so that the second bimodule extension ${}_B({}_A\mathcal{T}_A)_B$ of B-B bimodules in ${}_A\mathcal{T}_A$ is naturally isomorphic (monoidally equivalent) to the starting tensor category $\mathcal{T}$.

The author is greatful to A. Masuoka and M. Müger for helpful communications on the subject during the preparation of this article.

Convention: By a tensor category over a field $\mathbb{K}$, we shall mean a $\mathbb{K}$-linear category together with a compatible monoidal structure. If semisimplicty is involved, we assume that $\mathbb{K}$ is an algebraically closed field of zero characteristic.

Since we are primarily interested in the use for quantum symmetry, we shall not discriminate tensor categories as long as they provide the equivalent information; we shall implicitly assume the strictness of associativity as well as the saturation under taking direct sums and subobjects for example.

For basic categorical definitions, we refer to the standard text [8].

1 Monoidal algebras

Let $\mathcal{T}$ be a strict tensor category over a field $\mathbb{K}$ and assume that $\mathrm{End}(I) = \mathbb{K}1_I$ for the unit object I. Given an object X in $\mathcal{T}$, set

$$A_{m,n} = \mathrm{Hom}(X^{\otimes n}, X^{\otimes m})$$

for non-negative integers m, n. The family $\{A_{m,n}\}_{m,n\geq 0}$ is then a **block system of algebra** in the sense that $A = \oplus_{m,n\geq 0} A_{m,n}$ is an algebra satisfying $A_{k,l}A_{m,n} \subset \delta_{l,m}A_{k,n}$ and $A_{0,0} = \mathbb{K}$. Denote the unit of A_n by 1_n.

The tensor product in the category $\mathcal{T}$ defines a bilinear map

$$A_{k,l} \times A_{m,n} \ni f \times g \mapsto f \otimes g \in A_{k+m,l+n}$$

such that

1. the unit 1_0 of A_0 satisfies $1_0 \otimes f = f \otimes 1_0 = f$,
2. the tensor product is associative; $(f \otimes g) \otimes h = f \otimes (g \otimes h)$ and
3. compatible with the composition; $(f \otimes g)(f' \otimes g') = (ff') \otimes (gg')$.

A block system of algebra is called a **monoidal algebra** according to Kazhdan and Wenzl [4] (though they use this terminology in a more restricted meaning) if it is furnished with the operation of taking tensor products which satisfies the above conditions.

Conversely, given a monoidal algebra A, we define a tensor category $\mathcal{A}$ in the following way; objects in $\mathcal{A}$ are parametrized by non-negative integers and the hom-set $\mathrm{Hom}(m, n)$ is the vector space $A_{n,m}$ with the composition of morphisms given by the multiplication in the algebra A. The tensor product operation in $\mathcal{A}$ is the one naturally induced from that of monoidal algebra. If the monoidal algebra A is associated to an object X in a tensor category $\mathcal{T}$, the tensor category $\mathcal{A}$ associated to A is monoidally equivalent to the tensor category generated by X.

If the starting tensor category is semisimple, the monoidal algebra is **locally semisimple** in the sense that for any finite subset F of non-negative integers, the subalgebra $\oplus_{i,j \in F} A_{i,j}$ is semisimple. Conversely, a locally semisimple monoidal algebra A gives rise to a semisimple tensor category $\overline{\mathcal{A}}$ as the Karoubian envelope of $\mathcal{A}$: an object in $\overline{\mathcal{A}}$ is a pair (n, e) of an integer $n \geq 0$ and an idempotent e in A_n with hom-sets defined by

$$\mathrm{Hom}((m, e), (n, f)) = f A_{n,m} e.$$

The operation of tensor product is given by

$$(m, e) \otimes (n, f) = (m + n, e \otimes f)$$

on objects.

A similar construction works for bicategories as well; consider a (strict) bicategory of two objects $\{1, 2\}$ for example and choose objects X, Y in the hom-categories $\mathcal{H}om(2, 1)$, $\mathcal{H}om(1, 2)$ respectively. By using the tensor product notation for the composition in the bicategory, we have the four systems of block algebras

$$\begin{aligned} A_{m,n} &= \mathrm{Hom}((X \otimes Y)^{\otimes n}, (X \otimes Y)^{\otimes m}), \\ B_{m+1,n+1} &= \mathrm{Hom}((X \otimes Y)^{\otimes n} \otimes X, (X \otimes Y)^{\otimes m} \otimes X), \\ C_{m+1,n+1} &= \mathrm{Hom}(Y \otimes (X \otimes Y)^{\otimes n}, Y \otimes (X \otimes Y)^{\otimes m}), \\ D_{m,n} &= \mathrm{Hom}((Y \otimes X)^{\otimes n}, (Y \otimes X)^{\otimes m}) \end{aligned}$$

(note that $(X \otimes Y)^{\otimes n} \otimes X = X \otimes (Y \otimes X)^{\otimes n}$ are alternating tensor products of X and Y) with the operation of tensor product among them applied in a 2×2-matrix way,

$$\begin{aligned} A_{m,m'} \otimes B_{n,n'} &\subset B_{m+n,m''+n''}, \\ B_{m,m'} \otimes D_{n,n'} &\subset B_{m+n,m'+n'}, \\ C_{m,m'} \otimes A_{n,n'} &\subset C_{m+n,m'+n'}, \\ D_{m,m'} \otimes C_{n,n'} &\subset C_{m+n,m'+n'}, \\ B_{m,m'} \otimes C_{n,n'} &\subset A_{m+n-1,m'+n'-1}, \\ C_{m,m'} \otimes B_{n,n'} &\subset D_{m+n-1,m'+n'-1}, \end{aligned}$$

which satisfies the associativity and multiplicativity (and the unit condition for tensor products involving A_0 or D_0) exactly as in the definition of monoidal algebra.

Conversely, given such an algebraic system, we can recover a (two-object) bicategory together with off-diagonal objects X and Y in an obvious way.

We can also talk about isomorphisms of monoidal algebras or their bicategorical counterparts, which exactly correspond to isomorphisms between associated tensor categories or bicategories.

2 Frobenius algebras

It would be just a formal business to formulate axioms of algebraic systems in terms of categorical languages such as monoids or algebras, see [8] for example. Here is a bit more elaborate formulation of Frobenius algebra structure in tensor categories, which we shall describe here, following [3] and [9], mainly to fix the notation with some rewritings of axioms.

Let $\mathcal{T}$ be a tensor category. An **algebra** in $\mathcal{T}$ is a triplet (A, T, δ) with A an object in $\mathcal{T}$, $T \in \mathrm{Hom}(A \otimes A, A)$ and $\delta \in \mathrm{Hom}(I, A)$ satisfying $T(T \otimes 1_A) = T(1_A \otimes T)$, $T(\delta \otimes 1_A) = 1_A = T(1_A \otimes \delta)$, which are graphically denoted in the following way:

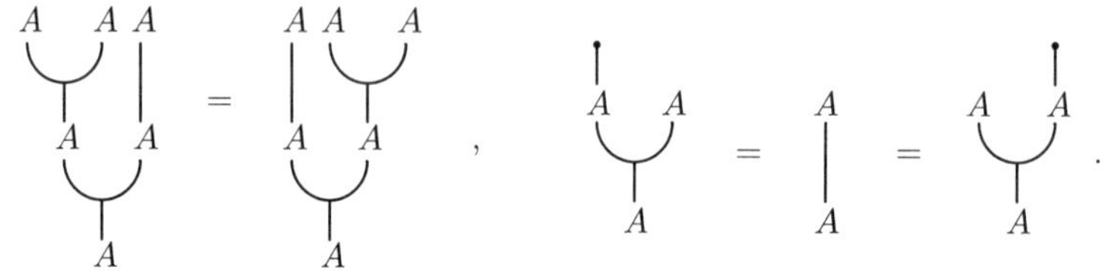

By reversing the direction of arrows, a **coalgebra** in $\mathcal{T}$ is a triplet (C, S, ϵ) with $S : C \to C \otimes C$ and $\epsilon : C \to I$ satisfying $(S \otimes 1_C)S = (1_C \otimes S)S$, $(\epsilon \otimes 1_C)S = 1_C = (1_C \otimes \epsilon)S$:

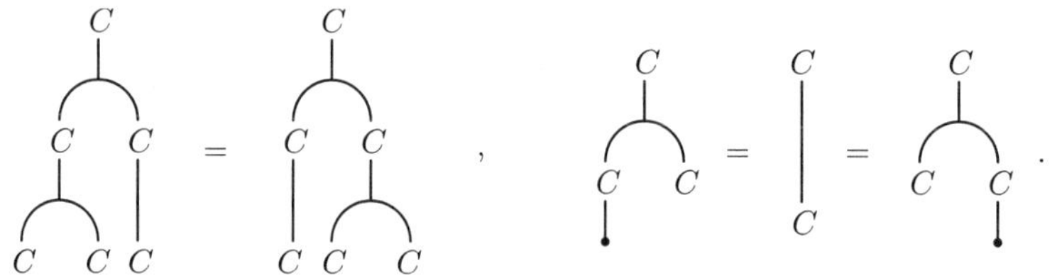

Note that δ and ϵ are uniquely determined by T and S respectively.

A **Frobenius algebra** in $\mathcal{T}$ is, by definition, a quintuplet $(A, S, T, \delta, \epsilon)$ with (A, T, δ) an algebra and (A, S, ϵ) a coalgebra, which satisfies the compatibility condition (st-duality), Fig. 1. The terminology is justified because the axioms turn out to be equivalent to those for ordinary Frobenius algebras if we work with the tensor category of finite-dimensional vector spaces. For an early appearance of categorical Frobenius structures, see [5].

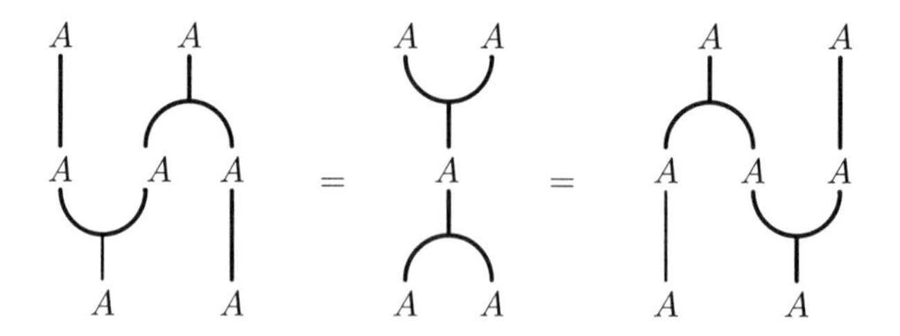

Figure 1

For a Frobenius algebra $(A, S, T, \delta, \epsilon)$, the object A is self-dual with the rigidity pair given by

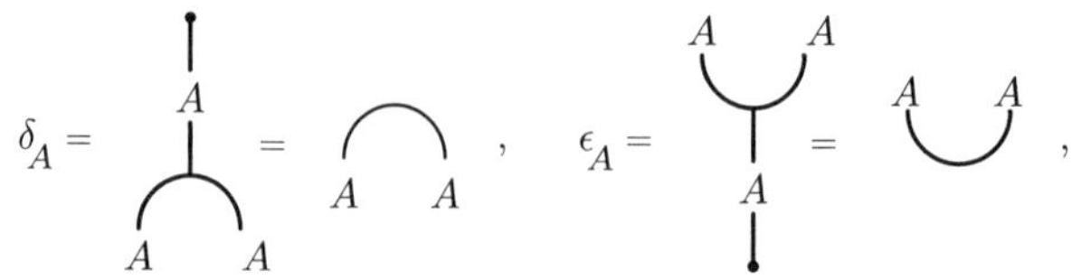

$\delta_A =$, $\epsilon_A =$,

which satisfies the conditions

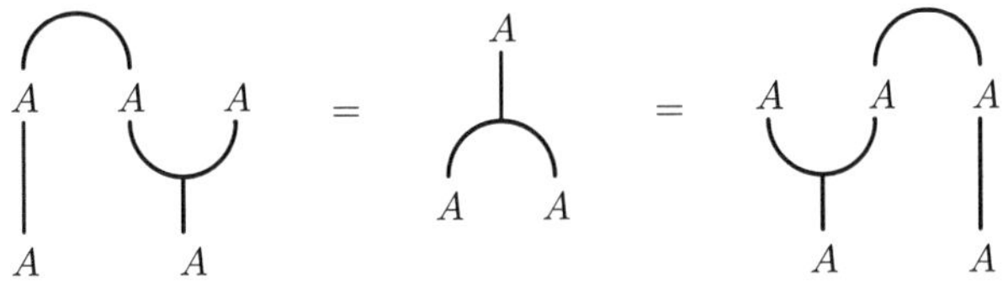

and

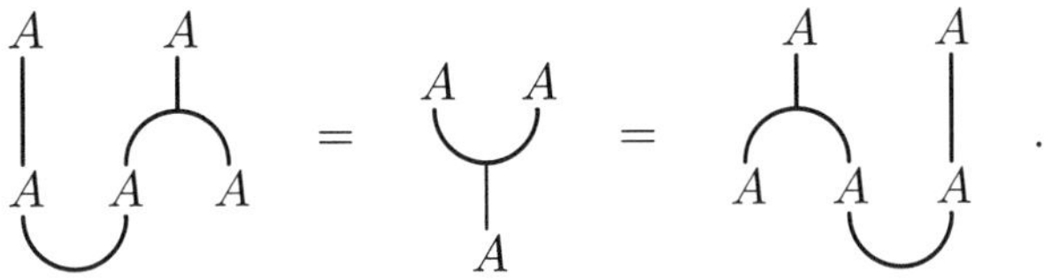

Conversely, given an algebra (A, T, δ) in $\mathcal{T}$ with A a self-dual object and a rigidity copairing $\delta_A : I \to A \otimes A$ fulfilling the **compatibility condition**

,

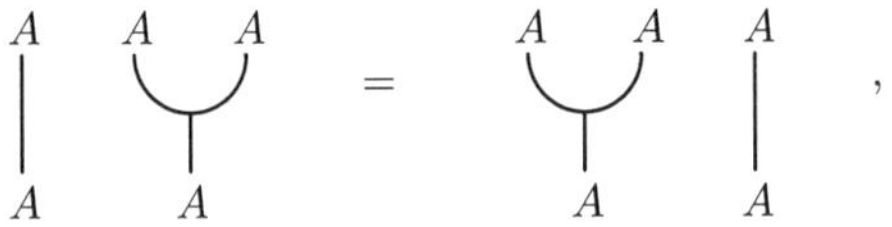

we can recover the Frobenius algebra by straightforward arguments so that the above common morphism $A \to A \otimes A$ serves as comultiplication. Since algebra and coalgebra structures are interchangeable with each other in the present context, we have the following characterizations of Frobenius algebra.

Proposition 2.1 *Let A be an object in a tensor category $\mathcal{T}$. Then the following data give the equivalent information on A.*

1. *A Frobenius algebra structure on A.*
2. *An algebra structure (T, δ) on A together with a rigidity copairing $\delta_A : I \to A \otimes A$ satisfying the compatibility condition (A being self-dual particularly).*
3. *A coalgebra structure (S, ϵ) on A together with a rigidity pairing $\epsilon_A : A \otimes A \to I$ satisfying the compatibility condition.*
4. *A pair of morphisms (S, T) satisfying the st-duality and the existence of units and counits.*

It would be worth pointing out here that, in a C*-tensor category $\mathcal{T}$, any coalgebra (A, S, ϵ) is canonically supplemented to a Frobenius algebra (with the coalgebra structure given by taking adjoints of S and ϵ) provided that S is a scalar multiple of an isometry [7].

In what follows, we shall assume that

$$TS = (\text{non-zero scalar})1_A \qquad \text{and} \qquad \epsilon\delta = (\text{non-zero scalar})1_I.$$

Note that the st-duality relation for the pair (S, T) uniquely determines ϵ and δ. For example, if we change (S, T) into $(\lambda S, \mu T)$, then (ϵ, δ) is modified into $(\mu^{-1}\epsilon, \lambda^{-1}\delta)$. Thus, by adjusting scalar multiplications, we may assume that the scalars appearing in TS and $\epsilon\delta$ coincide. If this is the case, we call the pair (S, T) an **algebraic Q-system** (see [6] for the original meaning of Q-systems) and denote the common scalar by d. The associated Frobenius algebra is then referred to as a **special Frobenius algebra** according to [3]. (In [9], the adjective 'strongly separable' is used instead of 'special'.)

A standard model for special Frobenius algebras is the following: Assume that we are given a (strict) bicategory of two objects $\{1,2\}$ and arrange the associated four hom-categories in the matrix form $\begin{pmatrix} \mathcal{H}_{11} & \mathcal{H}_{12} \\ \mathcal{H}_{21} & \mathcal{H}_{22} \end{pmatrix}$ with $\mathcal{H}_{ij} = \mathcal{H}om(j,i)$.

Choose off-diagonal objects $H \in \mathcal{H}_{12}$ and $H^* \in \mathcal{H}_{21}$ such that H^* is a left and right dual of H at the same time with a (right) rigidity pairing $\epsilon : H^* \otimes H \to I_2$ and a (left) copairing $\delta : I_2 \to H^* \otimes H$. Then $A = H \otimes H^*$ is a Frobenius algebra with multiplication and comultiplication given by $T = 1_H \otimes \epsilon \otimes 1_{H^*}$ and $S = 1_H \otimes \delta \otimes 1_{H^*}$ respectively.

If we further assume the irreducibility of H as well as the existence of unit objects $I_1 \in \mathcal{H}_{11}$ and $I_2 \in \mathcal{H}_{22}$, then A is a special Frobenius algebra.

Remark 2.2

1. If we consider the case of the tensor category of normal *-endomorphisms of an infinite factor, we are reduced to the situation of Q-systems in [6], [7].
2. See [9] for more information on the relationship with the notion of Q-system.

3 Dual systems

Given an algebraic Q-system and objects X, Y in $\mathcal{T}$, we introduce an idempotent operator $E = E_{Y,X} : \mathrm{Hom}(A \otimes X, A \otimes Y) \to \mathrm{Hom}(A \otimes X, A \otimes Y)$ by

$$E(f) = \frac{1}{d}(T \otimes 1_Y)(1_A \otimes f)(S \otimes 1_X),$$

where d is the non-zero scalar associated to the algebraic Q-system.

The following is an easy consequence of graphical computations.

Lemma 3.1 *For $f \in \mathrm{Hom}(A \otimes X, A \otimes Y)$, the following conditions are equivalent.*

1. $E(f) = f$.
2. $f(T \otimes 1_X) = (T \otimes 1_Y)(1_A \otimes f)$.
3. $(S \otimes 1_Y)f = (1_A \otimes f)(S \otimes 1_X)$.

Corollary 3.2 *The image of* $\mathrm{End}(A \otimes X)$ *under the map E, i.e., $\{f \in \mathrm{End}(A \otimes X); E(f) = f\}$, is a subalgebra of* $\mathrm{End}(A \otimes X)$.

Similarly we can introduce the idempotent operator F associated to the right tensoring of A. We consider the monoidal algebra $\{A_{m,n} = \mathrm{Hom}(A^{\otimes n}, A^{\otimes m})\}_{m,n \geq 0}$ associated with the object A. Set

$$D_{m,n} = \{f \in A_{m+1,n+1}; E(f) = f \text{ and } F(f) = f\}$$

for m, $n \geq 0$ and

$$B_{m,n} = \{f \in A_{m,n}; F(f) = f\},$$
$$C_{m,n} = \{f \in A_{m,n}; E(f) = f\}$$

for m, $n \geq 1$. Note here that $EF = FE$ on $A_{m+1,n+1}$ by the associativity of S and T.

The above corollary then shows that $\{D_{m,n}\}_{m,n \geq 0}$ is a block system of algebra, i.e., $D_{k,l} D_{m,n} \subset \delta_{l,m} D_{k,n}$, where the product is performed inside the block system of algebra $\bigoplus_{i,j \geq 0} A_{i,j}$. Similarly for $\{B_{m,n}\}_{m,n \geq 1}$ and $\{C_{m,n}\}_{m,n \geq 1}$.

We shall now make $\{D_{k,l}\}$ into a monoidal algebra. Let $f \in D_{k,l}$ and $g \in D_{m,n}$. We define $f\widehat{\otimes}g \in A_{k+m+1,l+n+1}$ by

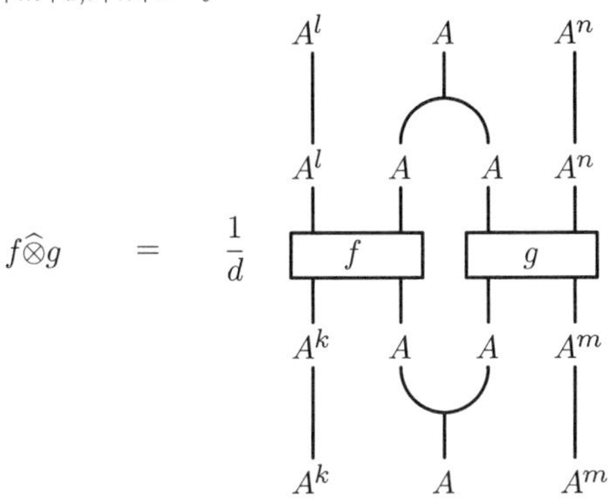

The following is easily checked by graphical computations.

Lemma 3.3

1. *We have* $f\widehat{\otimes}g \in D_{k+m,l+n}$.
2. *The unit* 1_A *of* $A_{0,0}$ *satisfies* $1_A\widehat{\otimes}f = f\widehat{\otimes}1_A = f$ *for* $f \in D_{m,n}$.
3. *For* $f : V \otimes A \to W \otimes A$, $h : A \otimes X \to A \otimes Y$ *and* $g \in D_{m,n}$ *with* $m,\ n \geq 0$, *we have* $(f\widehat{\otimes}g)\widehat{\otimes}h = f\widehat{\otimes}(g\widehat{\otimes}h)$.
4. *For* $f \in D_{m,m'}$, $f' \in D_{m',m''}$, $g \in D_{n,n'}$ *and* $g' \in D_{n',n''}$, *we have*

$$(f\widehat{\otimes}g)(f'\widehat{\otimes}g') = (ff')\widehat{\otimes}(gg').$$

The block system $\{D_{m,n}\}$ is now a monoidal algebra by the previous lemma. The construction can be obviously extended to the systems $\{B_{m,n}\}$ and $\{C_{m,n}\}$ so that they give rise to a 2×2-bicategory $\mathcal{B}$:

$$\begin{aligned}
A_{m,m'} \otimes B_{n,n'} &\subset B_{m+n,m'+n'}, \\
B_{m,m'}\widehat{\otimes}D_{n,n'} &\subset B_{m+n,m'+n'}, \\
C_{m,m'} \otimes A_{n,n'} &\subset C_{m+n,m'+n'}, \\
D_{m,m'}\widehat{\otimes}C_{n,n'} &\subset C_{m+n,m'+n'}, \\
B_{m,m'}\widehat{\otimes}C_{n,n'} &\subset A_{m+n-1,m'+n'-1}, \\
C_{m,m'} \otimes B_{n,n'} &\subset D_{m+n-1,m'+n'-1}
\end{aligned}$$

with analogous properties of tensor products for $\{D_{m,n}\}$.

If we denote by H and H^* objects associated to $B_{1,1}$ and $C_{1,1}$ respectively, then A is identified with $H \otimes H^*$ and $D_{m,n} = \mathrm{Hom}((H^* \otimes H)^{\otimes n}, (H^* \otimes H)^{\otimes m})$.

Proposition 3.4 *The bicategory* $\mathcal{B}$ *is rigid. More precisely, the generators* H *and* H^* *are rigid with rigidity pairs given by*

$$\begin{aligned}
&\delta : I \to H \otimes H^* = A, \qquad T \in D_{0,1} = \mathrm{Hom}(H^* \otimes H, J), \\
&S \in D_{1,0} = \mathrm{Hom}(J, H^* \otimes H), \qquad \epsilon : H \otimes H^* = A \to I.
\end{aligned}$$

(J denotes the unit object for D.)

Proof The hook identities for these pairs are nothing but the unit and counit identities for T and S respectively. □

The rigidity pairs then induce the Frobenius algebra structure on $H^* \otimes H$ by switching the roles of (δ, ϵ) and (S, T), which is referred to as the **dual Q-system**: the multiplication and comultiplication in $H^* \otimes H$ are given respectively by

$$1 \otimes \epsilon \otimes 1 : H^* \otimes H \otimes H^* \otimes H \to H^* \otimes H, \qquad 1 \otimes \delta \otimes 1 : H^* \otimes H \to H^* \otimes H \otimes H^* \otimes H.$$

Now the following duality for algebraic Q-systems, although obvious, generalizes an operator algebraic result in [6].

Proposition 3.5 *Given an algebraic Q-system (S, T), its bidual Q-system is canonically isomorphic to (S, T).*

4 Bicategory of bimodules

Recall that a morphism $f : X \to Y$ in a category is called a **monomorphism** (**epimorphism** respectively) if $g_j : Z \to X$ ($g_j : Y \to Z$) for $j = 1, 2$ satisfies $fg_1 = fg_2$ ($g_1 f = g_2 f$), then $g_1 = g_2$. A **subobject** of an object Y is a pair (X, j) of an object X and a monomorphism $j : X \to Y$. A subobject $j : X \to Y$ is called a **direct summand** if we can find a morphism $p : Y \to X$ such that $pj = 1_X$.

In what follows, categories are assumed to be linear, have splitting idempotents and be closed under taking direct sums. Given an idempotent $e \in \mathrm{End}(X)$, we denote the associated subobject of X by eX (with e regarded as a monomorphism in $\mathrm{Hom}(eX, X) = \mathrm{End}(X)e$), which is a direct summand of X and we have the obvious identification $eX \oplus (1 - e)X = X$.

Let A be a Frobenius algebra in a tensor category $\mathcal{T}$. By a **left A-module**, we shall mean an object M in $\mathcal{T}$ together with a morphism (called the action) $\lambda : A \otimes M \to M$ satisfying $\lambda(\epsilon \otimes 1_M) = 1_M$ and the commutative diagram

$$\begin{array}{ccc} A \otimes A \otimes M & \xrightarrow{1 \otimes \lambda} & A \otimes M \\ {\scriptstyle T \otimes 1}\downarrow & & \downarrow{\scriptstyle \lambda} \\ A \otimes M & \xrightarrow[\lambda]{} & M \end{array} .$$

The notion of **right A-module** is defined analogously. Let B be another Frobenius algebra. By an A-B bimodule, we shall mean a left A-module M (with the left action $\lambda : A \otimes M \to M$) which is a right B-module (with the right action $\mu : M \otimes B \to M$) at the same time and makes the diagram

$$\begin{array}{ccc} A \otimes M \otimes B & \xrightarrow{\lambda \otimes 1} & M \otimes B \\ {\scriptstyle 1 \otimes \mu}\downarrow & & \downarrow{\scriptstyle \mu} \\ A \otimes M & \xrightarrow[\lambda]{} & M \end{array}$$

commutative.

An A-B bimodule based on an object M in $\mathcal{T}$ is simply denoted by ${}_A M_B$. Given another A-B bimodule ${}_A N_B$, a morphism $f : M \to N$ in the category $\mathcal{T}$ is said to be A-B **linear** if the diagram

$$\begin{array}{ccc} A \otimes M \otimes B & \xrightarrow{1 \otimes f \otimes 1} & A \otimes N \otimes B \\ \downarrow & & \downarrow \\ M & \xrightarrow[f]{} & N \end{array}$$

commutes.

The totality of A-B bimodules $\{{}_AM_B\}$ forms a linear category ${}_A\mathcal{T}_B$ by

$$\mathrm{Hom}({}_AM_B, {}_AN_B) = \{f \in \mathrm{Hom}(M,N); f \text{ is } A\text{-}B \text{ linear}\}.$$

Recall here that we have assumed splitting idempotents in the category $\mathcal{T}$ and the same holds for ${}_A\mathcal{T}_B$: if $e \in \mathrm{End}({}_AM_B)$ is an idempotent, then the A-B action on M induces an A-B action on the subobject eM, i.e., ${}_A(eM)_B$.

From here on we exclusively deal with Frobenius algebras of algebraic Q-systems, i.e., special Frobenius algebras, and shall introduce the notion of tensor product for bimodules. A more general and categorical construction is available in [3] but we prefer the following less formal description, which enables us to easily check the associativity (the so-called pentagonal relation) of tensor products.

Let X_B and ${}_BY$ be right and left B-modules with action morphisms ρ and λ respectively. Let $e \in \mathrm{End}(X \otimes Y)$ be an idempotent defined by

$$e = d^{-1}(\rho \otimes \lambda)(1_X \otimes \delta_A \otimes 1_Y),$$

where d is the common scalar for TS and $\epsilon\delta$.

The **module tensor product** $X \otimes_B Y$ is, by definition, the subobject $e(X \otimes Y)$ of $X \otimes Y$ associated to the idempotent e. For bimodules ${}_AX_B$ and ${}_BY_C$, e belongs to $\mathrm{End}({}_AX \otimes Y_C)$ and hence it induces an A-C bimodule ${}_AX \otimes_B Y_C$.

Let ${}_CZ$ be another left C-module and $f \in \mathrm{End}(Y \otimes Z)$ be the idempotent associated to the inner action of C. Then it is immediate to show the commutativity $(e \otimes 1_Z)(1_X \otimes f) = (1_X \otimes f)(e \otimes 1_Z)$ by the compatibility of left and right actions on Y, which enables us to identify

$$(X \otimes_B Y) \otimes_C Z = (e \otimes 1_Z)(1_X \otimes f)(X \otimes Y \otimes Z) = X \otimes_B (Y \otimes_C Z).$$

Moreover, given morphisms $\varphi : {}_AX_B \to {}_AX'_B$ and $\psi : {}_BY_C \to {}_BY'_C$, $\varphi \otimes_B \psi :$ ${}_AX \otimes_B Y_C \to {}_AX' \otimes_B Y'_C$ is defined by

$$\varphi \otimes_B \psi = (\varphi \otimes \psi)e = e'(\varphi \otimes \psi),$$

where $e' \in \mathrm{End}(X' \otimes Y')$ denotes the idempotent associated to the inner action of B on $X' \otimes Y'$. It is also immediate to see the associativity for the tensor product of morphisms:

$$(\phi \otimes_A \varphi) \otimes_B \psi = \phi \otimes_A (\varphi \otimes_B \psi).$$

(More precisely, the identification is through the natural isomorphisms among module tensor products of objects.)

The Frobenius algebra A itself bears the structure of A-A bimodule by the multiplication morphism, which is denoted by ${}_AA_A$. Given a left A-module $\lambda :$ $A \otimes X \to X$, let $\lambda^* : X \to A \otimes X$ be the associated coaction: $\lambda^* = (1_A \otimes \lambda)(\delta_A \otimes 1_X)$.

Lemma 4.1 *Both of λ and λ^* are A-linear.*

Proof The A-linearity of λ is just the associativity of the action. To see the A-linearity of λ^*, we use the identity

$$(T \otimes 1_A)(1_A \otimes \delta_A) = S = (1_A \otimes T)(\delta_A \otimes 1_A).$$

□

Lemma 4.2 *Let $e \in \mathrm{End}(A \otimes X)$ be the idempotent associated to the inner action of A on $A \otimes X$. Then we have*

$$\lambda\lambda^* = d1_X, \qquad \lambda^*\lambda = de.$$

Proof These follow from simple graphical computations of $\lambda\lambda^*$ and $\lambda^*\lambda$. □

Lemma 4.3 *The action morphism $\lambda : A \otimes X \to X$ induces the A-linear isomorphism $l : A \otimes_A X \to X$ with the inverse given by $d^{-1}\lambda^*$. Likewise a right A-module $\rho : X \otimes A \to X$ induces the isomorphism $r : X \otimes_A A \to X$ with the inverse given by $d^{-1}\rho^*$.*

Here is another useful observation, which is an immediate consequence of definitions.

Lemma 4.4 *Let A be a Frobenius algebra. Then, by the correspondance $(\lambda : A \otimes X \to X) \iff (\lambda^* : X \to A \otimes X)$, there is an equivalence between the category of left A-modules and the category of left A-comodules.*

$$\begin{array}{ccc} A \otimes X & \xrightarrow{1\otimes f} & A \otimes Y \\ \downarrow & & \downarrow \\ X & \xrightarrow[f]{} & Y \end{array} \iff \begin{array}{ccc} X & \xrightarrow{f} & Y \\ \downarrow & & \downarrow \\ A \otimes X & \xrightarrow[1\otimes f]{} & A \otimes Y \end{array}.$$

Lemma 4.5 *Let X_A be a right A-module and ${}_AY$ be a left A module with the associated isomorphisms $r : X \otimes_A A \to X$ and $l : A \otimes_A Y \to Y$. Then r and l satisfy the triangle identity: $r \otimes_A 1_Y = 1_X \otimes_A l$ on $X \otimes_A A \otimes_A Y$.*

Proof Let $e_X \in \mathrm{End}(X \otimes A)$, $e_Y \in \mathrm{End}(A \otimes Y)$ and $e \in \mathrm{End}(X \otimes Y)$ be idempotents associated to the inner actions of A. We need to show the equality

$$e(\rho \otimes 1_Y)(e_X \otimes 1_Y)(1_X \otimes e_Y) = e(1_X \otimes \lambda)(e_X \otimes 1_Y)(1_X \otimes e_Y).$$

By a graphical computation, we see that

$$d^2(\rho\otimes 1_Y)(e_X\otimes 1_Y)(1_X\otimes e_Y) = d(\rho\otimes\lambda)(1_X\otimes S\otimes 1_Y) = d^2(1_X\otimes\lambda)(e_X\otimes 1_Y)(1_X\otimes e_Y).$$

□

Summarizing the discussions so far, we have

Proposition 4.6 *The family of categories $\{{}_A\mathcal{T}_B\}$ indexed by pairs of special Frobenius algebras forms a bicategory with unit constraints given by l and r in the previous lemma.*

The following is not needed in what follows but enables us to compare our definition with the one in [3].

Lemma 4.7 *The projection $e : X \otimes Y \to X \otimes_B Y$ gives the cokernel of*

$$(\rho \otimes 1_Y - 1_X \otimes \lambda) : X \otimes B \otimes Y \to X \otimes Y.$$

Proof By a graphical computation, we have

$$e(\rho \otimes 1_Y) = (\rho \otimes \lambda)(1_X \otimes S \otimes 1_Y) = e(1_X \otimes \lambda).$$

Conversely, given a morphism $f : X \otimes Y \to Z$ satisfying $f(\rho \otimes 1_Y) = f(1_X \otimes \lambda)$, we can show $ef = f$. □

5 Rigidity in bimodules

The rigidity of categorical modules is considered in [3] under the assumption of a certain 'commutativity' of Frobenius algebras. Although its general validity would be well-known for experts, we shall describe here the relevant points for completeness.

Let A and B be Frobenius algebras in a tensor category $\mathcal{T}$ and ${}_AX_B$ be an A-B bimodule in $\mathcal{T}$. Assume that the object X admits a (left) dual X^* in $\mathcal{T}$ with a rigidity pair given by $\epsilon : X \otimes X^* \to I$ and $\delta : I \to X^* \otimes X$. We can then define the B-A action on X^* as the transposed morphism: consider $B \otimes X^* \to X^*$ and $X^* \otimes A \to X^*$ defined by Fig. 2.

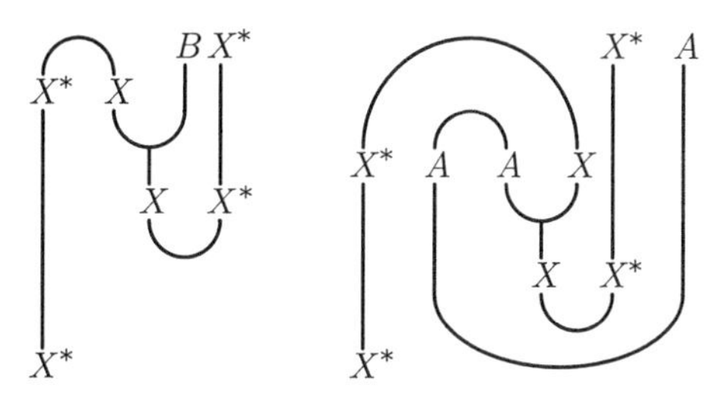

Figure 2

The following is immediate by easy graphical works.

Lemma 5.1 *These in fact define the left and right actions on X^*, which are compatible in the following sense.*

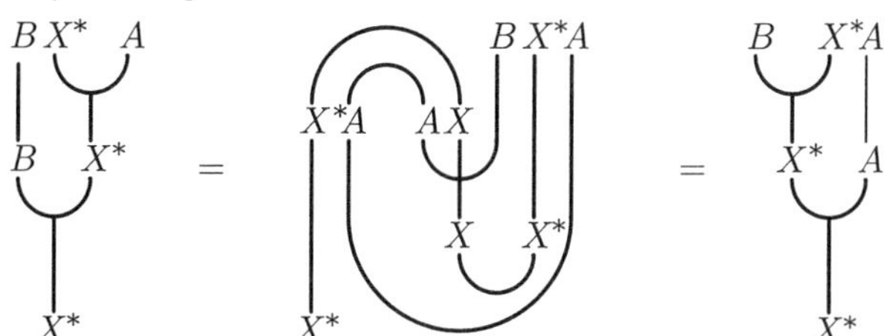

We shall show that the bimodule ${}_BX^*_A$ is a dual object of ${}_AX_B$. To this end, we first introduce morphisms $\epsilon : X \otimes X^* \to A$ and $\delta : B \to X^* \otimes X$ by Fig. 3

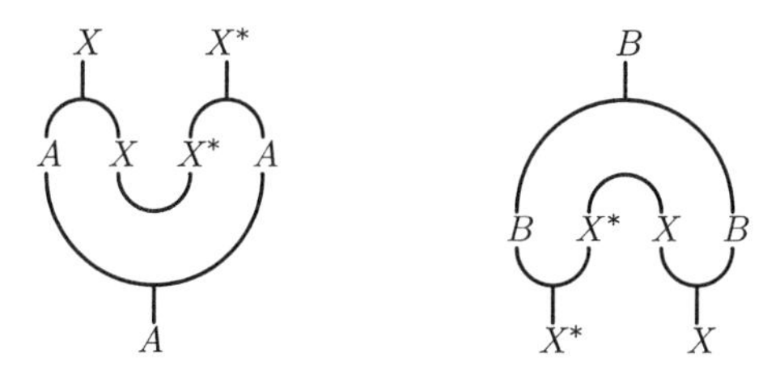

Figure 3

Lemma 5.2 *We have*

$$\epsilon = d_A(1_A \otimes \epsilon_X)((X \to A \otimes X) \otimes 1_{X^*}),$$
$$\delta = d_B(1_{X^*} \otimes (X \otimes B \to X))(\delta_X \otimes 1_B).$$

Proof Insert the definition of (co)actions on X^* and compute graphically. □

Lemma 5.3 *The morphism $\epsilon : X \otimes X^* \to A$ is A-A linear whereas the morphism $\delta : B \to X^* \otimes X$ is B-B linear.*

Proof We use the formulas in the above lemma together with the definition of (co)actions on X^*. □

Lemma 5.4 *Let $e_B \in \mathrm{End}(X \otimes X^*)$ and $e_A \in \mathrm{End}(X^* \otimes X)$ be idempotents associated with inner actions. Then ϵ and δ are supported by these idempotents: $\epsilon e_B = \epsilon$ and $e_A \delta = \delta$.*

Proof For example, the equality $e_A\delta = \delta$ is proved by

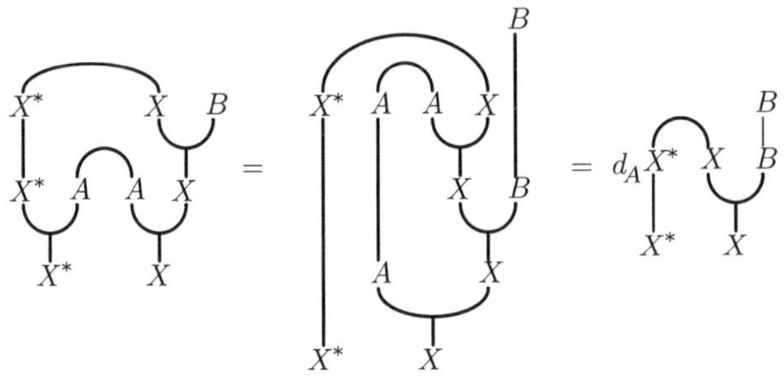

□

By the above lemmas, we can regard ϵ and δ as defining morphisms ${}_AX \otimes_B X^*_A \to {}_AA_A$ and ${}_BB_B \to {}_BX^* \otimes_A X_B$ respectively.

Lemma 5.5 *The compositions*

$$X \xrightarrow{1 \otimes_B \delta} X \otimes_B X^* \otimes_A X \xrightarrow{\epsilon \otimes_A 1} X,$$
$$X^* \xrightarrow{\delta \otimes_B 1} X^* \otimes_A X \otimes_B X^* \xrightarrow{1 \otimes_A \epsilon} X^*$$

are scalar multiplication of identities by the common scalar ${d_A}^2 d_B$.

Proof By the previous lemma together with Corollary 4.3, we need to compare compositions

$$X \xrightarrow{d_B^{-1}\rho^*} X \otimes B \xrightarrow{1_X \otimes \delta} X \otimes X^* \otimes X \xrightarrow{\epsilon \otimes 1_X} A \otimes X \xrightarrow{\lambda} X,$$
$$X^* \xrightarrow{d_B^{-1}\lambda^*} B \otimes X^* \xrightarrow{\delta \otimes 1_{X^*}} X^* \otimes X \otimes X^* \xrightarrow{1_{X^*} \otimes \epsilon} X^* \otimes A \xrightarrow{\rho} X^*,$$

where λ denotes one of the left actions $A \otimes X \to X$, $B \otimes X^* \to X^*$ and similarly for ρ, λ^* and ρ^*.

By multiplying d_A^{-1} on both of these compositions, the former is reduced to whereas the latter is given by

and turns out to be $d_A d_B 1_X$ from the relations in Fig. 4. □

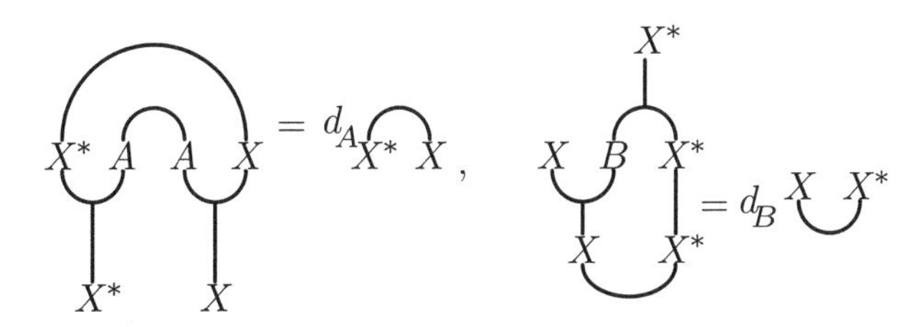

Figure 4

Proposition 5.6 *Let ${}_AX_B$ be a bimodule and assume that X is rigid in $\mathcal{T}$. Then the bimodule ${}_AX_B$ is rigid in the bicategory with the dual bimodule given by ${}_BX^*_A$.*

Proof This is just a paraphrase of the previous lemma. □

Definition 5.7 Given a Frobenius algebra A in a tensor category $\mathcal{T}$, we denote by ${}_A\mathcal{T}_A$ the tensor category of A-A bimodules.

Proposition 5.8 *Given a special Frobenius algebra A in a tensor category $\mathcal{T}$, let B be the dual Frobenius algebra of A. Then the bicategory connecting A and B is generated by the bimodule $H = {}_IA_A$ in $\mathcal{T}$: $H \otimes_A H^* \cong A$ while the Frobenius algebra ${}_AH^* \otimes H_A$ is isomorphic to B.*

Theorem 5.9 (Duality for Tensor Categories) *Given a special Frobenius algebra A in a tensor category $\mathcal{T}$, the dual Frobenius algebra B is canonically realized in the tensor category ${}_A\mathcal{T}_A$ and the tensor category ${}_B({}_A\mathcal{T}_A)_B$ of B-B bimodules in ${}_A\mathcal{T}_A$ is naturally monoidally equivalent to the starting tensor category $\mathcal{T}$.*

Proof By the identification $B = H^* \otimes H$, the object H has the structure of a right B-module in an obvious way and, if we regard this as defining an object M at an off-diagonal corner of a bicategory connecting $\mathcal{T}$ and ${}_B({}_A\mathcal{T}_A)_B$, then it satisfies the imprimitivity condition; $M \otimes_B M^* = I$ (the unit object in $\mathcal{T}$) and $M^* \otimes M = {}_B B_B$ (the unit object in ${}_B({}_A\mathcal{T}_A)_B$). Thus taking adjoint tensor multiplications by M gives rise to a monoidal equivalence of tensor categories in question. □

6 Semisimplicity

An object X is said to be **semisimple** if any subobject is a direct summand and said to be **simple** if there is no non-trivial subobject.

Note that, if End(X) is finite-dimensional for a semisimple object X, then X is isomorphic to a direct sum of simple objects.

A tensor category is **semisimple** if every object is semisimple.

The following is a direct and simplified version of the proof in [3, §5.4] (cf. also [10]).

Proposition 6.1 *Let A and B be special Frobenius algebras in a tensor category $\mathcal{T}$. An A-B bimodule ${}_A X_B$ is semisimple in ${}_A\mathcal{T}_B$ if the base object X is semisimple in $\mathcal{T}$.*

Proof Let $f : {}_A Y_B \to {}_A X_B$ be a monomorphism in ${}_A\mathcal{T}_B$. We first show that f is monomorphic as a morphism in $\mathcal{T}$.

In fact, given a morphism $h : Z \to Y$ in $\mathcal{T}$ such that $fh = 0$, the induced morphism $\widetilde{h} : A \otimes Z \otimes B \to Y$ defined by

$$\widetilde{h} = (A \otimes Y \otimes B \to Y)(1_A \otimes h \otimes 1_B)$$

is A-B linear and satisfies

$$f\widetilde{h} = (A \otimes Y \otimes B \to Y)(1_A \otimes fh \otimes 1_B) = 0$$

by the A-B linearity of f. Since f is assumed to be monomorphic in ${}_A\mathcal{T}_B$, this implies $\widetilde{h} = 0$ and hence

$$h = \widetilde{h}(\delta \otimes 1_Z \otimes \delta) = 0,$$

where δ denotes one of unit morphisms $I \to A$ and $I \to B$ in the Frobenius algebras.

So far, we have proved that $f : Y \to X$ gives a subobject of X. Since X is semisimple by our assumption, we can find a morphism $g : X \to Y$ satisfying $gf = 1_Y$. Let $\widetilde{g} : Y \to X$ be defined by

$$\widetilde{g} = (A \otimes X \otimes B \to X)(1_A \otimes g \otimes 1_B)(Y \to A \otimes Y \otimes B),$$

which is A-B linear as a composition of A-B linear morphisms.

Now the computation

$$\begin{aligned} \widetilde{g}f &= (A \otimes Y \otimes B \to Y)(1_A \otimes g \otimes 1_B)(Y \to A \otimes Y \otimes B)f \\ &= (A \otimes Y \otimes B \to Y)(1_A \otimes gf \otimes 1_B)(Y \to A \otimes Y \otimes B) \\ &= (A \otimes Y \otimes B \to Y)(Y \to A \otimes Y \otimes B) \\ &= d_A d_B 1_Y \end{aligned}$$

shows that ${}_A Y_B$ is a direct summand of ${}_A X_B$. □

Corollary 6.2 *Let A and B be special Frobenius algebras in a semisimpte tensor category $\mathcal{T}$. Then the category ${}_A\mathcal{T}_B$ of A-B bimodules in $\mathcal{T}$ is semisimple as well.*

7 Tannaka duals

By the Tannaka dual of a Hopf algebra H, we shall mean the tensor category of finite-dimensional (left) H-modules.

We shall here work with the Tannaka dual $\mathcal{A}$ of a semisimple Hopf algebra H which is realized in a tensor category $\mathcal{T}$, i.e., we are given a faithful monoidal functor $F : \mathcal{A} \to \mathcal{T}$. The notion of $\mathcal{A}$-modules is introduced in [13] in terms of the notion of trivializing isomorphisms.

Let $\mathbb{A}$ be the unit object in the tensor category of $\mathcal{A}$-$\mathcal{A}$ modules in $\mathcal{T}$. Recall that the object $\mathbb{A}$ is isomorphic to

$$\bigoplus_V F(V) \otimes V^*$$

as an object in $\mathcal{T}$. By interchanging left and right actions, the dual object $\mathbb{A}^*$ of $\mathbb{A}$ is an $\mathcal{A}$-$\mathcal{A}$ module in a canonical way, which is isomorphic to the unit object ${}_{\mathcal{A}}\mathbb{A}_{\mathcal{A}}$. We shall give an explicit formula for the isomorphism ${}_{\mathcal{A}}\mathbb{A}^*_{\mathcal{A}} \cong {}_{\mathcal{A}}\mathbb{A}_{\mathcal{A}}$.

Lemma 7.1 *The isomorphism* $\mathbb{A}^* \to \mathbb{A}$ *given by*

$$\mathbb{A}^* = \bigoplus\nolimits_V F(V^*) \otimes V \xrightarrow{\oplus_V d(V)1} \bigoplus\nolimits_V F(V^*) \otimes V = \mathbb{A}$$

is $\mathcal{A}$-$\mathcal{A}$ *linear.*

Proof Let us prove the left $\mathcal{A}$-linearity for example. To this end, we first recall that the left action $F(U) \otimes \mathbb{A}^* \to \mathbb{A}^* \otimes U$ on $\mathbb{A}^*$ is given by the composition

$$F(U)\otimes\mathbb{A}^* \to \mathbb{A}^*\otimes U\otimes U^*\otimes\mathbb{A}\otimes F(U)\otimes\mathbb{A}^* \to \mathbb{A}^*\otimes U\otimes\mathbb{A}\otimes F(U^*)\otimes F(U)\otimes\mathbb{A}^* \to \mathbb{A}^*\otimes U.$$

We can check this formula by working on vector spaces: Let

$$\{X \xrightarrow{\ \xi\ } W \otimes U \xrightarrow{\ \xi^*\ } X\}$$

be an irreducible decomposition of $W \otimes U$ and $\{x_l\}$, $\{w_k\}$, $\{u_i\}$ be bases of vector spaces X, W, U with the dual bases indicated by asterisk. Then

$$\begin{aligned}
&F(u) \otimes F(v^*) \otimes v \\
&\quad \mapsto \bigoplus_W \sum_{i,j,k} F(w_j^*) \otimes w_k \otimes u_i \otimes u_i^* \otimes F(w_j) \otimes w_k^* \otimes F(u) \otimes F(v^*) \otimes v \\
&\quad \mapsto \bigoplus_W \sum_{i,j,k} \sum_{X,\xi,l} F(w_j^*) \otimes w_k \otimes u_i \otimes u_i^* \otimes F(x_l) \otimes \xi \otimes w_k^* \otimes F(v^*) \\
&\qquad\qquad\qquad\qquad\qquad\qquad\qquad\qquad \otimes v\langle x_l^*, \xi^*(w_j \otimes u)\rangle \\
&\quad \mapsto \bigoplus_W \sum F(w_j^*) \otimes w_k \otimes u_i \otimes u_i^* \otimes F(x_l) \otimes \widetilde{\xi}(w_k^*) \otimes F(v^*) \\
&\qquad\qquad\qquad\qquad\qquad\qquad\qquad\qquad \otimes v\langle x_l^*, \xi^*(w_j \otimes u)\rangle
\end{aligned}$$

($F(x_l)$ and $\widetilde{\xi}(w_k^*)$ being coupled with $F(v^*)$ and $u_i^* \otimes v$ respectively)

$$\mapsto \bigoplus_W \sum F(w_j^*) \otimes w_k \otimes u_i \langle u_i^* \otimes v, \widetilde{\xi}(w_k^*)\rangle \langle v^*, \xi^*(w_j \otimes u)\rangle$$

(letting $X = V$)

$$= \bigoplus_W \sum F(w_j^*) \otimes \xi v \langle v^*, \xi^*(w_j \otimes u)\rangle$$

$$= \bigoplus_W \sum_\xi F(\widetilde{\xi^*}(u \otimes v^*)) \otimes \xi v$$

$$= \bigoplus_W \sum_{\eta: UV^* \to W^*} \frac{d(V)}{d(W)} F(\eta(u \otimes v^*)) \otimes \widetilde{\eta^*} v,$$

where the family $\{W^* \xrightarrow{\eta^*} U \otimes V^* \xrightarrow{\eta} W^*\}$ denotes an irreducible decomposition of $U \otimes V^*$.

Comparing the last expression with the definition of trivialization isomorphism $F(U) \otimes \mathbb{A} \to \mathbb{A} \otimes U$, we see that

$$\mathbb{A}^* = \bigoplus_V F(V^*) \otimes V \xrightarrow{\oplus_V d(V) 1} \bigoplus_V F(V^*) \otimes V = \mathbb{A}$$

is $\mathcal{A}$-linear. □

The object $A = \mathbb{A}^* \otimes_{\mathcal{A}} \mathbb{A}$ in $\mathcal{T}$ is a Frobenius algebra by the rigidity of ${}_{\mathcal{A}}\mathbb{A}$: the multiplication morphism is given by

$$A \otimes A = \mathbb{A}^* \otimes_{\mathcal{A}} \mathbb{A} \otimes \mathbb{A}^* \otimes_{\mathcal{A}} \mathbb{A} \xrightarrow{1 \otimes \epsilon \otimes 1} \mathbb{A}^* \otimes_{\mathcal{A}} \mathbb{A} \otimes_{\mathcal{A}} \mathbb{A} = \mathbb{A}^* \otimes_{\mathcal{A}} \mathbb{A} = A.$$

By the natural identification $\mathbb{A}^* \otimes_{\mathcal{A}} \mathbb{A} = \mathbb{A}$, this can be rewritten as

$$\mathbb{A} \otimes \mathbb{A} \to \mathbb{A} \otimes \mathbb{A}^* \xrightarrow{\epsilon} \mathbb{A},$$

where ϵ denotes a rigidity pair for ${}_{\mathcal{A}}\mathbb{A}$ (ϵ being $\mathcal{A}$-$\mathcal{A}$ linear) and is defined by the formula after Corollary 6.2:

$$\mathbb{A} \otimes \mathbb{A}^* \xrightarrow{\oplus_X d(X) 1 \otimes \epsilon_X \otimes 1} \bigoplus_X \mathbb{A} \otimes X \otimes X^* \otimes \mathbb{A}^*$$

$$\longrightarrow \bigoplus_X F(X) \otimes \mathbb{A} \otimes \mathbb{A}^* \otimes X^* \xrightarrow{\oplus_X 1 \otimes \epsilon_{\mathbb{A}} \otimes 1} \bigoplus_X F(X) \otimes X^* = \mathbb{A}$$

with ϵ_X the ordinary vector space pairing and $\epsilon_{\mathbb{A}}$ the rigidity pairing for the object $\mathbb{A}$ in $\mathcal{T}$ (with the trivial action).

Since ${}_{\mathcal{A}}\mathbb{A}_{\mathcal{A}}$ is identified with ${}_{\mathcal{A}}\mathbb{A}^*_{\mathcal{A}}$ by multiplying the weight $\{d(V)^{-1}\}_V$, the multiplication morphism $\mathbb{A} \otimes \mathbb{A} \to \mathbb{A}$ is given by the following process on vectors:

$$\begin{aligned}
&F(v) \otimes v^* \otimes F(w) \otimes w^* \\
&\quad \mapsto d(W)^{-1} F(v) \otimes v^* \otimes F(w) \otimes w^* \\
&\quad \mapsto \bigoplus_X \frac{d(X)}{d(W)} \sum_i F(v) \otimes v^* \otimes x_i \otimes x_i^* \otimes F(w) \otimes w^* \\
&\quad \mapsto \bigoplus_{X,U} \frac{d(X)}{d(W)} \sum_{\xi,i,j} F(v) \otimes \xi \otimes \langle u_j, \xi^*(v^* \otimes x_i)\rangle u_j^* \otimes x_i^* \otimes F(w) \otimes w^* \\
&\quad \mapsto \bigoplus_{X,U} \frac{d(X)}{d(W)} \sum_{\xi,i,j} F(\widetilde{\xi} v) \otimes \langle u_j, \xi^*(v^* \otimes x_i)\rangle u_j^* \otimes x_i^* \otimes F(w) \otimes w^*
\end{aligned}$$

(letting $U = W^*$ and $u_j = w_j^*$ for the pairing)

$$\mapsto \bigoplus_X \frac{d(X)}{d(W)} \sum F(\langle \widetilde{\xi} v \rangle_w) \otimes \langle w_j^*, \xi^*(v^* \otimes x_i) \rangle \langle w_j, w^* \rangle x_i^*$$

$$= \bigoplus_X \frac{d(X)}{d(W)} \sum_{\xi,i} F(\xi'(v \otimes w)) \otimes \langle w^*, \xi^*(v^* \otimes x_i) \rangle x_i^*$$

$$= \bigoplus_X \frac{d(X)}{d(W)} \sum_{\xi} F(\xi'(v \otimes w)) \otimes {}^t(\xi^*)'(w^* \otimes v^*),$$

where the family

$$\{U^* \xrightarrow{\ \xi\ } V^* \otimes X \xrightarrow{\ \xi^*\ } U^*\}$$

denotes an irreducible decomposition of $V^* \otimes X$ with $\widetilde{\xi}$ and ξ' Frobenius transforms of ξ.

Since

$$\frac{d(X)}{d(W)} \left(X \xrightarrow{(\xi^*)'} V \otimes W \xrightarrow{\ \xi'\ } X \right)$$

gives an irreducible decomposition of $V \otimes W$, we have the following.

Proposition 7.2 *The object $\mathbb{A}$ in $\mathcal{T}$ is an Frobenius algebra by the multiplication morphism*

$$F(v) \otimes v^* \otimes F(w) \otimes w^* \mapsto \bigoplus_U \sum_{\eta: U \to VW} F(\eta^*(v \otimes w)) \otimes {}^t\eta(w^* \otimes v^*)$$

(the family $\{U \xrightarrow{\ \eta\ } V \otimes W \xrightarrow{\ \eta^*\ } U\}$ being an irreducible decomposition of $V \otimes W$) with the compatible rigidity copairing $\delta_A : \mathbb{A} \otimes \mathbb{A} \to I$ given by the composition

$$\bigoplus_{V,W} F(V) \otimes V^* \otimes F(W) \otimes W^* \to \bigoplus_V F(V) \otimes V^* \otimes F(V^*) \otimes V \quad (\text{letting } W = V^*)$$

$$\xrightarrow{\oplus_V d(V) 1} \oplus_V F(V) \otimes V^* \otimes F(V^*) \otimes V \to I,$$

where the last morphism is the summation of the canonical pairing

$$F(V) \otimes F(V^*) \otimes V^* \otimes V \to I \otimes \mathbb{C} = I.$$

The associated unit (morphism) is given by the obvious imbedding

$$I \to F(\mathbb{C}) \otimes \mathbb{C} \subset \bigoplus_V F(V) \otimes V^*.$$

Corollary 7.3 *The multiplication morphism $\mathbb{A} \otimes \mathbb{A} \to \mathbb{A}$ is associative.*

We have seen so far that Tannaka duals give rise to a special class of Frobenius algebras in a canonical way (a depth two characterization of the class is possible in terms of factorization of Frobenius algebras, see [13]). It is worth pointing out here that a similar computation is carried out in [9, §6] based on the analysis of Hopf algebra strucutres. As can be recognized in the above arguments, our proof is purely categorical with the explicit use of fibre functors.

In what follows, we shall use calligraphic letters, say $\mathcal{A}$, to express Tannaka duals (realized in a tensor category $\mathcal{T}$) with the associated Frobenius algebras denoted by the corresponding boldface letters, say $\mathbb{A}$.

Recall here that Tannaka duals give rise to the bicategory of bimodules, whereas there is a natural notion of bimodule of Frobenius algebras which produces another bicategory.

We shall now construct a monoidal functor Φ, which associate an $\mathbb{A}$-$\mathbb{B}$ bimodule to each $\mathcal{A}$-$\mathcal{B}$ bimodule. For simplicity, consider a left $\mathcal{A}$-bimodule X with the trivialization isomorphisms $\{\phi_V : F(V) \otimes X \to X \otimes V\}$. We choose a representative family $\{V_j\}$ of simple objects in the relevant Tannaka dual and set $\phi_j = \phi_{V_j}$.

The action morphism $\phi : \mathbb{A} \otimes X \to X$ is then introduced by

$$\bigoplus_j \widetilde{\phi}_j : \bigoplus_j F(V_j) \otimes V_j^* \otimes X \to X,$$

where $\widetilde{\phi}_j : F(V_j) \otimes V_j^* \otimes X \to X$ corresponds to ϕ_j under the isomorphism

$$\mathrm{Hom}(F(V_j) \otimes V_j^* \otimes X, X) \cong \mathrm{Hom}(F(V_j) \otimes X, X) \otimes V_j = \mathrm{Hom}(F(V_j) \otimes X, X \otimes V_j).$$

Now the square diagram

$$\begin{array}{ccc} \mathbb{A} \otimes \mathbb{A} \otimes X & \xrightarrow{1 \otimes \phi} & \mathbb{A} \otimes X \\ {\scriptstyle \mu \otimes 1}\downarrow & & \downarrow{\scriptstyle \phi} \\ \mathbb{A} \otimes X & \xrightarrow[\phi]{} & X \end{array}$$

is commutative if and only if so is the diagram

$$\begin{array}{ccc} \bigoplus_{i,j} \mathrm{Hom}(Y, F(V_i) \otimes F(V_j) \otimes X) \otimes V_i^* \otimes V_j^* & \longrightarrow & \bigoplus_i \mathrm{Hom}(Y, F(V_i) \otimes X) \otimes V_i^* \\ \downarrow & & \downarrow \\ \bigoplus_k \mathrm{Hom}(Y, F(V_k) \otimes X) \otimes V_k^* & \longrightarrow & \mathrm{Hom}(Y, X) \end{array}$$

for any object Y. If we trace the morphisms starting from $f \otimes v_i^* \otimes v_j^*$ for $f : Y \to F(V_i) \otimes F(V_j) \otimes X$ and $v_i^* \in V_i^*$, then the commutativity is reduced to the identity

$$\sum_k \sum_\xi \langle \phi_k(\xi^* \otimes 1_X) f \rangle_{{}^t\xi(v_i^* \otimes v_j^*)} = \langle (\phi_i \otimes 1)(1 \otimes \phi_j) f \rangle_{v_i^* \otimes v_j^*},$$

where the family $\{V_k \xrightarrow{\xi} V_i \otimes V_j \xrightarrow{\xi^*} V_k\}$ denotes an irreducible decomposition.

Since the choice of $v_i^* \in V_i^*$ is arbitrary, the above relation is equivalent to

$$\sum_{k,\xi} (1 \otimes \xi) \phi_k (\xi^* \otimes 1_X) f = (\phi_i \otimes 1)(1 \otimes \phi_j) f$$

for any f or simply

$$\sum_{k,\xi} (1 \otimes \xi) \phi_k (\xi^* \otimes 1_X) = (\phi_i \otimes 1)(1 \otimes \phi_j),$$

which is exactly the $\mathcal{A}$-module property of X, i.e., the commutativity of the diagram

$$\begin{array}{ccc} F(V_i) \otimes F(V_j) \otimes X & \longrightarrow & F(V_i) \otimes X \otimes V_j \\ \downarrow & & \downarrow \\ \bigoplus_k F(V_k) \otimes X \otimes \begin{bmatrix} V_k \\ V_i\, V_j \end{bmatrix} & \longrightarrow & \bigoplus_k X \otimes V_k \otimes \begin{bmatrix} V_k \\ V_i\, V_j \end{bmatrix} = X \otimes V_i \otimes V_j \end{array}.$$

The unitality for the $\mathbb{A}$-action, which says that

$$X = I \otimes X \to A \otimes X \to X$$

is the identity, is reduced to that of the $\mathcal{A}$-action on X.

By summarizing the arguments so far, we have associated a left $\mathbb{A}$-module ${}_{\mathbb{A}}X$ to each $\mathcal{A}$-module ${}_{\mathcal{A}}X$ with the common base object X in $\mathcal{T}$. Moreover, given another ${}_{\mathcal{A}}Y$ with the associated ${}_{\mathbb{A}}Y$, we have the equality

$$\mathrm{Hom}({}_{\mathcal{A}}X, {}_{\mathcal{A}}Y) = \mathrm{Hom}({}_{\mathbb{A}}X, {}_{\mathbb{A}}Y)$$

as subsets of $\mathrm{Hom}(X, Y)$ from our construction.

Thus the correspondance ${}_{\mathcal{A}}X \mapsto {}_{\mathbb{A}}X$ defines a fully faithful functor $\Phi : {}_{\mathcal{A}}\mathcal{T} \to {}_{\mathbb{A}}\mathcal{T}$.

We shall now identify the tensor products. Given a right $\mathcal{A}$-module ${}_{\mathcal{A}}X$ and a left $\mathcal{A}$-module ${}_{\mathcal{A}}Y$ in $\mathcal{T}$with the trivialization isomorphisms $\phi_V : X \otimes F(V) \to V \otimes X$ and $\psi_V : F(V) \otimes Y \to Y \otimes V$, denote the associated action morphisms of $\mathbb{A}$ by $\phi : X \otimes \mathbb{A} \to X$ and $\psi : Y \otimes \mathbb{A} \to Y$ respectively.

Given a basis $\{v_i\}$ of V, we introduce morphisms $\phi_{V,i} : X \otimes F(V) \to X$ by the relation

$$\phi_V = \sum_i v_i \otimes \phi_{V,i}$$

in the vector space $\mathrm{Hom}(X \otimes F(V), V \otimes X) = V \otimes \mathrm{Hom}(F(V) \otimes X, X)$. Likewise, we define morphisms $\psi_{V,i} : F(V) \otimes Y \to Y$ so that

$$\psi_V = \sum_i \psi_{V,i} \otimes v_i.$$

From the definition of ϕ, $\phi \otimes 1_Y$ is identified with

$$\bigoplus_V \sum_i v_i \otimes \phi_{V,i} \quad \in \quad \bigoplus_V V \otimes \mathrm{Hom}(X \otimes F(V) \otimes Y, X \otimes Y)$$

in the vector space

$$\begin{aligned} \mathrm{Hom}(X \otimes \mathbb{A} \otimes Y, X \otimes Y) &= \bigoplus_V \mathrm{Hom}(X \otimes F(V) \otimes V^* \otimes Y, X \otimes Y) \\ &= \bigoplus_V V \otimes \mathrm{Hom}(X \otimes F(V) \otimes Y, X \otimes Y). \end{aligned}$$

Similarly we have the expression

$$1_X \otimes \psi = \bigoplus_V \sum_i v_i \otimes \psi_{V,i} \quad \text{in} \quad \bigoplus_V V \otimes \mathrm{Hom}(X \otimes F(V) \otimes Y, X \otimes Y).$$

Now the idempotent $p \in \mathrm{End}(X \otimes Y)$ producing the relative tensor product $X \otimes_{\mathbb{A}} Y$ is given by

$$\sum_V \frac{d_{F(V)}}{d_{\mathbb{A}}} \sum_i (\phi_{V,i} \otimes 1_Y)(1_X \otimes \psi^*_{V,i})$$

from the definition of p and the formula for δ_A. Here we denote by $\{\psi^*_{V,i} : Y \to F(V) \otimes Y\}$ the cosystem of $\{\psi_{V,i}\}_i$:

$$\psi_{V,i}\psi^*_{V,j} = \delta_{ij} 1_Y, \qquad \sum_i \psi^*_{V,i}\psi_{V,i} = 1_{F(V)\otimes Y}$$

and $d_{F(V)}$, $d_{\mathbb{A}}$ are quantum dimensions of the objects $F(V)$, $\mathbb{A}$ respectively.

We next derive an explicit formula for the idempotent $\pi(e)$ which is used to define $X \otimes_{\mathcal{A}} Y$. Recall here that π is an algebra homomorphism of the dual Hopf algebra H^* into $\mathrm{End}(X \otimes Y)$ and $e \in H^*$ denotes the counit functional of H.

By using the explicit definition of π in [13], we see that

$$\pi(e) = \sum_V \frac{\dim(V)}{\dim(H)} \sum_i (\phi_{V,i} \otimes 1_Y)(1_X \otimes \psi^*_{V,i}),$$

which is exactly the idempotent p because of $d_{F(V)} = \dim(V)$ and $d_{\mathbb{A}} = \dim(H)$.

Proposition 7.4 *The fully faithful functor $\Phi : {}_{\mathcal{A}}\mathcal{T}_{\mathcal{B}} \to {}_{\mathbb{A}}\mathcal{T}_{\mathbb{A}}$ is monoidal by the equality $X \otimes_{\mathcal{B}} Y = X \otimes_{\mathbb{B}} Y$ in $\mathcal{T}$.*

Proposition 7.5 *The monoidal functor $\Phi : {}_{\mathcal{A}}\mathcal{T}_{\mathcal{B}} \to {}_{\mathbb{A}}\mathcal{T}_{\mathbb{B}}$ is an equivalence of categories, i.e., any $\mathbb{A}$-$\mathbb{B}$ bimodule in $\mathcal{T}$ is isomorphic to $\Phi({}_{\mathcal{A}}X_{\mathcal{B}})$ with ${}_{\mathcal{A}}X_{\mathcal{B}}$ an $\mathcal{A}$-$\mathcal{B}$ bimodule in $\mathcal{T}$.*

Proof Let ${}_{\mathbb{A}}X_{\mathbb{B}}$ be an $\mathbb{A}$-$\mathbb{B}$ bimodule in $\mathcal{T}$. Since the $\mathbb{A}$-$\mathbb{B}$ bimodule ${}_{\mathbb{A}}\mathbb{A}\otimes X\otimes\mathbb{B}_{\mathbb{B}}$ is isomorphic to $\Phi({}_{\mathcal{A}}\mathbb{A}\otimes X\otimes\mathbb{B}_{\mathcal{B}})$ and since the functor is fully faithful, we can find an idempotent $p \in \mathrm{End}({}_{\mathcal{A}}\mathbb{A} \otimes X \otimes \mathbb{B}_{\mathcal{B}})$ such that $\Phi(p)$ induces the relative tensor product $\mathbb{A} \otimes_{\mathbb{A}} X \otimes_{\mathbb{B}} \otimes \mathbb{B}$. Thus ${}_{\mathbb{A}}X_{\mathbb{B}}$ is isomorphic to $\Phi({}_{\mathcal{A}}p(\mathbb{A} \otimes X \otimes \mathbb{B})_{\mathcal{B}})$. □

References

[1] P. Deligne, Catégories tannakiennes, 111–195, *The Grothendieck Festshrift, II* (Progress in Math. 87), Birkhäuser, 1990.

[2] S. Doplicher and J.E. Roberts, A new duality theory for compact groups, *Invent. Math.*, 98(1989), 157–218.

[3] J. Fuchs and C. Schweigert, Category theory for conformal boundary conditions, *Fields Institute Commun.*, to appear.

[4] D. Kazhdan and H. Wenzl, Reconstructing monoidal categories, *Adv. Soviet Math.*, 16(1993), 111–136.

[5] F.W. Lawvere, Ordinal sums and equational doctrines. Sem. on Triples and Categorical Homology Theory (ETH, Zürich, 1966/67). pp. 141–155. Springer, Berlin (1969).

[6] R. Longo, A duality for Hopf algebras and for subfactors. I, *Commun. Math. Phys.*, 159(1994), 133–150.

[7] R. Longo and J.E. Roberts, A theory of dimension, *K-Theory*, 11(1997), 103–159.

[8] S. Mac Lane, *Categories for the working mathematician*, 2nd ed., Springer Verlag, 1998.

[9] M. Müger, From subfactors to categories and topology I., *J. Pure and Applied Algebras*, 180(2003), 81–157.

[10] V. Ostrik, Module categories, weak Hopf algebras and modular invariants, *Transf. Groups*, to appear.

[11] D. Tambara, A duality for modules over monoidal categories of representations of semisimple Hopf algebras, *J. Algebra*, 241(2001), 515–547.

[12] S. Yamagami, Group symmetry in tensor categories and duality for orbifolds, *J. Pure and Applied Algebras*, 167(2002), 83–128.

[13] ___________, Tannaka duals in semisimple tensor categories, *J. Algebra*, 253(2002), 350–391.

Received December 4, 2002; in revised form May 20, 2003

Titles in This Series

18 **Panos M. Pardalos and Henry Wolkowicz, Editors,** Topics in semidefinite and interior-point methods, 1998

17 **Joachim J. R. Cuntz and Masoud Khalkhali, Editors,** Cyclic cohomology and noncommutative geometry, 1997

16 **Victor P. Snaith, Editor,** Algebraic K-theory, 1997

15 **Stephen P. Braham, Jack D. Gegenberg, and Robert J. McKellar, Editors,** Sixth Canadian conference on general relativity and relativistic astrophysics, 1997

14 **Mourad E. H. Ismail, David R. Masson, and Mizan Rahman, Editors,** Special functions, q-series and related topics, 1997

13 **Peter A. Fillmore and James A. Mingo, Editors,** Operator algebras and their applications, 1997

12 **Dan-Virgil Voiculescu, Editor,** Free probability theory, 1997

11 **Colleen D. Cutler and Daniel T. Kaplan, Editors,** Nonlinear dynamics and time series: Building a bridge between the natural and statistical sciences, 1997

10 **Jerrold E. Marsden, George W. Patrick, and William F. Shadwick, Editors,** Integration algorithms and classical mechanics, 1996

9 **W. H. Kliemann, W. F. Langford, and N. S. Namachchivaya, Editors,** Nonlinear dynamics and stochastic mechanics, 1996

8 **Larry M. Bates and David L. Rod, Editors,** Conservative systems and quantum chaos, 1996

7 **William F. Shadwick, Perinkulam Sambamurthy Krishnaprasad, and Tudor Stefan Ratiu, Editors,** Mechanics day, 1996

6 **Anna T. Lawniczak and Raymond Kapral, Editors,** Pattern formation and lattice gas automata, 1996

5 **John Chadam, Martin Golubitsky, William Langford, and Brian Wetton, Editors,** Pattern formation: Symmetry methods and applications, 1996

4 **William F. Langford and Wayne Nagata, Editors,** Normal forms and homoclinic chaos, 1995

3 **Anthony Bloch, Editor,** Hamiltonian and gradient flows, algorithms and control, 1994

2 **K. A. Morris, Editor,** Control of flexible structures, 1993

1 **Michael J. Enos, Editor,** Dynamics and control of mechanical systems: The falling cat and related problems, 1993

For a complete list of titles in this series, visit the AMS Bookstore at **www.ams.org/bookstore/**.

Titles in This Series

43 **George Janelidze, Bodo Pareigis, and Walter Tholen, Editors,** Galois theory, Hopf algebras, and semiabelian categories, 2004

42 **Saber Elaydi, Gerry Ladas, Jianhong Wu, and Xingfu Zou, Editors,** Difference and differential equations, 2004

41 **Alf van der Poorten and Andreas Stein, Editors,** High primes and misdemeanours: Lectures in honour of the 60th birthday of Hugh Cowie Williams, 2004

40 **Vlastimil Dlab and Claus Michael Ringel, Editors,** Representations of finite dimensional algebras and related topics in Lie theory and geometry, 2004

39 **Stephen Berman, Yuly Billig, Yi-Zhi Huang, and James Lepowsky, Editors,** Vector operator algebras in mathematics and physics, 2003

38 **Noriko Yui and James D. Lewis, Editors,** Calabi-Yau varieties and mirror symmetry, 2003

37 **Panos Pardalos and Henry Wolkowicz, Editors,** Novel approaches to hard discrete optimization, 2003

36 **Shigui Ruan, Gail S. K. Wolkowicz, and Jianhong Wu, Editors,** Dynamical systems and their applications in biology, 2003

35 **Yakov Eliashberg, Boris Khesin, and François Lalonde, Editors,** Symplectic and contact topology: Interactions and perspectives, 2003

34 **T. J. Lyons and T. S. Salisbury, Editors,** Numerical methods and stochastics, 2002

33 **Franz-Viktor Kuhlmann, Salma Kuhlmann, and Murray Marshall, Editors,** Valuation theory and its applications, Volume II, 2003

32 **Franz-Viktor Kuhlmann, Salma Kuhlmann, and Murray Marshall, Editors,** Valuation theory and its applications, Volume I, 2002

31 **A. Galves, J. K. Hale, and C. Rocha, Editors,** Differential equations and dynamical systems, 2002

30 **Roberto Longo, Editor,** Mathematical physics in mathematics and physics: Quantum and operator algebraic aspects, 2001

29 **Teresa Faria and Pedro Freitas, Editors,** Topics in functional differential and difference equations, 2001

28 **David R. McDonald and Stephen R. E. Turner, Editors,** Analysis of communication networks: Call centres, traffic and performance, 2000

27 **Shui Feng and Anna T. Lawniczak, Editors,** Hydrodynamic limits and related topics, 2000

26 **Neal Madras, Editor,** Monte Carlo methods, 2000

25 **A. G. Ramm, P. N. Shivakumar, and A. V. Strauss, Editors,** Operator theory and its applications, 2000

24 **Edward Bierstone, Boris Khesin, Askold Khovanskii, and Jerrold E. Marsden, Editors,** The Arnoldfest: Proceedings of a conference in honour of V. I. Arnold for his sixtieth birthday, 1999

23 **Myrna H. Wooders, Editor,** Topics in mathematical economics and game theory. Essays in honor of Robert J. Aumann, 1999

22 **Graciela Chichilnisky, Editor,** Topology and markets, 1999

21 **Shigui Ruan, Gail S. K. Wolkowicz, and Jianhong Wu, Editors,** Differential equations with applications to biology, 1998

20 **Peter A. Fillmore and James A. Mingo, Editors,** Operator algebras and their applications II, 1998

19 **William G. Dwyer, Stephen Halperin, Richard Kane, Stanley O. Kochman, Mark E. Mahowald, and Paul S. Selick (Editor-in-Chief), Editors,** Stable and unstable homotopy, 1998

18 **Panos M. Pardalos and Henry Wolkowicz, Editors,** Topics in semidefinite and interior-point methods, 1998

17 **Joachim J. R. Cuntz and Masoud Khalkhali, Editors,** Cyclic cohomology and noncommutative geometry, 1997

16 **Victor P. Snaith, Editor,** Algebraic K-theory, 1997

15 **Stephen P. Braham, Jack D. Gegenberg, and Robert J. McKellar, Editors,** Sixth Canadian conference on general relativity and relativistic astrophysics, 1997

14 **Mourad E. H. Ismail, David R. Masson, and Mizan Rahman, Editors,** Special functions, q-series and related topics, 1997

13 **Peter A. Fillmore and James A. Mingo, Editors,** Operator algebras and their applications, 1997

12 **Dan-Virgil Voiculescu, Editor,** Free probability theory, 1997

11 **Colleen D. Cutler and Daniel T. Kaplan, Editors,** Nonlinear dynamics and time series: Building a bridge between the natural and statistical sciences, 1997

10 **Jerrold E. Marsden, George W. Patrick, and William F. Shadwick, Editors,** Integration algorithms and classical mechanics, 1996

9 **W. H. Kliemann, W. F. Langford, and N. S. Namachchivaya, Editors,** Nonlinear dynamics and stochastic mechanics, 1996

8 **Larry M. Bates and David L. Rod, Editors,** Conservative systems and quantum chaos, 1996

7 **William F. Shadwick, Perinkulam Sambamurthy Krishnaprasad, and Tudor Stefan Ratiu, Editors,** Mechanics day, 1996

6 **Anna T. Lawniczak and Raymond Kapral, Editors,** Pattern formation and lattice gas automata, 1996

5 **John Chadam, Martin Golubitsky, William Langford, and Brian Wetton, Editors,** Pattern formation: Symmetry methods and applications, 1996

4 **William F. Langford and Wayne Nagata, Editors,** Normal forms and homoclinic chaos, 1995

3 **Anthony Bloch, Editor,** Hamiltonian and gradient flows, algorithms and control, 1994

2 **K. A. Morris, Editor,** Control of flexible structures, 1993

1 **Michael J. Enos, Editor,** Dynamics and control of mechanical systems: The falling cat and related problems, 1993

For a complete list of titles in this series, visit the AMS Bookstore at **www.ams.org/bookstore/**.